Finite Mathematics

Geoffrey C. Berresford
Long Island University

Andrew M. Rockett
Long Island University

Houghton Mifflin Company

Boston   New York

Editor-in-Chief: Charles Hartford
Sponsoring Editor: Paul Murphy
Editorial Assistant: Marika Hoe
Senior Project Editor: Maria Morelli
Editorial Assistants: Cecilia Molinari, Tanius Stamper
Senior Production/Design Coordinator: Carol Merrigan
Senior Manufacturing Coordinator: Priscilla J. Bailey
Marketing Manager: Michael Busnach

Cover design: Diana Coe/ko design
Cover image: Ian Lawrence/Photonica

Photo credits: Chapter 1 opener: PALM PRESS; Shroud of Turin: Corbis.
Chapter 2 opener: Alan Schein/The Stock Market.
Chapter 3 opener: Steve Leonard/Stone.
Chapter 4 opener: Digital Stock; stock quotron: Robert Brenner/Photo Edit.
Chapter 5 opener: Fontanon Nuñez/The Image Bank; baseball team: Brian Haimer/PhotoEdit; airport: Grafton Marshal Smith/The Stock Market; McGuire and Sosa: ©AFP/Corbis.
Chapter 6 opener: Spencer Grant/PhotoEdit; IBM Probability Machine: IBM.
Chapter 7 opener: Bruce Ayers/Stone.
Chapter 8 opener: John Newbauer/PhotoEdit.

Printed in the U.S.A.

Library of Congress Catalog Card Number: 00-133839

ISBN: 0-395-98574-9

123456789-DC-04 03 02 01 00

PREFACE

This book is an introduction to finite mathematics and its applications for students in the management, social, behavioral, and biomedical sciences, as well as other applied fields and general liberal arts studies. Chapter 1 provides a brief review of algebra and functions, and Chapter 2 applies exponential and logarithmic functions to the mathematics of finance. Chapters 3 and 4 explore matrices and linear programming problems, while Chapters 5 and 6 develop probability and the foundations of statistics. Chapter 7 is an extended application of matrices to the study of Markov chains, and Chapter 8 is an introduction to symbolic logic.

Our presentation endeavors to exhibit some of the many fascinating and useful applications of these areas of mathematics in business, the sciences, and everyday life. But beyond its utility, there is beauty to mathematics and we hope to convey some of its elegance and simplicity.

FEATURES

Realistic Applications Since courses using this book are by their basic nature very applied, every chapter contains an unusually large number of applications, many appearing in no other textbook, both in the development of the material and in the exercises. We explore learning curves in airplane production (pages 26–27 and 29), analyze long term trends in cigarette smoking (page 63), find the age of the Dead Sea Scrolls (page 85), model data flow in a computer network (pages 199–200), manage an investment portfolio (pages 346–348 and 367–368), analyze Mark McGuire's home run record (page 453), and make projections of farm size distributions (pages 543–544). These and many other applications convincingly show that mathematics is more than just the manipulation of abstract symbols and is deeply connected to everyday life.

Graphing Calculators (*Optional*)   Calculators with capabilities previously available only on computers have changed the way mathematics can be taught and used. Reading this book does not require a graphing calculator, but having one will enable you to do many problems more easily and at the same time deepen your understanding by allowing you to concentrate on concepts. Throughout the book are Graphing Calculator Explorations and Exercises, which

explore new topics (such as the graphs of $y = \sqrt[n]{x}$ on page 23 and then the "rule of 0.6" in Exercises 83–84 on page 29),

carry out otherwise "messy" calculations (such as the population growth comparison on page 71 and then Exercises 17–18 on page 73), and

show the advantages and limitations of technology (such as the differences between $\ln x^2$ and $2 \ln x$ on page 84).

While any graphing calculator (or a computer) may be used, the displays shown in the text are from the Texas Instruments TI-83 and TI-89. A discussion of the essentials of graphing calculators follows this preface. For those not using a graphing calculator, the Graphing Calculator Explorations have been carefully planned so that *most can be simply read for enrichment* (as with the histograms and box-and-whisker plots on pages 470–471 and 486).

Graphing Calculator Programs (*Optional*)   For certain topics, the computational effort may be reduced and the understanding enhanced by using the optional graphing calculator programs we created for use with this book. These programs are free and easy to obtain (see "How to Obtain Graphing Calculator Programs" later in this Preface). The topics covered are: amortization tables (pages 145–146), matrix row operations (pages 196–197), objective function values of a linear programming problem on the feasible region (pages 282–283), the pivot operation (pages 304–305), binomial probability distributions (page 450), the normal approximation to the binomial distribution (page 497), and the steady-state distribution for a regular Markov chain (pages 528–529).

Spreadsheets (*Optional*)   Conceptual understanding may also be gained from numerical explorations using computer spreadsheet programs. We have included spreadsheet examples of: the relation between revenues, costs, and profits (page 42), amortization tables (page 142), matrix arithmetic (page 213), matrix inverses (page 227), solving a transportation problem (page 334), the binomial theorem (page 404), graphs of statistical data (page 467), observations of a Markov chain and the number of times in a particular state (page 518), and truth values and logic functions (page 576). Ancillary materials for Microsoft Excel are also available (see "Supplements for Students" later in this Preface).

Enhanced Readability   An elegant four-color design increases the visual appeal and readability of this book. Color is frequently used for pedagogical purposes, such as in the examples of the Gauss-Jordan method (pages 194, 196, and 198), the construction of an initial simplex tableau (page 298), and the order of calculation in a truth table (pages 564–565). For the sake of continuity, references to earlier material are minimized by restating results whenever they are used. Where references are necessary, explicit page numbers are given.

Application Previews   Most sections begin with an Application Preview that presents an interesting application of the mathematics developed in the section. Each is self-contained (although some exercises may later refer to it) and serves to motivate interest in the coming section. The titles and page numbers of these thirty-eight essays are listed in the table of Contents.

Practice Problems   Learning mathematics requires your active participation—"mathematics is not a spectator sport." Throughout the readings are short pencil and paper Practice Problems designed to consolidate your understanding of one topic before moving ahead to another, such as constructing a dual linear programming problem (page 326) or identifying antecedents and consequents (page 577). Complete solutions to all Practice Problems are provided at the end of the book.

Annotations   Notes to the right of many mathematical formulas and manipulations state the results in words, assisting the important skill of "reading mathematics," as well as providing explanations and justifications for the steps in calculations and the interpretations of the results (for instance, see page 52). Notes and diagrams to the left of new concepts and results provide visual clarification (see the Gauss-Jordan method on page 193 and the pivot operation on page 302).

Extensive Exercises   Anyone who ever learned any mathematics did so by solving many many problems, and the exercises are the most essential part of the learning process. The exercises sets (see, for instance, pages 432–434) are graded from routine drills to significant applications and some conclude with Explorations and Excursions that extend and augment the material presented in the text. Exercises marked with the symbol ▦ require a business or scientific calculator with keys like $\boxed{\ln x}$ and $\boxed{y^x}$ for natural logarithms and powers. Exercises marked by ▦ require a graphing calculator. Answers to odd-numbered exercises and answers to all Chapter Summary exercises are given at the end of the book (full solutions are given in the Student Solutions Manual).

Levels of Reinforcement   Because there are many new ideas and techniques in this book, learning checks are provided at several different levels. As noted above, **Practice Problems** encourage mastery of new skills directly after they are demonstrated. **Section Summaries** briefly state both essential formulas and key concepts (see page 451). **Chapter Summaries** review the major developments of the chapter and are keyed to particular chapter review exercises (see page 456). **Hints and Suggestions** at the end of each chapter summary unify the chapter, give specific reminders of essential facts or "tricks" that sometimes are otherwise overlooked or forgotten, and list a selection of the review exercises for a **Practice Test** of the chapter material (see page 457).

Projects and Essays   Each chapter concludes with a collection of open-ended projects and topics for student essays that ask the student (or a group of students) to research a relevant person or an idea, to compare several different mathematical ideas, or to relate a concept to their lives (such as AIDS tests and false positives on page 461, the cost accounting problem as a Markov chain on page 555, or conversational implicature on page 606). These are in keeping with recent recommendations from the Mathematical Association of America and the National Council of Teachers of Mathematics. Other more challenging projects can be found in the highly recommended *MAA Notes* Numbers 27–30, available from the Mathematical Association of America in Washington, DC.

Explorations and Excursions   At the end of some exercise sets are optional problems of a more advanced nature that carry the development of certain topics beyond the level of the text, such as: round-off errors in systems of equations (page 173), Klee-Minty linear programming problems (page 321), Stirling's approximation (page 409), the mean of grouped data (pages 481–482), and the disjunctive normal form of a statement (page 573–574).

Accuracy and Proofs   All of the answers and other mathematics have been checked carefully by several mathematicians. The statements of definitions and theorem are mathematically accurate. Since the treatment is applied rather than theoretical, intuitive and geometric justifications have often been preferred to formal proofs. When proofs are given, however, they are correct and "honest."

Philosophy We wrote this book with several principles in mind. One is that to learn something, it is best to begin doing it as soon as possible: the preliminary material is brief, so you can begin doing useful mathematics immediately. An early start allows more time during the course for interesting applications and necessary review. Another principle is that the mathematics should be done together with the applications. Consequently every section contains applications (there are no "pure math" sections).

Prerequisites The only prerequisite for this course is some knowledge of algebra, functions, and graphing, and these are reviewed in Chapter 1. Other review material has been placed in relevant locations throughout the book.

HOW TO OBTAIN GRAPHING CALCULATOR PROGRAMS

The optional graphing calculator programs used in the text have been written for a variety of Texas Instruments graphing calculators (including the TI-82, TI-83, TI-85, TI-86, TI-89 and TI-92), and may be obtained for free by any of the following ways:

- If you know someone who already has the programs on a Texas Instruments graphing calculator of the same model as yours, you can easily transfer the programs from their calculator to yours using the black cable that came with the calculator and the LINK button. (Even if you use one of the following methods to get a "first" copy of the programs, you can use this method to transfer the programs to others in your class.)

- If you have access to the Internet, you may download the programs (together with descriptive materials) from the www.hmco.com Houghton Mifflin web site for this book. Then use the TI-GRAPH LINK™ software and cable (available for purchase in most stores that sell graphing calculators) to transfer the programs from your computer to your calculator.

- You may send a formatted $3\frac{1}{2}$-inch "floppy" disk, carefully packed, to the authors at the following address, specifying both the type of your computer (Windows or Macintosh) and your Texas Instruments calculator (TI-82, TI-83, TI-85, TI-86, TI-89, or TI-92). We will return a disk containing the appropriate programs and a packet of descriptive information. After you load the disk and programs into your computer, you will need the TI-GRAPH LINK™

software and cable (available for purchase in most stores that sell graphing calculators) to transfer the programs to your calculator.

■ You may send your calculator (TI-82, TI-83, TI-85, TI-86, TI-89 or TI-92), carefully packed, to the authors at the following address, and we will return it loaded with the appropriate programs, along with a packet of descriptive information.

Authors' Address   Dr. G. C. Berresford and Dr. A. M. Rockett, Department of Mathematics, C. W. Post Campus of Long Island University, 720 Northern Boulevard, Brookville, New York 11548–1300

SUPPLEMENTS FOR STUDENTS (available at college.hmco.com)

Student Solutions Manual, Jeanne Marie Draper, Solano Community College. Contains worked-out solutions to all odd-numbered exercises and all review exercises.

Graphing Calculator Guide, Frank Wilson, Green River Community College. Provides detailed instructions on how to use your graphing calculator as a tool for learning much of the mathematics developed in this book.

Excel Guide for Finite Math and Applied Calculus, Revathi Narasimhan, Kean University. Contains Microsoft Excel spreadsheet materials for many of the topics covered in this book.

SUPPLEMENTS FOR INSTRUCTORS

Instructor Solutions Manual, Jeanne Marie Draper, Solano Community College. This soft cover book contains full solutions for all exercises in the book.

Computerized Test Bank (*Windows and Macintosh versions*) The test bank contains more than 2000 test questions arranged by chapter and section, allowing instructors to create customized tests efficiently. Many of these test questions are applied problems. Test questions can be selected by section number as well as other criteria. Both versions have full editing capabilities and high quality graph reproduction. They produce scrambled and multiple test versions in multiple choice or free response format, and provide answer keys. The Windows version also provides on-line testing and gradebook functions, and allows importation of files from ASCII, Word, and WordPerfect.

Printed Test Bank with Chapter Tests, Jean Shutters, Harrisburg Area Community College, and Christi Verity. This is a printed version of the Computerized Test Bank for instructors who do not use computers. Also included are two comprehensive Chapter Tests for each chapter (one multiple choice and one free response). Answers to all test questions are included.

ACKNOWLEDGMENTS

We are indebted to numerous people for useful suggestions, conversations, and correspondence during the writing of this book, but most of all to our many "Math 5" students at the C.W. Post campus of Long Island University over the last several years for their questions, comments and enthusiasm for learning mathematics.

We have had the good fortune to have had wonderfully supportive and expert editors at Houghton Mifflin: Paul Murphy (Sponsoring Editor), Marika Hoe (Editorial Assistant), and Maria Morelli (Senior Project Editor). They made the most difficult tasks easy, and helped beyond words. Helen Medley (accuracy reviewer) and Jeanne Marie Draper (solutions manuals) saved us from many mistakes and confusions. We also express our gratitude to the many others at Houghton Mifflin who made many important contributions too numerous to mention.

The following reviewers have contributed greatly to the progression from first draft to finished book:

Richard Blecksmith, *Northern Illinois University*

Bob Bradshaw, *Ohlone College*

Roxanne Byrne, *University of Colorado at Denver*

Candy Giovanni, *Michigan State University*

W. Thomas Kiley, *George Mason University*

M. Maheswaran, *University of Wisconsin, Marathon County*

Gregory McColm, *University of South Florida*

Diana McCoy, *Florida International University*

Ho Kuen Ng, *San Jose State University*

Richard Pellerin, *Northern Virginia Community College*

Thomas Riedel, *University of Louisville*

Patty Schovanec, *Texas Tech University*

Jennifer Stevens Fowler, *University of Tennessee, Knoxville*

DEDICATION

We dedicate this book to our wives, Barbara and Kathryn, and our children, Lee, Chris, Justin and Joshua, for their understanding and patience, without which this book would not exist.

COMMENTS WELCOMED

With the knowledge that any book can always be improved, we welcome corrections, constructive criticisms and suggestions from every reader.

Graphing Calculator Terminology

The Graphing Calculator Explorations and Exercises have been kept as generic as possible for use with any of the popular graphing calculators. We assume that you or your teacher are familiar with the sequence of button pushes necessary to accomplish various basic operations on your calculator. Certain standard calculator terms are capitalized in this book and are described below. Your calculator may use slightly different terminology.

The viewing window or graphing **WINDOW** is the part of the Cartesian plane shown in the display screen of your graphing calculator. **XMIN** and **XMAX** are the smallest and largest x-values shown, and **YMIN** and **YMAX** are the smallest and largest y-values shown. These can be set by using the WINDOW or **RANGE** command and are changed automatically by using any of the **ZOOM** operations. **XSCALE** and **YSCALE** define the distance between tick marks on the x- and y-axes.

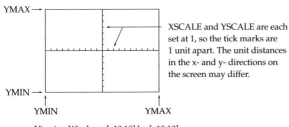

XSCALE and YSCALE are each set at 1, so the tick marks are 1 unit apart. The unit distances in the x- and y- directions on the screen may differ.

Viewing Windows [–10,10] by [–10,10]

The viewing window is always [XMIN, XMAX] by [YMIN, YMAX]. We will set XSCALE and YSCALE so that there are a reasonable number of tick marks (generally 2 to 20) on each axis. Either or both of the x- and y-axes may not be visible when the viewing window does not include the origin.

Pixel, an abbreviation for *pic*ture *el*ement, refers to a tiny rectangle on the screen that can be darkened to represent a dot on a graph. Pixels are arranged in a rectangular array on the screen. In the above window, the axes and tick marks are formed by darkened pixels. The size of the screen and the number of pixels vary with different calculators.

TRACE allows you to move a flashing pixel (or *cursor*) along a curve in the viewing window, with the x- and y-coordinates shown at the bottom of the screen.

Useful Hint To make the x-values in TRACE take simple values like 0.1, 0.2, and 0.3, choose XMIN and XMAX to be multiples of one less than the number of pixels across the screen. For example, on the TI-82 and TI-83, which have 95 pixels across the screen, using an x-window such as [−9.4, 9.4] or [−4.7, 4.7] or [−940, 940] will TRACE with simpler x-values than the standard windows used in this book.

ZOOM IN allows you to magnify any part of the viewing window to see finer detail around a chosen point. **ZOOM OUT** does the opposite, like stepping back to see a larger portion of a picture but with less detail. These and any of several other **ZOOM** commands change the viewing window.

VALUE or **EVALUATE** finds the value of a previously entered expression at a specified x-value.

SOLVE or **ROOT** or **ZERO** finds the x-value that solves $f(x) = 0$, or equivalently, the x-intercepts of a curve $y = f(x)$. When applied to a difference $f(x) - g(x)$, it finds the x-value where the two curves meet (as also done by the **INTERSECT** command).

MAX and **MIN** find the highest and lowest points of a (previously entered) curve between specified x-values.

In **CONNECTED MODE** your calculator will darken pixels to connect calculated points on a graph in an attempt to show it as a continuous or "unbroken" curve. However, this may lead to "false lines" in a graph that should have breaks or "jumps." Such false lines can be eliminated by using **DOT MODE**.

The **TABLE** command on some calculators lists in table form the values of a function, just as you have probably done when graphing a curve. The x-values may be chosen by you or by the calculator.

The **Order of Operations** used by most calculators evaluates operations in the following order: first powers and roots, then operations like LN and LOG, then multiplications and divisions, and finally additions and subtractions, always working from left to right within each level. For example, $5 \wedge 2x$ means $(5 \wedge 2)x$, *not* $5 \wedge (2x)$. Also, $1/x + 1$ means $(1/x) + 1$, *not* $1/(x + 1)$. See your calculator's instruction manual for further information. *Be careful*: some calculators evaluate $1/2x$ as $(1/2)x$ and some as $1/(2x)$. When in doubt, use parentheses to clarify the expression.

More information can be found in the manual for your graphing calculator and other features will be discussed later when they are needed.

CONTENTS

| Chapter 2 | Mathematics of Finance | 98 |

| Chapter 3 | Matrices and Systems of Equations | 157 |

Chapter 4   Linear Programming   261

Chapter 5   Probability   385

Chapter 6   Statistics   462

Chapter 7   Markov Chains   506

| **Chapter 8** | **Logic** | **556** |

1

Functions

Parabolas described by a bouncing ball

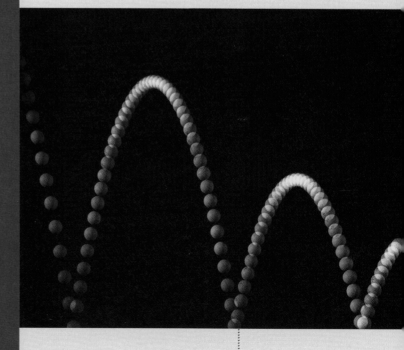

1.1 Real Numbers, Inequalities, and Lines

World Record Mile Runs

The dots on the graph below show the world record times for the mile run from 1865 to the 1999 world record of 3 minutes 43.13 seconds, set by the Moroccan runner Hicham el-Guerrouji. These points fall roughly along a line, called the **regression line.** The regression line is easily found by using a graphing calculator to apply a method called **least squares,** which is explained in Chapter 3. Notice that the times do not level off as you might expect, but continue to decrease.

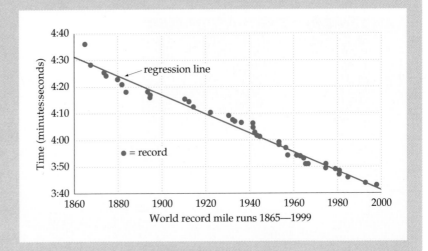

World record mile runs 1865—1999

HISTORY OF THE RECORD FOR THE MILE RUN

Time	Year	Athlete	Time	Year	Athlete
4:36.5	1865	Richard Webster	4:14.4	1913	John Paul Jones
4:29.0	1868	William Chinnery	4:12.6	1915	Norman Taber
4:28.8	1868	Walter Gibbs	4:10.4	1923	Paavo Nurmi
4:26.0	1874	Walter Slade	4:09.2	1931	Jules Ladoumegue
4:24.5	1875	Walter Slade	4:07.6	1933	Jack Lovelock
4:23.2	1880	Walter George	4:06.8	1934	Glenn Cunningham
4:21.4	1882	Walter George	4:06.4	1937	Sydney Wooderson
4:18.4	1884	Walter George	4:06.2	1942	Gunder Hägg
4:18.2	1894	Fred Bacon	4:06.2	1942	Arne Andersson
4:17.0	1895	Fred Bacon	4:04.6	1942	Gunder Hägg
4:15.6	1895	Thomas Conneff	4:02.6	1943	Arne Andersson
4:15.4	1911	John Paul Jones	4:01.6	1944	Arne Andersson

Time	Year	Athlete	Time	Year	Athlete
4:01.4	1945	Gunder Hägg	3:51.0	1975	Filbert Bayi
3:59.4	1954	Roger Bannister	3:49.4	1975	John Walker
3:58.0	1954	John Landy	3:49.0	1979	Sebastian Coe
3:57.2	1957	Derek Ibbotson	3:48.8	1980	Steve Ovett
3:54.5	1958	Herb Elliott	3:48.53	1981	Sebastian Coe
3:54.4	1962	Peter Snell	3:48.40	1981	Steve Ovett
3:54.1	1964	Peter Snell	3:47.33	1981	Sebastian Coe
3:53.6	1965	Michel Jazy	3:46.31	1985	Steve Cram
3:51.3	1966	Jim Ryun	3:44.39	1993	Noureddine Morceli
3:51.1	1967	Jim Ryun	3:43.13	1999	Hicham el-Guerrouji

Source: USA Track & Field

The equation of the regression line shown in the graph is $y = -0.356x + 257.44$, where x represents years after 1900 and y is in seconds. The regression line can be used to predict the world mile record in future years. Notice that the most recent world record would have been predicted quite accurately by this line, since the rightmost dot falls almost exactly on the line. Linear trends, however, must not be extended too far. The downward slope of this line means that it will eventually "predict" mile runs in a fraction of a second or even in *negative* time. Moral: In the real world, linear trends do not continue indefinitely. This and other topics in "linear" mathematics will be developed in Section 1.1.

Introduction

In this section we will study *linear* relationships between two quantities, that is, relationships that can be represented by *lines.* Later we will turn to *nonlinear* relationships, which can be represented by *curves.*

When reading this book, it will be helpful (but not necessary) to have a graphing calculator. The **Graphing Calculator Explorations** show how to use a graphing calculator to explore a concept more deeply or to analyze an application in more detail. The parts of the book that require graphing calculators are marked by the symbol ▦. Exercises that can be done with a graphing *or* scientific *or* business calculator (with keys like $\boxed{\ln x}$ and $\boxed{y^x}$) are marked by ▦.

Real Numbers and Inequalities

In this book the word "number" means *real* number, a number that can be represented by a point on the number line (also called the *real line*).

The *order* of the real numbers is expressed by inequalities, with $a < b$ meaning "a is to the *left* of b."

Inequalities

Inequality	In Words
$a < b$	a is less than (smaller than) b
$a \leq b$	a is less than or equal to b
$a > b$	a is greater than (larger than) b
$a \geq b$	a is greater than or equal to b

The inequalities $a < b$ and $a > b$ are called "strict" inequalities, and $a \leq b$ and $a \geq b$ are called "nonstrict" inequalities.

EXAMPLE 1   **Inequalities Between Numbers**

a. $3 \leq 5$　　　**b.** $6 > -2$　　　**c.** $-10 < -5$
$$\uparrow$$
-10 is less than (smaller than) -5

■

Throughout this book are many **Practice Problems,** short questions designed to help you check your understanding of a topic before moving on to new material. Full solutions are given at the back of the book. Solve the following Practice Problem and then check your answer.

PRACTICE PROBLEM 1   Which number is smaller: $\frac{1}{100}$ or $-1,000,000$?

Solution at the back of the book

Changing the signs in an inequality reverses its direction:

$$-3 < 2 \qquad \text{but} \qquad 3 > -2$$

A *double* inequality, such as $a < x < b$, means that *both* the inequalities $a < x$ and $x < b$ hold. The inequality $a < x < b$ can be interpreted graphically as "x is between a and b."

$a < x < b$

Sets and Intervals

Braces { } are read "the set of all" and a vertical bar $|$ is read "such that."

EXAMPLE 2   **Interpreting Sets**

The set of all

a. $\{x \mid x > 3\}$ means "the set of all x such that x is greater than 3."

Such that

b. $\{x \mid -2 < x < 5\}$ means "the set of all x such that x is between -2 and 5."

∎

PRACTICE PROBLEM 2

a. Write in set notation "the set of all x such that x is greater than or equal to -7."

b. Express in words: $\{x \mid x < -1\}$.   *Solutions at the back of the book*

The set $\{x \mid 2 \le x \le 5\}$ can be expressed in *interval* notation by enclosing the endpoints 2 and 5 in square brackets, [2, 5]. The *square* brackets indicate that the endpoints are *in*cluded. The set $\{x \mid 2 < x < 5\}$ can be written (2, 5). The *parentheses* indicate that the endpoints 2 and 5 are *ex*cluded. An interval is *closed* if it includes both endpoints; it is *open* if it includes neither endpoint. The four types of intervals are shown below: a *solid* dot • on the graph indicates that the point is *in*cluded in the interval; a *hollow* dot ○ indicates that the point is *ex*cluded.

Finite Intervals

Interval Notation	Set Notation	Graph	Type
[a, b]	$\{x \mid a \le x \le b\}$		Closed
(a, b)	$\{x \mid a < x < b\}$		Open
[a, b)	$\{x \mid a \le x < b\}$		Half-open
(a, b]	$\{x \mid a < x \le b\}$		or half-closed

An interval may extend infinitely far to the right (indicated by the symbol ∞ for "infinity") or infinitely far to the left (indicated by $-\infty$ for "negative infinity"). Note that ∞ and $-\infty$ are not numbers but are merely symbols to indicate that the interval extends endlessly in one direction or the other. The infinite intervals in the next box are said to be *closed* or *open* depending on whether they *include* or *exclude* their single endpoint.

Infinite Intervals

Interval Notation	Set Notation	Graph	Type
$[a, \infty)$	$\{x \mid x \geq a\}$		Closed
(a, ∞)	$\{x \mid x > a\}$		Open
$(-\infty, a]$	$\{x \mid x \leq a\}$		Closed
$(-\infty, a)$	$\{x \mid x < a\}$		Open

The interval $(-\infty, \infty)$ extends infinitely far in *both* directions (meaning the entire real line) and is denoted by the symbol $\mathbb{R}$ (the set of all real numbers).

$$\mathbb{R} = (-\infty, \infty)$$

Cartesian Plane

Two real lines or *axes*, one horizontal and one vertical, intersecting at their zero points, define the *Cartesian plane*.* The axes divide the plane into four *quadrants*, I through IV. Any point in the Cartesian plane can be specified uniquely by an ordered pair of numbers (x, y); x, called the *abscissa* or *x-coordinate*, is the number on the horizontal axis corresponding to the point; y, called the *ordinate* or *y-coordinate*, is the number on the vertical axis corresponding to the point.

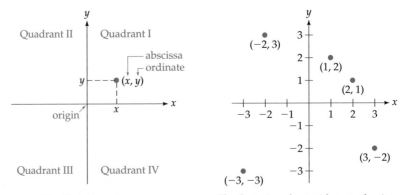

The Cartesian plane

The Cartesian plane with several points. Order matters: (1, 2) is not the same as (2, 1)

* So named because it was originated by the French philosopher and mathematician René Descartes (1596–1650). Following the custom of the day, Descartes signed his scholarly papers with his Latin name Cartesius—hence "Cartesian" plane.

Slope of a Line

The symbol Δ (read "delta," the Greek letter D) means "the change in." For any two points (x_1, y_1) and (x_2, y_2) we define

$\Delta x = x_2 - x_1$ The change in x is the difference in the x-coordinates.

$\Delta y = y_2 - y_1$ The change in y is the difference in the y-coordinates.

The *slope* of a nonvertical line measures the steepness of the line and is defined as *the change in y divided by the change in x* for any two points on the line.

Slope of Line Through (x_1, y_1) and (x_2, y_2)

$$m = \frac{\Delta y}{\Delta x} = \frac{y_2 - y_1}{x_2 - x_1}$$

Slope is the change in y over the change in x $(x_2 \neq x_1)$

The changes Δy and Δx are often called, respectively, the *rise* and the *run*, with the understanding that a negative "rise" means a "fall." Slope is then "rise over run."

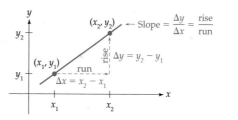

EXAMPLE 3   Finding Slopes and Graphing Lines

Find the slope of the line through each pair of points, and graph the line.

a. $(1, 3), (2, 5)$ **b.** $(2, 4), (3, 1)$ **c.** $(-1, 3), (2, 3)$ **d.** $(2, -1), (2, 3)$

Solution

We use the slope formula $m = \dfrac{y_2 - y_1}{x_2 - x_1}$ for each pair $(x_1, y_1), (x_2, y_2)$.

a. For $(1, 3)$ and $(2, 5)$ the slope is

$$\frac{5 - 3}{2 - 1} = \frac{2}{1} = 2$$

b. For $(2, 4)$ and $(3, 1)$ the slope is

$$\frac{1 - 4}{3 - 2} = \frac{-3}{1} = -3$$

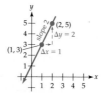

c. For $(-1, 3)$ and $(2, 3)$ the slope is $\dfrac{3 - 3}{2 - (-1)} = \dfrac{0}{3} = 0$

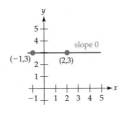

d. For $(2, -1)$ and $(2, 3)$ the slope is $\dfrac{3 - (-1)}{2 - 2} = \dfrac{4}{0}$ Undefined!

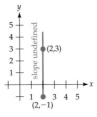

■

If $\Delta x = 1$, as in Examples 3a and 3b, then the slope is just the "rise." That is,

$$\text{Slope} = \begin{pmatrix} \text{the amount that the line rises} \\ \text{when } x \text{ increases by 1} \end{pmatrix}$$

PRACTICE PROBLEM 3

A company president is considering four different business strategies, called S_1, S_2, S_3, and S_4, each with different projected future profits. The graph on the right shows the annual projected profit for the first few years for each of the strategies. Which strategy yields:

a. The highest projected profit in year 1?

b. The highest projected profit in the long run?   *Solutions at the back of the book*

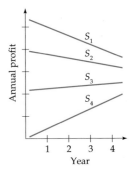

Equations of Lines

The point where a nonvertical line crosses the y-axis is called the *y-intercept* of the line. The y-intercept can be given either as the y-coordinate b or as the point $(0, b)$. Such a line can be expressed very simply in terms of its slope and y-intercept.

Slope–Intercept Form of a Line

$y = mx + b$
$m = \text{slope}$
$b = y\text{-intercept}$

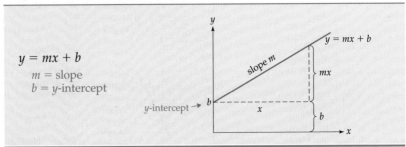

EXAMPLE 4   **Using the Slope-Intercept Form**

Find an equation of the line with slope -2 and y-intercept 4.

Solution

$$y = -2x + 4 \qquad\qquad y = mx + b \text{ with } m = -2 \text{ and } b = 4$$

We may graph this line by first plotting the y-intercept $(0, 4)$. Using the slope $m = -2$, we plot another point 1 unit over and 2 units *down* from the y-intercept. We then draw the line through these two points, as shown below.

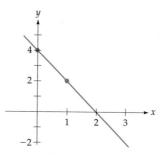

 Graphing Calculator Exploration

a. Use a graphing calculator to graph the lines $y_1 = x$, $y_2 = 2x$, and $y_3 = 3x$ simultaneously on the viewing window $[-10, 10]$ by $[-10, 10]$. How do the graphs change as the coefficient of x increases from 1 to 2 to 3?

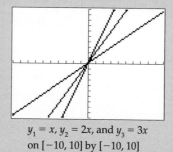

$y_1 = x$, $y_2 = 2x$, and $y_3 = 3x$
on $[-10, 10]$ by $[-10, 10]$

b. Predict what the graph of $y = 0.5x$ would look like. What about $y = -2x$? Check your predictions by graphing them.

c. Describe the graph of the line $y = mx$ for any number m.

Point–Slope Form of a Line

$$y - y_1 = m(x - x_1)$$

(x_1, y_1) = point on the line
m = slope

This form comes directly from the slope formula $m = \dfrac{y_2 - y_1}{x_2 - x_1}$ by dropping the subscript 2 and multiplying each side by $(x - x_1)$.

EXAMPLE 5   Using the Point–Slope Form

Find an equation of the line through $(6, -2)$ with slope $-\frac{1}{2}$.

Solution

$$y - (-2) = -\tfrac{1}{2}(x - 6)$$

$y - y_1 = m(x - x_1)$ with $y_1 = -2$, $m = -\frac{1}{2}$, and $x_1 = 6$

$$y + 2 = -\tfrac{1}{2}x + 3$$

Eliminating parentheses

$$y = -\tfrac{1}{2}x + 1$$

Subtracting 2 from each side

∎

Alternatively, we could have found this equation using $y = mx + b$, replacing m by the given slope $-\frac{1}{2}$ and substituting the given $x = 6$, $y = -2$ to evaluate b.

EXAMPLE 6   Finding an Equation for a Line Through Two Points

Find an equation for the line through the points $(4, 1)$ and $(7, -2)$.

Solution

The slope is not given, so we calculate it from the two points.

$$m = \frac{-2 - 1}{7 - 4} = \frac{-3}{3} = -1$$

$m = \dfrac{y_2 - y_1}{x_2 - x_1}$ with $(4, 1)$ and $(7, -2)$

Then we use the point–slope formula with this slope and either of the two points.

$$y - 1 = -1(x - 4)$$

$y - y_1 = m(x - x_1)$ with slope -1 and point $(4, 1)$

$$y - 1 = -x + 4$$

Eliminating parentheses

$$y = -x + 5$$

Adding 1 to each side

∎

PRACTICE PROBLEM 4

Find the slope–intercept form of the line through the points (2, 1) and (4, 7).

Solution at the back of the book

Vertical and horizontal lines have particularly simple equations: a variable equaling a constant.

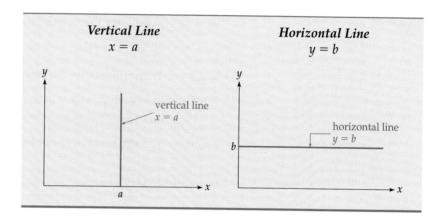

EXAMPLE 7   Graphing Vertical and Horizontal Lines

Graph the lines $x = 2$ and $y = 6$.

Solution

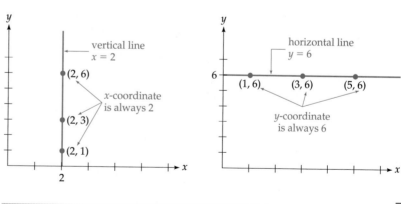

EXAMPLE 8   Finding Equations of Vertical and Horizontal Lines

a. Find an equation for the *vertical* line through the point (3, 2).

b. Find an equation for the *horizontal* line through the point (3, 2).

Solution

a. Vertical line $x = 3$ $x = a$, with a being the x-coordinate from $(3, 2)$

b. Horizontal line $y = 2$ $y = b$, with b being the y-coordinate from $(3, 2)$

■

PRACTICE PROBLEM 5

Find an equation for the vertical line through the point $(-2, 10)$.

Solution at the back of the book

Distinguish carefully between slopes of vertical and horizontal lines:

> Vertical line: slope is *undefined.*
> Horizontal line: slope *is* defined and is *zero.*

There is one form that covers *all* lines, vertical and nonvertical.

General Linear Equation

$$ax + by = c$$

For constants a, b, c, with a and b not both zero

Any equation that can be written in this form is called a *linear equation,* and the variables are said to *depend linearly* on each other.

EXAMPLE 9   Finding the Slope and the *y*-intercept from a Linear Equation

Find the slope and y-intercept of the line $2x + 3y = 12$.

Solution

We write the line in slope–intercept form. Solving for y:

$$3y = -2x + 12$$

$2x + 3y = 12$ after subtracting $2x$ from both sides

$$y = -\tfrac{2}{3}x + 4$$

Dividing each side by 3 gives the slope–intercept form $y = mx + b$

Therefore, the slope is $-\tfrac{2}{3}$ and the y-intercept is $(0, 4)$.

■

PRACTICE PROBLEM 6

Find the slope and y-intercept of the line $x - \dfrac{y}{3} = 2$.

Solution at the back of the book

Section Summary

An *interval* is a set of real numbers corresponding to a section of the real line. The interval is *closed* if it contains all of its endpoints; it is *open* if it contains none of its endpoints.

The nonvertical line through two points (x_1, y_1) and (x_2, y_2) has slope

$$m = \frac{\Delta y}{\Delta x} = \frac{y_2 - y_1}{x_2 - x_1} \qquad x_1 \neq x_2$$

There are five equations, or "forms," for lines:

$y = mx + b$	Slope–intercept form m = slope, b = y-intercept
$y - y_1 = m(x - x_1)$	Point–slope form (x_1, y_1) = point, m = slope
$x = a$	Vertical line (slope undefined) a = x-intercept
$y = b$	Horizontal line (slope zero) b = y-intercept
$ax + by = c$	General linear equation

EXERCISES 1.1

Write each interval in set notation and graph it on the real line.

1. $[0, 6)$ 2. $(-3, 5]$ 3. $(-\infty, 2]$ 4. $[7, \infty)$

5. Given the equation $y = 5x - 12$, how will y change if x:

   a. Increases by 3 units?
   b. Decreases by 2 units?

6. Given the equation $y = -2x + 7$, how will y change if x:

   a. Increases by 5 units?
   b. Decreases by 4 units?

Find the slope (if it is defined) of the line through each pair of points.

7. $(2, 3)$ and $(4, -1)$ 8. $(3, -1)$ and $(5, 7)$

9. $(-4, 0)$ and $(2, 2)$ 10. $(-1, 4)$ and $(5, 1)$

11. $(0, -1)$ and $(4, -1)$ 12. $\left(-2, \frac{1}{2}\right)$ and $\left(5, \frac{1}{2}\right)$

13. $(2, -1)$ and $(2, 5)$ 14. $(6, -4)$ and $(6, -3)$

For each equation, find the slope m and y-intercept $(0, b)$ (if they exist) and draw the graph.

15. $y = 3x - 4$ 16. $y = 2x$

17. $y = -\frac{1}{2}x$ 18. $y = -\frac{1}{3}x + 2$

19. $y = 4$ 20. $y = -3$

21. $x = 4$ 22. $x = -3$

23. $2x - 3y = 12$ 24. $3x + 2y = 18$

25. $x + y = 0$ 26. $x = 2y + 4$

27. $x - y = 0$ 28. $y = \frac{2}{3}(x - 3)$

29. $y = \dfrac{x + 2}{3}$ 30. $\dfrac{x}{2} + \dfrac{y}{3} = 1$

31. $\dfrac{2x}{3} - y = 1$ 32. $\dfrac{x + 1}{2} + \dfrac{y + 1}{2} = 1$

 Use a graphing calculator to graph each line. [*Note:* Your graph will depend on the viewing window you choose. Begin with a "standard" window such as $[-10, 10]$ by

[−10, 10], choosing a larger window (or "zooming out") if the line does not appear.]

33. $y = 2x - 8$ **34.** $y = 3x - 6$

35. $y = 7 - 3x$ **36.** $y = 5 - 2x$

37. $y = 50 - x$ **38.** $y = x - 40$

Find an equation of the line satisfying the following conditions. If possible, write your answer in the form $y = mx + b$.

39. Slope -2.25 and y-intercept 3

40. Slope $\frac{2}{3}$ and y-intercept -8

41. Slope 5 and passing through the point $(-1, -2)$

42. Slope -1 and passing through the point $(4, 3)$

43. Horizontal and passing through the point $(1.5, -4)$

44. Horizontal and passing through the point $\left(\frac{1}{2}, \frac{3}{4}\right)$

45. Vertical and passing through the point $(1.5, -4)$

46. Vertical and passing through the point $\left(\frac{1}{2}, \frac{3}{4}\right)$

47. Passing through the points $(5, 3)$ and $(7, -1)$

48. Passing through the points $(3, -1)$ and $(6, 0)$

49. Passing through the points $(1, -1)$ and $(5, -1)$

50. Passing through the points $(2, 0)$ and $(2, -4)$

Write an equation of the form $y = mx + b$ for each line graphed below. [*Hint:* Either find the slope and y-intercept or use any two points on the line.]

51.
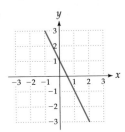

52.
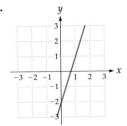

53.
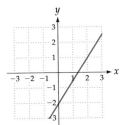

54.
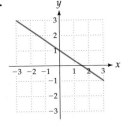

Find equations for the lines that make up the four-sided figures shown below.

55.
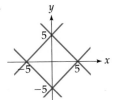

56.
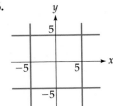

57. Show that $y - y_1 = m(x - x_1)$ simplifies to $y = mx + b$ if the point (x_1, y_1) is the y-intercept $(0, b)$.

58. Show that the linear equation $\frac{x}{a} + \frac{y}{b} = 1$ has x-intercept $(a, 0)$ and y-intercept $(0, b)$. (The x-*intercept* is the point where the line crosses the x-axis.)

59. Find the x-intercept $(a, 0)$ where the line $y = mx + b$ crosses the x-axis. Under what condition on m will a single x-intercept exist?

60. i. Show that the general linear equation $ax + by = c$ with $b \neq 0$ can be written as $y = -\frac{a}{b}x + \frac{c}{b}$, which is the equation of a line in slope–intercept form.

ii. Show that the general linear equation $ax + by = c$ with $b = 0$ but $a \neq 0$ can be written as $x = \frac{c}{a}$, which is the equation of a vertical line.

[*Note:* Because these steps are reversible, parts (i) and (ii) together show that the general linear equation $ax + by = c$ (for a and b not both zero) includes vertical and nonvertical lines.]

61. a. Graph the lines $y_1 = -x$, $y_2 = -2x$, and $y_3 = -3x$ on the window $[-5, 5]$ by $[-5, 5]$ (using the negation key $\boxed{(-)}$, not the *subtraction* key $\boxed{-}$). Observe how the coefficient of x changes the slope of the line.

b. Predict what the line $y = -9x$ would look like, and then check your prediction by graphing the line.

62. a. Graph the lines $y_1 = x + 2$, $y_2 = x + 1$, $y_3 = x$, $y_4 = x - 1$, and $y_5 = x - 2$ on the window $[-5, 5]$ by $[-5, 5]$. Observe how the constant changes the position of the line.

b. Predict what the lines $y = x + 4$ and $y = x - 4$ would look like, and then check your prediction by graphing them.

APPLIED EXERCISES

63. Business: Energy Usage A utility considers demand for electricity "low" if it is below 8 mkW (million kilowatts), "average" if it is at least 8 mkW but below 20 mkW, "high" if it is at least 20 mkW but below 40 mkW, and "critical" if it is 40 mkW or more. Express these demand levels in interval notation. [*Hint:* The interval for "low" is [0, 8).]

64. General: Grades If a grade of 90 through 100 is an A, at least 80 but less than 90 is a B, at least 70 but less than 80 a C, at least 60 but less than 70 a D, and below 60 an F, write these grade levels in interval form (ignoring rounding). [*Hint:* F would be [0, 60).]

65. General: Mile Run Read the Application Preview on pages 2–3.
a. Use the regression line $y = -0.356x + 257.44$ to predict the world record in the year 2010. [*Hint:* If x represents years after 1900, what value of x corresponds to the year 2010? The result will be in seconds and should be converted to minutes and seconds.]
b. According to this formula, when will the record be 3 minutes 30 seconds? [*Hint:* Set the formula equal to 210 seconds and solve. What year corresponds to this x-value?]

66. General: Mile Run Read the Application Preview on pages 2–3. Evaluate the regression line $y = -0.356x + 257.44$ at $x = 720$ and at $x = 724$ (corresponding to the years 2620 and 2624). Does the formula give reasonable times for the mile record in these years?

67. Business: Corporate Profit A company's profit increased linearly from $6 million at the end of year 1 to $14 million at the end of year 3.
a. Use the two (year, profit) data points (1, 6) and (3, 14) to find the linear relationship $y = mx + b$ between $x =$ year and $y =$ profit.
b. Find the company's profit at the end of 2 years.
c. Predict the company's profit at the end of 5 years.

68. Economics: Per Capita Personal Income In the short run, per capita personal income (PCPI) in the United States grows approximately linearly. In 1990 PCPI was 19.1, and in 1998 it had grown to 26.4 (both in thousands of dollars). (*Source: Bureau of Economic Analysis*)

a. Use the two (year, PCPI) data points (0, 19.1) and (8, 26.4) to find the linear relationship $y = mx + b$ between $x =$ years since 1990 and $y =$ PCPI.
b. Use your linear relationship to predict PCPI in 2010.

69. General: Temperature On the Fahrenheit temperature scale, water freezes at 32° and boils at 212°. On the Celsius (centigrade) scale, water freezes at 0° and boils at 100°.

a. Use the two (Celsius, Fahrenheit) data points (0, 32) and (100, 212) to find the linear relationship $y = mx + b$ between $x =$ Celsius temperature and $y =$ Fahrenheit temperature.
b. Find the Fahrenheit temperature that corresponds to 20° Celsius.

70. Ecology: Waste Disposal The amount of solid waste generated per person annually in the United States has increased approximately linearly, from 200 pounds in 1960 to 308 pounds in 1990.

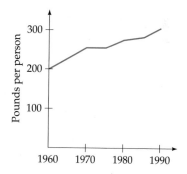

U.S. Solid Waste Generated Per Capita
(Source: U.S. Environmental Protection Agency)

a. Use the two (year, pounds) data points (0, 200) and (30, 308) to find the linear relationship $y = mx + b$ between $x =$ years since 1960 and $y =$ per capita waste.
b. Use your formula to predict the amount in the year 2010.

71–72: Business: Straight-Line Depreciation Straight-line depreciation is a method for estimating the value of an asset (such as a piece of machinery) as it loses value ("depreciates") through use. When we know the original *price* of an asset, its *useful lifetime*, and its *scrap* *value* (its value at the end of its useful lifetime), the value of the asset after t years is given by the formula

$$\text{Value} = (\text{price}) - \left(\frac{(\text{price}) - (\text{scrap value})}{(\text{useful lifetime})}\right) \cdot t$$

$$0 \leq t \leq (\text{useful lifetime})$$

71. a. A farmer buys a harvester for $50,000 and estimates its useful life to be 20 years, at the end of which its scrap value will be $6000. Use the given formula to find an expression for the value V of the harvester after t years, for $0 \leq t \leq 20$.
   b. Use your formula to find the value of the harvester after 5 years.
   c. Graph the function found in part (a) on a graphing calculator on the window [0, 20] by [0, 50,000]. [*Hint:* Use x instead of t.]

72. a. A newspaper buys a printing press for $800,000 and estimates its useful life to be 20 years, at the end of which its scrap value will be $60,000. Use the given formula to find an expression for the value V of the press after t years, for $0 \leq t \leq 20$.
   b. Use your formula to find the value of the press after 10 years.
   c. Graph the function found in part (a) on a graphing calculator on the window [0, 20] by [0, 800,000]. [*Hint:* Use x instead of t.]

73. Social Science: Age at First Marriage Americans are marrying later and later. Based on data from the U.S. Bureau of the Census for the years 1980 to 1998, the median age at first marriage for men is $y_1 = 24.8 + 0.13x$ and for women it is $y_2 = 22.1 + 0.17x$, where x is the number of years since 1980.
   a. Graph these lines on the window [0, 30] by [20, 30].
   b. Use these lines to predict the median marriage ages for men and women in the year 2005. [*Hint:* Which x-value corresponds to 2005? Then use TRACE, EVALUATE, or TABLE.]
   c. Predict the median marriage ages for men and women in the year 2010.

74. Social Science: Equal Pay for Equal Work Women's pay has often lagged behind men's, although Title VII of the Civil Rights Act requires equal pay for equal work. Based on data from 1980–1997, women's annual earning as a percentage of men's (for year round, full-time work) can be approximated by the formula $y = 0.794x + 60.9$, where x is the number of years since 1980. For example, $x = 10$ gives $y = 68.84$, so in 1990 women's wages were about 70% of men's wages. (*Source: U.S. Department of Labor, Women's Bureau*)
   a. Graph this line on the window [0, 30] by [0, 100].
   b. Use this line to predict the percentage in the year 2005. [*Hint:* Which x-value corresponds to 2005? Then use TRACE, EVALUATE, or TABLE.]
   c. Predict the percentage in the year 2010.

75–76: General: Life Expectancy The following tables give the life expectancy (years of life expected) for a newborn child born in the indicated year (Exercise 75 is for males, Exercise 76 for females). For each exercise:
   a. Enter the data into a graphing calculator and make a plot of the resulting points, with Years Since 1950 on the x-axis and Life Expectancy on the y-axis.
   b. Use the graphing calculator to find the linear regression line for these points. Enter the resulting formula as y_1, which then estimates life expectancy on the basis of the year of birth. Graph the points together with the regression line.
   c. Use your line y_1 to estimate the life expectancy of a child born in the year 2025. (This might be your child or grandchild.) [*Hint:* What x-value corresponds to 2025?]

75.

Birth Year (Years Since 1950)	Life Expectancy (Male)
1950	0	65.6
1960	10	66.6
1970	20	67.1
1980	30	70.0
1990	40	71.8

76.

Birth Year (Years Since 1950)	Life Expectancy (Female)
1950	0	71.1
1960	10	73.1
1970	20	74.7
1980	30	77.5
1990	40	78.8

1.2 Exponents

Size, Shape, and Exponents

The study of shape and size is called *allometry*, and many allometric relationships involve exponents that are fractions or decimals. For example, among all four-legged animals, from mice to elephants, (average) leg width and body length are governed (approximately) by the law

$$\left(\begin{array}{c}\text{Leg}\\\text{width}\end{array}\right) = (\text{constant}) \cdot \left(\begin{array}{c}\text{body}\\\text{length}\end{array}\right)^{3/2}$$

(hip to shoulder)

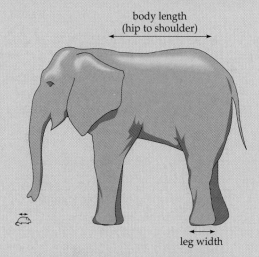

The constant depends on the units (inches, centimeters, etc.), but the exponent is always $\frac{3}{2}$. The precise meaning of the exponent $\frac{3}{2}$ will be explained later in this section, but the fact that it is greater than 1 means that for larger animals, leg width increases *faster* than body length. This is the reason why nature cannot build a land animal very much larger than an elephant—its oversized legs would get in each other's way. The preceding formula is an example of what is called the *law of simple allometry*, which states that two measurements x and y will be related by a *power law* of the form $y = a \cdot x^b$ for constants a and b.*

Another application of fractional exponents concerns maps and illustrations. Studies have shown that when people are asked to estimate the size of objects, the *perceived* size and the *actual* size are related by a power law

$$\left(\begin{matrix} \text{Perceived} \\ \text{size} \end{matrix}\right) = (\text{constant}) \cdot \left(\begin{matrix} \text{actual} \\ \text{size} \end{matrix}\right)^a$$

with exponent $a < 1$. The value of the exponent a depends on whether the object is one-, two-, or three-dimensional, but the exponent is always less than 1. This means that if you were using a geometric symbol on a map to indicate rainfall, to suggest that one country's rainfall was twice as great as another's, you should use a symbol *more* than twice as large. Many other uses of fractional exponents will be discussed in this section.

* For further information on allometry, see D'Arcy Wentworth Thompson, *On Growth and Form* (Cambridge University Press, 1942; Dover Publications, 1992); Stephen Jay Gould, "Allometry and Size in Ontogeny and Philogeny," *Biol. Rev.*, **41**:587–640, 1966; and Stefan Hildebrandt and Anthony Tromba, *The Parsimonious Universe* (Springer-Verlag, 1996)

Introduction

Not all variables are related linearly. In this section we will discuss exponents, which will enable us to express many nonlinear relationships.

Positive Integer Exponents

Numbers may be expressed with exponents, as in $2^3 = 2 \cdot 2 \cdot 2 = 8$. More generally, for any positive integer n, x^n means the product of n x's.

$$x^n = \overbrace{x \cdot x \cdots x}^{n}$$

The number being raised to the power is called the *base*: x^n

Exponent

Base

There are several *laws of exponents* for simplifying expressions. The first three are known, respectively, as the addition, subtraction, and multiplication laws of exponents.

Laws of Exponents

$x^m \cdot x^n = x^{m+n}$	To *multiply* powers of the same base, *add* the exponents
$\dfrac{x^m}{x^n} = x^{m-n}$	To *divide* powers of the same base, *subtract* the exponents (top exponent minus bottom exponent)
$(x^m)^n = x^{m \cdot n}$	To raise a power to a power, *multiply* the powers
$(xy)^n = x^n \cdot y^n$	To raise a product to a power, raise *each factor* to the power
$\left(\dfrac{x}{y}\right)^n = \dfrac{x^n}{y^n}$	To raise a fraction to a power, raise the numerator *and* denominator to the power

EXAMPLE 1   Simplifying Exponents

a. $x^2 \cdot x^3 = x^5$ $\overbrace{}^{2+3}$

b. $\dfrac{x^5}{x^3} = x^2$ $\overbrace{}^{5-3}$

c. $\left(x^2\right)^3 = x^6$ $\overbrace{}^{2 \cdot 3}$

d. $\dfrac{\left[(x^2)^3\right]^4}{x^5 \cdot x^7 \cdot x} = \dfrac{x^{24}}{x^{13}} = x^{11}$ $\overbrace{}^{2 \cdot 3 \cdot 4}$ $\overbrace{}^{24 - 13}$ $\underbrace{}_{5 + 7 + 1}$

e. $(2w)^3 = 2^3 w^3 = 8w^3$

f. $\left(\dfrac{x}{3}\right)^4 = \dfrac{x^4}{3^4} = \dfrac{x^4}{81}$

PRACTICE PROBLEM 1

Simplify: **a.** $\dfrac{x^5 \cdot x}{x^2}$ **b.** $\left[(x^3)^2\right]^2$ *Solutions at the back of the book*

Remember: For exponents in the form $x^2 \cdot x^3 = x^5$, *add* exponents.
For exponents in the form $(x^2)^3 = x^6$, *multiply* exponents.

Graphing Calculator Exploration

a. Use a graphing calculator to graph $y_1 = x$, $y_2 = x^2$, $y_3 = x^3$, and $y_4 = x^4$ on the viewing window $[0, 2]$ by $[0, 2]$. Use TRACE to identify which curve goes with which power.

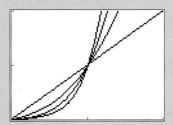

b. Which curve is highest for values of x between 0 and 1? Which is lowest?

c. Which curve is highest for values of x greater than 1? Which is lowest?

d. Predict what the curve $y = x^5$ will look like. Check your prediction by graphing it.

e. Predict which of these curves will be positive when x is negative. Check your prediction by changing the viewing window to $[-2, 2]$ by $[-2, 2]$.

Zero and Negative Exponents

For any number x other than zero, we define

$x^0 = 1$	x to the power zero is one
$x^{-1} = \dfrac{1}{x}$	x to the power -1 is one over x
$x^{-2} = \dfrac{1}{x^2}$	x to the power -2 is one over x squared
$x^{-n} = \dfrac{1}{x^n}$	x to a negative power is one over x to the positive power

EXAMPLE 2   **Simplifying Zero and Negative Exponents**

a. $5^0 = 1$

b. $7^{-1} = \dfrac{1}{7}$

c. $3^{-2} = \dfrac{1}{3^2} = \dfrac{1}{9}$

d. $(-2)^{-3} = \dfrac{1}{(-2)^3} = \dfrac{1}{-8} = -\dfrac{1}{8}$

e. 0^0 and 0^{-3} are undefined.

■

PRACTICE PROBLEM 2

Evaluate: **a.** 2^0 **b.** 2^{-4} *Solutions at the back of the book*

The definitions of x^0 and x^{-n} are motivated by the following calculations.

$$1 = \frac{x^2}{x^2} = x^{2-2} = x^0$$

The subtraction law of exponents leads to $x^0 = 1$

$$\frac{1}{x^n} = \frac{x^0}{x^n} = x^{0-n} = x^{-n}$$

$x^0 = 1$ and the subtraction law of exponents lead to $x^{-n} = 1/x^n$

A fraction to a negative power means *division* by the fraction, so we "invert and multiply."

$$\left(\frac{x}{y}\right)^{-1} = \frac{1}{\dfrac{x}{y}} = 1 \cdot \frac{y}{x} = \frac{y}{x}$$

└── Reciprocal of the original fraction

Therefore, for $x \neq 0$ and $y \neq 0$,

$$\left(\frac{x}{y}\right)^{-1} = \frac{y}{x}$$

A fraction to the power -1 is the reciprocal of the fraction

$$\left(\frac{x}{y}\right)^{-n} = \left(\frac{y}{x}\right)^n$$

A fraction to the negative power is the reciprocal of the fraction to the positive power

EXAMPLE 3   **Simplifying Fractions to Negative Exponents**

a. $\left(\dfrac{3}{2}\right)^{-1} = \dfrac{2}{3}$

b. $\left(\dfrac{1}{2}\right)^{-3} = \left(\dfrac{2}{1}\right)^3 = \dfrac{2^3}{1^3} = 8$

↑
Reciprocal of $\dfrac{3}{2}$

■

PRACTICE PROBLEM 3   Simplify: $\left(\dfrac{2}{3}\right)^{-2}$ *Solution at the back of the book*

Roots and Fractional Exponents

We may take the square root of any *nonnegative* number, and the cube root of *any* number.

EXAMPLE 4   **Evaluating Roots**

a. $\sqrt{9} = 3$ **b.** $\sqrt{-9}$ is undefined.   Square roots of negative numbers are not defined

c. $\sqrt[3]{8} = 2$ **d.** $\sqrt[3]{-8} = -2$ Cube roots of negative numbers *are* defined

e. $\sqrt[3]{\dfrac{27}{8}} = \dfrac{\sqrt[3]{27}}{\sqrt[3]{8}} = \dfrac{3}{2}$

There are *two* square roots of 9, namely 3 and -3, but the radical sign $\sqrt{}$ means just the *positive* one (the "principal" square root).

$\sqrt[n]{a}$ means the principal *n*th root of *a*.   Principal means the positive root if there are two

---

In general, we may take *odd* roots of *any* number, but *even* roots only if the number is positive or zero.

EXAMPLE 5   **Evaluating Roots of Positive and Negative Numbers**

Odd roots of negative numbers are defined

a. $\sqrt[4]{81} = 3$ **b.** $\sqrt[5]{-32} = -2$ Since $(-2)^5 = -32$

Graphing Calculator Exploration

a. Use a graphing calculator to graph $y_1 = x$, $y_2 = \sqrt{x}$, $y_3 = \sqrt[3]{x}$, and $y_4 = \sqrt[4]{x}$ simultaneously on the viewing window $[0, 3]$ by $[0, 2]$. Use TRACE to identify which curve goes with which root.

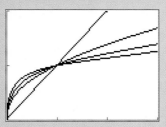

b. Which curve is highest for values of x between 0 and 1? Which is lowest?

c. Which curve is highest for values of x greater than 1? Which is lowest?

d. Predict what the curve $y = \sqrt[7]{x}$ would look like. Check your prediction by graphing it.

e. Which of these roots are defined for *negative* values of x? Check your answer by changing the window to $[-3, 3]$ by $[-2, 2]$ and using TRACE where x is negative.

Fractional Exponents

Fractional exponents are defined as follows:

$$x^{\frac{1}{2}} = \sqrt{x} \qquad \text{Power } \tfrac{1}{2} \text{ means the principal square root}$$

$$x^{\frac{1}{3}} = \sqrt[3]{x} \qquad \text{Power } \tfrac{1}{3} \text{ means the cube root}$$

$$x^{\frac{1}{n}} = \sqrt[n]{x} \qquad \text{Power } \tfrac{1}{n} \text{ means the principal } n\text{th root (for a positive integer } n)$$

EXAMPLE 6   **Evaluating Fractional Exponents**

a. $9^{\frac{1}{2}} = \sqrt{9} = 3$ **b.** $125^{\frac{1}{3}} = \sqrt[3]{125} = 5$

c. $81^{\frac{1}{4}} = \sqrt[4]{81} = 3$ **d.** $(-32)^{\frac{1}{5}} = \sqrt[5]{-32} = -2$

e. $\left(-\dfrac{27}{8}\right)^{\frac{1}{3}} = \sqrt[3]{-\dfrac{27}{8}} = -\dfrac{\sqrt[3]{27}}{\sqrt[3]{8}} = -\dfrac{3}{2}$ ∎

PRACTICE PROBLEM 4   Evaluate: **a.** $(-27)^{\frac{1}{3}}$ **b.** $\left(\dfrac{16}{81}\right)^{\frac{1}{4}}$ *Solutions at the back of the book*

The definition of $x^{1/2}$ is motivated by the multiplication law of exponents:

$$\left(x^{\frac{1}{2}}\right)^{2} = x^{\frac{1}{2}\cdot 2} = x^{1} = x$$

Taking square roots of each side of $\left(x^{\frac{1}{2}}\right)^{2} = x$ gives

$$x^{\frac{1}{2}} = \sqrt{x} \qquad\qquad \text{\small \(x\) to the half power means}$$
$$\text{\small the square root of \(x\)}$$

To define $x^{\frac{m}{n}}$ for positive integers m and n, we must fully reduce the exponent $\frac{m}{n}$ (for example, reduce $\frac{4}{6}$ to $\frac{2}{3}$). Then

$$x^{\frac{m}{n}} = \left(x^{\frac{1}{n}}\right)^{m} = (x^{m})^{\frac{1}{n}} \qquad\qquad \text{\small Since in both cases the exponents}$$
$$\text{\small multiply to \(\frac{m}{n}\)}$$

Therefore we define fractional exponents as follows:

Fractional Exponents

$$x^{\frac{m}{n}} = \left(\sqrt[n]{x}\right)^{m} = \sqrt[n]{x^{m}} \qquad \text{\small \(x^{m/n}\) means the \(m\)th power of the}$$
$$\text{\small \(n\)th root or, equivalently, the \(n\)th root}$$
$$\text{\small of the \(m\)th power}$$

Both expressions, $\left(\sqrt[n]{x}\right)^{m}$ and $\sqrt[n]{x^{m}}$, will give the same answer. In either case, the numerator determines the power and the denominator determines the root.

$$x^{\frac{m}{n}} \begin{array}{l} \nearrow \text{Power} \\ \searrow \text{Root} \end{array} \qquad\qquad \text{Power over root}$$

EXAMPLE 7 **Evaluating Fractional Exponents**

a. $8^{2/3} = \sqrt[3]{8^2} = \sqrt[3]{64} = 4$ First the power, then the root

> same

b. $8^{2/3} = \left(\sqrt[3]{8}\right)^2 = (2)^2 = 4$ First the root, then the power

c. $25^{3/2} = \left(\sqrt{25}\right)^3 = (5)^3 = 125$

d. $\left(\dfrac{-27}{8}\right)^{2/3} = \left(\sqrt[3]{\dfrac{-27}{8}}\right)^2 = \left(\dfrac{-3}{2}\right)^2 = \dfrac{9}{4}$

■

PRACTICE PROBLEM 5   Evaluate: **a.** $16^{3/2}$ **b.** $(-8)^{2/3}$ *Solutions at the back of the book*

Graphing Calculator Exploration

a. Use a graphing calculator to evaluate $25^{3/2}$. [On some calculators, press 25^(3 ÷ 2).] Your answer should agree with Exercise 7c above.

b. Evaluate $(-8)^{2/3}$. Use the $\boxed{(-)}$ key for negation, and place parentheses around the exponent. Your answer should be 4. If you get an "error," try evaluating the expression as $[(-8)^{1/3}]^2$ or $[(-8)^2]^{1/3}$. Whichever way works, remember it for evaluating negative numbers to fractional powers in the future.

EXAMPLE 8 **Evaluating Negative Fractional Exponents**

a. $8^{-2/3} = \dfrac{1}{8^{2/3}} = \dfrac{1}{\left(\sqrt[3]{8}\right)^2} = \dfrac{1}{2^2} = \dfrac{1}{4}$ A negative exponent means the reciprocal of the number to the positive exponent, which is then evaluated as before

b. $\left(\dfrac{9}{4}\right)^{-3/2} = \left(\dfrac{4}{9}\right)^{3/2} = \left(\sqrt{\dfrac{4}{9}}\right)^3 = \left(\dfrac{2}{3}\right)^3 = \dfrac{8}{27}$

———— Interpreting the power 3/2
———— Reciprocal to the positive exponent
———— Negative exponent

■

PRACTICE PROBLEM 6

Evaluate: **a.** $25^{-3/2}$ **b.** $\left(\dfrac{1}{4}\right)^{-1/2}$ **c.** $5^{1.3}$ [*Hint:* Use 📱.]

Solutions at the back of the book

Avoiding Pitfalls in Simplifying

The square root of a product is equal to the product of the square roots:

$$\sqrt{a \cdot b} = \sqrt{a} \cdot \sqrt{b}$$

However, the corresponding statement for *sums* is *not* true:

$$\sqrt{a + b} \quad \text{is } not \text{ equal to} \quad \sqrt{a} + \sqrt{b}$$

For example,

$$\underbrace{\sqrt{9 + 16}}_{\sqrt{25}} \neq \underbrace{\sqrt{9}}_{3} + \underbrace{\sqrt{16}}_{4}$$

The two sides are not equal: one is 5 and the other is 7

Therefore, do not "simplify" $\sqrt{x^2 + 9}$ into $x + 3$. The expression $\sqrt{x^2 + 9}$ *cannot be simplified.* Similarly,

$$(x + y)^2 \quad \text{is } not \text{ equal to} \quad x^2 + y^2$$

The expression $(x + y)^2$ means $(x + y)$ times itself:

$$(x + y)^2 = (x + y)(x + y) = x^2 + xy + yx + y^2 = x^2 + 2xy + y^2$$

This result will be useful in later sections.

$$(x + y)^2 = x^2 + 2xy + y^2$$

$(x + y)^2$ is the first number squared plus twice the product of the numbers plus the second number squared

---

Learning Curves in Airplane Production

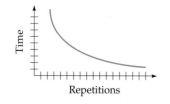

Time / Repetitions

It is a truism that the more you practice a task, the faster you can do it. Successive repetitions generally take less time, following a "learning curve" like that on the left. Learning curves are used in industrial production. For example, it took 150,000 work-hours to build the first Boeing 707 airliner, while later planes ($n = 2, 3, \ldots, 300$) took less time.*

* A work-hour is the amount of work that a person can do in one hour. For further information on learning curves in industrial production, see J. M. Dutton et al., "The History of Progress Functions as a Managerial Technology," Bus. Hist. Rev., 58:204–233, 1984.

$$\begin{pmatrix} \text{Time to build} \\ \text{plane number } n \end{pmatrix} = 150\,n^{-0.322} \qquad \text{thousand work-hours}$$

The time for the tenth Boeing 707 is found by substituting $n = 10$:

$$\begin{pmatrix} \text{Time to build} \\ \text{plane 10} \end{pmatrix} = 150(10)^{-0.322}$$
$$\approx 71.46 \text{ thousand work-hours}$$

$150n^{-0.322}$ with
$n = 10$; using
a calculator

This shows that building the tenth Boeing 707 took about 71,460 work-hours, which is less than half of the 150,000 work-hours needed for the first. For the 100th Boeing 707,

$$\begin{pmatrix} \text{Time to build} \\ \text{plane 100} \end{pmatrix} = 150(100)^{-0.322}$$
$$\approx 34.05 \text{ thousand work-hours}$$

$150n^{-0.322}$ with
$n = 100$

or about 34,050 work-hours, which is less than half the time needed to build the tenth. Such learning curves are used for determining the cost of a contract to build several planes.

Notice that the learning curve graphed on the previous page decreases *less steeply* as the number of repetitions increases. This means that although construction time continues to decrease, it does so more slowly for later planes. This behavior, called *diminishing returns*, is typical of learning curves.

Section Summary

We defined zero, negative, and fractional exponents as follows:

$$x^0 = 1 \qquad\qquad \text{for } x \neq 0$$

$$x^{-n} = \frac{1}{x^n} \qquad\qquad \text{for } x \neq 0$$

$$x^{\frac{m}{n}} = \left(\sqrt[n]{x}\right)^m = \sqrt[n]{x^m} \qquad m > 0, \quad n > 0, \quad \frac{m}{n} \text{ fully reduced}$$

With these definitions, the following laws of exponents hold for *all* exponents, whether integral or fractional, positive or negative.

$$x^m \cdot x^n = x^{m+n} \qquad (x^m)^n = x^{m\cdot n} \qquad \left(\frac{x}{y}\right)^n = \frac{x^n}{y^n}$$

$$\frac{x^m}{x^n} = x^{m-n} \qquad\qquad (xy)^n = x^n \cdot y^n$$

EXERCISES 1.2

Evaluate each expression *without* using a calculator:

1. $(2^2 \cdot 2)^2$

2. $(5^2 \cdot 4)^2$

3. 2^{-4}

4. 3^{-3}

5. $\left(\dfrac{1}{2}\right)^{-3}$

6. $\left(\dfrac{1}{3}\right)^{-2}$

7. $\left(\dfrac{5}{8}\right)^{-1}$

8. $\left(\dfrac{3}{4}\right)^{-1}$

9. $4^{-2} \cdot 2^{-1}$

10. $3^{-2} \cdot 9^{-1}$

11. $\left(\dfrac{3}{2}\right)^{-3}$

12. $\left(\dfrac{2}{3}\right)^{-3}$

13. $\left(\dfrac{1}{3}\right)^{-2} - \left(\dfrac{1}{2}\right)^{-3}$

14. $\left(\dfrac{1}{3}\right)^{-2} - \left(\dfrac{1}{2}\right)^{-2}$

15. $\left[\left(\dfrac{2}{3}\right)^{-2}\right]^{-1}$

16. $\left[\left(\dfrac{2}{5}\right)^{-2}\right]^{-1}$

17. $25^{1/2}$

18. $36^{1/2}$

19. $25^{3/2}$

20. $16^{3/2}$

21. $16^{3/4}$

22. $27^{2/3}$

23. $(-8)^{2/3}$

24. $(-27)^{2/3}$

25. $(-8)^{5/3}$

26. $(-27)^{5/3}$

27. $\left(\dfrac{25}{36}\right)^{3/2}$

28. $\left(\dfrac{16}{25}\right)^{3/2}$

29. $\left(\dfrac{27}{125}\right)^{2/3}$

30. $\left(\dfrac{125}{8}\right)^{2/3}$

31. $\left(\dfrac{1}{32}\right)^{2/5}$

32. $\left(\dfrac{1}{32}\right)^{3/5}$

33. $4^{-1/2}$

34. $9^{-1/2}$

35. $4^{-3/2}$

36. $9^{-3/2}$

37. $8^{-2/3}$

38. $16^{-3/4}$

39. $(-8)^{-1/3}$

40. $(-27)^{-1/3}$

41. $(-8)^{-2/3}$

42. $(-27)^{-2/3}$

43. $\left(\dfrac{25}{16}\right)^{-1/2}$

44. $\left(\dfrac{16}{9}\right)^{-1/2}$

45. $\left(\dfrac{25}{16}\right)^{-3/2}$

46. $\left(\dfrac{16}{9}\right)^{-3/2}$

47. $\left(-\dfrac{1}{27}\right)^{-5/3}$

48. $\left(-\dfrac{1}{8}\right)^{-5/3}$

Use a calculator to evaluate each expression. Round answers to 2 decimal places.

49. $7^{0.39}$

50. $5^{0.47}$

51. $8^{2.7}$

52. $5^{3.9}$

Use a graphing calculator to evaluate each expression.

53. $(-8)^{7/3}$

54. $(-8)^{5/3}$

55. $[(5/2)^{-1}]^{-2}$

56. $[(3/2)^{-2}]^{-1}$

57. $[(4)^{-1}]^{0.5}$

58. $[(0.25)^{-1}]^{0.5}$

59. $(0.4^{-7})^{-1/7}$

60. $[(0.5^{-1})^{-2}]^{-3}$

61. $[(0.1)^{0.1}]^{0.1}$

62. $\left(1 + \dfrac{1}{1000}\right)^{1000}$

63. $\left(1 - \dfrac{1}{1000}\right)^{-1000}$

64. $(1 + 10^{-6})^{10^6}$

Simplify.

65. $(x^3 \cdot x^2)^2$

66. $(x^4 \cdot x^3)^2$

67. $[z^2(z \cdot z^2)^2 z]^3$

68. $[z(z^3 \cdot z)^2 z^2]^2$

69. $[(x^2)^2]^2$

70. $[(x^3)^3]^3$

71. $\dfrac{(ww^2)^3}{w^3 w}$

72. $\dfrac{(ww^3)^2}{w^3 w^2}$

73. $\dfrac{(5xy^4)^2}{25x^3y^3}$

74. $\dfrac{(4x^3y)^2}{8x^2y^3}$

75. $\dfrac{(9xy^3z)^2}{3(xyz)^2}$

76. $\dfrac{(5x^2y^3z)^2}{5(xyz)^2}$

77. $\dfrac{(2u^2vw^3)^2}{4(uw^2)^2}$

78. $\dfrac{(u^3vw^2)^2}{9(u^2w)^2}$

79–80: Allometry: Dinosaurs The body measurements of most four-legged animals, from mice to elephants, obey (approximately) the following power law:

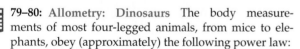

$$\left(\begin{array}{c}\text{Average body}\\ \text{thickness}\end{array}\right) = 0.4(\text{hip-to-shoulder length})^{3/2}$$

where body thickness is measured vertically and all measurements are in feet. Assuming that this same relationship held for dinosaurs, find the average body

thickness of the following dinosaurs, whose hip-to-shoulder lengths were measured from their skeletons:

79. Diplodocus, whose hip-to-shoulder length was 16 ft.

80. Triceratops, whose hip-to-shoulder length was 14 ft.

81–82: Business: The Rule of 0.6 in Industrial Production Many chemical and refining companies use "the Rule of Point Six" to estimate the cost of new equipment. According to this rule, if a piece of equipment (such as a storage tank) originally cost C dollars, then the cost of similar equipment that is x times as large will be approximately $x^{0.6}C$ dollars. For example, if the original equipment cost C dollars, then new equipment with twice the capacity ($x = 2$) costs $2^{0.6}C = 1.516C$ dollars—that is, about 1.5 times as much. Therefore, to increase capacity by 100% costs only about 50% more.*

81. Use the Rule of 0.6 to estimate how costs will change when a company quadruples ($x = 4$) its capacity.

82. Use the Rule of 0.6 to estimate how costs will change when a company triples ($x = 3$) its capacity.

83–84: Business: Continuation Use a graphing calculator to graph $y = x^{0.6}$, expressing y, the cost multiple for larger equipment, in terms of x, the size multiple. Use the viewing window [0, 5] by [0, 3].

83. By how much can a company multiply its capacity for twice the money? That is, find the value of x that satisfies $x^{0.6} = 2$. [*Hint:* Either use TRACE or find where $y_1 = x^{0.6}$ INTERSECTs $y_2 = 2$.]

84. Does the curve rise more steeply or less steeply as x increases? What does this mean about how rapidly cost increases as equipment size increases?

85–86: Allometry: Heart Rate It is well known that the hearts of smaller animals beat faster than the hearts of larger animals. The actual relationship is approximately

$$(\text{Heart rate}) = 250(\text{weight})^{-1/4}$$

where the heart rate is in beats per minute and the weight is in pounds. Use this relationship to estimate the heart rate of:

85. A 16-pound dog

86. A 625-pound grizzly bear

87–88: Allometry: Continuation Use a graphing calculator to graph $y = 250x^{-0.25}$, which expresses y, heartbeats per minute, in terms of x, the animal's weight in pounds. Use the viewing window [0, 200] by [0, 150].

87. Notice that the curve decreases less steeply for larger values of x. Explain what this means about how rapidly heart rate decreases as body weight increases.

88. Evaluate this formula at your own weight, x, to find your predicted heart rate. Then take your pulse and see if the numbers (roughly) agree.

89–90: Business and Psychology: Learning Curves in Airplane Production Recall (pages 26–27) that the learning curve for the production of Boeing 707 airplanes is $150x^{-0.322}$ (thousand work-hours). Find how many work-hours it took to build:

89. The 50th Boeing 707

90. The 250th Boeing 707

91. General: Richter Scale The Richter scale (developed by Charles Richter in 1935) is widely used to measure the strength of earthquakes. Every increase of one on the Richter scale corresponds to a tenfold increase in ground motion. Therefore, an increase on the Richter scale from A to B means that ground motion increases by a factor of 10^{B-A} (for $B > A$). Find the increase in ground motion between the following earthquakes:

a. The 1994 Northridge, California, earthquake, measuring 6.8 on the Richter scale, and the 1906 San Francisco earthquake, measuring 8.3. (The San Francisco earthquake resulted in 500 deaths and a 3-day fire that destroyed 4 square miles of San Francisco.)

b. The 1995 Kobe (Japan) earthquake, measuring 7.2 on the Richter scale, and the 1933 Miyagi earthquake, measuring 8.1. (The Miyagi earthquake caused a 90-foot-high tsunami, or "tidal wave," that killed 3064 people. The death toll in the Kobe earthquake was more than 5000.)

*Although the Rule of 0.6 is only a rough "rule of thumb," it can be somewhat justified on the basis that the equipment of such industries consists mainly of containers, and the cost of a container depends on its surface area (square units), which increases more slowly than its capacity (cubic units).

92. Continuation Every increase of one on the Richter scale corresponds to an approximately *thirtyfold* increase in *energy released*. Therefore, an increase on the Richter scale from A to B means that the energy released increases by a factor of 30^{B-A} (for $B > A$).

a. Find the increase in *energy released* between the earthquakes in Exercise 91a.

b. Find the increase in *energy released* between the earthquakes in Exercise 91b.

93–94: General: Waterfalls Water falling from a waterfall that is x feet high will hit the ground with speed $\frac{60}{11}x^{0.5}$ miles per hour (neglecting air resistance).

93. Find the speed of the water at the bottom of the highest waterfall in the world, Angel Falls in Venezuela (3281 feet high).

94. Find the speed of the water at the bottom of the highest waterfall in the United States, Ribbon Falls in Yosemite, California (1650 feet high).

95–96: Environmental Science: Biodiversity Larger land areas can support larger numbers of species. According to one study,* multiplying the land area by a factor of x multiplies the number of species by a factor of $x^{0.239}$. Use a graphing calculator to graph $y = x^{0.239}$. Use the viewing window [0, 100] by [0, 4].

95. Find the multiple x for the land area that leads to *double* the number of species. That is, find the value of x such that $x^{0.239} = 2$. [*Hint:* Either use TRACE or find where $y_1 = x^{0.239}$ INTERSECTs $y_2 = 2$.]

96. Find the multiple x for the land area that leads to triple the number of species. That is, find the value of x such that $x^{0.239} = 3$. [*Hint:* Either use TRACE or find where $y_1 = x^{0.239}$ INTERSECTs $y_2 = 3$.]

*The Theory of Island Biogeography, by Robert H. MacArthur and Edward O. Wilson (Princeton University Press, 1967).

97. Business: Learning Curves A manufacturer of supercomputers finds that the numbers of work-hours required to build the first, the tenth, the twentieth, and the thirtieth supercomputers are as follows:

Supercomputer Number	Work-Hours Required
1	3200
10	1900
20	1400
30	1300

a. Enter these numbers into a graphing calculator and make a plot of the resulting points (Supercomputer Number on the x axis and Work-Hours Required on the y axis).

b. Have the calculator find the power regression formula for these data, fitting a curve of the form $y = ax^b$ to the points. Enter the results as y_1. Plot the points together with the regression line. Observe that the line fits the points rather well.

c. Evaluate y_1 at $x = 50$ to predict the number of work-hours required to build the fiftieth supercomputer.

98. General: Paper Stacking Suppose that you take an ordinary piece of paper (about $\frac{1}{250}$ of an inch thick, cut it in half, stack the two halves, and repeat this cutting and stacking operation many times. Each operation doubles the height of the stack, so that after a total of 25 such operations, the stack will be $2^{25} \cdot \dfrac{1}{250} \cdot \dfrac{1}{12} \cdot \dfrac{1}{5280}$ miles high.

a. Evaluate this height.

b. Use a graphing calculator to make a TABLE showing the height of the stack when the number of operations is 15 or more. [*Hint:* Use $y_1 = 2^x/(250 \cdot 12 \cdot 5280)$ for values of x beginning at 15.] For what number of operations is the stack over a mile high? over 10 miles high? over 100 miles high?

1.3 Functions

Functional Drunkenness

When you drink liquor, the alcohol begins to enter your blood-stream almost immediately and your blood-alcohol level rises rapidly. After you stop drinking, your natural metabolic processes slowly eliminate the alcohol and your blood-alcohol level begins to fall. How long must you wait after drinking before you can drive safely?

Experiments show that after drinking 6 beers, 4 glasses of wine, or 4 shots of liquor in an hour, your blood-alcohol level typically rises to 0.11 gram per deciliter of blood. Thereafter, alcohol is eliminated at the rate of 0.02 gram per hour. (Many substances are eliminated from the bloodstream *non*linearly, but alcohol is elimi-nated *linearly*.) The following graph shows the resulting blood-al-cohol level over time, first rising rapidly and then falling linearly. By calculating where the graph falls below the legal limit (0.08 in most states), you can find how long you must wait—about an hour and a half. Incidentally, coffee and exercise have no effect; only time helps.

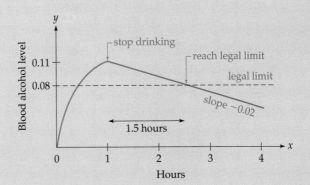

Mathematical relationships like this, in which each *x* (time, in hours) gives exactly one *y* (blood-alcohol level) are called *functions*.

Introduction

In the previous section we saw that the time required to build a Boeing 707 airliner depends on the number that have already been built. Mathematical relationships like this, in which one number depends on another, are called *functions*, and are central to the study of mathematics. In this section we define and give some applications of functions.

Functions

A *function* is a rule or procedure for finding, from a given number, a new number.* If the function is denoted by f and the given number by x, then the resulting number is written $f(x)$ (read "f of x") and is called *the value of the function f at x.* The set of numbers x for which a function f is defined is called the *domain* of f, and the set of all function values $f(x)$ is called the *range* of f. For any x in the domain, $f(x)$ must be a *single* number. Formally,

Function

A *function* f is a rule that assigns to each number x in a set D exactly one number $f(x)$. The set D is called the *domain*, and the set of all values $f(x)$ for x in the domain is called the *range*.

For example, recording the temperature at a given location throughout a particular day would define a *temperature* function:

$$f(x) = \begin{pmatrix} \text{Temperature at} \\ \text{time } x \text{ hours} \end{pmatrix} \qquad \text{Domain would be } [0, 24]$$

A function f may be thought of as a numerical procedure or "machine" that takes an "input" number x and produces an "output" number $f(x)$. The permissible input numbers form the domain, and the resulting output numbers form the range.

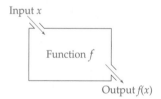

Input x

Function f

Output $f(x)$

We will be mostly concerned with functions that are defined by *formulas* for calculating $f(x)$ from x. If the domain of such a function is

---

* In this chapter the word "function" will mean *function of one variable.* In Chapter 4 we will discuss functions of more than one variable.

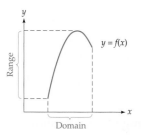

Range

$y = f(x)$

Domain

The domain of a function $y = f(x)$ is the set of all possible x-values, and the range is the set of all corresponding y-values.

not stated, then it is always taken to be the *largest* set of numbers for which the function is defined, called the *natural domain* of the function. To *graph* a function f, we plot all points (x, y) such that x is in the domain and $y = f(x)$. We call x the *independent variable* and y the *dependent variable*, because y *depends on* (is calculated from) x. The domain and range can be illustrated graphically, as shown on the left.

PRACTICE PROBLEM 1

Find the domain and range of the function graphed on the right.
Solution at the back of the book

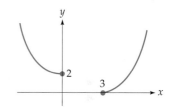

EXAMPLE 1 **Finding the Domain and Range of a Rational Function**

For the function $f(x) = \dfrac{1}{x - 1}$, find:

a. $f(5)$, **b.** the domain, **c.** the range.

Solution

a. $f(5) = \dfrac{1}{5 - 1} = \dfrac{1}{4}$

b. Domain $= \{x \mid x \neq 1\}$

c. The graph of the function (from a graphing calculator) is shown on the right. From it, and realizing that the curve continues upward and downward (as may be verified by zooming out), it is clear that every y-value is taken on except for $y = 0$ (because the curve does not touch the x-axis). Therefore,

Range $= \{y \mid y \neq 0\}$

$f(x) = \dfrac{1}{x\,2\,1}$ with $x = 5$

$f(x) = \dfrac{1}{x\,2\,1}$ is defined for all x *except* $x = 1$

$f(x) = \dfrac{1}{x - 1}$ on $[-5, 5]$ by $[-5, 5]$

May also be written $\{z \mid z \neq 0\}$ or with any other letter

The range could also be found by solving $y = \dfrac{1}{x-1}$ for x to find

$x = \dfrac{1}{y} + 1$, again showing that all values of y are possible except $y = 0$.

EXAMPLE 2 **Finding the Domain and Range of a Polynomial**

For $f(x) = 2x^2 + 4x - 5$, find:

a. $f(-3)$,   **b.** the domain,   **c.** the range.

Solution

a. $f(-3) = 2(-3)^2 + 4 \cdot (-3) - 5$
$\qquad = 18 - 12 - 5 = 1$

$f(x) = 2x^2 + 4x - 5$ with each x replaced by -3

b. Domain $= \mathbb{R}$

$2x^2 + 4x - 5$ is defined for *all* real numbers

c. From the graph of $f(x) = 2x^2 + 4x - 5$ on the right, the lowest y-value is -7 (as can be found from TRACE or MIN), and all higher y-values are taken on (because the curve is a parabola opening upward). Therefore:

$$\text{Range} = \{y \mid y \geq -7\}$$

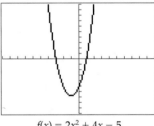

$f(x) = 2x^2 + 4x - 5$
on $[-10, 10]$ by $[-10, 10]$

Any letters may be used for defining a function or describing the domain and the range.

PRACTICE PROBLEM 2

For $g(z) = \sqrt{z - 2}$, find: **a.** $g(27)$,   **b.** the domain,   **c.** the range.

Solutions at the back of the book

For each x in the domain of a function, there must be a *single* number $y = f(x)$, and so the graph of a function cannot have two points (x, y) with the same x-value but different y-values. This leads to the following *graphical* test for functions.

Vertical Line Test for Functions

> A curve in the Cartesian plane is the graph of a *function* if and only if every vertical line intersects the curve at most once.

EXAMPLE 3 **Using the Vertical Line Test**

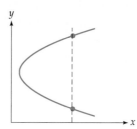

This is *not* the graph of a function of x because there is a vertical line (shown dashed) that intersects the curve twice.

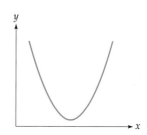

This *is* the graph of a function of x because no vertical line intersects the curve more than once.

A graph that has two or more points (x, y) with the same x-value but different y-values, like the one on the left above, defines a *relation* rather than a function. We will be concerned exclusively with *functions,* and so we will use the terms "function," "graph," and "curve" interchangeably.

Linear Function

A *linear function* is a function that can be expressed in the form

$$f(x) = mx + b$$

with constants m and b. Its graph is a line with slope m and y-intercept b.

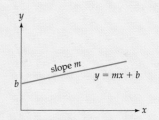

EXAMPLE 4 **Finding a Company's Cost Function**

An electronics company manufactures pocket calculators at a cost of $9 each, and the company's fixed costs (such as rent) amount to $400 per day. Find a function $C(x)$ that gives the total cost of producing x pocket calculators in a day.

Solution

Each calculator costs $9 to produce, and so x calculators will cost $9x$ dollars, to which we must add the fixed costs of $400.

$$C(x) \quad = \quad 9x \quad + \quad 400$$

Total cost   Unit cost Number of units Fixed cost

The graph of $C(x) = 9x + 400$ is a line with slope 9 and y-intercept 400, as shown below. Notice that the *slope* is the same as the *rate of change* of cost (costs increase at the rate of $9 per additional calculator), which is also the company's *marginal cost* (the cost of producing one more calculator is $9). The *slope*, the *rate of change*, and the *marginal cost* are always the same for a linear function.

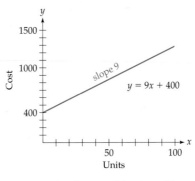

Cost function $C(x) = 9x + 400$

PRACTICE PROBLEM 3

A trucking company will deliver furniture for a charge of $25 plus 5% of the purchase price of the furniture. Find a function $D(x)$ that gives the delivery charge for a piece of furniture that cost x dollars.

Solution at the back of the book

A mathematical description of a real-world situation is called a *mathematical model*. For instance, the cost function $C(x) = 9x + 400$ from Example 4 is a mathematical model for the cost of manufacturing calculators. In this model, x, the number of calculators, should take only whole-number values (0, 1, 2, 3, . . .), and the graph should consist of discrete dots rather than a continuous curve. Instead, we will find it easier to let x take *continuous* values and to round up or down as necessary at the end.

Quadratic Function

A *quadratic function* is a function that can be expressed in the form

$$f(x) = ax^2 + bx + c$$

with constants ("coefficients") $a \neq 0$, b, and c. Its graph is called a *parabola*.

The condition $a \neq 0$ keeps the function from becoming $f(x) = bx + c$, which would be linear. Many familiar curves are parabolas.

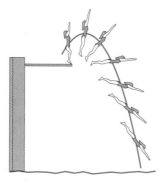

The center of gravity of a diver describes a parabola.

A stream of water from a hose takes the shape of a parabola.

The parabola $f(x) = ax^2 + bx + c$ opens *upward* if the constant a is *positive* and opens *downward* if the constant a is *negative*. The *vertex* of a parabola is its "central" point. The vertex is the *lowest* point on the parabola if it opens *up*, and the *highest* point if it opens *down*.

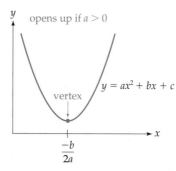

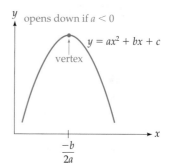

Graphing Calculator Exploration

a. Graph the parabolas $y_1 = x^2$, $y_2 = 2x^2$, and $y_3 = 4x^2$ on the graphing window $[-10, 10]$ by $[-10, 10]$. Use TRACE to identify which curve goes with which formula. How does the shape of the parabola change when the coefficient of x^2 increases?

b. Graph $y_4 = -x^2$. What did the negative sign do to the parabola?

c. Predict the shape of the parabolas $y_5 = -2x^2$ and $y_6 = \frac{1}{3}x^2$. Then check your predictions by graphing the functions.

The x-coordinate of the vertex of a parabola may be found by a formula, which will be derived in Exercises 94–96 on page 49.

Vertex Formula for a Parabola

The x-coordinate of the vertex of the parabola $f(x) = ax^2 + bx + c$ is

$$x = \frac{-b}{2a}$$

EXAMPLE 5   **Graphing a Quadratic Function**

Graph the quadratic function $f(x) = 2x^2 - 40x + 104$.

Solution

Graphing using a graphing calculator is largely a matter of finding an appropriate viewing window, as the following three unsatisfactory windows show.

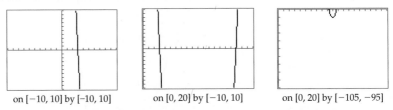

on $[-10, 10]$ by $[-10, 10]$ on $[0, 20]$ by $[-10, 10]$ on $[0, 20]$ by $[-105, -95]$

To find an appropriate viewing window, we use the vertex formula:

$$x = \frac{-b}{2a} = \frac{-(-40)}{2 \cdot 2} = \frac{40}{4} = 10 \qquad \begin{array}{l} x \text{ coordinate of the vertex, from} \\ x = -b/2a \text{ with } a = 2 \text{ and } b = -40 \end{array}$$

We move a few units, say 5, to either side of $x = 10$, making the x window $[5, 15]$. Using the calculator to EVALUATE the given function at $x = 10$ (or evaluating by hand) gives $y(10) = -96$. Since the parabola opens upward (the coefficient of x^2 is positive), the curve rises up from its vertex, so we select a y interval from -96 upward, say $[-96, -70]$. Graphing the function on the window $[5, 15]$ by $[-96, -70]$ gives the result shown below. (Some other graphing windows are just as good.)

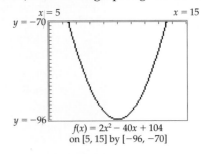

$f(x) = 2x^2 - 40x + 104$
on $[5, 15]$ by $[-96, -70]$

Solving Quadratic Equations

A value of x that solves an equation $f(x) = 0$ is called a *root* of the equation, a *zero* of the function, or an *x-intercept* on the graph of $y = f(x)$. The roots of a quadratic equation can often be found by factoring.

EXAMPLE 6   **Solving a Quadratic Equation by Factoring**

Solve $2x^2 - 4x = 6$.

Solution

$$2x^2 - 4x - 6 = 0$$
Subtracting 6 from each side to get zero on the right

$$2(x^2 - 2x - 3) = 0$$
Factoring out a 2

$$2(x - 3) \cdot (x + 1) = 0$$
Factoring $x^2 - 2x - 3$

Equals 0   Equals 0
at $x = 3$ at $x = -1$

Finding x-values that make each factor zero

$$x = 3, x = -1$$
Solutions

Graphing Calculator Exploration

You can find the solutions to the equation in Example 6 by graphing the function $f(x) = 2x^2 - 4x - 6$ and using ZERO or TRACE to find where the curve crosses the x-axis. Your answers should agree with those found in Example 6.

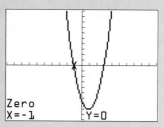

PRACTICE PROBLEM 4

Solve by factoring or graphing: $9x - 3x^2 = -30$

Solution at the back of the book

Quadratic equations can often be solved by the "quadratic formula." A derivation of this formula is given on page 44.

Quadratic Formula

The solutions to $ax^2 + bx + c = 0$ are

$$x = \frac{-b \pm \sqrt{b^2 - 4ac}}{2a}$$

The "plus or minus" sign $\pm$ means calculate both ways, first using the $+$ sign and then using the $-$ sign

In a business, it is often important to find a company's *break-even points*, the numbers of units of production where a company's costs are equal to its revenue.

EXAMPLE 7   **Finding Break-Even Points**

A company that installs automobile compact disc (CD) players finds that if it installs x CD players per day, then its costs will be $C(x) = 120x + 4800$ and its revenue will be $R(x) = -2x^2 + 400x$ (both in dollars). Find the company's break-even points.

Solution

$$120x + 4800 = -2x^2 + 400x$$

Setting $C(x) = R(x)$

$$2x^2 - 280x + 4800 = 0$$

Combining all terms on one side

$$x = \frac{280 \pm \sqrt{(-280)^2 - 4 \cdot 2 \cdot 4800}}{2 \cdot 2}$$

Quadratic formula with $a = 2, b = -280$, and $c = 4800$

$$= \frac{280 \pm \sqrt{40,000}}{4} = \frac{280 \pm 200}{4}$$

Working out the formula on a calculator

$$= \frac{480}{4} \text{ or } \frac{80}{4} = 120 \text{ or } 20$$

The company will break even when it makes either 20 units or 120 units.

Although it is important for a company to know where its break-even points are, most companies want to do better than break even—they want to maximize their profits. Profit is defined as *revenue minus cost* (since profit is what is left over after subtracting expenses from income).

Profit = Revenue − Cost

EXAMPLE 8   **Maximizing Profit**

For the CD installer whose daily revenue and cost functions were given in Example 7, find the number of units that maximizes profit, and find the maximum profit.

Solution

The profit function is the revenue function minus the cost function.

$$P(x) = \underbrace{-2x^2 + 400x}_{R(x)} - \underbrace{(120x + 4800)}_{C(x)}$$

$P(x) = R(x) - C(x)$ with $R(x) = -2x^2 + 400x$ and $C(x) = 120x + 4800$

$$= -2x^2 + 280x - 4800$$

Simplifying

This function represents a parabola opening downward (because of the -2), so it is maximized at its vertex, which is found using the vertex formula.

$$x = \frac{-280}{2(-2)} = \frac{-280}{-4} = 70$$

$x = \frac{-b}{2a}$ with $a = -2$ and $b = 280$

Thus profit is maximized when 70 units are installed. For the maximum profit, we substitute $x = 70$ into the profit function:

$$P(70) = -2(70)^2 + 280 \cdot 70 - 4800$$

$P(70) = -2x^2 + 280x - 4800$ with $x = 70$

$$= 5000$$

Multiplying and combining

Therefore, the company will maximize its profit when it installs 70 CD players per day. Its maximum profit will be $5000 per day.

SPREADSHEET EXPLORATION

The spreadsheet below shows the graphs of the functions $R(x)$, $C(x)$, and $P(x)$ from Examples 7 and 8. The values for x were made by entering 0 in A3 and then copying the formula =A3+10 for A4 into cells A5 through A18. The values for $R(x)$ were found by entering the formula =-2*A3*A3+400*A3 in B3 and then copying it into cells B4 through B18. Then $C(x)$ was similarly found by starting with C3 being =120*A3+4800. $P(x)$, the difference $R(x) - C(x)$, was found with D3 being =B3-C3. The chart was made by plotting the values for the three columns corresponding to the revenue, cost, and profit.

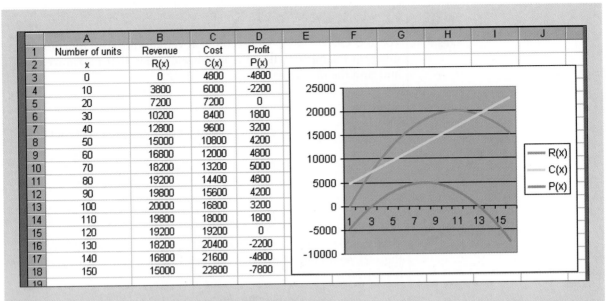

	A	B	C	D	E	F	G	H	I	J
1	Number of units	Revenue	Cost	Profit						
2	x	R(x)	C(x)	P(x)						
3	0	0	4800	-4800						
4	10	3800	6000	-2200						
5	20	7200	7200	0						
6	30	10200	8400	1800						
7	40	12800	9600	3200						
8	50	15000	10800	4200						
9	60	16800	12000	4800						
10	70	18200	13200	5000						
11	80	19200	14400	4800						
12	90	19800	15600	4200						
13	100	20000	16800	3200						
14	110	19800	18000	1800						
15	120	19200	19200	0						
16	130	18200	20400	-2200						
17	140	16800	21600	-4800						
18	150	15000	22800	-7800						
19										

Notice that the break-even points from Example 7 (at $x = 20$ and $x = 120$) correspond to a profit of zero, and that the maximum profit (at $x = 70$) occurs halfway between the two break-even points.

Why doesn't a company make more profit the more it sells? Because to increase its sales it must lower its prices, which eventually leads to lower profits. The relationship among the cost, revenue, and profit functions can be represented graphically as follows.

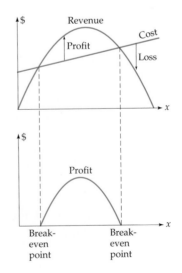

Not all quadratic equations have (real) solutions.

EXAMPLE 9   **Using the Quadratic Formula**

Solve $\frac{1}{2}x^2 - 3x + 5 = 0$.

Solution

The quadratic formula with $a = \frac{1}{2}$, $b = -3$, and $c = 5$ gives

$$x = \frac{3 \pm \sqrt{9 - 4\left(\frac{1}{2}\right)(5)}}{2\left(\frac{1}{2}\right)} = \frac{3 \pm \sqrt{9 - 10}}{1} = 3 \pm \sqrt{-1} \qquad \text{Undefined!}$$

Therefore, the equation $\frac{1}{2}x^2 - 3x + 5 = 0$ has *no real solutions* (because of the undefined $\sqrt{-1}$). The geometrical reason why there are no solutions can be seen in the graph below: The curve never reaches the x-axis, so the function never equals zero.

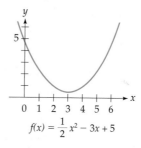

$$f(x) = \frac{1}{2}x^2 - 3x + 5$$

The quantity $b^2 - 4ac$, whose square root appears in the quadratic formula, is called the *discriminant*. If the discriminant is *positive* (as in Example 7), the equation $ax^2 + bx + c = 0$ has *two* solutions (because the square root is added and subtracted). If the discriminant is *zero*, there is only *one* root (because adding and subtracting zero gives the same answer). If the discriminant is *negative* (as in Example 9), then the equation has *no* real roots. Therefore, the discriminant being positive, zero, or negative corresponds to the parabola meeting the x-axis at 2, 1, or 0 points, as shown below.

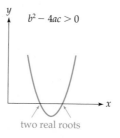

two real roots

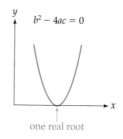

one real root

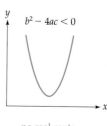

no real roots

Derivation of the Quadratic Formula

$$ax^2 + bx + c = 0$$ The quadratic set equal to zero

$$ax^2 + bx = -c$$ Subtracting c

$$4a^2x^2 + 4abx = -4ac$$ Multiplying by $4a$

$$4a^2x^2 + 4abx + b^2 = b^2 - 4ac$$ Adding b^2

$$(2ax + b)^2 = b^2 - 4ac$$ Since $4a^2x^2 + 4abx + b^2 = (2ax + b)^2$

$$2ax + b = \pm \sqrt{b^2 - 4ac}$$ Taking square roots

$$2ax = -b \pm \sqrt{b^2 - 4ac}$$ Subtracting b

$$x = \frac{-b \pm \sqrt{b^2 - 4ac}}{2a}$$ Dividing by $2a$ gives the quadratic formula

Section Summary

In this section we defined and gave examples of *functions* and saw how to find their domains and ranges. The most important characteristic of a function f is that for any given "input" number x in the domain there is exactly one "output" number $f(x)$. This requirement is stated geometrically in the *vertical line test:* No vertical line can intersect the graph of a function more than once. We then defined *linear functions* (whose graphs are lines) and *quadratic functions* (whose graphs are parabolas). We solved quadratic equations by factoring, graphing, and using the quadratic formula. We maximized and minimized quadratic functions using the vertex formula.

EXERCISES 1.3

Determine whether each graph defines a function of x.

1.

2.

3.

4.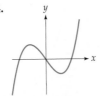

5.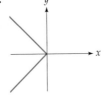

6.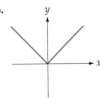

7.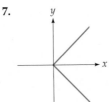

8.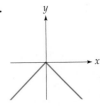

Find the domain and range of each function graphed below.

9.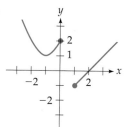

10.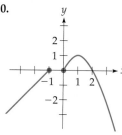

11–22: For each function:

   a. Evaluate the given expression.
   b. Find the domain of the function.
   c. Find the range.
     [*Hint:* Use a graphing calculator.]

11. $f(x) = \sqrt{x - 1}$; find $f(10)$

12. $f(x) = \sqrt{x - 4}$; find $f(40)$

13. $h(z) = \dfrac{1}{z + 4}$; find $h(-5)$

14. $h(z) = \dfrac{1}{z + 7}$; find $h(-8)$

15. $h(x) = x^{1/4}$; find $h(81)$

16. $h(x) = x^{1/6}$; find $h(64)$

17. $f(x) = x^{2/3}$; find $f(-8)$

[*Hint for Exercises 17 and 18:* You may need to enter $x^{m/n}$ as $(x^m)^{1/n}$ or as $(x^{1/n})^m$, as discussed on page 25.]

18. $f(x) = x^{4/5}$; find $f(-32)$

19. $f(x) = \sqrt{4 - x^2}$; find $f(0)$

20. $f(x) = \dfrac{1}{\sqrt{x}}$; find $f(4)$

21. $f(x) = \sqrt{-x}$; find $f(-25)$

22. $f(x) = -\sqrt{-x}$; find $f(-100)$

23–30: Graph each function "by hand." (*Note:* Even if you have a graphing calculator, it is important to be able to sketch simple curves by finding a few important points.)

23. $f(x) = 3x - 2$

24. $f(x) = 2x - 3$

25. $f(x) = -x + 1$

26. $f(x) = -3x + 5$

27. $f(x) = 2x^2 + 4x - 16$

28. $f(x) = 3x^2 - 6x - 9$

29. $f(x) = -3x^2 + 6x + 9$

30. $f(x) = -2x^2 + 4x + 16$

31–34: For each quadratic function:

   a. Find the vertex using the vertex formula.
   b. Graph the function on an appropriate viewing window. (Answers may differ.)

31. $f(x) = x^2 - 40x + 500$

32. $f(x) = x^2 + 40x + 500$

33. $f(x) = -x^2 - 80x - 1800$

34. $f(x) = -x^2 + 80x - 1800$

35–52: Solve each equation by factoring or by using the quadratic formula, as appropriate.

35. $x^2 - 6x - 7 = 0$

36. $x^2 - x - 20 = 0$

37. $x^2 + 2x = 15$

38. $x^2 - 3x = 54$

39. $2x^2 + 40 = 18x$

40. $3x^2 + 18 = 15x$

41. $5x^2 - 50x = 0$

42. $3x^2 - 36x = 0$

43. $2x^2 - 50 = 0$

44. $3x^2 - 27 = 0$

45. $4x^2 + 24x + 40 = 4$

46. $3x^2 - 6x + 9 = 6$

47. $-4x^2 + 12x = 8$

48. $-3x^2 + 6x = -24$

49. $2x^2 - 12x + 20 = 0$

50. $2x^2 - 8x + 10 = 0$

51. $3x^2 + 12 = 0$

52. $5x^2 + 20 = 0$

53–62: Solve each equation using a graphing calculator. [*Hint:* Begin with the viewing window $[-10, 10]$ by $[-10, 10]$ or another of your choice (see Useful Hint in Graphing Calculator Terminology following the Preface) and use ZERO, SOLVE, or TRACE and ZOOM IN.] (In Exercises 61 and 62, round answers to two decimal places.)

53. $x^2 - x - 20 = 0$

54. $x^2 + 2x = 15$

55. $2x^2 + 40 = 18x$

56. $3x^2 + 18 = 15x$

57. $4x^2 + 24x + 45 = 9$

58. $3x^2 - 6x + 5 = 2$

59. $3x^2 + 7x + 12 = 0$

60. $5x^2 + 14x + 20 = 0$

61. $2x^2 + 3x - 6 = 0$

62. $3x^2 + 5x - 7 = 0$

63. Use a graphing calculator to graph the following four equations simultaneously on the viewing window $[-10, 10]$ by $[-10, 10]$.

$$y_1 = 2x + 6$$

$$y_2 = 2x + 2$$

$$y_3 = 2x - 2$$

$$y_4 = 2x - 6$$

a. What do the lines have in common and how do they differ?

b. Write the equation of another line with the same slope that lies 2 units below the lowest line. Then check your answer by graphing it with the others.

64. Use a graphing calculator to graph the following four equations simultaneously on the viewing window $[-10, 10]$ by $[-10, 10]$.

$$y_1 = 3x + 4$$

$$y_2 = 1x + 4$$

$$y_3 = -1x + 4 \quad \text{(Use } \boxed{(-)} \text{ to get } -1x.)$$

$$y_4 = -3x + 4$$

a. What do the lines have in common and how do they differ?

b. Write the equation of a line through this y-intercept with slope $\frac{1}{2}$. Then check your answer by graphing it with the others.

APPLIED EXERCISES

65. Business: Cost Functions   A lumberyard will deliver wood for $4 per board foot plus a delivery charge of $20. Find a function $C(x)$ for the cost of having x board feet of lumber delivered.

66. Business: Cost Functions   A company manufactures bicycles at a cost of $55 each. If the company's fixed costs are $900, express the company's costs as a linear function of x, the number of bicycles produced.

67. Business: Salary   An employee's weekly salary is $500 plus $15 per hour of overtime. Find a function $P(x)$ giving his pay for a week in which he worked x hours of overtime.

68. Business: Salary   A sales clerk's weekly salary is $300 plus 2% of her total week's sales. Find a function $P(x)$ for her pay for a week in which she sold x dollars of merchandise.

69. General: Water Pressure   At a depth of d feet underwater, the water pressure is $p(d) = 0.45d + 15$ pounds per square inch. Find the pressure at:

a. The bottom of a swimming pool 6 feet deep

b. The maximum ocean depth of 35,000 feet

70. General: Boiling Point   At higher altitudes water boils at lower temperatures. This is why at high altitudes foods must be boiled for longer times—the lower boiling point imparts less heat to the food. At an altitude of h thousand feet above sea level, water

boils at a temperature of $B(h) = -1.8h + 212$ degrees Fahrenheit. Find the altitude at which water boils at 98.6 degrees Fahrenheit. (Your answer will show that at a high enough altitude, water boils at normal body temperature. This is why airplane cabins must be pressurized—at high enough altitudes one's blood would boil.)

71–72: General: Stopping Distance   According to data from the National Transportation Safety Board, a car traveling at speed v miles per hour should be able to come to a full stop in a distance of

$$D(v) = 0.055v^2 + 1.1v \quad \text{feet}$$

Find the stopping distance required for a car traveling at:

71. 40 mph   **72.** 60 mph

73. Biomedical: Cell Growth   The number of cells in a culture after t days is given by $N(t) = 200 + 50t^2$. Find the size of the culture after:

a. 2 days

b. 10 days

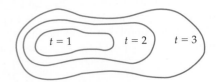

74. General: Juggling If you toss a ball h feet straight up, it will return to your hand after $T(h) = 0.5\sqrt{h}$ seconds. This leads to the "juggler's dilemma." Juggling more balls means tossing them higher. However, the square root in the above formula means that tossing them twice as high does not gain twice as much time, but only $\sqrt{2} \approx 1.4$ times as much time. Because of this, there is a limit to the number of balls that a person can juggle, which seems to be about ten. Use this formula to find:

a. How long will a ball spend in the air if it is tossed to a height of 4 ft? 8 ft?

b. How high must it be tossed to spend 2 seconds in the air? 3 seconds in the air?

75. General: Impact Velocity If a marble is dropped from a height of x feet, it will hit the ground with velocity $v(x) = \frac{60}{11}\sqrt{x}$ miles per hour (neglecting air resistance). Use this formula to find the velocity with which a marble will strike the ground if it is dropped from the top of the tallest building in the United States, the 1454-foot Sears Tower in Chicago.

76. General: Tsunamis The speed of a tsunami (popularly known as a tidal wave, although it has nothing whatever to do with tides) depends on the depth of the water through which it is traveling. At a depth of d feet, the speed of a tsunami will be $s(d) = 3.86\sqrt{d}$ miles per hour. Find the speed of a tsunami in the Pacific basin, where the average depth is 15,000 feet.

77–78: Impact Time of a Projectile If an object is thrown upward so that its height (in feet) above the ground t seconds after it is thrown is given by the function $h(t)$ below, find when the object hits the ground. That is, find the positive value of t such that $h(t) = 0$. Give the answer correct to two decimal places. [*Hint:* Enter the function in terms of x rather than t. Use the ZERO operation, or use TRACE and ZOOM IN, or use similar operations.]

77. $h(t) = -16t^2 + 45t + 5$

78. $h(t) = -16t^2 + 40t + 4$

79. Break-Even Points and Maximum Profit A company that produces devices for computer disk drives finds that if it produces x devices per week, then its costs will be $C(x) = 180x + 16,000$ and its revenue will be $R(x) = -2x^2 + 660x$ (both in dollars).

a. Find the company's break-even points.

b. Find the number of devices that maximizes profit, and find the maximum profit.

80. Break-Even Points and Maximum Profit A bicycle store finds that if it sells x racing bicycles per month, then its costs will be $C(x) = 420x + 72,000$ and its revenue will be $R(x) = -3x^2 + 1800x$ (both in dollars).

a. Find the store's break-even points.

b. Find the number of bicycles that maximizes profit, and find the maximum profit.

81. Break-Even Points and Maximum Profit A sporting goods store finds that if it sells x exercise machines per day, then its costs will be $C(x) = 100x + 3200$ and its revenue will be $R(x) = -2x^2 + 300x$ (both in dollars).

a. Find the store's break-even points.

b. Find the number of sales that will maximize profit, and find the maximum profit.

82. Break-Even Points and Maximum Profit A company that installs car alarm systems finds that if it installs x systems per week, then its costs will be $C(x) = 210x + 72,000$ and its revenue will be $R(x) = -3x^2 + 1230x$ (both in dollars).

a. Find the company's break-even points.

b. Find the number of installations that will maximize profit, and find the maximum profit.

83. General: Smoking and Education According to a recent study,* the probability that a smoker will quit smoking increases with the smoker's educational level. The probability (expressed as a percent) that a smoker with x years of education will

---

* William Sander, "Schooling and Quitting Smoking," *The Review of Economics and Statistics* LXXVII(1):191–199, February 1995.

quit is approximately $y = 0.831x^2 - 18.1x + 137.3$ (for $10 \le x \le 16$).

a. Graph this curve on the window [10, 16] by [0, 100].

b. Find the probability that a smoker who is a high school graduate ($x = 12$) will quit.

c. Find the probability that a smoker who is a college graduate ($x = 16$) will quit.

84. **Environmental Science: Wind Energy** After a slow start, the use of wind power is growing rapidly, especially in Europe, where it is seen as an efficient and renewable source of energy. Global wind-power-generating capacity for the years 1990 to 1998 is given approximation by $y = 125.62x^2 - 69.9x + 2037$ megawatts, where x is the number of years since 1990. (One megawatt would supply the electrical needs of approximately 100 homes.) (*Source: Worldwatch Institute*)

a. Graph this curve on the window [0, 20] by [0, 51000].

b. Use this curve to predict the global wind-power-generating capacity in the year 2005. [*Hint:* Which x-value corresponds to 2005? Then use TRACE, EVALUATE, or TABLE.]

c. Predict the global wind-power-generating capacity in the year 2010.

85. **Biomedical: Muscle Contraction** The fundamental equation of muscle contraction is of the form $(w + a)(v + b) = c$, where w is the weight placed on the muscle, v is the velocity of the contraction of the muscle, and a, b, and c are constants that depend on the muscle and the units of measurement. Solve this equation for v as a function of w, a, b, and c.

86. **General: Longevity** According to insurance data, the probability that a 65-year-old person will live for another x decades is approximated by the function $f(x) = -0.077x^2 - 0.057x + 1$ (for $0 \le x \le 3$). Find the probability that such a person will live for another:

a. Decade
b. Two decades
c. Three decades

Explorations and Excursions

The following problems extend and augment the material presented in the text.

Shifts of Parabolas

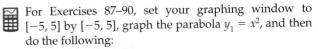 For Exercises 87–90, set your graphing window to [−5, 5] by [−5, 5], graph the parabola $y_1 = x^2$, and then do the following:

87. **Horizontal Shifts**

a. Graph $y_2 = (x - 4)^2$ (a parabola in factored form) together with the original parabola. How is the new parabola shifted compared to the original one? What are the coordinates of the vertex of the new parabola?

b. Return to y_2, change it to $y_2 = (x + 3)^2$, and graph it together with the original parabola $y_1 = x^2$. Now how is the new parabola shifted compared to the original one? What are the coordinates of its vertex?

c. In general, for any number $a > 0$, how would you describe the parabola $y = (x - a)^2$ compared to the original parabola? What are the coordinates of its vertex? What if the sign is plus instead of minus?

88. **Vertical Shifts**

a. Graph $y_2 = x^2 + 2$ together with the original parabola. How is the new parabola shifted compared to the original one? What are the coordinates of the vertex of the new parabola?

b. Return to y_2, change it to $y_2 = x^2 - 3$, and graph it together with the original parabola $y_1 = x^2$. Now how is the new parabola shifted compared to the original one? What are the coordinates of its vertex?

c. In general, for any number $b > 0$, how would you describe the parabola $y = x^2 + b$ compared to the original parabola? What if the sign is minus instead of plus?

89. **Horizontal and Vertical Shifts**

a. Graph $y_2 = (x - 3)^2 + 2$ together with the original parabola $y_1 = x^2$. How is the new parabola shifted compared to the original one? What are the coordinates of the vertex of the new parabola?

b. Return to y_2, change it to $y_2 = (x + 2)^2 - 5$, and graph it together with the original parabola $y_1 = x^2$. Now how is the new parabola shifted compared to the original one? What are the coordinates of its vertex?

c. In general, for any numbers a and b, how would you describe the parabola $y = (x - a)^2 + b$ compared to the original parabola?

90. The General Case

a. In the Graphing Calculator Exploration on page 37 you investigated the effect of the coefficient of x^2 on the shape of the parabola. From that, together with the results of the previous exercise, describe the parabola $y = 5(x - 2)^2 - 1$ as compared to the original parabola. Then check your description by graphing the new parabola together with the original one.

b. Describe the parabola $y = -5(x - 2)^2 - 1$ as compared to the original parabola. Check your description by graphing it.

c. Describe the parabola $y = c(x - a)^2 + b$ (for any numbers $a \neq 0$, b, and c) as compared with the original parabola $y_1 = x^2$.

The Vertex of a Parabola

91. a. Sketch $y = x^2 + 1$ and explain why 1 is the smallest possible value for y. [*Hint:* Can x^2 ever be negative?]

b. Sketch $y = -x^2 + 1$ and explain why 1 is the largest possible value for y.

92. a. Sketch $y = (x - 1)^2 + 2$ and explain why 2 is the smallest possible value for y. [*Hint:* Can $(x - 1)^2$ ever be negative?]

b. Sketch $y = (x + 2)^2 - 1$ and explain why -1 is the smallest possible value for y.

93. a. Verify that $(x + 3)^2 + 1 = x^2 + 6x + 10$ by "multiplying out" the expression on the left and collecting "like terms."

b. Verify that $-(x - 2)^2 + 3 = -x^2 + 4x - 1$ by "multiplying out" the expression on the left and collecting "like terms."

94. Verify that $a\left(x + \dfrac{b}{2a}\right)^2 + c - \dfrac{b^2}{4a} = ax^2 + bx + c$ by "multiplying out" the expression on the left and collecting "like terms."

95. a. For $a > 0$, show that the smallest possible value of $f(x) = a\left(x + \dfrac{b}{2a}\right)^2 + c - \dfrac{b^2}{4a}$ is obtained when $x = \dfrac{-b}{2a}$. [*Hint:* Can $a\left(x + \dfrac{b}{2a}\right)^2$ ever be negative? When is it zero?]

b. For $a < 0$, show the largest possible value of $f(x) = a\left(x + \dfrac{b}{2a}\right)^2 + c - \dfrac{b^2}{4a}$ is obtained when $x = \dfrac{-b}{2a}$. [*Hint:* Can $a\left(x + \dfrac{b}{2a}\right)^2$ ever be positive? When is it zero?]

96. Vertex Formula Explain why Exercise 94 and 95 show that the vertex of the parabola $f(x) = ax^2 + bx + c$ occurs at $x = \dfrac{-b}{2a}$. [*Hint:* Recall that the vertex is the lowest point on the parabola if it opens up and is the highest point if the parabola opens down.]

1.4 Functions, continued

APPLICATION PREVIEW

Automobile Efficiency and Rational Functions

Automobile efficiency is measured by miles per gallon (mpg), the number of miles that a car can drive on one gallon of gas (under specified conditions). How much money will you actually save if you change from a car that gets 10 mpg to one that gets 20 mpg?

Suppose you drive 12,000 miles in a typical year, and gas costs $1.50 per gallon. Your savings will be

$$\underbrace{\frac{12,000}{10}\,(1.50)}_{\substack{\text{Cost of driving}\\ \text{12,000 miles}\\ \text{at 10 mpg}}} - \underbrace{\frac{12,000}{20}\,(1.50)}_{\substack{\text{Cost of driving}\\ \text{12,000 miles}\\ \text{at 20 mpg}}} = 1800 - 900 = \underbrace{\$900}_{\substack{\text{Annual}\\ \text{savings}}}$$

Instead of asking about improving efficiency from 10 to 20 mpg, let us ask the more general question: How much is saved by improving efficiency from x to $(x + 10)$ mpg? Based on the preceding calculation, the savings $f(x)$ in changing from x to $(x + 10)$ mpg is

$$f(x) = \frac{12,000}{x}\,(1.50) - \frac{12,000}{x + 10}\,(1.50)$$

After some algebra, this simplifies to

$$f(x) = \frac{180,000}{x(x + 10)}$$

Such functions are called *rational functions* and will be discussed in this section. The graph of this rational function is shown below, although for our purposes the graph is meaningful only for positive values of x. Note that the graph becomes arbitrarily high and low near the values $x = 0$ and $x = -10$, respectively, where the denominator becomes zero.

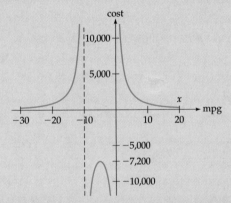

The rightmost "branch" of the curve shows that as x increases, the function takes values closer and closer to zero. This means that at high mpg efficiencies, another 10 mpg yields only a very small saving. For example, improving from 10 mpg to 20 mpg saves $900, but improving from 50 mpg to 60 mpg would save only $60 for the year, with negligible savings thereafter.

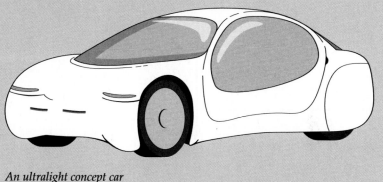

An ultralight four-passenger car achieving 66 mpg, called the Prius, is now available in Japan, and hybrid gas–electric cars achieving up to 300 mpg now seem possible. (*Source: Rocky Mountain Institute*)

An ultralight concept car

Introduction

In this section we will define other useful types of functions and an important operation, the *composition* of two functions.

Polynomial Functions

A *polynomial function* (or simply a *polynomial*) is a function that can be written in the form

$$f(x) = a_n x^n + a_{n-1}x^{n-1} + \cdots + a_2 x^2 + a_1 x + a_0$$

where n is a nonnegative integer and $a_0, a_1, \ldots, a_n$ are (real) numbers, called *coefficients*. The *domain* of a polynomial is $\mathbb{R}$, the set of all (real) numbers. The *degree* of a polynomial is the highest power of the variable. The following are polynomials.

$$f(x) = 2x^8 - 3x^7 + 4x^5 - 5$$

A polynomial of degree 8 (since the highest power of x is 8)

$$f(x) = -4x^2 - \tfrac{1}{3}x + 19$$

A polynomial of degree 2 (a quadratic function)

$$f(x) = x - 1$$

A polynomial of degree 1 (a linear function)

$$f(x) = 6$$

A polynomial of degree 0 (a constant function)

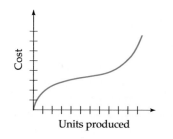

A cost function may increase at different rates at different production levels.

Polynomials are used to model many situations in which change occurs at different rates. For example, the polynomial in the graph on the left might represent the total cost of manufacturing x units of a product. At first costs rise quite steeply, as a result of high start-up expenses, then they rise more slowly as the economies of mass production come into play, and finally they rise more steeply as new production facilities need to be built.

Polynomial equations can often be solved by factoring (just as with quadratic equations).

EXAMPLE 1   **Solving a Polynomial Equation**

Solve $3x^4 - 6x^3 = 24x^2$.

Solution

$$3x^4 - 6x^3 - 24x^2 = 0$$
Rewritten with all the terms on the left side

$$3x^2(x^2 - 2x - 8) = 0$$
Factoring out $3x^2$

$$3x^2 \ (x - 4) \ (x + 2) = 0$$
Factoring further

Equals zero at $x = 0$ Equals zero at $x = 4$ Equals zero at $x = -2$
Finding the zeros of each factor

$$x = 0, \quad x = 4, \quad x = -2$$
Solutions

As in this example, if a positive power of x can be factored out of a polynomial, then $x = 0$ is one of the roots.

PRACTICE PROBLEM 1

Solve $2x^3 - 4x^2 = 48x$.   *Solution at the back of the book*

Rational Functions

The word "ratio" means fraction or quotient, and the quotient of two polynomials is called a *rational function*. The following are rational functions.

$$f(x) = \frac{4x^3 + 3x^2}{x^2 - 2x + 1} \qquad g(x) = \frac{1}{x^2 + 1}$$
A rational function is a polynomial over a polynomial

The domain of a rational function is the set of all numbers for which the denominator is not zero. For example, the domain of the function on the left above is $\{x \mid x \neq 1\}$ (because $x = 1$ makes the denominator zero), and the domain of the function on the right is $\mathbb{R}$ (because $x^2 + 1$ is never zero).

PRACTICE PROBLEM 2

What are the domain and range of the rational function graphed on page 50?   *Solution at the back of the book*

Simplifying a rational function by canceling a common factor from the numerator and the denominator can change the domain of the function, so that the "simplified" and "original" versions may not be equal (because they have different domains). For example, the following rational function is not defined at $x = 1$, whereas the simplified version on its right *is* defined at $x = 1$, so the two functions are technically not equal.

$$\underbrace{\frac{x^2 - 1}{x - 1}} = \underbrace{\frac{(x + 1)(x - 1)}{x - 1}} \neq \underbrace{x + 1}$$

Not defined at $x = 1$,   *Is* defined at $x = 1$,
so the domain is $\{x \mid x \neq 1\}$ so the domain is $\mathbb{R}$

However, the functions *are* equal at every x-value *except* $x = 1$, and the graphs are the same except that the rational function omits the point at $x = 1$.

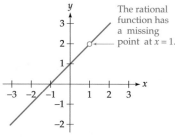

The rational function has a missing point at $x = 1$.

Graph of $y = \dfrac{x^2 - 1}{x - 1}$

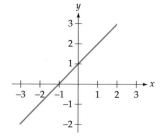

Graph of $y = x + 1$

Piecewise Linear Functions

The rule for calculating the values of a function may be given in several parts. If each part is linear, the function is called a *piecewise linear function*, and its graph consists of "pieces" of straight lines.

EXAMPLE 2 **Graphing a Piecewise Linear Function**

Graph

$$f(x) = \begin{cases} 5 - 2x & \text{if } x \geq 2 \\ x + 3 & \text{if } x < 2 \end{cases}$$

This notation means: Use the top formula for $x \geq 2$ and the bottom formula for $x < 2$

Solution We graph one "piece" at a time.

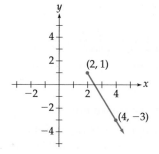

Step 1: To graph the first part, $f(x) = 5 - 2x$ if $x \geq 2$, we use the "endpoint" $x = 2$ and also $x = 4$ (or any other x-value satisfying $x \geq 2$). The points are $(2, 1)$ and $(4, -3)$, with the y-coordinates calculated from $f(x) = 5 - 2x$. Draw the line through these two points, but only for $x \geq 2$ (from $x = 2$ to the *right*), as shown in the left margin.

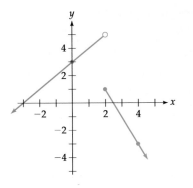

Step 2: For the second part, $f(x) = x + 3$ if $x < 2$, the restriction $x < 2$ means that the line ends just *before* $x = 2$. We mark this "missing point" (2, 5) by an "open circle" (○) to indicate that it is *not* included in the graph (the *y*-coordinate comes from $f(x) = x + 3$). For a second point, choose $x = 0$ (or any other $x < 2$), giving (0, 3). Draw the line through these two points, but only for $x < 2$ (to the *left* of $x = 2$), completing the graph of the function. ∎

An important piecewise linear function is the *absolute value* function.

EXAMPLE 3 Graphing the Absolute Value Function

The absolute value function is defined as follows:

$$f(x) = \begin{cases} x & \text{if } x \geq 0 \\ -x & \text{if } x < 0 \end{cases}$$

The lower line, for *negative x,* attaches a *second* negative sign to make the result *positive*

For example, when applied to either 3 or −3, the function gives *positive* 3:

$$f(3) = 3$$

Using the top formula (since $3 \geq 0$)

$$f(-3) = -(-3) = 3$$

Using the bottom formula (since $-3 < 0$)

Its graph, drawn as in Example 2, is shown below. ∎

Absolute Value Function

$$f(x) = |x| = \begin{cases} x & \text{if } x \geq 0 \\ -x & \text{if } x < 0 \end{cases}$$

The absolute value function $f(x) = |x|$ has a corner at the origin.

Examples 2 and 3 show that the "pieces" of a piecewise linear function may or may not be connected.

EXAMPLE 4 **Graphing an Income Tax Function**

Federal income taxes are "progressive," which means that they take a larger percentage of higher incomes. For example, the 2000 federal income tax for a single taxpayer whose taxable income is less than $115,000 is determined by a three-part rule: 15% of income up to $22,100, plus 28% of any amount over $22,100 up to $53,500, plus 31% of any amount over $53,500 up to $115,000. For an income of x dollars, the tax $f(x)$ may then be expressed as follows:

$$f(x) = \begin{cases} 0.15x & \text{if } 0 \le x \le 22{,}100 \\ 3315 + 0.28(x - 22{,}100) & \text{if } 22{,}100 < x \le 53{,}500 \\ 12{,}107 + 0.31(x - 53{,}500) & \text{if } 53{,}500 < x \le 115{,}000 \end{cases}$$

Graphing this by the same technique as before leads to the following graph. The slopes 0.15, 0.28, and 0.31 are called the *marginal tax rates.*

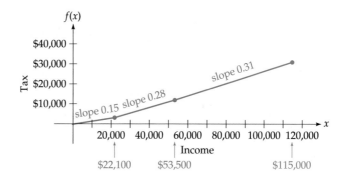

Composite Functions

Just as we substitute a *number* into a function, we may substitute a *function* into a function giving a *composite* function. For two functions f and g, evaluating f at $g(x)$ gives $f(g(x))$, called the *composition of f with g* evaluated at x.*

Composite Functions

> The composition of f with g evaluated at x is $f(g(x))$.

The *domain* of $f(g(x))$ is the set of all numbers x in the domain of g such that $g(x)$ is in the domain of f. If we think of the functions f and g

---

* The composite function $f(g(x))$ may also be written $(f \circ g)(x)$, although we will not use this notation.

as "numerical machines," then the composition $f(g(x))$ may be thought of as a *combined* machine in which the output of g is connected to the input of f.

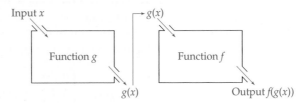

A "machine" for generating the composition of f with g. A number x is fed into the function g, and the output $g(x)$ is then fed into the function f, resulting in $f(g(x))$.

EXAMPLE 5   Finding a Composite Function

If $f(x) = x^7$ and $g(x) = x^3 - 2x$, find the composition $f(g(x))$.

Solution

$$f(g(x)) \quad = \quad \underbrace{[g(x)]^7}_{\substack{f(x) = x^7 \text{ with } x \\ \text{replaced by } g(x)}} \quad = \quad \underbrace{(x^3 - 2x)^7}_{\substack{\text{Using} \\ g(x) = x^3 - 2x}}$$

EXAMPLE 6   Finding Both Composite Functions

If $f(x) = \dfrac{x + 8}{x - 1}$ and $g(x) = \sqrt{x}$, find $f(g(x))$ and $g(f(x))$.

Solution

$$f(g(x)) = \frac{g(x) + 8}{g(x) - 1} = \frac{\sqrt{x} + 8}{\sqrt{x} - 1} \qquad f(x) = \frac{x + 8}{x - 1} \text{ with } x \text{ replaced by}$$
$$g(x) = \sqrt{x}$$

$$g(f(x)) = \sqrt{f(x)} = \sqrt{\frac{x + 8}{x - 1}} \qquad g(x) = \sqrt{x} \text{ with } x \text{ replaced by}$$
$$f(x) = \frac{x + 8}{x - 1}$$

The order of composition makes a difference: $f(g(x))$ is *not* the same as $g(f(x))$. To show this, we evaluate the preceding $f(g(x))$ and $g(f(x))$ at $x = 4$:

$$f(g(4)) = \frac{\sqrt{4} + 8}{\sqrt{4} - 1} = \frac{2 + 8}{2 - 1} = \frac{10}{1} = 10 \qquad f(g(x)) = \frac{\sqrt{x} + 8}{\sqrt{x} - 1} \text{ at } x = 4$$

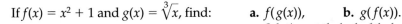

Different
answers

$$g(f(4)) = \sqrt{\frac{4 + 8}{4 - 1}} = \sqrt{\frac{12}{3}} = \sqrt{4} = 2 \qquad g(f(x)) = \sqrt{\frac{x + 8}{x - 1}} \text{ at } x = 4$$

PRACTICE PROBLEM 3   If $f(x) = x^2 + 1$ and $g(x) = \sqrt[3]{x}$, find: **a.** $f(g(x))$,   **b.** $g(f(x))$.

Solutions at the back of the book

EXAMPLE 7   **Predicting Water Usage**

A planning commission estimates that if a city's population is p thousand people, its daily water usage will be $W(p) = 30p^{1.2}$ thousand gallons. The commission further predicts that the population in t years will be $p(t) = 60 + 2t$ thousand people. Express the water usage W as a function of t, the number of years from now, and find the water usage 10 years from now.

Solution

Water usage W as a function of t is the *composition* of $W(p)$ and $p(t)$.

$$W(p(t)) = 30[p(t)]^{1.2} = 30(60 + 2t)^{1.2} \qquad \begin{array}{l} W = 30p^{1.2} \text{ with } p \\ \text{replaced by } p(t) = 60 + 2t \end{array}$$

To find water usage in 10 years, we evaluate $W(p(t))$ at $t = 10$.

$$\begin{aligned} W(p(10)) &= 30(60 + 2 \cdot 10)^{1.2} \qquad &30(60 + 2t)^{1.2} \text{ with } t = 10 \\ &= 30(80)^{1.2} \approx 5765 \qquad &\text{Using a calculator} \end{aligned}$$

Thousand gallons

Therefore, in 10 years the city will need about 5,765,000 gallons of water per day.   ∎

Shifts of Graphs

In certain cases, the graph of a composite function is simply a "shift" of an original graph. That is, adding a number to or subtracting it from a variable or a function results in a horizontal or vertical shift of the graph. The following diagram shows the graph of $y = x^2$ in the center, along with *horizontal* shifts (adding to or subtracting from x) and *vertical* shifts (adding to or subtracting from the *function*).

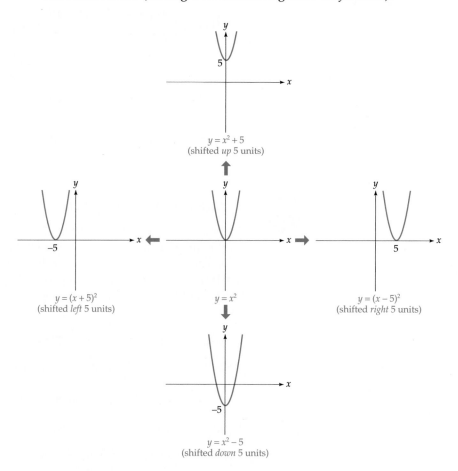

$y = x^2 + 5$
(shifted *up* 5 units)

$y = (x + 5)^2$
(shifted *left* 5 units)

$y = x^2$

$y = (x - 5)^2$
(shifted *right* 5 units)

$y = x^2 - 5$
(shifted *down* 5 units)

In general, if the addition or subtraction is done to the *x-value*, then the shift is *horizontal*, while if it is done to the *function*, then the shift is *vertical*. That is, for any function $y = f(x)$ and positive numbers a and b:

The graph of	is the graph of $y = f(x)$ shifted
$y = f(x + a)$	*left* by a units
$y = f(x - a)$	*right* by a units
$y = f(x) + b$	*up* by b units
$y = f(x) - b$	*down* by b units

Of course, a graph can be shifted both horizontally and vertically:

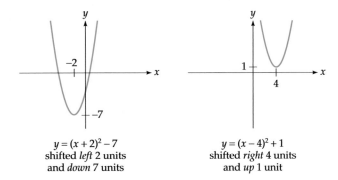

$y = (x + 2)^2 - 7$
shifted *left* 2 units
and *down* 7 units

$y = (x - 4)^2 + 1$
shifted *right* 4 units
and *up* 1 unit

Such shifts apply to *any* function $f(x)$: The graph of $y = f(x + a) + b$ is shifted *left a* units and *up b* units (with the understanding that a *negative a* or *b* means that the direction is reversed).

   Be careful: Remember that adding a *positive* number to x means a *left* shift.

Graphing Calculator Exploration

The absolute value function $y = |x|$ may be graphed on some graphing calculators as $y_1 = \text{ABS}(x)$

a. Graph $y_1 = \text{ABS}(x - 2) - 6$ and observe that the absolute value function is shifted *right* 2 units and *down* 6 units. (The graph shown is drawn using ZOOM ZSquare.)

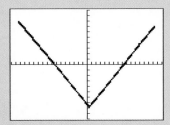

b. Predict the shift of $y_1 = \text{ABS}(x + 4) + 2$ and then verify your prediction by graphing the function on your calculator.

Section Summary

We introduced a variety of functions: polynomials (which include linear and quadratic functions), rational functions, and piecewise linear functions. Examples of these are shown below. You should be able to identify these basic types of functions from their algebraic forms. We also added constants to perform horizontal and vertical *shifts* of graphs of functions, and we combined functions by using the "output" of one as the "input" of the other, resulting in *composite* functions.

A Gallery of Functions

POLYNOMIALS

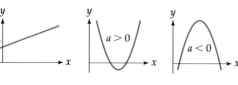

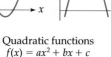

Linear function
$f(x) = mx + b$

Quadratic functions
$f(x) = ax^2 + bx + c$

$f(x) = ax^4 + bx^3 + cx^2 + dx + e$

RATIONAL FUNCTIONS

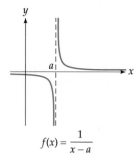

$$f(x) = \frac{1}{x - a}$$

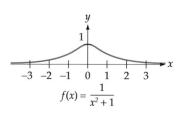

$$f(x) = \frac{1}{x^2 + 1}$$

PIECEWISE LINEAR FUNCTIONS

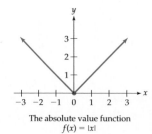

The absolute value function
$f(x) = |x|$

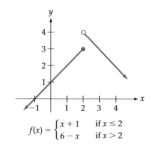

$$f(x) = \begin{cases} x + 1 & \text{if } x \le 2 \\ 6 - x & \text{if } x > 2 \end{cases}$$

EXERCISES 1.4

Find the domain and range of each function graphed below.

1.

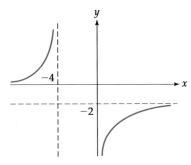

2.

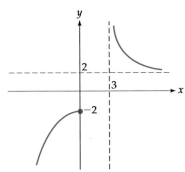

For each function in Exercises 3–10:

   a. Evaluate the given expression.
   b. Find the domain of the function.
   c. Find the range. [*Hint:* Use a graphing calculator. You may have to ignore some false connecting lines on the graph. Graphing in "dot mode" will eliminate false lines.]

3. $f(x) = \dfrac{1}{x + 4}$; find $f(-3)$

4. $f(x) = \dfrac{1}{(x - 1)^2}$; find $f(-1)$

5. $f(x) = \dfrac{x^2}{x - 1}$; find $f(-1)$

6. $f(x) = \dfrac{x^2}{x + 2}$; find $f(2)$

7. $f(x) = \dfrac{12}{x(x + 4)}$; find $f(2)$

8. $f(x) = \dfrac{16}{x(x - 4)}$; find $f(-4)$

9. $g(x) = |x + 2|$; find $g(-5)$

10. $g(x) = |x| + 2$; find $g(-5)$

Solve each equation by factoring.

11. $x^5 + 2x^4 - 3x^3 = 0$ **12.** $x^6 - x^5 - 6x^4 = 0$

13. $5x^3 - 20x = 0$ **14.** $2x^5 - 50x^3 = 0$

15. $2x^3 + 18x = 12x^2$ **16.** $3x^4 + 12x^2 = 12x^3$

17. $6x^5 = 30x^4$ **18.** $5x^4 = 20x^3$

19. $3x^{5/2} - 6x^{3/2} = 9x^{1/2}$ **20.** $2x^{7/2} + 8x^{5/2} = 24x^{3/2}$
[*Hint for Exercises 19 and 20:* First factor out a fractional power.]

Solve each equation using a graphing calculator. (For Exercises 31 and 32, round answers to two decimal places.)

21. $x^3 - 2x^2 - 8x = 0$ **22.** $x^3 + 2x^2 - 8x = 0$

23. $x^5 - 2x^4 - 3x^3 = 0$ **24.** $x^6 + x^5 - 6x^4 = 0$

25. $2x^3 = 12x^2 - 18x$

26. $3x^4 + 12x^3 + 12x^2 = 0$

27. $6x^5 + 30x^4 = 0$ **28.** $5x^4 + 20x^3 = 0$

29. $2x^{5/2} + 4x^{3/2} = 6x^{1/2}$

30. $3x^{7/2} - 12x^{5/2} = 36x^{3/2}$

31. $x^5 - x^4 - 5x^3 = 0$

32. $x^6 + 2x^5 - 5x^4 = 0$

Graph each function.

33. $f(x) = |x - 3| - 3$

34. $f(x) = |x + 2| - 2$

35. $f(x) = \begin{cases} 2x - 7 & \text{if } x \geq 4 \\ 2 - x & \text{if } x < 4 \end{cases}$

36. $f(x) = \begin{cases} 2 - x & \text{if } x \geq 3 \\ 2x - 4 & \text{if } x < 3 \end{cases}$

37. $f(x) = \begin{cases} 8 - 2x & \text{if } x \geq 2 \\ x + 2 & \text{if } x < 2 \end{cases}$

38. $f(x) = \begin{cases} 2x - 4 & \text{if } x > 3 \\ 5 - x & \text{if } x \leq 3 \end{cases}$

Identify each function as a polynomial, a rational function, a piecewise linear function, or none of these. (Do not graph the functions; just identify their types.)

39. $f(x) = x^5$

40. $f(x) = 3|x|$

41. $f(x) = 1 - |x|$

42. $f(x) = x^4$

43. $f(x) = x + 2$

44. $f(x) = \begin{cases} 3x - 1 & \text{if } x \geq 2 \\ 1 - x & \text{if } x < 2 \end{cases}$

45. $f(x) = \dfrac{1}{x + 2}$

46. $f(x) = x^2 + 9$

47. $f(x) = \begin{cases} x - 2 & \text{if } x < 3 \\ 7 - 4x & \text{if } x \geq 3 \end{cases}$

48. $f(x) = \dfrac{x}{x^2 + 9}$

49. $f(x) = 3x^2 - 2x$

50. $f(x) = x^3 - x^{2/3}$

51. $f(x) = x^2 + x^{1/2}$

52. $f(x) = 5$

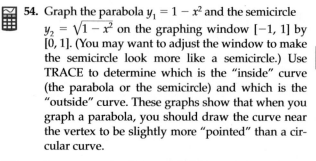

53. For the functions $y_1 = 100x$, $y_2 = 10x^2$, $y_3 = x^3$:
 a. Predict which curve will be the highest for large values of x.
 b. Predict which curve will be the lowest for large values of x.
 c. Check your prediction by graphing the functions on the window [0, 12] by [0, 2000].
 d. From your graph, where do all these curves meet?

54. Graph the parabola $y_1 = 1 - x^2$ and the semicircle $y_2 = \sqrt{1 - x^2}$ on the graphing window [−1, 1] by [0, 1]. (You may want to adjust the window to make the semicircle look more like a semicircle.) Use TRACE to determine which is the "inside" curve (the parabola or the semicircle) and which is the "outside" curve. These graphs show that when you graph a parabola, you should draw the curve near the vertex to be slightly more "pointed" than a circular curve.

55. For any x, the function INT(x) is defined as the greatest integer less than or equal to x. For example, INT(3.7) = 3 and INT(−4.2) = −5.
 a. Use a graphing calculator to graph the function $y_1 = $ INT(x). (You may need to graph it in DOT mode to eliminate false connecting lines.)
 b. From your graph, what are the domain and range of this function?

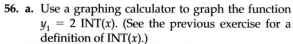

56. a. Use a graphing calculator to graph the function $y_1 = 2$ INT(x). (See the previous exercise for a definition of INT(x).)
 b. From your graph, what are the domain and range of this function?

For each pair $f(x)$ and $g(x)$, find **a.** $f(g(x))$, **b.** $g(f(x))$.

57. $f(x) = x^5$, $g(x) = 7x - 1$

58. $f(x) = x^8$, $g(x) = 2x + 5$

59. $f(x) = \frac{1}{x}$, $g(x) = x^2 + 1$

60. $f(x) = \sqrt{x}$, $g(x) = x^3 - 1$

61. $f(x) = x^3 - x^2$, $g(x) = \sqrt{x} - 1$

62. $f(x) = x - \sqrt{x}$, $g(x) = x^2 + 1$

63. $f(x) = \dfrac{x^3 - 1}{x^3 + 1}$, $g(x) = x^2 - x$

64. $f(x) = \dfrac{x^4 + 1}{x^4 - 1}$, $g(x) = x^3 + x$

65. a. Find the composition $f(g(x))$ of the two linear functions $f(x) = ax + b$ and $g(x) = cx + d$ (for constants a, b, c, and d).
 b. Is the composition of two linear functions always a linear function?

66. a. Is the composition of two quadratic functions always a quadratic function? [*Hint:* Find the composition of $f(x) = x^2$ and $g(x) = x^2$.]
 b. Is the composition of two polynomials always a polynomial?

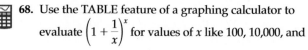

67. Find, rounding to five decimal places:
 a. $\left(1 + \dfrac{1}{100}\right)^{100}$
 b. $\left(1 + \dfrac{1}{10,000}\right)^{10,000}$
 c. $\left(1 + \dfrac{1}{1,000,000}\right)^{1,000,000}$
 d. Do the resulting numbers seem to be approaching a limiting value? Estimate the limiting value to five decimal places. The number that you have approximated is denoted e. You will use it in the next section and again in Chapter 2.

68. Use the TABLE feature of a graphing calculator to evaluate $\left(1 + \dfrac{1}{x}\right)^x$ for values of x like 100, 10,000, and

1,000,000 and higher. Do the resulting numbers seem to be approaching a limiting value? Estimate the limiting value to five decimal places. The number that you have approximated is denoted e. You will use it in the next section and again in Chapter 2.

 69. How will the graph of $y = (x + 3)^3 + 6$ differ from the graph of $y = x^3$? Check by graphing both functions together.

70. How will the graph of $y = -(x - 4)^2 + 8$ differ from the graph of $y = -x^2$? Check by graphing both functions together.

APPLIED EXERCISES

 71. Economics: Income Tax   The function below expresses an income tax that is 10% for incomes up to $5000, and otherwise is $500 plus 30% of income in excess of $5000.

$$f(x) = \begin{cases} 0.10x & \text{if } 0 \le x \le 5000 \\ 500 + 0.30(x - 5000) & \text{if } x > 5000 \end{cases}$$

a. Calculate the tax on an income of $3000.
b. Calculate the tax on an income of $5000.
c. Calculate the tax on an income of $10,000.
d. Graph the function.

72. Economics: Income Tax   The following function expresses an income tax that is 15% for incomes up to $6000, and otherwise is $900 plus 40% of incomes in excess of $6000.

$$f(x) = \begin{cases} 0.15x & \text{if } 0 \le x \le 6000 \\ 900 + 0.40(x - 6000) & \text{if } x > 6000 \end{cases}$$

a. Calculate the tax on an income of $3000.
b. Calculate the tax on an income of $6000.
c. Calculate the tax on an income of $10,000.
d. Graph the function.

 73. Business: Insurance Reserves   An insurance company keeps reserves (money to pay claims) of $R(v) = 2v^{0.3}$, where v is the value of all of its policies, and the value of its policies is predicted to be $v(t) = 60 + 3t$, where t is the number of years from now. (Both R and v are in millions of dollars.) Express the reserves R as a function of t, and evaluate the function at $t = 10$.

 74. Business: Research Expenditures   An electronics company's research budget is $R(p) = 3p^{0.25}$, where p is the company's profit, and the profit is predicted to be $p(t) = 55 + 4t$, where t is the number of years from now. (Both R and p are in millions of dollars.) Express the research expenditure R as a function of t, and evaluate the function at $t = 5$.

75. General: Oil Imports and Fuel Efficiency   The average fuel efficiency of all cars in America, called the *fleet mpg*, is 21.6 mpg. The amount of crude oil that would be saved if the fleet mpg were increased to a value of x is

$$S(x) = 3208 - \frac{69{,}300}{x} \quad \begin{array}{l}\text{million barrels}\\\text{annually}\end{array}$$

a. Graph the function $S(x)$ on the viewing window [21.6, 40] by [0, 2000].
b. For what fleet mpg x will the savings reach 720 million barrels, which is the amount annually imported from the Middle East OPEC nations? [*Hint:* Either ZOOM IN around the point at $y = 720$ or, better, use INTERSECT with $y_2 = 720$.] (*Note:* New cars average 29 mpg, but fleet mpg always lags. Ten manufacturers have already built and tested prototype cars getting from 67 to 138 mpg.) (*Source: Information Please Almanac and the Rocky Mountain Institute.*)

 76. General: Long-Term Trends in Cigarette Smoking   The table below gives the annual cigarette consumption in the United States per adult from 1910 to 1990.

	Years since 1900	Cigarettes per Adult
1910	10	200
1930	30	1500
1950	50	3400
1970	70	3900
1990	90	3300

a. Enter these data into a graphing calculator and make a plot of the resulting points (Years since 1900 on the x-axis and Cigarettes per Adult on the y-axis).

b. Have your calculator find the cubic (third-order polynomial) regression formula for these data. Then enter the result as y_1, which gives a formula for cigarette consumption for each year. Plot the points together with the regression curve.

c. Use the regression formula to predict the cigarette consumption in the years 2000 ($x = 100$) and 2005 ($x = 105$).

d. Use the regression formula to predict the cigarette consumption in the year 2020 ($x = 120$). The answer shows that polynomial predictions, if extended too far, give nonsensical results. *(Source: Worldwatch Institute.)*

Exponential Functions

APPLICATION PREVIEW

Exponential Functions and the World's Worst Currency

Most economies of the world suffer from some degree of inflation, which drives up prices over the years. For example, if the inflation rate is 5%, prices rise by 5% each year, so that a $100 item rises in a year to $105. This is equivalent to multiplying the original price by 1.05, and so another year of 5% inflation increases the already inflated prices by multiplying them by 1.05 again, to $100(1.05)^2$. After x years, the price would rise to $100(1.05)^x$. This sort of function, with the variable in the exponent, is called an *exponential function* and is the subject of this section. Exponential functions are used extensively for calculating bank interest and the effects of inflation.

What country suffers the worst inflation? At the end of 1993, the inflation rate in Yugoslavia exceeded *1 million percent per year*. (Yugoslavia was under United Nations sanctions for supporting a war in neighboring Bosnia.) This rate of inflation rendered its currency, the Yugoslavian dinar, almost worthless within days of issue. Near the end of 1993, the government issued a 500-billion-dinar note. At the time of issue it was worth about $5. Within a month it was worth one-thousandth of a cent. Such is the power of exponential functions.

From $5 to 0.001¢ in a month

Introduction

In this section we introduce *exponential* functions, in which the variable is in the exponent. We will use exponential functions extensively when we discuss compound interest. We will also introduce the very important mathematical constant e.

  Exponential functions are often used to model the processes of growth and decay.

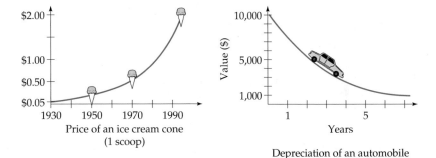

Price of an ice cream cone
(1 scoop)

Depreciation of an automobile

Exponential Functions

A function that has a variable in an exponent, such as $f(x) = 2^x$, is called an *exponential function*.

$$f(x) = 2^x$$

Exponent

Base

The table below shows some values of the exponential function $f(x) = 2^x$, and its graph (based on these points) is shown on the right.

x	$y = 2^x$
-3	$2^{-3} = \frac{1}{8}$
-2	$2^{-2} = \frac{1}{4}$
-1	$2^{-1} = \frac{1}{2}$
0	$2^0 = 1$
1	$2^1 = 2$
2	$2^2 = 4$
3	$2^3 = 8$

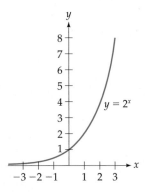

Domain of 2^x is $\mathbb{R} = (-\infty, \infty)$ and range is $(0, \infty)$.

Clearly, the graph of the exponential function 2^x is quite different from the parabola x^2.

The exponential function $f(x) = \left(\frac{1}{2}\right)^x$ has base $\frac{1}{2}$. The following table shows some of its values, and its graph is shown to the right of the table. Notice that it is the mirror image of the curve $y = 2^x$.

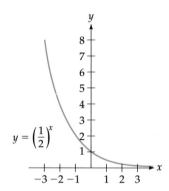

x	$y = \frac{1}{2}^x$
-3	$\frac{1}{2}^{-3} = 8$
-2	$\frac{1}{2}^{-2} = 4$
-1	$\frac{1}{2}^{-1} = 2$
0	$\frac{1}{2}^{0} = 1$
1	$\frac{1}{2}^{1} = \frac{1}{2}$
2	$\frac{1}{2}^{2} = \frac{1}{4}$

$y = \left(\frac{1}{2}\right)^x$

Domain of $\left(\frac{1}{2}\right)^x$ is $\mathbb{R} = (-\infty, \infty)$ and range is $(0, \infty)$.

We can define an exponential function $f(x) = a^x$ for any positive base a. We always take the base to be positive, so for the rest of this section *the letter a will stand for a positive constant.*

Exponential functions with bases $a > 1$ are used to model *growth*, as in populations or savings accounts, and exponential functions with bases $a < 1$ are used to model *decay*, as in depreciation. (For base $a = 1$ the graph is a horizontal line, since $1^x = 1$ for any x.)

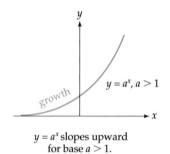

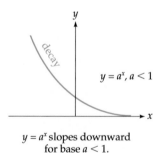

$y = a^x$ slopes upward for base $a > 1$.

$y = a^x$ slopes downward for base $a < 1$.

Value Appreciation

Suppose that you make an investment of P dollars that increases in value ("appreciates") by 8% each year. Then after 1 year the value will be

$$\left(\begin{array}{c}\text{Value after} \\ \text{1 year}\end{array}\right) = P + 0.08P = P(1 + 0.08)$$

Notice that increasing a quantity by 8% is the same as multiplying it by $(1 + 0.08)$. Therefore, to find the value of the investment after a *second* year, we would multiply the previous answer by $(1 + 0.08)$, giving $P(1 + 0.08)(1 + 0.08)$, or $P(1 + 0.08)^2$. For the value after t years we simply multiply P by $(1 + 0.08)$ t times:

$$\left(\begin{matrix}\text{Value after}\\ t \text{ years}\end{matrix}\right) = \overbrace{P(1 + 0.08) \cdot (1 + 0.08) \cdots (1 + 0.08)}^{t \text{ times}}$$

$$= P(1 + 0.08)^t$$

Clearly, the 8% can be replaced by *any* growth rate r (written in decimal form).

Value Appreciation

For an investment of P dollars that increases at rate r each year,

$$\left(\begin{matrix}\text{Value after}\\ t \text{ years}\end{matrix}\right) = P(1 + r)^t$$

EXAMPLE 1 **Finding the Value of an Investment**

A coin collection recently appraised at $25,000 is expected to appreciate in value by 8.5% each year. Predict its value 6 years from now.

Solution

$$25,000(1 + 0.085)^6 = 25,000 \cdot 1.085^6 \qquad \begin{matrix}P(1 + r)^t \text{ with } P = 25,000,\\ r = 0.085, \text{ and } t = 6\end{matrix}$$

$$\approx 40,786.69 \qquad \text{Using a calculator}$$

The coin collection will be worth approximately $40,787.

■

PRACTICE PROBLEM 1

Find the value of a 30-year Treasury bond (called a "long bond") at the end of its 30-year term if it is now selling for $10,000 and if the interest rate is 7%. *Solution at the back of the book*

Depreciation by a Fixed Percentage

Depreciation by a fixed percentage means that a piece of equipment loses a fixed percentage (say 30%) of its value each year. Losing a percentage of value is like value appreciation but with a *negative* growth rate. Therefore, we use the formula $P(1 + r)^t$, but for depreciation r is *negative*.

EXAMPLE 2   **Depreciating an Asset**

A car worth $15,000 depreciates in value by 40% each year. How much is it worth after 3 years?

Solution

The car loses 40% of its value each year, which is equivalent to an interest rate of *negative* 40%. The value appreciation formula with $P = 15{,}000$, $r = -0.40$, and $t = 3$ gives

$$P(1 + r)^t = 15{,}000(1 - 0.40)^3 = 15{,}000(0.60)^3 = \$3240$$

Using a
calculator

Its value after 3 years is $3240.

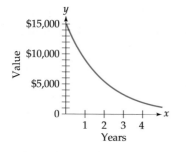

The exponential function $f(x) = 15{,}000(0.60)^x$, giving the value of the car after x years of depreciation, is graphed on the left.

PRACTICE PROBLEM 2

A printing press, originally worth $50,000, loses 20% of its value each year. What is its value after 4 years?   *Solution at the back of the book*

The graph above shows that depreciation by a fixed percentage is quite different from "straight-line" depreciation. Under straight-line depreciation the same *dollar* value is lost each year, whereas under fixed-percentage depreciation the same *percentage* of value is lost each year, resulting in larger dollar losses in the early years and smaller dollar losses in later years. Depreciation by a fixed percentage (also called the "declining balance" method) is one type of "accelerated" depreciation. The method of depreciation that one uses depends on how one chooses to estimate value and, in practice, is often determined by the tax laws.

The Number *e*

Which is better: a one-time 100% increase or two successive 50% increases? You might think that they are the same, but let's apply both to $1 and compare the answers. If $1 is increased by 100%, the result is obviously $2. But if $1 is first increased by 50% to $1.50, and then this $1.50 is increased by 50% (adding $0.75), the total is $2.25. Therefore, the single 100% increase gave $2.00, while two successive 50% increases gave $2.25, or 25¢ more. (An additional 25¢ may seem insignificant, but it *is* significant based on $1, and if instead we began with $1000, the difference would be $250.)

There is a faster way to carry out this calculation: Increasing something by 50% means keeping the original and adding half more, which is equivalent to multiplying it by $\left(1+\frac{1}{2}\right)$. Increasing by 50% again means multiplying a *second* time—that is, finding $\left(1+\frac{1}{2}\right)^2 = 2.25$, the same answer as before.

Even better is to have four successive 25% increases, because $\left(1+\frac{1}{4}\right)^4 \approx 2.44$, which is better than \$2.25 by 19¢.

How about 100 successive increases of 1% each, or 200 increases of $\frac{1}{2}$%, and so on? The general formula for the result of performing n successive equal increases that add up to 100% is $\left(1+\frac{1}{n}\right)^n$. The following graphing calculator display evaluates this expression for values up to a million (replacing n by x).

	X	Y₁	
a single 100% increase :	1	2	
4 successive 25% increases :	4	2.4414	
100 successive 1% increases :	100	2.7048	
	1000	2.7169	
	10000	2.7181	
	100000	2.7183	
1E6 means $1 \cdot 10^6 = 1000000$:	**1E6**	2.7183	
	X=1000000		

$\left. \right\}$ last two (rounded) y-values agree

Notice that the y-values do not become arbitrarily large but are "settling down" to a result approximately equal to 2.7183. That is, breaking 100% up into smaller and smaller successive increases does *not* result in an arbitrarily large increase but, rather, results in a multiple of approximately 2.7183 (which is still much better than the multiple of 2 resulting from a single 100% increase). This number is so important in mathematics that it is given its own symbol (like π)—it is called e. More formally, using the notation $\lim\limits_{n\to\infty}$ to mean the value *approached* by a quantity as n becomes arbitrarily large, we can represent e as follows:

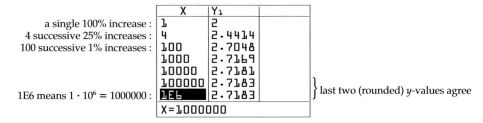

$$e = \lim_{n\to\infty}\left(1+\frac{1}{n}\right)^n = 2.71828 \ldots$$

$\lim\limits_{n\to\infty}\left(1+\dfrac{1}{n}\right)^n$ is read "the limit as n approaches infinity of 1 plus $1/n$ to the nth power"

The number e has an infinitely long decimal representation and has been calculated to several million decimal places. This same e appears on the e^x key of most calculators (including business calculators) and is used in statistics to define the famous "bell-shaped" or "normal" curve (see Chapter 6). The value of e may be found on most calculators by pressing 2nd LN 1 or simply the "e" key to give $e \approx 2.718281828$ (rounded to nine decimal places), as you should check.

The Function $y = e^x$

The number e gives us a new exponential function $f(x) = e^x$. This function is used extensively in business, economics, and all areas of science. The table below shows the values of e^x for various values of x. These values lead to the graph of $f(x) = e^x$ shown on the right.

x	$y = e^x$
-3	$e^{-3} \approx 0.05$
-2	$e^{-2} \approx 0.14$
-1	$e^{-1} \approx 0.37$
0	$e^0 \approx 1$
1	$e^1 \approx 2.7$
2	$e^2 \approx 7.4$
3	$e^3 \approx 20.1$

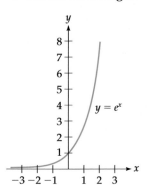

Domain of e^x is $\mathbb{R} = (-\infty, \infty)$ and range is $(0, \infty)$.

Notice that e^x is never zero and is positive for all values of x, even when x is negative. We restate this important observation as follows:

---

e **to any power is positive.**

The function $f(x) = e^{kx}$ for various values of the constant k is graphed below. For positive values of k the curve rises, and for negative values of k the function falls (as you move to the right). For higher values of k the curve rises more steeply.

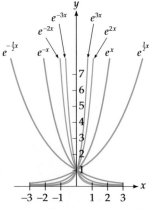

$f(x) = e^{kx}$ for various values of k

EXAMPLE 3 **Predicting the Population of the United States**

As the following graph shows, the population of the United States since 1900 has grown approximately exponentially, closely matching the curve $y = 80.1e^{0.013x}$, where x is the number of years since 1900. Use this exponential function to predict the U.S. population in the year 2010.

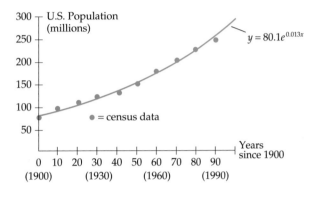

Solution

$$80.1e^{0.013(110)} = 80.1e^{1.43} \approx 335$$

$80.1e^{0.013x}$ with $x = 110$
(2010 is 110 years after 1900)

Using a calculator

In the year 2010 the population of the United States will be approximately 335 million (based on the last nine censuses).

Graphing Calculator Exploration

The most populous states are California and Texas, with New York third and Florida fourth but gaining. According to data from the Census Bureau, x years after 1990 the population of New York will be $17.99e^{0.0025x}$ and the population of Florida will be $12.94e^{0.0283x}$ (both in millions).

a. Graph these two functions on a calculator on the window [0, 20] by [0, 25]. (Use the $\boxed{e^x}$ key above the $\boxed{\text{LN}}$ key for entering the power of e.)

b. Use INTERSECT to find the x-value where the curves intersect.

c. From your answer to part (b), in which year is Florida projected to overtake New York as the third-largest state? [*Hint:* x is years after 1990.]

Exponential Growth

All exponential growth has one common characteristic: The amount of growth is proportional to the size of the quantity. For example, the interest that a bank account earns is proportional to the size of the account, and the growth of a population is proportional to the size of the population. This is in contrast, for example, to a person's height, which does not increase exponentially. That is, exponential growth occurs in those situations when a quantity grows *in proportion to its size*.

Section Summary

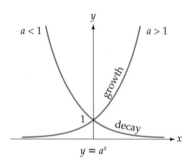

Exponential functions have exponents that involve variables. The exponential functions $f(x) = a^x$ slope *upward* or *downward* depending on whether the base a (which must be positive) satisfies $a > 1$ or $a < 1$.

The formula $P(1 + r)^t$ gives the value after t years of an investment originally worth P dollars that increases at rate r each year. This same formula, with a *negative* growth rate r, governs depreciation by a fixed percentage.

By considering a 100% increase not given all at once but divided up into smaller and smaller successive increases, we defined a new constant e:

$$e = \lim_{n \to \infty}\left(1 + \frac{1}{n}\right)^n \approx 2.71828$$

Exponential functions with base e are used extensively in modeling many types of growth, such as the growth of populations and investments.

EXERCISES 1.5 (Most require .)

Graph each function. If you are using a graphing calculator, make a hand-drawn sketch from the screen.

1. $y = 3^x$ **2.** $y = 5^x$

3. $y = \left(\frac{1}{3}\right)^x$ **4.** $y = \left(\frac{1}{5}\right)^x$

Calculate each value of e^x using a calculator.

5. $e^{1.74}$ **6.** $e^{-0.09}$

 7. e^x **versus** x^n Which curve is eventually higher, x to a power or e^x?

   a. Graph x^2 and e^x on the window [0, 5] by [0, 20]. Which curve is higher?

b. Graph x^3 and e^x on the window [0, 6] by [0, 200]. Which curve is higher for large values of x?

c. Graph x^4 and e^x on the window [0, 10] by [0, 10,000]. Which curve is higher for large values of x?

d. Graph x^5 and e^x on the window [0, 15] by [0, 1,000,000]. Which curve is higher for large values of x?

e. Do you think that e^x will exceed x^6 for large values of x? On the basis of these observations, can you make a conjecture about e^x and *any* power of x?

8. Geometric versus Exponential Growth

    a. Graph $y_1 = x$ and $y_2 = e^{0.01x}$ on the window [0, 10] by [0, 10]. Which curve is higher for x near 10?

    b. Then graph the same curves on the window [0, 1000] by [0, 1000]. Which curve is higher for x near 1000?

A linear function like y_1 represents *geometric* growth, and y_2 represents *exponential* growth, and the result here is true in general: Exponential growth always beats geometric growth (eventually, no matter what the constants).*

APPLIED EXERCISES

9. General: Art Appreciation A Picasso etching, *The Frugal Repast*, bought for $250 in 1944, has been appreciating in value by 15% per year. Predict its value in the year 2004.

10. General: Art Appreciation A Van Gogh painting, *Intérieur d'un Restaurant*, was auctioned in 1996 for $10,342,500. Assuming that it continues to appreciate in value by 12% per year, predict its value in the year 2016.

11. General: Asset Appreciation A mint condition 1910 Honus Wagner** baseball card was sold at an auction in 1996 for $640,500, a record price for an item of sports memorabilia. Assuming that it will continue to appreciate in value by 12% per year, predict its value in the year 2006.

12. General: Asset Appreciation In 1626, Peter Minuit purchased Manhattan Island from the Shinnecock Indians for trinkets and beads worth 60 guilders, or approximately $24. If this $24 had been invested to grow by 5% annually, what would it be worth now?

13. General: Retirement Funds At the birth of your child you put $1000 into an investment that will grow by 8% per year. Find its value if your child holds it until retirement, 60 years later.

14. General: College Costs In an effort to defray the cost of a college education, at the birth of your child you put $15,000 in an investment that will increase by 6% per year. What will be the investment's value when your child is ready for college at age 18?

15. General: Depreciation A $25,000 automobile depreciates by 35% per year. Find its value after:

    a. 4 years   **b.** 6 months

16. Business: Depreciation A $22 million corporate jet depreciates by 18% per year. Find its value after:

    a. 10 years   **b.** 6 months

17. General: Population According to the United Nations Fund for Population Activities, the population of the world x years after the year 2000 will be $P(x) = 5.89e^{0.0125x}$ billion people (for $0 \leq x \leq 20$). Use this formula to predict the world population in the year 2015.

18. General: Population The most populous country is China, with a (2000) population of 1.26 billion, increasing by 0.8% per year. The 2000 population of India was 1.02 billion, increasing by 1.7% per year. Assuming that these growth rates continue, which population will be larger in the year 2025?

19. General: Nuclear Meltdown According to the Nuclear Regulatory Commission, the probability of a "severe core meltdown accident" at a nuclear reactor in the United States within the next n years is

$$P(n) = 1 - (0.9997)^{100n}$$

Find the probability of a meltdown within:

    a. 25 years   **b.** 40 years
(The 1986 core meltdown in the Chernobyl reactor in the Soviet Union spread radiation over much of Eastern Europe, leading to an undetermined number of fatalities.)

* The realization that populations grow exponentially while food supplies grow only geometrically caused the great nineteenth-century essayist Thomas Carlyle to dub economics the "dismal science." He was commenting not on how interesting economics is, but on the grim conclusions that follow from populations outstripping their food supplies.

** Honus Wagner, regarded as the greatest player of his time, played for the Pittsburgh Pirates. Baseball cards were then sold with cigarettes, and Wagner became even more famous when he refused to accept payment for the use of his picture as a protest against the use of tobacco.

20. General: Mosquitoes   Female mosquitoes (*Culex pipiens*) feed on blood (only the females drink blood) and then lay several hundred eggs. In this way each mosquito can, on the average, breed another 300 mosquitoes in about 9 days. Find the number of great-grandchildren mosquitoes that will be descended from one female mosquito, assuming that all eggs hatch and mature.

21. Environmental Science: Light   According to the Bouguer–Lambert law, the proportion of light that penetrates ordinary seawater to a depth of x feet is $e^{-0.44x}$. Find the proportion of light that penetrates to a depth of:

   a. 3 feet   **b.** 10 feet

22–23: Biomedical: Drug Dosage   If a dosage d of a drug is administered to a patient, the amount of the drug remaining in the tissues t hours later will be $f(t) = d \cdot e^{-kt}$, where k (the "absorption constant") depends on the drug.*

22. For the immunosuppresent cyclosporine, the absorption constant is $k = 0.012$. For a dose of $d = 400$ mg, use the above formula to find the amount of cyclosporine remaining in the tissues after:

   a. 24 hours   **b.** 48 hours

23. For the cardioregulator Digoxin, the absorption constant is $k = 0.018$. For a dose of $d = 2$ mg, use the above formula to find the amount remaining in the tissues after:

   a. 24 hours   **b.** 48 hours

24. Biomedical: Bacterial Growth   A colony of bacteria in a Petri dish doubles in size every hour. At noon the Petri dish is just covered with bacteria. At what time was the Petri dish

   a. 50% covered [*Hint:* No calculation needed.]
   b. 25% covered

25. Business: Advertising   A company finds that x days after the conclusion of an advertising campaign the daily sales of a new product are $S(x) = 100 + 800e^{-0.2x}$. Find the daily sales 10 days after the advertising campaign.

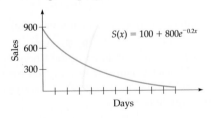

$$S(x) = 100 + 800e^{-0.2x}$$

26. Business: Quality Control   A company finds that the proportion of its light bulbs that will burn continuously for longer than t weeks is $p(t) = e^{-0.01t}$. Find the proportion of bulbs that burn for longer than 10 weeks.

27. General: Temperature   A covered mug of coffee originally at 200 degrees, if left for t hours in a 70-degree room will cool to a temperature of $T(t) = 70 + 130e^{-1.8t}$ degrees. Find the temperature of the coffee after:

   a. 15 minutes   **b.** half an hour

28. Behavioral Science: Learning   In certain experiments the percentage of items that are remembered after t time units is

$$p(t) = 100 \frac{1 + e}{1 + e^{t+1}}$$

Such curves are called "forgetting" curves. Find the percentage remembered after:

   a. 0 time units   **b.** 2 time units

29. Biomedical: Epidemics   The Reed–Frost model for an epidemic predicts that the number I of newly infected people is $I = S(1 - e^{-rx})$, where S is the number of susceptible people, r is the effective contact rate, and x is the number of infectious people. Suppose that a school reports an outbreak of measles with $x = 10$ cases and that the effective contact rate is $r = 0.01$. If the number of susceptibles is $S = 400$, use the Reed–Frost model to estimate how many students will be newly infected during this stage of the epidemic.

30. Social Science: Election Cost   The cost of winning a seat in the House of Representatives in recent years has been approximately $C(x) = 655e^{0.0728x}$ thousand dollars, where x is the number of years since 1996. Estimate the cost of winning a House seat in the year 2004. (*Source: Center for Responsive Politics*)

31. General: Rate of Return   An investment of $8000 grows to $10,291.73 in 4 years. Find the annual rate of return. [*Hint:* Use $P(1 + r)^t$ and solve for r (rounded).]

* For further details, see T. R. Harrison (ed), *Principles of Internal Medicine*, 11th ed. (New York: McGraw-Hill, 1987), pp. 342–352.

32. General: Rate of Return An investment of $9000 grows to $10,380.65 in 2 years. Find the annual rate of return [*Hint:* Use $P(1 + r)^t$ and solve for r (rounded).]

33. General: Population As noted earlier, the most populous states are California and Texas, with New York third and Florida fourth but gaining. According to the Census Bureau, x years after 1990 the population of New York will be $17.99e^{0.0025x}$, the population of Texas will be $16.99e^{0.0177x}$, and the population of Florida will be $12.94e^{0.0283x}$ (all in millions).

a. Graph these three functions on a calculator on the window [0, 30] by [0, 30].
b. Note that the order of these states (largest to smallest) in 1995 was Texas, New York, Florida. What will be the (projected) order in 2005?
c. What will be the (projected) order in 2020?
d. In which year is Florida projected to overtake Texas as the second-largest state? [*Hint:* Use INTERSECT.]

34. General: St. Louis Arch The Gateway Arch in St. Louis is built around a mathematical curve called a "catenary." The height of this catenary above the ground at a point x feet from the center line is

$$y = 688 - 31.5(e^{0.01033x} + e^{-0.01033x})$$

a. Graph this curve on a calculator on the window [−400, 400] by [0, 700].
b. Find the height of the Gateway Arch at its highest point, using the fact that the top of the arch is 5 feet higher than the top of the central catenary.

35. Athletics: Olympic Games When the Olympic Games were held near Mexico City in the summer of 1968, many athletes were concerned that the high elevation would affect their performance. If air pressure decreases exponentially by 0.4% for each 100 feet of altitude, by what percentage did the air pressure decrease in moving from Tokyo (the site of the 1964 Summer Olympics, at altitude 30 feet) to Mexico City (altitude 7347 feet)?

1.6 Logarithmic Functions

APPLICATION PREVIEW

Carbon-14 Dating and the Shroud of Turin

Carbon-14 dating is a method for estimating the age of ancient plants and animals. While a plant is alive, it absorbs carbon dioxide, and with it both "ordinary" carbon and carbon-14, a radioactive form of carbon produced by cosmic rays in the upper atmosphere. When the plant dies, it stops absorbing both types of carbon, and the carbon-14 decays exponentially at a known rate, as shown in the following graph.

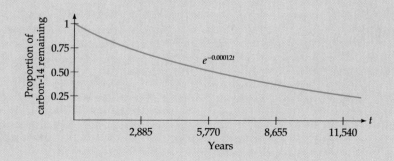

The time since the plant died can then be estimated by comparing the amount of carbon-14 that remains to the amount of ordinary carbon remaining, since they were originally in a known proportion. The comparison technique involves using *logarithms*, as explained in this section. Animal remains can also be dated in this way because the diets of animals contain plant matter.

In 1988, the Roman Catholic Church used carbon-14 dating to determine the authenticity of the Shroud of Turin, believed by many to be the burial cloth of Jesus. This was done by estimating the age of the flax plants from which the linen shroud was woven. In Exercise 32 on page 88, you will find that the shroud is only a few hundred years old and therefore cannot be the burial cloth of Christ.

A few of the technical details of carbon-14 dating are as follows: The amount of carbon-14 remaining in a sample can be found by burning the sample near a Geiger counter to measure the radioactivity. For the shroud, this would have required a handkerchief-sized sample, so a more advanced method, employing a tandem accelerator–mass spectrometer and requiring only a postage-stamp-sized sample, was used instead.

Carbon-14 dating assumes, however, that the ratio of carbon-14 to ordinary carbon in the atmosphere has remained steady over the centuries. This assumption seemed difficult to verify until the discovery in 1955 of a bristlecone pine tree in the White Mountains of California that was over 2000 years old (according to its growth rings). Each annual ring had absorbed carbon and carbon-14 from the air and so provided a year-by-year record of their ratio. The analysis showed that the ratio has remained relatively steady over the centuries, and where variations did occur, scientists were able to construct a table for making corrections in the original carbon-14 dates. (Incidentally, bristlecone pine trees that are 4900 years old have been found, making them the oldest living things on earth.) For extremely old remains, such as dinosaur fossils, archeologists use longer-lasting radioactive elements, such as potassium-40.

Carbon-14 dating, for which its inventor Willard Libby received a Nobel Prize in 1960, has become an invaluable tool in the social and biological sciences.

Introduction

In this section we introduce logarithmic functions, concentrating on *common* (base 10) and *natural* (base *e*) logarithms, and then apply them to the doubling of investments and carbon-14 dating. Both types of logarithms will be used in Chapter 2.

Common Logarithms

The word "logarithm" (abbreviated "log") means *power* or *exponent*. The number being raised to the power is called the *base* and is written as a subscript. For example, the expression

$$\log_{10} 1000 \qquad \text{Read: log (base 10) of 1000}$$

$$\overset{\curvearrowleft}{} \text{base}$$

means the *exponent* of 10 that gives 1000. Since $10^3 = 1000$, the exponent is 3, so the *logarithm* is 3.

$$\log_{10} 1000 = 3 \qquad \text{Since } 10^3 = 1000$$

Logarithms with base 10 are called *common logarithms*. For common logarithms we often omit the subscript, with base 10 understood.

Common Logarithms

$$\log x = y \quad \text{is equivalent to} \quad 10^y = x \qquad \log x \text{ means } \log_{10} x \text{ (base 10)}$$

Since $10^y = x$ is positive for every value of y, $\log x$ is defined only for *positive* values of x. The common logarithm of a number can often be found by expressing the number as a power of 10 and then finding the exponent.

EXAMPLE 1   **Finding a Common Logarithm**

Find log 100.

Solution

$$\log 100 = y \qquad \text{is equivalent to} \qquad 10^y = 100 \qquad \begin{array}{l} y = 2 \text{ works, so } 2 \\ \text{is the logarithm} \end{array}$$

The logarithm y is the exponent that solves

Therefore,

$$\log 100 = 2 \qquad\qquad \text{Since } 10^2 = 100$$

■

EXAMPLE 2   **Finding a Common Logarithm**

Find $\log \dfrac{1}{10}$.

Solution

$$\log \frac{1}{10} = y \qquad \text{is equivalent to} \qquad 10^y = \frac{1}{10} \qquad \begin{array}{l} y = -1 \text{ works, so } -1 \\ \text{is the logarithm} \end{array}$$

The logarithm y is the exponent that solves

Therefore,

$$\log \frac{1}{10} = -1 \qquad\qquad \text{Since } 10^{-1} = \frac{1}{10}$$

■

PRACTICE PROBLEM 1   Find log 10,000.   *Solution at the back of the book*

Graphing Calculator Exploration

On a graphing calculator, common logarithms are found using the $\boxed{\text{LOG}}$ key. Verify that the last three common logarithms we considered can be found in this way.

log(100)
2
log(1/10)
-1
log(10000)
4

Properties of Common Logarithms

Because logarithms are exponents, the properties of exponents can be restated as properties of logarithms. For positive numbers M and N:

Properties of Common Logarithms

1. $\log 1 = 0$	The log of 1 is 0 (since $10^0 = 1$)
2. $\log 10 = 1$	The log of 10 is 1 (since $10^1 = 10$)
3. $\log 10^x = x$	The log of 10 to a power is just the power (since $10^x = 10^x$)
4. $10^{\log x} = x$	10 raised to the log of a number is just the number ($x > 0$)
5. $\log (M \cdot N) = \log M + \log N$	The log of a product is the sum of the logs
6. $\log \left(\dfrac{1}{N}\right) = -\log N$	The log of 1 over a number is minus the log of the number
7. $\log \left(\dfrac{M}{N}\right) = \log M - \log N$	The log of a quotient is the difference of the logs
8. $\log (M^N) = N \cdot \log M$	The log of a number to a power is the power times the log

The first two properties are simply special cases of the third (with $x = 0$ and $x = 1$). Because logs are exponents, the third property simply says that the exponent of 10 that gives 10^x is x, which is obvious when you think about it. Because $\log x$ is the power of 10 that gives x, raising 10 to that power must give x, which is the fourth property. Justifications for properties 5–8 are given on page 86. Property 8 will be particularly useful in applications, and can be summarized: *Logarithms bring down exponents.*

EXAMPLE 3   **Using the Properties of Common Logarithms**

a. $\log 10^7 = 7$ Property 3

b. $10^{\log 13} = 13$ Property 4

c. $\log (3 \cdot 4) = \log 3 + \log 4$ Property 5

d. $\log \left(\dfrac{1}{4}\right) = -\log 4$ Property 6

e. $\log \left(\dfrac{3}{4}\right) = \log 3 - \log 4$ Property 7

f. $\log (5^3) = 3 \log 5$ Property 8: $\ln(5^3) = 3 \ln 5$

EXAMPLE 4   **Finding the Doubling Time of an Investment**

An investment grows in value by 7% per year. How many years will it take to double?

Solution

Let P stand for the size of the investment (since we are not told the amount). As we saw on page 67, an investment grows according to the formula $P(1 + r)^t$, and for the investment to double it must reach $2P$. This gives the following equation, which we solve for t.

$$P(1 + 0.07)^t = 2P \qquad \text{$P(1 + r)^t$ with $r = 0.07$ set equal to $2P$}$$

$$1.07^t = 2 \qquad \text{Canceling Ps and simplifying}$$

$$\log 1.07^t = \log 2 \qquad \text{Taking log of each side}$$

$$t \cdot \log 1.07 = \log 2 \qquad \text{Bringing down t (property 8)}$$

$$t = \frac{\log 2}{\log 1.07} \qquad \text{Dividing by log 1.07}$$

$$t \approx 10.24 \qquad \text{Using a calculator}$$

The investment will have doubled in value in 11 years.

Note that we round *up* to the next whole year to ensure that the value actually doubles, by which time the value will have somewhat *more* than doubled. This same technique can be used to find the doubling time for a population or for any other quantity that grows exponentially.

Graphs of Logarithmic and Exponential Functions

If a point (x, y) lies on the graph of $y = \log x$ or, equivalently, $x = 10^y$, then, reversing x and y, the point (y, x) lies on the graph of $y = 10^x$. That is, the curves $y = \log x$ and $y = 10^x$ are related by having their x- and y-coordinates *reversed*, so the curves are *mirror images* of each other in the line $y = x$.

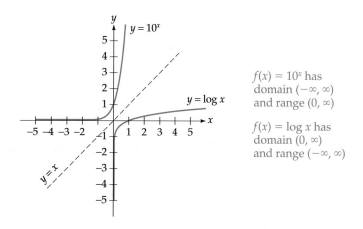

$f(x) = 10^x$ has domain $(-\infty, \infty)$ and range $(0, \infty)$

$f(x) = \log x$ has domain $(0, \infty)$ and range $(-\infty, \infty)$

This graphical relationship is equivalent to the fact that the functions $\log x$ and 10^x "undo" or "reverse" each other, as is shown by properties 3 and 4 on page 79. Such functions are called *inverse functions*:

$$\log x \quad \text{and} \quad 10^x \quad \text{are inverse functions}$$

Logarithms to Other Bases

We may calculate logarithms to bases other than 10. In fact, *any* positive number other than 1 may be used as a base for logs, using the following definition.

Base a Logarithms

$$\log_a x = y \quad \text{is equivalent to} \quad a^y = x \quad\quad (x > 0)$$

For example,

$$\log_2 8 = 3 \quad\quad \text{Since } 2^3 = 8$$

and

$$\log_9 3 = \frac{1}{2} \quad\quad \text{Since } 9^{1/2} = \sqrt{9} = 3$$

Natural Logarithms

We will use only one other base, the number e (approximately 2.718) that we defined on page 69. Logarithms to the base e are called *natural* or *Napierian* logarithms.* The natural logarithm of a positive number x is written ln x ("n" for "natural") and is found using the $\boxed{\text{LN}}$ key on a calculator.

Natural Logarithms

$\ln x = y$	is equivalent to	$e^y = x$ ln x means $\log_e x$ (base e)

PRACTICE PROBLEM 2   Use a calculator to find ln 8.34.   *Solution at the back of the book*

The properties of natural logarithms are similar to those of common logarithms but with ln instead of log and e instead of 10. For positive numbers M and N:

Properties of Natural Logarithms

1. $\ln 1 = 0$	The natural log of 1 is 0 (since $e^0 = 1$)
2. $\ln e = 1$	The natural log of e is 1 (since $e^1 = e$)
3. $\ln e^x = x$	The natural log of e to a power is just the power (since $e^x = e^x$)
4. $e^{\ln x} = x$	e raised to the natural log of a number is just the number ($x > 0$)
5. $\ln (M \cdot N) = \ln M + \ln N$	The natural log of a product is the sum of the logs
6. $\ln \left(\dfrac{1}{N} \right) = -\ln N$	The natural log of 1 over a number is minus the log of the number
7. $\ln \left(\dfrac{M}{N} \right) = \ln M - \ln N$	The natural log of a quotient is the difference of the logs
8. $\ln (M^N) = N \cdot \ln M$	The natural log of a number to a power is the power times the log

As with common logs, the first two properties are special cases of the third (with $x = 0$ and $x = 1$). The third and fourth properties have interpretations analogous to their "common" counterparts (for example, the third says that the exponent of e that gives e^x is x). As before, Property 8 can be summarized: *Logarithms bring down exponents.*

---

* After John Napier, a seventeenth-century Scottish mathematician and, incidentally, the inventor of the decimal point.

EXAMPLE 5 **Using the Properties of Natural Logarithms**

a. $\ln e^7 = 7$ Property 3

b. $e^{\ln 13} = 13$ Property 4

c. $\ln (3 \cdot 4) = \ln 3 + \ln 4$ Property 5

d. $\ln \left(\dfrac{1}{4}\right) = -\ln 4$ Property 6

e. $\ln \left(\dfrac{3}{4}\right) = \ln 3 - \ln 4$ Property 7

f. $\ln (5^3) = 3 \ln 5$ Property 8: $\ln(5^3) = 3 \ln 5$

Properties 3 and 4 show that $y = \ln x$ and $y = e^x$ are *inverse functions*, so their graphs are reflections of each other in the diagonal line $y = x$.

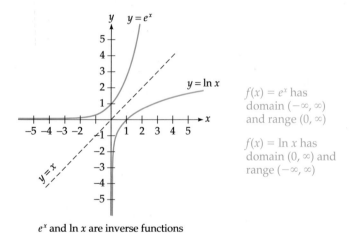

$f(x) = e^x$ has domain $(-\infty, \infty)$ and range $(0, \infty)$

$f(x) = \ln x$ has domain $(0, \infty)$ and range $(-\infty, \infty)$

e^x and $\ln x$ are inverse functions

The properties of natural logarithms are helpful for simplifying functions.

EXAMPLE 6 **Simplifying a Function**

$f(x) = \ln (2x) - \ln 2$

$= \ln 2 + \ln x - \ln 2$ Since $\ln (2x) = \ln 2 + \ln x$ by property 5

$= \ln x$ Canceling

EXAMPLE 7 **Simplifying a Function**

$$f(x) = \ln\left(\frac{x}{e}\right) + 1 = \ln x - \ln e + 1$$

Since $\ln (x/e) = \ln x - \ln e$ by property 7

$$= \ln x - 1 + 1$$

Since $\ln e = 1$ by property 2

$$= \ln x$$

Canceling

EXAMPLE 8 **Simplifying a Function**

$$f(x) = \ln (x^5) - \ln (x^3) = 5 \ln x - 3 \ln x$$

Bringing down exponents by property 8

$$= 2 \ln x$$

Combining

Graphing Calculator Exploration

Some advanced graphing calculators have computer algebra systems that can simplify algebraic expressions. For example, the Texas Instruments TI-89 graphing calculator simplifies logarithmic expressions, but only if the condition $x > 0$ is included so that the logarithms are defined.

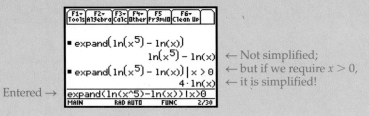

← Not simplified;
← but if we require $x > 0$,
← it is simplified!

Entered →

Graphing Calculator Exploration

a. Graph the function $y = \ln x^2$ on the window $[-5, 5]$ by $[-5, 5]$.
b. Change the function to $y = 2 \ln x$ and explain why the two graphs are different. (Doesn't property 8 say that the two functions should be the same?)

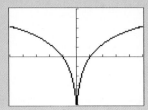
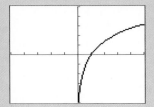

Carbon-14 Dating

All living things absorb small amounts of radioactive carbon-14 from the atmosphere. When they die, the carbon-14 stops being absorbed and decays exponentially into ordinary carbon. Therefore, the proportion of carbon-14 still present in a fossil or other ancient remain can be used to estimate how old it is. The proportion of the original carbon-14 that will be present after t years is

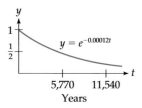

$y = e^{-0.00012t}$

5,770   11,540
Years

$$\begin{pmatrix} \text{Proportion of carbon-14} \\ \text{remaining after } t \text{ years} \end{pmatrix} = e^{-0.00012t}$$

EXAMPLE 9 **Dating by Carbon-14**

The Dead Sea Scrolls, discovered in a cave near the Dead Sea in Jordan, are among the earliest documents of Western civilization. Estimate the age of the Dead Sea Scrolls if the animal skins on which some were written contain 78% of their original carbon-14.

Solution The proportion of carbon-14 remaining after t years is $e^{-0.00012t}$. We equate this formula to the actual proportion (expressed as a decimal).

$e^{-0.00012t} = 0.78$	Equating the proportions
$\ln e^{-0.00012t} = \ln 0.78$	Taking natural logs
$-0.00012t = \ln 0.78$	$\ln e^{-0.00012t} = -0.00012t$ by property 3
$t = \dfrac{\ln 0.78}{-0.00012} \approx \dfrac{-0.24846}{-0.00012} \approx 2071$	Solving for t using a calculator

Therefore, the Dead Sea Scrolls are approximately 2070 years old. ∎

Graphing Calculator Exploration

Solve Example 9 by graphing $y_1 = e^{-0.00012x}$ and $y_2 = 0.78$ on the window [0, 10000] by [0, 1] and using INTERSECT to find where they meet.

Both here and in Example 4 on page 80 we solved for the variable in the exponent by taking logarithms, but here we used *natural* logarithms and there we used *common* logarithms. Is there a difference?

Not really—you can use logarithms to *any* base to solve for exponents, and the final answer will be the same (see Exercise 41). However, when *e* is involved, it is usually easier to use *natural* logs.

Justification of Properties 5–8 of Logarithms

Properties 1–4 of logarithms (both common and natural) were justified earlier. Properties 5–8 of common logarithms are justified below, with the "natural" justifications obtained simply by replacing 10 by *e*.

The addition law of exponents, $10^x \cdot 10^y = 10^{x+y}$, can be stated in words: "The exponent of a product is the sum of the exponents." Because logs are exponents, this can be restated, "The log of a product is the sum of the logs," which is just property 5.

The subtraction law of exponents, $10^x/10^y = 10^{x-y}$ can be stated, "The exponent of a quotient is the difference of the exponents." This translates into "The log of a quotient is the difference of the logs," which is just property 7. Property 6 is simply a special case of property 7 with $M = 1$ (and using property 1).

The law of exponents $(10^x)^y = 10^{x \cdot y}$ says that the exponent *y* can be "brought down" and multiplied by the *x*. Because logs are exponents, this says that in $\log(M^N)$, the exponent *N* can be brought down and multiplied by the logarithm $\log M$, giving $N \cdot \log M$, which is just property 8.

Section Summary

Logarithms are exponents: Common logs are exponents of 10, and natural logs are exponents of *e*. That is,

$y = \log x$ is equivalent to $x = 10^y$ $\log x$ means $\log_{10} x$ (base 10)

$y = \ln x$ is equivalent to $x = e^y$ $\ln x$ means $\log_e x$ (base *e*)

Each property of logarithms is equivalent to a property of exponents.

Logarithmic Property	Exponential Property
$\log 1 = 0$	$10^0 = 1$
$\log(M \cdot N) = \log M + \log N$	$10^x \cdot 10^y = 10^{x+y}$
$\log\left(\dfrac{1}{N}\right) = -\log N$	$\dfrac{1}{10^y} = 10^{-y}$
$\log\left(\dfrac{M}{N}\right) = \log M - \log N$	$\dfrac{10^x}{10^y} = 10^{x-y}$
$\log M^N = N \log M$	$(10^x)^y = 10^{x \cdot y}$

In practice, logarithms are found using the ⎡LOG⎤ and ⎡LN⎤ keys on a calculator. The properties of exponents lead to properties of logarithms, which are listed on pages 79 and 82. Property 8, that *logs bring down exponents*, is particularly useful in applications that require solving for a variable in the exponent.

EXERCISES 1.6 (Most require or ▦.)

Find each logarithm *without* using a calculator.

1. **a.** $\log 100{,}000$ **b.** $\log \frac{1}{100}$ **c.** $\log \sqrt{10}$

2. **a.** $\log 1000$ **b.** $\log \frac{1}{1000}$ **c.** $\log \sqrt[3]{10}$

3. **a.** $\ln e^5$ **b.** $\ln \frac{1}{e}$ **c.** $\ln \sqrt[3]{e}$

4. **a.** $\ln e^3$ **b.** $\ln \frac{1}{e^2}$ **c.** $\ln \sqrt{e}$

5. **a.** $\ln 1$ **b.** $\ln (\ln e^e)$ **c.** $\ln \sqrt[3]{e^2}$

6. **a.** $\ln e$ **b.** $\ln (\ln e)$ **c.** $\ln \sqrt{e^3}$

7. **a.** $\log_4 16$ **b.** $\log_4 \frac{1}{4}$ **c.** $\log_4 2$

8. **a.** $\log_9 81$ **b.** $\log_9 3$ **c.** $\log_9 \frac{1}{9}$

▦ For each logarithm:

   i. Find the logarithm using a calculator, rounding your answer to three decimal places.

   ii. Raise the appropriate base (either 10 or *e*) to the power you found in part (i) and check that the result agrees with the number in the original problem. (*Note:* For the power, include *all* of the digits that your calculator showed for part (i) to minimize the error.)

9. **a.** $\log 22.3$ **b.** $\ln 22.3$

10. **a.** $\log 44.9$ **b.** $\ln 44.9$

Use the properties of natural logarithms to simplify each function.

11. $f(x) = \ln (9x) - \ln 9$ 12. $f(x) = \ln \left(\frac{x}{2}\right) + \ln 2$

13. $f(x) = \ln (x^3) - \ln x$ 14. $f(x) = \ln (4x) - \ln 4$

15. $f(x) = \ln \left(\frac{x}{4}\right) + \ln 4$

16. $f(x) = \ln (x^5) - 3 \ln x$

17. $f(x) = \ln (e^{5x}) - 2x - \ln 1$

18. $f(x) = \ln (e^{-2x}) + 3x + \ln 1$

19. $f(x) = 8x - e^{\ln x}$

20. $f(x) = e^{\ln x} + \ln (e^{-x})$

21. Without using a calculator, sketch the graph of $f(x) = \log (x + 1)$.

22. Without using a calculator, sketch the graph of $f(x) = \ln (x + e)$.

▦ 23. Find the domain and range and graph the function $f(x) = \ln (x^2 - 1)$.

▦ 24. Find the domain and range and graph the function $f(x) = \ln (1 - x^2)$.

APPLIED EXERCISES

25. **Business: Investments**   A real estate investment increases in value by 10% per year. How soon will it double in value?

26. **Business: Investments**   An art collection increases in value by 4% each year. How soon will it double in value?

27. **General: Population**   During the years 1990–1998, the population of Georgia increased by 2.1% annually. At this rate, how soon will the population double?

28. **General: Population**   During the years 1990–1998, the population of Nevada increased by 4.8% annually. At this rate, how soon will the population double?

29. **Economics: Energy Output**   The world's output of primary energy (petroleum, natural gas, coal, hydroelectricity, and nuclear electricity) is increasing by 2.4% annually. How soon will it double? (*Source: Worldwatch Institute.*)

30. Environmental Science: Carbon Emissions Between 1995 and 1997, the global carbon emissions increased by 1.5%. At this rate, how soon will carbon emissions double? (*Note:* Carbon dioxide is the most prevalent "greenhouse gas" that many scientists believe is destabilizing the Earth's climate.) (*Source: Worldwatch Institute.*)

31–32: General: Carbon-14 Dating The proportion of carbon-14 still present in a sample after t years is $e^{-0.00012t}$.

31. Estimate the age of the cave paintings discovered in 1994 in the Ardèche region of France if the carbon with which they were drawn contains only 2.3% of its original carbon-14. They are the oldest known paintings in the world.

32. Estimate the age of the Shroud of Turin, believed by many to be the burial cloth of Christ (see the Application Preview on pages 75–77), from the fact that its linen fibers contained only 92.3% of their original carbon-14.

33–34: General: Potassium-40 Dating The radioactive isotope potassium-40 is used to date very old remains. The proportion of potassium-40 that remains after t million years is $e^{-0.00054t}$. Use this function to estimate the age of the following fossils.

33. The most nearly complete skeleton of an early human ancestor ever found was discovered in Kenya in 1984. Use the above formula to estimate the age of the remains if they contained 99.91% of their original potassium-40.

34. Dating Older Women Use the above formula to estimate the age of the partial skeleton of *Australopithecus afarensis* known as "Lucy" which was found in Ethiopia in 1977, if it had 99.82% of its original potassium-40.

35. Social Science: Education and Income According to a 1992 study,* each additional year of education increases one's income by 16%. How many additional years of schooling would then be required to double one's income?

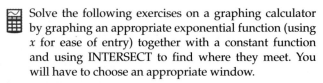

Solve the following exercises on a graphing calculator by graphing an appropriate exponential function (using x for ease of entry) together with a constant function and using INTERSECT to find where they meet. You will have to choose an appropriate window.

36. General: Value of an Automobile An antique automobile increases in value by 7.5% annually. How soon will it double in value?

37. General: Value of a Home A vacation home increases in value by 5.5% annually. How soon will it double in value?

38. General: Carbon-14 Dating In 1991 two hikers in the Italian Alps found the frozen but well-preserved body of the most ancient human ever found, dubbed "Iceman." Estimate the age of Iceman if his grass cape contained 53% of its original carbon-14. (Use the carbon-14 decay function stated in the directions for Exercises 31–32.)

39. General: Potassium-40 Dating Estimate the age of the oldest known dinosaur, a dog-sized creature called Herrerasaurus found in Argentina in 1988, if volcanic material found with it contained 88.4% of its original potassium-40. (Use the potassium-40 decay function given in the directions for Exercises 33–34.)

40. Environmental Science: Nuclear Waste More than half a century after the beginning of the nuclear age, not a single country has found a safe or permanent way of disposing of long-lived radioactive waste. Among the most hazardous radioactive waste is irradiated fuel from nuclear power plants, which totaled 84,000 tons in 1990 and is growing by 11.3% annually. At this rate, how long will it take for this amount to double? (*Source: Worldwatch Institute.*)

---

* Orley Ashenfelter and Alan Krueger, "Estimate of the Economic Return to Schooling from a New Sample of Twins," *The American Economic Review,* pp. 1157–1173, Vol. 84, No. 5, December 1994.

41. **Change-of-Base Formula for Logarithms** Let a and b be any two bases, and let x be any positive number.

a. Give a justification for each numbered equals sign.

$$\log_a x \overset{1}{=} \log_a b^{\log_b x} \overset{2}{=} (\log_b x) \cdot (\log_a b)$$

b. Show that the result can be written as

$$\log_a x = (\log_a b) \cdot (\log_b x)$$

This is the "change-of-base" formula for logarithms, enabling one to express $\log_b x$ in terms of $\log_a x$.

c. Use the change-of-base formula in the numerator and denominator of the following fraction to justify each numbered equals sign.

$$\frac{\log_a x}{\log_a y} \overset{3}{=} \frac{(\log_a b) \cdot (\log_b x)}{(\log_a b) \cdot (\log_b y)} \overset{4}{=} \frac{\log_b x}{\log_b y}$$

This equation shows that when you are finding a *ratio* of logarithms, using one base gives the same result as using any other base. Because solving for a variable in the exponent involves calculating *ratios* of logarithms, you may do so using logarithms to *any* base.

Chapter Summary with Hints and Suggestions

Reading the text and doing the exercises in this chapter have helped you to master the following skills, which are listed by section (in case you need to review them) and are keyed to particular Review Exercises. Answers for all Review Exercises are given at the back of the book, and full solutions can be found in the Student Solutions Manual.

1.1 REAL NUMBERS, INEQUALITIES, AND LINES

- Translate an interval into set notation and graph it on the real line. *(Review Exercises 1–4.)*

$$[a, b] \quad (a, b) \quad [a, b) \quad (a, b] \quad (-\infty, b]$$

$$(-\infty, b) \quad [a, \infty) \quad (a, \infty) \quad (-\infty, \infty)$$

- Express given information in interval form. *(Review Exercises 5–6.)*

- Find an equation for a line that satisfies certain conditions. *(Review Exercises 7–12.)*

$$m = \frac{y_2 - y_1}{x_2 - x_1} \qquad y = mx + b$$

$$y - y_1 = m(x - x_1) \qquad x = a \qquad y = b$$

$$ax + by = c$$

- Find an equation of a line from its graph. *(Review Exercises 13–14.)*

- Use straight-line depreciation to find the value of an asset. *(Review Exercises 15–16.)*

- Use 📇 on real-world data to find a regression line and make a prediction. *(Review Exercises 17–18.)*

1.2 EXPONENTS

- Evaluate negative and fractional exponents without using a calculator. *(Review Exercises 19–26.)*

$$x^0 = 1 \qquad x^{-n} = \frac{1}{x^n} \qquad x^{m/n} = \sqrt[n]{x^m} = \left(\sqrt[n]{x}\right)^m$$

- Evaluate an exponential expression using a calculator. *(Review Exercises 27–28.)*

1.3 FUNCTIONS

- Evaluate and find the domain and range of a function. *(Review Exercises 29–32.)*

  A function f is a rule that assigns to each number x in a set (the domain) a (single) number $f(x)$. The range is the set of all resulting values $f(x)$.

- Use the vertical line test to see whether a graph defines a function. *(Review Exercises 33–34.)*

- Graph a linear function.   *(Review Exercises 35–36.)*

$$f(x) = mx + b$$

- Graph a quadratic function.   *(Review Exercises 37–38.)*

$$f(x) = ax^2 + bx + c$$

- Solve a quadratic equation by factoring and by the quadratic formula.   *(Review Exercises 39–42.)*

vertex: $x = \dfrac{-b}{2a}$

x-intercepts: $x = \dfrac{-b \pm \sqrt{b^2 - 4ac}}{2a}$

- Use ⊞ to graph a quadratic function.   *(Review Exercises 43–44.)*

- Construct a linear function from an applied problem or from real-life data, and then use the function in an application.   *(Review Exercises 45–48.)*

- For given cost and revenue functions, find the break-even points and maximum profit.   *(Review Exercises 49–50.)*

1.4   FUNCTIONS, CONTINUED

- Evaluate and find the domain and range of a more complicated function.   *(Review Exercises 51–54.)*
- Solve a polynomial equation by factoring.   *(Review Exercises 55–58.)*
- Graph a "shifted" function.   *(Review Exercises 59–60.)*
- Graph a piecewise linear function.   *(Review Exercises 61–62.)*
- Given two functions $f(x)$ and $g(x)$, find their composition.   *(Review Exercises 63–68.)*

$$f(g(x))   \quad g(f(x))$$

- Solve an applied problem involving the composition of functions.   *(Review Exercise 69.)*
- Use factoring or ⊞ to solve a polynomial equation.   *(Review Exercises 70–71.)*
- Use ⊞ to fit a curve to real-life data and make a prediction.   *(Review Exercise 72.)*

1.5   EXPONENTIAL FUNCTIONS

- Sketch the graph of an exponential function.   *(Review Exercises 73–74.)*
- Find the appreciated value of an investment.   *(Review Exercises 75–76.)*

$$P(1 + r)^t$$

- Depreciate an asset by a fixed percentage.   *(Review Exercises 77–78.)*

$$P(1 + r)^t \qquad \text{(for depreciation, } r \text{ is negative)}$$

- Use exponential functions to estimate the growth of populations or other quantities.   *(Review Exercises 79–80.)*

1.6   LOGARITHMIC FUNCTIONS

- Evaluate common and natural logarithms without using a calculator.   *(Review Exercises 81–82.)*

$$\log x = y \quad \text{is equivalent to } 10^y = x$$

$$\ln x = y \quad \text{is equivalent to } e^y = x$$

- Use the properties of natural logarithms to simplify a function.   *(Review Exercises 83–84.)*

$$\ln 1 = 0 \qquad \ln e = 1 \qquad \ln e^x = x$$

$$e^{\ln x} = x \qquad \ln(M \cdot N) = \ln M + \ln N$$

$$\ln\left(\frac{1}{N}\right) = -\ln N \qquad \ln\left(\frac{M}{N}\right) = \ln M - \ln N$$

$$\ln(M^N) = N \cdot \ln M$$

- Find the doubling time for an investment, a population, or some other quantity.   *(Review Exercises 85–88.)*

$$P(1 + r)^t = 2P$$

- Estimate the age of a fossil.   *(Review Exercises 89–90.)*

Hints and Suggestions

- **Overview:** In reviewing this chapter, try to understand the difference between two different kinds of mathematical objects, *geometric* objects (points, curves, etc.) and *analytic* objects (numbers, functions, etc.), and the connections between them. Descartes first made this connection: By drawing axes in the plane, he saw that points could be specified by numerical coordinates, and so *curves* could be specified by *equations* governing their coordinates. This idea connected geometry to algebra, previously distinct subjects. You should be able to express geometric objects analytically, and vice versa. For example, given a *graph* of a line, you should be able to find an *equation* for it, and given a quadratic *function*, you should be able to *graph* it. As you do the Review Exercises, try to be aware of the difference between geometric and analytic objects and their interrelations.

- A graphing calculator or a computer with appropriate software can help you to *explore* a concept more fully (for example, seeing how a curve changes as a coefficient or exponent changes) and also to *solve* a problem (for example, eliminating the point-plotting aspect of graphing, or finding a regression line).

- If you don't have a graphing calculator, you should have a scientific or business calculator to carry out calculations, especially in later chapters.

- The Practice Problems help you to check your mastery of the skills presented. Complete solutions are given at the back of the book.

- The Student Solutions Manual, available separately from your bookstore, provides fully worked-out solutions to selected exercises.

- **Practice for Test:** Review Exercises 1, 9, 11, 13, 15, 17, 19, 31, 33, 35, 37, 43, 47, 49, 55, 61, 65, 71, 75, 77, 81, 83, and 85.

Review Exercises for Chapter 1   *Practice test exercise numbers are in blue.*

1.1   Real Numbers, Inequalities, and Lines

For each interval, write it in set notation and graph it on the real line.

1. $(2, 5]$ 2. $[-2, 0)$ 3. $[100, \infty)$ 4. $(-\infty, 6]$

5. **General: Wind Speed** The United States Coast Guard defines a "hurricane" as winds of at least 74 mph, a "storm" as winds of at least 55 mph but less than 74 mph, a "gale" as winds of at least 38 mph but less than 55 mph, and a "small craft warning" as winds of at least 21 mph but less than 38 mph. Express each of these wind conditions in interval form. [*Hint*: A small craft warning is $[21, 38)$.]

6. State in interval form:
   a. The set of all positive numbers
   b. The set of all negative numbers
   c. The set of all nonnegative numbers
   d. The set of all nonpositive numbers

Find an equation of the line satisfying the following conditions. If possible, write your answer in the form $y = mx + b$.

7. Slope 2 and passing through the point $(1, -3)$

8. Slope -3 and passing through the point $(-1, 6)$

9. Vertical and passing through the point $(2, 3)$

10. Horizontal and passing through the point $(2, 3)$

11. Passing through the points $(-1, 3)$ and $(2, -3)$

12. Passing through the points $(1, -2)$ and $(3, 4)$

Write an equation of the form $y = mx + b$ for each line graphed below.

13.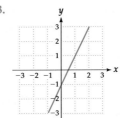

14.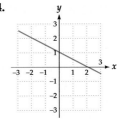

15. Business: Straight-line Depreciation   A contractor buys a backhoe for $25,000 and estimates its useful life to be 8 years, at the end of which its scrap value will be $1,000.

   a. Use straight-line depreciation to find a formula for the value V of the backhoe after t years, for $0 \leq t \leq 8$.

   b. Use your formula to find the value of the backhoe after 4 years.

16. Business: Straight-line Depreciation   A trucking company buys a satellite communication system for $78,000 and estimates its useful life to be 15 years, at the end of which its scrap value will be $3,000.

   a. Use straight-line depreciation to find a formula for the value V of the system after t years, for $0 \leq t \leq 15$.

   b. Use your formula to find the value of the system after 8 years.

17. Ecology: Sulfur Oxide Pollution   Sulfur oxide pollution has decreased significantly in the United States during the last two decades, mostly because of antipollution devices on automobiles and on coal- and oil-fired power plants. The following table shows sulfur oxide emissions (in millions of tons) in the United States from 1975 to 1995. To avoid large numbers, years are listed in the table as years since 1975.

	Years Since 1975	Sulfur Oxide Emissions
1975	0	26.0
1980	5	23.5
1985	10	21.6
1990	15	19.3
1995	20	18.2

Source: Worldwatch Institute and U.S. Environmental Protection Agency.

   a. Enter the table numbers into a graphing calculator and make a plot of the resulting points (Years Since 1975 on the x-axis and Sulfur Oxide Emissions on the y-axis).

   b. Have your calculator find the linear regression formula for this data. Then enter the result as y_1, which gives a formula for Sulfur Oxide Emissions in each year. Plot the points together with the regression line. How well does the line fit the data?

   c. Use your formula to predict the sulfur oxide pollution in the years 2005 and 2010 (assuming that the current trend continues).

18. Social Science: Gap Between Rich and Poor   During the last few decades, the richest 20% of the world's people have been growing richer, while the poorest 20% have become poorer. The probable consequences of this growing gap are not only social instability but also environmental decline, because the richest consume more wastefully while the poorest must cut down rainforest and overgraze land just to survive. The following table shows the ratio of income of the richest 20% to that of the poorest 20% from 1960 to 1990. To avoid large numbers, years are listed in the table as years since 1960.

	Years Since 1960	Ratio of Richest to Poorest
1960	0	30 to 1
1970	10	32 to 1
1980	20	45 to 1
1990	30	60 to 1

Source: United Nations Development Programme.

   a. Enter the table numbers into a graphing calculator and make a plot of the resulting points [Years Since 1960 on the x-axis and the larger number in the Ratio column (the 30, 32, etc.) on the y-axis].

   b. Have your calculator find the linear regression formula for these data. Then enter the result as y_1, which gives a formula for the ratio of richest to poorest in each year. Plot the points together with the regression line. How well does the line fit the data?

   c. Use your formula to predict the ratio in the years 2005 and 2010 (assuming that the current trend continues).

1.2   Exponents

Evaluate each expression *without* using a calculator.

19. $\left(\frac{1}{6}\right)^{-2}$ **20.** $\left(\frac{4}{3}\right)^{-1}$ **21.** $64^{1/2}$

22. $1000^{1/3}$ **23.** $81^{-3/4}$ **24.** $100^{-3/2}$

25. $\left(-\frac{8}{27}\right)^{-2/3}$ **26.** $\left(\frac{9}{16}\right)^{-3/2}$

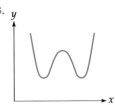

Use a calculator to evaluate each expression. Round answers to two decimal places.

27. $3^{2.4}$ **28.** $12^{1.9}$

1.3 Functions

For each function in Exercises 29–32:

    a. Evaluate the given expression.
    b. Find the domain.
    c. Find the range.

29. $f(x) = \sqrt{x - 7}$; $f(11)$ **30.** $g(t) = \dfrac{1}{t + 3}$; $g(-1)$

31. $h(w) = w^{-3/4}$; $h(16)$ **32.** $w(z) = z^{-4/3}$; $w(8)$

Determine whether each graph defines a function of x.

33.
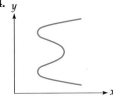

34.

Graph each function.

35. $f(x) = 4x - 8$ **36.** $f(x) = 6 - 2x$

37. $f(x) = -2x^2 - 4x + 6$ **38.** $f(x) = 3x^2 - 6x$

Solve each equation by: **a.** factoring, **b.** the quadratic formula.

39. $3x^2 + 9x = 0$ **40.** $2x^2 - 8x - 10 = 0$

41. $3x^2 + 3x + 5 = 11$ **42.** $4x^2 - 2 = 2$

For each quadratic function in Exercises 43–44:

    a. Find the vertex using the vertex formula.
    b. Graph the function on an appropriate viewing window. (Answers may vary.)

43. $f(x) = x^2 - 10x - 25$ **44.** $f(x) = x^2 + 14x - 15$

45. Business: Car Rentals A rental company rents cars for $45 per day and $0.12 per mile. Find a function $C(x)$ for the cost of a rented car driven for x miles in a day.

46. Business: Simple Interest If money is borrowed for a short period of time, generally less than a year, the interest is often calculated as *simple* interest, according to the formula Interest $= P \cdot r \cdot t$, where P is the principal, r is the rate (expressed as a decimal), and t is the time (in years). Find a function $I(t)$ for the interest charged on a loan of $10,000 at an interest rate of 8% for t years. Simplify your answer.

47. General: Air Temperature The air temperature decreases by about 1 degree Fahrenheit for each 300 feet of altitude. Find a function $T(x)$ for the temperature at an altitude of x feet if the sea level temperature is 70°.

48. Ecology: Carbon Dioxide Pollution The burning of fossil fuels (such as oil and coal) added 25 billion tons of carbon dioxide to the atmosphere during 1997, and this annual amount grows by 0.58 billion tons per year. Find a function $C(t)$ for the amount of carbon dioxide added during the year t years after 1997, and use the formula to find how soon this annual amount will reach 30 billion tons. (*Note:* Carbon dioxide traps solar heat, increasing the earth's temperature, and may lead to flooding of lowland areas by melting the polar ice.) (*Source: U.S. Environmental Protection Agency.*)

49. Break-Even Points and Maximum Profit A store that installs satellite TV receivers finds that if it installs x receivers per week, then its costs will be $C(x) = 80x + 1950$ and its revenue will be $R(x) = -2x^2 + 240x$ (both in dollars).

    a. Find the store's break-even points.
    b. Find the number of receivers it should install to maximize profit, and find the maximum profit.

50. Break-Even Points and Maximum Profit An air conditioner outlet finds that if it sells x air conditioners per month, then its costs will be $C(x) = 220x + 202,500$ and its revenue will be $R(x) = -3x^2 + 2020x$ (both in dollars).

    a. Find the break-even points.
    b. Find the number of air conditioners it should sell to maximize profit, and find the maximum profit.

1.4   Functions, continued

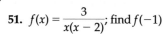

For each function in Exercises 51–54:

    a. Evaluate the given expression.
    b. Find the domain.
    c. Find the range.

51. $f(x) = \dfrac{3}{x(x-2)}$; find $f(-1)$

52. $f(x) = \dfrac{16}{x(x+4)}$; find $f(-8)$

53. $g(x) = |x+2| - 2$; find $g(-4)$

54. $g(x) = x - |x|$; find $g(-5)$

Solve each equation by factoring.

55. $5x^4 + 10x^3 = 15x^2$ **56.** $4x^5 + 8x^4 = 32x^3$

57. $2x^{5/2} - 8x^{3/2} = 10x^{1/2}$ **58.** $3x^{5/2} + 3x^{3/2} = 18x^{1/2}$

Graph each function.

59. $f(x) = (x+1)^2 - 1$ **60.** $f(x) = (x-2)^2 - 4$

61. $f(x) = \begin{cases} 3x - 7 & \text{if } x \ge 2 \\ -x - 1 & \text{if } x < 2 \end{cases}$

If you use a graphing calculator for Exercises 61 and 62, be sure to indicate any missing points.

62. $f(x) = \begin{cases} 6 - 2x & \text{if } x > 2 \\ 2x - 1 & \text{if } x \le 2 \end{cases}$

For each pair $f(x)$ and $g(x)$, find **a.** $f(g(x))$, **b.** $g(f(x))$.

63. $f(x) = x^2 + 1$, $g(x) = \dfrac{1}{x}$

64. $f(x) = \sqrt{x}$, $g(x) = 5x - 4$

65. $f(x) = \dfrac{x+1}{x-1}$, $g(x) = x^3$

66. $f(x) = |x|$, $g(x) = x + 2$

67. $f(x) = \sqrt[3]{x+1}$, $g(x) = x^3 - 1$

68. $f(x) = 5$, $g(x) = 12$

69. Business: Advertising Budget A company's advertising budget is $A(p) = 2p^{0.15}$, where p is the company's profit, and the profit is predicted to be $p(t) = 18 + 2t$, where t is the number of years from now. (Both A and p are in millions of dollars.) Ex-

press the advertising budget A as a function of t, and evaluate the function at $t = 4$.

70. a. Solve the equation $x^4 - 2x^3 - 3x^2 = 0$ by factoring.
    b. Use a graphing calculator to graph $y = x^4 - 2x^3 - 3x^2$ and find the x-intercepts of the graph. Be sure that you understand why your answers to parts (a) and (b) agree.

71. a. Solve the equation $x^3 + 2x^2 - 3x = 0$ by factoring.
    b. Use a graphing calculator to graph $y = x^3 + 2x^2 - 3x$ and find the x-intercepts of the graph. Be sure that you understand why your answers to parts (a) and (b) agree.

72. Business: Revenue The following table gives a company's annual revenue (in millions of dollars) from its overseas operations during its first 5 years.

Year	Revenue
1	2.0
2	1.8
3	1.9
4	2.1
5	2.8

    a. Enter the table numbers into a graphing calculator and make a plot of the resulting points (Years on the x-axis and Revenue on the y-axis). What kind of curve do the points suggest?
    b. Have your calculator fit such a curve to the data. Then enter the result as y_1, which gives a formula for the annual revenue each year. Plot the points together with the regression curve.
    c. Use your formula to predict the revenue in years 6 and 7.

1.5   Exponential Functions

Graph each function.

73. $f(x) = 4^x$ **74.** $f(x) = \left(\frac{1}{4}\right)^x$

75. General: Appreciation A fifteenth-century chalk drawing by Raphael, *Study for the Head and Hand of an Apostle*, was sold in 1996 for \$8.7 million. Assuming that it continues to appreciate in value by 6.5% per year, estimate its value in the year 2005.

76. General: Appreciation   A 1949 Scuderia Ferrari (166MM Barchetta) automobile was auctioned in 1996 for $1.65 million. If it continues to appreciate in value by 5.5% per year, find its value in the year 2010.

77. Business: Depreciation   An $800,000 computer depreciates by 20% each year.

  a. Give a formula for its value after t years.
  b. Find its value after 4 years.

78. General: Depreciation   A $5.4 million off-shore oil drilling platform depreciates by 12% each year.

  a. Give a formula for its value after t years.
  b. Find its value after 10 years.

 79. General: Population   The largest city in the world is Tokyo, followed by Mexico City and São Paulo (Brazil). According to the Census Bureau, x years after 1995 the population of Tokyo will be $27e^{0.0155x}$, the population of Mexico City will be $16.9e^{0.0194x}$, and the population of São Paulo will be $16.1e^{0.0249x}$ (all in millions). Graph these three functions on the window $[0, 100]$ by $[0, 100]$. Assuming that this growth continues:

  a. When will São Paulo overtake Mexico City as the second largest city?
  b. When will São Paulo overtake Tokyo as the largest city?

80. Computers: Moore's Law   The amount of information that can be stored on a computer chip can be measured in megabits (a "bit" is a binary digit, 0 or 1, and a "megabit" is a million bits). The first 1-megabit chips became available in 1987, and 4-megabit chips became available in 1990. This quadrupling of capacity every three years is expected to continue, so that t years after 1987, chip capacity will be $C(t) = 4^{t/3}$ megabits. Use this formula (known as Moore's law, after Gordon Moore, a founder of the Intel Corporation) to predict chip capacity in the year 2005. [*Hint:* What value of t corresponds to 2005?]

1.6   Logarithmic Functions

81. Find each logarithm *without* using a calculator.

  a. $\log 1000$ **b.** $\log \frac{1}{1000}$ **c.** $\ln e^3$ **d.** $\ln \sqrt[4]{e}$

82. Find each logarithm *without* using a calculator.

  a. $\log \sqrt{10}$ **b.** $\log 10^8$ **c.** $\ln \frac{1}{e}$ **d.** $\ln e^{3/2}$

Use the properties of natural logarithms to simplify each function.

83. $f(x) = \ln (x^4) - \ln (x^3) - \ln 1$

84. $f(x) = \ln (e^{7x}) - 5x - \ln e$

85. Business: Demand   Demand for a computer memory chip is increasing by 18% per year. How soon will demand double?

86. Economics: Geothermal Power   Geothermal electricity-generating power (using heat from the center of the Earth) has been increasing worldwide by 5.5% annually. How soon will it double? (*Source: Worldwatch.*)

87. General: Population   According to the Census Bureau, the population of Arizona is increasing by 3.1% annually. At this rate, how soon will Arizona double in population?

88. General: Population   According to the Census Bureau, the population of Colorado is increasing by 2.4% annually. At this rate, how soon will Colorado double in population?

89–90: General: Fossils   In the following exercises, use the fact that the proportion of potassium-40 remaining after t million years is $e^{-0.00054t}$.

89. In 1984 in the Wind River Basin of Wyoming, scientists discovered a fossil of a small, three-toed horse, an ancestor of the modern horse. Estimate the age of this fossil if it contained 97.3% of its original potassium-40.

90. Estimate the age of a skull found in 1959 in Tanzania (dubbed "Nutcracker Man" because of its huge jawbone) that contained 99.9% of its original potassium-40.

Projects and Essays

The following projects and essays are based on Chapter 1. There are no right or wrong answers—the results depend only on your imagination and resourcefulness.

1. Write a page about the mile record example on page 2–3, discussing the following questions: What factors might make modern runners faster than earlier runners? Would you expect this approximately linear trend to continue indefinitely? For how long do you think it might continue? What would you expect the trend to look like *eventually*, and why?

2. Discuss the relationships between the various forms of straight lines.
   For example:
   a. Explain how the slope–intercept form comes from the diagram beside it on page 8.
   b. Show how the point–slope form leads to the slope–intercept form if the point is an intercept.
   c. Show how the point–slope and slope–intercept forms lead to the horizontal line form.
   d. Show how the general linear equation $ax + by = c$ includes both vertical and nonvertical lines.

3. Look up René Descartes in a book on the history of mathematics, and write an essay about his discovery of Cartesian geometry and his other contributions to mathematics.

4. Give two examples of real-life situations that can be modeled by *linear* functions, and explain why they are linear. Then give and discuss two examples of situations that can be modeled by *quadratic* functions. Finally, give and discuss two examples of situations that are more complicated than linear or quadratic functions.

5. For each of the following two stories, find the graph (in the next column) that best matches it, and explain (briefly) why it matches. Then make up a story to match the remaining graph. Finally, make up two new stories and graphs that match them. (Be imaginative.)
   Story a. I drove away from home in a hurry but got a speeding ticket, after which I continued on more slowly.

Story b. I went out for a jog around town, returning home too exhausted to go out again.

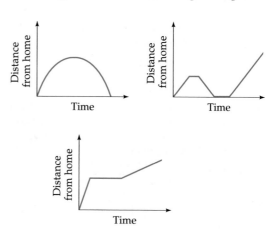

6. Define a "function" on people as follows: For any person x, let

$$f(x) = \text{the } father \text{ of person } x$$

$$m(x) = \text{the } mother \text{ of person } x$$

Then the maternal grandmother of x would be $m(m(x))$. Define all grandparents of x, and state how many of them there are. Do the same for great-grandparents. Can you generalize even further?

7. The constant e is defined as $e = \lim\limits_{n \to \infty} \left(1 + \frac{1}{n}\right)^n \approx 2.71828$. Use a calculator to evaluate $\left(1 + \frac{1}{n}\right)^n$ for large values of n. Compare e to $\lim\limits_{n \to \infty} \left(1 + \frac{1}{100}\right)^n$ and $\lim\limits_{n \to \infty} \left(1 + \frac{1}{n}\right)^{100}$ where, in each case, one of the n's is replaced by a constant. Estimate $\lim\limits_{n \to \infty} \left(1 + \frac{1}{n}\right)^{2n}$ and $\lim\limits_{n \to \infty} \left(1 + \frac{1}{2n}\right)^n$ and try to determine how their values are related to e. What if 2 is replaced by another number?

8. The "Money Angles" column in *Time* magazine (May 17, 1993) recommended buying wine by the case, saving an estimated 10% every 12 weeks. The column then suggests that repeating this purchase every 12 weeks will lead to a saving of "more than 40% per year." Questions: If you did this for two years, would you save 80%? For three years, would you save 120% (so the wine would be free, or even better)? Explain what is wrong with *Time's* reasoning.

9. Find out how logarithms (base 10) were used for calculating products and quotients before the invention of pocket calculators. Which properties of logs enable multiplication problems to be changed into addition problems, and division problems into subtraction problems? Look up "slide rule" and explain how slide rules were used for calculations, and how they are based on the addition and subtraction of logarithms.

10. Write about the difference between straight-line depreciation and depreciation by a fixed percentage, describing the advantages of each. For example, if the line (from straight-line depreciation) and the curve (from fixed-percentage depreciation) both begin at the same point and end at the same point, which method of depreciation provides the greater dollar drop in value in the first year? In the last year? If you had to pay income tax on the value each year, which method would you prefer? If you had to sell the asset midway, which method would you prefer? If you had to buy the asset midway?

11. Write about logarithms and their connection with exponents. Include discussions of the following questions: Why is the logarithm of 1 always zero (for any base)? Why can't 1 be a base for logarithms? Why can't a negative number be a base? Why can't you find logarithms of zero and negative numbers? What does the graph of the logarithm function look like for different bases (including bases between zero and one)? Look up John Napier and include information about the history of logarithms. Why do we use the letter e to represent the constant defined by the limit on page 69. [*Hint:* Look up Leonhard Euler.]

12. Reread the first item under "Hints and Suggestions" on pages 90–91. Give other examples of geometric and analytic objects, and discuss the relationship between them. Be sure to include a discussion of curves and functions and the relationships between them, including the vertical line test.

13. Look over your notes, your homework, and the text, and write a page about how a graphing calculator has helped you in this chapter. Include examples of how it has helped you to *explore concepts* and how it has helped to *simplify your work*. What was its *most* helpful or interesting use? What was its *least* helpful or interesting use? Are there problems that can be done on a graphing calculator but that are easier to do "by hand"?

2

Mathematics of Finance

Lenders offer borrowers many choices. This chapter explains how loans work and how to compare them.

6 MONTH   4.49%

1 YEAR   4.59% A

2 YEAR   4.70%

5 YEAR   5.01%

Simple Interest

APPLICATION PREVIEW

Benjamin Franklin's Will

In 1789, the eighty-three-year-old Benjamin Franklin added a codicil to his will creating a long-term bequest of £1000 (about $4500) to the Town of Boston. For the first hundred years, the money was to start a loan fund to help young married tradesmen start their own businesses. Each year the borrowers would repay one-tenth of the principal together with 5% interest, and this money would then be lent to fresh borrowers. Franklin hoped ". . . that no part of the money will at any time lie dead or be diverted to other purposes . . ." and calculated that after one hundred years the fund would grow to £131,000 (about $500,000). This money would then be divided, with £100,000 for public works in Boston and the rest to continue the loan fund for another hundred years. At the end of the second hundred years, Franklin estimated the fund would total over £4,061,000 (more than $18 million) and he directed that it then be divided between Boston and the state government, ". . . not presuming to carry my views farther."

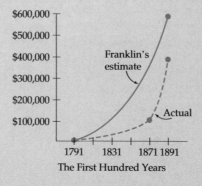

The First Hundred Years

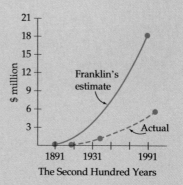

The Second Hundred Years

After Franklin's death in 1790, the Town of Boston accepted his offer, and the money arrived in March 1791. Despite the fund's early popularity, by the 1830s the changes brought about by the Industrial Revolution forced many workers into factories instead of starting up businesses on their own, and demand for loans all but disappeared. The managers of the fund placed some of the money in other investments, and by the hundredth anniversary in July 1891, its value reached $391,000. After various legal squabbles over the division of this considerable sum, a technical training school was established that has now become the Franklin Institute of Boston. In 1994, the second part of the Franklin Fund was ended and its $4.6 million assets were turned over to the school. The 200-year history of Franklin's £1000 gift provides an example of the relation between money and time, a subject that will be explored further in this chapter.

Introduction

In the modern credit world, the old adage "time is money" has become a basic fact of economic life. When you open a savings account or take out a car loan, you directly experience the "time value of money." This chapter covers the basic financial properties of a loan between a lender and a borrower and the calculation of the interest and payments. We begin with simple interest.

Simple Interest

The *principal* of a loan is the amount of money borrowed from the lender, the *term* is the time the borrower has the money, and the *interest* is the additional money paid by the borrower for the use of the lender's money. The interest is called *simple interest* if it is calculated as a fixed percentage of the principal and is paid at the end of the term. The *interest rate* of the loan is the dollars of interest per $100 of principal per year (or "per annum"). It is usually stated as a percentage but is always written as a decimal in calculations. Simple interest is calculated as follows:

Simple Interest Formula

The interest I on a loan of P dollars at simple interest rate r for t years is

$$I = Prt$$

EXAMPLE 1   **Finding Simple Interest**

Find the interest on a loan of two million dollars at 7.2% simple interest for 4 months.

Solution

Writing the interest rate in decimal form and changing the term to years, we find that

$$I = (2{,}000{,}000)(0.072)\left(\frac{4}{12}\right) = 48{,}000$$

$I = Prt$ with
$P = \$2{,}000{,}000$,
$r = 0.072$, and $t = 4/12$

The interest is $48,000.

■

Be Careful! When using a calculator, *round off only your final answer.* For example, if you use $t = 0.33$ in the previous example instead of $\frac{4}{12}$ or $\frac{1}{3}$, you would get the value $(2{,}000{,}000)(0.072)(0.33) = \$47{,}520$, which is wrong by $480.

Banker's Rule

Loan agreements for short terms (that is, less than a year) often add the phrase *using the Banker's rule* after stating the interest rate. This means that the term is calculated as the number of days divided by 360. At first glance, this seems merely to simplify the banker's arithmetic by making each year into 12 months of 30 days each, but in fact it sometimes gives a slight advantage to the lender.

EXAMPLE 2   **Using the Banker's Rule**

Find the interest on a loan of $2,000,000 at 7.2% simple interest from June 15 to October 15 using the Banker's rule.

Solution

Since the term is 4 months, this appears to be the same as Example 1. But the particular 4 months from June 15 to October 15 total 122 days, and by the Banker's rule, the interest is

$$I = (2{,}000{,}000)(0.072)\left(\frac{122}{360}\right) = 48{,}800 \qquad\qquad I = Prt$$

In this case, the Banker's rule resulted in an $800 advantage to the lender.

■

PRACTICE PROBLEM 1

Find the interest on a loan of $50,000 at 19.8% simple interest from March 3 to June 3 using the Banker's rule. How does this compare to a loan of $50,000 at 19.8% simple interest for 3 months?

Solution at the back of the book

If you have values for any three unknowns in the interest formula $I = Prt$, you can solve for the fourth one.

EXAMPLE 3   Finding the Simple Interest Rate

What is the interest rate of a loan charging $18 simple interest on a principal of $150 after 2 years?

Solution

We solve $I = Prt$ for interest rate r:

$$r = \frac{I}{Pt}$$ $I = Prt$ divided by Pt

Then

$$r = \frac{18}{(150)(2)} = \frac{18}{300} = 0.06$$ Substituting $I = 18$, $P = 150$, and $t = 2$

The interest rate is 6%.

■

Total Amount Due on a Loan

When the term of a loan is over, the borrower repays the principal and interest, so the total amount due is

$$\underbrace{P}_{\text{Principal}} + \underbrace{Prt}_{\text{Interest}} = \underbrace{P(1 + rt)}_{\text{Total amount}}$$ Factoring out the common term

This gives the following formula:

Total Amount Due for Simple Interest

> The total amount A due at the end of a loan of P dollars at simple interest rate r for t years is
>
> $$A = P(1 + rt)$$

The amount due may also be regarded as the *accumulated value* of an investment or the *future value* of the principal. As before, this formula may be used as it is or it may be solved for any one of the other variables, as the next few examples will show.

EXAMPLE 4 **Finding the Total Amount Due**

What it the total amount due on a loan of $3000 at 6% simple interest for 4 years?

Solution

$$A = 3000(1 + (0.06)(4))$$

$$= 3000(1.24) = 3720$$

```
3000→P
                         3000
.06→R
                          .06
4→T
                            4
P(1+RT)
                         3720
```

The total due on the loan is $3720.

The amount of the loan in Example 4 grew from $3000 to $3720 in 4 years. The $3720 is sometimes called the *future value* of the original $3000. Reversing our viewpoint, we say that the principal of $3000 is the *present value* of the later $3720. To find a formula for the present value, we solve the "total amount due" formula (on the previous page) for P by dividing by $1 + rt$ and rename the result PV:

$$PV = \frac{A}{1 + rt}$$

PV means **P**resent **V**alue of the future amount A

EXAMPLE 5 **Finding a Present Value**

How much should be invested now at 8.6% simple interest if $10,000 is needed in 6 years?

Solution

The amount to invest now means the *present value* of $10,000 in 6 years.

$$PV = \frac{10,000}{1 + (0.086)(6)} = \frac{10,000}{1.516} \approx 6596.31$$

$PV = A/(1 + rt)$ with $A = 10,000$, $r = 0.086$, and $t = 6$

The amount required is $6596.31.

We emphasize that the *present value* is the amount that will grow to the required sum in the given time period (at the stated interest rate). For this reason, it gives the actual value *now* of an amount to be paid later.

PRACTICE PROBLEM 2

Find the present value of a "promissory note" that will pay $5000 in 4 years at 12% simple interest.   *Solution at the back of the book*

We found the present value by deriving a formula for it. We could instead have substituted the given numbers into the "total amount due" formula and then solved for P. We will do the next example in this way, substituting numbers into the original formula and then solving for the remaining variable. You can use either method to solve these problems.

EXAMPLE 6   Finding the Term of a Simple Interest Loan

What is the term of a loan of $2000 at 4% simple interest if the amount due is $2400?

Solution

Substituting the given numbers for the appropriate variables in the "total amount due" formula (page 102) gives

$2400 = 2000(1 + 0.04t)$	$A = P(1 + rt)$ with $A = 2400$, $P = 2000$, and $r = 0.04$
$1.2 = 1 + 0.04t$	Dividing by 2000
$0.2 = 0.04t$	Subtracting 1
$t = \dfrac{0.2}{0.04} = 5$	Dividing by 0.04 and reversing sides

The term is 5 years.

■

Graphing Calculator Exploration

The formula $A = P(1 + rt)$ may be written as $A = (Pr)t + P$ so that it has the familiar $y = mx + b$ form of a straight line with slope Pr and y-intercept P (but with t and A instead of x and y).

a. Using the values $P = 2000$ and $P \cdot r = (2000)(0.04) = 80$ from Example 6, graph the line $y_1 = 80x + 2000$ on the window [0, 8] by [0, 4000], so that y gives the amount of the loan for any term x.

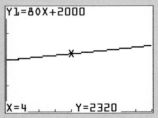

b. Use TRACE to estimate the term x that gives $y = 2400$. How does your "graphical" answer compare to the answer for Example 6? Try ZOOMing IN to improve your estimate.

c. Reset the window to [0, 8] by [0, 4000] and add the horizontal line $y_2 = 2400$ to your graph. Use INTERSECT to find the intersection point. Does the x-value of this point exactly match the answer for Example 6?

Discounted Loans and Effective Interest Rates

A *discounted* loan is a loan in which the lender deducts the interest from the amount the borrower receives at the start. How different is this from a simple interest loan? It must be better for the lender because getting the money early (so it can earn interest somewhere else) is always better than later. Since the borrower actually is receiving a smaller amount, we may recalculate the interest rate as a "standard" loan on this smaller amount. The resulting rate is called the *effective simple interest rate* of the loan.

EXAMPLE 7   **Finding the Effective Simple Interest Rate**

Find the effective simple interest rate on a discounted loan of $1000 at 6% simple interest for 2 years.

Solution

The interest is ($1000)(0.06)(2) = $120, so as a simple interest loan, the borrower receives only $P = 1000 - 120 = \$880$ and agrees to pay back $1000 at the end of 2 years. If we write r_s for the simple interest rate of this loan, we see that

$$1000 = 880(1 + r_s 2)$$ $A = P(1 + rt)$ with $A = 1000$, $P = 880$, and $t = 2$

Solving for r_s,

$$r_s = \frac{1}{2}\left(\frac{1000}{880} - 1\right) \approx 0.0682$$ Dividing by 880, subtracting 1, and dividing by 2

The effective simple interest rate on this discounted loan is 6.82%, which is significantly greater than the stated rate of 6%.

Using these same steps, we can find a formula for the effective rate. For a discounted loan of amount A at interest rate r for term t, the borrower receives $A - Art = A(1 - rt)$ dollars. Therefore, the simple interest equation with effective simple interest rate r_s gives

$$A = A(1 - rt)(1 + r_s t)$$

$A = P(1 + rt)$ with $A(1 - rt)$ for P

$\underbrace{}_{\text{Principal}}$

Solving for r_s:

$$\frac{1}{1 - rt} = 1 + r_s t$$

Canceling the As and dividing by $(1 - rt)$

$$r_s t = \frac{1}{1 - rt} - 1 = \frac{1 - (1 - rt)}{1 - rt} = \frac{rt}{1 - rt}$$

Reversing sides, subtracting 1, and simplifying

$$r_s = \frac{1}{t} \frac{rt}{1 - rt} = \frac{r}{1 - rt}$$

Dividing by t and then canceling

This gives the following formula for r_s.

Effective Simple Interest Rate for a Discounted Loan

For a discounted loan at interest rate r for t years, the effective simple interest rate r_s is

$$r_s = \frac{r}{1 - rt}$$

Notice that the effective rate r_s will be larger than r (since the denominator is less than 1) and that it depends only on the *rate* and *term* of the loan, and not on its amount. For the loan in Example 7, our formula gives the answer that we found:

$$r_s = \frac{0.06}{1 - (0.06)(2)} = \frac{0.06}{0.88} \approx 0.0682$$

Using $r = 0.06$ and $t = 2$

PRACTICE PROBLEM 3

What is the effective simple interest rate of a discounted loan at 6% interest for 3 years? for 5 years?   *Solution at the back of the book*

Section Summary

The simple interest formula is

$$I = Prt$$

I = simple interest
P = principal
r = interest rate
t = term in years

We can solve for any one of the variables if the others are known. Using the *Banker's rule*, we calculate the term as the number of days divided by 360. The total amount due at the end of the loan (principal plus interest) is

$$A = P(1 + rt)$$ A = total amount due

This equation may be solved for any one of the variables, keeping in mind that the amount A at the end of the loan is the *future value* of the principal P, and P is the *present value* of the future amount A.

For a *discounted* loan, the interest is subtracted from the principal at the beginning of the loan. The effective simple interest rate r_s for a discounted loan at rate r for t years is

$$r_s = \frac{r}{1 - rt}$$

EXERCISES 2.1 (Most require ▦ or ▦.)

Find the simple interest on each loan.

1. $1500 at 7% for 10 years.

2. $2000 at 9% for 7 years.

3. $6000 at 6.5% for 8 years.

4. $4500 at 4.25% for 9 years.

5. $825 at 6.58% for 5 years 6 months.

6. $950 at 5.87% for 6 years 3 months.

7. $1280 at 4.8% for 3 months.

8. $5275 at 5.3% for 2 months.

9. $1280 at 4.8% from March 8 to June 8 using the Banker's rule.

10. $5275 at 5.3% from October 14 to December 14 using the Banker's rule.

Find the total amount due for each simple interest loan.

11. $1500 at 7% for 10 years.

12. $2000 at 9% for 7 years.

13. $6100 at 5.7% for 4 years 9 months.

14. $4500 at 6.3% for 3 years 6 months.

15. $3125 at 4.81% for 10 months.

16. $8775 at 13.11% for 7 months.

APPLIED EXERCISES

17. Interest Rate   Find the interest rate on a loan charging $704 simple interest on a principal of $2750 after 4 years.

18. Interest Rate   Find the interest rate on a loan charging $1127 simple interest on a principal of $4900 after 5 years.

19. Principal   Find the principal of a loan at 8.4% if the simple interest after 5 years 6 months is $1155.

20. Principal   Find the principal of a loan at 7.6% if the simple interest after 9 years 3 months is $2109.

21. Term   Find the term of a loan of $175 at 9% if the simple interest is $63.

22. Term   Find the term of a loan of $225 at 7% if the simple interest is $94.50.

23. Present Value   How much should be invested now at 5.2% simple interest if $8670 is needed in 3 years?

24. Present Value   How much should be invested now at 4.8% simple interest if $4530 is needed in 4 years 4 months?

Zero Coupon Bonds   A *zero coupon bond* pays only its *face value* on maturity (getting its name because it has no "coupons" for interest payments before that date). The *fair market price* is the present value of the face value at the current interest rate.

25. What is the fair market price of a $10,000 zero coupon bond due in 1 year if today's long-term simple interest rate is 5.81%?

26. What is the fair market price of a $15,000 zero coupon bond due in 1 year if today's long-term simple interest rate is 7.23%?

27. What is the fair market price of a $5000 zero coupon bond due in 2 years if today's long-term simple interest rate is 3.54%?

28. How Interest Rates Affect Present Value   The present value formula $PV = A/(1 + rt)$ can be viewed on your graphing calculator as a function y of the simple interest rate x by rewriting it as $y_1 = A/(1 + xt)$. Using $A = \$10{,}000$ and $t = 1$ year, graph this expression with window [0, 1] by [0, 12000] to see the present value of a $10,000 zero coupon bond due in 1 year as a function of the interest rate. Use TRACE or VALUE to find the present value of the bond for interest rates of 4%, 5%, 6%, 7%, 8%, and 9%. Does a 1% increase in the interest rate always cause the same decrease in the present value of the bond?

29. Term   What should be the term for a loan of $6500 at 7.3% simple interest if the lender wants to receive $9347 when the loan is paid off?

30. Term   What should be the term for a loan of $5400 at 5.8% simple interest if the lender wants to receive $6966 when the loan is paid off?

31. Term   How long will it take an investment at 8% simple interest to increase by 70%?

32. Term   How long will it take an investment at 4.2% simple interest to increase by 26.6%?

Doubling Time   The *doubling time* of an investment is the number of years it takes for the value to double. This is the same as the number of years for the value to increase by 100%.

33. What is the doubling time of a 5% simple interest investment?

34. Show that the doubling time of a simple interest investment is $1/r$ where r is the simple interest rate.

35. Effective Rate   What is the effective simple interest rate of a discounted loan at 4.6% interest for 3 years 6 months?

36. Effective Rate   What is the effective simple interest rate of a discounted loan at 7.2% interest for 2 years 10 months?

37. Discounted Loan   Would you agree to borrow $1000 as a discounted loan at 5% for 20 years? Explain.

38. How the Term Affects the Discounted Rate   The effective rate formula $r_s = r/(1 - rt)$ can be viewed on your graphing calculator as a function of the term t by rewriting it as $y_1 = r/(1 - rx)$. Using $r = 0.05$, graph this expression with window [0, 20] by [0, 1] to see the effective rate y of the discounted loan as a function of the term x. Use TRACE or VALUE to find the effective rate of the loan for terms of 5, 10, and 15 years. Does a 1-year increase in the term always cause the same increase in the effective rate?

39. Bridge Loan   You have a buyer for your condominium, but the seller of your dream house wants to close now. In order to get enough money to go to the closing, you take out a 2-month bridge loan for $72,000 at 7.92% simple interest. You assume that you will close on the sale of your condominium before the 2 months are up. How much interest will you have to pay on the loan?

40. Banker's Rule   Suppose you wanted a loan for the month of February. Would the Banker's rule work in your favor or the bank's favor? [*Hint:* Calculate t both ways. Which will give a larger value of I?] What about a 3-month loan starting in January and ending in March?

41. Lawyer's Fees   After successfully defending a client against a driving while intoxicated (DWI) charge, a lawyer reluctantly accepted as payment a promissory note for $5000 plus 20% simple interest due in 270 days. Learning subsequently of the client's "deadbeat" reputation, she sold the note 90 days later to an experienced bill collector for $4600.

Using the Banker's rule and assuming the collector will actually obtain the full value of the note, find the simple interest rate the collector will earn for his trouble.

42. Home Improvements Because the contractor was short of cash when the job was finished, a plumber accepted as payment a promissory note for $2000 plus 18% simple interest due in 60 days. Needing cash himself, the plumber sold the note 15 days later for $2000 to a local loan agency. Using the Banker's rule, find the simple interest rate the agency will earn on its investment.

Brokerage Commissions An Internet discount brokerage firm charges a commission of 10% on the first $20,000 plus 5% of the excess over $20,000 on each buy or sell transaction. Find the simple interest rate earned by each of the following investments after including the commissions paid to the brokerage firm.

43. Purchase 900 shares of American WebWide Education at $18.50 per share and sell them 4 months later at $26.75 per share.

44. Purchase 1500 shares of Well Care Deluxe at $11.90 per share and sell them 10 months later at $14.80 per share. *under 20,000 so only 10%*

45. Purchase 800 shares of DucoFood Services at $28.70 per share and sell them 7 months later at $37.80 per share.

2.2 Compound Interest

Musical Instruments as Investments*

No two guitars sound alike—good ones, anyway. Over the last 30 years, prices for high-quality acoustic instruments have soared. In 1993, a John D'Angelico New Yorker 18-inch-wide cutaway archtop ("jazz") guitar sold for $150,000, setting the record for the price of an acoustic fretted instrument not previously owned by a deceased superstar. Many pre-World War II Gibson and Martin instruments are valued not only for their craftsmanship but also for their distinctive sound. A mint condition Martin D-45 made in 1932–1942 that might have sold for $3500 in 1971 could be sold for $125,000 in 1996, while a Gibson 1932–1939 Mastertone Granada 5-string (replacement neck) banjo selling for $18,000 in 1996 might have sold for just $1200 in 1971.

How can these price increases be compared to other investments such as stocks, bonds, or bank savings accounts? The dots on

* Much of this material was generously provided by Stanley M. Jay and Larry Wexer of *mandolin bros. Ltd.* (Staten Island, New York), dealers in fine mandolins, guitars, and banjos.

the graph show the typical selling prices over the 25-year period from 1971 to 1996 for a signed 1922 Gibson F-5 mandolin, together with a curve showing the increasing value of a $2000 investment earning 12% compound interest.

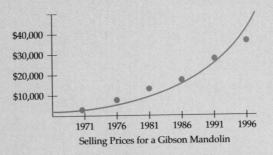

Selling Prices for a Gibson Mandolin

The dots closely match the curve, showing that this mandolin has increased in value by about 12% annually, better than many bond and stock investments. The values of fine musical instruments and other investments can be found by the methods of this section.

Introduction

When a simple interest loan is not paid off but "rolled over" into another loan, the new principal includes the unpaid interest from the old loan. The interest on this new loan is called *compound interest* because it combines interest both on the original principal and on the unpaid interest. Naturally, this "interest on interest" is to the benefit of the lender. For example, if you lend $1000 at 6% simple interest for 3 years, your return will be $A = \$1000(1 + (0.06)(3)) = \$1180$. Now suppose you lend it for the first year only—the amount due will be $\$1000(1 + 0.06) = \$1060$. You then lend this $1060 for the next year, at the end of which the amount due is $\$1060(1 + 0.06) = \$1123.60$. For the third and final year, you lend this $1123.60 and receive $\$1123.60(1 + 0.06) = \$1191.02$. This amount exceeds the simple interest amount by $\$1191.02 - \$1180 = \$11.02$, the difference coming from *interest on interest*. For a term longer than 3 years, you would continue multiplying the debt by (1 + 0.06)—that is, by 1 plus the interest rate—each year.

Compound Interest

In general, the amount A due on a compound interest loan of P dollars at interest rate r per year compounded annually for t years is found by repeatedly multiplying the principal P by $(1 + r)$, once for each year:

$$A = \underbrace{P(1 + r)(1 + r) \cdots (1 + r)}_{\substack{t \text{ multiplications} \\ \text{by } (1 + r)}} = P(1 + r)^t$$

Graphing Calculator Exploration

To see that compound interest eventually surpasses simple interest (even with a higher rate and principal), compare $500 invested at 4% compounded annually with $2500 invested at 8% simple interest.

a. Graph the simple interest amount $A = P(1 + rt)$ as $y_1 = 2500(1 + .08x)$ on the window [0, 150] by [0, 35000] so that y_1 is the amount of the investment after x years.

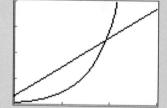

b. Graph the compound interest amount $A = P(1 + r)^t$ as $y_2 = 500(1 + .04)^x$ on the same window.

c. Use TRACE or INTERSECT to find when the compound interest amount equals the simple interest amount. What happens after this intersection point? Try your graphs with the larger windows [0, 300] by [0, 70000] and by [0, 500000].

Compounding can be done more than once a year. Some standard compounding periods are described in the following table.

Compounding Frequency	Periods per Year
Annually	1
Semiannually	2
Quarterly	4
Bimonthly	6
Monthly	12
Biweekly	26
Weekly	52
Daily	365

With m compounding periods per year, the interest rate for each period becomes r/m while the number of compoundings in t years is mt. Compound interest is then calculated as follows:

Compound Interest Formula

> The amount A due on a loan of P dollars at yearly interest rate r compounded m times per year for t years is
>
> $$A = P\left(1 + \frac{r}{m}\right)^{mt}$$

The stated yearly interest rate r is called the *nominal rate* of the loan.

EXAMPLE 1 **Finding the Amount Due on a Compound Interest Loan**

Find the amount due on a loan of $1500 at 4.8% compounded monthly for 5 years.

Solution

The nominal rate of 4.8% expressed as a decimal is 0.048, and monthly compounding means that $m = 12$, so we have

$$A = 1500\left(1 + \frac{0.048}{12}\right)^{(12)(5)}$$

$A = P(1 + r/m)^{mt}$
with $P = 1500$, $r = 0.048$, $m = 12$, and $t = 5$

$$= 1500(1.004)^{60} \approx 1905.96$$

The amount due is $1905.96.

PRACTICE PROBLEM 1

Find the amount that is due on a loan of $3500 at 5.1% compounded bimonthly for 8 years.   *Solution at the back of the book*

A loan is really an amount that a bank *invests* in a borrower, so any question about a loan can be rephrased as a question about an *investment*. Example 1 could have asked for the value of an investment of $1500 that grows by 4.8% compounded monthly for 5 years—the answer would still be $1905.96. In this section we will speak interchangeably of *loans* and *investments*, since they are the same but from different viewpoints.

In Example 1, the amount $1905.96 may be called the *future value* of the original $1500, and conversely, the $1500 is the *present value* of the later amount $1905.96. As before, the present value gives the value *today* of a payment that will be received at some time in the future. To find a formula for the present value, we solve the compound interest formula (page 112) for P, using the familiar $1/x^n = x^{-n}$ rule of exponents and renaming the result PV.

$$PV = \frac{A}{\left(1 + \dfrac{r}{m}\right)^{mt}} = A\left(1 + \frac{r}{m}\right)^{-mt} \qquad PV \text{ means Present Value}$$

EXAMPLE 2   Finding the Present Value

How much should be invested now at 8.6% compounded weekly if $10,000 is needed in 6 years?

Solution

$$PV = 10,000\left(1 + \frac{0.086}{52}\right)^{-(52)(6)} \approx 5971.58 \qquad \begin{array}{l} PV = A(1 + r/m)^{-mt} \\ \text{with } A = 10,000, r = 0.086, \\ m = 52, \text{ and } t = 6 \end{array}$$

The amount required is $5971.58.   ∎

Growth Times

We solved Example 2 by using the formula that we had derived for present value. Instead, we could have substituted the given numbers into the compound interest formula and *then* solved for the remaining variable. We will do the next example in this way, finding a "growth time," the time for a loan to reach a given value. We will also use the rule of logarithms $\log(x^n) = n \log(x)$ to "bring down the power." Logarithms with *any* base may be used in these calculations.

EXAMPLE 3   **Finding the Term of a Compound Interest Loan**

What is the term of a loan of $2000 at 6% compounded monthly that will have an amount due of $2400?

Solution

We substitute the given numbers into the compound interest formula.

$$2400 = 2000 \left(1 + \frac{0.06}{12}\right)^{12 \cdot t} \qquad \begin{array}{l} A = P(1 + r/m)^{mt} \text{ with} \\ A = 2400, P = 2000, \\ r = 0.06, \text{ and } m = 12 \end{array}$$

$$1.2 = 1.005^{12t} \qquad \begin{array}{l} \text{Dividing by 2000 and} \\ \text{simplifying} \end{array}$$

$$\log 1.2 = \log 1.005^{12t} \qquad \text{Taking logarithms}$$

$$\log 1.2 = 12t \log 1.005 \qquad \begin{array}{l} \text{Using } \log (x^n) = n \log (x) \\ \text{to bring down the power} \end{array}$$

$$12t = \frac{\log 1.2}{\log 1.005} \approx 36.555 \qquad \begin{array}{l} \text{Dividing by } \log 1.005, \\ \text{reversing sides, and} \\ \text{calculating logarithms} \end{array}$$

Months

This amount of time, $12t = 36.555$, is in *months* (t is years, but $12t$ is *months*). We round *up* to the nearest month (because a shorter term will not reach the needed amount), so the term needed is 37 months, or 3 years 1 month.

∎

Hint: Instead of solving for t, if the compounding is monthly stop when you find $12t$ (the number of months), if quarterly stop when you find $4t$ (the number of quarters), if weekly stop when you find $52t$ (the number of weeks), and so on. Then round *up* to the next whole number of periods to ensure that the needed amount will actually be reached.

Notice that the actual amounts $2400 and $2000 did not matter in this calculation, only their *ratio* $\frac{2400}{2000} = 1.2$. Thus, a loan of $2000 grows to $2400 in the same amount of time that a loan of 20 *million* dollars would grow to 24 million dollars (at the stated interest rate) since the ratios are the same. This is true in general because in the compound interest formula (page 112), the total amount and the principal are *proportional*. In Example 3 we could have omitted the dollar amounts and simply asked for the term of a loan that would *multiply the principal by 1.2*. Furthermore, multiplying by 1.2 means *increasing* by 0.2, or 20%, so we could have asked for the term of a loan that would *increase the principal by 20%*—all three formulations are equivalent.

Be careful! Distinguish carefully between the *increase* and the *multiplier*. Increasing by 20% means *multiplying by 1.2* (the "1" keeps the original amount and the ".2" increases it by 20%). For example, to *increase* an amount by 35% you would *multiply* it by 1.35.

EXAMPLE 4 **Finding the Term to Increase the Principal**

What is the term of a loan at 6% compounded weekly that will increase the principal by 40%?

Solution

Increasing by 40% means multiplying the amount by 1.4. Using P for the unknown principal, we must solve

$$1.4P = P\left(1 + \frac{0.06}{52}\right)^{52 \cdot t} \qquad \begin{array}{l} P \text{ (for principal)} \\ \text{on both sides} \end{array}$$

$$1.4 = (1 + 0.06/52)^{52t} \qquad \begin{array}{l} \text{Canceling the } Ps \text{ and} \\ \text{simplifying} \end{array}$$

$$\log 1.4 = 52t \log (1 + 0.06/52) \qquad \begin{array}{l} \text{Taking logarithms and} \\ \text{bringing down the power} \end{array}$$

$$52t = \frac{\log 1.4}{\log (1 + 0.06/52)} \approx 291.8 \qquad \begin{array}{l} \text{Dividing by } \log (1 + 0.06/52), \\ \text{reversing sides, and} \\ \text{calculating logarithms} \end{array}$$

Weeks

Rounding *up*, the term needed is 292 weeks, or 5 years 32 weeks.

∎

PRACTICE PROBLEM 2

What is the term of a loan at 6% compounded quarterly that multiplies the principal by 2?   *Solution at the back of the book*

Rule of 72

The *doubling time* of an investment or a loan is the number of years it takes for the value to double. Doubling times are often estimated by using the *rule of 72:*

$$\left(\begin{array}{c} \text{Doubling} \\ \text{time} \end{array}\right) \approx \frac{72}{r \times 100} \qquad \text{Divide 72 by the rate times 100}$$

The rule of 72, however, gives only an *approximation* for the doubling time, but it is often quite accurate. (A justification of this rule is given in Exercise 60.) For example, for a loan at 6% compounded quarterly the rule of 72 gives

$$\left(\begin{array}{c} \text{Doubling} \\ \text{time} \end{array}\right) \approx \frac{72}{0.06 \times 100} = \frac{72}{6} = 12 \text{ years} \qquad \text{Rule of 72 estimate}$$

In Practice Problem 2 we found the correct doubling time to be $11\frac{3}{4}$ years, so the rule of 72 is not far off. The following Graphing Calculator Exploration shows that the rule of 72 is reasonably accurate for many "everyday" interest rates (but see also Exercise 59).

Graphing Calculator Exploration

Compare the "rule of 72" with the exact formula for the doubling time by graphing both as functions of the interest rate.

a. Graph $y_1 = 72/(x \times 100)$ on the window $[0, .5]$ by $[0, 20]$ to show the "rule of 72" estimate for the number of years y at interest rate x.

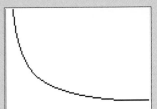

b. Graph the curve $y_2 = \log(2)/(12 \log(1 + x/12))$ on the same window to show the doubling time of a compounded monthly loan with interest rate x.

c. Verify that the graph shows *both* curves.

Effective Rates

How different is a loan charging 6.2% compounded *quarterly* from a loan charging 6.2% compounded *monthly?* In general, how can we compare different compound interest rates? We can compare them by calculating the *actual percentage increase* that each will generate during 1 year. More formally, we calculate the *effective rate* r_e, which is the simple interest rate that will return the same amount on a 1-year loan. The effective rate is also called the *annual percentage rate* or *APR* and by law must be stated in consumer loan agreements. The effective rate r_e is found by solving

$$P(1 + r_e) = P\left(1 + \frac{r}{m}\right)^m$$

Finding the "simple" rate that gives the "compound" rate for $t = 1$

Canceling the P's and subtracting the 1 on the left gives

Effective Rate for a Compound Interest Loan

For a compound interest loan at interest rate r compounded m times per year, the effective rate r_e is

$$r_e = \left(1 + \frac{r}{m}\right)^m - 1$$

The effective rate can be interpreted as the interest earned by $1 for 1 year.

EXAMPLE 5 **Finding the Effective Rate**

What is the effective rate of a loan at 6.2% compounded quarterly?

Solution

$$r_e = \left(1 + \frac{0.062}{4}\right)^4 - 1 \approx 0.0635 \qquad \begin{array}{l} r_e = (1 + r/m)^m - 1 \\ \text{with } r = 0.062 \text{ and } m = 4 \end{array}$$

The effective rate is 6.35%. That is, 6.2% compounded quarterly results in an annual gain of 6.35%.

PRACTICE PROBLEM 3

What is the effective rate of a loan at 6.2% compounded monthly?

Solution at the back of the book

Graphing Calculator Exploration

The effective rate should be larger for more frequent compounding since the sooner the interest is credited, the sooner it begins to earn more interest.

a. Using the value $r = 0.062$ from Example 5, graph the curve $y_1 = (1 + .062/x)^x - 1$ on the window [0, 100] by [.061, .065] so that y is the effective rate of a 6.2% loan compounded x times a year.

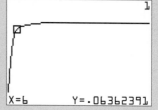

b. Use TRACE or VALUE to find y when $x = 4$ and when $x = 12$. How do your answers compare to the answers to Example 5 and Practice Problem 3?

c. What happens for large values of x? Does the effective rate become arbitrarily large or does it "settle down"?

d. Can the effective rate be as large as 6.395674%? (You may need a larger window!) Can it be as large as 6.3963%?

Choices between different loans or investments are often made on the basis of their effective rates.

EXAMPLE 6   Using Effective Rates to Compare Investments

A self-employed carpenter is setting up a Keogh retirement plan (which defers taxes on interest until withdrawals are made) and has decided to purchase either a certificate of deposit (CD) with the First & Federal Bank at 6.2% compounded semiannually or a CD with the Chicago Nationswide Bank at 6.15% compounded monthly. Which bank offers the greater effective rate?

Solution

The effective rate for 6.2% compounded semiannually (First & Federal) is

$$r_e = \left(1 + \frac{0.062}{2}\right)^2 - 1 \approx 0.0630 \qquad\qquad r_e \approx 6.30\%$$

whereas for 6.15% compounded monthly (Chicago Nationswide), the effective rate is

$$r_e = \left(1 + \frac{0.0615}{12}\right)^{12} - 1 \approx 0.0633 \qquad\qquad r_e \approx 6.33\%$$

The CD at the Chicago Nationswide Bank is better.

■

The effective rate of return for an investment of P dollars that returns an amount A after t years can be found from the compound interest formula (page 112) with $m = 1$:

$$A = P(1 + r_e)^t \qquad\qquad \begin{array}{l} A = P(1 + r/m)^{mt} \\ \text{with } m = 1 \text{ and } r = r_e \end{array}$$

An example of this is a *zero coupon bond*, which pays its *face value* on maturity and sells now for a lower price. This kind of bond makes no payments before maturity and so has no interest "coupons."

EXAMPLE 7   Effective Rate for a Zero Coupon Bond

What is the effective rate of return of a $10,000 zero coupon bond maturing in 6 years and offered now for sale at $5500?

Solution

We solve

$$10,000 = 5500(1 + r_e)^6$$

$A = P(1 + r_e)^t$ with $A = 10,000$, $P = 5500$, and $t = 6$

$$\frac{10,000}{5500} = (1 + r_e)^6$$

Dividing by 5500

$$(100/55)^{1/6} = 1 + r_e$$

Raising to power $\frac{1}{6}$

$$r_e = (100/55)^{1/6} - 1 \approx 0.1048$$

Solving for r_e

The effective rate is 10.48%.

■

Continuous Compounding

How much does the amount due on a loan change as the compounding changes? Clearly, if the term and interest rate are fixed, the amount due will increase if the compounding is more frequent, because the sooner the interest is charged, the sooner it will begin generating more interest. But will more frequent compounding increase the amount arbitrarily or is there a limit? To answer these questions we look at the compound interest formula and let the number of compounding periods m approach infinity, $m \to \infty$. We will use the fact from page 69 that as $n \to \infty$, the quantity $\left(1 + \dfrac{1}{n}\right)^n$ approaches the number $e \approx 2.71828$.

$$A = P\left(1 + \frac{r}{m}\right)^{mt} = P\left(1 + \frac{r}{m}\right)^{\frac{m}{r} rt}$$

<u>Compound interest formula</u>   Multiplying and dividing by r in the exponent

$$= P\left(1 + \frac{1}{n}\right)^{nrt} = P\left[\left(1 + \frac{1}{n}\right)^{n}\right]^{rt} \xrightarrow[\text{As } n \to \infty]{} Pe^{rt}$$

Letting $n = \frac{m}{r}$ $\to e$ as $n \to \infty$
so $\frac{1}{n} = \frac{r}{m}$

The arrow on the right means that the amount A approaches the quantity Pe^{rt} as n (or, equivalently, m) approaches infinity. That is, as the number m of compounding periods per year gets larger, the amount A approaches the limiting value Pe^{rt}. This "infinitely frequent" compounding is called *continuous compounding*.

Interest Compounded Continuously

> The amount A due on a loan of P dollars at annual interest rate r compounded *continuously* for t years is
>
> $$A = Pe^{rt}$$

To evaluate e to a power, use the ⎡2nd⎤ and ⎡LN⎤ keys on your calculator.

EXAMPLE 8   **Finding Amount Due with Continuous Compounding**

Find the amount due on a loan of $1500 at 4.8% compounded continuously for 5 years.

Solution

$$A = 1500e^{0.048 \cdot 5} = 1500e^{0.24} \approx 1906.87$$

```
1500→P
              1500
.048→R
               .048
5→T
                 5
Pe^(RT)
        1906.873725
```

The amount due is $1906.87.   ∎

Intuitive Meaning of Continuous Compounding

In Example 1 on page 112, we found that the amount due on a similar loan but compounded *monthly* was $1905.96. Notice that the amount found here for *continuous* compounding is slightly greater. Why does continuous compounding grow faster? With monthly compounding, the interest is not charged until the *end* of the month. Under *continuous* compounding the interest is charged *continuously, without delay,* so the interest starts generating interest immediately. The extra growth in continuous compounding comes from this "instant charging" of interest, generating more interest from the start.

For "present value under continuous compounding," we simply solve the continuous compounding formula $A = Pe^{rt}$ for P and rename it PV.

$$PV = Ae^{-rt}$$ Present value with
                            continuous compounding

EXAMPLE 9 **Finding Present Value with Continuous Compounding**

How much should be invested now at $6\frac{1}{2}\%$ compounded continuously if $5000 is needed in 4 years?

Solution

$$PV = 5000e^{-0.065 \cdot 4} = 5000e^{-0.26} \approx 3855.26$$

$PV = Ae^{-rt}$
with $A = 5000$,
$r = 0.065$, and $t = 4$

The amount needed now is $3855.26.

■

Growth Times Under Continuous Compounding

To find the time needed to reach a particular amount, we solve the continuous compounding formula $A = Pe^{rt}$ for t. It is easiest to use *natural* logarithms, which have the property that $\ln e^x = x$ (see page 82).

EXAMPLE 10 **Finding the Doubling Time with Continuous Compounding**

What is the doubling time of a loan at 6% compounded continuously?

Solution

We solve

$$2P = Pe^{0.06t}$$ $A = Pe^{rt}$ with A equal to $2P$

$$2 = e^{0.06t}$$ Canceling Ps

$$\ln 2 = \ln e^{0.06t}$$ Taking natural logs of both sides

$$\ln 2 = 0.06t$$ Since $\ln e^{0.06t} = 0.06t$

$$t = \frac{\ln 2}{0.06} \approx 11.6$$ Dividing by 0.06, reversing sides, and evaluating using the $\boxed{\text{LN}}$ key

The term is about 11.6 years.

■

In Practice Problem 2 (page 115), you found that the doubling time for a similar loan but with *quarterly* compounding was $11\frac{3}{4}$ years. Do you understand why continuous compounding should give a slightly shorter doubling time?

Be careful! For continuous compounding there is no "rounding to the next period," because there is no "period"—the term is simply an amount of years. How do you know when to use which formula? Use $A = Pe^{rt}$ when the compounding is *continuous* and $A = P(1 + r/m)^{mt}$ (the "discrete" formula) in all other cases.

Other types of continuous growth problems can be done just as Examples 3 and 4 (pages 114–115), but using the "continuous" formula $A = Pe^{rt}$ and then taking *natural* logarithms. That is, if a loan with continuous compounding is to increase by 75%, then we would solve $1.75P = Pe^{rt}$ for t. If specific dollar amounts are given, we would substitute them for P and A and then solve for t.

Effective Rates for Continuous Compounding

For *continuous* compounding, the effective rate satisfies

$$P(1 + r_e) = Pe^r$$

r_e is the "simple" rate that gives the continuous compound rate for $t = 1$

Canceling the Ps and subtracting 1 gives

$$r_e = e^r - 1$$

Effective rate for continuous compounding at rate r

EXAMPLE 11   Finding the Effective Rate with Continuous Compounding

What is the effective rate of a loan at 6.2% compounded continuously?

Solution

$$r_e = e^{0.062} - 1 \approx 0.0640 \qquad\qquad r_e = e^r - 1 \text{ with } r = 0.062$$

The effective rate is 6.40%. That is, 6.2% compounded continuously results in an annual gain of 6.4%.   ∎

Section Summary

The *compound interest formula* is

$$A = P\left(1 + \frac{r}{m}\right)^{mt}$$

A = amount due
P = principal
r = interest rate
m = compoundings per year
t = term in years

A is the *future value* of P, and P is the *present value* of A. Given values for four of the variables, we can solve for the fifth. Remember, however, to round the term *up* to the next whole number of compounding periods. The *doubling time*, the amount of time for an amount to double ($A = 2P$), can be approximated by the *rule of 72*: $t \approx 72/(r \times 100)$ years. Remember the difference between *increasing* by a percentage and *multiplying* by a number: Increasing by 50% means multiplying by 1.50.

The *effective rate* or *APR* of a loan is

$$r_e = \left(1 + \frac{r}{m}\right)^m - 1$$

r_e = effective rate of interest at rate r compounded m times each year

The effective rate of return for *any* investment of P dollars that returns A dollars after t years (such as a zero coupon bond) is found from the compound interest formula with $m = 1$, $A = P(1 + r_e)^t$.

For continuous compounding, the formula is

$$A = Pe^{rt}$$

A = amount
r = rate
t = years

For the present value under continuous compounding we solve this formula for P, whereas for the term we solve for t using natural logarithms. The effective rate for nominal rate r compounded continuously is $r_e = e^r - 1$.

EXERCISES 2.2 (Most require 📷 or 📷.)

Find the amount due on each compound interest loan.

1. $15,000 at 8% for 10 years if the interest is compounded:
   a. Annually   **b.** Quarterly **c.** Weekly.

2. $7500 at 6.5% for 8 years 6 months if the interest is compounded:
   a. Annually   **b.** Quarterly **c.** Weekly.

3. $17,500 at 8.5% for 6 years 9 months if the interest is compounded:
   a. Annually   **b.** Quarterly **c.** Weekly.

4. $25,000 at 9% for 5 years if the interest is compounded:
   a. Semiannually **b.** Monthly   **c.** Continuously

5. $12,000 at 7.5% for 4 years 3 months if the interest is compounded:
   a. Semiannually **b.** Monthly   **c.** Continuously

6. $16,750 at 4.5% for 6 years 6 months if the interest is compounded:
   a. Semiannually **b.** Monthly   **c.** Continuously

Find the present value of each compound interest loan.

7. $25,000 after 7 years at 12% if the interest is compounded:
   a. Annually   **b.** Quarterly **c.** Weekly.

8. $9500 after 4 years 6 months at 8.4% if the interest is compounded:
   a. Annually   **b.** Quarterly **c.** Weekly.

9. $19,500 after 6 years 9 months at 7.3% if the interest is compounded:
   a. Annually   **b.** Quarterly **c.** Weekly.

10. $16,500 after 5 years at 9% if the interest is compounded:
   a. Semiannually **b.** Monthly   **c.** Continuously

11. $11,500 after 4 years 3 months at 8.4% if the interest is compounded:
   a. Semiannually **b.** Monthly   **c.** Continuously

12. $2200 after 2 years 6 months at 7.5% if the interest is compounded:
   a. Semiannually **b.** Monthly   **c.** Continuously

Find the term of each compound interest loan.

13. 8.2% compounded quarterly to obtain $8400 from a principal of $2000.

14. 8.2% compounded continuously to obtain $8400 from a principal of $2000.

15. 6.8% compounded quarterly to multiply the principal by 1.8.

16. 5.48% compounded continuously to multiply the principal by 1.5.

17. 8.5% compounded monthly to increase the principal by 65%.

18. 8.5% compounded continuously to increase the principal by 65%.

Use the "rule of 72" to estimate the doubling time (in years) for each interest rate, and then calculate it exactly.

19. 9% compounded annually.

20. 6% compounded quarterly.

21. 7.9% compounded weekly.

22. 9% compounded continuously.

23. 6.1% compounded continuously.

24. 5.9% compounded continuously.

Find the effective rate of each compound interest rate or investment.

25. 4.3% compounded weekly.

26. 18% compounded monthly. [*Note:* This is a typical credit card interest rate, often stated as 1.5% per month.]

27. 8.57% compounded semiannually.

28. 7.5% compounded continuously.

29. A $50,000 zero coupon bond maturing in 8 years and selling now for $23,500.

30. A $10,000 zero coupon bond maturing in 12 years and selling now for $6400.

APPLIED EXERCISES

31. Bond Funds During the mid-1990s, the T. Rowe Price International Bond fund returned 10.43% compounded monthly. How much would a $5000 investment in this fund have been worth after 3 years?

32. Mutual Funds During the first half of the 1990s, the Twentieth Century Ultra aggressive growth mutual fund returned 19.83% compounded quarterly. How much would a $10,000 investment in this fund have been worth after 5 years?

33. College Savings How much would your parents have needed to set aside 17 years ago at 7.3% compounded weekly to give you $50,000 for college expenses today?

34. College Savings How much would your parents have needed to set aside 16 years ago at 6.7% compounded continuously to give you $60,000 for college expenses today?

35. Bond Funds You have just received $125,000 from the estate of a long-lost rich uncle. If you invest all of your inheritance in a tax-free bond fund earning 6.9% compounded quarterly, how long do you have to wait to become a millionaire?

36. Home Buying You and your new spouse have decided to use all $6500 of your wedding present monies as a nest egg for a house down payment. Investing at 11.47% compounded monthly, how long must you wait to have enough to put 10% down on a $140,000 house?

37. Mutual Funds A $10,000 investment in the Fidelity Blue Chip Growth mutual fund in 1987 would have been worth $30,832 seven years later. What was the effective rate of this investment?

38. Rate Comparisons The First Federal Bank offers 4.7% compounded weekly passbook savings accounts while Consolidated Nationwide Savings offers 4.73% compounded quarterly. Which bank offers the better rate?

39. Rate Comparisons The Second Peoples National Bank offers a long-term certificate of deposit earning 6.43% compounded monthly. Your broker locates a $20,000 zero coupon bond rated AA by Standard & Poor's for $7965 and maturing in 14 years. Which investment will give the greater rate of return?

40. Horse Trading In March 1995, a descendant of an early Texas settler sent $100 to Sam Houston IV to make good on a $100 debt (possibly from the sale of a horse) owed for 160 years to Sam Houston, the hero of Texas independence. The check was donated to the Sam Houston Museum and the debt considered as settled. However, newspapers reported that a banker had estimated the accumulated interest on the loan to be $420 million. What yearly interest rate did the banker use in her calculations?

41. **Lottery Winnings** You have just won $100,000 from a lottery. If you invest all this amount in a tax-free money market fund earning 7% compounded continuously, how long do you have to wait to become a millionaire?

42. **Becoming a Millionaire** How much would you have to invest now at 6.5% interest compounded continuously to have a million dollars in 40 years?

43. **Rate Comparisons** The People's State Bank offers 4.2% compounded quarterly, while Statewide Federal offers 4.1% compounded continuously. Which bank offers the better rate?

44. **Rate Comparisons** The Southwestern Savings and Loan Bank offers a 10-year certificate of deposit earning 6.4% compounded continuously. Your broker offers you a $10,000 zero coupon bond costing $5325 and maturing in 10 years. Which gives the greater rate of return?

Explorations and Excursions

The following problems extend and augment the material presented in the text.

45. Show that after two years, the (yearly) compound interest amount $A = P(1 + r)^2$ exceeds the simple interest amount $A = P(1 + 2r)$ by $P(r^2)$.

46. Show that after three years, the (yearly) compound interest amount $A = P(1 + r)^3$ exceeds the simple interest amount $A = P(1 + 3r)$ by $P(3r^2 + r^3)$.

47. Show that after n years ($n > 2$), the (yearly) compound interest amount $A = P(1 + r)^n$ exceeds the simple interest amount $A = P(1 + nr)$ by more than $P(nr^{n-1})$.

48. Solve any of Exercises 1–3 as follows: Enter the compound interest formula as $y = P(1 + r/m)^{mt}$. Then STORE the values for P, r, m, and t, and evaluate y to find the amount due.

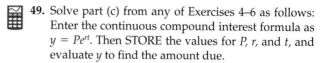

49. Solve part (c) from any of Exercises 4–6 as follows: Enter the continuous compound interest formula as $y = Pe^{rt}$. Then STORE the values for P, r, and t, and evaluate y to find the amount due.

50. Solve any of Exercises 7–9 as follows: Enter the present value formula as $y = A(1 + r/m)^{-mt}$. Then STORE the values A, r, m, and t, and evaluate y to find the present value.

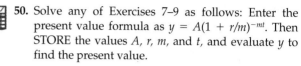

51. Solve part (c) from any of Exercises 10–12 as follows: Enter the continuous present value formula as $y = Ae^{-rt}$. Then STORE the values for A, r, and t, and evaluate y to find the amount due.

52. Solve the compound interest formula (page 112) for the number of compounding periods mt, and show that the result is $mt = \log(A/P)/\log(1 + r/m)$. Then show that if this number is not an integer, rounding it *up* is the same as using the formula $n = \lfloor\log(A/P)/\log(1 + r/m)\rfloor + 1$, where $\lfloor x \rfloor$ denotes the greatest integer less than or equal to the number x.

53. The formula for the number y of compounding periods from Exercise 52 can be viewed on your graphing calculator as a function of the interest rate x by entering it as $y = \text{int}(\log(A/P)/\log(1 + x/m)) + 1$. STORE the values 2400 for A, 2000 for P, and 12 for m (these are the numbers from Example 3 on page 114), and then graph y on the window [0, .1] by [0, 100]. Does this curve have the same shape as the curve in the "Rule of 72" Graphing Calculator Exploration on page 116? Use TRACE to explore the y-values for various values of x. Is y always a whole number? Use VALUE to find the number of compounding periods when the interest rate is 6%. Does this agree with the answer to Example 3?

54. Solve Exercise 13 as follows: Enter the formula for the number of compounding periods from Exercise 52 as $y = \text{int}(\log(A/P)/\log(1 + r/m)) + 1$. Then STORE the values for A, P, r and m, and evaluate y to find the number of compounding periods. Can you modify the expression for y to solve Exercise 15 in a similar manner?

55. Show that $P(1 + r/m)^{mt} = P(1 + r_e)^t$.

56. Show that the *effective m-compound rate* r_m for an investment compounded m times per year and returning A dollars from P dollars after t years is

$r_m = m((A/P)^{1/mt} - 1)$. Then check this formula by showing that the rate r_m compounded m times for one year gives the effective rate $r_e = (A/P)^{1/t} - 1$.

57. Suppose you found a bank that offered 100% interest compounded continuously. Being duly cautious, you decided to "test" them with a $1 deposit for one year. Show that your deposit would be worth exactly e dollars after 1 year.

58. Find the value of $100 at 20% compounded quarterly for the following terms: 1 year, 5 years, 25 years. Repeat these calculations for the same amount, rate, and terms, but with *continuous* compounding. Is continuous compounding significantly better over one year? over 25 years?

More About the Rule of 72

 59. This problem continues the "Rule of 72" Graphing Calculator Exploration on page 116. Use the same $[0, .5]$ by $[0, 20]$ window for all your graphs.

a. For interest compounded quarterly instead of monthly, graph $y_1 = \log(2)/(4 \log(1 + x/4))$ together with $y_2 = 72/(x \times 100)$. Do these two curves still closely match?

b. For interest compounded weekly instead of monthly, graph $y_1 = \log(2)/(52 \log(1 + x/52))$ together with $y_2 = 72/(x \times 100)$. Do these two curves still closely match?

c. Using your monthly, quarterly, or weekly compounding curves, replace $y_2 = 72/(x \times 100)$ with $y_2 = 70(x \times 100)$. Try it again with $y_2 = 71/(x \times 100)$. Do these curves still closely match the compound interest curve? Which "rule" is easier to use in mental calculations: 70, 71, or 72?

d. Using your monthly, quarterly, or weekly compounding curves, replace $y_2 = 72/(x \times 100)$ with $y_2 = 60/(x \times 100)$ and $y_3 = 80/(x \times 100)$. Is 60 much too small and 80 much too big?

60. a. Show that the doubling time for interest rate r compounded continuously is $\frac{1}{r} \ln 2$.

b. How does the numerical value of $\ln 2$ explain the "rule of 72"?

c. Show that the time it takes for an amount at interest rate r compounded continuously to be multiplied by a factor of m is $\frac{1}{r} \ln m$.

d. Use the result of part (c) and the numerical value of $\ln 3$ to devise your own rule to estimate tripling times under compound interest.

2.3 Annuities

APPLICATION PREVIEW

Retirement Planning

Although buying your first car and your first house will give you the most immediate satisfaction when you take possession, planning and saving for your retirement will extend over more years and involve more money than perhaps anything else. Most financial planners will tell you that the amount you set aside is less important than the fact that you are saving *something* regularly, and the earlier you start, the greater the results. An individual retirement account, or IRA, allows almost anyone to set aside up to $2000 each year to earn interest tax free until retirement age. The power of compound interest is so great that if you fund an IRA *only* during your twenties, you can end up with more money than if you wait until your thirties and then invest every year until retirement.

But where should you invest your IRA money? From 1926 to 1995, the average annual return was 3.72% on Treasury bills, 5.69% on long-term bonds, and 10.54% on common stocks. Although stocks are the most volatile, they do have the most growth potential over time, and investment in a mutual fund can spread your money over a wide variety of companies. Some mutual funds charge annual fees for IRA accounts, while others do not. Can a $30 annual fee make much difference over the 40 years before you retire? The graph shows the difference between the values of two IRA accounts earning 10% with annual $2000 deposits when one charges a $30 annual fee and the other does not—after 40 years, the difference is $13,278.

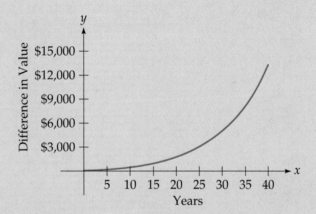

Difference between no-fee and annual-fee IRA accounts.

Such accounts are called *annuities* and are the subject of this section.

Introduction

An *annuity* is a scheduled sequence of payments. Some annuities, such as retirement pensions, provide money to you, while others, such as car loans and home mortgages, require that you pay someone else. In this section we will consider an *ordinary annuity*, which is an annuity with equal payments at regular intervals, a fixed interest rate, and the interest compounded at the end of each payment period. For example, a car loan for 4 years with monthly payments of $200 at 6% compounded monthly is an ordinary annuity.

Although having a new car is nice, when the first payment comes due at the end of the first month, you might wish that instead you were saving that $200 for yourself. If you were, then 4 years later your first $200 would have grown to 200(1 + 0.06/12)^{47}$ (the exponent is 47 rather than 48 because the first payment was made at the *end* of the first month and so earns interest for only 47 months). Similarly, your second $200 would grow to 200(1 + 0.06/12)^{46}$, your third to 200(1 + 0.06/12)^{45}$, and so on. The total amount you could have saved would be the sum of the 48 payments with the interest earned on each one:

$$200\left(1 + \frac{0.06}{12}\right)^{47} + 200\left(1 + \frac{0.06}{12}\right)^{46} + 200\left(1 + \frac{0.06}{12}\right)^{45}$$

$$+ \cdots + 200\left(1 + \frac{0.06}{12}\right)^{2} + 200\left(1 + \frac{0.06}{12}\right)^{1} + 200$$

Notice that your last $200 earns no interest because you pay it at the end of the 4 years. With patience and careful attention to detail, you could add up the 48 values to find this sum. However, there is an easier way.

Geometric Series

The above sum is an example of a *geometric series,* where each value in the sum is a fixed multiple of the previous value. The simplest geometric series is of the form

$$x^{n-1} + x^{n-2} + x^{n-3} + \cdots + x^2 + x + 1$$

If we multiply this sum by $x - 1$, we obtain

$$(x^{n-1} + x^{n-2} + x^{n-3} + \cdots + x^2 + x + 1)(x - 1)$$

$$= x^{n-1}(x - 1) + x^{n-2}(x - 1) + x^{n-3}(x - 1)$$

$$+ \cdots + x^2(x - 1) + x^1(x - 1) + 1(x - 1)$$

Multiplying each by $x - 1$

$$= x^n - x^{n-1} + x^{n-1} - x^{n-2} + x^{n-2} - x^{n-3} + \cdots$$

Cancel   Cancel   Cancels with next term

All cancel except $x^n - 1$

$$\cdots + x^3 - x^2 + x^2 - x + x - 1$$

Cancels with previous term   Cancel   Cancel

$$= x^n - 1$$

Setting the first and last expressions equal to each other, we have

$$(x^{n-1} + x^{n-2} + x^{n-3} + \cdots + x^2 + x + 1)(x - 1) = x^n - 1$$

or

$$x^{n-1} + x^{n-2} + x^{n-3} + \cdots + x^2 + x + 1 = \frac{x^n - 1}{x - 1} \qquad \text{Dividing by } x - 1$$

Multiplying both sides by a gives a formula for the sum of a geometric series:

Sum of a Geometric Series

$$ax^{n-1} + ax^{n-2} + ax^{n-3} + \cdots + ax^2 + ax + a = a\frac{x^n - 1}{x - 1}$$

For instance, the geometric series $48 + 24 + 12 + 6 + 3$, in which $a = 3$, $x = 2$, and $n = 5$, has the value $3(2^5 - 1)/(2 - 1) = 93$, as you should check by adding up the five numbers. Applying this formula to our 4-year savings account, we find that

$$200\left(1 + \frac{0.06}{12}\right)^{47} + 200\left(1 + \frac{0.06}{12}\right)^{46} + \cdots$$

$$+ 200\left(1 + \frac{0.06}{12}\right)^{2} + 200\left(1 + \frac{0.06}{12}\right)^{1} + 200$$

$$= 200\frac{\left(1 + \dfrac{0.06}{12}\right)^{48} - 1}{\left(1 + \dfrac{0.06}{12}\right) - 1} \approx 10{,}819.57 \qquad \begin{aligned} & a\frac{x^n - 1}{x - 1} \text{ with } a = 200, \\ & x = (1 + 0.06/12), \\ & \text{and } n = 48 \end{aligned}$$

Therefore, regular $200 deposits at the end of each month for 4 years into a savings account earning 6% interest compounded monthly will total $10,819.57. (This is significantly more than we get when we simply add the 48 payments of $200, which total only $9600.)

Accumulated Amount Formula

Applying the same method to any annuity with regular payments of P dollars repeated m times per year for t years with the interest rate r

compounded at each payment, we find that the final accumulated amount A will be

$$A = P\left(1 + \frac{r}{m}\right)^{mt-1} + P\left(1 + \frac{r}{m}\right)^{mt-2} + \cdots + P\left(1 + \frac{r}{m}\right)^{1} + P$$

$$= P\frac{\left(1 + \frac{r}{m}\right)^{mt} - 1}{\left(1 + \frac{r}{m}\right) - 1}$$

Canceling the 1 and the -1 in the denominator, we have the following formula.

Accumulated Amount of an Annuity

> The accumulated amount A of an ordinary annuity with payments of P dollars m times per year for t years at interest rate r compounded at the end of each payment period is
>
> $$A = P\frac{(1 + r/m)^{mt} - 1}{r/m}$$

This formula is simple to evaluate using a calculator, but compound interest and annuity calculations have been common features of finance for many centuries. Writing i for r/m and n for mt, the archaic symbol $s_{\overline{n}|i}$ (pronounced "s sub n angle i") is sometimes still used to denote the value of $((1 + i)^n - 1)/i$ in business textbooks and tables. Using this notation, the above formula takes the form $A = Ps_{\overline{n}|i}$ with $n = mt$ and $i = r/m$.

EXAMPLE 1   Finding the Accumulated Amount of an Annuity

What is the final balance of a retirement account earning 4% interest compounded weekly if $50 is deposited at the end of every week for 40 years? How much of this final balance is interest?

Solution

$$A = 50\,\frac{(1 + 0.04/52)^{(52)(40)} - 1}{0.04/52} \approx 256{,}749.15$$

$A = P\dfrac{(1 + r/m)^{mt} - 1}{r/m}$
with $P = 50$, $r = 0.04$, $m = 52$, and $t = 40$

The final balance is $256,749.15. The deposits total only $50 × 52 × 40 = $104,000, so the account has earned $152,749.15 in interest during the 40 years.

PRACTICE PROBLEM 1

What is the final balance of a retirement account earning 5% interest compounded weekly if $40 is deposited at the end of every week for 35 years? How much of this final balance is interest?

Solution at the back of the book

Sinking Funds

A *sinking fund* is a regular savings plan designed to provide a given amount after a certain number of years. For example, suppose you decide to save up for a $18,000 car instead of taking out a loan. What amount should you save each month to have the $18,000 in 3 years? If you just put your money in a shoe box, you will need to set aside $18,000 ÷ (12 × 3) = $500 each month. It would make more sense to put your money into a savings account each month and let it earn compound interest. This is the same as setting up an annuity with a given accumulated amount and asking what the regular payment should be. Solving the accumulated amount formula (on previous page) for P by dividing gives the following formula.

Sinking Fund Payment

> The payment P to make m times per year for t years at interest rate r compounded at each payment to accumulate amount A is
>
> $$P = A \frac{r/m}{(1 + r/m)^{mt} - 1}$$

This formula may also be written $P = A/s_{\overline{n}|i}$ with $n = mt$ and $i = r/m$.

EXAMPLE 2 **Finding a Sinking Fund Payment**

What amount should be deposited at the end of each month for 3 years in a savings account earning 4.5% interest compounded monthly to accumulate $18,000 to buy a new car?

Solution

$$P = 18,000 \frac{0.045/12}{(1 + 0.045/12)^{(12)(3)} - 1} \approx 467.945$$

$P = A \dfrac{r/m}{(1 + r/m)^{mt} - 1}$
with $A = 18,000$,
$r = 0.045$, $m = 12$,
and $t = 3$

Rounding up to the next penny, $467.95 should be saved each month. (If we rounded down to $467.94, the savings account would accumulate slightly less than the amount needed. Notice that this amount is significantly lower than the $500 that would be needed *without* compound interest.)

This example could also have been done using the accumulated amount formula (page 130) by substituting $A = 18{,}000$, $r = 0.045$, $m = 12$, and $t = 3$, and then solving for the remaining variable to find $P = \$467.95$.

PRACTICE PROBLEM 2

For wage earners paid every other week instead of monthly, Example 2 becomes: "What amount should be deposited biweekly for 3 years in a savings account earning 4.5% interest compounded biweekly to accumulate $18,000 for a new car?" Find this amount.

Solution at the back of the book

How Long Will It Take?

Continuing with our sinking fund example, suppose you could save only $200 each month. How long would it take to accumulate the $18,000?

We solve the accumulated amount formula (page 130) for the number of years t.

$$\frac{A}{P}\frac{r}{m} + 1 = \left(1 + \frac{r}{m}\right)^{mt}$$

$A = P\,\dfrac{(1 + r/m)^{mt} - 1}{r/m}$
Dividing by P, multiplying by r/m, and adding 1

$$mt \log\left(1 + \frac{r}{m}\right) = \log\left(\frac{A}{P}\frac{r}{m} + 1\right)$$

Switching sides and taking logs of both to bring down the exponent

Dividing by $\log(1 + r/m)$, we have:

Number of Periods

$$mt = \frac{\log\left(\dfrac{A}{P}\dfrac{r}{m} + 1\right)}{\log\left(1 + \dfrac{r}{m}\right)} \quad \text{periods}$$

As usual, we round *up* to the next whole number of compounding periods and then express the answer as years plus any extra periods. Notice that the time t depends only on the ratio $\frac{A}{P}$ between the accumulated amount A and the regular payment P.

EXAMPLE 3 Finding the Term for a Sinking Fund

How long will it take to accumulate $18,000 by depositing $200 at the end of each month in a savings account earning 4.5% interest compounded monthly?

Solution

$$12t = \frac{\log\left(\dfrac{18{,}000}{200}\dfrac{0.045}{12} + 1\right)}{\log\left(1 + \dfrac{0.045}{12}\right)} \approx 77.7$$

$$mt = \frac{\log\left(\dfrac{A}{P}\dfrac{r}{m} + 1\right)}{\log\left(1 + \dfrac{r}{m}\right)}$$

with $A = 18{,}000$, $P = 200$, $r = 0.045$, and $m = 12$

The time is 77.7 months, which we round *up* to 78 months. Therefore, it will take 6 years 6 months to accumulate the $18,000.

■

This example could also have been done using the accumulated amount formula from page 130 by substituting $A = 18{,}000$, $P = 200$, $r = 0.045$, and $m = 12$, and then solving (using logarithms) for the remaining variable to find that $12t$ is 78 months.

PRACTICE PROBLEM 3

How long will it take to accumulate $18,000 by depositing $250 each month in a savings account earning 4.5% interest compounded monthly?   *Solution at the back of the book*

Graphing Calculator Exploration

Depending on the size of the monthly deposit, the time needed to accumulate a given amount may vary greatly.

a. Using the values $r = 0.045$, $m = 12$, and $A = 18{,}000$ from Example 3, and replacing the payment amount P by x, graph the curve $y_1 = \log((18000/x)$ $(.045/12) + 1)/\log(1 + .045/12)$ on the window [0, 1000] by [0, 240]. The graph shows how y, the number of months needed to accumulate $18,000, decreases as the monthly payment x increases.

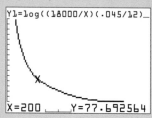

b. Use TRACE or VALUE to find y when $x = 200$ and when $x = 250$. How do your answers compare to the answers to Example 3 and Practice Problem 3?

c. Use TRACE or VALUE to find the years y needed when the monthly deposits x are $100, $150, $200, $250, and $300. Does each additional $50 reduce the time needed by the same amount? How does the graph show this?

Internal Rate of Return

Suppose you make regular investments into a stock market fund, and after several years you want to find the "rate of return" on your money. One measure of this is the *internal rate of return*, which is the interest rate r of the annuity having the same payments, payment frequency, accumulated amount, and term. That is, in the accumulated amount formula from page 130,

$$A = P \frac{(1 + r/m)^{mt} - 1}{r/m}$$

we know the values of P, m, t, and A, and we want to solve for the interest rate r. Unfortunately, there is no simple algebraic solution of this problem. However, the solution can be approximated using a graphing calculator by replacing r by x and graphing $y = P((1 + x/m)^{mt} - 1)/(x/m)$ using the known values of P, m, and t. You can then use TRACE or INTERSECT to find an interest rate x giving a y-value near the given amount A. Other, more tedious methods include searching a $s_{\overline{n}|i}$ table for values near $\frac{A}{P}$ or by trial and error using guesses for the interest rate in the annuity formula.

Graphing Calculator Exploration

What is the internal rate of return of $500 quarterly investments in a mutual fund account valued at $27,660 after 8 years?

Solution

We have the values $P = 500$, $m = 4$, $t = 8$, and $A = 27,660$. The exponent mt is $(4)(8) = 32$.

a. Graph the curve $y = 500((1 + x/4)^{32} - 1)/(x/4)$ on the window [0, .2] by [0, 40000] to show the relation between the amount y and the internal rate of return x.

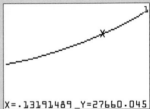

X=.13191489 Y=27660.045

b. Use TRACE or INTERSECT to estimate the value for x that corresponds to a y-value of 27,660.

The internal rate of return of this mutual fund is 13.19% over the last 8 years. You can verify this estimate by checking that this value for r with the given values for P, m, and t in the annuity formula gives a value for A near the actual value of $27,660.

Overview

In the previous section we found the eventual value of an investment, the initial amount necessary to reach that later amount, the time it takes, and the internal rate of return, just as we did in this section. What's the difference between the two sections? *There* we were considering investments of a *single* payment of P dollars, while *here* we are considering *multiple and regular payments* of P dollars.

Section Summary

For an *ordinary annuity* with regular payments of P dollars m times per year for t years at interest rate r compounded at the end of each payment, the accumulated amount A is

$$A = P\frac{(1 + r/m)^{mt} - 1}{r/m}$$

A = accumulated amount, P = payment, r = interest rate, m = times per year, t = years

This formula may be solved for the other variables. Solving for P finds the regular payment to make into a *sinking fund* to accumulate the amount A after t years:

$$P = A\frac{r/m}{(1 + r/m)^{mt} - 1}$$

Solving for t finds the number of years required for the payments into a sinking fund to accumulate to a final amount A:

$$mt = \frac{\log\left(\dfrac{A}{P}\dfrac{r}{m} + 1\right)}{\log\left(1 + \dfrac{r}{m}\right)}$$

mt = number of periods (other variables as above)

Round *up* to find the number of payments and then express your answer as a number of years plus any additional periods.

There is no simple algebraic way to solve for the interest rate r. However, the *internal rate of return* of an investment with known regular payments and accumulated amount can be approximated using a graphing calculator.

EXERCISES 2.3 (Most exercises require 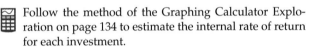 or 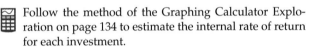 .)

In the following ordinary annuities, the interest is compounded with each payment, and the payment is made at the end of the compounding period.

Find the accumulated amount of each annuity.

1. $1500 annually at 7% for 10 years.

2. $1750 annually at 8% for 7 years.

3. $2000 semiannually at 6% for 12 years.

4. $950 quarterly at 10.7% for 6 years.

5. $1000 monthly at 6.9% for 20 years.

6. $200 weekly at 11.3% for 5 years.

Find the required payment for each sinking fund.

7. Monthly deposits earning 5% to accumulate $5000 after 10 years.

8. Monthly deposits earning 4% to accumulate $6000 after 15 years.

9. Monthly deposits earning 11.7% to accumulate $14,000 after 5 years.

10. Quarterly deposits earning 6.4% to accumulate $50,000 after 20 years.

11. Yearly deposits earning 12.3% to accumulate $8500 after 12 years.

12. Weekly deposits earning 9.8% to accumulate $15,000 after 6 years.

Find the amount of time needed for each sinking fund to reach the given accumulated amount.

13. $1500 yearly at 8% to accumulate $100,000.

14. $2000 yearly at 9% to accumulate $125,000.

15. $800 semiannually at 6.6% to accumulate $80,000.

16. $500 quarterly at 8.2% to accumulate $19,500.

17. $235 monthly at 5.9% to accumulate $25,000.

18. $70 weekly at 10.9% to accumulate $7500.

Follow the method of the Graphing Calculator Exploration on page 134 to estimate the internal rate of return for each investment.

19. $350 quarterly to accumulate $21,600 after 10 years.

20. $300 quarterly to accumulate $16,940 after 9 years.

21. $50 monthly to accumulate $5270 after 6 years.

22. $180 monthly to accumulate $52,800 after 14 years.

23. $10 weekly to accumulate $9572 after 12 years.

24. $1.50 daily to accumulate $106,795 after 50 years.

APPLIED EXERCISES

25. **Retirement Savings**   An individual retirement account, or IRA, earns tax-deferred interest and allows the owner to invest up to $2000 each year. Joe and Jill both will make IRA deposits for 30 years (from age 35 to 65) into stock mutual funds yielding 9.8%. Joe deposits $2000 once each year while Jill has $38.46 (which is 2000/52) withheld from her weekly paycheck and deposited automatically. How much will each have at age 65?

26. **Retirement Savings**

   a. Steve opens a retirement account yielding 9% and deposits $150 each month for the next 30 years (from age 35 to 65). How much will he have at age 65?

   b. Sue opens her own retirement account yielding 8% and deposits $150 each month for the 10 years from age 25 to 35. She then makes no further deposits but lets the accumulated amount earn

compound interest for the next 30 years. How much will she have at age 65?

   c. Who has more money for retirement?

27. **Mutual Funds**   How much must you invest each month in a mutual fund yielding 13.4% compounded monthly to become a millionaire in 10 years?

28. **Bond Funds**   How much must you invest each quarter in a bond fund yielding 9.7% compounded quarterly to become a millionaire in 20 years?

29. **Lifetime Savings**   The Oseola McCarty Scholarship Fund at the University of Southern Mississippi was established by a $150,000 gift from an 87-year-old black woman who had dropped out of sixth grade and worked for most of her life as a washerwoman. How much must be saved each week in a bank account earning 3.9% compounded weekly to have $150,000 after 75 years?

30. Lifetime Savings If each day your fairy god-mother put $1 into a savings account at 8.37% compounded daily, how much would the account be worth when you were 65 years old?

31. Home Buying You and your new spouse each bring home $1500 each month after taxes and other payroll deductions. By living frugally, you intend to live on just one paycheck and save the other in a bond fund yielding 7.86% compounded monthly. How long will it take to have enough for a 20% down payment on a $165,000 condo in the city?

32. Boat Buying How long will it take to save $16,000 for a new motorboat if you deposit $375 each month into a money market fund yielding 5.73% compounded monthly?

33. Stock Market According to the *Wall Street Journal*, the $16\frac{1}{2}$ years from February 1966 to August 1982 was perhaps the most brutal period for stock market investors since the Great Depression in the 1930s. Ibbotson Associates estimates that if you had invested $100 each month into the stocks that make up the S&P 500 during this period, by the market low at mid-1982 your portfolio would have been worth almost $32,600 (with all dividends reinvested). What would the internal rate of return have been on such an investment?

34. Technology Stocks What is the internal rate of return for $500 quarterly investments in a science and technology specialty stock fund valued at $17,375 after 6 years?

27441.91

Explorations and Excursions

The following problems extend and augment the material presented in the text.

35. Solve any three of Exercises 1–6 as follows: Enter the accumulated amount formula as $y = P((1 + r/m)^{mt} - 1)/(r/m)$. Then for each exercise, STORE the values for P, r, m, and t, and evaluate y to find the amount due.

36. Solve any three of Exercises 7–12 as follows: Enter the sinking-fund payment formula as $y = A(r/m)/((1 + r/m)^{mt} - 1)$. Then for each problem, STORE the values for A, r, m, and t, and evaluate y to find the payment.

37. Show that if the number of deposits (or, equivalently, the number of periods) $mt = \log((A/P)(r/m) + 1)/(\log(1 + r/m))$ from page 132 is not a whole number, then rounding up is the same as using the formula $n = \lfloor \log((A/P)(r/m) + 1)/\log(1 + r/m)\rfloor + 1$, where $\lfloor x \rfloor$ denotes the greatest integer less than or equal to the number x.

38. The formula for the number y of deposits from Exercise 37 can be viewed on your graphing calculator as a function of the deposit x by entering it as $y = \text{int}(\log((A/x)(r/m) + 1)/\log(1 + r/m)) + 1$. STORE the values 18,000 for A, 0.045 for r, and 12 for m (these are the numbers from Example 3 on pages 132–133), and then graph y on the window [0, 1000] by [0, 500]. Does this curve have the same shape as the curve in the Graphing Calculator Exploration on page 133 after Practice Problem 3? Use TRACE to explore the y-values for various values of x. Is y always a whole number? Use VALUE to find the number y of deposits when the deposit x is 200. Does this agree with the answer to Example 3?

39. Solve any three of Exercises 13–18 as follows: Enter the formula for the number of deposits from Exercise 37 as $y = \text{int}(\log((A/P)(r/m) + 1)/\log(1 + r/m)) + 1$. For each exercise, STORE the values for A, P, r, and m, and evaluate y to find the number of deposits needed.

The following exercises provide further information about geometric series.

40. For each choice of values for a, x, and n, write out the geometric series $ax^{n-1} + ax^{n-2} + ax^{n-3} + \cdots + ax^2 + ax + a$, find the sum, and then evaluate the corresponding $a\dfrac{x^n - 1}{x - 1}$ expression to check that both have the same value.

a. $a = 2, x = 3, n = 5$
b. $a = 162, x = \frac{1}{3}, n = 5$. How do parts (a) and (b) differ?
c. $a = 3, x = 10, n = 6$

41. Check the formula $x^{n-1} + x^{n-2} + x^{n-3} + \cdots + x^2 + x + 1 = \dfrac{x^n - 1}{x - 1}$ for the case $n = 5$ by carrying out the long division $x - 1\overline{)x^5 - 1}$.

42. Proof by Induction Show that

$$x^{n-1} + x^{n-2} + x^{n-3} + \cdots + x^2 + x + 1 = \frac{x^n - 1}{x - 1}$$

by *induction* as follows:

a. Check that $(x + 1)(x - 1) = x^2 - 1$ by multiplying out the left side. Then divide by $x - 1$ to show that $x + 1 = \dfrac{x^2 - 1}{x - 1}$.

b. Factor out an x to check that $x^2 + x + 1 = x(x + 1) + 1$. Using (a), substitute $\dfrac{x^2 - 1}{x - 1}$ for $x + 1$ on the right side, get a common denominator, and collect like terms to show that $x^2 + x + 1 = \dfrac{x^3 - 1}{x - 1}$.

c. Factor out an x to check that $x^3 + x^2 + x + 1 = x(x^2 + x + 1) + 1$. Using (b), substitute $\dfrac{x^3 - 1}{x - 1}$ for $x^2 + x + 1$ on the right side, get a common denominator, and collect like terms to show that $x^3 + x^2 + x + 1 = \dfrac{x^4 - 1}{x - 1}$. Do you now see how to use this method to show that $x^4 + x^3 + x^2 + x + 1 = \dfrac{x^5 - 1}{x - 1}$?

d. We can now show that $x^{n-1} + x^{n-2} + x^{n-3} + \cdots + x^2 + x + 1 = \dfrac{x^n - 1}{x - 1}$ for every value of $n \geq 2$.

Part (a) showed that this formula is true for $n = 2$, part (b) showed it true for $n = 3$ because it is true for $n = 2$, and part (c) showed that this formula is true for $n = 4$ because it is true for $n = 3$. A *proof by induction* concludes that the formula is true for every $n \geq 2$ if we know that it is true for $n = 2$ (the "starting" value) and that the formula is true for $n + 1$ if it is true for n (that is, we can always prove the "next" version of the formula from the one we just proved). To check the *induction step* to complete the proof, show that

$$x^n + x^{n-1} + x^{n-2} + \cdots + x^2 + x + 1$$
$$= x(x^{n-1} + x^{n-2} + x^{n-3} + \cdots + x^2 + x + 1) + 1$$

Substitute $\dfrac{x^n - 1}{x - 1}$ for $(x^{n-1} + x^{n-2} + x^{n-3} + \cdots + x^2 + x + 1)$, get a common denominator, and

collect like terms to show that

$$x^n + x^{n-1} + x^{n-2} + \cdots + x^2 + x + 1 = \frac{x^{n+1} - 1}{x - 1}$$

e. By the way, is the formula true for $n = 1$?

43. *Requires a calculator with series operations.* Solve Exercise 40 as follows. Your calculator can LIST the numbers making up a geometric series as a *sequence* of values by using SEQUENCE. For the values of a, x, and n in Exercise 40a, list the numbers as the SEQUENCE $2(3^n)$ for n starting at 0, ending at 4, and increasing by 1. This list of numbers can be added using SUM. In this way, check each of the SUMs in Exercise 40. Then check that the geometric SEQUENCE $200(1 + 0.06/12)^n$ on page 129 for n from 0 to 47 SUMs to \$10,819.57.

44. Infinite Geometric Series *Requires a calculator with series operations.* With the values $a = 1$ and $x = \frac{1}{2}$, the geometric series formula becomes

$$\left(\tfrac{1}{2}\right)^{n-1} + \left(\tfrac{1}{2}\right)^{n-2} + \left(\tfrac{1}{2}\right)^{n-3} + \cdots + \left(\tfrac{1}{2}\right)^2 + \tfrac{1}{2} + 1 = \frac{\left(\tfrac{1}{2}\right)^n - 1}{\tfrac{1}{2} - 1}$$

a. Check this formula by evaluating both sides for $n = 5$, $n = 10$, and $n = 15$. As n gets larger, do these values get closer to 2?

b. Graph the SUM y of the first x terms of this geometric SEQUENCE on the window $[0, 25]$ by $[0, 3]$. Use TRACE to explore the graph. As x gets larger, do these values get closer to 2?

An *infinite geometric series* is a sum of the form $a + ax + ax^2 + ax^3 + ax^4 + \cdots$, where the $+ \cdots$ indicates that the sum continues on to include every power of x. For $a = 1$ and $x = \frac{1}{2}$, we suspect from parts (a) and (b) that

$$1 + \tfrac{1}{2} + \left(\tfrac{1}{2}\right)^2 + \left(\tfrac{1}{2}\right)^3 + \left(\tfrac{1}{2}\right)^4 + \cdots = 2$$

c. Check this arithmetic: Because $\left(\tfrac{1}{2}\right)^n$ gets closer to 0 as n gets larger, the geometric series formula $\dfrac{\left(\tfrac{1}{2}\right)^n - 1}{\tfrac{1}{2} - 1}$ gets closer to $\dfrac{0 - 1}{\tfrac{1}{2} - 1}$ as n gets larger.

d. If $1 + \tfrac{1}{2} + \left(\tfrac{1}{2}\right)^2 + \left(\tfrac{1}{2}\right)^3 + \left(\tfrac{1}{2}\right)^4 + \cdots$ really is 2, it should behave just like 2. Show that $\tfrac{1}{2}\left(1 + \tfrac{1}{2} + \left(\tfrac{1}{2}\right)^2 + \left(\tfrac{1}{2}\right)^3 + \left(\tfrac{1}{2}\right)^4 + \cdots\right) = 1$ by multiplying out the left side and using the fact that $1 + \tfrac{1}{2} + \left(\tfrac{1}{2}\right)^2 + \left(\tfrac{1}{2}\right)^3 + \left(\tfrac{1}{2}\right)^4 + \cdots = 2$.

(continues)

If $|x| < 1$, then x^n gets closer to 0 as n increases and we have the following formula for an *infinite* geometric series:

$$a + ax + ax^2 + \cdots = a \cdot \frac{1}{1-x}$$

e. Find the value of $1 + \frac{1}{3} + \left(\frac{1}{3}\right)^2 + \left(\frac{1}{3}\right)^3 + \left(\frac{1}{3}\right)^4 + \cdots$.

f. Find the value of $9 + 9\left(\frac{1}{10}\right) + 9\left(\frac{1}{10}\right)^2 + 9\left(\frac{1}{10}\right)^3 + 9\left(\frac{1}{10}\right)^4 + \cdots$. Can you write this geometric series as a decimal?

Amortization

Loan Repayment and the Rule of 78

Many car loans include the clause that early repayment of the loan will be calculated using the "rule of 78." While most buyers expect to satisfy their loans by making the agreed monthly payments, unforeseen circumstances could make it necessary to end the loan early. The rule of 78 is a method of dividing the remaining payments between the interest owed and the principal borrowed in order to find the amount yet to be repaid. As in the Banker's rule, the simplicity of the rule of 78 hides the advantage it gives the lender. For a 1-year loan with monthly payments, the rule of 78 states that the last payment contains one part of the total interest, the second to last contains two parts of the total interest, and so on back to the first monthly payment, which then contains 12 parts of the total interest. Thus the interest for the year has been divided into $1 + 2 + 3 + \cdots + 11 + 12 = 78$ equal parts, which gives this rule its name. This means that the first payment pays $\frac{12}{78}$ of the interest and that $\frac{66}{78}$ remain. For a longer term, the rule of 78 similarly sums the number of payments.

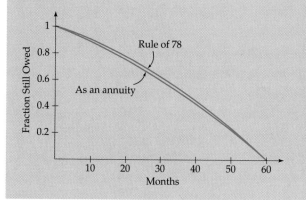

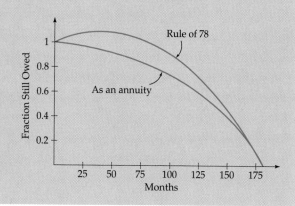

Is this fair to the borrower? The left graph shows the amount still owed on a 5-year car loan at 18% calculated both by the rule of 78 and as an annuity formed by the remaining payments. The curve for the rule of 78 is significantly higher for the first two-thirds of the loan period. The right graph shows the amount still owed on a 15-year home improvement loan at 18%. Not only is the rule of 78 amount larger, but for the first 7 years it is even larger than the amount of the loan! This "negative amortization" provides an extreme yet possible demonstration of the advantage the rule of 78 gives the lender.

Introduction

In this section we will find the *present value of an annuity*—the total value now of all the future payments. Such calculations are important for buying and selling annuities and for using them to pay off debts (amortization). For example, on page 129 we found the accumulated amount for monthly payments of $200 for 4 years at 6% compounded monthly. But how much would a loan company give right now for that promise of future payments?

Present Value of an Annuity

We know that for an ordinary annuity with payments of P dollars m times per year for t years at interest rate r compounded at each payment, the accumulated amount A is

$$A = P\frac{(1 + r/m)^{mt} - 1}{r/m}$$

From page 130

Because the present value PV of this annuity must grow to match the accumulated amount of the annuity at the end of the t years, we must also have, by the compound interest formula,

$$A = PV\,(1 + r/m)^{mt}$$

From page 112

To find the present value PV, we set these two expressions for A equal to each other and divide by $(1 + r/m)^{mt}$.

$$PV = P\frac{(1 + r/m)^{mt} - 1}{(r/m)(1 + r/m)^{mt}}$$

Equating As and dividing by $(1 + r/m)^{mt}$

$$= P\frac{1 - (1 + r/m)^{-mt}}{r/m}$$

Dividing numerator and denominator by $(1 + r/m)^{mt}$

Therefore:

Present Value of an Annuity

> The present value PV of an ordinary annuity with payments of P dollars m times per year for t years at interest rate r compounded at each payment is
>
> $$PV = P\frac{1 - (1 + r/m)^{-mt}}{r/m}$$

Writing i for r/m and n for mt, the archaic symbol $a_{\overline{n}|i}$ (pronounced "a sub n angle i") is sometimes still used in business textbooks and tables to denote the value of $(1 - (1 + i)^{-n})/i$. Using this notation, the above formula takes the form $PV = Pa_{\overline{n}|i}$ with $n = mt$ and $i = r/m$.

EXAMPLE 1   Finding the Present Value of an Annuity

What is the present value of a 6% car loan for 4 years with monthly payments of $200?

Solution

$$PV = 200\frac{1 - (1 + 0.06/12)^{-(12)(4)}}{0.06/12} \approx 8516.06$$

$PV = P\dfrac{1 - (1 + r/m)^{-mt}}{r/m}$
with $P = 200$, $r = 0.06$,
$m = 12$, and $t = 4$

The present value is $8516.06.

⬛

What interpretation can we give to this present value? If we put it into a savings account earning the same 6% compounded monthly for 4 years, it would grow to $8516.06 $(1 + 0.06/12)^{48}$ = $10,819.57. But we saw on page 129 that 4 years of monthly $200 payments into a savings account earning the same interest will accumulate this same final value. Thus the return from depositing this present value into a bank account for 4 years is exactly the same as that from the monthly annuity payments over the same period, and the lender will not care which she receives.

PRACTICE PROBLEM 1

What is the present value of a 20-year retirement annuity paying $850 per month if the current long-term interest rate is 7.53%?

Solution at the back of the book

Amortization

A debt is *amortized* (or "killed off") if it is repaid by a regular sequence of payments. How do we calculate the payments to amortize a debt, since the unpaid portion continues to accumulate compound interest?

SPREADSHEET EXPLORATION

For example, the spreadsheet below shows an "amortization table" for the repayment of a $1000 debt at 10% compounded yearly with five annual payments. Notice that the last payment is adjusted to *exactly* pay off the remaining debt and interest.

	A	B	C	D	E	F	G
1	Debt		Payment				
2	1000		($263.80)				
3							
4	Annual Rate						
5	10%		Year	Payment	Interest	Debt Reduction	Outstanding Debt
6			0	0	0	0	1000
7	Term in Years		1	($263.80)	100	($163.80)	$836.20
8	5		2	($263.80)	83.62	($180.18)	$656.02
9			3	($263.80)	65.6	($198.20)	$457.82
10	Compoundings per Year		4	($263.80)	45.78	($218.02)	$239.80
11	1		5	($263.78)	23.98	($239.80)	$0.00
12							
13			Totals:	($1,318.98)	$318.98	($1,000.00)	

The payment amount shown in cell C2 was found using the spreadsheet PMT (. . .) function. Our first task is to explain how such amounts are actually calculated.

The payments necessary to amortize a debt form an annuity with present value PV equal to the debt D. To find the payments P to amortize debt D, we solve for P in the "present value of an annuity" formula on the previous page.

Amortization Payment

The payment P to make m times per year for t years at interest rate r compounded at each payment to amortize a debt of D dollars is

$$P = D\frac{r/m}{1 - (1 + r/m)^{-mt}}$$

This formula may also be written $P = D/a_{\overline{n}|i}$ with $n = mt$ and $i = r/m$.

EXAMPLE 2 **Finding the Payment to Amortize a Debt**

What monthly payment will amortize a $150,000 home mortgage at 8.6% in 30 years?

Solution

$$P = 150{,}000 \frac{0.086/12}{1 - (1 + 0.086/12)^{-(12)(30)}} \approx 1164.02$$

$$P = D\frac{r/m}{1 - (1 + r/m)^{-mt}}$$
with $D = 150{,}000$, $r = 0.086$, $m = 12$, and $t = 30$

The required payment is $1164.02 each month.

■

Notice that the borrower will pay a total of $1164.02 \times 12 \times 30 = \$419{,}047.20$ during the 30 years of the loan to pay off a $150,000 debt.

PRACTICE PROBLEM 2

What monthly payment will amortize a $150,000 home mortgage at 8.6% in 25 years? What is the total amount the borrower will pay?

Solution at the back of the book

Graphing Calculator Exploration

When amortizing a loan, a longer term means smaller payments, but the total amount the borrower pays is much larger.

a. To find the monthly payment y to amortize a debt over x years using the values $D = 150{,}000$, $r = 0.086$, and $m = 12$ from Practice Problem 2, graph the curve $y_1 = 150000(.086/12)/(1 - (1 + .086/12)^{-12x})$ on the window $[0, 50]$ by $[0, 5000]$.

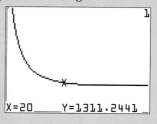

b. Use TRACE or VALUE to find the payments for mortgage terms of 15, 20, 25, and 30 years. How do your answers compare to the answers to Example 2 and Practice Problem 2?

c. To find the total amount the borrower pays, multiply the payment by m and t (here replaced by x). Graph the new curve $y_1 = 12x(150000)(.086/12)/(1 - (1 + .086/12)^{-12x})$ on the window $[0, 50]$ by $[0, 600000]$ to find the total amount paid y on a loan for x years.

d. Use TRACE or VALUE to find the total amount paid for mortgage terms of 15, 20, 25, and 30 years. Does a small change in the payment make a large difference in both the term and the total amount paid on a debt?

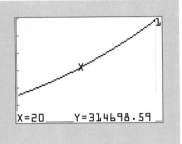

X=20   Y=314698.59

Unpaid Balance

How much does the borrower still owe partway through an agreed amortization schedule? Suppose you have made mortgage payments for 10 years on a 30-year loan and now you must move. How much should you pay to settle your debt? Your remaining payments form another annuity and are worth precisely the present value of this new annuity.

EXAMPLE 3 **Finding the Unpaid Balance**

What is the unpaid balance after 10 years of monthly payments on a 30-year mortgage of $150,000 at 8.6%?

Solution

First we must calculate the monthly payments. But for this problem, we know from Example 2 (on the previous page) that the monthly payment to amortize the original mortgage is $1164.02. The remaining payments thus form an annuity with $P = \$1164.02$, $r = 0.086$, $m = 12$, and $t = 20$ years. The amount still owed on the mortgage is the present value of this annuity, which we can calculate by the formula on page 141.

$$PV = 1164.02 \, \frac{1 - (1 + 0.086/12)^{-(12)(20)}}{0.086/12}$$

$PV = P\dfrac{1 - (1 + r/m)^{-mt}}{r/m}$
with $P = 1164.02$, $r = 0.086$, $m = 12$, and $t = 20$

$$\approx 133{,}158.27$$

The unpaid balance after 10 years of payments on this $150,000 mortgage is $133,158.27.

Notice that after the first 10 of 30 years, much less than one-third of the mortgage has been paid. The early payments of any loan pay mostly interest and only a little of the principal.

PRACTICE PROBLEM 3

How much is still owed on a 4-year car loan of $12,000 at 4.7% after 1 year of monthly payments?   *Solution at the back of the book*

Graphing Calculator Exploration

The program* AMORTABL constructs an *amortization table* that shows how each payment is allocated to paying the interest and reducing the outstanding debt. To explore the table for the situation used in Examples 2 and 3, run this program by selecting it from the program menu and proceed as follows:

a. Enter 150000 for the debt, .086 for the interest rate, 30 for the number of years, and 12 for the number of compoundings per year (because the payments are monthly).

```
DEBT (DOLLARS)
  150000
RATE (DECIMAL)
  .086
NUMBER OF YEARS
  30
COMPOUNDINGS/YR
  12
```

b. After the table of values has been calculated by finding the payment rounded to the nearest penny and then applying each payment to the rounded interest due on the debt for that period and reducing the debt by the excess, you may choose which part of the table you wish to see.

```
   VIEW TABLE
1:FROM START
2:JUMP TO YEAR
3:AT END
4:FINISHED
```

c. The amortization table contains five columns: X is the number of the payment (negative X's or X's larger than the number of payments display zeros in the other columns), Y_1 is the payment, Y_2 is the interest part of the payment, Y_3 is the remaining part used to reduce the debt, and Y_4 is the debt remaining after this reduction. Use the arrow keys to scroll right and left through these columns.

X	Y_1	Y_2	
0	0	0	
1	1164	1075	
2	1164	1074.4	
3	1164	1073.7	
4	1164	1073.1	
5	1164	1072.4	
6	1164	1071.8	
X=0			

X	Y_3	Y_4	
0	0	150000	
1	89.02	149911	
2	89.66	149821	
3	90.3	149731	
4	90.95	149640	
5	91.6	149548	
6	92.26	149456	
Y_4=150000			

*See the Preface for information on how to obtain this and other graphing calculator programs.

X	Y₁	Y₂	⋮
356	1164	40.8	
357	1164	32.75	
358	1164	24.64	
359	1164	16.48	
360	1160	8.25	
361	0	0	
362	419043	269043	
X=361			

X	Y₃	Y₄	⋮
356	1123.2	4570	
357	1131.3	3438.7	
358	1139.4	2299.3	
359	1147.5	1151.8	
360	1151.8	0	
361	0	**0**	
362	150000	0	
Y₄=0			

The second line after the end of the amortization contains the total payments, interest, and debt reduction. Notice that the final payment is adjusted to correct for rounding errors and the debt is reduced to zero.

d. To use the table to find the remaining debt after a given number of years, "jump" to that position in the table, and scroll to the Y_4 column. After ten years of payments, the table shows $133,157.56 remaining.

X	Y₃	Y₄	⋮
118	205.28	133573	
119	206.75	133366	
120	208.23	**133158**	
121	209.72	132948	
122	211.23	132737	
123	212.74	132524	
124	214.27	132310	
Y₄=133157.56			

Section Summary

The *present value* of an ordinary annuity compounded at each payment is

$$PV = P\frac{1 - (1 + r/m)^{-mt}}{r/m}$$

PV = present value, P = payment, r = rate, m = times per year, and t = years

The payment to *amortize a debt* of D dollars is found by solving this formula for P.

$$P = D\frac{r/m}{1 - (1 + r/m)^{-mt}}$$

D = debt (other variables as before)

The amount still owed after making amortization payments for part of the term is the present value of a new annuity formed by the remaining payments and is found by the first formula.

EXERCISES 2.4 (Most require ▦ or ▦.)

Find the present value of each annuity.

1. $15,000 annually at 7% for 10 years.

2. $18,000 annually at 6% for 15 years.

3. $10,000 semiannually at 11% for 7 years.

4. $3000 quarterly at 8% for 8 years.

5. $1400 monthly at 6.9% for 30 years.

6. $25 weekly at 7.7% for 6 years.

Find the payment to amortize each debt.

7. Monthly payments on $100,000 at 5% for 25 years.

8. Monthly payments on $125,000 at 5.4% for 20 years.

9. Monthly payments on $6000 at 6% for 5 years.

10. Annual payments on $50,000 at 8.2% for 10 years.

11. Quarterly payments on $14,500 at 12.7% for 6 years.

12. Weekly payments on $2500 at 18% for 3 years.

Find the unpaid balance on each debt (you should already know the payments from Exercises 7–12).

13. After 6 years of monthly payments on $100,000 at 5% for 25 years.

14. After 15 years of monthly payments on $125,000 at 5.4% for 20 years.

15. After 2 years 11 months of monthly payments on $6000 at 6% for 5 years.

16. After 3 years of annual payments on $50,000 at 8.2% for 10 years.

17. After 2 years 3 months of quarterly payments on $14,500 at 12.7% for 6 years.

18. After 2 years of weekly payments on $2500 at 18% for 3 years.

APPLIED EXERCISES

19. **Contest Prizes** The super prize in a contest is $10 million. This prize will be paid out in equal yearly payments over the next 25 years. If the prize money is guaranteed by U.S. Treasury bonds yielding 7.9% and is placed into an escrow account when the contest is announced 1 year before the first payment, how much do the contest sponsors have to deposit in the escrow account?

20. **Grant Funding** In September 1994, the New York City public schools system was awarded a grant of $50 million over 5 years from the Annenberg Foundation. If the grant paid equal yearly amounts for the next 5 years and was financed at 6.8%, how much did this grant cost the Foundation when it was announced?

21. **NFL Contracts** When Michael Westbrook signed a 7-year, $18 million contract with the Washington Redskins, it included a $6.5 million signing bonus that was the largest for a wide receiver in NFL history. If the $18 million was paid out in equal quarterly payments for the 7 years and the current long-term interest rate was 6.85%, how much was the contract worth when it was signed?

22. **Escrow Accounts** On September 1, 1994, Judge Sam C. Pointer, Jr., of the Federal District Court in Birmingham approved a $4.25 billion settlement against the manufacturers of silicon breast implants to be paid out over the next 30 years. If this money was to be paid out in equal semiannual amounts from an escrow account earning 8.7% interest, how much would the manufacturers have to deposit into the account (six months before the first payment) to meet the terms of the settlement?

23. **College Savings** When I graduated from high school last spring, my dear Aunt Sallie gave me a savings account passbook for an account earning 5.61%. She told me that starting next fall, there would be enough money for me to take out $10,000 every 6 months for the next 4 years to pay for my college expenses. I thought "Wow! Thanks! That's $80,000!" but she just smiled, shook her head "No," and told me to look in the passbook. How much was in the account when she gave it to me?

24. **Car Buying** A Cadillac dealer offers the following terms on a 2-year lease of a new Eldorado: either $1995 down and $339 per month for 24 months or

$a = 7,711.38 + 1995 = 9706.38$ $b = $9,599$
$10,599.63$

$9599 down and nothing more to pay for 24 months. (a) If the current interest rate were 5.2%, which option would be cheaper? (b) Since the car dealer is an expert money manager, what must the current interest rate be if both offers are equally good for the dealer?

25. **Life Insurance** Just before his first attempt at bungee jumping, John decides to buy a life insurance policy. His annual income at age 30 is $35,000, so he figures he should get enough insurance to provide his wife and new baby with that amount each year for the next 35 years. If the long-term interest rate is 6.7%, what is the present value of John's future annual earnings? Rounding up to the next $50,000, how much life insurance should he buy?

26. **Life Insurance** A 60-year-old grandmother wants a life insurance policy that could replace her annual $55,000 earnings for the next 10 years. If the long-term interest rate is now 6.2%, how much life insurance (rounded up to the next $10,000) should she buy?

27. **Mortgages** Real estate prices are so high in Tokyo that some mortgages are written for 99 years in an attempt to keep down the monthly payments. (a) What is the monthly payment on a $500,000 mortgage at 8% for 99 years? (b) How much does this save the borrower each month compared to a 30-year mortgage at 8% for the same amount? (c) Which term results in the higher total payment?

28. **Apartment Rents** The *New York Times* reported that the Associates for International Research had estimated the following monthly rents for a two-bedroom apartment in major cities around the world.

Two-Bedroom Apartment Monthly Rents Around the World			
Sydney	$1100	London	$2950
Cairo	$1200	Moscow	$5000
Buenos Aires	$1500	Shanghai	$6300
Paris	$2150	Tokyo	$7100
New York	$2300	Hong Kong	$7200

Assuming that the monthly rent represents the payment on the current market value of the apartment as a 30-year investment and that the long-term interest rate is 6.52%, estimate the current market value for an apartment in (a) Sydney, (b) Paris, (c) Moscow, and (d) Tokyo.

29. **Credit Cards** A MasterCard statement shows a balance of $560 at 13.9% compounded monthly. What monthly payment will pay off this debt in 1 year 8 months?

30. **Credit Cards** The MasterCard statement in Exercise 29 also states that the "minimum payment" is $15. How long will it take to pay off this debt by making this minimum payment each month?
   a. To estimate the number of years required, STORE the values 15 for P, 0.139 for r, and 12 for m before graphing the present value formula $y = P(1 - (1 + r/m)^{-mx})/(r/m)$ on the window [0, 10] by [0, 1000]. Use TRACE or INTERSECT to estimate the number of years x that corresponds to the debt's present value $y = 560. Be sure to change this number of years into the number of months the payments must be made.
   b. To estimate the number of years required, STORE the values 560 for D, 0.139 for r, and 12 for m before graphing the amortization payment formula $y = D(r/m)/(1 - (1 + r/m)^{-mx})$ on the window [0, 10] by [0, 20]. Use TRACE or INTERSECT to estimate the number of years x that corresponds to the payment $y = 15. Be sure to change this number of years into the number of months the payments must be made.
   c. Why are the answers from parts (a) and (b) exactly the same?

31. **Credit Cards** What monthly payment should you make on your Visa card to pay off a new $1575 stereo in 15 months if the interest rate is 14.8% compounded monthly?

32. **Mortgages** The *Town Gossip* local newspaper lost the libel suit filed by the village's former mayor. What quarterly payment will pay off the $150,000 judgment in 5 years at 11.6% compounded quarterly?

33. **College Costs** When Jill graduated from college, her loans with interest totaled $58,720. Because her last co-op employer invited her to stay on as a full-time employee, she was able to handle the monthly payments even though the interest rate was 9.4% and the bank expected everything to be paid back in 8 years. How much debt remained after 3 years of payments?

34. Car Buying Tom bought a new turbo Ford Thunderbird and financed it with a new-car loan of $18,000 at 13.78% for 4 years. After making the monthly payments for 7 months, he decided he couldn't study for his business courses at college and put in enough hours at his part-time job to keep up his grades, so he offered to give the car to his sister if she would take over the payments. What purchase price would she be paying for the car?

35. Automobile Financing A car dealer's newspaper ad offers "0% financing with guaranteed credit approval" and includes an example: "Finance $10,000 for 48 months—normal payment is $262.84 while 0% financing is $208.33—you save $2616." What is the "normal" finance rate? To estimate the answer, graph the payment formula $y = 10000(x/12)/(1 - (1 + x/12)^{-48})$ on the window $[0, .2]$ by $[0, 500]$ to see the monthly payment y for a 4-year loan of $10,000 at interest rate x compounded monthly. Use TRACE or INTERSECT to find the rate x that gives a y-value near 262.84. (By the way, the small print at the bottom of the ad contains the statement "0% financing may affect selling price of car.")

36. Computer Financing A Radio Shack back-to-school sale ad offers a "complete PC system for $39 per month after no payments for six months." The ad explains that "if you do not pay the full amount of purchase [$1270] by the end of the deferred period," then "finance charges will accrue from the date of purchase and will be added to the purchase balance" and states that the APR is 22.55%.

a. Show that the nominal interest rate for this loan is 20.508%.

b. How much will the balance be after six months if you make no payments?

c. How long will it then take to pay off the debt by paying the $39 each month? To estimate the number of years required, STORE the values 39 for P, 0.20508 for r, and 12 for m before graphing the present value formula $y = P(1 - (1 + r/m)^{-mx})/(r/m)$ on the window $[0, 20]$ by $[0, 2000]$. Use TRACE or INTERSECT to estimate the number of years x that corresponds to the present value y from part (b). Be sure to change this number of years into the number of months the payments must be made.

d. Using the advertised financing, what is the buyer's total cost?

37. Yearly Interest A home owner finances a new luxury car using a "home equity" loan so that she can deduct the interest each year when calculating her taxable income. If she plans to repay the $40,000 in equal monthly payments over 4 years and the interest rate is fixed at 9.35%, how much interest will she pay in the second year of the loan?

38. Home Refinancing A widow needs cash and decides to refinance the home she and her husband bought 20 years ago for $35,000 with a monthly payment, 30-year mortgage at 6.8% on 80% of the purchase price. The Home Sweet Home Mortgage Corporation has appraised her house at $125,000 and has agreed to lend her 70% of this value. How much cash will she have after paying off the old mortgage?

Explorations and Excursions

The following problems extend and augment the material presented in the text.

39. Solve any three of Exercises 1–6 as follows: Enter the "present value of an annuity" formula as $y = P(1 - (1 + r/m)^{-mt})/(r/m)$. Then for each exercise, STORE the values for P, r, m, and t, and evaluate y to find the present value.

40. The unpaid balance y of the mortgage in Example 3 on page 144 can be viewed as a function of the number of years x that payments have been made by graphing the function $y = (1164.02)(1 - (1 + .086/12)^{-12(30-x)})/(.086/12)$ on the window $[0, 35]$ by $[0, 160000]$. The "t" in the annuity present value formula has been replaced by "$30 - x$" since the number of years remaining is 30 minus the number of years x already paid. Use TRACE to explore the curve. Does it start at $(0, 150000)$ and end at $(30, 0)$? Notice that these points correspond to the initial value of the mortgage and the length of the loan. Does the unpaid balance decrease by the same amount each year? How long does it take to reduce the unpaid balance to half the initial amount?

41. Solve any three of Exercises 7–12 as follows: Enter the amortization payment formula as $y = D(r/m)/(1 - (1 + r/m)^{-mt})$. Then for each exercise, STORE the values for m, D, r, and t, and evaluate y to find the required payment.

42. Unpaid Balance Formula Combine the "present value of an annuity" formula with the amortization payment formula to show that the unpaid balance after x years of payments made m times each year for t years on a debt of D dollars at interest rate r is

$$D(1 + r/m)^{mx} \frac{(1 + r/m)^{m(t-x)} - 1}{(1 + r/m)^{mt} - 1}$$

43. Solve any three of Exercises 13–18 as follows: Enter the unpaid balance formula from Exercise 42 as $y = D(1 + r/m)^{mx}((1 + r/m)^{m(t-x)} - 1)/((1 + r/m)^{mt} - 1)$. Then, for each exercise, STORE the values for x, m, t, D, and r, and evaluate y to find the unpaid balance after x years of payments. Why does this unpaid balance formula give slightly different answers for these exercises?

44. Present Value of an Annuity Show that the present value of an annuity formula may be rewritten in the form

$$PV = P \frac{1 - (1 + i)^{-n}}{i}$$

where $i = r/m$ and $n = mt$.

45. Show that the present value of an annuity is the sum of the present values of the payments. [*Hint:* Use the notation of Exercise 44 and show that the present value of the first payment is $P(1 + i)^{-1}$, that of the second is $P(1 + i)^{-2}$, and so on for the n payments. Factor out the common terms from the sum and use the geometric series formula on page 129.]

46. Use the program AMORTABL to solve Exercise 13. How does the amortization table solution differ from your previous solution to Exercise 13? How much does the final payment in the table differ from the payments made during the rest of the table?

47. Use the program AMORTABL to solve Exercise 16. Be sure to enter "1" for the "compoundings per year." How does the amortization table solution differ from your previous solution to Exercise 16? How much does the final payment in the table differ from the payments made during the rest of the table?

48. Use the program AMORTABL to solve Exercise 18. Be sure to enter "52" for the "compoundings per year." How does the amortization table solution differ from your previous solution to Exercise 18? How much does the final payment in the table differ from the payments made during the rest of the table?

Chapter Summary with Hints and Suggestions

Reading the text and doing the exercises in this chapter have helped you to master the following skills, which are listed by section (in case you need to review them) and are keyed to particular Review Exercises. Answers for all Review Exercises are given at the back of the book, and full solutions can be found in the Student Solutions Manual.

2.1 SIMPLE INTEREST

- Find the interest due on a simple interest loan. (*Review Exercises 1–4.*)

$$I = Prt$$

- Use the Banker's rule in a simple interest calculation. (*Review Exercises 5–8.*)

- Find the total amount due on a simple interest loan. (*Review Exercises 9–12.*)

$$A = P(1 + rt)$$

- Solve a simple interest situation for the interest, the interest rate, the principal, or the term. (*Review Exercises 13–17.*)

- Find a simple interest future or present value. (*Review Exercises 18–20.*)

$$A = P(1 + rt) \qquad PV = \frac{A}{1 + rt}$$

- Find the effective simple interest rate of a discounted loan. (*Review Exercises 21–22.*)

$$r_s = \frac{r}{1 - rt}$$

- Solve an applied simple-interest problem. *(Review Exercises 23–25.)*

2.2 COMPOUND INTEREST

- Find the amount due on a compound interest loan. *(Review Exercises 26–30.)*

$$A = P\,(1 + r/m)^{mt} \qquad A = Pe^{rt}$$

- Find a future or present value. *(Review Exercises 31–35.)*

$$A = P(1 + r/m)^{mt} \qquad PV = \frac{A}{(1 + r/m)^{mt}}$$

- Find the term needed for a given principal to grow to a future value. *(Review Exercises 36–40.)*

- Use the "rule of 72" to estimate a doubling time. *(Review Exercises 41–45.)*

$$\left(\begin{array}{c}\text{Doubling}\\\text{time}\end{array}\right) \approx \frac{72}{r \times 100}$$

- Find the effective rate of a loan or investment. *(Review Exercises 46–50.)*

$$r_e = (1 + r/m)^m - 1 \qquad r_e = e^r - 1$$

2.3 ANNUITIES

- Find the accumulated amount of an ordinary annuity. *(Review Exercises 51–55.)*

$$A = P\,\frac{(1 + r/m)^{mt} - 1}{r/m}$$

- Find the regular payment to make into a sinking fund to accumulate a given amount. *(Review Exercises 56–60.)*

$$P = A\,\frac{r/m}{(1 + r/m)^{mt} - 1}$$

- Find the number of periods to make payments into a sinking fund to accumulate a given amount. *(Review Exercises 61–65.)*

$$mt = \frac{\log\,((A/P)\,(r/m) + 1)}{\log\,(1 + r/m)}$$

- Estimate the internal rate of return of regular investments using a graphing calculator. *(Review Exercises 66–70.)*

2.4 AMORTIZATION

- Find the present value of an ordinary annuity. *(Review Exercises 71–75.)*

$$PV = P\,\frac{1 - (1 + r/m)^{-mt}}{r/m}$$

- Find the regular payment to amortize a debt. *(Review Exercises 76–80.)*

$$P = D\,\frac{r/m}{1 - (1 + r/m)^{-mt}}$$

- Find the amount still owed after making amortization payments for part of the term. *(Review Exercises 81–85.)*

HINTS AND SUGGESTIONS

- **Overview:** Compound interest is repeated simple interest with the interest added to the principal in each successive period. The *future value* is the amount the principal will become after all the interest is included. Reversing the point of view, the *present value* is the principal needed now that will grow to the final amount. The four basic amount formulas are: $P(1 + rt)$ for simple interest, $P(1 + r/m)^{mt}$ for compound interest, Pe^{rt} for continuous compound interest, and $P((1 + r/m)^{mt} - 1)/(r/m)$ for an annuity (multiple payments). The other formulas all follow from these basic formulas.

- The rate is usually stated as a percent but is always used in decimal form for calculations: 6.7% in decimal form is 0.067.

- Round off only your final answer when using your calculator. Don't use the decimal 0.33 for the fraction $\frac{1}{3}$.

- Don't confuse a *percent increase* with the *multiplier*. *Increasing* an amount by 25% means *multiplying* it by 1.25.

- Round *up* to find the whole number of compoundings needed to reach a future value because a smaller number won't reach the stated goal.

- The "rule of 72" is only an approximation but is a helpful check that can catch "button pushing" errors when using your calculator.

- **Practice for Test:** Review Exercises 1, 5, 9, 13, 17, 22, 28, 31, 36, 38, 44, 47, 51, 54, 58, 62, 66, 72, 77, 82.

Review Exercises for Chapter 2   *Practice test exercise numbers are in blue.*

Round all dollar amounts to the nearest penny, all times in years to two decimal places, and all terms to the appropriate number of compounding periods. Most exercises require 🖩 or 🖩.

2.1   Simple Interest

Find the simple interest on each loan.

1. $1875 at 5.8% for 2 years.

2. $1150 at 9.2% for 6 months.

3. $8000 at $7\frac{1}{2}$% for 3 years 9 months.

4. $2385 at 11.3% for 1 year 3 months.

Use the Banker's rule to find the simple interest on each loan.

5. $1575 at 8.6% from January 10 to April 10 (not a leap year).

6. $2835 at 4.7% from March 15 to August 15.

7. $800 at 12% from July 21 to September 21.

8. $10,000 at 11.1% from September 9 to November 9.

Find the total amount due on each simple interest loan.

9. $8900 at 5.9% for 2 years 6 months.

10. $1375 at 11.3% for 9 months.

11. $1795 at 6.38% for 3 years.

12. $3700 at 8.3% for 1 year 11 months.

Solve each problem.

13. Find the interest rate on a loan charging $272 simple interest on a principal of $2000 after 2 years.

14. Find the principal of a loan at 9.3% if the simple interest after 1 year 8 months is $279.

15. Find the term of a loan of $7600 at 9% if the simple interest is $1026.

16. What should be the term for a loan of $5500 at 8.7% simple interest if the lender wants to receive $7414 when the loan is paid off?

17. How long will it take an investment at 10% simple interest to increase by 50%?

18. How much should be invested now at 5.9% simple interest if $5208 is needed in 2 years 8 months?

19. What would be the fair market price of a $10,000 zero coupon bond due in 1 year if today's long-term simple interest rate is 6.45%?

20. What would be the fair market price of a $50,000 zero coupon bond due in 1 year if today's long-term simple interest rate is 5.78%?

21. What is the effective simple interest rate of a discounted loan at 12% interest for 6 years?

22. What is the effective simple interest rate of a discounted loan at 15% interest for 2 years?

23. Treasury Bonds   *Forbes* magazine ranked John W. Kluge as the richest American, with a net worth of $5.2 billion, at the time of his divorce from Patricia Kluge. A newspaper reported that besides a 45-room Georgian mansion, Mrs. Kluge ". . . received $1 billion that invested conservatively in 30-year Treasury bonds would throw off $66 million a year." What was the interest rate when the article was written?

24. Insurance Settlements   How large an insurance settlement does an accident victim need for a yearly income of $29,000 from long-term bond investments paying 5.8%?

25. Furniture Buying   A furniture store offers a complete living room suite for $999 and will finance the entire price as a discounted loan at 10% for 1 year. What is the effective simple interest rate for this loan and how much will the buyer need to pay at the end of the year?

2.2   Compound Interest

Find the amount due on each compound interest loan.

26. $15,000 at 7.5% compounded quarterly for 10 years.

27. $65,000 at 8.25% compounded continuously for 17 years.

28. **Mutual Funds** From 1970 to 1995, the AIM Charter Fund posted an average annual return of 15.55% including sales charges. How much would a $10,000 investment in this fund have been worth after 25 years?

29. **Mutual Funds** During the mid-1990s, the Brandywine long-term growth fund returned 19.6% compounded quarterly. How much would a $5000 investment in this fund have been worth after 4 years?

30. **National Debt** From 1835 to 1837, the United States not only was free of debt but actually had a surplus in the Treasury. On January 1, 1837, after $5 million had been set aside as a reserve fund, there remained $37,468,859. If that money had not been distributed to the 26 states but instead had been invested at 5% compounded continuously, would there have been enough by 1997 to pay off a $4.3 trillion national debt?

Find the present value of each compound interest loan.

31. $25,000 after 10 years at 9% compounded monthly.

32. $30,000 after 8 years at 6% compounded continuously.

33. $175,000 after 16 years 6 months at 5.5% compounded quarterly.

34. **Certificates of Deposit** From 1968 to 1995, the average annual return on 6-month CDs was 7.79% compounded semiannually. How much would one have to have invested in 1968 in order to have $100,000 in 1995?

35. **Bond Funds** During the mid-1990s, the Northeast Investors Trust general-term bond fund returned 15.7% compounded monthly. How much would you need to invest in this fund to have $15,000 after 3 years 8 months?

Find the term of each compound interest loan.

36. 8.9% compounded quarterly to obtain $10,000 from a principal of $2000.

37. 11.5% compounded monthly to multiply the principal by 1.60.

38. How long will it take the value of an apartment building to increase by 80% if real estate prices are increasing 8% continuously?

39. **Treasury Bonds** How long will you have to wait to become a millionaire if you invest all of your $850,000 lottery winnings in Treasury bonds paying 5.6% compounded semiannually?

40. **Art Appreciation** How long will it take for the value of an early Picasso pencil sketch to triple if its market value is increasing 6.5% annually?

Use the "rule of 72" to estimate the doubling time (in years) for each interest rate and then calculate it exactly.

41. 8% compounded annually.

42. 6% compounded monthly.

43. 11.9% compounded weekly.

44. **Real Estate** How long will it take for the value of a house to double if real estate prices are increasing 12% each year?

45. **Index Funds** How long will it take an investment in a stock market index fund yielding 10% annually to double?

46. Find the effective rate of 13.25% compounded quarterly and compounded continuously.

47. **Baseball Cards** A mint condition 1955 Topps "Sandy Koufax" baseball card selling for $500 in 1989 was worth $1350 in 1993. What is the effective rate of this price increase?

48. **Stocks** One of Wall Street's most cherished buys in the 1950s was Texas Instruments stock, which rose spectacularly from $72\frac{1}{8}$ to $214\frac{1}{4}$ in 18 months. What is the effective rate of this price increase?

49. **Baseball Teams** The Haas family bought the Oakland Athletics from Charles O. Finley for $12.7 million in 1980 and sold it in 1995 for $85 million. Neglecting all other expenses, what is the effective rate of return on their investment?

50. **Pianos** A properly maintained 90-year-old Steinway grand piano is now worth 13.6 times its initial purchase price. What is the effective rate of this increase?

2.3 Annuities

Find the accumulated amount of each annuity.

51. $1500 annually at 11% for 20 years.

52. $500 semiannually at 8% for 12 years.

53. Home Buying How much will you have for a vacation home if you save $25 each week for 16 years in a 5.5% passbook savings account?

54. Retirement Savings How much will an IRA stock fund earning 11.2% be worth if you deposit $180 each month for 35 years?

55. Apartments When André-François Raffray died at age 77 in Arles, France, he had been making $500 monthly payments to Jeanne Calment for 30 years for the right to take over her apartment when she died. Mrs. Calment, who at 120 was then the world's oldest person with the records to prove it, remarked that "In life, one sometimes makes bad deals." How much would Mr. Raffray's money be worth if instead he had deposited it into a 7.8% bond fund?

Find the required payment for each sinking fund.

56. Yearly deposits earning 7% to accumulate $16,000 after 8 years.

57. Quarterly deposits earning 4.9% to accumulate $40,000 after 25 years.

58. How much must you pay biweekly to pay off a 13.2% car loan of $19,500 in 5 years?

59. Stock Funds How much must you deposit each month into a stock fund earning 11.2% to accumulate $150,000 in 20 years?

60. Home Buying How much must you save each week in a 5.5% passbook savings account to have $26,000 for a vacation home in 16 years?

Find the number of years needed for each sinking fund to reach the given accumulated amount.

61. $1000 yearly at 10% to accumulate $100,000.

62. $250 quarterly at 8.1% to accumulate $75,000.

63. Money Market Funds How long must you save $275 each month in a 6.2% money market fund to accumulate $19,500 for a new car?

64. Stock Funds How long must you deposit $300 each month into a stock fund earning 11.2% to accumulate $150,000?

65. Home Buying How long must you save $25 each week in a 5.5% passbook savings account to have $26,000 for a vacation home?

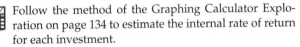 Follow the method of the Graphing Calculator Exploration on page 134 to estimate the internal rate of return for each investment.

66. $150 quarterly to accumulate $10,000 after 10 years.

67. $225 monthly to accumulate $100,000 after 25 years.

68. $500 monthly to accumulate $1,000,000 after 30 years.

69. $1.25 weekly to accumulate $50,000 after 60 years.

70. $10 daily to accumulate $100,000 after 20 years.

2.4 Amortization

71. Find the present value of an annuity of $25,000 annually at 6% for 25 years.

72. Football Contracts Deion Sanders became the highest-paid defensive player in football when he signed a $25 million, 5-year contract with the Dallas Cowboys. If the $25 million was paid out in equal weekly payments over the 5 years and the current interest rate was 7.32%, how much was the contract worth when it was signed?

73. Grant Funding A foundation is funding a new program that will award twenty $250,000 grants each year for the next 8 years to promote art and music instruction in elementary schools. How much must be placed in a money market account paying 6.75% to fund this initiative?

74. Contest Prizes The grand prize in a lottery drawing is $14 million to be paid out in equal quarterly payments over the next 20 years. If the prize money is guaranteed by U.S. Treasury bonds yielding 7.6% and is placed into an escrow account when the lottery is announced 1 year before the first payment, how much do the lottery sponsors have to deposit in the escrow account?

75. Home Buying Before looking for their first house, the Jones family calculates that they could afford a $1250 monthly mortgage payment. How large a mortgage could they afford at 8.75% for 30 years?

76. Find the weekly payments to amortize a debt of $45,000 at 9.15% over 15 years.

77. Credit Cards   A Discover card statement shows a balance of $945 at 15.7% compounded monthly. What monthly payment will pay off this debt in 2 years?

78. Credit Cards   What monthly payment should you make on your Optima card to pay off $2784 of spring break vacation charges in 7 months if the interest rate is 18.9% compounded monthly?

79. Home Buying   The Jones family has found a nice house in the suburbs. What is the monthly payment on their $150,000 mortgage at 8.75% for 30 years?

80. Stock Funds   How much must you save each week in a stock fund yielding 10.1% to become a millionaire in 15 years?

Find the unpaid balance on each debt.

81. After 5 years of annual payments on $85,000 at 9.1% for 20 years.

82. After 3 years 8 months of monthly payments on $150,000 at 8.7% for 25 years.

83. Car Buying   After 1 year 5 months of $297.92 monthly car payments on a 3-year loan at 4.6%, Karen wants to get rid of her sedan and get a red sports car. How much should she pay to settle her debt?

84. Contract Buyout   The Middleville Central School Board made a mistake when they hired the new superintendent and agreed to a 6-year contract paying $125,000 per year. With current interest rates at 6.2%, how much should they offer to break the contract after the first year?

85. Home Buying   After 4 years 9 months of monthly mortgage payments on their 30-year, $150,000 mortgage at 8.75%, the Jones family has to relocate to Atlanta and must sell their nice house in the suburbs. How much do they still owe on their mortgage?

Projects and Essays

The following projects and essays are based on Chapter 2. There are no right or wrong answers—the results depend only on your imagination and resourcefulness.

1. Discuss the differences between two $1000 loans for 8 years: one at 5% simple interest and the other at 5% compounded annually.

2. Make a list of the basic financial concepts from Section 2.1 that were used in the later sections of this chapter and discuss how your understanding of each changed as you mastered the subsequent material.

3. Visit your local bank and copy down the nominal and effective interest rates on several CDs or loans. How accurately did the bank do its calculations?

4. Ask your parents, grandparents, or other relatives for stories about how much things cost when they were your age. Use the rule of 72 or your calculator to estimate the yearly inflation rate since the time of these stories. Are these rates approximately the same for all the stories you can collect?

5. Find a car loan contract and read the "fine print." Does it contain a reference to the rule of 78 in the early repayment section? If there is an example, compare it to the actual amount using the formulas from Section 2.4.

6. Look up the article "A Hidden Case of Negative Amortization" by Bert K. Waits and Franklin Demana in the March 1990 issue of *The College Mathematics Journal* and write a page about their example. Would you ever agree to use the rule of 78 with a loan longer than 5 years?

7. Ask your parents or other relatives if you may read their home mortgage contract. Check the bank's calculations and write a page about what you find and how much of the total mortgage payments is interest.

8. Ask your grandparents or other relatives to show you their monthly Social Security checks. Look up the current long-term interest rate in today's newspaper and calculate the present value of an annuity with the same monthly payment for the next 20 years.

9. If your state runs a lottery game, read the rules and regulations to find out how the payments are made. Work out an example showing when and how much a $10 million winner would actually receive, and find the present value of an equivalent annuity using the current long-term interest rates from today's newspaper.

10. Look over your notes, your homework, and the text, and write a page about how your graphing calculator has helped you in this chapter. Include examples of how it has helped you to explore concepts and how it has helped to simplify your work. What was the *most* helpful or interesting use? What was the *least* helpful or interesting use? Are there problems that are easier to do "by hand" that you tried to solve on your graphing calculator?

3

Matrices and Systems of Equations

Each number in a grid has a special meaning given by its row and column position. This chapter uses matrices to represent and solve problems that depend on many numbers.

Systems of Two Linear Equations in Two Variables

A "Taxing" Problem

To be acceptable, a tax law must be perceived as fair by the taxpayers. When several government divisions each attempt to tax the same resource, abhorrence of "double jeopardy" suggests that the part paid as tax to one government authority should not be subjected to further taxation by another.

Consider a simplified situation in which the federal income tax is 21% of the taxable income after the state income tax has first been deducted from the taxable income, and the state income tax is 4% of the taxable income after the federal income tax has first been deducted. How could a taxpayer with a taxable income of $29,748 find the correct taxes due? There is a widespread misconception that the resulting federal and state taxes can only be approximated but not found exactly. This confusion starts with the observation that because neither tax amount is obvious, it is reasonable to estimate the state tax, calculate the corresponding federal tax, and then readjust the state tax estimate and try again. Of course, this "solution method" will be even worse for an example involving federal, state, and city taxes. Perhaps as a result of this general confusion, state income tax codes do not allow for the deduction of federal taxes, although the federal income tax Schedule A form allows the deduction of state income taxes. This chapter develops the mathematical tools that directly and simply solve the kinds of problems represented by this tax situation.

Form **W-4**	**Employee's Withholding Allowance Certificate**	OMB No. 1545-0010
Department of the Treasury Internal Revenue Service	▶ For Privacy Act and Paperwork Reduction Act Notice, see reverse.	**2001**

1 Type or print your first name and middle initial	Last name		2 Your social security number

Home address (number and street or rural route)		3 ☐ Single ☐ Married ☐ Married, but withhold at higher Single rate. Note: If married, but legally separated, or spouse is a nonresident alien, check the Single box.

City or town, state, and ZIP code		4 If your last name differs from that on your social security card, check here and call 1-800-772-1213 for a new card . . . . ▶ ☐

5	Total number of allowances you are claiming (from line G above or from the worksheets on page 2 if they apply) .	**5**	
6	Additional amount, if any, you want withheld from each paycheck	**6** $	
7	I claim exemption from withholding for 1995 and I certify that I meet **BOTH** of the following conditions for exemption:		

• Last year I had a right to a refund of ALL federal income tax withheld because I had NO tax liability; AND
• This year I expect a refund of ALL Federal income tax withheld because I expect to have NO tax liability.

Introduction

Neither of the statements "Bob and Sue together have $100" and "Bob has $20 less than Sue" tells us how much either has. But taken together, they force the conclusion that Bob has $40 and Sue has $60. Many methods of solving such simple problems have been used since antiquity, but as the problem situations become more complicated and the statements more numerous, many methods that succeed for simple problems become unwieldy and often fail. This chapter explores a solution method that solves many such problems and has the pleasing property that the method for large situations is an easy extension of the method for the simplest.

Systems of Equations

We begin with the simplest form of these problems.

Systems of Two Linear Equations in Two Variables

> A system of two linear equations in two variables is any problem expressible in the form
>
> $$\begin{cases} ax + by = h \\ cx + dy = k \end{cases}$$
>
> where x and y are the variables and the constants a, b, c, d, h, k are such that at least one of a, b, c, d is not zero.

EXAMPLE 1   System of Two Linear Equations in Two Variables

Express the statements "Bob and Sue together have $100" and "Bob has $20 less than Sue" as a system of two linear equations in two variables.

Solution

Let x represent the amount of money that Bob has and y represent the amount that Sue has. The first statement may be written "$x + y = 100$" and the second as "$x = y - 20$". Rearranging this second equation by subtracting y from both sides, we obtain the system of equations

$$\begin{cases} x + y = 100 \\ x - y = -20 \end{cases}$$

$ax + by = h$ with
$a = 1, b = 1, h = 100$

$cx + dy = k$ with
$c = 1, d = -1, k = -20$

This is not the only possible representation of this situation. Other possible representations are

$$\begin{cases} x - y = -20 \\ x + y = 100 \end{cases}$$

Switch the order of the equations

or

$$\begin{cases} x + y = 100 \\ -x + y = 20 \end{cases}$$

Multiply $x - y = -20$ through by -1

or

$$\begin{cases} 2x + 2y = 200 \\ x - y = -20 \end{cases}$$

Multiply $x + y = 100$ through by 2

■

Although we have already noted that $x = 40$, $y = 60$ is the solution of this example, it is important to notice that this solution is itself a system of two linear equations in two variables:

$$\begin{cases} 1x + 0y = 40 \\ 0x + 1y = 60 \end{cases}$$

$x = 40, y = 60$ is a system of equations

PRACTICE PROBLEM 1

Express the statements "a jar of pennies and nickels contains 80 coins" and "the coins in the jar are worth \$1.60" as a system of two linear equations in two variables.   *Solution at the back of the book*

A *solution* of a system of equations in two variables is a pair of values for the variables that satisfy all the equations (such as $x = 40$, $y = 60$ for Example 1). The *solution set* is the collection of all solutions. *Solving the system of equations* means finding this solution set.

The two statements "Bob has \$20 less than Sue" and "Sue has \$20 more than Bob" really express the same relationship between Bob's and Sue's wealth. Equations representing the same information are *dependent*, while equations representing different facts are *independent*.

The two statements "Bob has \$20 less than Sue" and "Sue has \$50 more than Bob" can not both be true. Equations representing conflicting statements are *inconsistent*, while equations representing statements that do not conflict are *consistent*. An inconsistent system has no solution, but a consistent system always has at least one solution.

Graphical Representations of Lines

An equation of the form $ax + by = h$ (with at least one of a and b not zero) is the equation of a line written in *general form*. If both a and b are not zero, the line $ax + by = h$ has x-intercept $\left(\frac{h}{a}, 0\right)$ and y-intercept $\left(0, \frac{h}{b}\right)$.

If $a = 0$, the line is the horizontal line $y = \frac{h}{b}$. If $b = 0$, the line is the vertical line $x = \frac{h}{a}$. If $h = 0$, the line passes through the origin $(0, 0)$ and the points $(b, -a)$ and $(-b, a)$.

EXAMPLE 2   **Sketching a Linear Equation in Two Variables**

Sketch the graph of each linear equation.

a. $2x + 3y = 12$

b. $0x + 3y = 12$ Which is just $3y = 12$

c. $2x + 0y = 12$ Which is just $2x = 12$

d. $2x + 3y = 0$

Solution

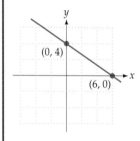

a. $2x + 3y = 12$

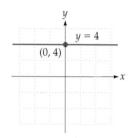

b. $0x + 3y = 12$

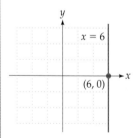

c. $2x + 0y = 12$

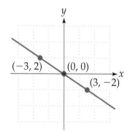

d. $2x + 3y = 0$

■

If the system $\begin{cases} ax + by = h \\ cx + dy = k \end{cases}$ is *independent*, it represents two *distinct* lines, whereas if it is *dependent*, the equations represent the *same* line. If it does represent two distinct lines, they are *parallel* if the system is *inconsistent* and *intersecting* if it is *consistent*.

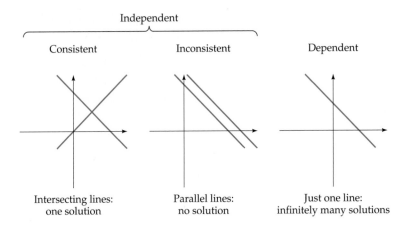

Independent

Consistent   Inconsistent   Dependent

Intersecting lines: Parallel lines: Just one line:
one solution   no solution   infinitely many solutions

The nature of the solution of a system of equations (and any values) can be determined from the graph if it is drawn with sufficient accuracy.

EXAMPLE 3   **Solving by Graphing**

Solve each system of equations by sketching the graphs of the lines.

a. $\begin{cases} 2x + 3y = 12 \\ x - y = 1 \end{cases}$ b. $\begin{cases} 2x + 3y = 12 \\ 4x + 6y = 12 \end{cases}$ c. $\begin{cases} 2x + 3y = 12 \\ 4x + 6y = 24 \end{cases}$

Solution

The line $2x + 3y = 12$ was sketched in Example 2a. For part (a), the second line has intercepts $(1, 0)$ and $(0, -1)$; for (b), the second line has intercepts $(3, 0)$ and $(0, 2)$; and for (c), the second line has intercepts $(6, 0)$ and $(0, 4)$.

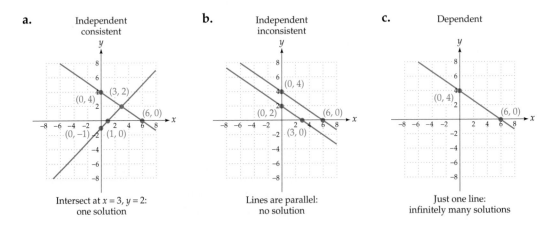

a.   Independent
consistent

Intersect at $x = 3$, $y = 2$:
one solution

b.   Independent
inconsistent

Lines are parallel:
no solution

c.   Dependent

Just one line:
infinitely many solutions

Notice that the solution to (a) checks because $2(3) + 3(2) = 12$ and $(3) - (2) = 1$ as required by the equations. In (c), *every* point on the line $2x + 3y = 12$ is a solution. Rewriting $2x + 3y = 12$ as $x = (12 - 3y)/2 = 6 - \frac{3}{2}y$, we can *parameterize* these solutions as $x = 6 - \frac{3}{2}t$, $y = t$, where t is any real number. Choosing $t = 8$, we have the particular solution $x = -6$, $y = 8$ and similarly for any other value of t. This parameterized solution for (c) checks because $2(6 - \frac{3}{2}t) + 3(t) = 12 - 3t + 3t = 12$ and $4(6 - \frac{3}{2}t) + 6(t) = 24 - 6t + 6t = 24$ as required by the equations. ∎

Graphing Calculator Exploration

You can view a nonvertical line $ax + by = h$ by entering it as $y = (h - ax)/b$. To see the system of equations $\begin{cases} 2x + 3y = 12 \\ x - y = 1 \end{cases}$ from Example 3a:

a. Enter the first equation as $y_1 = (12 - 2x)/3$ and the second as $y_2 = (1 - 1x)/(-1)$. Graph them on the window $[-3, 7]$ by $[-3, 5]$.

b. Use TRACE or EVALUATE to check the x- and y-intercepts.

c. Use TRACE or INTERSECT to find the intersection point of these two lines.

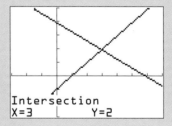

Substitution Method

The *substitution method* quickly finds the solution of an independent and consistent system when one of the equations represents a vertical or horizontal line.

EXAMPLE 4 **Solving by the Substitution Method**

Solve $\begin{cases} x + 3y = 15 \\ x = 9 \end{cases}$ by the substitution method.

Solution

Since the second equation states that $x = 9$, we may substitute 9 for x in the first equation.

$$(9) + 3y = 15 \qquad\qquad x + 3y = 15 \text{ with } x = 9$$
$$3y = 6 \qquad\qquad \text{Subtracting 9 from both sides}$$
$$y = 2 \qquad\qquad \text{Dividing by 3}$$

The solution is $x = 9$, $y = 2$.

■

The substitution method also can be used when one of the equations can be solved for a variable appearing in the other equation.

EXAMPLE 5   Solving by the Substitution Method

Solve $\begin{cases} x + 3y = 15 \\ 2x - 5y = 8 \end{cases}$ by the substitution method.

Solution

Since the first equation can easily be solved for x as $x = 15 - 3y$, we may substitute $15 - 3y$ for x in the second equation.

$$2(15 - 3y) - 5y = 8 \qquad\qquad \text{Replacing } x \text{ with } 15 - 3y$$
$$30 - 6y - 5y = 8 \qquad\qquad \text{Multiplying out}$$
$$-11y = -22 \qquad\qquad \text{Collecting like terms}$$
$$y = 2 \qquad\qquad \text{Dividing by } -11$$

and then

$$x = 15 - 3(2) = 15 - 6 = 9 \qquad\qquad x = 15 - 3y \text{ with } y = 2$$

The solution is $x = 9$, $y = 2$ and the original system is independent and consistent.

■

PRACTICE PROBLEM 2

Solve $\begin{cases} 2x + y = 10 \\ x + 2y = 8 \end{cases}$ by the substitution method.

Solution at back of the book

An *inconsistent* system of equations always leads to an "impossible" equation stating that two different numbers are equal. For instance, if we attempt to solve $\begin{cases} x + y = 10 \\ x + y = 20 \end{cases}$ by using the first equation to find that $x = 10 - y$ and then substituting this for x in the second equation, we find $(10 - y) + y = 20$ so that $10 = 20$. But this is contradictory, so there is no solution and the equations are inconsistent.

A *dependent* system of equations always results in a "useless" equation stating that a number equals itself. For instance, if we attempt to solve $\begin{cases} x + y = 10 \\ 2x + 2y = 20 \end{cases}$ by using the first equation to find that $x = 10 - y$ and then substituting this for x in the second equation, we find $2(10 - y) + 2y = 20$ so that $20 - 2y + 2y = 20$. But $20 = 20$ is just uninformative but not contradictory, giving as solutions *all* the points on the line $x + y = 10$, so the system is dependent.

Equivalent Systems of Equations

Two systems of equations are *equivalent* if they have the same solution. Returning to Example 1 (pages 159–160), we have already remarked that the following systems are all equivalent:

$$\begin{cases} x + y = 100 \\ x - y = -20 \end{cases} \qquad \begin{cases} x - y = -20 \\ x + y = 100 \end{cases} \qquad \begin{cases} x + y = 100 \\ -x + y = 20 \end{cases}$$

$$\begin{cases} 2x + 2y = 200 \\ x - y = -20 \end{cases} \quad \text{and} \quad \begin{cases} 1x + 0y = 40 \\ 0x + 1y = 60 \end{cases}$$

An equivalent system can be obtained from a given system by doing one or more of the following:

1. Switch the order of the equations.

$$\begin{cases} x + y = 100 \\ x - y = -20 \end{cases} \xrightarrow[\text{equations}]{\text{Switch}} \begin{cases} x - y = -20 \\ x + y = 100 \end{cases}$$

2. Multiply or divide one of the equations by a nonzero number.

$$\begin{cases} x + y = 100 \\ x - y = -20 \end{cases} \xrightarrow[\text{equation by } -1]{\text{Multiply second}} \begin{cases} x + y = 100 \\ -x + y = 20 \end{cases}$$

3. Replace an equation with its sum or difference with the other equation.

$$\begin{cases} x + y = 100 \\ x - y = -20 \end{cases} \xrightarrow[\text{to the second}]{\text{Add first equation}} \begin{cases} x + y = 100 \\ 2x + 0y = 80 \end{cases}$$

Elimination Method

The *elimination method* solves $\begin{cases} ax + by = h \\ cx + dy = k \end{cases}$ by attempting to change it into an equivalent system of the form $\begin{cases} 1x + 0y = p \\ 0x + 1y = q \end{cases}$ so that the y variable has been removed from the first equation and the x variable has been removed from the second. If this can be done, the solution is $x = p$, $y = q$ and the system of equations is independent and consistent. Even if this cannot be done, the attempt will still identify the system as dependent or inconsistent.

EXAMPLE 6   **Solving by the Elimination Method**

Solve $\begin{cases} x + 3y = 15 \\ 2x - 5y = 8 \end{cases}$ by the elimination method.

Solution

We hope to remove the x variable from the second equation and the y variable from the first. The following sequence of equivalent systems is one possible way of reaching the solution. First, we eliminate the x variable from the second equation by subtracting an appropriate multiple of the first equation.

$$\begin{cases} 2x + 6y = 30 \\ 2x - 5y = 8 \end{cases}$$

Multiplying the first equation by 2 to get a second $2x$

$$\begin{cases} 2x + 6y = 30 \\ 0x - 11y = -22 \end{cases}$$

Subtracting the first equation from the second to get a zero

Because the coefficients of the first equation are all even and those of the second equation are all multiples of -11, we can simplify the numbers by dividing:

$$\begin{cases} 1x + 3y = 15 \\ 0x + 1y = 2 \end{cases}$$

Dividing the first equation by 2
Dividing the second equation by -11

Next we eliminate the y-variable from the first equation by subtracting an appropriate multiple of the second equation:

$$\begin{cases} 1x + 3y = 15 \\ 0x + 3y = 6 \end{cases}$$

Multiplying the second equation by 3 to get another $3y$

$$\begin{cases} 1x + 0y = 9 \\ 0x + 3y = 6 \end{cases}$$

Subtracting the second equation from the first to get $0y$

We finish by dividing to get just one y in the second equation:

$$\begin{cases} 1x + 0y = 9 \\ 0x + 1y = 2 \end{cases}$$

Dividing the second equation by 3

The solution is $x = 9$, $y = 2$ and the original system is independent and consistent.

You may have noticed in the preceding example that after we found $y = 2$ (in the third step), we then could have solved for x by "back substituting" $y = 2$ into the other equation, $1x + 3 \cdot 2 = 15$, to find $x = 9$. But our solution using *only* elimination demonstrates what will become a general method to solve problems with more equations and more variables.

An inconsistent system of equations always leads to the impossible equation that zero equals a nonzero number. For instance, solving Example 3b by the elimination method could be done as follows:

$$\begin{cases} 2x + 3y = 12 \\ 4x + 6y = 12 \end{cases} \xrightarrow[\text{equation by 2}]{\text{Divide second}} \begin{cases} 2x + 3y = 12 \\ 2x + 3y = 6 \end{cases} \xrightarrow[\text{from second}]{\substack{\text{Subtract first} \\ \text{equation}}} \begin{cases} 2x + 3y = 12 \\ 0x + 0y = -6 \end{cases}$$

Since the second equation says $0 = -6$, which is contradictory, there is no solution and the equations are inconsistent.

A dependent system of equations always results in one equation becoming $0x + 0y = 0$. For instance, solving Example 3c by the elimination method could be done as follows:

$$\begin{cases} 2x + 3y = 12 \\ 4x + 6y = 24 \end{cases} \xrightarrow[\text{equation by 2}]{\text{Divide second}} \begin{cases} 2x + 3y = 12 \\ 2x + 3y = 12 \end{cases} \xrightarrow[\text{from second}]{\substack{\text{Subtract first} \\ \text{equation}}} \begin{cases} 2x + 3y = 12 \\ 0x + 0y = 0 \end{cases}$$

Since the second equation says $0 = 0$, which is uninformative but not contradictory, the solutions are all the points on the line $2x + 3y = 12$ and the system is dependent.

PRACTICE PROBLEM 3

Solve $\begin{cases} x + y = 100 \\ x - y = -20 \end{cases}$ by the elimination method.

Solution at the back of the book

We conclude this section with the solution of the tax problem posed in the Application Preview.

EXAMPLE 7   **A Taxing Problem**

What are the federal and state taxes on an income of $29,748 if the federal tax is 21% of the income after first deducting the state tax and the state tax is 4% of the income after first deducting the federal tax?

Solution

Let x represent the federal tax and y represent the state tax. The statements

"the federal tax is 21% of ($29,748 less the state tax)"

"the state tax is 4% of ($29,748 less the federal tax)"

become the equations

$$x = 0.21(29{,}748 - y) \quad \text{and} \quad y = 0.04(29{,}748 - x)$$

Multiplying out to remove the parentheses yields

$$x = 6247.08 - 0.21y \quad \text{and} \quad y = 1189.92 - 0.04x$$

Moving the variables to the left sides of the equals signs, we can represent the given situation as a system of two linear equations in two variables:

$$\begin{cases} x + 0.21y = 6247.08 \\ 0.04x + y = 1189.92 \end{cases}$$

Solving by the elimination method, we create the following sequence of equivalent systems of equations. First, we eliminate the x-variable from the second equation by subtracting an appropriate multiple of the first equation.

$$\begin{cases} 0.04x + 0.0084y = 249.8832 \\ 0.04x + y = 1189.92 \end{cases} \qquad \text{Multiplying the first equation by 0.04}$$

$$\begin{cases} 0.04x + 0.0084y = 249.8832 \\ 0x + 0.9916y = 940.0368 \end{cases} \qquad \text{Subtracting the first equation from the second}$$

Simplifying the numbers by dividing, we obtain

$$\begin{cases} 1x + 0.21y = 6247.08 \\ 0x + 1y = 948 \end{cases} \qquad \begin{array}{l} \text{Dividing the first equation by 0.04} \\ \text{Dividing the second equation by 0.9916} \end{array}$$

Next we eliminate the y-variable from the first equation by subtracting an appropriate multiple of the second equation.

$$\begin{cases} 1x + 0.21y = 6247.08 \\ 0x + 0.21y = 199.08 \end{cases} \qquad \text{Multiplying the second equation by 0.21}$$

$$\begin{cases} 1x + 0y = 6048 \\ 0x + 0.21y = 199.08 \end{cases} \qquad \text{Subtracting the second equation from first}$$

We finish by dividing the second equation to get just one y.

$$\begin{cases} 1x + 0y = 6048 \\ 0x + 1y = 948 \end{cases} \qquad \text{Dividing the second equation by 0.21}$$

The solution is $x = 6048$, $y = 948$. The federal tax is \$6048 and the state tax is \$948. You should check that each of these tax amounts is the correct percentage of the original income once the other tax is subtracted.

Section Summary

A *system of two linear equations in two variables* can be written in the form

$$\begin{cases} ax + by = h \\ cx + dy = k \end{cases}$$

Since the graphs of these linear equations are lines, the system of equations can be solved by sketching the lines.

$\begin{cases} ax + by = h \\ cx + dy = k \end{cases}$	Graph	Number of Solutions
Independent and consistent	Lines intersect	One
Independent and inconsistent	Parallel lines	None
Dependent	Just one line	Infinitely many

The *substitution method* solves one of the equations for a variable and uses this new expression to rewrite the other equation with just one variable. Solving this new equation and then finding the first variable solves the system of equations.

Equivalent systems of equations have the same solution and can be found by

1. Switching the order of the equations,
2. Multiplying or dividing one equation by a nonzero number, or
3. Replacing an equation with its sum or difference with the other.

The *elimination method* solves the system by attempting to change it into an equivalent system of the form $\begin{cases} 1x + 0y = p \\ 0x + 1y = q \end{cases}$. If this can be done, the solution is $x = p, y = q$, and the system of equations is independent and consistent. If an equation of the form $0x + 0y = 1$ is obtained, there is *no solution*, and the system is *independent* and *inconsistent*. If an equation of the form $0x + 0y = 0$ is obtained, there are *infinitely many solutions* and the system is *dependent*.

EXERCISES 3.1

Represent each pair of statements as a system of two linear equations in two variables using the given definitions of x and y. Verify that the values given for x and y are a solution for the system of equations.

1. "The sum of two numbers is eighteen" and "the first number is two more than the second number." Let x be the first number and y be the second number. $x = 10, y = 8$

2. "The sum of two numbers is twenty-five" and "twice the first number added to the second number totals thirty-two." Let x be the first number and y be the second number. $x = 7, y = 18$

3. "Tom has $6 more than Alice" and "together, they have $40." Let x be the amount of money that Tom has and y be the amount Alice has. $x = 23, y = 17$

4. "Bill and Jessica together have $25" and "Jessica has $12." Let x be the amount of money that Bill has and y be the amount that Jessica has. $x = 13, y = 12$

5. "An envelope of $1 and $5 bills contains thirty bills" and "the money in the envelope is worth $70." Let x be the number of $1 bills and y be the number of $5 bills. $x = 20, y = 10$

6. "An envelope of $10 and $20 bills contains eight bills" and "the money in the envelope is worth $110." Let x be the number of $10 bills and y be the number of $20 bills. $x = 5, y = 3$

7. "A small theater sold tickets for all one hundred seats" and "the box office receipts of $650 came from adult tickets at $10 and child tickets at $5." Let x be the number of adult tickets sold and y be the number of child tickets sold. $x = 30, y = 70$

8. "A movie theater sold tickets for three hundred seats" and "the box office receipts of $2400 came from adult tickets at $9 and child tickets at $6." Let x be the number of adult tickets sold and y be the number of child tickets sold. $x = 200, y = 100$

9. "A corn and beet farmer planted 225 acres of crops" and "he planted twice as many acres of corn as acres of beets." Let x be the number of acres of corn and y be the number of acres of beets. $x = 150, y = 75$

10. "A stock and bond speculator invested $10,000 in the market" and "she invested three times as much in stocks as in bonds." Let x be the amount in stocks and y be the amount in bonds. $x = 7500, y = 2500$

Solve each of the following systems of equations by graphing. Identify each system as "independent and consistent," "independent and inconsistent," or "dependent."

You may use a graphing calculator if permitted by your instructor.

11. $\begin{cases} x + y = 6 \\ x - y = 2 \end{cases}$

12. $\begin{cases} -x + y = 2 \\ x + y = 4 \end{cases}$

13. $\begin{cases} 2x + y = 8 \\ x = 3 \end{cases}$

14. $\begin{cases} x + 2y = 10 \\ y = 4 \end{cases}$

15. $\begin{cases} x - y = 4 \\ -x + 2y = -6 \end{cases}$

16. $\begin{cases} 2x - y = 2 \\ x + 2y = 6 \end{cases}$

17. $\begin{cases} x + y = 10 \\ -x - y = 10 \end{cases}$

18. $\begin{cases} -2x + 4y = -16 \\ x - 2y = 4 \end{cases}$

19. $\begin{cases} x + y = 10 \\ -x - y = -10 \end{cases}$

20. $\begin{cases} 2x - 4y = 16 \\ -x + 2y = -8 \end{cases}$

Solve each of the following systems of equations by the substitution method. Identify each system as "independent and consistent," "independent and inconsistent," or "dependent."

21. $\begin{cases} x + 2y = 10 \\ y = 3 \end{cases}$ **22.** $\begin{cases} 2x + y = 8 \\ x = 2 \end{cases}$

23. $\begin{cases} 2x + y = 20 \\ x + y = 15 \end{cases}$ **24.** $\begin{cases} x + y = 12 \\ x + 2y = 14 \end{cases}$

25. $\begin{cases} 5x + 2y = 30 \\ 2x + y = 10 \end{cases}$ **26.** $\begin{cases} 2x + 3y = 25 \\ x - y = 5 \end{cases}$

27. $\begin{cases} 3x + 2y = 30 \\ x - y = -5 \end{cases}$ **28.** $\begin{cases} 2x + y = 20 \\ x + 3y = 15 \end{cases}$

29. $\begin{cases} -2x + 2y = -20 \\ x - y = 10 \end{cases}$ **30.** $\begin{cases} 3x - 2y = 30 \\ -6x + 4y = 30 \end{cases}$

Solve each of the following systems of equations by the elimination method. Identify each system as "independent and consistent," "independent and inconsistent," or "dependent."

31. $\begin{cases} x + y = 11 \\ 2x + 3y = 30 \end{cases}$ **32.** $\begin{cases} 3x + 2y = 30 \\ x + y = 13 \end{cases}$

33. $\begin{cases} 3x + y = 15 \\ x + 2y = 10 \end{cases}$ **34.** $\begin{cases} x + 3y = 30 \\ 2x + y = 10 \end{cases}$

35. $\begin{cases} x + 2y = 14 \\ 3x + 4y = 36 \end{cases}$ **36.** $\begin{cases} 2x + 3y = 30 \\ x - y = 10 \end{cases}$

37. $\begin{cases} 2x + 5y = 60 \\ 2x + 3y = 48 \end{cases}$ **38.** $\begin{cases} 4x + 3y = 36 \\ x + 3y = 18 \end{cases}$

39. $\begin{cases} 3x + 4y = -24 \\ 6x + 8y = 24 \end{cases}$ **40.** $\begin{cases} x - 4y = 20 \\ -2x + 8y = -40 \end{cases}$

APPLIED EXERCISES

Formulate each situation as a system of two linear equations in two variables. Be sure to state clearly the meaning of your x- and y-variables. Solve the system of equations by the elimination method. Be sure to state your final answer in terms of the original question.

41. Coins in a Jar   A jar contains 60 nickels and dimes worth $4.30. How many of each are in the jar?

42. Bills in an Envelope   An envelope found in a safe deposit box contains a total of ninety $5 and $20 bills worth $1200. How many of each are in the envelope?

43. Financial Planning   A retired couple wish to invest their nest egg of $10,000 in a money market account paying 6% and in a stock mutual fund returning 11%. If their income tax and Social Security situation requires that they earn $1000 from these investments, how much should they invest in each?

44. Home Financing   A young couple needs to borrow $168,000 to finance their first house. They could borrow the whole amount at 12% from their bank, but her father offers to lend them enough of the money at 5% (with the same terms as the bank loan) to reduce the overall interest rate to just 8%. How much does he lend them and how much do they borrow from the bank?

45. Ice Hockey Concession Receipts   The concession stand at an ice hockey rink had receipts of $7200 from selling a total of 3000 sodas and hot dogs. If each soda sold for $2 and each hot dog sold for $3, how many of each were sold?

46. Baseball Tickets   A college baseball game generated box office receipts of $4800 from 600 ticket sales. If general admission tickets were $12 and student tickets were half-price, how many of each were sold?

47. Income Taxes   Find the federal and state taxes on a taxable income of $49,900 if the federal tax is 10% of the taxable income after first deducting the state tax and the state tax is 2% of the taxable income after first deducting the federal tax.

48. Estate Division   A will specifies that each of two sons receive one-half of the $3 million estate after first deducting the other's share, and that any remainder is then to be given to their sister. How much does each son receive? How much is left for the sister?

49. Sports Nutrition   The dietician at a sports training facility has determined that one of her athletes needs an additional 600 mg each of calcium and phosphorus daily. Two supplements are available containing the milligrams of calcium and phosphorus per tablet as given by the table. How many tablets of each supplement will provide the required calcium and phosphorus?

	Calcium	Phosphorus
Supplement A	150	100
Supplement B	120	120

50. Bicycle Shop Management A bicycle shop has $10,500 to spend on new bikes and 390 hours of assembly time to put them together. Each mountain bike costs $50 wholesale and takes two hours to assemble. Each racing bike costs $70 wholesale and takes two-and-one-half hours to assemble. How many of each can the shop buy and assemble to use all the available money and time?

Explorations and Excursions

The following problems extend and augment the material presented in the text.

Systems of Three Linear Equations in Two Variables. If we have a system that represents three lines instead of just two, there are more possibilities. The second line could be parallel to the first and the third might be the same as the first, so the system could be both inconsistent and dependent.

Sketch each of the following systems and identify them as "consistent" or "inconsistent" and as "dependent" or "independent." If the system is consistent, find the solution.

You may use a graphing calculator if permitted by your instructor.

51. $\begin{cases} x + y = 10 \\ x + y = 5 \\ 2x + 2y = 20 \end{cases}$ **52.** $\begin{cases} x + y = 10 \\ x - y = 0 \\ 2x + 2y = 20 \end{cases}$

53. $\begin{cases} x + 3y = 27 \\ 2x + 5y = 50 \\ x + y = 19 \end{cases}$ **54.** $\begin{cases} 2x + 3y = 12 \\ x - y = 1 \\ 4x + y = 14 \end{cases}$

55. $\begin{cases} 2x + y = 24 \\ x - y = 6 \\ x + y = 16 \end{cases}$

More About Parameterizations

56. In Example 2a we sketched the line $2x + 3y = 12$, and in Example 3c we parameterized this line as $x = 6 - \frac{3}{2}t, y = t$. To graph these *parametric equations* on your calculator, change the MODE from FUNCtion to PARametric and enter the pair of equations $x_{1T} = 6 - (3/2)T, y_{1T} = T$. Set the WINDOW with Tmin $= -2$, Tmax $= 2$, Tstep $= 0.1$, Xmin $= -10$, Xmax $= 10$, Ymin $= -10$, and Ymax $= 10$, and watch as your calculator draws the graph. Change Tmin to -3, Tmax to 3, and GRAPH it again. Experiment with different values for Tmin and Tmax until you see how to get a "complete" picture of this line in this window.

57. Check that the line $2x - 5y = 8$ can be parameterized as $x = 4 + \frac{5}{2}t, y = t$. Set the WINDOW with Tmin $= -2$, Tmax $= 2$, Tstep $= 0.1$, Xmin $= -10$, Xmax $= 10$, Ymin $= -10$, and Ymax $= 10$. GRAPH these equations (be sure to first change the MODE from FUNCtion to PARametric) and use TRACE to explore this line segment. Experiment with different values for Tmin and Tmax until you see how to get a "complete" picture of this line in this window.

58. Find a parameterization for the line $x + 3y = 15$ and use your calculator to verify that your parametric equations determine a line with x-intercept $(15, 0)$ and y-intercept $(0, 5)$.

59. Use the parametric equations from Exercises 57 and 58 to GRAPH Example 5 on the WINDOW with Tmin $= -5$, Tmax $= 10$, Tstep $= 0.1$, Xmin $= -5$, Xmax $= 20$, Ymin $= -10$, and Ymax $= 10$. Use TRACE to explore both lines and verify that the intersection point is $(9, 2)$.

60. Using the WINDOW with Tmin $= -5$, Tmax $= 10$, Tstep $= 0.1$, Xmin $= -5$, Xmax $= 5$, Ymin $= -5$, and Ymax $= 5$, explore each of the following pairs of parametric equations. Can you tell in advance which give lines and which give curves? Why does part (c) give just some of the line?

a. $\begin{cases} x = T - 1 \\ y = T \end{cases}$ **b.** $\begin{cases} x = T^2 - 1 \\ y = T \end{cases}$

c. $\begin{cases} x = T^2 - 1 \\ y = T^2 \end{cases}$ **d.** $\begin{cases} x = T^2 - 1 \\ y = T^3 \end{cases}$

e. $\begin{cases} x = T^3 - 1 \\ y = T^2 \end{cases}$ **f.** $\begin{cases} x = T^3 - 1 \\ y = T^3 \end{cases}$

Change Tstep from 0.1 to 0.05 and redo (a) and (f). Watch while your calculator draws each line. How do these parameterizations of the line $x - y = -1$ differ?

Round-off errors can completely misrepresent the true nature of a system of equations. (If you wish to remove fractions from a problem, multiply each equation through by the least common denominator instead of rounding off.)

61. Use the elimination method to show that the system $\begin{cases} x + \frac{1}{3}y = 39 \\ 2x + \frac{2}{3}y = 84 \end{cases}$ is inconsistent.

62. Rounding to one decimal place, the system in Exercise 61 becomes $\begin{cases} x + 0.3y = 39 \\ 2x + 0.7y = 84 \end{cases}$. Use the elimination method to show that this new system is consistent and independent with solution $x = 21$, $y = 60$.

63. Use the elimination method to show that the system $\begin{cases} x + \frac{2}{5}y = 9.79 \\ 4x + \frac{8}{9}y = 39.16 \end{cases}$ is dependent.

64. Rounding to two decimal places, the system in Exercise 63 becomes $\begin{cases} x + 0.22y = 9.79 \\ 4x + 0.89y = 39.16 \end{cases}$. Use the elimination method to show that this new system is consistent and independent with solution $x = 9.79$, $y = 0$.

65. Use the elimination method to show that the system $\begin{cases} 2.1x + \frac{1}{7}y = 157 \\ 3x + \frac{1}{5}y = 224 \end{cases}$ is consistent and independent with solution $x = 70$, $y = 70$. Rounding to two decimal places, this system becomes $\begin{cases} 2.10x + 0.14y = 157 \\ 3.00x + 0.20y = 224 \end{cases}$. Use the elimination method to show that this new system is inconsistent.

3.2 Matrices and Linear Equations in Two Variables

APPLICATION PREVIEW

Spreadsheets

For hundreds of years the business spreadsheet has been a standard accounting tool. Originally written on slates or paper, it organizes data and calculated values into rows and columns containing similar quantities. Because even simple business situations may lead to large collections of numbers linked by complicated formulas, carrying out the required calculations can be a formidable, time-consuming task. Even worse, should one of the data values change, the effect on the rest of the spreadsheet can be found only by recalculating the entire grid of numbers.

When Dan Bricklin was a graduate student at the Harvard Business School, he had the idea that business calculations could be greatly simplified if the spreadsheet was really a computer program with the formulas embedded in the positions where the calculated values should appear. Together with Robert Frankston, a friend from their undergraduate days at the Massachusetts Institute of Technology, he created the first personal computer spreadsheet program in 1979. Written for the Apple II computer with 32K memory, VisiCalc had only 254 rows and 63 columns and a crude user interface. But it worked and became the original "killer application" of the revolution that changed the personal computer from an interesting toy to an essential business tool.

C1 = A1 + B1		
	A	**B**	**C**
1	300	800	1100
2	200	700	900

The formula in cell C1 is A1 + B1, so C1 displays the sum of the current values in A1 and B1. Can you find a similar formula for the value in cell C2?

Introduction

The solutions of many problems are greatly simplified when expressed in appropriate notations. In this section, we will use matrix notation to streamline the elimination method used in the previous section to solve systems of two linear equations in two variables. In the next section, we shall extend this method to the solutions of systems of many linear equations in many variables.

Matrices

A *matrix* is a rectangular array of numbers called *elements.* This rectangular array may be viewed as consisting of *rows* (with the first at the top, the second below the first, and so on) or *columns* (with the first on the left, the second to the right of the first, and so on). The *dimension* of a matrix is the size expressed as "rows by columns," and we will write $m \times n$ for the dimension of a matrix with m rows and n columns. Thus a 5×2 matrix is "tall and thin," while a 2×5 matrix is "short and wide." A *square matrix* has the same number of rows as columns. A *row matrix* has just one row, and a *column matrix* has just one column.

$$
(1\ 2\ 3) \qquad \begin{pmatrix} 1 \\ 2 \\ 3 \end{pmatrix} \qquad \begin{pmatrix} 1 & 2 & 3 \\ 4 & 5 & 6 \\ 7 & 8 & 9 \end{pmatrix}
$$

Row matrix Column matrix Square matrix
(dimension 1×3) (dimension 3×1) (dimension 3×3)

Each element of a matrix has both a *value* and a *position* given by its row and column address. We name matrices with capital letters (*A, B, C, . . .*) and then the elements are named by the corresponding lower-case letter together with the appropriate row and column address. For instance, if

$$A = \begin{pmatrix} 1 & 2 & 3 & 4 \\ 5 & 6 & 7 & 8 \\ 9 & 10 & 11 & 12 \end{pmatrix}$$

then the dimension of A is 3×4 and $a_{2,3} = 7$ because 7 is the value of the element of A in the second row and third column. This *double subscript* notation for the elements of the matrix is sometimes used without the comma between the row and column addresses, so that a_{23} means $a_{2,3}$. The "L" shape of the letter and subscripts suggests the "follow the L" nickname sometimes given this notation: Just as you make the letter L by a downstroke then a right-stroke, the element $a_{2,3}$ is found by moving *down* to row 2 and then *right* to column 3. The elements on the *main diagonal* are those with the same row address as column address. For the matrix A above, the main diagonal consists of the elements $a_{1,1} = 1$, $a_{2,2} = 6$, and $a_{3,3} = 11$.

Augmented Matrices from Systems of Equations

An *augmented matrix* is a matrix created from two "smaller" matrices having the same number of rows by placing them beside each other and joining them into one "larger" matrix. The system of equations $\begin{cases} ax + by = h \\ cx + dy = k \end{cases}$ naturally gives rise to a *coefficient matrix* $\begin{pmatrix} a & b \\ c & d \end{pmatrix}$ and a *constant term matrix* $\begin{pmatrix} h \\ k \end{pmatrix}$. Taken together, these form the augmented matrix $\begin{pmatrix} a & b & h \\ c & d & k \end{pmatrix}$, which represents the system of equations. The first column gives the coefficients of the *x*-variable, the second gives those of the *y*-variable, and the last column contains the constant terms after the equals signs.

Augmented Matrix of a System of Equations

Sometimes written as $\begin{pmatrix} a & b & | & h \\ c & d & | & k \end{pmatrix}$

The augmented matrix $\begin{pmatrix} a & b & h \\ c & d & k \end{pmatrix}$ represents the system of equations $\begin{cases} ax + by = h \\ cx + dy = k \end{cases}$.

EXAMPLE 1 **Augmented Matrices and Systems of Equations**

a. Find the augmented matrix representing the system $\begin{cases} 2x + 3y = 24 \\ 4x + 5y = 60 \end{cases}$.

b. Find the system of equations represented by the augmented matrix

$$\begin{pmatrix} 6 & 8 & 84 \\ 4 & 5 & 60 \end{pmatrix}.$$

Solution

a. The system of equations $\begin{cases} 2x + 3y = 24 \\ 4x + 5y = 60 \end{cases}$ is represented by $\begin{pmatrix} 2 & 3 & 24 \\ 4 & 5 & 60 \end{pmatrix}$.

b. The augmented matrix $\begin{pmatrix} 6 & 8 & 84 \\ 4 & 5 & 60 \end{pmatrix}$ represents $\begin{cases} 6x + 8y = 84 \\ 4x + 5y = 60 \end{cases}$.

■

PRACTICE PROBLEM 1

a. Find the augmented matrix representing the system of equations $\begin{cases} 2x - y = 14 \\ x + 3y = 21 \end{cases}$.

b. Find the system of equations represented by the augmented matrix

$$\begin{pmatrix} 3 & 2 & 35 \\ 1 & 3 & 21 \end{pmatrix}.$$

Solutions at the back of the book

Row Operations

In order to solve the system of equations represented by an augmented matrix, we transform it into *equivalent matrices* using the following *row operations* that correspond to the manipulations we used to solve the system by the elimination method on page 165.

Matrix Row Operations

1. Switch any two rows.
2. Multiply or divide one of the rows by a nonzero number.
3. Replace a row by its sum or difference with another row.

We have phrased the above definition to apply to matrices of any size. We shall indicate the row operation used by writing next to the changed row a short formula explaining what was done. We shall write R for "row" and R' (read "R prime") for the "new row."

EXAMPLE 2   **Performing Matrix Row Operations**

Carry out the indicated row operation on the given matrix.

a. $R_1' = R_2$ and $R_2' = R_1$ on $\begin{pmatrix} 2 & 3 & 25 \\ 3 & 4 & 36 \end{pmatrix}$

b. $R_2' = 3R_2$ on $\begin{pmatrix} 2 & 3 & 30 \\ 1 & 1 & 13 \end{pmatrix}$

c. $R_1' = R_1 - R_2$ on $\begin{pmatrix} 2 & 3 & 30 \\ 2 & -2 & 20 \end{pmatrix}$

Solution

a. $R_1' = R_2$ says that "the new row 1 is the (old) row 2" and $R_2' = R_1$ says that "the new row 2 is the (old) row 1," so this row operation switches rows 1 and 2:

$$\begin{pmatrix} 3 & 4 & 36 \\ 2 & 3 & 25 \end{pmatrix} \begin{matrix} R_1' = R_2 \\ R_2' = R_1 \end{matrix} \qquad \text{From } \begin{pmatrix} 2 & 3 & 25 \\ 3 & 4 & 36 \end{pmatrix}$$

b. $R_2' = 3R_2$ says that "the new row 2 is 3 times the (old) row 2," so the second row is multiplied by 3:

$$\begin{pmatrix} 2 & 3 & 30 \\ 3 & 3 & 39 \end{pmatrix} R_2' = 3R_2 \qquad \text{From } \begin{pmatrix} 2 & 3 & 30 \\ 1 & 1 & 13 \end{pmatrix}$$

c. $R_1' = R_1 - R_2$ says that "the new row 1 is the (old) row 1 minus the (old) row 2," so the first row is replaced by its difference with the second row:

$$\begin{pmatrix} 0 & 5 & 10 \\ 2 & -2 & 20 \end{pmatrix} R_1' = R_1 - R_2 \qquad \text{From } \begin{pmatrix} 2 & 3 & 30 \\ 2 & -2 & 20 \end{pmatrix}$$

■

PRACTICE PROBLEM 2

a. Carry out $R_1' = 3R_1$ on $\begin{pmatrix} 2 & 1 & 14 \\ 1 & -3 & 21 \end{pmatrix}$.

b. Carry out $R_1' = R_1 + R_2$ on $\begin{pmatrix} 6 & 3 & 42 \\ 1 & -3 & 21 \end{pmatrix}$.

c. Carry out $R_1' = \frac{1}{7}R_1$ on $\begin{pmatrix} 7 & 0 & 63 \\ 1 & -3 & 21 \end{pmatrix}$.

d. Is $R_1' = 0R_1$ a valid row operation?

e. Is $R_2' = 5R_1$ a valid row operation?   *Solutions at the back of the book*

Solving Equations by Row Reduction

Two matrices are *equivalent* if there is a sequence of row operations that transforms one into the other. Since the augmented matrix represents the system of equations and the row operations correspond to the steps used to solve the system, we can "solve" the augmented matrix by *row-reducing* it to an equivalent matrix that displays the solution. If we can obtain the form

$$\begin{pmatrix} 1 & 0 & p \\ 0 & 1 & q \end{pmatrix} \qquad \begin{cases} 1x + 0y = p \\ 0x + 1y = q \end{cases} \text{ or } \begin{cases} x = p \\ y = q \end{cases}$$

then the solution of the corresponding system of equations is $x = p$, $y = q$, and the system is independent and consistent. However, if we obtain a row of zeros ending with a nonzero number:

$$0 \quad 0 \quad m \qquad\qquad m \neq 0$$

then the system is *inconsistent* and has *no solution* because the equation $0x + 0y = m$ makes the impossible claim that zero equals a nonzero number. On the other hand, if there is no such "inconsistent" row but there is a row consisting entirely of zeros:

$$0 \quad 0 \quad 0$$

then the system is *dependent,* and there are *infinitely many solutions* because the equation $0x + 0y = 0$ is *always* true and represents no restriction at all. These important observations will be extended in the next section.

EXAMPLE 3 **Solving Equations by Row Reduction**

Solve $\begin{cases} x + 3y = 15 \\ 2x - 5y = 8 \end{cases}$ by row reducing the corresponding augmented matrix.

Solution

The augmented matrix for this system of equations is

$$\begin{pmatrix} 1 & 3 & 15 \\ 2 & -5 & 8 \end{pmatrix}$$

We hope to change the first column from $\frac{1}{2}$ to $\frac{1}{0}$ (which is the same as removing the x-variable from the second equation) and then the second column to $\frac{0}{1}$ (to remove the y-variable from the first equation).

If we are successful, we will have the matrix $\begin{pmatrix} 1 & 0 & p \\ 0 & 1 & q \end{pmatrix}$ and the solution $x = p$, $y = q$. If not, we will find that the system is inconsistent or dependent. The following sequence of row operations and equivalent matrices is one possible way of reaching the solution. First, we get a zero at the bottom of the first column by using the row 1, column 1 element.

$$\begin{pmatrix} 2 & 6 & 30 \\ 2 & -5 & 8 \end{pmatrix} R_1' = 2R_1$$

To get a 2 above the other 2

$$\begin{pmatrix} 2 & 6 & 30 \\ 0 & -11 & -22 \end{pmatrix} R_2' = R_2 - R_1$$

Subtracting to get the zero

Since the first row has a common factor of 2 and the second row has a common factor of -11, we can simplify the numbers by dividing.

$$\begin{pmatrix} 1 & 3 & 15 \\ 0 & -11 & -22 \end{pmatrix} R_1' = \tfrac{1}{2}R_1$$

Dividing the first row by 2

$$\begin{pmatrix} 1 & 3 & 15 \\ 0 & 1 & 2 \end{pmatrix} R_2' = \tfrac{1}{-11}R_2$$

Dividing the second row by -11

Next we get a zero at the top of the second column by getting a 3 in the second row and then subtracting.

$$\begin{pmatrix} 1 & 3 & 15 \\ 0 & 3 & 6 \end{pmatrix} R_2' = 3R_2$$

Multiplying the second row by 3

$$\begin{pmatrix} 1 & 0 & 9 \\ 0 & 3 & 6 \end{pmatrix} R_1' = R_1 - R_2$$

Subtracting to get the zero

Since we now have zeros at the desired locations, we conclude by changing the 3 in the second row back into a 1.

$$\begin{pmatrix} 1 & 0 & 9 \\ 0 & 1 & 2 \end{pmatrix} R_2' = \tfrac{1}{3}R_2$$

Dividing the second row by 3 to get a 1

The solution is $x = 9$, $y = 2$, and the original system is consistent and independent. [We solved this same system in the previous section by the elimination method in Example 6 (page 166) and found the same solution.] ∎

PRACTICE PROBLEM 3

Solve $\begin{cases} 2x + y = 14 \\ x - 3y = 21 \end{cases}$ by row reducing the corresponding augmented matrix. [*Hint:* Use your work from Practice Problem 2.]

Solution at the back of the book

Graphing Calculator Exploration

Some graphing calculators can row-reduce matrices. If your calculator has a RREF command (for "reduced row-echelon form," as explained in the next section), you can easily check your row reduction. For the system of equations in Example 3:

a. Enter $\begin{pmatrix} 1 & 3 & 15 \\ 2 & -5 & 8 \end{pmatrix}$ as matrix [A] using MATRIX EDIT.

```
MATRIX[A] 2 X3
[1    3    15  ]
[2   -5     8  ]



2,3=8
```

b. QUIT and select the RREF command from the MATRX MATH menu.

```
NAMES  MATH  EDIT
0↑cumSum(
A:ref(
B:rref(
C:rowSwap(
D:row+(
E:*row(
F:*row+(
```

c. Apply RREF to the matrix [A].

```
rref([A])
         [[1 0 9]
          [0 1 2]]


```

Although this serves as a useful check of your answer, do not rely on it completely because the calculator sometimes returns an answer with round-off errors. Furthermore, this calculator command may not work for matrices with more rows than columns. We will be interested in such matrices in the next section.

EXAMPLE 4 **Solving Equations by Row Reduction**

Solve $\begin{cases} 6x - 3y = 30 \\ -8x + 4y = -40 \end{cases}$ by row-reducing the corresponding augmented matrix.

Solution

The augmented matrix is

$$\begin{pmatrix} 6 & -3 & 30 \\ -8 & 4 & -40 \end{pmatrix}$$

When the numbers in a row have an obvious common factor, removing that factor can sometimes simplify the reduction. Because the first row is divisible by 3, we begin with

$$\begin{pmatrix} 2 & -1 & 10 \\ -8 & 4 & -40 \end{pmatrix} \quad R_1' = \tfrac{1}{3}R_1$$

Similarly for the second row:

$$\begin{pmatrix} 2 & -1 & 10 \\ -2 & 1 & -10 \end{pmatrix} \quad R_2' = \tfrac{1}{4}R_2$$

Adding the first row to the second, we get a zero at the bottom of the first column.

$$\begin{pmatrix} 2 & -1 & 10 \\ 0 & 0 & 0 \end{pmatrix} \quad R_2' = R_2 + R_1$$

To finish, we divide the first row by 2 to make the row begin with a 1 on the left.

$$\begin{pmatrix} 1 & -\tfrac{1}{2} & 5 \\ 0 & 0 & 0 \end{pmatrix} \quad R_1' = \tfrac{1}{2}R_1$$

The zero row means that the system is *dependent*, so there are infinitely many solutions. The first row of this final matrix says that

$$x - \tfrac{1}{2}y = 5$$

or, solving for x,

$$x = 5 + \tfrac{1}{2}y$$

We may let y be *any* number t and then determine x from the above equation. That is, the solutions may be parameterized as $x = 5 + \tfrac{1}{2}t$, $y = t$, where t is *any* number. The following table lists some of these solutions for various values of the parameter t.

t	$x = 5 + \tfrac{1}{2}t$	$y = t$
-20	-5	-20
-10	0	-10
0	5	0
10	10	10
20	15	20

There are many other sequences of row operations to reduce this augmented matrix and all reach the same conclusion. ∎

PRACTICE PROBLEM 4

Verify that $x = -5$, $y = -20$, and $x = 15$, $y = 20$ from the above table solve $\begin{cases} 6x - 3y = 30 \\ -8x + 4y = -40 \end{cases}$.

Solution at the back of the book

EXAMPLE 5 Production Management

A worker in a plastics factory breaks apart sheets of component A and strips of component B and then snaps one of each together to make a finished item. If the worker can break off 20 A's or 30 B's per minute and snap together 10 pairs of A's and B's per minute, how should the worker's 440-minute workday be divided so as to complete as many items as possible with no unused pieces left over?

Solution

Let x be the number of minutes spent breaking apart sheets of component A and y be the number of minutes breaking apart strips of component B. The remainder of the worker's time, $440 - (x + y)$, will then be spent snapping A's and B's together. Because the worker needs as many A's as B's,

$$20x = 30y$$

Rate × time gives number finished

As many finished items as A's will be completed:

$$20x = 10(440 - (x + y))$$

Rewriting these as a system of equations, the problem becomes

$$\begin{cases} 20x - 30y = 0 \\ 30x + 10y = 4400 \end{cases}$$

The augmented matrix is

$$\begin{pmatrix} 20 & -30 & 0 \\ 30 & 10 & 4400 \end{pmatrix}$$

We begin the row reduction by removing from both rows the common factor of 10, writing both results on one matrix.

$$\begin{pmatrix} 2 & -3 & 0 \\ 3 & 1 & 440 \end{pmatrix} \begin{matrix} R_1' = \frac{1}{10} R_1 \\ R_2' = \frac{1}{10} R_2 \end{matrix}$$

To get a one in the upper lefthand corner:

$$\begin{pmatrix} 1 & 4 & 440 \\ 3 & 1 & 440 \end{pmatrix} R_1' = R_2 - R_1$$

To get a zero below the 1 in the upper lefthand corner:

$$\begin{pmatrix} 3 & 12 & 1320 \\ 3 & 1 & 440 \end{pmatrix} \quad R_1' = 3R_1$$

$$\begin{pmatrix} 3 & 12 & 1320 \\ 0 & 11 & 880 \end{pmatrix} \quad R_2' = R_1 - R_2$$

Removing common factors again:

$$\begin{pmatrix} 1 & 4 & 440 \\ 0 & 1 & 80 \end{pmatrix} \quad \begin{aligned} R_1' &= \tfrac{1}{3}R_1 \\ R_2' &= \tfrac{1}{11}R_2 \end{aligned}$$

To get a zero above the 1 at the bottom of the second column:

$$\begin{pmatrix} 1 & 4 & 440 \\ 0 & 4 & 320 \end{pmatrix} \quad R_2' = 4R_2$$

$$\begin{pmatrix} 1 & 0 & 120 \\ 0 & 4 & 320 \end{pmatrix} \quad R_1' = R_1 - R_2$$

Now that we have zeros in the proper positions, we finish by returning the column 2, row 2 entry to 1:

$$\begin{pmatrix} 1 & 0 & 120 \\ 0 & 1 & 80 \end{pmatrix} \quad R_2' = \tfrac{1}{4}R_2 \qquad\qquad \begin{aligned} x &= 120 \\ y &= 80 \end{aligned}$$

The solution is $x = 120$, $y = 80$. In terms of the original question, the worker should break apart sheets of component A for 120 minutes, break apart strips of component B for 80 minutes, and snap A's and B's together for the remaining 240 minutes.

 If you have a graphing calculator with a RREF command, you may use it to check the result of the row reduction in Example 5.

Because we will often employ sequences of row operations similar to the last three in Example 5 to use a 1 to "zero out" another entry in the same column, we shall find it useful to compress them into one operation:

$$\begin{pmatrix} 1 & 4 & 440 \\ 0^0 & 1^4 & 80^{320} \end{pmatrix} \rightarrow \begin{pmatrix} 1 & 0 & 120 \\ 0 & 1 & 80 \end{pmatrix} \quad R_1' = R_1 - 4R_2$$

We have written the temporary multiple of R_2 as superscripts on the second row entries so that we can see the values while carrying out the arithmetic. Thus we may combine the second and third basic row operations (see page 176) to "replace a row by its sum or difference with a multiple of another row."

Section Summary

Each element of a matrix has both a *value* and a *position*. The *augmented matrix* $\begin{pmatrix} a & b & h \\ c & d & k \end{pmatrix}$ represents the system of equations $\begin{cases} ax + by = h \\ cx + dy = k \end{cases}$. The rows correspond to the equations, and the first column corresponds to the x-variable, the second column to the y-variable, and the third column to the constant terms of the equations.

The three basic row operations are

1. Switch any two rows.
2. Multiply or divide one of the rows by a nonzero number.
3. Replace a row by its sum or difference with another row.

We may also use the *combined* row operation:

4. Replace a row by its sum or difference with a multiple of another row.

Any 2×3 augmented matrix corresponding to a system of equations is equivalent to exactly one of the matrices in the following table. These special matrices display the solutions of the original systems of equations.

The augmented matrix representation $\begin{pmatrix} a & b & h \\ c & d & k \end{pmatrix}$ of the system $\begin{cases} ax + by = h \\ cx + dy = k \end{cases}$ is equivalent to exactly one of the following matrices (where p and q stand for any numbers).		
Matrix	**Equations**	**Solution**
$\begin{pmatrix} 1 & 0 & p \\ 0 & 1 & q \end{pmatrix}$	Independent and consistent	One solution: $x = p, y = q$
$\begin{pmatrix} 1 & p & 0 \\ 0 & 0 & 1 \end{pmatrix}$	Independent and inconsistent	No solution
$\begin{pmatrix} 0 & 1 & 0 \\ 0 & 0 & 1 \end{pmatrix}$	Independent and inconsistent	No solution
$\begin{pmatrix} 1 & p & q \\ 0 & 0 & 0 \end{pmatrix}$	Dependent	Infinitely many solutions: $x = q - pt, y = t$
$\begin{pmatrix} 0 & 1 & p \\ 0 & 0 & 0 \end{pmatrix}$	Dependent	Infinitely many solutions: $x = t, y = p$

EXERCISES 3.2

Find the dimension of each matrix and the value(s) of the specified element(s).

1. $\begin{pmatrix} 1 & 4 \\ 2 & 5 \\ 3 & 6 \end{pmatrix}$; $a_{1,1}, a_{3,2}$

2. $\begin{pmatrix} 6 & 3 \\ 5 & 2 \\ 4 & 1 \end{pmatrix}$; $a_{1,1}, a_{3,2}$

3. $\begin{pmatrix} 1 & -1 & 2 & -2 \\ -3 & 3 & -4 & 4 \\ 5 & -5 & 6 & -6 \end{pmatrix}$; $a_{2,2}, a_{3,4}$

4. $\begin{pmatrix} 1 & -2 & -1 \\ -3 & 4 & 3 \\ 5 & -6 & -5 \\ -7 & 8 & 7 \end{pmatrix}$; $a_{2,2}, a_{4,3}$

5. $\begin{pmatrix} 1 & 0 & 0 & 0 \\ 0 & 1 & 0 & 0 \\ 0 & 0 & 1 & 0 \\ 0 & 0 & 0 & 1 \end{pmatrix}$; $a_{1,1}, a_{2,2}, a_{3,3}, a_{4,4}, a_{1,4}$

6. $\begin{pmatrix} 1 & 0 & 0 \\ 0 & 1 & 0 \\ 0 & 0 & 1 \end{pmatrix}$; $a_{1,1}, a_{2,2}, a_{3,3}, a_{2,1}$

7. $(4 \quad 5 \quad 6 \quad 7)$; $a_{1,3}$

8. $(2 \quad 3 \quad 4 \quad 5 \quad 6)$; $a_{1,4}$

9. $\begin{pmatrix} 9 \\ 8 \\ 7 \\ 6 \\ 5 \end{pmatrix}$; $a_{2,1}, a_{4,1}$

10. $\begin{pmatrix} 2 \\ 8 \\ 3 \\ 7 \end{pmatrix}$; $a_{2,1}, a_{3,1}$

Find the augmented matrix representing the system of equations.

11. $\begin{cases} x + 2y = 2 \\ 3x + 4y = 12 \end{cases}$

12. $\begin{cases} 2x - y = 10 \\ -3x + 4y = 60 \end{cases}$

13. $\begin{cases} -4x + 3y = 84 \\ 5x - 2y = 70 \end{cases}$

14. $\begin{cases} -x + 2y = 2 \\ 2x - 3y = 6 \end{cases}$

15. $\begin{cases} 3x - 2y = 24 \\ x = 6 \end{cases}$

16. $\begin{cases} 4x - 3y = 24 \\ y = 8 \end{cases}$

17. $\begin{cases} 5x - 15y = 30 \\ -4x + 12y = 24 \end{cases}$

18. $\begin{cases} 12x - 4y = 36 \\ -15x + 5y = 45 \end{cases}$

19. $\begin{cases} x = 20 \\ y = 30 \end{cases}$

20. $\begin{cases} y = 18 \\ x = 12 \end{cases}$

Find the system of equations represented by the augmented matrix.

21. $\begin{pmatrix} 1 & 1 & 9 \\ 0 & 1 & 4 \end{pmatrix}$

22. $\begin{pmatrix} 1 & 0 & -3 \\ 1 & 1 & 5 \end{pmatrix}$

23. $\begin{pmatrix} -4 & 3 & -60 \\ 1 & -2 & 20 \end{pmatrix}$

24. $\begin{pmatrix} 3 & 4 & 24 \\ 1 & 2 & 6 \end{pmatrix}$

25. $\begin{pmatrix} 1 & -3 & -70 \\ 1 & 1 & 10 \end{pmatrix}$

26. $\begin{pmatrix} 1 & 1 & 5 \\ 2 & 3 & 7 \end{pmatrix}$

27. $\begin{pmatrix} 2 & 1 & 6 \\ 1 & 2 & -6 \end{pmatrix}$

28. $\begin{pmatrix} 3 & 1 & -24 \\ 1 & 3 & 24 \end{pmatrix}$

29. $\begin{pmatrix} 20 & -15 & 60 \\ -16 & 12 & -48 \end{pmatrix}$

30. $\begin{pmatrix} 20 & -15 & 60 \\ -16 & 12 & 48 \end{pmatrix}$

Carry out the row operation on the matrix.

31. $R_1' = R_2$ and $R_2' = R_1$ on $\begin{pmatrix} 3 & 4 & 24 \\ 5 & 6 & 30 \end{pmatrix}$

32. $R_1' = R_2$ and $R_2' = R_1$ on $\begin{pmatrix} 8 & 7 & 56 \\ 6 & 5 & 60 \end{pmatrix}$

33. $R_1' = R_1 - R_2$ on $\begin{pmatrix} 8 & 7 & 56 \\ 6 & 5 & 60 \end{pmatrix}$

34. $R_1' = 2R_1$ on $\begin{pmatrix} 3 & 4 & 24 \\ 5 & 6 & 30 \end{pmatrix}$

35. $R_2' = 5R_2$ on $\begin{pmatrix} 5 & 6 & 30 \\ 1 & 2 & 18 \end{pmatrix}$

36. $R_2' = 3R_2$ on $\begin{pmatrix} 6 & 5 & 60 \\ 2 & 2 & -4 \end{pmatrix}$

37. $R_2' = R_1 - R_2$ on $\begin{pmatrix} 6 & 6 & -12 \\ 6 & 5 & 60 \end{pmatrix}$

38. $R_1' = R_1 - R_2$ on $\begin{pmatrix} 5 & 6 & 30 \\ 4 & 8 & 72 \end{pmatrix}$

39. $R_2' = \frac{1}{8}R_2$ on $\begin{pmatrix} 1 & -2 & -42 \\ 0 & 8 & 120 \end{pmatrix}$

40. $R_1' = \frac{1}{6}R_1$ on $\begin{pmatrix} 6 & 0 & 420 \\ 0 & 1 & -72 \end{pmatrix}$

Interpret each reduced row-echelon form matrix as the solution of a system of equations. Identify each system as "independent and consistent," "independent and inconsistent," or "dependent."

41. $\begin{pmatrix} 1 & 0 & 7 \\ 0 & 1 & -3 \end{pmatrix}$ **42.** $\begin{pmatrix} 1 & 0 & -5 \\ 0 & 1 & 8 \end{pmatrix}$

43. $\begin{pmatrix} 1 & 1 & 0 \\ 0 & 0 & 1 \end{pmatrix}$ **44.** $\begin{pmatrix} 1 & -1 & 0 \\ 0 & 0 & 1 \end{pmatrix}$

45. $\begin{pmatrix} 0 & 1 & 0 \\ 0 & 0 & 1 \end{pmatrix}$ **46.** $\begin{pmatrix} 1 & 0 & 0 \\ 0 & 1 & 0 \end{pmatrix}$

47. $\begin{pmatrix} 1 & 2 & 3 \\ 0 & 0 & 0 \end{pmatrix}$ **48.** $\begin{pmatrix} 1 & -4 & 6 \\ 0 & 0 & 0 \end{pmatrix}$

49. $\begin{pmatrix} 0 & 1 & -3 \\ 0 & 0 & 0 \end{pmatrix}$ **50.** $\begin{pmatrix} 0 & 1 & 9 \\ 0 & 0 & 0 \end{pmatrix}$

Solve each system of equations by row-reducing the corresponding augmented matrix. Identify each system as "independent and consistent," "independent and inconsistent," or "dependent."

 If you have a graphing calculator with a RREF command, use it to check your row reduction.

51. $\begin{cases} x + y = 5 \\ x = 3 \end{cases}$ **52.** $\begin{cases} x + y = 8 \\ y = 6 \end{cases}$

53. $\begin{cases} x + y = 4 \\ x - y = 2 \end{cases}$ **54.** $\begin{cases} x - 2y = 6 \\ x + y = 3 \end{cases}$

55. $\begin{cases} 2x + y = 4 \\ x + y = 3 \end{cases}$ **56.** $\begin{cases} x + y = 4 \\ 3x + y = 6 \end{cases}$

57. $\begin{cases} x + y = 5 \\ 2x + 3y = 12 \end{cases}$ **58.** $\begin{cases} 3x + y = 9 \\ 2x + y = 4 \end{cases}$

59. $\begin{cases} 2x + y = 20 \\ x + 3y = 15 \end{cases}$ **60.** $\begin{cases} x + y = 8 \\ 3x + 5y = 30 \end{cases}$

61. $\begin{cases} -x + 2y = 4 \\ x - 2y = 6 \end{cases}$ **62.** $\begin{cases} x - 3y = 12 \\ -x + 3y = 9 \end{cases}$

63. $\begin{cases} 6x + 2y = 18 \\ 5x + 2y = 10 \end{cases}$ **64.** $\begin{cases} 5x + 3y = 15 \\ 4x + 2y = 8 \end{cases}$

65. $\begin{cases} -3x + y = 3 \\ 5x - 2y = -10 \end{cases}$ **66.** $\begin{cases} 3x - y = 3 \\ -2x + y = 2 \end{cases}$

67. $\begin{cases} 2x - 6y = 18 \\ -3x + 9y = -27 \end{cases}$ **68.** $\begin{cases} 3x - 6y = 24 \\ -5x + 10y = -40 \end{cases}$

69. $\begin{cases} 4x + 7y = 56 \\ 2x + 3y = 30 \end{cases}$ **70.** $\begin{cases} 4x + 3y = 24 \\ 6x + 5y = 30 \end{cases}$

APPLIED EXERCISES

Formulate each situation as a system of two linear equations in two variables. Be sure to state clearly the meaning of your x- and y-variables. Solve the system of equations by row-reducing the corresponding augmented matrix. State your final answer in terms of the original question.

 If you have a graphing calculator with a RREF command, use it to check your row reduction.

71. Commodities   A corn and soybean commodities speculator invested $15,000 yesterday with twice as much in soybean futures as in corn futures. How much did she invest in each?

72. Stamps and Coins   A stamp and coin dealer spent $8000 at a numismatics auction last weekend. If he spent three times as much on coins as on stamps, how much did he spend on each?

73. Estate Division   A will specifies that the older brother is to receive one-half of the $12 million estate after first deducting the younger brother's share, the younger brother is to receive one-third of the estate after first deducting the older brother's share, and the remainder is to be given to their sister. How much does each brother receive? How much is left over for their sister?

74. Real Estate Taxes   Find the state and city property taxes on an apartment building assessed at $833,000 if the state tax is 4% of the assessed value after first deducting the city tax and the city tax is 1% of the assessed value after first deducting the state tax.

75. Coins in a Jar   A jar contains nickels and quarters worth $36.75. If there were twice as many nickels and half as many quarters, they would be worth $31.50 instead. How many of each are in the jar?

76. Financial Planning Last year, a retired couple's investments in a money market account yielding 5% and in a stock mutual fund yielding 14% paid a total of $16,800. If the stock mutual fund can return 17% this year and the money market account can continue to yield 5%, the couple will receive $17,850. How much is invested in each?

77. Geriatrics Nutrition The dietician at a senior care facility has decided to supplement the weekly menu with "Fountain of Youth" NutraDrink and "Get Up & Go" VitaPills, which contain calcium and vitamin C as listed in the table. If each resident needs an additional 300 mg of calcium and 240 mg of vitamin C per week, how many cans of NutraDrink and tablets of VitaPills should each resident be given per week?

	Calcium (mg)	Vitamin C (mg)
NutraDrink	25	20
VitaPills	20	16

78. Plant Fertilizer A garden field needs 308 pounds of potash and 330 pounds of nitrogen. Two brands of fertilizer, GrowRite and GreatGreen, are available and contain the amounts of potash and nitrogen per bag listed in the table. How many bags of each brand should be used to provide the required potash and nitrogen?

(*Pounds per Bag*)	GrowRite	GreatGreen
Potash	4	7
Nitrogen	5	6

79. Political Advertising For the final days before the election, the campaign manager has set aside $36,000 to spend on TV and radio campaign advertisements. Each TV ad costs $3000 and is seen by 10,000 voters, while each radio ad costs $500 and is heard by 2000 voters. Ignoring repeated exposures to the same voter, how many TV and radio ads will contact 130,000 voters using the allocated funds?

80. Consumer Preferences A marketing company wants to gather information on consumer preferences for laundry detergents using telephone interviews and direct mail questionnaires. Each attempted telephone interview costs $2.25 whether successful or not, and each mailed questionnaire costs $0.75 whether returned or not. The research director estimates that 21% of the telephone calls will result in usable interviews while only 7% of the questionnaires will be returned in usable condition. If the director's budget is $15,750 and she needs 1470 usable responses, how many telephone calls should she attempt and how many questionnaires should she mail out?

Explorations and Excursions

The following problems extend and augment the material presented in the text.

More about 2 × 3 Row Reduced Matrices The table on page 184 lists exactly five possible final matrices for the solution of a system of equations by reducing the corresponding augmented matrix. The following problems provide examples of each possibility.

Sketch each system of equations and then solve it by row-reducing the corresponding augmented matrix. Identify each system as "independent and consistent," "independent and inconsistent," or "dependent."

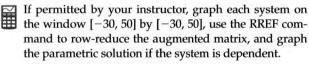 If permitted by your instructor, graph each system on the window [−30, 50] by [−30, 50], use the RREF command to row-reduce the augmented matrix, and graph the parametric solution if the system is dependent.

81. $\begin{cases} x - 2y = -14 \\ 2x + 3y = 84 \end{cases}$

82. $\begin{cases} x + 2y = 56 \\ 2x - 3y = 42 \end{cases}$

83. $\begin{cases} 3x - 6y = 18 \\ -2x + 4y = 16 \end{cases}$

84. $\begin{cases} 4x - 6y = 36 \\ -2x + 3y = 30 \end{cases}$

85. $\begin{cases} 2y = 8 \\ 3y = 18 \end{cases}$

86. $\begin{cases} 5y = 20 \\ 3y = 30 \end{cases}$

87. $\begin{cases} 4x + 14y = 56 \\ 6x + 21y = 84 \end{cases}$

88. $\begin{cases} 6x + 14y = 42 \\ 9x + 21y = 63 \end{cases}$

89. $\begin{cases} 5y = 30 \\ 3y = 18 \end{cases}$

90. $\begin{cases} 2y = 18 \\ 3y = 27 \end{cases}$

Row Operations are Reversible Carry out the row operations on the augmented matrix $\begin{pmatrix} 3 & -10 & -65 \\ -4 & 13 & 84 \end{pmatrix}$ and then find row operations to perform on your new matrix that will return it to $\begin{pmatrix} 3 & -10 & -65 \\ -4 & 13 & 84 \end{pmatrix}$.

91. $R_1' = R_2$ and $R_2' = R_1$

92. $R_2' = R_1$ and $R_1' = R_2$

93. $R_1' = 5R_1$

94. $R_2' = 3R_2$

95. $R_2' = \frac{1}{4}R_2$

96. $R_1' = \frac{1}{3}R_1$

97. $R_1' = R_1 + R_2$

98. $R_2' = R_2 + R_1$

99. $R_1' = R_1 - R_2$

100. $R_2' = R_2 - R_1$

Systems of Linear Equations and the Gauss–Jordan Method

APPLICATION PREVIEW

Ancient Chinese Mathematics*

The earliest known instance of matrix notation used for the systematic solution of linear equations appears in the eighth section of the ancient Chinese mathematics treatise *Arithmetic in Nine Sections*, written about 200 B.C. The method is explained by solving example problems, the first of which may be rephrased as follows:

> A farmer grows three kinds of corn, and each kind has been harvested and gathered into bundles. Three bundles of the first kind, two of the second, and one of the third make 39 bushels. Two of the first, three of the second, and one of the third make 34 bushels. And one of the first, two of the second, and three of the third make 26 bushels. How many bushels of corn are contained in one bundle of each kind of corn?

The rule for finding the solution is to write the 3, 2, 1 bundles of the three kinds and the 39 bushels as a column on the right, and then write the other conditions in the middle and on the left. The problem leads to the following table, whose columns correspond to our rows:

1	2	3	First kind
2	3	2	Second kind
3	1	1	Third kind
26	34	39	Total bushels

* Based on material in Yoshio Mikami, *The Development of Mathematics in China and Japan*, 2nd ed. (New York: Chelsea Publishing Company, 1974).

The calculation begins by multiplying the middle column by the top number of the righthand column.

1	6	3	First kind
2	9	2	Second kind
3	3	1	Third kind
26	102	39	Total bushels

The righthand column of numbers is then taken away as many times as possible from the middle column. In this case, it can be subtracted two times to leave

1	0	3	First kind
2	5	2	Second kind
3	1	1	Third kind
26	24	39	Total bushels

We then continue the process by multiplying columns by numbers and subtracting other columns as many times as possible until each column describes only one kind of corn and gives a number of bushels at the bottom.

0	0	540	First kind
0	180	0	Second kind
36	0	0	Third kind
99	765	4995	Total bushels

The solution is finished by dividing each lower number by the number of the kind to find the number of bushels: $\frac{4995}{540} = 9\frac{1}{4}$ for the first kind, $\frac{765}{180} = 4\frac{1}{4}$ for the second, and $\frac{99}{36} = 2\frac{3}{4}$ for the third.

Although this method was used in China over 2000 years ago, it is interesting how similar the problem statement and the column manipulations are to the augmented matrices and row operations used in the last section.

Introduction

In this section, we use augmented matrices to solve systems of many linear equations in many variables. We shall simply enlarge the size of the augmented matrix to allow for more equations (rows) and more variables (columns) and then apply row operations to find an equivalent matrix that will display the solution.

Names for Many Variables

To deal with many variables, we shall now distinguish them by subscripts instead of different letters. Our first variable will be named x_1 ("x sub one" or "the first x"), the second x_2 ("x sub two" or "the second x"), and so on for as many as we need (possibly x_{10}, x_{20}, or even x_{100}). With this new notation, we can rewrite the problem from the Application Preview as

$$\begin{cases} 3x_1 + 2x_2 + x_3 = 39 \\ 2x_1 + 3x_2 + x_3 = 34 \\ x_1 + 2x_2 + 3x_3 = 26 \end{cases}$$

We form the augmented matrix exactly as before, so for this system of equations we have

$$\begin{array}{ccc} x_1 & x_2 & x_3 \end{array}$$
$$\begin{pmatrix} 3 & 2 & 1 & 39 \\ 2 & 3 & 1 & 34 \\ 1 & 2 & 3 & 26 \end{pmatrix}$$

← Variables corresponding to columns (last column represents constant terms)

← Each row represents an equation

Reduced Row-Echelon Form

We shall continue to use row operations to "solve" the augmented matrix by finding an equivalent matrix that displays the solution. With our list of 2×3 matrices on page 184 as a guide, we make the following definition for matrices of any dimension. In this definition, a *zero row* is a row containing only zeros, and a *nonzero row* is a row with at least one nonzero element.

Reduced Row-Echelon Form

A matrix is in *reduced row-echelon form* if it satisfies the following four conditions.

1. All zero rows are below every nonzero row.
2. The first nonzero element of every nonzero row is a one (we will call these special ones "leftmost ones").

3. The column of each leftmost one [from (2)] contains only zeros in the other positions.

4. Each leftmost one [from (2)] appears to the right of the leftmost ones in the rows above it.

The following matrices are in reduced row-echelon form:

$$\begin{pmatrix} 0 & 1 & 2 & 0 & 1 \\ 0 & 0 & 0 & 1 & 4 \end{pmatrix}, \begin{pmatrix} 1 & 2 & 3 & 4 & 0 \\ 0 & 0 & 0 & 0 & 1 \\ 0 & 0 & 0 & 0 & 0 \end{pmatrix}, \text{ and } \begin{pmatrix} 1 & 0 & 0 & 0 & 1 \\ 0 & 1 & 0 & 0 & 2 \\ 0 & 0 & 1 & 0 & 3 \\ 0 & 0 & 0 & 1 & 4 \end{pmatrix}$$

Leftmost ones   Zero row

Notice that the ones in the upper right corners of the first and last matrices above are not *leftmost* ones and do not need to have zeros below them. The following matrix is *not* in reduced row-echelon form because although conditions (1), (2), and (3) are satisfied, condition (4) fails for the third row.

$$\begin{pmatrix} 1 & 0 & 0 & 2 & 3 \\ 0 & 0 & 1 & 4 & 5 \\ 0 & 1 & 0 & 6 & 7 \\ 0 & 0 & 0 & 0 & 0 \end{pmatrix}$$

The 1 in the third row is not to the right of the 1 in the second row

Can you think of a row operation that will correct this defect and result in a matrix in reduced row-echelon form?

Solutions from Augmented Matrices

Just as in the previous section, if a reduced row-echelon form matrix has a zero row, then the system of equations from which it comes is *dependent*; otherwise, the system is *independent*. If a reduced row-echelon form matrix has a row of zeros ending in a 1 (such as "0 0 0 1"), then the system of equations is *inconsistent*; otherwise, the system is *consistent*. The following are examples of each possibility.

	Independent	Dependent
Consistent	$\begin{pmatrix} 1 & 0 & 0 & 2 \\ 0 & 1 & 0 & 3 \\ 0 & 0 & 1 & 4 \end{pmatrix}$	$\begin{pmatrix} 1 & 0 & 0 & 2 \\ 0 & 1 & 0 & 3 \\ 0 & 0 & 0 & 0 \end{pmatrix}$
Inconsistent	$\begin{pmatrix} 1 & 0 & 0 & 0 \\ 0 & 1 & 0 & 0 \\ 0 & 0 & 0 & 1 \end{pmatrix}$	$\begin{pmatrix} 1 & 0 & 0 & 0 \\ 0 & 0 & 0 & 1 \\ 0 & 0 & 0 & 0 \end{pmatrix}$

In a consistent system, a variable is *determined* or *dependent* if its column in the reduced row-echelon form matrix has a leftmost one, and otherwise it is *free* or *independent*. The free variables may take any values, and then the values of the determined variables follow from the equations represented by their rows. The free variables give a *parameterization* of the solution by determining the values of the other variables. For instance, for the matrix

$$\begin{pmatrix} 1 & 2 & 3 \\ 0 & 0 & 0 \end{pmatrix} \qquad \begin{aligned} &1x_1 + 2x_2 = 3 \\ &\text{so } x_1 = 3 - 2x_2 \end{aligned}$$

the x_1 is *determined* (its column has a leftmost one) while the x_2 is *free*. Since x_2 is "free" to take any value t and then $x_1 = 3 - 2t$ is "determined," the solution is $x_1 = 3 - 2t$, $x_2 = t$, where t may take any value.

EXAMPLE 1 Finding a Parameterized Solution

Find the solution of a system of equations with augmented matrix equivalent to the reduced row-echelon form matrix

$$\begin{pmatrix} 1 & 0 & 2 & 3 & 0 & 4 \\ 0 & 1 & 5 & 6 & 0 & 7 \\ 0 & 0 & 0 & 0 & 1 & 8 \end{pmatrix}$$

Solution

There were five variables in the system of equations because the augmented matrix has six columns. The variables x_1, x_2, and x_5 are determined, while x_3 and x_4 are free. Letting the free variables take the values $x_3 = t_1$ and $x_4 = t_2$, we can use the rows to find the determined variables.

$$\begin{aligned} 1x_1 + 0x_2 + 2x_3 + 3x_4 + 0x_5 = 4 & \qquad \text{so} \qquad & x_1 = 4 - 2t_1 - 3t_2 \\ 0x_1 + 1x_2 + 5x_3 + 6x_4 + 0x_5 = 7 & \qquad \text{so} \qquad & x_2 = 7 - 5t_1 - 6t_2 \\ 0x_1 + 0x_2 + 0x_3 + 0x_4 + 1x_5 = 8 & \qquad \text{so} \qquad & x_5 = 8 \end{aligned}$$

The solution is: For example, if $t_1 = 10$ and $t_2 = -10$:

$$x_1 = 4 - 2t_1 - 3t_2 \qquad\qquad x_1 = 14$$
$$x_2 = 7 - 5t_1 - 6t_2 \qquad\qquad x_2 = 17$$
$$x_3 = t_1 \qquad\qquad x_3 = 10$$
$$x_4 = t_2 \qquad\qquad x_4 = -10$$
$$x_5 = 8 \qquad\qquad x_5 = 8$$

∎

PRACTICE PROBLEM 1

Find the solution of a system of equations with augmented matrix equivalent to the reduced row-echelon form matrix

$$\begin{pmatrix} 1 & 2 & 0 & 0 & 0 \\ 0 & 0 & 1 & 0 & 3 \\ 0 & 0 & 0 & 1 & 4 \\ 0 & 0 & 0 & 0 & 0 \end{pmatrix}$$

Solution at the back of the book

Row-Reducing a Large Matrix: The Gauss–Jordan Method

Row-reducing a large matrix is essentially the same process as reducing a 2 × 3 matrix. The *Gauss–Jordan method* is a systematic approach to reducing a matrix with many rows and columns. It is named after Carl F. Gauss (1777–1855), the "prince of mathematicians," and Wilhelm Jordan (1842–1899), the German geodesist. This method uses row operations to transform the matrix to reduced row-echelon form one column at a time, starting from the left and working to the right and down.

Gauss–Jordan Method

For instance, you might get:

$$\begin{pmatrix} * & * & * \\ * & * & * \\ * & * & * \end{pmatrix} \rightarrow \begin{pmatrix} 1 & * & * \\ * & * & * \\ * & * & * \end{pmatrix}$$

$$\rightarrow \begin{pmatrix} 1 & * & * \\ 0 & * & * \\ 0 & * & * \end{pmatrix} \rightarrow \begin{pmatrix} 1 & * & * \\ 0 & 1 & * \\ 0 & * & * \end{pmatrix}$$

$$\rightarrow \begin{pmatrix} 1 & 0 & * \\ 0 & 1 & * \\ 0 & 0 & * \end{pmatrix} \rightarrow \cdots$$

To row-reduce a matrix:

1. If the first row begins with one or more zeros and there is a row below it that begins with *fewer* zeros, interchange those two rows. Repeat this step until *no* row below the first row has fewer leading zeros.

2. Find the leftmost non-zero entry in the first row and divide the row through by that number, obtaining a *leftmost one* in that row. If there is no non-zero entry (that is, the entire row is zero), then stop—the matrix is in reduced row-echelon form.

3. Add (or subtract) multiples of the first row to (or from) the other rows to make the entries above and below that leftmost one into zeros.

4. Repeat steps 1 to 3 but replace "first row" by "second row," then by "third row," and so on. Stop when you have performed these steps on the last row.

EXAMPLE 2 **The Gauss–Jordan Method**

Row reduce the matrix $\begin{pmatrix} 2 & 6 & 10 & 8 \\ 3 & 9 & 15 & 12 \\ 2 & 5 & 8 & 7 \end{pmatrix}$ by the Gauss–Jordan method.

Solution

Because the first entry in the first row is non-zero, we do not need to switch any rows, and step 1 takes us immediately to step 2. To make the first entry in the first row into 1 (step 2), we divide by 2.

$$\begin{pmatrix} 1 & 3 & 5 & 4 \\ 3 & 9 & 15 & 12 \\ 2 & 5 & 8 & 7 \end{pmatrix} \begin{array}{l} R_1' = \frac{1}{2}R_1 \end{array}$$

Get a 1 in the first column of the first row

To make the rest of the first column zero (step 3), we subtract appropriate multiples of the first row.

$$\begin{pmatrix} 1 & 3 & 5 & 4 \\ 0 & 0 & 0 & 0 \\ 0 & -1 & -2 & -1 \end{pmatrix} \begin{array}{l} \\ R_2' = R_2 - 3R_1 \\ R_3' = R_3 - 2R_1 \end{array}$$

"Zero out" the rest of the first column

The first column is now finished, and we move on to the second row (step 4). Because the second row is a zero row and the third row begins with fewer zeros, we switch the second and third rows (step 1).

$$\begin{pmatrix} 1 & 3 & 5 & 4 \\ 0 & -1 & -2 & -1 \\ 0 & 0 & 0 & 0 \end{pmatrix} \begin{array}{l} \\ R_2' = R_3 \\ R_3' = R_2 \end{array}$$

Move row with many leading zeros to the bottom

To make the leftmost non-zero entry of the (new) second row into 1 (step 2), we multiply (or divide) by −1.

$$\begin{pmatrix} 1 & 3 & 5 & 4 \\ 0 & 1 & 2 & 1 \\ 0 & 0 & 0 & 0 \end{pmatrix} \begin{array}{l} \\ R_2' = -R_2 \end{array}$$

Get a 1 in the second column of the second row

To make the rest of the second column zero (step 3), we subtract an appropriate multiple of the second row from the first row and do nothing to the third row.

$$\begin{pmatrix} 1 & 0 & -1 & 1 \\ 0 & 1 & 2 & 1 \\ 0 & 0 & 0 & 0 \end{pmatrix} \begin{array}{l} R_1' = R_1 - 3R_2 \end{array}$$

"Zero out" the rest of the second column

The second column is now finished and we move on to the third row (step 4). Because the third row is a zero row, there is nothing more to be done (step 2), and the matrix is in reduced row-echelon form.

∎

When row-reducing matrices "by hand," it is often possible to make the problem easier to solve by first removing common factors of a row by dividing, and by getting any "easy zeros" before beginning the Gauss–Jordan method.

EXAMPLE 3   **Row-Reducing a Large Matrix**

Row-reduce the matrix $\begin{pmatrix} 5 & 5 & 0 & 5 & 50 \\ 2 & 3 & 1 & 0 & 17 \\ 2 & 2 & 1 & -1 & 9 \\ 2 & 3 & 1 & 1 & 22 \end{pmatrix}$.

Solution

Before setting to work, always look for simple row operations that may make the problem easier. Since the elements of the first row have 5 as a common factor, we begin by simplifying that row.

$$\begin{pmatrix} 1 & 1 & 0 & 1 & 10 \\ 2 & 3 & 1 & 0 & 17 \\ 2 & 2 & 1 & -1 & 9 \\ 2 & 3 & 1 & 1 & 22 \end{pmatrix} \begin{matrix} R'_1 = \frac{1}{5}R_1 \\ \\ \\ \\ \end{matrix}$$ Remove common factor

Next, notice that the second and fourth rows have the same first three elements, so we can get "many zeros" by subtracting.

$$\begin{pmatrix} 1 & 1 & 0 & 1 & 10 \\ 2 & 3 & 1 & 0 & 17 \\ 2 & 2 & 1 & -1 & 9 \\ 0 & 0 & 0 & 1 & 5 \end{pmatrix} \begin{matrix} \\ \\ \\ R'_4 = R_4 - R_2 \end{matrix}$$ Get "easy" zeros

In the same way, the second and third rows can give us zeros and a one where we would wish in the second row.

$$\begin{pmatrix} 1 & 1 & 0 & 1 & 10 \\ 0 & 1 & 0 & 1 & 8 \\ 2 & 2 & 1 & -1 & 9 \\ 0 & 0 & 0 & 1 & 5 \end{pmatrix} \begin{matrix} \\ R'_2 = R_2 - R_3 \\ \\ \\ \end{matrix}$$ Get easy "0 1 0" pattern

We can now easily finish the reduction by the Gauss–Jordan method.

$$\begin{pmatrix} 1 & 1 & 0 & 1 & 10 \\ 0 & 1 & 0 & 1 & 8 \\ 0 & 0 & 1 & -3 & -11 \\ 0 & 0 & 0 & 1 & 5 \end{pmatrix} R_3' = R_3 - 2R_1$$

"Zero out" the rest of the first column

$$\begin{pmatrix} 1 & 0 & 0 & 0 & 2 \\ 0 & 1 & 0 & 1 & 8 \\ 0 & 0 & 1 & -3 & -11 \\ 0 & 0 & 0 & 1 & 5 \end{pmatrix} R_1' = R_1 - R_2$$

"Zero out" the rest of the second column (the third column needs no work)

$$\begin{pmatrix} 1 & 0 & 0 & 0 & 2 \\ 0 & 1 & 0 & 0 & 3 \\ 0 & 0 & 1 & 0 & 4 \\ 0 & 0 & 0 & 1 & 5 \end{pmatrix} \begin{matrix} \\ R_2' = R_2 - R_4 \\ R_3' = R_3 + 3R_4 \\ \end{matrix}$$

"Zero out" the rest of the fourth column.
Reduced row-echelon form

■

PRACTICE PROBLEM 2

Solve the system of equations $\begin{cases} 5x_1 + 5x_2 + 5x_4 = 50 \\ 2x_1 + 3x_2 + 1x_3 = 17 \\ 2x_1 + 2x_2 + x_3 - x_4 = 9 \\ 2x_1 + 3x_2 + x_3 + x_4 = 22 \end{cases}$. Be sure to check your answer in the equations. [*Hint:* No work necessary—just look at the reduced row-echelon form matrix from Example 3.]

Solution at the back of the book

Graphing Calculator Exploration

The program* ROWOPS carries out the arithmetic for the type of row operation you select from a menu. To carry out on your calculator the row operations used in Example 3 (above), proceed as follows:

a.   Enter the augmented matrix as matrix [A]. Because this matrix is too large to fit on the calculator screen, it will scroll from left to right and back again as you enter your numbers.

```
MATRIX[A] 4 x5
 5    0    5   _
 3    1    0   _
 2    1   -1   _
 3    1    1   _

4,4=1
```

* See the Preface for information on how to obtain this and other programs.

b.   Run the program ROWOPS. It will display the current values in matrix [A] and you can use the arrows to scroll the screen to see the rest of it. Press ENTER to select the type of row operation you would like to perform.

```
┌────────────────────┐
│ ROW OPERATIONS     │
│1: SWITCH           │
│2:  MULTIPLY        │
│3:  DIVIDE          │
│4: ADD TOGETHER     │
│5: SUBTRACT FROM    │
│6:  ADD MULTIPLE    │
│7: QUIT             │
└────────────────────┘
```

c.   Choose the type of row operation you want by using the arrows to move to its number and pressing ENTER, or just press the number. Enter the specific details for the particular operation you want and the program will carry out your request.

```
┌─────────────────────┐
│        DIVIDE    ┊  │
│DIVIDE ROW 1         │
│  BY 5               │
│  [[1  1  0  1   10] │
│   [2  3  1  0   17] │
│   [2  2  1 -1  9 ]  │
│   [2  3  1  1   22]]│
└─────────────────────┘
```

d.   Press ENTER to select another row operation or to choose 7 and QUIT the program.

The program ROWOPS allows you to multiply rows by fractions and displays fractions in the usual 3/5 notation. The reduction of this matrix took eight row operations to complete in Example 3. Can you reduce it in just seven?

EXAMPLE 4   **Managing Production**

A hand-thrown pottery shop manufactures plates, cups, and vases. Each plate requires 4 ounces of clay, 6 minutes of shaping, and 5 minutes of painting; each cup requires 4 ounces of clay, 5 minutes of shaping, and 3 minutes of painting; and each vase requires 3 ounces of clay, 4 minutes of shaping, and 4 minutes of painting. This week the shop has 165 pounds of clay, 59 hours of skilled labor for shaping, and 46 hours of skilled labor for painting. If the shop manager wishes to use all these resources fully, how many of each product should the shop produce?

Solution

Let x_1, x_2, and x_3 be the numbers of plates, cups, and vases produced. The number of ounces of clay required is then $4x_1 + 4x_2 + 3x_3$ and this must match the 165 pounds available:

$$4x_1 + 4x_2 + 3x_3 = 2640$$

Use ounces on both sides of the equation

Similarly, for the time in minutes required for shaping and painting,

$$6x_1 + 5x_2 + 4x_3 = 3540$$
$$5x_1 + 3x_2 + 4x_3 = 2760$$

Use minutes on both sides of the equation

Therefore, the augmented matrix is

$$\begin{pmatrix} 4 & 4 & 3 & 2640 \\ 6 & 5 & 4 & 3540 \\ 5 & 3 & 4 & 2760 \end{pmatrix}$$

We now solve this problem by row-reducing the augmented matrix using the Gauss–Jordan method.

$$\begin{pmatrix} 1 & 1 & \frac{3}{4} & 660 \\ 6 & 5 & 4 & 3540 \\ 5 & 3 & 4 & 2760 \end{pmatrix} R_1' = \frac{1}{4}R_1$$

Get a 1 in the first column of the first row

$$\begin{pmatrix} 1 & 1 & \frac{3}{4} & 660 \\ 0 & -1 & -\frac{1}{2} & -420 \\ 0 & -2 & \frac{1}{4} & -540 \end{pmatrix} \begin{matrix} \\ R_2' = R_2 - 6R_1 \\ R_3' = R_3 - 5R_1 \end{matrix}$$

"Zero out" the rest of the first column

$$\begin{pmatrix} 1 & 1 & \frac{3}{4} & 660 \\ 0 & 1 & \frac{1}{2} & 420 \\ 0 & -2 & \frac{1}{4} & -540 \end{pmatrix} R_2' = -R_2$$

Get a 1 in the second column of the second row

$$\begin{pmatrix} 1 & 0 & \frac{1}{4} & 240 \\ 0 & 1 & \frac{1}{2} & 420 \\ 0 & 0 & \frac{5}{4} & 300 \end{pmatrix} \begin{matrix} R_1' = R_1 - R_2 \\ \\ R_3' = R_3 + 2R_2 \end{matrix}$$

"Zero out" the rest of the second column

$$\begin{pmatrix} 1 & 0 & \frac{1}{4} & 240 \\ 0 & 1 & \frac{1}{2} & 420 \\ 0 & 0 & 1 & 240 \end{pmatrix} R_3' = \frac{4}{5}R_3$$

Get a 1 in the third column of the third row

$$\begin{pmatrix} 1 & 0 & 0 & 180 \\ 0 & 1 & 0 & 300 \\ 0 & 0 & 1 & 240 \end{pmatrix} \begin{matrix} R_1' = R_1 - \frac{1}{4}R_3 \\ R_2' = R_2 - \frac{1}{2}R_3 \\ \end{matrix}$$

"Zero out" the rest of the third column

The system of equations is independent and consistent with solution $x_1 = 180$, $x_2 = 300$, $x_3 = 240$. You should check that these values satisfy the original equations and that the clay and skilled labor resources are fully used. In terms of the original question, the shop should produce 180 plates, 300 cups, and 240 vases this week.

EXAMPLE 5   **Modeling a Computer Network**

The office manager for the accounting division of a large company is writing a recommendation report to her supervisor about the demands placed on the computers serving her division. Besides meeting the needs of the accounting division, her four "file servers" are expected to accept and relay messages to and from other areas. The diagram shows the four file servers (I, II, III, IV), and the arrows and numbers indicate the data packets per minute passing from the senders to the receivers. Computers A through H are outside the accounting division. Connections passing an unknown number of data packets per minute are marked with variables (for instance, x_1 is the number of data packets per minute sent from server I to server II, while x_5 is the number leaving server IV to outside computer G). In order for the network to function, the number of data packets per minute arriving at each computer must match the number leaving it. How many data packets per minute must leave IV for G? Is it possible for each connection within the accounting division to carry no more than 1500 data packets per minute?

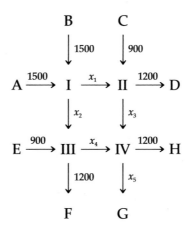

Solution

In terms of the notations from the diagram, the requirements that the data packets per minute arriving at each file server match the number leaving are represented by the equations

"In" equals "out"

$$1500 + 1500 = x_1 + x_2 \qquad \text{Server I}$$

$$x_1 + 900 = x_3 + 1200 \qquad \text{Server II}$$

$$900 + x_2 = x_4 + 1200 \qquad \text{Server III}$$

$$x_3 + x_4 = 1200 + x_5 \qquad \text{Server IV}$$

Rewriting these to put the variables on the left, we have the system

$$\begin{cases} x_1 + x_2 = 3000 \\ x_1 - x_3 = 300 \\ x_2 - x_4 = 300 \\ x_3 + x_4 - x_5 = 1200 \end{cases}$$

Because there are fewer equations than variables, the solution will not consist of unique values for all the variables. If the system is consistent, we will find a parameterized solution, whereas if it is inconsistent, there will be no solution. The augmented matrix representing these equations is shown on the left below, with its reduced row-echelon form shown on the right. (You may wish to carry out the reduction—it can be done easily because of the many zeros and ones—or use your calculator's RREF command to check the reduction.)

$$\begin{pmatrix} 1 & 1 & 0 & 0 & 0 & 3000 \\ 1 & 0 & -1 & 0 & 0 & 300 \\ 0 & 1 & 0 & -1 & 0 & 300 \\ 0 & 0 & 1 & 1 & -1 & 1200 \end{pmatrix} \xrightarrow{\text{Leads to}} \begin{pmatrix} 1 & 0 & 0 & 1 & 0 & 2700 \\ 0 & 1 & 0 & -1 & 0 & 300 \\ 0 & 0 & 1 & 1 & 0 & 2400 \\ 0 & 0 & 0 & 0 & 1 & 1200 \end{pmatrix}$$

The system is consistent, and x_4 is a free variable (its column does *not* have a leftmost one). We may parameterize the solution as follows:

$$x_1 = 2700 - t$$

$$x_2 = 300 + t$$

$$x_3 = 2400 - t$$

$$x_4 = t \qquad \text{The free variable}$$

$$x_5 = 1200$$

To prevent a "back flow" along the connections, each variable must stay nonnegative, which means that $t \geq 0$ (since $x_4 = t$) and $t \leq 2400$ (since $x_3 = 2400 - t$). For example, the choice $t = 1200$ gives the solution

$$x_1 = 1500, x_2 = 1500, x_3 = 1200, x_4 = 1200, x_5 = 1200$$

In terms of the original questions, 1200 data packets per minute must leave computer IV for G, and the example with $t = 1200$ shows that it *is* possible to arrange the transmissions so that each connection within the accounting division carries no more than 1500 data packets per minute.

Section Summary

The solution of any system of linear equations can be found by representing it as an augmented matrix, applying row operations (page 184) using the Gauss–Jordan method (page 193) to find the equivalent reduced row-echelon form matrix (pages 190–191), and interpreting this final matrix as a statement about the original system of equations (pages 191–192). Each step in this method is a direct extension of the method for solving systems of two linear equations in two variables presented in the previous section.

For three equations in three unknowns with a unique solution, this would look like:

$$\begin{pmatrix} * & * & * & * \\ * & * & * & * \\ * & * & * & * \end{pmatrix} \xrightarrow{\text{Row–reduces to}} \begin{pmatrix} 1 & 0 & 0 & * \\ 0 & 1 & 0 & * \\ 0 & 0 & 1 & * \end{pmatrix}$$

$\underset{\text{solution}}{\uparrow}$

The numbers in the last column are the solution.

EXERCISES 3.3

Find the augmented matrix representing the system of equations.

1. $\begin{cases} x_1 + x_2 + x_3 = 4 \\ x_1 + 2x_2 + x_3 = 3 \\ x_1 + 2x_2 + 2x_3 = 5 \end{cases}$

2. $\begin{cases} x_1 + 5x_2 + 4x_3 = 6 \\ x_1 + x_2 + x_3 = 4 \\ 2x_1 + 3x_2 + 3x_3 = 9 \end{cases}$

3. $\begin{cases} 2x_1 - x_2 + 2x_3 = 11 \\ -x_1 + x_2 - 3x_3 = -12 \\ 2x_1 - 2x_2 + 7x_3 = 27 \end{cases}$

4. $\begin{cases} 4x_1 + 3x_2 - x_3 = 2 \\ 3x_1 + 3x_2 + 2x_3 = 9 \\ 2x_1 + x_2 - 3x_3 = -6 \end{cases}$

5. $\begin{cases} 2x_1 + x_2 + 5x_3 + 4x_4 + 5x_5 = 2 \\ x_1 + x_2 + 3x_3 + 3x_4 + 3x_5 = -1 \end{cases}$

6. $\begin{cases} 5x_1 + 2x_2 - 4x_3 + x_4 + 5x_5 = 7 \\ 3x_1 + x_2 - 3x_3 + x_4 + 3x_5 = 5 \end{cases}$

7. $\begin{cases} 6x_1 + 3x_2 + 5x_3 = 8 \\ x_1 + 2x_2 + 2x_3 = 1 \\ 4x_1 + 3x_2 + 4x_3 = 5 \\ 5x_1 + x_2 + 3x_3 = 7 \end{cases}$

8. $\begin{cases} 5x_1 + 9x_2 + 9x_3 = 11 \\ 4x_1 + 7x_2 + 6x_3 = 9 \\ 3x_1 + 5x_2 + 3x_3 = 8 \\ 4x_1 + 7x_2 + 5x_3 = 10 \end{cases}$

9. $\begin{cases} 3x_1 + 4x_2 + 2x_3 + 4x_4 = 12 \\ x_1 + 2x_2 + x_3 + x_4 = 4 \\ 4x_1 + 5x_2 + 2x_3 + 5x_4 = 14 \\ 6x_1 + 6x_2 + x_3 + 6x_4 = 15 \end{cases}$

10. $\begin{cases} 5x_1 + 4x_2 + 7x_3 + 6x_4 = 18 \\ 2x_1 + 2x_2 + 3x_3 + 3x_4 = 9 \\ 4x_1 + 3x_2 + 5x_3 + 5x_4 = 16 \\ 3x_1 + 2x_2 + 3x_3 + 3x_4 = 11 \end{cases}$

Find the system of equations represented by the augmented matrix.

11. $\begin{pmatrix} 4 & 3 & 2 & 11 \\ 3 & 3 & 1 & 6 \\ 1 & -2 & 3 & 13 \end{pmatrix}$

12. $\begin{pmatrix} 3 & -2 & 5 & 23 \\ -1 & 1 & -3 & -12 \\ 2 & -2 & 7 & 27 \end{pmatrix}$

13. $\begin{pmatrix} 2 & 1 & 1 & 7 \\ 2 & 2 & 1 & 6 \\ 3 & 3 & 2 & 10 \end{pmatrix}$ **14.** $\begin{pmatrix} 5 & 1 & 3 & 20 \\ 1 & 1 & 2 & 6 \\ 4 & 1 & 3 & 17 \end{pmatrix}$

15. $\begin{pmatrix} 8 & 3 & -2 & 19 & 15 \\ 3 & 1 & -1 & 7 & 6 \end{pmatrix}$

16. $\begin{pmatrix} 6 & -2 & -4 & -2 & 36 \\ 2 & -1 & -10 & 5 & 6 \end{pmatrix}$

17. $\begin{pmatrix} 2 & 3 & 2 & 5 \\ 3 & 5 & 3 & 8 \\ 1 & 2 & 2 & 2 \\ 4 & 7 & 5 & 9 \end{pmatrix}$ **18.** $\begin{pmatrix} 3 & 1 & 3 & 5 \\ 2 & 2 & 1 & 1 \\ 3 & 2 & 2 & 3 \\ 5 & 3 & 4 & 6 \end{pmatrix}$

19. $\begin{pmatrix} 3 & 3 & 5 & 4 & 11 \\ 2 & 2 & 3 & 3 & 9 \\ 2 & 1 & 2 & 2 & 7 \\ 3 & 2 & 3 & 3 & 11 \end{pmatrix}$ **20.** $\begin{pmatrix} 2 & 1 & -1 & 1 & 1 \\ 1 & 1 & 0 & 1 & 2 \\ -2 & 1 & 4 & 1 & 8 \\ 1 & 2 & 1 & 1 & 4 \end{pmatrix}$

Interpret each reduced row-echelon form matrix as the solution of a system of equations. Identify each system as "independent" or "dependent" and as "consistent" or "inconsistent."

21. $\begin{pmatrix} 1 & 0 & 0 & 4 \\ 0 & 1 & 0 & 5 \\ 0 & 0 & 1 & -4 \end{pmatrix}$ **22.** $\begin{pmatrix} 1 & 0 & 0 & 0 \\ 0 & 1 & 0 & 0 \\ 0 & 0 & 1 & 1 \end{pmatrix}$

23. $\begin{pmatrix} 1 & 0 & 0 & 0 & 2 \\ 0 & 1 & 0 & 0 & -1 \\ 0 & 0 & 1 & 0 & 3 \\ 0 & 0 & 0 & 1 & 1 \end{pmatrix}$ **24.** $\begin{pmatrix} 1 & 0 & 7 \\ 0 & 1 & 3 \\ 0 & 0 & 0 \end{pmatrix}$

25. $\begin{pmatrix} 1 & 0 & 1 & 0 \\ 0 & 1 & 0 & 0 \\ 0 & 0 & 0 & 1 \end{pmatrix}$ **26.** $\begin{pmatrix} 1 & 1 & 0 \\ 0 & 0 & 1 \end{pmatrix}$

27. $\begin{pmatrix} 1 & 0 & -1 & -5 \\ 0 & 1 & 1 & 5 \\ 0 & 0 & 0 & 0 \end{pmatrix}$ **28.** $\begin{pmatrix} 1 & 1 & 0 & 0 & 2 \\ 0 & 0 & 1 & 0 & -1 \\ 0 & 0 & 0 & 1 & 3 \\ 0 & 0 & 0 & 0 & 0 \end{pmatrix}$

29. $\begin{pmatrix} 1 & -1 & 0 & 1 & 8 \\ 0 & 0 & 1 & -1 & 4 \\ 0 & 0 & 0 & 0 & 0 \end{pmatrix}$ **30.** $\begin{pmatrix} 1 & 0 & 0 & 6 \\ 0 & 0 & 1 & 3 \\ 0 & 0 & 0 & 0 \end{pmatrix}$

Use an appropriate row operation or sequence of row operations to find the equivalent reduced row-echelon form matrix.

If your instructor permits, you may carry out the calculations on a graphing calculator using the ROWOPS program.

31. $\begin{pmatrix} 0 & 1 & 0 & 2 \\ 1 & 0 & 0 & 1 \\ 0 & 0 & 1 & 3 \end{pmatrix}$ **32.** $\begin{pmatrix} 1 & 0 & 0 & 1 \\ 0 & 0 & 1 & 3 \\ 0 & 1 & 0 & 2 \end{pmatrix}$

33. $\begin{pmatrix} 1 & 0 & 1 & 4 \\ 0 & 1 & 0 & 2 \\ 0 & 0 & 1 & 3 \end{pmatrix}$ **34.** $\begin{pmatrix} 1 & 0 & 0 & 1 \\ 1 & 1 & 0 & 3 \\ 0 & 0 & 1 & 3 \end{pmatrix}$

35. $\begin{pmatrix} 2 & 4 & 0 & 6 \\ 0 & 0 & 1 & -3 \\ 0 & 0 & 0 & 0 \end{pmatrix}$ **36.** $\begin{pmatrix} 1 & 0 & 2 & 5 \\ 0 & 3 & -6 & 3 \\ 0 & 0 & 0 & 0 \end{pmatrix}$

37. $\begin{pmatrix} 2 & 0 & 4 & 0 & 6 \\ 0 & 1 & 0 & -1 & 1 \\ 0 & 0 & 1 & 1 & 1 \\ 0 & 0 & 1 & 1 & 2 \end{pmatrix}$ **38.** $\begin{pmatrix} 1 & 0 & -1 & 0 & 1 \\ 0 & 1 & 2 & 1 & 3 \\ 0 & 2 & 4 & 2 & 8 \\ 0 & 0 & 0 & 1 & -2 \end{pmatrix}$

39. $\begin{pmatrix} 1 & 1 & 0 & 0 & 4 \\ 0 & 1 & 0 & 0 & 2 \\ 0 & 0 & 0 & 1 & 1 \\ 0 & 0 & 1 & 0 & 3 \end{pmatrix}$ **40.** $\begin{pmatrix} 0 & 1 & 0 & 1 & 4 \\ 1 & 1 & 0 & 1 & 2 \\ 0 & 0 & 1 & 1 & 3 \\ 0 & 0 & 1 & 0 & 2 \end{pmatrix}$

Solve each system of equations by row-reducing the corresponding augmented matrix using the Gauss–Jordan method. Identify each system as "independent" or "dependent" and as "consistent" or "inconsistent."

If you have a graphing calculator with a RREF command, use it to check your row reduction.

41. $\begin{cases} x_1 + x_2 + x_3 = 2 \\ x_1 + 2x_2 + 2x_3 = 3 \\ x_1 + 3x_2 + 2x_3 = 1 \end{cases}$

42. $\begin{cases} 2x_1 + 3x_2 + x_3 = 4 \\ 3x_1 + 3x_2 + 2x_3 = 12 \\ x_1 + 2x_2 + x_3 = 4 \end{cases}$

43. $\begin{cases} 2x_1 + 2x_2 - x_3 = -5 \\ -2x_1 - x_2 + x_3 = 3 \\ 3x_1 + 4x_2 - x_3 = -8 \end{cases}$

44. $\begin{cases} 3x_1 - 4x_2 + 2x_3 = -15 \\ -x_1 + 2x_2 - x_3 = 6 \\ 4x_1 - 3x_2 + 2x_3 = -16 \end{cases}$

45. $\begin{cases} 2x_1 + x_2 - 2x_3 + x_4 = 2 \\ x_1 + x_2 + 2x_3 + x_4 = 5 \\ x_1 + x_2 + x_3 + x_4 = 4 \\ 2x_1 + 2x_2 + 3x_3 + x_4 = 8 \end{cases}$

46. $\begin{cases} x_1 - x_2 + x_3 + 2x_4 = 7 \\ x_1 - 2x_2 + x_3 + 2x_4 = 8 \\ 2x_1 + 2x_2 + x_3 + 3x_4 = 7 \\ x_1 - x_2 + x_3 + x_4 = 6 \end{cases}$

47. $\begin{cases} 2x_1 + 3x_2 + x_3 = 4 \\ 3x_1 + 5x_2 + 2x_3 = 12 \\ x_1 + 2x_2 + x_3 = 3 \end{cases}$

48. $\begin{cases} x_1 + 3x_2 + 2x_3 = 6 \\ x_1 + 2x_2 + 2x_3 = 3 \\ 2x_1 + 5x_2 + 4x_3 = 8 \end{cases}$

49. $\begin{cases} 4x_1 + 3x_2 + 2x_3 = 24 \\ x_1 + x_2 + 3x_3 = 7 \\ 5x_1 + 4x_2 + 5x_3 = 31 \end{cases}$

50. $\begin{cases} 5x_1 - 7x_2 + 3x_3 = -9 \\ -x_1 + x_2 - x_3 = 1 \\ 4x_1 - 5x_2 + 3x_3 = -6 \end{cases}$

51. $\begin{cases} x_1 + x_2 + 2x_3 + x_4 = 2 \\ 2x_1 + 2x_2 + 3x_3 + 3x_4 = 9 \\ 2x_1 + x_2 + 2x_3 + 2x_4 = 7 \\ x_1 + x_2 + x_3 + x_4 = 4 \end{cases}$

52. $\begin{cases} 3x_1 + 7x_2 + 4x_3 + 4x_4 = -7 \\ 7x_1 + 7x_2 + 4x_3 + 7x_4 = 3 \\ 4x_1 + 3x_2 + 2x_3 + 3x_4 = 2 \\ 3x_1 + 2x_2 + x_3 + 3x_4 = 4 \end{cases}$

53. $\begin{cases} 2x_1 + 2x_2 - 2x_3 + x_4 = -6 \\ -3x_1 - x_2 + x_3 - 2x_4 = 8 \\ 2x_1 + x_2 - x_3 + x_4 = -5 \\ -x_1 - x_2 + 2x_3 - x_4 = 6 \end{cases}$

54. $\begin{cases} 2x_1 - 5x_2 + 3x_3 + 6x_4 = -16 \\ 5x_1 - 11x_2 + 5x_3 + 14x_4 = -41 \\ -4x_1 + 9x_2 - 3x_3 - 13x_4 = 37 \\ -3x_1 + 7x_2 - 4x_3 - 8x_4 = 23 \end{cases}$

55. $\begin{cases} 4x_1 - x_2 + 3x_3 - x_4 + x_5 = -2 \\ 2x_1 + 2x_2 + x_3 + 4x_4 = 0 \\ 4x_1 - x_2 + 2x_3 - 2x_4 + x_5 = 1 \\ 2x_1 + x_2 + x_3 + 2x_4 = 0 \\ 3x_1 - x_2 + x_3 - 2x_4 + x_5 = 2 \end{cases}$

56. $\begin{cases} x_1 + x_3 + x_5 = 4 \\ -x_1 + 2x_2 - x_3 + x_4 - x_5 = -9 \\ 4x_1 + 2x_2 + 4x_3 + 2x_4 + 3x_5 = 9 \\ -x_1 + 4x_2 - 2x_3 + 2x_4 - 2x_5 = -17 \\ 3x_1 + x_2 + 2x_3 + x_4 + x_5 = 5 \end{cases}$

57. $\begin{cases} x_1 - x_2 + x_4 = 1 \\ 2x_1 - 2x_2 + 2x_4 = 2 \\ x_1 - x_2 - x_3 - x_4 = 1 \\ 2x_1 - 2x_2 - x_3 = 1 \end{cases}$

58. $\begin{cases} x_1 - x_3 + x_4 = 2 \\ x_2 + 2x_3 + x_4 = 0 \\ 2x_1 - 2x_2 - 6x_3 = 5 \\ x_1 + x_2 + x_3 + 2x_4 = 2 \end{cases}$

59. $\begin{cases} x_1 + x_2 + x_3 + 2x_4 = 3 \\ x_1 - x_3 + x_4 = 2 \\ x_1 + 2x_2 + 3x_3 + 3x_4 = 4 \\ x_2 + 2x_3 + x_4 = 1 \end{cases}$

60. $\begin{cases} x_1 + 2x_2 + x_3 - 2x_4 = -3 \\ 2x_1 + 4x_2 + 2x_3 - x_4 = 0 \\ x_1 + 2x_2 + x_3 + x_4 = 3 \\ x_1 + 2x_2 + x_3 = 1 \end{cases}$

Formulate each situation as a system of linear equations in an appropriate number of variables and be sure to state clearly the meaning of each. Solve the system of equations by row-reducing the corresponding augmented matrix using the Gauss–Jordan method. State your final answer in terms of the original question.

If you have a graphing calculator with a RREF command, use it to check your row reduction.

61. Plant Fertilizer A backyard garden needs 35 pounds of potash, 68 pounds of nitrogen, and 25 pounds of phosphoric acid. Three brands of fertilizer, GrowRite, MiracleMix, and GreatGreen, are available and contain the amounts of potash, nitrogen, and phosphoric acid per bag listed in the table. How many bags of each brand should be used to provide the required potash, nitrogen, and phosphoric acid?

(Pounds per Bag)	GrowRite	MiracleMix	GreatGreen
Potash	4	6	7
Nitrogen	5	10	16
Phosphoric acid	3	4	5

62. Nutrition A student athlete has decided to "bulk up" before the wrestling season by supplementing his weekly diet with an additional 325 grams of protein, 185 grams of fiber, and 110 grams of fat. If a hamburger contains 20 grams of protein, 10 grams of fiber, and 5 grams of fat; a cheeseburger contains 25 grams of protein, 10 grams of fiber, and 5 grams of fat; and a "sloppy-joe" contains 20 grams of protein, 15 grams of fiber, and 10 grams of fat, how many of each should he eat this week to meet his goal?

63. Coins in a Jar A jar contains a total of 700 nickels, dimes, and quarters worth $60. (a) How many of each are in the jar? (b) What is the greatest possible number of quarters in the jar?

64. Financial Planning An international investment banker wishes to invest $150,000 in U.S. and German stocks and bonds. Stocks are not as secure as bonds, so he plans to spread his investments so that he has three times as much in U.S. bonds as in U.S.

stocks and twice as much in German bonds as in German stocks. Furthermore, he intends to invest just $45,000 in stocks altogether. How much does he invest in each?

65. Income Taxes (a) Find the federal, state, and city income taxes on a taxable income of $180,000 if the federal tax is 50% of the taxable income after first deducting the state and city taxes, the state tax is 17% of the taxable income after first deducting the federal and city taxes, and the city tax is 3% of the taxable income after first deducting the federal and state taxes. (b) Although it appears that the taxpayer is facing a 70% nominal tax rate, use the actual taxes to determine the effective combined tax rate for this situation.

66. Estate Division As the old patriarch lay dying, he spoke to his four sons: "I leave you my fortune of 11,100 pieces of gold, which you must divide as I command. The oldest is to take one half, the second oldest is to take one third, the third oldest is to take one quarter, and the youngest is to take one fifth. But each of you must take your part from what remains after the others have taken theirs and then you must give any remainder to your mother and sister." The old woman was frantic: "How can there be any left for our daughter when you have given away more than everything?" "Don't worry, Mother," said the daughter, who had gotten an MBA while her brothers had been off fighting at Troy. How much did each son receive and how much was left over for the mother and daughter?

67. Apparel Production (a) A "limited edition" ladies fashion shop has a 240-yard supply of a silk fabric suitable for scarves, dresses, blouses, and skirts. Each scarf requires 1 yard of material, 3 minutes of cutting, and 5 minutes of sewing; each dress requires 3 yards of material, 14 minutes of cutting, and 40 minutes of sewing; each blouse requires 1.5 yards of material, 9 minutes of cutting, and 30 minutes of sewing; and each skirt requires 2 yards of material, 8 minutes of cutting, and 20 minutes of sewing. If the shop has 17 hours of skilled pattern cutter labor and 45 hours of skilled seamstress labor available, how many of each can be made? (b) If 20 blouses and 10 skirts are made, how many scarves and dresses can be made?

68. Agriculture Management   A farmer grows wheat, barley, and oats on his 360-acre farm. The labor (for planting, tending, and harvesting) and capital (for seeds, fertilizers, and pesticides) requirements for each crop are given in the table. If the farmer can contract for 1050 days of migrant worker labor and has $18,600 set aside for capital expenses, how many acres of each crop can he grow?

(Per Acre)	Wheat	Barley	Oats
Days of labor	2	4	3
Capital expenses	$50	$40	$60

69. Advertising   The promotional director for a new movie has a total of $110,000 to spend on TV, radio, and newspaper advertisements in the metropolitan area. Each TV ad costs $2500 and is seen by 10,000 moviegoers, each radio ad costs $500 and is heard by 2000 moviegoers, and each newspaper ad costs $1000 and is read by 5000 moviegoers. Ignoring repeated exposures to the same person, how many of each ad will contact 500,000 moviegoers using the allocated funds?

70. Mass Transit   Part of a subway system is shown in the diagram. The numbers of subway cars per hour arriving at and leaving stations I, II, III, and IV are indicated by arrows with numbers or variables. In order for the subway to function, the numbers of cars arriving and leaving per hour must match at each station. To prevent collisions, the subway cars must travel in the directions indicated by the arrows, and so none of the variables may be negative. A group of concerned citizens has petitioned the city council to build an express service from station I to IV (indicated on the diagram as x_5) capable of handling 40 cars per hour. Is this a good idea or is it a waste of money?

Explorations and Excursions

The following problems extend and augment the material presented in the text.

Row-echelon form has the same requirements as reduced row-echelon form (pages 190–191) except that condition (3) is relaxed to require only that *the column of each leftmost one contain zeros below it.* Such a matrix is also called a *triangular matrix*, because the nonzero elements can only be on and above the main diagonal.

Find the system of equations represented by each row-echelon form matrix:

71. $\begin{pmatrix} 1 & 2 & 3 \\ 0 & 1 & 1 \end{pmatrix}$

72. $\begin{pmatrix} 1 & 3 & 9 \\ 0 & 1 & 2 \end{pmatrix}$

73. $\begin{pmatrix} 1 & 1 & 1 & 4 \\ 0 & 1 & 1 & 3 \\ 0 & 0 & 1 & 2 \end{pmatrix}$

74. $\begin{pmatrix} 1 & 0 & 1 & 0 & 7 \\ 0 & 1 & 0 & 1 & 3 \\ 0 & 0 & 1 & 0 & 4 \\ 0 & 0 & 0 & 1 & 1 \end{pmatrix}$

Unlike reduced row-echelon form matrices, there are many row-echelon form matrices equivalent to a given augmented matrix.

Row-reduce each group of matrices to show that they are all equivalent because they are all equivalent to the same reduced row-echelon form matrix.

75. $\begin{pmatrix} 1 & 2 & 3 \\ 2 & 3 & 5 \end{pmatrix}, \begin{pmatrix} 1 & 2 & 3 \\ 0 & 1 & 1 \end{pmatrix}, \begin{pmatrix} 1 & 3 & 4 \\ 0 & 1 & 1 \end{pmatrix}$

76. $\begin{pmatrix} 2 & 5 & 16 \\ 1 & 4 & 11 \end{pmatrix}, \begin{pmatrix} 1 & 3 & 9 \\ 0 & 1 & 2 \end{pmatrix}, \begin{pmatrix} 1 & -1 & 1 \\ 0 & 1 & 2 \end{pmatrix}$

77. $\begin{pmatrix} 1 & 2 & 2 & 7 \\ 1 & 1 & -1 & 0 \\ 2 & 2 & 1 & 6 \end{pmatrix}, \begin{pmatrix} 1 & 1 & 1 & 4 \\ 0 & 1 & 1 & 3 \\ 0 & 0 & 1 & 2 \end{pmatrix},$

$\begin{pmatrix} 1 & 2 & 2 & 7 \\ 0 & 1 & 2 & 5 \\ 0 & 0 & 1 & 2 \end{pmatrix}$

78. $\begin{pmatrix} 1 & 1 & 1 & 1 & 10 \\ 1 & 0 & 2 & 0 & 11 \\ 1 & 1 & 0 & 1 & 6 \\ 1 & 0 & 1 & 1 & 8 \end{pmatrix}, \begin{pmatrix} 1 & 0 & 1 & 0 & 7 \\ 0 & 1 & 0 & 1 & 3 \\ 0 & 0 & 1 & 0 & 4 \\ 0 & 0 & 0 & 1 & 1 \end{pmatrix},$

$\begin{pmatrix} 1 & -1 & 1 & -1 & 4 \\ 0 & 1 & -1 & 1 & -1 \\ 0 & 0 & 1 & -1 & 3 \\ 0 & 0 & 0 & 1 & 1 \end{pmatrix}$

Find an equivalent matrix in row-echelon form using the REF matrix command.

79. $\begin{pmatrix} 1 & 2 & 3 \\ 2 & 3 & 5 \end{pmatrix}$

80. $\begin{pmatrix} 1 & 2 & 2 & 7 \\ 1 & 1 & -1 & 0 \\ 2 & 2 & 1 & 6 \end{pmatrix}$

Back Substitution If the augmented matrix for a system of equations is in row-echelon form, the last equation gives the value for the last variable. Substituting this value into the second-to-last equation, the value of the second-to-last variable is easy to find. Continuing upward through the equations, the values of all the remaining variables are easily found.

Solve each system of equations by back substitution.

81. $\begin{cases} x_1 + 2x_2 = 3 \\ x_2 = 1 \end{cases}$

82. $\begin{cases} x_1 + 3x_2 = 9 \\ x_2 = 2 \end{cases}$

83. $\begin{cases} x_1 + 3x_2 = 4 \\ x_2 = 1 \end{cases}$

84. $\begin{cases} x_1 - x_2 = 1 \\ x_2 = 2 \end{cases}$

85. $\begin{cases} x_1 + x_2 + x_3 = 4 \\ x_2 + x_3 = 3 \\ x_3 = 2 \end{cases}$

86. $\begin{cases} x_1 + 2x_2 + 2x_3 = 7 \\ x_2 + 2x_3 = 5 \\ x_3 = 2 \end{cases}$

87. $\begin{cases} x_1 + x_3 = 7 \\ x_2 + x_4 = 3 \\ x_3 = 4 \\ x_4 = 1 \end{cases}$

88. $\begin{cases} x_1 - x_2 + x_3 - x_4 = 4 \\ x_2 - x_3 + x_4 = -1 \\ x_3 - x_4 = 3 \\ x_4 = 1 \end{cases}$

Gauss–Jordan Elimination solves a system of equations by row-reducing the augmented matrix. It consists of two steps: First use row operations to find an equivalent row-echelon form matrix, and then use more row operations to back-substitute and complete the solution. Of course, if the row-echelon form matrix contains a row of zeros ending in a one, the corresponding system of equations is inconsistent, there can be no solution, and the method stops *without* attempting to carry out the back substitution.

Solve each system of equations by Gauss–Jordan elimination.

89. $\begin{cases} x_1 + 3x_2 + x_3 = 2 \\ 2x_1 + 5x_2 + 2x_3 = 5 \\ x_1 + 2x_2 + 2x_3 = 5 \end{cases}$

90. $\begin{cases} 2x_1 + 2x_2 + x_3 + 3x_4 = 8 \\ x_1 + 2x_2 + x_3 + x_4 = 4 \\ 2x_1 + 3x_2 + x_3 + 2x_4 = 6 \\ 3x_1 + 2x_2 - x_3 + 2x_4 = 3 \end{cases}$

3.4 Matrix Arithmetic

Matrices and Computers

Because matrices and their arithmetic are naturally suited to representing and manipulating vast quantities of numerical information, it is hardly surprising that even some of the earliest programming languages included commands to define and process matrices. BASIC, perhaps the most widely used computer programming language, was created at Dartmouth College in May 1964 by John G. Kemeny and Thomas E. Kurtz. This very first version contained an "array" data structure so that whole grids of values could be conveniently represented. A variety of BASIC introduced during the summer of 1964 called CARD-BASIC included the now-standard MAT instructions, so that single commands could specify arithmetic operations on entire matrices. For instance, if the manufacturing costs of a company's entire product line are entered in a matrix named Costs and the corresponding product revenues are stored in a matrix named Revenues, the single program statement

```
MAT Profits = Revenues - Costs
```

will subtract each cost from the corresponding revenue and place the result at the correct position in a matrix named Profits. From the beginning, BASIC has stored arrays in row-major order, in which the values of the first column are placed in adjacent computer memory locations, with those of the next column immediately following, to permit the fastest possible access to the matrix elements during MAT calculations. The inclusion of matrix commands in virtually all computer languages has stimulated many useful applications of matrix methods.

Introduction

Our focus thus far has been on the augmented matrix representation of a system of equations. The augmented matrix representation is "unnatural" in the sense that it somehow "contains" but does not show the equals signs from the equations. We shall now define the arithmetic of matrices so that we may write an entire system of equations as a single *matrix equation*. The solution in the next section of such an equation of matrices will deepen our understanding of what it means to solve a system of equations.

Some Basic Terms

We write $A = (a_{i,j})$ to indicate that the matrix A is composed of the elements $a_{i,j}$. Two matrices are *equal* if they have the same dimension and if elements in corresponding locations are equal; that is, $A = B$ if $a_{i,j} = b_{i,j}$ for every row i and column j. For instance,

$$\begin{pmatrix} 1 & 2 & 3 \\ 4 & 5 & 6 \end{pmatrix} = \begin{pmatrix} 1 & 1+1 & 4-1 \\ 2^2 & 10/2 & 2\cdot3 \end{pmatrix}$$

Equal because same dimension and corresponding values are equal but

$$\begin{pmatrix} 1 & 2 \\ 3 & 4 \end{pmatrix} \neq \begin{pmatrix} 1 \\ 2 \\ 3 \\ 4 \end{pmatrix}$$

Not equal because different dimensions

and

$$\begin{pmatrix} 1 & 2 \\ 3 & 4 \end{pmatrix} \neq \begin{pmatrix} 1 & 2 \\ 4 & 3 \end{pmatrix}$$

Not equal because some corresponding values differ.

The *transpose* of a matrix is formed by turning each row into a column (or, equivalently, each column into a row). More precisely, the transpose (denoted by a superscript t) of the $m \times n$ matrix $A = (a_{i,j})$ is the $n \times m$ matrix $A^t = (a_{j,i})$. For instance,

$$\begin{pmatrix} 1 & 2 & 3 \\ 4 & 5 & 6 \end{pmatrix}^t = \begin{pmatrix} 1 & 4 \\ 2 & 5 \\ 3 & 6 \end{pmatrix}$$

First row becomes first column
Second row becomes second column

A matrix is *symmetric* if it is equal to its transpose. Notice that a symmetric matrix must be square and that the elements "above" the main diagonal must mirror those "below" it. For instance,

$$\begin{pmatrix} 1 & 5 & \\ 5 & 2 & 7 \\ & 7 & 3 \end{pmatrix} \text{ is symmetric but } \begin{pmatrix} 1 & 2 & 3 \\ 2 & 1 & 2 \\ 2 & 3 & 1 \end{pmatrix} \text{ is not}$$

An *identity matrix* is a square matrix with ones on the main diagonal and zeros elsewhere. We write $I = I_n$ for the $n \times n$ identity matrix. Notice that identity matrices are symmetric. The identity matrices $I_1, I_2, I_3,$ and I_4 are written below.

$$(1), \begin{pmatrix} 1 & 0 \\ 0 & 1 \end{pmatrix}, \begin{pmatrix} 1 & 0 & 0 \\ 0 & 1 & 0 \\ 0 & 0 & 1 \end{pmatrix}, \begin{pmatrix} 1 & 0 & 0 & 0 \\ 0 & 1 & 0 & 0 \\ 0 & 0 & 1 & 0 \\ 0 & 0 & 0 & 1 \end{pmatrix}$$

Scalar multiplication of a matrix simply means multiplying each element of a matrix by the same number. More precisely, a number (or "scalar") s times a matrix A is defined as $sA = (sa_{i,j})$. In particular, the *negative* of a matrix is $-A = (-a_{i,j})$. For example, if

$$A = \begin{pmatrix} 1 & 2 & 3 \\ 4 & 5 & 6 \end{pmatrix},$$

then

$$3A = \begin{pmatrix} 3 & 6 & 9 \\ 12 & 15 & 18 \end{pmatrix} \text{ and } -A = \begin{pmatrix} -1 & -2 & -3 \\ -4 & -5 & -6 \end{pmatrix}$$

Matrix Addition

For two matrices with the same dimensions, *matrix addition* means adding elements in corresponding locations; that is, $A + B = (a_{i,j} + b_{i,j})$. Matrices with *different* dimensions cannot be added.

$$\begin{pmatrix} 1 & 4 \\ 2 & 5 \\ 3 & 6 \end{pmatrix} + \begin{pmatrix} 11 & 12 \\ 10 & 9 \\ 7 & 8 \end{pmatrix} = \begin{pmatrix} 12 & 16 \\ 12 & 14 \\ 10 & 14 \end{pmatrix}$$

but $\quad\begin{pmatrix} 1 & 4 \\ 2 & 5 \\ 3 & 6 \end{pmatrix} + \begin{pmatrix} 7 & 8 & 9 \\ 10 & 11 & 12 \end{pmatrix}\quad$ is not possible.

Similarly, two matrices with the same dimensions can be *subtracted* by subtracting corresponding elements (or equivalently, adding the negative of the second matrix).

$$\begin{pmatrix} 11 & 12 \\ 10 & 9 \\ 7 & 8 \end{pmatrix} - \begin{pmatrix} 1 & 4 \\ 2 & 5 \\ 3 & 6 \end{pmatrix} = \begin{pmatrix} 10 & 8 \\ 8 & 4 \\ 4 & 2 \end{pmatrix}$$

A *zero matrix* has all elements equal to zero. We write 0 for the zero matrix of whatever size is appropriate for the situation. If two matrices are equal, their difference is a zero matrix: that is, if $A = B$, then $A - B = 0$ means that 0 is the zero matrix of the same dimension as A and B.

$$\begin{pmatrix} 1 & 2 \\ 3 & 4 \\ 5 & 6 \end{pmatrix} - \begin{pmatrix} 1 & 2 \\ 3 & 4 \\ 5 & 6 \end{pmatrix} = 0 \quad \text{means} \quad \begin{pmatrix} 1 & 2 \\ 3 & 4 \\ 5 & 6 \end{pmatrix} - \begin{pmatrix} 1 & 2 \\ 3 & 4 \\ 5 & 6 \end{pmatrix} = \begin{pmatrix} 0 & 0 \\ 0 & 0 \\ 0 & 0 \end{pmatrix}$$

Graphing Calculator Exploration

The basic matrix operations are found in the MATRX MATH menu. To have the calculator demonstrate the preceding concepts, proceed as follows:

a. Use MATRX EDIT to enter $\begin{pmatrix} 1 & 2 & 3 \\ 4 & 5 & 6 \end{pmatrix}$ in [A].

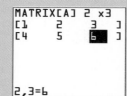

b.   Find the transpose of A.

```
[A]
        [[1 2 3]
         [4 5 6]]
[A]ᵀ
        [[1 4]
         [2 5]
         [3 6]]
```

c.   Find the scalar multiples $3A$ and $(1/2)A$. Check that attempting the division [A]/2 gives a "data type" error.

```
3[A]
    [[3  6  9 ]
     [12 15 18]]

(1/2)[A]
    [[.5 1    1.5]
     [2  2.5 3   ]]
```

d.   Find $A + A$ and $A - A$.

```
[A]+[A]
        [[2 4  6 ]
         [8 10 12]]

[A]-[A]
        [[0 0 0]
         [0 0 0]]
```

e.   Find the 4×4 identity matrix.

```
identity(4)
        [[1 0 0 0]
         [0 1 0 0]
         [0 0 1 0]
         [0 0 0 1]]
```

EXAMPLE 1   **Using Matrix Arithmetic**

A McBurger restaurant sells hamburgers for $1, cheeseburgers for $1.50, and fries for 75¢, while the BurgerQueen across the street sells hamburgers for $1.25, cheeseburgers for $1.75, and fries for 50¢. At McBurger the preparation costs are 30¢ per hamburger, 40¢ per cheeseburger, and 25¢ per order of fries, while at BurgerQueen the costs are 35¢ per hamburger, 45¢ per cheeseburger, and 20¢ per order of fries.

a.   Represent the selling prices and preparation costs as matrices.

b.   Using the matrices from (a), find the profit margin matrix for the items at the restaurants.

c.   Using the matrix from (b), find the franchise fee charged on each item at each restaurant if this fee is 30% of the profit margin.

Solution

a. Let R be the revenue matrix of selling prices and let C be the cost matrix of preparation costs. We have 3 items from 2 restaurants, so we can choose the matrices to be either 2×3 or 3×2. We choose the matrices to be 2×3, with the rows corresponding to the different restaurants and the columns to the different menu items:

$$
\begin{array}{ccc}
\text{Hamburger} & \text{Cheeseburger} & \text{Fries} \\
\downarrow & \downarrow & \downarrow
\end{array}
$$

$$
\begin{array}{c}
\text{McBurger} \rightarrow \\
\text{BurgerQueen} \rightarrow
\end{array}
\left(
\begin{array}{ccc}
\underline{\quad} & \underline{\quad} & \underline{\quad} \\
\underline{\quad} & \underline{\quad} & \underline{\quad}
\end{array}
\right)
$$

From the given information, we obtain the matrices

$$
R = \begin{pmatrix} 1.00 & 1.50 & 0.75 \\ 1.25 & 1.75 & 0.50 \end{pmatrix} \quad \text{and} \quad C = \begin{pmatrix} 0.30 & 0.40 & 0.25 \\ 0.35 & 0.45 & 0.20 \end{pmatrix}
$$

b. Let P be the profit margin matrix, which is revenue R minus cost C.

$$
P = R - C = \begin{pmatrix} 1.00 & 1.50 & 0.75 \\ 1.25 & 1.75 & 0.50 \end{pmatrix} - \begin{pmatrix} 0.30 & 0.40 & 0.25 \\ 0.35 & 0.45 & 0.20 \end{pmatrix}
$$

$$
= \begin{pmatrix} 0.70 & 1.10 & 0.50 \\ 0.90 & 1.30 & 0.30 \end{pmatrix}
\qquad \text{\small Using matrix}\\ \text{\small subtraction}
$$

c. Let F be the franchise fee matrix, which is 30% of the profit matrix P.

$$
F = 0.30P = 0.30 \cdot \begin{pmatrix} 0.70 & 1.10 & 0.50 \\ 0.90 & 1.30 & 0.30 \end{pmatrix}
$$

$$
= \begin{pmatrix} 0.21 & 0.33 & 0.15 \\ 0.27 & 0.39 & 0.09 \end{pmatrix}
\qquad \text{\small Using scalar}\\ \text{\small multiplication}
$$

∎

Matrix Multiplication as Evaluation

To evaluate the expression $5x_1 + 6x_2 + 7x_3$ at $x_1 = 2$, $x_2 = 3$, and $x_3 = 4$, we substitute the values for the variables, multiply them out, and add them up: $5 \cdot 2 + 6 \cdot 3 + 7 \cdot 4 = 56$. With this kind of evaluation in mind, we define *matrix multiplication* as multiplying in order the numbers from a row by the numbers from a column and adding the results. Thus, *we link matrix arithmetic to systems of equations by defining matrix multiplication as "evaluation" of the row "expression" on the left by the column "list of values" on the right.* The matrix product of a row matrix by a column matrix is a 1×1 matrix:

$$
\begin{pmatrix} 5 & 6 & 7 \end{pmatrix} \cdot \begin{pmatrix} 2 \\ 3 \\ 4 \end{pmatrix} = (5 \cdot 2 + 6 \cdot 3 + 7 \cdot 4) = (56)
$$

Of course, there must be exactly as many elements in the row on the left as in the column on the right.

PRACTICE PROBLEM 1

Find $(1 \quad 2) \cdot \begin{pmatrix} 3 \\ 4 \end{pmatrix}.$

Solution at the back of the book

If there are several rows on the left, and several columns on the right, then we multiply *each row* times *each column*, with each answer placed in the product matrix at the row and column position from whence it came:

First row times second column goes here, in first row, second column

$$\begin{pmatrix} 3 & 2 & 1 \\ 2 & 0 & -2 \end{pmatrix}\begin{pmatrix} 2 & 7 \\ 3 & 6 \\ 4 & 5 \end{pmatrix} = \begin{pmatrix} (3 \quad 2 \quad 1)\begin{pmatrix} 2 \\ 3 \\ 4 \end{pmatrix} & (3 \quad 2 \quad 1)\begin{pmatrix} 7 \\ 6 \\ 5 \end{pmatrix} \\ (2 \quad 0 \quad -2)\begin{pmatrix} 2 \\ 3 \\ 4 \end{pmatrix} & (2 \quad 0 \quad -2)\begin{pmatrix} 7 \\ 6 \\ 5 \end{pmatrix} \end{pmatrix} = \begin{pmatrix} 16 & 38 \\ -4 & 4 \end{pmatrix}$$

Multiply each row times column "in your head"

Go from here

Directly to here

In general, to multiply two matrices, the row length of the first must match the column length of the second, with the other numbers giving the dimension of the product:

$$\underset{m \times p}{A} \quad \cdot \quad \underset{p \times n}{B} \quad = \quad \underset{m \times n}{C}$$

Inside numbers get "absorbed"

Formally, matrix multiplication is defined as follows:

Matrix Multiplication

The product of A with dimension $m \times p$ and B with dimension $q \times n$ is defined only if $p = q$, in which case $A \cdot B$ is the matrix C with dimension $m \times n$ with elements

$$c_{i,j} = a_{i,1}b_{1,j} + a_{i,2}b_{2,j} + \cdots + a_{i,p}b_{p,j}.$$

That is, the element of $A \cdot B$ in the ith row and jth column is found by multiplying each element of the ith row of A by the corresponding element of the jth column of B and adding them all together.

Not every matrix product can be found; for instance, the product

$$\begin{pmatrix} 2 & 7 \\ 3 & 6 \\ 4 & 5 \end{pmatrix}\begin{pmatrix} 5 & 6 & 7 \\ 1 & -1 & 1 \\ 2 & 0 & -2 \\ 3 & 2 & 1 \end{pmatrix} \text{ is not defined.}$$

The row length (2) of the first does not match the column length (4) of the second

Notice that while $(5 \quad 6 \quad 7)\begin{pmatrix} 2 \\ 3 \\ 4 \end{pmatrix} = (56)$ is a 1×1 matrix, the product

in the reverse order is a very different 3×3 matrix:

$$\begin{pmatrix} 2 \\ 3 \\ 4 \end{pmatrix}(5 \quad 6 \quad 7) = \begin{pmatrix} 2 \cdot 5 & 2 \cdot 6 & 2 \cdot 7 \\ 3 \cdot 5 & 3 \cdot 6 & 3 \cdot 7 \\ 4 \cdot 5 & 4 \cdot 6 & 4 \cdot 7 \end{pmatrix} = \begin{pmatrix} 10 & 12 & 14 \\ 15 & 18 & 21 \\ 20 & 24 & 28 \end{pmatrix}$$

SPREADSHEET EXPLORATION

The spreadsheet below shows the sum and product of the matrices

$$A = \begin{pmatrix} 1 & 2 & 3 \\ 4 & -2 & 4 \\ 3 & 2 & 1 \end{pmatrix} \text{ and } B = \begin{pmatrix} 6 & 3 & 8 \\ 2 & 4 & 1 \\ 5 & 7 & 9 \end{pmatrix}.$$ The sum $A + B$ is found by

adding the range of cells a3:c5 (containing the entries of matrix A) to the range of cells e3:g5 (containing the entries of matrix B, while the product $A \cdot B$ uses the matrix multiplication function MMULT(,) applied to the two ranges of cells representing the matrices.

	A	B	C	D	E	F	G
1		Matrix A				Matrix B	
2							
3	1	2	3		6	3	8
4	4	-2	4		2	4	1
5	3	2	1		5	7	9
6							
7		Sum A+B				Product A*B	
8							
9	7	5	11		25	32	37
10	6	2	5		40	32	66
11	8	9	10		27	24	35

Can you modify this spreadsheet to calculate $B + A$ and $B \cdot A$? Which answer will change and which will remain the same?

Unlike number multiplication, *matrix multiplication is not commutative: A · B may not be the same as B · A.* In fact, $A \cdot B$ and $B \cdot A$ may have different dimensions. However, if we change rows into columns and columns into rows (that is, transpose each matrix) and then reverse the order, we will be multiplying and adding the same collections of numbers, with only their positions changed. That is, *the transpose of a product is the product of the transposes in the opposite order.*

If $A \cdot B$ is possible, then $(A \cdot B)^t = B^t \cdot A^t$.

EXAMPLE 2   Matrix Multiplication

Use the selling price ("revenue") matrix from Example 1 (pages 210–211) and matrix multiplication to find the total selling price for a meal of three hamburgers, a cheeseburger, and two fries at each of the restaurants.

Solution

The revenue matrix R from Example 1 gives the selling prices in rows, one row for each restaurant. If we write the three hamburgers, one cheeseburger, and two fries as a *column*, we can use matrix multiplication to find the total price at each restaurant.

$$\begin{pmatrix} 1.00 & 1.50 & 0.75 \\ 1.25 & 1.75 & 0.50 \end{pmatrix} \begin{pmatrix} 3 \\ 1 \\ 2 \end{pmatrix} = \begin{pmatrix} 1.00 \cdot 3 + 1.50 \cdot 1 + 0.75 \cdot 2 \\ 1.25 \cdot 3 + 1.75 \cdot 1 + 0.50 \cdot 2 \end{pmatrix}$$

$\qquad\qquad\uparrow\qquad\qquad\quad\uparrow$
R (from Example 1)   Order (3 hamburgers, 1 cheeseburger, 2 fries)

$$= \begin{pmatrix} 6.00 \\ 6.50 \end{pmatrix} \begin{matrix} \leftarrow \text{Price at McBurger} \\ \leftarrow \text{Price at BurgerQueen} \end{matrix}$$

Notice that we could have found the same numbers by multiplying the transposes in the opposite order:

$$(3 \quad 1 \quad 2) \begin{pmatrix} 1.00 & 1.25 \\ 1.50 & 1.75 \\ 0.75 & 0.50 \end{pmatrix} = (6.00 \quad 6.50)$$

PRACTICE PROBLEM 2

Modify Example 2 to find the total selling price for a meal of four hamburgers, two cheeseburgers, and five fries at each of the restaurants.

Solution at the back of the book

Identity Matrices

Multiplication by an identity matrix (of the proper size) does not change the matrix:

$$\begin{pmatrix} 1 & 0 & 0 \\ 0 & 1 & 0 \\ 0 & 0 & 1 \end{pmatrix} \begin{pmatrix} 2 & 7 \\ 3 & 6 \\ 4 & 5 \end{pmatrix} = \begin{pmatrix} 2 & 7 \\ 3 & 6 \\ 4 & 5 \end{pmatrix}$$

Same matrix on the right as on the left

The first row (1 0 0) picks out just the first value, the second (0 1 0) picks out the second value, and so on. (You should carefully verify this multiplication to see how the identity works.) To multiply by the identity matrix on the other side, we need in this case to use an identity matrix of a different size:

$$\begin{pmatrix} 2 & 7 \\ 3 & 6 \\ 4 & 5 \end{pmatrix} \begin{pmatrix} 1 & 0 \\ 0 & 1 \end{pmatrix} = \begin{pmatrix} 2 & 7 \\ 3 & 6 \\ 4 & 5 \end{pmatrix}$$

Again, the matrix is duplicated

Writing I for an identity matrix of the appropriate size, we have for any matrix A

$$I \cdot A = A \cdot I = A$$

That is, the matrix I plays the role in matrix arithmetic that the number 1 plays in ordinary arithmetic—multiplying by it gives back exactly what you started with.

Matrix Multiplication and Systems of Equations

Because we defined matrix multiplication in order to evaluate linear expressions (page 211), we may use it to write an entire system of equations as a single matrix equation. Consider the system of equations

$$\begin{cases} 2x_1 + 3x_2 + 3x_3 = 15 \\ 3x_1 + 4x_2 + 4x_3 = 22 \\ 3x_1 + 4x_2 + 5x_3 = 30 \end{cases}$$

The "coefficient" matrix A and the "constant term" matrix B are

$$A = \begin{pmatrix} 2 & 3 & 3 \\ 3 & 4 & 4 \\ 3 & 4 & 5 \end{pmatrix} \quad \text{and} \quad B = \begin{pmatrix} 15 \\ 22 \\ 30 \end{pmatrix}$$

Let X be the column matrix of the variables:

$$X = \begin{pmatrix} x_1 \\ x_2 \\ x_3 \end{pmatrix}$$

Then the matrix product $A \cdot X$ is precisely the left sides of the equations (as you should check):

$$\underbrace{\begin{pmatrix} 2 & 3 & 3 \\ 3 & 4 & 4 \\ 3 & 4 & 5 \end{pmatrix}}_{A} \underbrace{\begin{pmatrix} x_1 \\ x_2 \\ x_3 \end{pmatrix}}_{X} = \begin{pmatrix} 2x_1 + 3x_2 + 3x_3 \\ 3x_1 + 4x_2 + 4x_3 \\ 3x_1 + 4x_2 + 5x_3 \end{pmatrix}$$

The system of equations is the same as the matrix equation $AX = B$:

$$\begin{cases} 2x_1 + 3x_2 + 3x_3 = 15 \\ 3x_1 + 4x_2 + 4x_3 = 22 \\ 3x_1 + 4x_2 + 5x_3 = 30 \end{cases} \text{ is the same as } \underbrace{\begin{pmatrix} 2 & 3 & 3 \\ 3 & 4 & 4 \\ 3 & 4 & 5 \end{pmatrix}}_{A} \underbrace{\begin{pmatrix} x_1 \\ x_2 \\ x_3 \end{pmatrix}}_{X} = \underbrace{\begin{pmatrix} 15 \\ 22 \\ 30 \end{pmatrix}}_{B}$$

That is, the system of equations on the left can be written as one matrix equation $A \cdot X = B$. Such matrix representations will be particularly useful for larger systems of equations.

Matrix Multiplication and Row Operations

Because matrix multiplication of A by I picks out the elements of A in the correct order, switching the rows of I before multiplying will pick out the elements of A in a different order:

I with rows 1 and 2 switched $\longrightarrow$ $\begin{pmatrix} 0 & 1 & 0 \\ 1 & 0 & 0 \\ 0 & 0 & 1 \end{pmatrix}\begin{pmatrix} 2 & 7 \\ 3 & 6 \\ 4 & 5 \end{pmatrix} = \begin{pmatrix} 3 & 6 \\ 2 & 7 \\ 4 & 5 \end{pmatrix}$ $\longleftarrow$ A with rows 1 and 2 switched

Replacing one of the 1s in I with a different value before multiplying will pick out the corresponding element of A that many times:

I with row 3 multiplied by 8 $\longrightarrow$ $\begin{pmatrix} 1 & 0 & 0 \\ 0 & 1 & 0 \\ 0 & 0 & 8 \end{pmatrix}\begin{pmatrix} 2 & 7 \\ 3 & 6 \\ 4 & 5 \end{pmatrix} = \begin{pmatrix} 2 & 7 \\ 3 & 6 \\ 32 & 40 \end{pmatrix}$ $\longleftarrow$ A with row 3 multiplied by 8

Replacing one of the 0s in I with a 1 before multiplying will pick out the elements of another row and add them to the current row of A:

I with row 3 added to row 1 $\longrightarrow$ $\begin{pmatrix} 1 & 0 & 1 \\ 0 & 1 & 0 \\ 0 & 0 & 1 \end{pmatrix}\begin{pmatrix} 2 & 7 \\ 3 & 6 \\ 4 & 5 \end{pmatrix} = \begin{pmatrix} 6 & 12 \\ 3 & 6 \\ 4 & 5 \end{pmatrix}$ $\longleftarrow$ A with row 3 added to row 1

Thus the row operations of switching two rows, multiplying a row by a constant, and adding a row to another can be accomplished by matrix multiplications. Furthermore, to find the matrix that performs a given row operation, we need only apply the same row operation to the identity matrix (of the appropriate size).

> Any row operation can be accomplished by
> a matrix multiplication.

This observation is the basis for our discussion of "inverse" matrices in the next section, and provides a second reason why matrix multiplication is not commutative: Changing the order of a sequence of row operations usually changes the result, so changing the order of the corresponding matrix multiplications must similarly change the result.

Any *sequence* of row operations can be carried out by multiplying by a single matrix. This is because multiplying by several matrices on the left is equivalent to multiplying just once by the *product* of all of the matrices. Moreover, the matrix that we multiply by on the left is found by applying the same sequence of row operations to the identity matrix. Therefore, if a matrix A is equivalent to the identity matrix, then there is a matrix R (corresponding to the row operations performed on the identity matrix) such that $R \cdot A = I$. This observation will be very important in the next section.

Section Summary

The elements of a matrix have both values and positions. Matrix arithmetic uses the following terms and operations.

Equal matrices	$A = B$	Same dimension and same values in same positions
Transpose of a matrix	A^t	Row and column positions reversed
Symmetric matrix	$A^t = A$	"Above" diagonal mirrors "below" diagonal
Identity matrix	I	Square, zeros except for ones on diagonal
Scalar multiple	sA	Multiply every element by s
Matrix addition	$A + B$	Add elements in same positions
Matrix subtraction	$A - B$	Subtract elements in same positions
Zero matrix	0	All elements are zeros
Matrix multiplication	$A \cdot B$	Rows on left times columns on right (row length of A = column length of B)

Associative *Commutative* (handwritten margin note)

Any system of linear equations may be written as a matrix equation $A \cdot X = B$, where A is the coefficient matrix, X is the column matrix of variables, and B is the constant term matrix.

Any row operation can be accomplished by a matrix multiplication, and this "row operation matrix" can be found by applying the row operation to the identity matrix. Furthermore, a succession of such row operations can be carried out by multiplication on the left by a *single* matrix, the product of all of the matrices corresponding to the row operations.

EXERCISES 3.4

Use the given matrices to find each matrix expression.

$$A = \begin{pmatrix} 1 & 2 & 3 \\ 4 & 5 & 6 \\ 7 & 8 & 9 \end{pmatrix} \quad B = \begin{pmatrix} 9 & 8 & 7 \\ 6 & 5 & 4 \\ 3 & 2 & 1 \end{pmatrix} \quad C = \begin{pmatrix} 1 & 6 & 8 \\ 4 & 2 & 7 \\ 9 & 5 & 3 \end{pmatrix}$$

1. A^t

2. B^t

3. $3C$

4. $2A$

5. $-B$

6. $-C$

7. $A + C$

8. $B + C$

9. $C - (A + I)$

10. $(A + I) - B$

Find each matrix product.

11. $(1 \quad -1 \quad 1)\begin{pmatrix} 4 \\ 3 \\ 5 \end{pmatrix}$

12. $(1 \quad 1 \quad -1 \quad -1)\begin{pmatrix} 3 \\ 2 \\ 4 \\ 1 \end{pmatrix}$

13. $\begin{pmatrix} 1 \\ 2 \end{pmatrix} \cdot (3 \quad 4)$

14. $\begin{pmatrix} 2 \\ -1 \end{pmatrix} \cdot (4 \quad 0)$

15. $\begin{pmatrix} 1 & 2 & 1 \\ 2 & 1 & 2 \end{pmatrix}\begin{pmatrix} 2 \\ -3 \\ 2 \end{pmatrix}$

16. $\begin{pmatrix} 1 & 3 & 1 \\ 3 & 1 & 3 \end{pmatrix}\begin{pmatrix} 4 \\ -2 \\ 4 \end{pmatrix}$

17. $\begin{pmatrix} 1 & 3 & 1 \\ 2 & 1 & 2 \end{pmatrix}\begin{pmatrix} 1 & 2 \\ 1 & 1 \\ 2 & 1 \end{pmatrix}$

18. $\begin{pmatrix} 1 & 2 & 1 \\ 3 & 1 & 3 \end{pmatrix}\begin{pmatrix} 2 & 3 \\ 3 & 1 \\ 2 & -1 \end{pmatrix}$

19. $\begin{pmatrix} 2 & 3 \\ 3 & 1 \\ 2 & -1 \end{pmatrix}\begin{pmatrix} 1 & 2 & 1 \\ 3 & 1 & 3 \end{pmatrix}$

20. $\begin{pmatrix} 1 & 2 \\ 1 & 1 \\ 2 & 1 \end{pmatrix}\begin{pmatrix} 1 & 3 & 1 \\ 2 & 1 & 2 \end{pmatrix}$

Use the given matrices to find each matrix expression.

 If your instructor permits, you may carry out the calculations on a graphing calculator.

$$A = \begin{pmatrix} 2 & -1 & 1 \\ 1 & 0 & 1 \end{pmatrix} \quad B = \begin{pmatrix} 1 & 2 & 1 \\ 3 & -2 & 0 \end{pmatrix}$$

$$C = \begin{pmatrix} 2 & -1 & 2 \\ 1 & 1 & 1 \\ 0 & 2 & 1 \end{pmatrix}$$

21. $A \cdot C$

22. $B \cdot C$

23. $C \cdot B^t$

24. $C \cdot A^t$

25. $(A - B) \cdot C$

26. $(B - A) \cdot C$

27. $A^t \cdot B + C$

28. $B^t \cdot A - C$

29. $B \cdot (C + I)$

30. $A \cdot (C - I)$

Rewrite each system of linear equations as a matrix equation $A \cdot X = B$.

31. $\begin{cases} x_1 + 5x_2 + 4x_3 = 6 \\ x_1 + x_2 + x_3 = 4 \\ 2x_1 + 3x_2 + 3x_3 = 9 \end{cases}$

32. $\begin{cases} x_1 + x_2 + x_3 = 4 \\ x_1 + 2x_2 + x_3 = 3 \\ x_1 + 2x_2 + 2x_3 = 5 \end{cases}$

33. $\begin{cases} 4x_1 + 3x_2 - x_3 = 2 \\ 3x_1 + 3x_2 + 2x_3 = 9 \\ 2x_1 + x_2 - 3x_3 = -6 \end{cases}$

34. $\begin{cases} 2x_1 - x_2 + 2x_3 = 11 \\ -x_1 + x_2 - 3x_3 = -12 \\ 2x_1 - 2x_2 + 7x_3 = 27 \end{cases}$

35. $\begin{cases} 5x_1 + 2x_2 - 4x_3 + x_4 + 5x_5 = 7 \\ 3x_1 + x_2 - 3x_3 + x_4 + 3x_5 = 5 \end{cases}$

36. $\begin{cases} 2x_1 + x_2 + 5x_3 + 4x_4 + 5x_5 = 2 \\ x_1 + x_2 + 3x_3 + 3x_4 + 3x_5 = -1 \end{cases}$

Rewrite each matrix equation $A \cdot X = B$ as a system of linear equations.

37. $\begin{pmatrix} 5 & 9 & 9 \\ 4 & 7 & 6 \\ 3 & 5 & 3 \\ 4 & 7 & 5 \end{pmatrix} \cdot \begin{pmatrix} x_1 \\ x_2 \\ x_3 \end{pmatrix} = \begin{pmatrix} 11 \\ 9 \\ 8 \\ 10 \end{pmatrix}$

38. $\begin{pmatrix} 6 & 3 & 5 \\ 1 & 2 & 2 \\ 4 & 3 & 4 \\ 5 & 1 & 3 \end{pmatrix} \cdot \begin{pmatrix} x_1 \\ x_2 \\ x_3 \end{pmatrix} = \begin{pmatrix} 8 \\ 1 \\ 5 \\ 7 \end{pmatrix}$

39. $\begin{pmatrix} 5 & 4 & 7 & 6 \\ 2 & 2 & 3 & 3 \\ 4 & 3 & 5 & 5 \\ 3 & 2 & 3 & 3 \end{pmatrix} \cdot \begin{pmatrix} x_1 \\ x_2 \\ x_3 \\ x_4 \end{pmatrix} = \begin{pmatrix} 18 \\ 9 \\ 16 \\ 11 \end{pmatrix}$

40. $\begin{pmatrix} 3 & 4 & 2 & 4 \\ 1 & 2 & 1 & 1 \\ 4 & 5 & 2 & 5 \\ 6 & 6 & 1 & 6 \end{pmatrix} \cdot \begin{pmatrix} x_1 \\ x_2 \\ x_3 \\ x_4 \end{pmatrix} = \begin{pmatrix} 12 \\ 4 \\ 14 \\ 15 \end{pmatrix}$

For each row operation (or sequence of row operations), find a 4×4 matrix R such that the matrix product $R \cdot A$ is the same as the result of carrying out the row operation(s) on the matrix

$$A = \begin{pmatrix} 3 & 4 & 3 & 2 \\ 1 & 6 & 2 & 5 \\ 2 & -3 & 1 & -4 \\ 1 & 3 & 1 & 2 \end{pmatrix}$$

41. $R'_2 = R_4$ and $R'_4 = R_2$

42. $R'_3 = R_1$ and $R'_1 = R_3$

43. $R'_4 = 3R_4$

44. $R'_2 = 2R_2$

45. $R'_1 = R_1 - R_3$

46. $R'_4 = R_2 - R_4$

47. $R'_3 = R_3 - 2R_4$

48. $R'_1 = R_1 - 3R_2$

49. $R'_1 = R_1 - 3R_3,$
    $R'_2 = R_2 - 2R_3,$ and
    $R'_4 = R_4 - R_3$

50. $R'_1 = R_1 - 3R_2,$
    $R'_3 = R_3 - 2R_2,$ and
    $R'_4 = R_4 - R_2$

APPLIED EXERCISES

Formulate each situation in matrix form. Be sure to indicate the meaning of your rows and columns. Find the requested quantities using the appropriate matrix arithmetic.

51. **Sales Commissions** A salesman at a furniture store sells bed mattresses manufactured by SlumberKing, DreamOn, and RestEasy. The selling prices for the three models of SlumberKing mattresses are $300 for the economy, $350 for the best, and $500 for the deluxe; DreamOn mattresses are $350 for the economy, $400 for the best, and $550 for the deluxe; and RestEasy mattresses are $400 for the economy, $500 for the best, and $700 for the deluxe. Represent these selling prices as a price matrix. Use this matrix to find the salesperson's commission matrix for these mattresses from these manufacturers if the commission is 15% of the selling price.

52. **Sales Taxes** The ToysForYou stores in Rockland and Martinville sell "Little Tykes" baseballs, bats, gloves, and caps. At the Rockland store, the retail prices are $1 per baseball, $8 per bat, $7 per glove, and $10 per cap while at the Martinville store, the prices are $1 per baseball, $9 per bat, $8 per glove, and $11 per cap. Represent these retail prices as a price matrix. Use this matrix to find the sales tax matrix for these items at these stores if the state sales tax is 5% of the retail price.

53. Car Sales A car dealer sells sedans, station wagons, vans, and pickup trucks at sales lots in Oakdale and Roanoke. The "dealer markup" is the difference between the sticker price and the dealer invoice price. The dealer invoice prices at both locations are the same: $15,000 per sedan, $19,000 per wagon, $23,000 per van, and $25,000 per pickup. The sticker prices at the Oakdale lot are $18,900 per sedan, $22,900 per wagon, $26,900 per van, and $29,900 per pickup, while at the Roanoke lot the sticker prices are $19,900 per sedan, $21,900 per wagon, $27,900 per van, and $28,900 per pickup. Represent these prices as a dealer invoice matrix and a sticker price matrix. Use these matrices to find the dealer markup matrix for these vehicles at these sales lots.

54. Fuel Prices An oil refinery in Louisiana produces gasoline, kerosene, and diesel fuel for sale at service stations in Tennessee, Alabama, and Florida. In Tennessee, the pump prices per gallon are $1.18 for gasoline, $0.87 for kerosene, and $1.09 for diesel fuel while the combined federal and state taxes per gallon are 52¢ for gasoline, 35¢ for kerosene, and 46¢ for diesel fuel. In Alabama, the pump prices per gallon are $1.15 for gasoline, $0.88 for kerosene, and $1.04 for diesel fuel, while the combined federal and state taxes per gallon are 48¢ for gasoline, 33¢ for kerosene, and 41¢ for diesel fuel. And in Florida, the pump prices per gallon are $1.27 for gasoline, $0.93 for kerosene, and $1.16 for diesel fuel while the combined federal and state taxes per gallon are 56¢ for gasoline, 41¢ for kerosene, and 48¢ for diesel fuel. Represent these prices and taxes as matrices. Use these matrices to find the pretax price matrix for these fuels in these states.

55. Overseas Manufacturing A sports apparel company manufactures shorts, tee shirts, and caps in Costa Rica and Honduras for importation and sale in the United States. In Costa Rica, the labor costs per item are 75¢ per pair of shorts, 25¢ per tee shirt, and 45¢ per cap, while the costs of the necessary materials are $1.60 per pair of shorts, 95¢ per tee shirt, and $1.15 per cap. In Honduras, the labor costs per item are 80¢ per pair of shorts, 20¢ per tee shirt, and 55¢ per cap, while the costs of the necessary materials are $1.50 per pair of shorts, 80¢ per tee shirt, and $1.10 per cap. Represent these costs as a labor cost matrix and a materials cost matrix. Use these matrices to find the total cost matrix for these products in these countries.

56. Retirement Income A study of retired Chicago municipal workers now living in the Ozark Plateau found that in Missouri the average monthly pension benefits were $2700 for former police officers, $2500 for former mass transit workers, and $2800 for former firefighters, while in Arkansas the averages were $2750 for former police officers, $2300 for former mass transit workers, and $2900 for former firefighters. The average monthly Social Security benefits received by the same individuals in Missouri were $1100 for former police officers, $800 for former mass transit workers, and $1300 for former firefighters, while in Arkansas the averages were $1000 for former police officers, $900 for former mass transit workers, and $1200 for former firefighters. Represent these average monthly incomes as a Chicago pension matrix and a Social Security matrix. Use these matrices to find the average monthly retirement income matrix for these groups of retired employees in these states.

57. Picnic Supplies A supermarket advertises a "summer picnic sale" with 2-liter bottles of soda for 89¢, a large bottle of pickles for $1.29, packages of hot dogs for $2.39, and large bags of chips for $1.69. The Culbert family wants 12 sodas, 2 bottles of pickles, 3 packages of hot dogs, and 4 bags of chips. Represent the sale prices as a row matrix and the Culbert's shopping list as a column matrix. Use these matrices to find the total cost of these items at these prices.

58. Part-Time Jobs A college student makes money during the semester shelving books in the library at $5.50 per hour, tutoring freshmen at $20 per hour, and pumping gas on weekends at $4.75 per hour. Last week this student shelved books for 10 hours, tutored for 3 hours, and pumped gas for 16 hours. Represent the hourly pay rates as a row matrix and the hours worked as a column matrix. Use these matrices to find the total amount this student earned last week.

59. Furniture Production A furniture company manufactures pine tables, chairs, and desks at factories in Wytheville and Andersen. Each table requires

2 hours of cutting and milling, 1 hour of assembly, and 2 hours of finishing; each chair requires 1.5 hours of cutting and milling, 1 hour of assembly, and 0.5 hours of finishing; and each desk requires 3 hours of cutting and milling, 2 hours of assembly, and 3 hours of finishing. At the Wytheville factory, the per-hour labor costs are $9 for cutting and milling work, $14 for assembly work, and $13 for finishing work, while at the Andersen factory, the per-hour labor costs are $10 for cutting and milling work, $13 for assembly work, and $12 for finishing work. Represent the table, chair, and desk labor requirements as a time matrix and the factory per-hour labor costs as a cost matrix. Use these matrices to find the production costs of these pieces of furniture at these factories.

60. Municipal Management The Hollins County business manager has received a request from the police department for 3 new cars, 5 new motorcycles, and 1 van as well as a request from the rescue squad for 1 new car, 2 new vans, and 4 new ambulances. The local Ford dealer's prices are $20,000 per car, $8000 per motorcycle, $27,000 per van, and $52,000 per ambulance, while the local GM dealer's prices are $19,000 per car, $6000 per motorcycle, $29,000 per van, and $58,000 per ambulance. Represent the police and rescue requests as a vehicle matrix and the Ford and GM bids as a price matrix. Use these matrices to find the costs of these requests at these dealers.

Explorations and Excursions

The following problems extend and augment the material presented in the text.

Symmetric Matrices Find $A^t \cdot A$ and $A \cdot A^t$ for each matrix A and identify these matrix products as "symmetric" or "not symmetric."

 If your instructor permits, you may carry out the calculations on a graphing calculator.

61. $\begin{pmatrix} 1 & 2 & 3 \\ 4 & 5 & 6 \end{pmatrix}$

62. $\begin{pmatrix} 1 & 0 & 1 & 0 \\ 0 & -1 & 0 & -1 \end{pmatrix}$

63. $\begin{pmatrix} 1 & -1 & 2 \\ 0 & 2 & -1 \\ 0 & 0 & 1 \end{pmatrix}$

64. $\begin{pmatrix} 1 & 0 & 1 & -1 \\ 0 & 2 & -1 & 1 \\ 0 & 1 & 2 & 0 \\ 1 & 0 & 0 & 1 \end{pmatrix}$

65. Show that the matrix products $A^t \cdot A$ and $A \cdot A^t$ are symmetric for any matrix A by showing that $(A^t \cdot A)^t = A^t \cdot A$ and $(A \cdot A^t)^t = A \cdot A^t$. [*Hint:* Use the fact that $(A \cdot B)^t = B^t \cdot A^t$.]

Matrix Multiplication and Function Composition A *fractional linear transformation* is a function $f(x) = \dfrac{ax + b}{cx + d}$ with at least one of c and d not zero. The composition of two fractional linear transformations $f(x)$ and $g(x)$ is the fractional linear transformation $f(g(x))$. Each fractional linear transformation $f(x) = \dfrac{ax + b}{cx + d}$ corresponds to a 2×2 matrix $F = \begin{pmatrix} a & b \\ c & d \end{pmatrix}$.

Find the composition $f(g(x))$ and matrix product $F \cdot G$ for each pair of fractional linear transformations $f(x)$ and $g(x)$ and corresponding matrices F and G.

66. $f(x) = \dfrac{2x + 3}{x + 2}$ and $g(x) = \dfrac{x - 2}{-x + 1}$

$F = \begin{pmatrix} 2 & 3 \\ 1 & 2 \end{pmatrix}$ and $G = \begin{pmatrix} 1 & -2 \\ -1 & 1 \end{pmatrix}$

67. $f(x) = \dfrac{3x + 2}{2x + 1}$ and $g(x) = \dfrac{2x + 1}{x - 2}$

$F = \begin{pmatrix} 3 & 2 \\ 2 & 1 \end{pmatrix}$ and $G = \begin{pmatrix} 2 & 1 \\ 1 & -2 \end{pmatrix}$

68. $f(x) = \dfrac{x + 1}{2x - 1}$ and $g(x) = \dfrac{1}{x}$

$F = \begin{pmatrix} 1 & 1 \\ 2 & -1 \end{pmatrix}$ and $G = \begin{pmatrix} 0 & 1 \\ 1 & 0 \end{pmatrix}$

69. $f(x) = \dfrac{2x - 1}{x + 1}$ and $g(x) = 2x$

$F = \begin{pmatrix} 2 & -1 \\ 1 & 1 \end{pmatrix}$ and $G = \begin{pmatrix} 2 & 0 \\ 0 & 1 \end{pmatrix}$

70. Show that the composition $f(g(x))$ of $f(x) = \dfrac{ax + b}{cx + d}$ and $g(x) = \dfrac{px + q}{rx + s}$ corresponds to the matrix product

$F \cdot G$ of $F = \begin{pmatrix} a & b \\ c & d \end{pmatrix}$ and $G = \begin{pmatrix} p & q \\ r & s \end{pmatrix}$.

Zero Divisors   A familiar property of the real numbers is that if the product of two numbers is zero, then at least one of the numbers must have been zero as well. This is *not* true for matrix multiplication. A *zero divisor* is a nonzero matrix A such that $A \cdot B = 0$ for some nonzero matrix B.

Show that the first of each pair of matrices is a zero divisor by finding the product of the two matrices in the order in which they are given.

71. $\begin{pmatrix} 1 & 0 \\ 1 & 0 \end{pmatrix}, \begin{pmatrix} 0 & 0 \\ 1 & 1 \end{pmatrix}$

72. $\begin{pmatrix} 1 & 1 & 0 \\ 1 & 1 & 0 \end{pmatrix}, \begin{pmatrix} 0 & 1 & -1 \\ 0 & -1 & 1 \\ 1 & 1 & 1 \end{pmatrix}$

73. $\begin{pmatrix} 1 & 0 & 1 \\ 0 & 1 & 0 \\ 1 & 0 & 1 \end{pmatrix}, \begin{pmatrix} 1 & -2 & 1 \\ 0 & 0 & 0 \\ -1 & 2 & -1 \end{pmatrix}$

74. $\begin{pmatrix} 0 & 1 & 1 & 0 \\ 1 & 1 & 1 & 1 \\ 1 & 1 & 1 & 1 \\ 0 & 1 & 1 & 0 \end{pmatrix}, \begin{pmatrix} 1 & 0 & 0 & -1 \\ 0 & -1 & 1 & 0 \\ 0 & 1 & -1 & 0 \\ -1 & 0 & 0 & 1 \end{pmatrix}$

75. $\begin{pmatrix} 2 & 4 \\ 3 & 6 \end{pmatrix}, \begin{pmatrix} 2 & -2 \\ -1 & 1 \end{pmatrix}$

76. $\begin{pmatrix} 1 & 2 & 3 \\ 4 & 5 & 6 \end{pmatrix}, \begin{pmatrix} 7 & -8 \\ -14 & 16 \\ 7 & -8 \end{pmatrix}$

77. $\begin{pmatrix} 1 & 2 & 3 \\ 4 & 5 & 6 \\ 7 & 8 & 9 \end{pmatrix}, \begin{pmatrix} 1 & -1 & 1 \\ -2 & 2 & -2 \\ 1 & -1 & 1 \end{pmatrix}$

78. $\begin{pmatrix} 1 & 2 & 3 & 4 \\ 5 & 6 & 7 & 8 \\ 9 & 10 & 11 & 12 \\ 13 & 14 & 15 & 16 \end{pmatrix}, \begin{pmatrix} -1 & -1 & 1 & 1 \\ 1 & 3 & -3 & -1 \\ 1 & -3 & 3 & -1 \\ -1 & 1 & -1 & 1 \end{pmatrix}$

79. $\begin{pmatrix} 1 & 5 & 9 & 13 \\ 2 & 6 & 10 & 14 \\ 3 & 7 & 11 & 15 \\ 4 & 8 & 12 & 16 \end{pmatrix}, \begin{pmatrix} 12 & -11 & 10 & -9 \\ -16 & 15 & -14 & 13 \\ -4 & 3 & -2 & 1 \\ 8 & -7 & 6 & -5 \end{pmatrix}$

80. $\begin{pmatrix} 2 & -1 & 2 & -1 & 2 \\ -1 & 3 & -1 & 3 & -1 \\ 3 & -2 & 3 & -2 & 3 \end{pmatrix}, \begin{pmatrix} 1 & 2 & 1 \\ -2 & -2 & 1 \\ 1 & -1 & -2 \\ 2 & 2 & -1 \\ -2 & -1 & 1 \end{pmatrix}$

The Pivot Operation   The *pivot operation* is a sequence of row operations that changes a nonzero element of the matrix into a one and then "zeros out" the column above and below this "pivot element." Let PE be the pivot element, R_{pivot} be the pivot row, R_{other} be any other row, and PCE be the entry in this other row in the same column as the pivot element. The pivot operation is $R'_{pivot} = \frac{1}{PE}R_{pivot}$ ("divide the pivot row by the pivot element") followed by $R'_{other} = R_{other} - PCE \cdot R^{new}_{pivot}$ for every other row ("subtract multiples of the new pivot row from the other rows to get zeros in the rest of the pivot element's column").

For each matrix, carry out the given sequence of row operations to pivot on the given pivot element.

81. Pivot on the 2 in row 1 and column 1 of
$\begin{pmatrix} 2 & 1 & 1 \\ 1 & 1 & 2 \\ 1 & 1 & 1 \end{pmatrix}$ by doing:
$R'_1 = \frac{1}{2}R_1$
$R'_2 = R_2 - R_1$ and
$R'_3 = R_3 - R_1$

82. Pivot on the 3 in row 1 and column 1 of
$\begin{pmatrix} 3 & 2 & 2 \\ 1 & 1 & 2 \\ 2 & 2 & 3 \end{pmatrix}$ by doing:
$R'_1 = \frac{1}{3}R_1,$
$R'_2 = R_2 - R_1,$ and
$R'_3 = R_3 - 2R_1$

83. Pivot on the $\frac{1}{3}$ in row 2 and column 2 of
$\begin{pmatrix} 1 & 2/3 & 2/3 \\ 0 & 1/3 & 4/3 \\ 0 & 2/3 & 5/3 \end{pmatrix}$ by doing:
$R'_2 = 3R_2,$
$R'_1 = R_1 - \frac{2}{3}R_2,$ and
$R'_3 = R_3 - \frac{2}{3}R_2$

84. Pivot on the -1 in row 3 and column 3 of
$\begin{pmatrix} 1 & 0 & -2 \\ 0 & 1 & 4 \\ 0 & 0 & -1 \end{pmatrix}$ by doing:
$R'_3 = -R_3,$
$R'_1 = R_1 + 2R_3,$ and
$R'_2 = R_2 - 4R_3$

Determinants   The *determinant* of a square matrix that is row equivalent to the identity matrix is the product of any sequence of pivot elements that reduce the matrix to the identity matrix. If the square matrix is not row equivalent to the identity matrix, the determinant is zero. The determinant of a non-square matrix is not defined. The basic properties of determinants were first published in 1750 by Gabriel Cramer (1704–1752) in an investigation of the solutions of systems of linear equations. By the mid-nineteenth century, James Joseph Sylvester (1814–1897) and Arthur Cayley (1821–1895) had realized that determinants were just part of a larger "theory of invariants" and Sylvester's invention of the term "matrix" (based on the Latin *mater* for "mother") was meant to indicate the larger context within which determinants should be understood. The following century saw determinants completely supplanted by the matrix methods presented in this chapter.

Find the determinant of each matrix by pivoting at row 1 and column 1, row 2 and column 2, and then row 3 and column 3.

 If your calculator has a DETERMINANT command, use it to check your answers.

85. $\begin{pmatrix} 2 & 1 & 1 \\ 1 & 1 & 2 \\ 1 & 1 & 1 \end{pmatrix}$ **86.** $\begin{pmatrix} 3 & 2 & 2 \\ 1 & 1 & 2 \\ 2 & 2 & 3 \end{pmatrix}$

87. $\begin{pmatrix} 3 & -2 & 4 \\ 1 & -7 & -2 \\ -3 & -7 & -9 \end{pmatrix}$ **88.** $\begin{pmatrix} 6 & 5 & -3 \\ -5 & 5 & -4 \\ 0 & -4 & 3 \end{pmatrix}$

Properties of Determinants The determinant is multiplied by -1 if two rows are switched; is multiplied by a number if a row is multiplied by that same number; and is unchanged if one row is added to (or subtracted from) another row. Furthermore, the determinant of the transpose of a matrix is the same as the determinant of the original matrix.

Use the properties of the determinant to predict the determinant of the second matrix of each pair, given that the determinant of the first is -1.

 Verify your prediction using the DET calculator matrix math command.

89. $\begin{pmatrix} 3 & 2 & 2 \\ 1 & 1 & 2 \\ 2 & 2 & 3 \end{pmatrix}$ and $\begin{pmatrix} 1 & 1 & 2 \\ 3 & 2 & 2 \\ 2 & 2 & 3 \end{pmatrix}$

90. $\begin{pmatrix} 2 & 1 & 1 \\ 1 & 1 & 2 \\ 1 & 1 & 1 \end{pmatrix}$ and $\begin{pmatrix} 2 & 1 & 1 \\ 1 & 1 & 1 \\ 1 & 1 & 2 \end{pmatrix}$

91. $\begin{pmatrix} 3 & 2 & 2 \\ 1 & 1 & 2 \\ 2 & 2 & 3 \end{pmatrix}$ and $\begin{pmatrix} 3 & 2 & 2 \\ 5 & 5 & 10 \\ 2 & 2 & 3 \end{pmatrix}$

92. $\begin{pmatrix} 2 & 1 & 1 \\ 1 & 1 & 2 \\ 1 & 1 & 1 \end{pmatrix}$ and $\begin{pmatrix} 2 & 1 & 1 \\ 1 & 1 & 2 \\ 6 & 6 & 6 \end{pmatrix}$

93. $\begin{pmatrix} 3 & 2 & 2 \\ 1 & 1 & 2 \\ 2 & 2 & 3 \end{pmatrix}$ and $\begin{pmatrix} 3 & 2 & 2 \\ 16 & 11 & 12 \\ 2 & 2 & 3 \end{pmatrix}$

94. $\begin{pmatrix} 2 & 1 & 1 \\ 1 & 1 & 2 \\ 1 & 1 & 1 \end{pmatrix}$ and $\begin{pmatrix} 2 & 1 & 1 \\ 6 & 6 & 7 \\ 1 & 1 & 1 \end{pmatrix}$

95. $\begin{pmatrix} 3 & 2 & 2 \\ 1 & 1 & 2 \\ 2 & 2 & 3 \end{pmatrix}$ and $\begin{pmatrix} 3 & 1 & 2 \\ 2 & 1 & 2 \\ 2 & 2 & 3 \end{pmatrix}$

96. $\begin{pmatrix} 2 & 1 & 1 \\ 1 & 1 & 2 \\ 1 & 1 & 1 \end{pmatrix}$ and $\begin{pmatrix} 2 & 1 & 1 \\ 1 & 1 & 1 \\ 1 & 2 & 1 \end{pmatrix}$

97. A square matrix is a *diagonal matrix* if every element not on the main diagonal is zero. Use the definition and properties of the determinant to show that the determinant of any diagonal matrix is the product of the elements on the main diagonal.

98. A square matrix is an *upper triangular matrix* if every element "below" the main diagonal is zero. Use the definition and properties of the determinant to show that the determinant of any upper triangular matrix is the product of the elements on the main diagonal.

99. A square matrix is a *lower triangular matrix* if every element "above" the main diagonal is zero. Use the definition and properties of the determinant to show that the determinant of any lower triangular matrix is the product of the elements on the main diagonal.

100. Show that the determinant of any 2×2 matrix $\begin{pmatrix} a & b \\ c & d \end{pmatrix}$ is $ad - bc$ by *symbolically* pivoting at row 1 and column 1 and then again at row 2 and column 2 or by *symbolically* pivoting at row 1 and column 1 and using the result stated in Exercise 98.

Inverse Matrices and Systems of Linear Equations

Sensitivity Analysis

Although every dawn does greet a new day, one day's business problems are usually not entirely new but rather just variations on the problems of the previous day. But if yesterday's problems were understood and solved yesterday, there is little reason to begin today's work by solving similar problems all over again. For example, a factory that manufactures products in response to orders received each day should require only minor changes in its production lines to fulfill the new day's orders. *Sensitivity analysis* is the study of how small changes in a problem affect the solution. Consider the following system of equations and the solution.

$$\begin{cases} x_1 + x_2 + x_3 = 15 \\ 2x_1 + 2x_2 + x_3 = 23 \\ x_1 + 2x_2 + x_3 = 18 \end{cases} \longrightarrow \begin{array}{l} x_1 = 5 \\ x_2 = 3 \\ x_3 = 7 \end{array}$$

Problem   Solution

If we change the 15 in the first equation to 16, the solution becomes $x_1 = 5$, $x_2 = 2$, $x_3 = 9$. Notice that this change to the first equation did not change the value of the first variable. If we change the 15 to 17, the solution becomes $x_1 = 5$, $x_2 = 1$, $x_3 = 11$. Is it possible that every increase of the 15 results in a decrease of the x_2 value by the same amount, together with an increase of the x_3 value by twice that same amount? The answers to this and other similar questions are simply found using the inverse matrix discussed in this section.

Introduction

We can write a system of linear equations as a single matrix equation $AX = B$, so we will solve the system by solving this same matrix equation. Just as the simple equation $2x = 10$ is easily solved by multiplying each side by $\frac{1}{2}$ (the inverse of 2) to obtain $1x = 5$, we will solve the matrix equation $AX = B$ by multiplying by an "inverse" matrix to transform the left side from AX to IX or simply X (because the identity matrix I plays the role of 1 in matrix arithmetic).

Inverse Matrices

The inverse of the number 2 is $\frac{1}{2}$ because $\frac{1}{2} \cdot 2 = 1$ and $2 \cdot \frac{1}{2} = 1$. Similarly, the *inverse* of a matrix is defined as another matrix whose product with the first (in either order) gives the identity matrix. The inverse of the matrix A is denoted A^{-1} just as the inverse of the number 2 is $2^{-1} = \frac{1}{2}$.

Inverse Matrix

The square matrix A has an *inverse matrix* A^{-1} such that

$$A^{-1} \cdot A = I \quad \text{and} \quad A \cdot A^{-1} = I$$

if and only if A is row equivalent to the identity matrix.

EXAMPLE 1 **Checking Inverse Matrices**

For $A = \begin{pmatrix} \frac{1}{2} & 3 \\ 1 & 5 \end{pmatrix}$ and $A^{-1} = \begin{pmatrix} -10 & 6 \\ 2 & -1 \end{pmatrix}$, verify that A^{-1} is indeed the inverse of A by showing that $A^{-1} \cdot A = I$ and $A \cdot A^{-1} = I$.

Solution

$$A^{-1} \cdot A = \begin{pmatrix} -10 & 6 \\ 2 & -1 \end{pmatrix}\begin{pmatrix} \frac{1}{2} & 3 \\ 1 & 5 \end{pmatrix} = \begin{pmatrix} 1 & 0 \\ 0 & 1 \end{pmatrix} = I$$

and

$$A \cdot A^{-1} = \begin{pmatrix} \frac{1}{2} & 3 \\ 1 & 5 \end{pmatrix}\begin{pmatrix} -10 & 6 \\ 2 & -1 \end{pmatrix} = \begin{pmatrix} 1 & 0 \\ 0 & 1 \end{pmatrix} = I$$

as required. (You should check the arithmetic in both multiplications.)

∎

To verify that two matrices are inverses of each other, it is enough to multiply them together in one order to get I (see Exercise 69 on page 237).

How to Find Inverse Matrices

Not all matrices have inverses. A square matrix is *invertible* if it has an inverse, and it is *singular* if it does not. We saw on page 217 that if a matrix A is equivalent to the identity matrix I, then the row reduction can be accomplished by multiplying A by the product R of the row operation matrices: $R \cdot A = I$, making R the inverse of A. Thus a matrix is invertible if it is equivalent to the identity matrix. Furthermore, the inverse of an invertible matrix A can be found by adjoining an identity matrix, $(A\,|\,I)$, and row reducing to obtain $(I\,|\,A^{-1})$, with the inverse matrix on the right. This means we can calculate A^{-1} as follows:

Calculating the Inverse Matrix

> If the square matrix A is equivalent to the identity matrix, then row reducing the augmented matrix $(A\,|\,I)$ gives $(I\,|\,A^{-1})$, so that A^{-1} appears in the right half.

Of course, if A cannot be row-reduced to the identity matrix, this will be discovered during the attempt to row-reduce $(A\,|\,I)$ and the reduction process can be stopped immediately with the conclusion that A is singular.

EXAMPLE 2 **Calculating an Inverse Matrix**

Find the inverse of $A = \begin{pmatrix} 1 & 0 & 2 \\ 1 & 1 & 1 \\ 1 & 1 & 2 \end{pmatrix}$.

Solution

A is 3×3, so we augment it by I_3 and row-reduce $(A\,|\,I)$.

Sometimes written as
$$\begin{pmatrix} 1 & 0 & 2 & | & 1 & 0 & 0 \\ 1 & 1 & 1 & | & 0 & 1 & 0 \\ 1 & 1 & 2 & | & 0 & 0 & 1 \end{pmatrix}$$

$$\begin{pmatrix} 1 & 0 & 2 & 1 & 0 & 0 \\ 1 & 1 & 1 & 0 & 1 & 0 \\ 1 & 1 & 2 & 0 & 0 & 1 \end{pmatrix} \qquad (A\,|\,I)$$

$\underbrace{}_{A} \quad \underbrace{}_{I}$

There are many different sequences of row operations to reduce this augmented matrix, and all reach the same conclusion. One way is as follows:

$$\begin{pmatrix} 1 & 0 & 2 & 1 & 0 & 0 \\ 1 & 1 & 1 & 0 & 1 & 0 \\ 0 & 0 & 1 & 0 & -1 & 1 \end{pmatrix} R_3' = R_3 - R_2$$

$$\begin{pmatrix} 1 & 0 & 0 & 1 & 2 & -2 \\ 1 & 1 & 0 & 0 & 2 & -1 \\ 0 & 0 & 1 & 0 & -1 & 1 \end{pmatrix} \begin{matrix} R_1' = R_1 - 2R_3 \\ R_2' = R_2 - R_3 \end{matrix}$$

$$\begin{pmatrix} 1 & 0 & 0 & 1 & 2 & -2 \\ 0 & 1 & 0 & -1 & 0 & 1 \\ 0 & 0 & 1 & 0 & -1 & 1 \end{pmatrix} R_2' = R_2 - R_1 \qquad \begin{matrix} \text{Since } I \text{ is on} \\ \text{the left, } A^{-1} \text{ is} \\ \text{on the right} \end{matrix}$$

$\underbrace{}_{I} \quad \underbrace{}_{A^{-1}}$

Therefore, $A^{-1} = \begin{pmatrix} 1 & 2 & -2 \\ -1 & 0 & 1 \\ 0 & -1 & 1 \end{pmatrix}$ is the inverse of the original matrix

$A = \begin{pmatrix} 1 & 0 & 2 \\ 1 & 1 & 1 \\ 1 & 1 & 2 \end{pmatrix}$.

PRACTICE PROBLEM 1

For the matrices A and A^{-1} in Example 2, verify that A^{-1} *is* the inverse of A by showing that $A^{-1} \cdot A = I$.   *Solution at the back of the book*

SPREADSHEET EXPLORATION

The spreadsheet below uses the matrix inverse function
MINVERSE(   ) to calculate the inverses of the matrices

$\begin{pmatrix} 1 & 0 & 2 \\ 1 & 1 & 1 \\ 1 & 1 & 2 \end{pmatrix}$ and $\begin{pmatrix} 1 & -6 & 4 & 6 \\ 2 & 5 & 4 & 0 \\ 2 & 2 & 3 & 2 \\ 1 & 4 & 2 & -1 \end{pmatrix}$.

	A	B	C	D	E	F	G	H	I
1		Matrix A					Inverse of A		
2									
3	1	0	2			1	2	-2	
4	1	1	1			-1	0	1	
5	1	1	2			0	-1	1	
6									
7	1	-6	4	6		-5	34	-8	-46
8	2	5	4	0		2	-15	4	20
9	2	2	3	2		-3.25E-17	2	-1	-2
10	1	4	2	-1		3	-22	6	29

Spreadsheet programs sometimes produces round-off errors (such as the small number -3.25×10^{-17} instead of zero). The correct inverse

of $\begin{pmatrix} 1 & -6 & 4 & 6 \\ 2 & 5 & 4 & 0 \\ 2 & 2 & 3 & 2 \\ 1 & 4 & 2 & -1 \end{pmatrix}$ is $\begin{pmatrix} -5 & 34 & -8 & -46 \\ 2 & -15 & 4 & 20 \\ 0 & 2 & -1 & -2 \\ 3 & -22 & 6 & 29 \end{pmatrix}$ as you can easily check.

Be careful! Because apparent round-off errors might also indicate that A is invertible when it is in fact singular, *always* verify that the spreadsheet (or computer or calculator) answer does indeed satisfy $A^{-1} \cdot A = I$.

Solving $A \cdot X = B$ Using A^{-1}

Just as the equation $2x = 10$ is solved simply by multiplying by 2^{-1} or $\frac{1}{2}$, we can solve $AX = B$ by multiplying both sides (on the left) by A^{-1}.

$$AX = B \qquad \text{Original equation}$$

$$A^{-1} \cdot A \cdot X = A^{-1} \cdot B \qquad \text{Left-multiplying by } A^{-1}$$

$$I \cdot X = A^{-1} \cdot B \qquad \text{Since } A^{-1} \cdot A = I$$

$$X = A^{-1}B \qquad \text{Since } I \cdot X = X$$

Notice that because matrix multiplication is *not* cummutative, we *must* multiply both sides of $A \cdot X = B$ on the *left* by A^{-1}. (And besides, the product $B \cdot A^{-1}$ is not defined because B is a column matrix and A^{-1} is a square matrix.)

Solving $A \cdot X = B$ Using A^{-1}

The solution of $A \cdot X = B$ is $X = A^{-1} \cdot B$, provided the square matrix A is invertible.

If A is singular, the system of equations is *dependent* and we must use regular row reduction (as in Section 3.3) to find the solution if the equations are consistent or to establish that there is no solution because they are inconsistent.

EXAMPLE 3 **Solving a System of Equations Using the Inverse Matrix**

Use the inverse matrix to solve $\begin{cases} x_1 + 2x_3 = 22 \\ x_1 + x_2 + x_3 = 11 \\ x_1 + x_2 + 2x_3 = 20 \end{cases}$.

Solution

Writing this system of equations as the matrix equation $A \cdot X = B$, we have

$$\begin{pmatrix} 1 & 0 & 2 \\ 1 & 1 & 1 \\ 1 & 1 & 2 \end{pmatrix} \cdot \begin{pmatrix} x_1 \\ x_2 \\ x_3 \end{pmatrix} = \begin{pmatrix} 22 \\ 11 \\ 20 \end{pmatrix} \qquad A \cdot X = B$$

$$\underbrace{}_{A} \quad \underbrace{}_{X} \; \underbrace{}_{B}$$

Ordinarily, we would now find the inverse of matrix A by row-reducing $(A \mid I)$. However, this is exactly the matrix that we just row-reduced in

Example 2, so we will simply use the A^{-1} found on page 227 and write the solution as $X = A^{-1} \cdot B$.

$$\begin{pmatrix} x_1 \\ x_2 \\ x_3 \end{pmatrix} = \begin{pmatrix} 1 & 2 & -2 \\ -1 & 0 & 1 \\ 0 & -1 & 1 \end{pmatrix} \cdot \begin{pmatrix} 22 \\ 11 \\ 20 \end{pmatrix} = \begin{pmatrix} 4 \\ -2 \\ 9 \end{pmatrix} \qquad \text{Using } A^{-1} \text{ from Example 2}$$

$$\underbrace{}_{X} \quad \underbrace{}_{A^{-1}} \quad \underbrace{}_{B}$$

That is, the solution is $x_1 = 4$, $x_2 = -2$, $x_3 = 9$. We check this solution by substituting into the original system of equations.

$$4 + 2 \cdot 9 \qquad = 22$$

$$4 - 2 + 9 \qquad = 11$$

$$4 - 2 + 2 \cdot 9 = 20 \qquad \text{It checks!}$$

Graphing Calculator Exploration

If A is invertible, you can solve $A \cdot X = B$ by entering the matrices A and B and then calculating $A^{-1} \cdot B$. To solve the problem from Example 3, enter $\begin{pmatrix} 1 & 0 & 2 \\ 1 & 1 & 1 \\ 1 & 1 & 2 \end{pmatrix}$ in [A], enter $\begin{pmatrix} 22 \\ 11 \\ 20 \end{pmatrix}$ in [B], and calculate $[A]^{-1}[B]$:

```
[A]
      [[1 0 2]
       [1 1 1]
       [1 1 2]]
```

```
[B]
        [[22]
         [11]
         [20]]
```

```
[A]⁻¹[B]
         [[4 ]
          [-2]
          [9 ]]
```

Now enter your choices for any three integers in [B] and find $[A]^{-1}[B]$. Do these values satisfy your new equations?

PRACTICE PROBLEM 2

Solve $\begin{cases} x_1 + 2x_3 = -5 \\ x_1 + x_2 + x_3 = 10. \\ x_1 + x_2 + 2x_3 = 0 \end{cases}$ *Solution at the back of the book*

[*Hint:* Write this system as $AX = B$ and notice that A is the same as in Example 3 but B is different and so the solution can be found by matrix multiplication using the inverse found in Example 2.]

Solving $A \cdot X = B$ by finding the inverse and then using $X = A^{-1}B$ is only slightly more difficult than row-reducing the augmented matrix for the system of equations. However, if you need to solve $A \cdot X = B$ with the same A but different values in B, using the inverse A^{-1} means that you only do the row reductions *once* and then the solution for any new matrix B is given by a simple matrix multiplication. Such problems, where A remains the same but B changes, occur frequently in applications.

Solving $AX = B$ for Many Different B's

If A is invertible, the solutions of the matrix equations

$$A \cdot X = B_1, \quad A \cdot X = B_2, \ldots, A \cdot X = B_k$$

may all be found by calculating A^{-1} once and then finding the solutions as the products

$$X = A^{-1} \cdot B_1, \quad X = A^{-1} \cdot B_2, \ldots, X = A^{-1} \cdot B_k$$

EXAMPLE 4 **Jewelry Production**

An employee-owned jewelry company fabricates enameled gold rings, pendants, and bracelets. Each ring requires 3 grams of gold, 1 gram of enameling compound, and 2 hours of labor; each pendant requires 6 grams of gold, 2 grams of enameling compound, and 3 hours of labor; and each bracelet requires 8 grams of gold, 3 grams of enameling compound, and 2 hours of labor. Each of the five employee–owners works 160 hours each month, and the company has contracts guaranteeing the delivery of the grams of gold and enameling compound shown in the table on the first day of the months of March, April, May, and June. How many rings, pendants, and bracelets should the company fabricate each month to use all the available materials and time?

	March	April	May	June
Gold	1720	2620	2460	2220
Enamel	600	960	900	800

Solution

Let x_1, x_2, and x_3 be the numbers of rings, pendants, and bracelets produced in one month. Then the required grams of gold, grams of enameling compound, and hours of labor are

Gold	$3x_1 + 6x_2 + 8x_3$
Enamel	$x_1 + 2x_2 + 3x_3$
Labor	$2x_1 + 3x_2 + 2x_3$

For the month of March, these quantities must match the amounts available:

$$\begin{cases} 3x_1 + 6x_2 + 8x_3 = 1720 \\ x_1 + 2x_2 + 3x_3 = 600 \\ 2x_1 + 3x_2 + 2x_3 = 800 \end{cases}$$

From the table and 5 workers at 160 hours each

The systems of equations for the other months follow in a similar manner, and we have four problems to solve:

March	April
$\begin{cases} 3x_1 + 6x_2 + 8x_3 = 1720 \\ x_1 + 2x_2 + 3x_3 = 600 \\ 2x_1 + 3x_2 + 2x_3 = 800 \end{cases}$	$\begin{cases} 3x_1 + 6x_2 + 8x_3 = 2620 \\ x_1 + 2x_2 + 3x_3 = 960 \\ 2x_1 + 3x_2 + 2x_3 = 800 \end{cases}$
May	**June**
$\begin{cases} 3x_1 + 6x_2 + 8x_3 = 2460 \\ x_1 + 2x_2 + 3x_3 = 900 \\ 2x_1 + 3x_2 + 2x_3 = 800 \end{cases}$	$\begin{cases} 3x_1 + 6x_2 + 8x_3 = 2220 \\ x_1 + 2x_2 + 3x_3 = 800 \\ 2x_1 + 3x_2 + 2x_3 = 800 \end{cases}$

We could solve these four problems by row-reducing the corresponding augmented matrix for the March problem and then repeating the same sequence of row operations to solve the augmented matrices for the April, May, and June problems. The coefficient matrices for these four problems are the same, so if we find the inverse matrix for the March problem by row-reducing $(A \mid I)$, we will only have to carry out the row operations to reduce A once, and then the four problems can be solved with just four matrix multiplications.

We first find A^{-1} by row-reducing $(A \mid I)$. There are many different sequences of row operations to reduce this augmented matrix, and all reach the same conclusion. One possible way is as follows:

$$(A\,|\,I) = \begin{pmatrix} 3 & 6 & 8 & 1 & 0 & 0 \\ 1 & 2 & 3 & 0 & 1 & 0 \\ 2 & 3 & 2 & 0 & 0 & 1 \end{pmatrix}$$

$$\begin{pmatrix} 0 & 0 & 1 & -1 & 3 & 0 \\ 1 & 2 & 3 & 0 & 1 & 0 \\ 0 & 1 & 4 & 0 & 2 & -1 \end{pmatrix} \begin{matrix} R_1' = 3R_2 - R_1 \\ \\ R_3' = 2R_2 - R_3 \end{matrix}$$

Use 1 in the first column to zero out the rest of that column

$$\begin{pmatrix} 0 & 0 & 1 & -1 & 3 & 0 \\ 1 & 0 & -5 & 0 & -3 & 2 \\ 0 & 1 & 4 & 0 & 2 & -1 \end{pmatrix} R_2' = R_2 - 2R_3$$

Use 1 in second column to zero out the rest of that column

$$\begin{pmatrix} 0 & 0 & 1 & -1 & 3 & 0 \\ 1 & 0 & 0 & -5 & 12 & 2 \\ 0 & 1 & 0 & 4 & -10 & -1 \end{pmatrix} \begin{matrix} R_2' = R_2 + 5R_1 \\ R_3' = R_2 - 4R_1 \end{matrix}$$

Use 1 in third column to zero out the rest of that column

$$\begin{pmatrix} 1 & 0 & 0 & -5 & 12 & 2 \\ 0 & 1 & 0 & 4 & -10 & -1 \\ 0 & 0 & 1 & -1 & 3 & 0 \end{pmatrix} \begin{matrix} R_1' = R_2 \\ R_2' = R_3 \\ R_3' = R_1 \end{matrix}$$

Switch rows to achieve the correct order

Since the left side is I, the right side is the inverse matrix

$$A^{-1} = \begin{pmatrix} -5 & 12 & 2 \\ 4 & -10 & -1 \\ -1 & 3 & 0 \end{pmatrix}$$

The solution for each month is simply the product of this inverse matrix with the column matrix of the constant terms for that month.

Month	$X = A^{-1} \cdot B$				Solution
March	$\begin{pmatrix} x_1 \\ x_2 \\ x_3 \end{pmatrix} =$	$\begin{pmatrix} -5 & 12 & 2 \\ 4 & -10 & -1 \\ -1 & 3 & 0 \end{pmatrix} \cdot$	$\begin{pmatrix} 1720 \\ 600 \\ 800 \end{pmatrix} =$	$\begin{pmatrix} 200 \\ 80 \\ 80 \end{pmatrix}$	200 rings 80 pendants 80 bracelets
April	$\begin{pmatrix} x_1 \\ x_2 \\ x_3 \end{pmatrix} =$	$\begin{pmatrix} -5 & 12 & 2 \\ 4 & -10 & -1 \\ -1 & 3 & 0 \end{pmatrix} \cdot$	$\begin{pmatrix} 2620 \\ 960 \\ 800 \end{pmatrix} =$	$\begin{pmatrix} 20 \\ 80 \\ 260 \end{pmatrix}$	20 rings 80 pendants 260 bracelets
May	$\begin{pmatrix} x_1 \\ x_2 \\ x_3 \end{pmatrix} =$	$\begin{pmatrix} -5 & 12 & 2 \\ 4 & -10 & -1 \\ -1 & 3 & 0 \end{pmatrix} \cdot$	$\begin{pmatrix} 2460 \\ 900 \\ 800 \end{pmatrix} =$	$\begin{pmatrix} 100 \\ 40 \\ 240 \end{pmatrix}$	100 rings 40 pendants 240 bracelets
June	$\begin{pmatrix} x_1 \\ x_2 \\ x_3 \end{pmatrix} =$	$\begin{pmatrix} -5 & 12 & 2 \\ 4 & -10 & -1 \\ -1 & 3 & 0 \end{pmatrix} \cdot$	$\begin{pmatrix} 2220 \\ 800 \\ 800 \end{pmatrix} =$	$\begin{pmatrix} 100 \\ 80 \\ 180 \end{pmatrix}$	100 rings 80 pendants 180 bracelets

Section Summary

A square matrix A is *invertible* if there is an *inverse matrix* A^{-1} such that $A^{-1} \cdot A = I$ and $A \cdot A^{-1} = I$.   A square matrix is *singular* if it is not invertible.

The inverse of an invertible matrix A may be found by row-reducing the augmented matrix $(A \mid I)$ to obtain $(I \mid A^{-1})$.

The solution of the matrix equation $A \cdot X = B$ is $X = A^{-1} \cdot B$, provided the matrix A is invertible.

EXERCISES 3.5

Find each matrix product. Identify each pair of matrices as "a matrix and its inverse" or "not a matrix and its inverse."

1. $\begin{pmatrix} 1 & 2 \\ -1 & -1 \end{pmatrix}$ and $\begin{pmatrix} -1 & -2 \\ 1 & 1 \end{pmatrix}$

2. $\begin{pmatrix} 5 & 3 \\ -3 & -2 \end{pmatrix}$ and $\begin{pmatrix} 2 & 3 \\ -3 & -5 \end{pmatrix}$

3. $\begin{pmatrix} 1 & 1 & 0 \\ 2 & 1 & 1 \\ 1 & 0 & 0 \end{pmatrix}$ and $\begin{pmatrix} 0 & 0 & 1 \\ 1 & 0 & -1 \\ -1 & 1 & -1 \end{pmatrix}$

4. $\begin{pmatrix} 2 & 1 & -1 \\ -2 & 0 & 1 \\ -3 & -1 & 2 \end{pmatrix}$ and $\begin{pmatrix} 1 & -1 & 1 \\ 1 & 1 & 0 \\ 2 & -1 & 2 \end{pmatrix}$

5. $\begin{pmatrix} 4 & 6 & 3 \\ 3 & 4 & 1 \\ 5 & 7 & 3 \end{pmatrix}$ and $\begin{pmatrix} -5 & -3 & 6 \\ 4 & 3 & -5 \\ -1 & -2 & 2 \end{pmatrix}$

6. $\begin{pmatrix} 3 & 2 & 3 \\ 5 & 2 & 6 \\ 2 & 3 & 1 \end{pmatrix}$ and $\begin{pmatrix} 16 & -7 & -6 \\ -7 & 3 & 3 \\ -11 & 5 & 4 \end{pmatrix}$

7. $\begin{pmatrix} 10 & -4 & -7 \\ -7 & 3 & 5 \\ 4 & -1 & -3 \end{pmatrix}$ and $\begin{pmatrix} 4 & 5 & -1 \\ 1 & 2 & 1 \\ 5 & 6 & 2 \end{pmatrix}$

8. $\begin{pmatrix} 4 & 1 & 5 \\ -1 & 1 & -2 \\ 3 & 4 & 2 \end{pmatrix}$ and $\begin{pmatrix} 10 & 18 & -7 \\ -4 & -7 & -3 \\ -7 & -13 & 5 \end{pmatrix}$

9. $\begin{pmatrix} 2 & 0 & 1 & 0 \\ 1 & 1 & 1 & 0 \\ -2 & 0 & -1 & 1 \\ 1 & 0 & 0 & 1 \end{pmatrix}$ and $\begin{pmatrix} -1 & 0 & -1 & 1 \\ -2 & 1 & -1 & 1 \\ 3 & 0 & 2 & -2 \\ 1 & 0 & 1 & 0 \end{pmatrix}$

10. $\begin{pmatrix} 1 & -1 & 0 & 0 \\ -1 & 1 & 0 & 1 \\ -3 & 2 & 1 & 2 \\ 1 & 0 & 0 & -1 \end{pmatrix}$ and $\begin{pmatrix} 1 & 1 & 0 & 1 \\ 0 & 1 & 0 & 1 \\ 1 & -1 & 1 & 1 \\ 1 & 1 & 0 & 0 \end{pmatrix}$

Row-reduce $(A \mid I)$ for each matrix A to find the inverse matrix A^{-1} or to identify A as a singular matrix.

11. $\begin{pmatrix} 1 & 3 \\ 0 & 1 \end{pmatrix}$

12. $\begin{pmatrix} 1 & 4 \\ 0 & 1 \end{pmatrix}$

13. $\begin{pmatrix} 11 & 2 \\ 6 & 1 \end{pmatrix}$

14. $\begin{pmatrix} 5 & 7 \\ 2 & 3 \end{pmatrix}$

15. $\begin{pmatrix} 1 & 1 & 0 \\ 3 & 0 & 2 \\ 1 & 0 & 1 \end{pmatrix}$

16. $\begin{pmatrix} 1 & 0 & 1 \\ 1 & 1 & 0 \\ 5 & 0 & 4 \end{pmatrix}$

17. $\begin{pmatrix} 1 & 1 & 0 \\ 0 & 1 & 1 \\ 1 & 2 & 1 \end{pmatrix}$

18. $\begin{pmatrix} 1 & 0 & 1 \\ 2 & 1 & 3 \\ 0 & 1 & 1 \end{pmatrix}$

19. $\begin{pmatrix} 1 & 1 & 0 & 1 \\ 0 & 1 & 0 & 0 \\ 1 & 0 & 1 & 0 \\ 0 & 1 & 0 & 1 \end{pmatrix}$

20. $\begin{pmatrix} 1 & 0 & 0 & 1 \\ 0 & 1 & 1 & 1 \\ 2 & 0 & 0 & 1 \\ 0 & 1 & 0 & 1 \end{pmatrix}$

21. $\begin{pmatrix} 2 & 3 & 1 \\ 1 & 2 & 1 \\ 2 & 3 & 2 \end{pmatrix}$

22. $\begin{pmatrix} 1 & 2 & 2 \\ 1 & 1 & 1 \\ 1 & 3 & 2 \end{pmatrix}$

23. $\begin{pmatrix} 3 & -4 & 2 \\ -1 & 2 & -1 \\ 4 & -3 & 2 \end{pmatrix}$

24. $\begin{pmatrix} 2 & 2 & -1 \\ -2 & -1 & 1 \\ 3 & 4 & -1 \end{pmatrix}$

25. $\begin{pmatrix} 1 & -2 & 1 & 2 \\ 1 & -1 & 1 & 2 \\ 2 & 2 & 1 & 3 \\ 1 & -1 & 1 & 1 \end{pmatrix}$

26. $\begin{pmatrix} 2 & 1 & -2 & 1 \\ 1 & 1 & 2 & 1 \\ 1 & 1 & 1 & 1 \\ 2 & 2 & 3 & 1 \end{pmatrix}$

27. $\begin{pmatrix} 1 & 0 & -1 & 1 \\ 1 & 1 & 1 & 2 \\ 2 & -2 & -6 & 0 \\ 1 & 0 & 2 & 1 \end{pmatrix}$

28. $\begin{pmatrix} 1 & -1 & 0 & 1 \\ 2 & -2 & -1 & 0 \\ 1 & -1 & -1 & -1 \\ 2 & -2 & 0 & 2 \end{pmatrix}$

 29. $\begin{pmatrix} 1 & 0 & 1 & 0 & 1 \\ -1 & 2 & -1 & 1 & -1 \\ 4 & 2 & 4 & 2 & 3 \\ -1 & 4 & -2 & 2 & -2 \\ 3 & 1 & 2 & 1 & 1 \end{pmatrix}$

 30. $\begin{pmatrix} 4 & -1 & 3 & -1 & 1 \\ 2 & 2 & 1 & 4 & 0 \\ 4 & -1 & 2 & -2 & 1 \\ 2 & 1 & 1 & 2 & 0 \\ 3 & -1 & 1 & -2 & 1 \end{pmatrix}$

Rewrite each system of equations as a matrix equation $A \cdot X = B$ and use the inverse of A to find the solution. Be sure to check your solution in the original system of equations.

31. $\begin{cases} 11x_1 + 2x_2 = 9 \\ 6x_1 + x_2 = 5 \end{cases}$

32. $\begin{cases} 5x_1 + 7x_2 = 2 \\ 2x_1 + 3x_2 = 1 \end{cases}$

33. $\begin{cases} x_1 + x_2 = 2 \\ 3x_1 + 2x_3 = 5 \\ x_1 + x_3 = 2 \end{cases}$

34. $\begin{cases} x_1 + x_3 = 1 \\ x_1 + x_2 = 3 \\ 5x_1 + 4x_3 = 6 \end{cases}$

35. $\begin{cases} 2x_1 + 3x_2 + x_3 = 6 \\ x_1 + 2x_2 + x_3 = 4 \\ 2x_1 + 3x_2 + 2x_3 = 7 \end{cases}$

36. $\begin{cases} x_1 + 2x_2 + 2x_3 = 5 \\ x_1 + x_2 + x_3 = 3 \\ x_1 + 3x_2 + 2x_3 = 6 \end{cases}$

37. $\begin{cases} 3x_1 - 4x_2 + 2x_3 = 12 \\ -x_1 + 2x_2 - x_3 = -4 \\ 4x_1 - 3x_2 + 2x_3 = 15 \end{cases}$

38. $\begin{cases} 2x_1 + 2x_2 - x_3 = 1 \\ -2x_1 - x_2 + x_3 = 0 \\ 3x_1 + 4x_2 - x_3 = 2 \end{cases}$

39. $\begin{cases} x_1 - 2x_2 + x_3 + 2x_4 = 8 \\ x_1 - x_2 + x_3 + 2x_4 = 7 \\ 2x_1 + 2x_2 + x_3 + 3x_4 = 7 \\ x_1 - x_2 + x_3 + x_4 = 6 \end{cases}$

40. $\begin{cases} 2x_1 + x_2 - 2x_3 + x_4 = 2 \\ x_1 + x_2 + 2x_3 + x_4 = 5 \\ x_1 + x_2 + x_3 + x_4 = 4 \\ 2x_1 + 2x_2 + 3x_3 + x_4 = 8 \end{cases}$

41. $\begin{cases} x_1 - 2x_2 + x_3 + 2x_4 = 10 \\ x_1 - x_2 + x_3 + 2x_4 = 8 \\ 2x_1 + 2x_2 + x_3 + 3x_4 = 5 \\ x_1 - x_2 + x_3 + x_4 = 6 \end{cases}$

42. $\begin{cases} 2x_1 + x_2 - 2x_3 + x_4 = 8 \\ x_1 + x_2 + 2x_3 + x_4 = 3 \\ x_1 + x_2 + x_3 + x_4 = 4 \\ 2x_1 + 2x_2 + 3x_3 + x_4 = 5 \end{cases}$

43. $\begin{cases} x_1 + x_3 + x_5 = 4 \\ -x_1 + 2x_2 - x_3 + x_4 - x_5 = -9 \\ 4x_1 + 2x_2 + 4x_3 + 2x_4 + 3x_5 = 9 \\ -x_1 + 4x_2 - 2x_3 + 2x_4 - 2x_5 = -17 \\ 3x_1 + x_2 + 2x_3 + x_4 + x_5 = 5 \end{cases}$

44. $\begin{cases} 4x_1 - x_2 + 3x_3 - x_4 + x_5 = -2 \\ 2x_1 + 2x_2 + x_3 + 4x_4 = 0 \\ 4x_1 - x_2 + 2x_3 - 2x_4 + x_5 = 1 \\ 2x_1 + x_2 + x_3 + 2x_4 = 0 \\ 3x_1 - x_2 + x_3 - 2x_4 + x_5 = 2 \end{cases}$

 45. $\begin{cases} x_1 + x_3 + x_5 = 3 \\ -x_1 + 2x_2 - x_3 + x_4 - x_5 = -9 \\ 4x_1 + 2x_2 + 4x_3 + 2x_4 + 3x_5 = 3 \\ -x_1 + 4x_2 - 2x_3 + 2x_4 - 2x_5 = -17 \\ 3x_1 + x_2 + 2x_3 + x_4 + x_5 = 2 \end{cases}$

 46. $\begin{cases} 4x_1 - x_2 + 3x_3 - x_4 + x_5 = 12 \\ 2x_1 + 2x_2 + x_3 + 4x_4 = -9 \\ 4x_1 - x_2 + 2x_3 - 2x_4 + x_5 = 13 \\ 2x_1 + x_2 + x_3 + 2x_4 = -3 \\ 3x_1 - x_2 + x_3 - 2x_4 + x_5 = 11 \end{cases}$

 47. $\begin{cases} x_1 + 2x_2 + x_3 = 11 \\ x_1 + 4x_2 + x_3 = 19 \\ 2x_1 + 2x_2 + x_3 = 13 \end{cases}$

 48. $\begin{cases} 2x_1 + 3x_2 + x_3 = 4 \\ 3x_1 + 3x_2 + 2x_3 = 12 \\ x_1 + 2x_2 + x_3 = 4 \end{cases}$

 49. $\begin{cases} 4x_1 + 6x_2 + 5x_3 + 9x_4 = 75 \\ 4x_1 + x_2 + 2x_3 + 4x_4 = 10 \\ x_1 + 4x_2 + 3x_3 + 6x_4 = 65 \\ 4x_1 + 3x_2 + 3x_3 + 4x_4 = 10 \end{cases}$

 50. $\begin{cases} 3x_1 + 2x_2 + x_3 + 3x_4 = 10 \\ 4x_1 + 3x_2 + 3x_3 + 7x_4 = 15 \\ 6x_1 + 7x_2 + 3x_3 + 8x_4 = 50 \\ 5x_1 + 3x_2 + 3x_3 + 7x_4 = 10 \end{cases}$

Formulate each situation as a collection of systems of linear equations. Be sure to state clearly the meaning of each variable. Solve each collection by finding the inverse of the coefficient matrix and then using matrix multiplications. State your final answers in terms of the original questions.

If permitted by your instructor, you may use a graphing calculator to find the necessary inverse matrices and matrix products.

51. Movie Tickets A five-screen multiplex cinema charges $10 for adults and $5 for children under twelve. The number of tickets sold for each of today's shows and the corresponding gross receipts are given in the table. How many tickets of each kind were sold for each film?

	Film No. 1	Film No. 2	Film No. 3	Film No. 4	Film No. 5
Tickets sold	500	400	450	500	600
Gross receipts	$3250	$3000	$3500	$4500	$6000

52. Summer Day Care An inner-city antipoverty foundation staffs summer day care sites serving children aged 6 to 12 at Hollis Avenue, Beaverton Boulevard, Gramson Park, and Riverside Street. Certified instructors earn $350 per week and supervise 8 children, and college student group leaders earn $250 per week and supervise 6 children. The number of children and the weekly payroll at each site are given in the table. How many instructors and group leaders work at each site?

	Hollis Avenue	Beaverton Boulevard
Children	92	130
Payroll	$3900	$5500

	Gramson Park	Riverside Street
Children	124	152
Payroll	$5300	$6500

53. Coins in Jars Three glass jars (red, green, and blue) each contain pennies, nickels, and dimes. The value and number of coins in each jar are given in the table, together with the "altered value" of the coins when the pennies are replaced by the same number of nickels. How many of each coin are in each jar?

	Red Jar	Green Jar	Blue Jar
Value	$20	$39	$25
Number of coins	500	700	600
"Altered" value	$30	$45	$38

54. Plant Fertilizer The manager of a garden supply store has received the soil test results for the gardens of Mr. Smith, Mrs. Jones, Miss Roberts, and Mr. Wheeler. Using the size of each garden plot, the manager calculated the ounces of nutrients needed for each garden as listed in the first table. The store sells three brands of fertilizer, GreatGreen, MiracleMix, and GrowRite, which contain the amounts of potash, nitrogen, and phosphoric acid per box listed in the second table. How many boxes of each fertilizer should be sold to each customer?

Ounces Needed	Mr. Smith	Mrs. Jones	Miss Roberts	Mr. Wheeler
Potash	26	31	25	25
Nitrogen	31	33	31	29
Phosphoric acid	19	20	19	18

Ounces per Box	GreatGreen	MiracleMix	GrowRite
Potash	4	5	2
Nitrogen	5	5	3
Phosphoric acid	3	3	2

55. Infant Nutrition   A pediatric dietician at an inner-city foundling hospital needs to supplement each bottle of baby formula given to three infants in her ward with the units of vitamin A, vitamin D, calcium, and iron given in the first table. If four diet supplements are available with the nutrient content per drop given in the second table, how many drops of each supplement per bottle of formula should each infant receive?

Units Needed	Vitamin A	Vitamin D	Calcium	Iron
Billy	48	26	19	49
Susie	46	26	27	65
Jimmy	47	26	19	49

Units per Drop	Vitamin A	Vitamin D	Calcium	Iron
Supplement No. 1	5	3	0	1
Supplement No. 2	0	1	3	6
Supplement No. 3	4	2	3	7
Supplement No. 4	4	2	2	5

56. International Investments   An international investment advisor recommends industrial stocks in Nigeria, Bolivia, Thailand, and Hungary, with four times as much in Thailand as in Bolivia and three times as much in Hungary as in Nigeria. Furthermore, he suggests that more than 20% but less than 25% of the investor's portfolio be Bolivian and Nigerian stocks. Acting on this advice, four investors commit the amounts given in the table. How much does each invest in each country?

	Mr. Croft	Mrs. Fredericks
Total investment	$100,000	$150,000
Bolivia and Nigeria	$ 23,000	$ 33,000

	Mr. Spencer	Ms. Winpeace
Total investment	$90,000	$135,000
Bolivia and Nigeria	$22,000	$ 28,000

57. Financial Planning   A retirement planning counselor recommends investing in a stock fund yielding 18%, a money market fund returning 6%, and a bond fund paying 8%, with twice as much in the bond and money market funds together as in the stock fund. Mr. and Mrs. Jordan have $300,000 to invest and need an annual return of $31,000; Mr. and Mrs. French have $234,900 to invest and need an annual return of $25,600; and Mr. and Mrs. Daimen have $270,000 to invest and need an annual return of $28,500. How much should each elderly couple place in each investment to receive their desired income?

58. Year-End Bonuses   At the end of each year, the owner of a small company gives himself, his salesman, and his secretary a bonus by dividing up whatever remains in the "office supplies" account. He takes one third for himself, gives one quarter to his salesman, and gives one fifth to his secretary, but each share is taken after the others have been given out. Any remaining money is then spent on the company's New Year's Eve dinner at the best restaurant in town. The ending balances in the office supplies account for four years in the 1990s are given in the table. How much did each person receive each year and how much was left over each year for the dinner party?

1994	1995	1996	1997
$2500	$2000	$3000	$2800

59. Mass Transit   The metropolitan area mass transit manager is revising the subway, bus, and jitney service to the suburbs of Brighton, Conway, Longwood, and Oakley. To meet federal clean air mandates, the mayor's office demands that twice as many electric subway cars as diesel buses and jitneys combined be used for each suburb. The transit workers' union contract requires that like the buses and jitneys, each subway car must have a driver/ticket taker whether or not that subway car is part of a longer train. The number of commuters from each suburb using mass transit and the number of transit workers assigned to each area are given in the table. If each subway car carries 70 commuters, each bus carries 60 commuters, and each jitney carries 10 commuters, how many of each vehicle must be assigned to each suburb?

	Brighton	Conway
Commuters	11,500	9000
Transit workers	180	150

	Longwood	Oakley
Commuters	9500	10,250
Transit workers	150	165

60. Community Food Pantry A town-wide food drive to aid the interfaith ministries' food pantry for the needy was supported by the Boy Scouts, the Girl Scouts, and the Lions Club. Both dry food (in 10-ounce boxes) and canned food, in small (15-ounce) and large (40-ounce) sizes, were collected. The Boy Scouts collected 240 items weighing 355 pounds, of which 315 pounds were canned goods; the Girl Scouts collected 240 items weighing 320 pounds, of which 260 pounds were canned goods; and the Lions Club collected 416 items weighing 565 pounds, of which 465 pounds were canned goods. How many boxes, small cans, and large cans did each group collect?

Explorations and Excursions

The following problems extend and augment the material presented in the text.

Solve each matrix equation *symbolically* for the unknown matrix X. Evaluate this expression for X using the given matrices and verify that it satisfies the original equation.

If permitted by your instructor, store the matrices A, B, C, and D in your calculator and use your calculator to evaluate each expression for X.

$$A = \begin{pmatrix} 3 & 1 & 2 \\ 1 & 2 & 1 \\ 4 & 2 & 3 \end{pmatrix} \quad B = \begin{pmatrix} 1 & -1 & 2 \\ 4 & 1 & 1 \\ 6 & 1 & 2 \end{pmatrix}$$

$$C = \begin{pmatrix} 280 \\ 240 \\ 320 \end{pmatrix} \quad D = \begin{pmatrix} 200 \\ 120 \\ 280 \end{pmatrix}$$

61. $A \cdot X + X = C$

62. $A \cdot X - X = D$

63. $A \cdot X + B \cdot X = C + D$

64. $A \cdot X - B \cdot X = C - D$

65. Show that the inverse of an inverse matrix is the original matrix; that is, show that $(A^{-1})^{-1} = A$ for any invertible matrix A.

66. Show that the inverse of a transposed matrix is the transpose of the inverse matrix; that is, show that $(A^t)^{-1} = (A^{-1})^t$ for any invertible matrix A.

67. Show that the inverse of a product is the product of the inverses in the reverse order; that is, show that $(A \cdot B)^{-1} = (B^{-1}) \cdot (A^{-1})$ for any invertible matrices A and B of the same dimension.

68. Show that the $n \times n$ identity matrix is unique; that is, given that $J \cdot A = A$ for every A, establish that $J = I$. [*Hint:* consider $A = I$.]

69. Show that if $B \cdot A = I$, then $A \cdot B = I$ and $B = A^{-1}$.

70. Show that if the determinant $ad - bc$ of the matrix $\begin{pmatrix} a & b \\ c & d \end{pmatrix}$ is nonzero, then the inverse of the matrix $\begin{pmatrix} a & b \\ c & d \end{pmatrix}$ is the matrix $\dfrac{1}{ad - bc} \begin{pmatrix} d & -b \\ -c & a \end{pmatrix}$.

Matrix Inverses and Geometric Series

Exercises 71–80 develop material that will be used in Chapter 7 on Markov Chains.

71. Show that the geometric series formula on page 129 can be written as

$$1 + x + x^2 + \cdots + x^{n-1} = \frac{1 - x^n}{1 - x}$$

and that the corresponding matrix equation is then

$$I + A + A^2 + \ldots + A^{n-1} = (I - A^n) \cdot (I - A)^{-1}$$

where A is a square matrix such that $I - A$ is invertible.

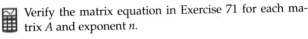 Verify the matrix equation in Exercise 71 for each matrix A and exponent n.

72. $A = \begin{pmatrix} 1 & 2 \\ 3 & 4 \end{pmatrix}$ for (a) $n = 2$ and (b) $n = 3$.

73. $A = \begin{pmatrix} 1 & 2 & 0 \\ 0 & 3 & 2 \\ 2 & 0 & 1 \end{pmatrix}$ for (a) $n = 2$ and (b) $n = 4$.

A square matrix with non-negative entries is a *substochastic* matrix of size w (where $0 \le w < 1$) if the sum of the entries in each row is no more than w.

74. Show that for square matrices A and B of the same dimesion, if A is substochastic of size u and B is substochastic of size v, then $A \cdot B$ is substochastic of size $u \cdot v$.

75. Show that if A is a substochastic matrix of size w, then A^n is a substochastic matrix of size w^n for any positive integer n.

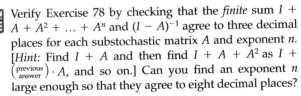 Verify Exercise 75 for each substochastic matrix A and exponent n.

76. $A = \begin{pmatrix} 0.2 & 0.4 \\ 0.3 & 0.5 \end{pmatrix}$ for (a) $n = 2$ and (b) $n = 3$. [*Hint:* Find w first.]

77. $A = \begin{pmatrix} 0.1 & 0.2 & 0 & 0.3 \\ 0.2 & 0.1 & 0.3 & 0.1 \\ 0 & 0.2 & 0.4 & 0.3 \\ 0.5 & 0.1 & 0.1 & 0.1 \end{pmatrix}$ for (a) $n = 2$ and (b) $n = 4$.

78. Use Exercises 75 and 71 to show that if A is a substochastic matrix of size w, then the inverse matrix $(I - A)^{-1}$ is the same as the infinite geometric series $I + A + A^2 + \cdots$.

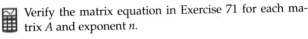 Verify Exercise 78 by checking that the *finite* sum $I + A + A^2 + \ldots + A^n$ and $(I - A)^{-1}$ agree to three decimal places for each substochastic matrix A and exponent n. [*Hint:* Find $I + A$ and then find $I + A + A^2$ as $I + \left(\substack{previous \\ answer} \right) \cdot A$, and so on.] Can you find an exponent n large enough so that they agree to eight decimal places?

79. $A = \begin{pmatrix} 0.2 & 0.1 \\ 0.1 & 0.2 \end{pmatrix}$ with $n = 6$

80. $A = \begin{pmatrix} 0.10 & 0.05 & 0.10 \\ 0.05 & 0.10 & 0.05 \\ 0.10 & 0.05 & 0.10 \end{pmatrix}$ with $n = 4$

3.6   Two Applications

Introduction

We conclude this chapter with two applications of matrix methods: the Leontief input–output model of an economy and least squares estimation. Because of the computational nature of these topics, the use of a graphing calculator or a computer for matrix calculations is appropriate throughout this section.

Leontief "Open" Input–Output Models

Input–output analysis was developed by Wassily Leontief (who received the 1973 Noble Memorial Prize for economics) to study the flow of goods and services among different sectors of an economy. In a "closed" model, all the goods produced are used by the producers, while the economy of an "open" model produces more than is needed by the producers, with the extra output available to consumers. In the global economy of today, each country may be considered an open system. Given the relationships among the different sectors of an economy, it should be possible to calculate the extra production from the amount of economic activity within each sector. And, conversely, it should be possible to determine the level of economic activity necessary within each sector to achieve a desired level of excess production. The Leontief "open" input–output economic model performs these two computational tasks by expressing the relation between the activity levels of the sectors and the extra production as a matrix equation.

We begin by considering the simple economy shown in the diagram, consisting of a blacksmith (B), who makes nails, plows, and other tools, a carpenter (C), who builds barns and other useful buildings, and a farmer (F), who grows food. Each arrow connects a producer with a consumer and indicates that goods and services from the producer are a part of the *input* needed by the consumer to manufacture *output*, which in turn becomes the input used by the others. Because producers also take as input some of their own outputs, the circular arrows indicate their consumption of parts of their own production. The number near the head of each arrow indicates the value of that input needed to produce one dollar of output from the target. Thus the topmost arrow from B to C labeled 35¢ means that each dollar of value produced by the carpenter requires the input of 35¢ worth of the blacksmith's output. The connections between the sectors of the economy are the values of materials needed to produce each dollar of output from each, and are the basic data needed to construct the Leontief model of the economy.

Let x_1, x_2, and x_3 be the values of the goods and services produced by the blacksmith, the carpenter, and the farmer, and let y_1, y_2, and y_3 be the values of the *excess* production from each that can be sold outside this economic system. The value produced by the blacksmith is the same as the values of the blacksmith's products used by the blacksmith, the carpenter, and the farmer, together with the amount sold outside this economy. Using the numbers from the diagram, we have

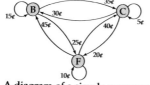

A diagram of a simple economy.

$$x_1 = \quad 0.15x_1 \quad + \quad 0.35x_2 \quad + \quad 0.25x_3 \quad + \quad y_1$$

| Value of blacksmith's products used by the blacksmith | Value of blacksmith's products used by the carpenter | Value of blacksmith's products used by the farmer | Value of blacksmith's products sold to the "outside" |

Similarly, for the carpenter and the blacksmith,

$$x_2 = \underbrace{0.30x_1}_{\substack{\text{Value of carpenter's} \\ \text{products used} \\ \text{by the blacksmith}}} + \underbrace{0.05x_2}_{\substack{\text{Value of carpenter's} \\ \text{products used} \\ \text{by the carpenter}}} + \underbrace{0.20x_3}_{\substack{\text{Value of carpenter's} \\ \text{products used} \\ \text{by the farmer}}} + \underbrace{y_2}_{\substack{\text{Value of carpenter's} \\ \text{products sold} \\ \text{to the "outside"}}}$$

$$x_3 = \underbrace{0.45x_1}_{\substack{\text{Value of farmer's} \\ \text{products used} \\ \text{by the blacksmith}}} + \underbrace{0.40x_2}_{\substack{\text{Value of farmer's} \\ \text{products used} \\ \text{by the carpenter}}} + \underbrace{0.10x_3}_{\substack{\text{Value of farmer's} \\ \text{products used} \\ \text{by the farmer}}} + \underbrace{y_3}_{\substack{\text{Value of farmer's} \\ \text{products sold} \\ \text{to the "outside"}}}$$

In matrix form, this system of equations becomes

$$\begin{pmatrix} x_1 \\ x_2 \\ x_3 \end{pmatrix} = \begin{pmatrix} 0.15 & 0.35 & 0.25 \\ 0.30 & 0.05 & 0.20 \\ 0.45 & 0.40 & 0.10 \end{pmatrix} \begin{pmatrix} x_1 \\ x_2 \\ x_3 \end{pmatrix} + \begin{pmatrix} y_1 \\ y_2 \\ y_3 \end{pmatrix}$$

The numerical matrix in this equation was called the "interindustry matrix of technical coefficients" by Leontief and is now usually referred to as the "technology matrix" of the economy. The columns of this matrix are the input values taken from each sector as shown in the economy diagram. In general,

Leonteiff "Open" Input–Output Model

A Leontief "open" input–output model is a matrix equation

$$X = A \cdot X + Y$$

where the $n \times 1$ column matrix X lists the values produced by each of the n economic sectors, the $n \times 1$ column matrix Y lists the values of the excess production of these same sectors, and the element a_{ij} of the $n \times n$ technology matrix A represents the value of sector i goods and services needed to produce one dollar of sector j output.

Solving this matrix equation for the excess productions Y in terms of the sector production values X, we have $X - A \cdot X = Y$ so that $Y = (I - A) \cdot X$. Alternatively, solving for the sector production values X required for given excess productions Y, we have $X = (I - A)^{-1} \cdot Y$. That is,

Excess and Sector Productions

In a Leontief "open" input–output model economy, the excess productions Y from given sector production values X is

$$Y = (I - A) \cdot X$$

and the sector production values X necessary to provide required excess productions Y is

$$X = (I - A)^{-1} \cdot Y$$

Returning to our example economy, if a government survey of economic activity establishes that both the blacksmith and the farmer each produce \$200 of value each year and the carpenter produces \$160, so that X is

$$\begin{pmatrix} x_1 \\ x_2 \\ x_3 \end{pmatrix} = \begin{pmatrix} 200 \\ 160 \\ 200 \end{pmatrix} \qquad \begin{array}{l} \text{Blacksmith} \\ \text{Carpenter} \\ \text{Farmer} \end{array}$$

Then the amount of extra production is given by $Y = (I - A) \cdot X$.

$$\begin{pmatrix} y_1 \\ y_2 \\ y_3 \end{pmatrix} = \left(\begin{pmatrix} 1 & 0 & 0 \\ 0 & 1 & 0 \\ 0 & 0 & 1 \end{pmatrix} - \begin{pmatrix} 0.15 & 0.35 & 0.25 \\ 0.30 & 0.05 & 0.20 \\ 0.45 & 0.40 & 0.10 \end{pmatrix} \right) \cdot \begin{pmatrix} 200 \\ 160 \\ 200 \end{pmatrix} = \begin{pmatrix} 64 \\ 52 \\ 26 \end{pmatrix}$$

Therefore, the amount of extra production will be \$64 from the blacksmith, \$52 from the carpenter, and \$26 from the farmer, for a total of \$142.

How much will the production of each sector have to change in order to provide \$151 of excess production, specifically \$59 from the blacksmith, \$48 from the carpenter, and \$44 from the farmer? So Y is

$$\begin{pmatrix} y_1 \\ y_2 \\ y_3 \end{pmatrix} = \begin{pmatrix} 59 \\ 48 \\ 44 \end{pmatrix}$$

The formula $X = (I - A)^{-1} \cdot Y$ becomes (using a calculator)

$$\begin{pmatrix} x_1 \\ x_2 \\ x_3 \end{pmatrix} = \left(\begin{pmatrix} 1 & 0 & 0 \\ 0 & 1 & 0 \\ 0 & 0 & 1 \end{pmatrix} - \begin{pmatrix} 0.15 & 0.35 & 0.25 \\ 0.30 & 0.05 & 0.20 \\ 0.45 & 0.40 & 0.10 \end{pmatrix} \right)^{-1} \cdot \begin{pmatrix} 59 \\ 48 \\ 44 \end{pmatrix} = \begin{pmatrix} 200 \\ 160 \\ 220 \end{pmatrix}$$

Thus if the blacksmith and the carpenter continue to produce the same value as before but the farmer increases production by \$20 to \$220, the economy can generate an additional \$9 in excess production (but notice that the respective sources of this excess change drastically).

Graphing Calculator Exploration

Leontief "open" input–output model calculations are simple to do on your calculator. To verify the above results for the black-smith–carpenter–farmer economy example:

a. Store the technology matrix A in [A].

```
[A]
[[.15 .35 .25]
 [.3  .05 .2 ]
 [.45 .4  .1 ]]
```

b. To find the excess productions Y for productions $X = \begin{pmatrix} 200 \\ 160 \\ 200 \end{pmatrix}$, store these values in [B].

```
[B]
       [[200]
        [160]
        [200]]
```

c. Then use the MATRX MATH IDENTITY(3) command to find $Y = (I - A) \cdot X$.

```
(identity(3)-[A]
)*[B]
        [[64]
         [52]
         [26]]
```

d. To find the sector productions X necessary to provide $Y = \begin{pmatrix} 59 \\ 48 \\ 44 \end{pmatrix}$ excess productions, store these values in [C] and use the $\boxed{x^{-1}}$ button to find $X = (I - A)^{-1} \cdot Y$.

```
(identity(3)-[A]
)⁻¹*[C]
        [[200]
         [160]
         [220]]
```

There are many reasons to alter the production levels or the excess production levels of the sectors, such as raising the standard of living or supporting government programs through taxes. The Leontief "open" input–output model provides a tool to evaluate the effects of such possible changes.

Least Squares

The method of least squares was invented in 1794 by Carl Friedrich Gauss (1777–1855) to find the best "compromise solution" to an inconsistent system of linear equations. He became world famous as a scientist in 1801 when he used his method to predict when and where the asteroid Ceres could next be seen after it had been lost in the sun's glare on February 11, 1801 shortly after its discovery on January 1, 1801 by the Italian astronomer Piazzi. The sensation created by the accuracy of Gauss's prediction ultimately led to his appointment as astronomer at the Gottingen Observatory, assuring him the financial security that allowed him to pursue his many other ideas.

We have seen that a system of linear equations written in the matrix form $A \cdot X = B$ may be solved as $X = A^{-1} \cdot B$ only in the special case that A is invertible (and hence square). But for any matrix A, the product $A^t \cdot A$ is square and symmetric. Let us now attempt to solve any matrix equation $A \cdot X = B$ by multiplying both sides by A^t and then trying to solve $A^t A \cdot X = A^t B$ as $X = (A^t A)^{-1} \cdot (A^t B)$. We begin with two examples in two variables.

The three lines $2x + y = 12$, $x + y = 8$, and $x + 2y = 12$ intersect in the common point $x = 4$, $y = 4$. Writing these equations as a single matrix equation $A \cdot X = B$, we have

$$\begin{pmatrix} 2 & 1 \\ 1 & 1 \\ 1 & 2 \end{pmatrix} \begin{pmatrix} x \\ y \end{pmatrix} = \begin{pmatrix} 12 \\ 8 \\ 12 \end{pmatrix}$$

Multiplying by A^t, we obtain $A^t A \cdot X = A^t B$:

$$\begin{pmatrix} 2 & 1 & 1 \\ 1 & 1 & 2 \end{pmatrix} \begin{pmatrix} 2 & 1 \\ 1 & 1 \\ 1 & 2 \end{pmatrix} \begin{pmatrix} x \\ y \end{pmatrix} = \begin{pmatrix} 2 & 1 & 1 \\ 1 & 1 & 2 \end{pmatrix} \begin{pmatrix} 12 \\ 8 \\ 12 \end{pmatrix} \qquad A^t A \cdot X = A^t B$$

Simplifying,

$$\begin{pmatrix} 6 & 5 \\ 5 & 6 \end{pmatrix} \begin{pmatrix} x \\ y \end{pmatrix} = \begin{pmatrix} 44 \\ 44 \end{pmatrix}$$

so

$$\begin{pmatrix} x \\ y \end{pmatrix} = \begin{pmatrix} 6 & 5 \\ 5 & 6 \end{pmatrix}^{-1} \begin{pmatrix} 44 \\ 44 \end{pmatrix}$$

$$= \begin{pmatrix} 6/11 & -5/11 \\ -5/11 & 6/11 \end{pmatrix} \begin{pmatrix} 44 \\ 44 \end{pmatrix} = \begin{pmatrix} 4 \\ 4 \end{pmatrix} \qquad \text{The known solution}$$

Thus, in this particular example, we were able to solve a consistent system of equations with a unique solution using an inverse matrix after multiplying both sides by the transpose of the coefficient matrix.

What happens if the equations are *inconsistent?* The diagram below shows that the four lines $x + y = 10$, $x + y = 2$, $x - y = 6$, and $x - y = -6$ have no point in common.

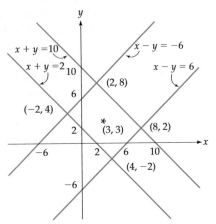

An inconsistent system of lines

These lines can be expressed as a matrix equation $A \cdot X = B$:

$$\begin{pmatrix} 1 & 1 \\ 1 & 1 \\ 1 & -1 \\ 1 & -1 \end{pmatrix} \begin{pmatrix} x \\ y \end{pmatrix} = \begin{pmatrix} 10 \\ 2 \\ 6 \\ -6 \end{pmatrix}$$

Multiplying by A^t, we obtain $A^t A \cdot X = A^t B$:

$$\begin{pmatrix} 1 & 1 & 1 & 1 \\ 1 & 1 & -1 & -1 \end{pmatrix} \begin{pmatrix} 1 & 1 \\ 1 & 1 \\ 1 & -1 \\ 1 & -1 \end{pmatrix} \begin{pmatrix} x \\ y \end{pmatrix} = \begin{pmatrix} 1 & 1 & 1 & 1 \\ 1 & 1 & -1 & -1 \end{pmatrix} \begin{pmatrix} 10 \\ 2 \\ 6 \\ -6 \end{pmatrix}$$

Simplifying,

$$\begin{pmatrix} 4 & 0 \\ 0 & 4 \end{pmatrix} \begin{pmatrix} x \\ y \end{pmatrix} = \begin{pmatrix} 12 \\ 12 \end{pmatrix}$$

so

$$\begin{pmatrix} x \\ y \end{pmatrix} = \begin{pmatrix} 4 & 0 \\ 0 & 4 \end{pmatrix}^{-1} \begin{pmatrix} 12 \\ 12 \end{pmatrix}$$

$$= \begin{pmatrix} 1/4 & 0 \\ 0 & 1/4 \end{pmatrix} \begin{pmatrix} 12 \\ 12 \end{pmatrix} = \begin{pmatrix} 3 \\ 3 \end{pmatrix} \qquad \begin{aligned} x &= 3 \\ y &= 3 \end{aligned}$$

However, $x = 3$, $y = 3$ fails to satisfy any of the original equations. But looking back at the diagram, we see that the point $(3, 3)$ is "in the middle" of the four lines. In a clear geometrical sense, this "solution" $(3, 3)$ is the "best compromise" for an answer in that it is as close as possible to each of the lines at the same time. A common measure of the error in an estimate is the sum of the squares of the vertical distances from the point to the lines. For this example, the sum of the squares of the vertical distances from the point (3,3) to the lines is $(4)^2 + (-4)^2 + (-6)^2 + (6)^2 = 104$ (the minus signs mean the lines were below the point). Any other point will give a larger sum of squares; for instance, repeating the previous calculation for the point (3, 4) gives the larger value $(3)^2 + (-5)^2 + (-5)^2 + (7)^2 = 108$. Thus, in this special case, $A \cdot X = B$ is inconsistent but $A^t \cdot A$ is invertible and the "solution" $X = (A^t A)^{-1} \cdot (A^t B)$ is *the best compromise answer* in that it minimizes the sum of the squares of the vertical errors.*

We now turn to prediction problems similar to those solved by Gauss with such spectacular results. Suppose we know that some quantity y should depend on some other quantity x in such a way that y is a linear function of x (that is, $y = mx + b$). If we could take *perfect* measurements of x and y, we might obtain the following data.

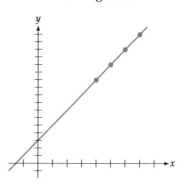

x	4	5	6	7
y	11	13	15	17

The above graph shows the points together with the line $y = 2x + 3$ passing through them (the values $m = 2$ and $b = 3$ are easily found by the usual slope and intercept methods).

However, in the "real world," measurements are not exact, and data often have errors. Suppose that the data were as follows:

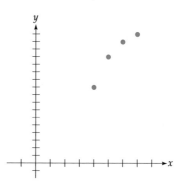

x	4	5	6	7
y	10	14	16	17

* For a proof, see pages 961–962 of *Calculus with Finite Mathematics* by the same authors.

Clearly, these four data points are *not* collinear, so it is impossible to find a line $y = mx + b$ passing through all four of them. If we were to write the problems of finding m and b as a system of linear equations,

$$\begin{cases} 4m + b = 10 \\ 5m + b = 14 \\ 6m + b = 16 \\ 7m + b = 17 \end{cases} \longrightarrow \begin{pmatrix} 4 & 1 \\ 5 & 1 \\ 6 & 1 \\ 7 & 1 \end{pmatrix}\begin{pmatrix} m \\ b \end{pmatrix} = \begin{pmatrix} 10 \\ 14 \\ 16 \\ 17 \end{pmatrix} \qquad AX = B$$

then the system would be *inconsistent* since there is no solution. However, instead of an *exact* solution we can look for the best *compromise* line passing *closest* to the points. As before, we multiply both sides of the above matrix equation by A^t.

$$\begin{pmatrix} 4 & 5 & 6 & 7 \\ 1 & 1 & 1 & 1 \end{pmatrix}\begin{pmatrix} 4 & 1 \\ 5 & 1 \\ 6 & 1 \\ 7 & 1 \end{pmatrix}\begin{pmatrix} m \\ b \end{pmatrix} = \begin{pmatrix} 4 & 5 & 6 & 7 \\ 1 & 1 & 1 & 1 \end{pmatrix}\begin{pmatrix} 10 \\ 14 \\ 16 \\ 17 \end{pmatrix} \qquad A^tAX = A^tB$$

Multiplying out yields

$$\begin{pmatrix} 126 & 22 \\ 22 & 4 \end{pmatrix}\begin{pmatrix} m \\ b \end{pmatrix} = \begin{pmatrix} 325 \\ 57 \end{pmatrix}$$

so

$$\begin{pmatrix} m \\ b \end{pmatrix} = \begin{pmatrix} 126 & 22 \\ 22 & 4 \end{pmatrix}^{-1}\begin{pmatrix} 325 \\ 57 \end{pmatrix} = \begin{pmatrix} 1/5 & -11/10 \\ -11/10 & 63/10 \end{pmatrix}\begin{pmatrix} 325 \\ 57 \end{pmatrix} = \begin{pmatrix} 2.3 \\ 1.6 \end{pmatrix}$$

These values, $m = 2.3$ and $b = 1.6$, give the line $y = 2.3x + 1.6$, which is called the "least squares" line, providing the best fit to the four points in that it minimizes the sum of the squared vertical distances. The following graph shows the four points together with the least squares line.

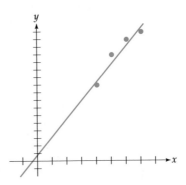

If the four points *did* lie on a line, then the m and b given by this procedure would be the *exact* values for the correct line. The least squares line is used widely in business and the sciences to predict future trends from current but imperfect data.

Graphing Calculator Exploration

The least squares line calculation is simple to do on your calculator. To verify the above results for the data with errors:

a. Store the x data values and the 1s in [A].

```
[A]
            [[4 1]
             [5 1]
             [6 1]
             [7 1]]
```

b. Store the y data values in [B].

```
[B]
           [[10]
            [14]
            [16]
            [17]]
```

c. Then use the TRANSPOSE command and the $\boxed{x^{-1}}$ button to find $X = (A^tA)^{-1} \cdot (A^tB)$.

```
([A]ᵀ[A])⁻¹([A]ᵀ[
B])
           [[2.3]
            [1.6]]
```

The method of least squares can be summarized as follows:

Least Squares Approximation

The slope m and y-intercept b of the least squares best approximation line $y = mx + b$ for the data points (x_1, y_1), (x_2, y_2), . . . , (x_n, y_n) are the unique solution of the system of linear equations

$$\begin{pmatrix} x_1 & x_2 \cdots x_n \\ 1 & 1 \cdots 1 \end{pmatrix} \begin{pmatrix} x_1 & 1 \\ x_2 & 1 \\ \vdots & \vdots \\ x_n & 1 \end{pmatrix} \begin{pmatrix} m \\ b \end{pmatrix} = \begin{pmatrix} x_1 & x_2 \cdots x_n \\ 1 & 1 \cdots 1 \end{pmatrix} \begin{pmatrix} y_1 \\ y_2 \\ \vdots \\ y_n \end{pmatrix}$$

provided the values $x_1, x_2, \ldots, x_n$ are distinct.

Section Summary

A *Leontief "open" input–output model* is a matrix equation $X = AX + Y$, where the column matrix X lists the values produced by each of the n economic sectors, the column matrix Y lists the values of the excess production of these same sectors, and the element $a_{i,j}$ of the $n \times n$ technology matrix A represents the value of sector i goods and services needed to produce one dollar of sector j output. Solving this model for the excess productions Y in terms of the sector production values X gives that $Y = (I - A) \cdot X$, while the sector production values X necessary to provide excess productions Y is $X = (I - A)^{-1} \cdot Y$.

For a system of linear equations $A \cdot X = B$, if A is not invertible but $A^t \cdot A$ is, the "solution" $X = (A^t A)^{-1} \cdot (A^t B)$ has the smallest possible sum of the squares of the errors $B - A \cdot X$. For data points (x_1, y_1), $(x_2, y_2), \ldots, (x_n, y_n)$ with distinct $x_1, x_2, \ldots, x_n$ values, the slope m and y-intercept b of the best approximation *least squares line* $y = mx + b$ are the unique solution of the system of linear equations

$$\begin{pmatrix} x_1 & x_2 \cdots x_n \\ 1 & 1 \cdots 1 \end{pmatrix} \begin{pmatrix} x_1 & 1 \\ x_2 & 1 \\ \vdots & \vdots \\ x_n & 1 \end{pmatrix} \begin{pmatrix} m \\ b \end{pmatrix} = \begin{pmatrix} x_1 & x_2 \cdots x_n \\ 1 & 1 \cdots 1 \end{pmatrix} \begin{pmatrix} y_1 \\ y_2 \\ \vdots \\ y_n \end{pmatrix}$$

This least squares line can be used to predict values for y from values of x.

EXERCISES 3.6 (⊞ will be helpful.)

Leontief "Open" Input–Output Models

Let A denote agriculture, C denote construction, E denote electronics, F denote fishing, H denote heavy industry, L denote light industry, M denote mining, R denote railroads, and T denote tourism.

Find the technology matrix for each economy diagram.

1.

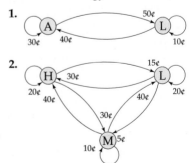

2.

3.

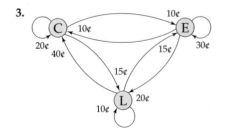

4.

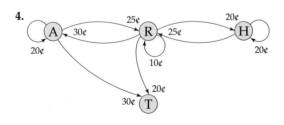

Draw an economy diagram for each technology matrix.

5. $\begin{pmatrix} 0.15 & 0.10 \\ 0.25 & 0.30 \end{pmatrix}$ for sectors M and H

6. $\begin{pmatrix} 0.10 & 0.30 & 0.10 \\ 0.20 & 0.40 & 0.20 \\ 0.10 & 0.30 & 0.10 \end{pmatrix}$ for sectors A, R, and M

7. $\begin{pmatrix} 0.20 & 0.30 & 0.10 \\ 0.50 & 0.20 & 0.20 \\ 0.30 & 0.10 & 0.20 \end{pmatrix}$ for sectors C, A, and L

8. $\begin{pmatrix} 0.10 & 0 & 0.20 \\ 0 & 0.20 & 0.10 \\ 0.10 & 0.40 & 0 \end{pmatrix}$ for sectors A, C, and R

Find the excess production Y of each economy with technology matrix A and economic activity level X.

9. $A = \begin{pmatrix} 0.20 & 0.30 \\ 0.35 & 0.25 \end{pmatrix}$ and $X = \begin{pmatrix} 130 \\ 110 \end{pmatrix}$

10. $A = \begin{pmatrix} 0.45 & 0.20 \\ 0.30 & 0.35 \end{pmatrix}$ and $X = \begin{pmatrix} 140 \\ 100 \end{pmatrix}$

11. $A = \begin{pmatrix} 0.05 & 0.15 & 0.20 \\ 0.15 & 0.05 & 0.15 \\ 0.10 & 0.10 & 0.05 \end{pmatrix}$ and $X = \begin{pmatrix} 150 \\ 170 \\ 140 \end{pmatrix}$

12. $A = \begin{pmatrix} 0.10 & 0.05 & 0.10 & 0.15 \\ 0.15 & 0.10 & 0.10 & 0.05 \\ 0 & 0.05 & 0.10 & 0 \\ 0.10 & 0 & 0.10 & 0.05 \end{pmatrix}$ and $X = \begin{pmatrix} 120 \\ 100 \\ 110 \\ 80 \end{pmatrix}$

Find the economic activity level X for each economy with technology matrix A necessary to generate excess production Y.

13. $A = \begin{pmatrix} 0.20 & 0.30 \\ 0.30 & 0.20 \end{pmatrix}$ and $Y = \begin{pmatrix} 84 \\ 51 \end{pmatrix}$

14. $A = \begin{pmatrix} 0.25 & 0.35 \\ 0.45 & 0.15 \end{pmatrix}$ and $Y = \begin{pmatrix} 33 \\ 57 \end{pmatrix}$

15. $A = \begin{pmatrix} 0.10 & 0.20 & 0 \\ 0 & 0.15 & 0.20 \\ 0.30 & 0.10 & 0.20 \end{pmatrix}$ and $Y = \begin{pmatrix} 60 \\ 31 \\ 50 \end{pmatrix}$

16. $A = \begin{pmatrix} 0.10 & 0.15 & 0.10 & 0.05 \\ 0.15 & 0.15 & 0.05 & 0.10 \\ 0 & 0.20 & 0.10 & 0.10 \\ 0.10 & 0 & 0.15 & 0.05 \end{pmatrix}$ and $Y = \begin{pmatrix} 60 \\ 55 \\ 60 \\ 70 \end{pmatrix}$

Represent each situation as a Leontief "open" input–output model by constructing an economy diagram

and the corresponding technology matrix. Find the required excess production or level of economic activity and be sure to state your final answer in terms of the original question.

17. Industrial Production The heavy and light industry sectors of the Birmingham economy depend on each other in the following way: Each dollar of production from the heavy industry sector requires $0.25 of heavy industry produce and $0.15 of light industry produce, while each dollar of production from the light industry sector requires $0.35 of heavy industry produce and $0.05 of light industry produce. How much must each type of industry produce to yield an excess production of $127 million of heavy industry produce and $221 million of light industry produce?

18. County Production Planning An analysis of the mining, railroad, construction, and light industry sectors of the Hanover County economy revealed that each dollar produced by the mining sector requires $0.20 of mining products, $0.10 of railroad services, $0.10 of construction, and $0.10 of light industry products. Each dollar produced by the railroad sector requires $0.10 of mining products, $0.20 of railroad services, and $0.10 of light industry products. Each dollar produced by the construction sector requires $0.30 of mining products, $0.20 of construction, and $0.10 of light industry products. Each dollar produced by the light industry sector requires $0.10 of mining products, $0.20 of railroad services, and $0.10 of light industry products. If these industries in Hanover County presently produce excess productions valued at $8 million from mining, $30 million from railroads, $32 million from construction, and $31 million from light industry, what is the current production level of each of these economic sectors?

19. Third-World Productivity The new government of a third-world country wants to increase its excess production by stimulating the heavy industry sector of the national economy. An analysis of the relationships between its heavy industry, light industry, and railroad sectors found that the current production levels for these sectors are $100 million from heavy industry, $150 million from light industry, and $100 million from the railroads, with the technology matrix given in the table. Find the current excess production from these sectors of the economy. If the light industry and railroad productions remain the same, how much does this excess production increase with each $10 million increase in heavy industry production? What is the greatest heavy industry production level that this part of the national economy can tolerate?

	Heavy industry	Light industry	Railroads
Heavy industry	0.20	0.20	0.30
Light industry	0.50	0.20	0.20
Railroads	0.20	0.10	0.20

20. Island Economy An economist is studying the agriculture-, fishing-, and tourism-based economy of a small Pacific island nation. Although the country's excess production was $158 million last year, she has found that the economy actually produced $300 million, of which $142 million was consumed in the course of production. Her breakdown of each sector's production is given in the table. (The first line shows that the total value of agriculture production was $100 million, with $20 million consumed by the agriculture sector, $8 million consumed by the fishing sector, and $36 million consumed by the tourism sector, leaving an excess of $36 million.) If the same relations persist among the sectors of the economy, what will happen to the excess production next year if the fishing sector declines to $60 million while the other sectors produce the same amounts as before?

in Millions of $	Agriculture	Fishing	Tourism	Excess Production
Agriculture	$20	$8	$36	$36
Fishing	$10	$8	$36	$26
Tourism	$0	$0	$24	$96

Least Squares

Solve each system of equations by row-reducing the corresponding augmented matrix. Rewrite the system of equations in the matrix form $A \cdot X = B$, calculate $X = (A^tA)^{-1} \cdot (A^tB)$, and verify that this is the same solution.

21. $\begin{cases} x + y = 1 \\ y = -4 \\ x - y = 9 \end{cases}$

22. $\begin{cases} x - y = 1 \\ x = 4 \\ x + y = 7 \end{cases}$

23. $\begin{cases} x_1 + x_2 = 5 \\ x_2 + x_3 = 7 \\ x_1 + x_3 = 6 \\ x_1 - x_2 + x_3 = 3 \end{cases}$

24. $\begin{cases} x_1 - x_2 = 2 \\ -x_2 + x_3 = 1 \\ x_1 - x_3 = 1 \\ x_1 - x_2 + x_3 = 5 \end{cases}$

Row-reduce the corresponding augmented matrix for each system of equations to verify that the system is inconsistent. Rewrite the system of equations in the matrix form $A \cdot X = B$, calculate $X = (A^tA)^{-1} \cdot (A^tB)$, and verify that this compromise solution "almost" satisfies the original equations.

25. $\begin{cases} x + y = 12 \\ x = 3 \\ y = 3 \end{cases}$

26. $\begin{cases} 2x - y = 6 \\ x + y = 6 \\ -x + 2y = 6 \end{cases}$

27. $\begin{cases} x_1 + x_2 = 5 \\ x_2 + x_3 = 7 \\ x_1 + x_3 = 1 \\ x_1 - x_2 + x_3 = 3 \end{cases}$

28. $\begin{cases} x_1 - x_2 = 2 \\ -x_2 + x_3 = 4 \\ x_1 - x_3 = 4 \\ x_1 - x_2 + x_3 = 5 \end{cases}$

29. $\begin{cases} x_1 + x_3 = 6 \\ x_2 + x_4 = 3 \\ x_2 = 4 \\ x_3 = 3 \\ x_1 + x_4 = 7 \end{cases}$

30. $\begin{cases} x_1 = 4 \\ x_2 + x_3 = 6 \\ x_2 - x_3 = 2 \\ x_4 = 3 \\ x_1 + x_4 = 4 \end{cases}$

Find the least squares best approximation line $y = mx + b$ for each collection of x and y data pairs.

31.

x	−1	0	1
y	24	36	42

32.

x	−1	0	1
y	8	8	14

33.

x	−1	0	1	2
y	120	90	70	30

34.

x	−1	0	1	3
y	761	656	691	551

35.

x	1	4	5	8	12
y	185	195	195	245	275

36.

x	1	4	5	7	8	11
y	50	80	110	140	170	200

Use the least squares best approximation line to make each prediction. Be sure to state your final answer in terms of the original question.

37. Real Estate The new salesman at Abbott Associates, Real Estate Brokers, had an impressive first three months with sales of $300,000, $480,000, and $600,000. If he can keep improving this much every month, how much will he sell in his fifth month?

38. Commodity Futures A soybean speculator has been nervously watching the December delivery price per bushel drop from $6.59 on Monday to $6.41 on Tuesday and then to $6.29 on Wednesday. If this trend continues, what will the price be on Friday?

39. Price–Demand The market research division of a large candy manufacturer has test-marketed the new treat YummieCrunchies in five different markets at five different prices per eight-ounce bag to determine the relation between the selling price and the sales volume. The results of its study are presented in the table and are normalized to give the weekly sales per 20,000 consumers. How many weekly sales per 20,000 consumers can the manufacturer expect when it begins national distribution

next week with an introductory price of 79¢ per eight-ounce bag?

Selling price	Sales volume
70¢	2050
55¢	2675
75¢	1975
90¢	1250
85¢	1425

40. TV Advertising The accounts manager at television station WXXB claims that every dollar spent on commercials by local retailers generates $23 of sales. As proof, she shows prospective advertisers the following table of advertising budgets and gross receipts for five area companies that aired commercials last month. Is her claim justified?

Dollars spent on WXXB commercials	Gross receipts
$8000	$245,000
$5000	$215,000
$2000	$125,000
$9000	$305,000
$6000	$230,000

Chapter Summary with Hints and Suggestions

Reading the text and doing the exercises in this chapter have helped you to master the following skills, which are listed by section (in case you need to review them) and are keyed to particular Review Exercises. Answers for all Review Exercises are given at the back of the book, and full solutions can be found in the Student Solutions Manual.

3.1 SYSTEMS OF TWO LINEAR EQUATIONS IN TWO VARIABLES

- Represent a pair of statements as a system of two linear equations in two variables after making an appropriate choice for the meaning of the x- and y-variables. *(Review Exercises 1–2.)*

$$\begin{cases} ax + by = h \\ cx + dy = k \end{cases}$$

- Solve a system of two linear equations in two variables by graphing and identify the system as "independent and consistent," "independent and inconsistent" or "dependent." *(Review Exercises 3–4.)*

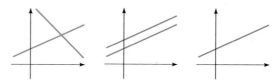

- Solve a system of two linear equations in two variables by the substitution method and identify the system as "independent and consistent," "independent and inconsistent" or "dependent." *(Review Exercises 5–6.)*

- Solve a system of two linear equations in two variables by the elimination method and identify the system as "independent and consistent," "independent and inconsistent" or "dependent." *(Review Exercises 7–8.)*

- Formulate an application as a system of two linear equations in two variables after making an appropriate choice for the meaning of the x- and y-variables. Solve the system of equations by the elimination method, and then state the final answer in terms of the original question. *(Review Exercises 9–10.)*

3.2   MATRICES AND LINEAR EQUATIONS IN TWO VARIABLES

- Find the dimension of a matrix and identify a particular element using double subscript notation. *(Review Exercises 11–12.)*

$$A = \begin{pmatrix} a_{1,1} & \cdots & a_{1,n} \\ \vdots & & \\ a_{m,1} & \cdots & a_{m,n} \end{pmatrix}$$

- Carry out a given row operation on an augmented matrix. *(Review Exercises 13–14.)*

$$\begin{cases} ax + by = h \\ cx + dy = k \end{cases} \longrightarrow \begin{pmatrix} a & b & h \\ c & d & k \end{pmatrix}$$

- Solve a system of two linear equations in two variables by row-reducing the corresponding augmented matrix, interpreting the final reduced row-echelon form matrix as the solution of the system of equations, and identifying the system as "independent and consistent," "independent and inconsistent," or "dependent." See the table on page 184. *(Review Exercises 15–18.)*

- Formulate an application as a system of two linear equations in two variables after making an appropriate choice for the meaning of the x- and y-variables. Solve the system of equations by row-reducing the corresponding augmented matrix, and then state the final answer in terms of the original question. *(Review Exercises 19–20.)*

3.3   SYSTEMS OF LINEAR EQUATIONS AND THE GAUSS–JORDAN METHOD

- Interpret a reduced row-echelon form matrix as the solution of a system of equations and identify the system as "independent" or "dependent" and as "consistent" or "inconsistent." *(Review Exercises 21–22.)*

   $0 \quad 0 \quad \dots \quad 0 \quad 0$ means dependent.

   $0 \quad 0 \quad \dots \quad 0 \quad 1$ means inconsistent.

- Use the Gauss–Jordan method to find the equivalent reduced row-echelon form matrix. *(Review Exercises 23–24.)*

- Solve a system of linear equations by row-reducing the corresponding augmented matrix via the Gauss–Jordan method, interpreting the final reduced row-echelon form matrix as the solution of the system of equations, and identifying the system as "independent" or "dependent" and as "consistent" or "inconsistent." *(Review Exercises 25–28.)*

- Formulate an application as a system of linear equations in an appropriate number of variables after making an appropriate choice for the meaning of each. Solve the system of equations by row-reducing the corresponding augmented matrix via the Gauss–Jordan method, and then state the final answer in terms of the original question. *(Review Exercises 29–30.)*

3.4   MATRIX ARITHMETIC

- Find a matrix product or determine that the product is not defined. *(Review Exercises 31–32.)*

$$\underset{m \times p}{A} \cdot \underset{p \times n}{B} = \underset{m \times n}{C}$$

- Find the value of a matrix expression involving scalar multiplication, matrix addition, subtraction, transposition, or multiplication, and the zero and identity matrices. *(Review Exercises 33–34.)*

   $A \cdot I = I \cdot A = A$ \qquad $A + 0 = 0 + A = A$

- Rewrite a system of linear equations as a matrix equation $A \cdot X = B$. *(Review Exercises 35–36.)*

- Find a matrix R so that the matrix product $R \cdot A$ is the same as the result of carrying out a given row operation or sequence of row operations on the matrix A.   *(Review Exercises 37–38.)*

- Formulate an application in matrix form after making an appropriate choice for the meaning of each row and column, and then find the requested quantity using the appropriate matrix arithmetic. *(Review Exercises 39–40.)*

3.5   INVERSE MATRICES AND SYSTEMS OF LINEAR EQUATIONS

- Find the product of a pair of matrices to identify the pair as "a matrix and its inverse" or "not a matrix and its inverse."   *(Review Exercises 41–42.)*

$$A \cdot A^{-1} = A^{-1} \cdot A = I$$

- Row-reduce $(A \mid I)$ for a square matrix A to find the inverse matrix A^{-1} or to identify A as a singular matrix.   *(Review Exercises 43–44.)*

$$(A \mid I) \to (I \mid A^{-1})$$

- Rewrite a system of equations as a matrix equation $A \cdot X = B$ and use the inverse of A to find the solution as $X = A^{-1} \cdot B$.   *(Review Exercises 45–48.)*
- Formulate an application as a collection of systems of linear equations in an appropriate number of variables after making an appropriate choice for the meaning of each. Solve the collection of systems of equations by finding the inverse of the common coefficient matrix and using matrix multiplications of this inverse times the various constant term matrices. State the final answer in terms of the original question.   *(Review Exercises 49–50.)*

3.6   TWO APPLICATIONS

Leontief "Open" Input–Output Models

- Find the technology matrix from an economy diagram.   *(Review Exercises 51–52.)*
- Draw an economy diagram from a technology matrix.   *(Review Exercises 53–54.)*
- Find the excess production of an economy with a given technology matrix and economic activity level.   *(Review Exercises 55–56.)*

$$Y = (I - A) \cdot X$$

- Find, for an economy with a given technology matrix, the economic activity level necessary to generate a specified excess production.   *(Review Exercises 57–58.)*

$$X = (I - A)^{-1} \cdot Y$$

- Represent an application as a Leontief "open" input–output model by constructing an economy diagram and the corresponding technology matrix. Find the required excess production or level of economic activity, and state the final answer in terms of the original question.   *(Review Exercises 59–60.)*

Least Squares

- Solve a consistent and independent system of linear equations by row-reducing the corresponding augmented matrix. Rewrite the system of equations in the matrix form $A \cdot X = B$, calculate $X = (A^t A)^{-1} \cdot (A^t B)$, and verify that this is the same solution.   *(Review Exercises 61–62.)*
- Row-reduce the corresponding augmented matrix for an inconsistent system of linear equations to verify that the system is inconsistent. Rewrite the system of equations in the matrix form $A \cdot X = B$, calculate $X = (A^t A)^{-1} \cdot (A^t B)$, and verify that this compromise solution "almost" satisfies the original equations.   *(Review Exercises 63–64.)*
- Find the least squares best approximation line $y = mx + b$ for a collection of x and y data pairs such that all the x-values are distinct.   *(Review Exercises 65–68.)*

$$\begin{pmatrix} x_1 & x_2 \cdots x_n \\ 1 & 1 \cdots 1 \end{pmatrix} \begin{pmatrix} x_1 & 1 \\ x_2 & 1 \\ \vdots & \vdots \\ x_n & 1 \end{pmatrix} \begin{pmatrix} m \\ b \end{pmatrix} = \begin{pmatrix} x_1 & x_2 \cdots x_n \\ 1 & 1 \cdots 1 \end{pmatrix} \begin{pmatrix} y_1 \\ y_2 \\ \vdots \\ y_n \end{pmatrix}$$

- Use the least squares best approximation line to make a prediction from information presented in an application.   *(Review Exercises 69–70.)*

HINTS AND SUGGESTIONS

- **Overview:** Row operations on matrices are a generalization of the elimination method of finding the intersection of two lines, extending the technique to problems with many equations in many variables. Matrix arithmetic is similar to real number operations except that matrix multiplication is not commutative and many matrices do not have inverses. However, if a square matrix A *does* have an inverse, then the equation $AX = B$ can be solved as $X = A^{-1}B$, just as a real number linear equation can be solved by dividing. Matrices are used to represent and solve many large and complicated problems important to both science and society.

- Although not every system of linear equations has a solution, they can all be identified as "consistent" or "inconsistent" and as "dependent" or "independent" by row-reducing the corresponding augmented matrix. Only "consistent" equations have a solution, which may be a single collection of values for the variables or many values for the variables given in terms of one or more parameters.

- When setting up a word problem, look first for the questions "how many" or "how much" to help identify the variables. Be sure that finding values for your variables will answer the question stated in the problem. Use the rest of the given information to build equations describing facts about your variables.

- Many row reduction problems require many steps to solve, so don't give up; keep improving the matrix until it meets all the requirements to be row-reduced. Make sure that each row operation you choose will move you toward your goal without undoing the parts you already have gotten the way you need.

- **Practice for Test:** Review Exercises 4, 6, 8, 11, 17, 20, 25, 30, 33, 35, 40, 43, 49, 51, 53, 57, 59, 65, and 69.

Review Exercises for Chapter 3   *Practice test exercise numbers are in blue.*

3.1   Systems of Two Linear Equations in Two Variables

Represent each pair of statements as a system of two linear equations in two variables. Be sure to state clearly the meaning of your x- and y-variables.

1. "A small commuter airplane has thirty passengers" and "the ticket receipts of $3970 come from 30-day advance sale tickets at $79 and full fare tickets at $159."

2. "A cow and horse rancher has four hundred twenty animals" and "there are twice as many cows as horses."

Solve each system of equations by graphing. Identify each system as "independent and consistent," "independent and inconsistent," or "dependent."

3. $\begin{cases} x + y = 18 \\ x - y = 8 \end{cases}$ 4. $\begin{cases} 2x - 4y = 36 \\ -3x + 6y = -54 \end{cases}$

Solve each system of equations by the substitution method. Identify each system as "independent and consistent," "independent and inconsistent," or "dependent."

5. $\begin{cases} 5x - 2y = 10 \\ -2x + y = 2 \end{cases}$ 6. $\begin{cases} -4x + 2y = 12 \\ 2x - y = 12 \end{cases}$

Solve each system of equations by the elimination method. Identify each system as "independent and consistent," "independent and inconsistent," or "dependent."

7. $\begin{cases} x + y = 12 \\ x + 3y = 18 \end{cases}$ 8. $\begin{cases} 4x + 5y = 60 \\ 2x + 3y = 42 \end{cases}$

Formulate each situation as a system of two linear equations in two variables. Be sure to state clearly the meaning of your x- and y-variables. Solve the system of equations by the elimination method. State your final answer in terms of the original question.

9. Garden Plants   A retired investment broker has rosebushes in his flower garden and tomato plants in his vegetable garden. He spends one hour each day tending his twenty-five plants. If each rosebush takes three minutes of care and each tomato plant takes two minutes, how many of each does he have?

10. Fraternity Convention   Twenty-six members of the Alpha Alpha Alpha fraternity want to go to the national convention in Orlando and each has put in $10 for gas. If each car holds 5 people and uses $45 worth of gas and each van holds 8 people and uses $85 worth of gas, how many of each vehicle do they need for the trip?

3.2 Matrices and Linear Equations in Two Variables

Find the dimension of each matrix and the values of the specified elements.

11. $\begin{pmatrix} 8 & 3 & 4 \\ 1 & 5 & 9 \\ 6 & 7 & 2 \end{pmatrix}; a_{2,2}, a_{3,1}, a_{1,3}$

12. $\begin{pmatrix} 1 & 15 & 14 & 4 \\ 12 & 6 & 7 & 9 \\ 8 & 10 & 11 & 5 \\ 13 & 3 & 2 & 16 \end{pmatrix}; a_{2,3}, a_{3,2}, a_{4,1}$

Carry out the row operation on the matrix.

13. $R_2' = 3R_2$ on $\begin{pmatrix} 3 & 4 & 12 \\ 1 & 2 & 2 \end{pmatrix}$

14. $R_1' = R_1 + R_2$ on $\begin{pmatrix} -1 & 2 & 2 \\ 2 & -3 & 6 \end{pmatrix}$

Solve each system of equations by row-reducing the corresponding augmented matrix and interpreting the final reduced row-echelon form matrix as the solution. Identify each system as "independent and consistent," "independent and inconsistent," or "dependent."

15. $\begin{cases} x + 3y = 63 \\ 4x + 5y = 140 \end{cases}$

16. $\begin{cases} -3x + 4y = 60 \\ 2x - y = 10 \end{cases}$

17. $\begin{cases} 12x - 4y = 36 \\ -15x + 5y = 45 \end{cases}$

18. $\begin{cases} -16x + 12y = -48 \\ 20x - 15y = 60 \end{cases}$

Formulate each situation as a system of two linear equations in two variables. Be sure to state clearly the meaning of your x- and y-variables. Solve the system of equations by row-reducing the corresponding augmented matrix. State your final answer in terms of the original question.

19. Pharmaceuticals The pharmacist at the Charter Drug Shop filled ninety-two prescriptions today for antibiotics and cough suppressants. If there were thirty-four more prescriptions for antibiotics than for cough suppressants, how many prescriptions for each were filled?

20. Classic Magazines A used book store offers grab bag packages of old *Life* and *The New Yorker* magazines from the 1940s containing 4 copies of *Life* and 3 copies of *The New Yorker* for $39 and larger bags of 12 copies of *Life* and 10 copies of *The New Yorker* for $122. What is the price of one copy of each old magazine?

3.3 Systems of Linear Equations and the Gauss–Jordan Method

Interpret each reduced row-echelon form matrix as the solution of a system of equations. Identify each system as "independent" or "dependent" and as "consistent" or "inconsistent."

21. $\begin{pmatrix} 1 & 0 & 0 & 3 \\ 0 & 1 & 0 & -3 \\ 0 & 0 & 1 & 6 \end{pmatrix}$

22. $\begin{pmatrix} 1 & 1 & 0 & 4 \\ 0 & 0 & 1 & 2 \\ 0 & 0 & 0 & 0 \end{pmatrix}$

Use the Gauss–Jordan method to find the equivalent reduced row-echelon form matrix.

23. $\begin{pmatrix} 0 & 1 & 1 & -1 \\ 1 & 0 & 1 & 6 \\ 2 & 0 & 1 & 10 \end{pmatrix}$

24. $\begin{pmatrix} 1 & 0 & 1 & 1 & 4 \\ 5 & 1 & 1 & 4 & 12 \\ 2 & 1 & 0 & 1 & 5 \\ 3 & 0 & 1 & 3 & 8 \end{pmatrix}$

Solve each system of equations by row-reducing the corresponding augmented matrix using the Gauss–Jordan method. Identify each system as "independent" or "dependent" and as "consistent" or "inconsistent."

If permitted by your instructor, you may use a graphing calculator.

25. $\begin{cases} x_1 - 2x_3 = 2 \\ -x_1 + x_2 + 2x_3 = 1 \\ -x_1 + 2x_2 + 3x_3 + x_4 = 7 \\ x_1 - 2x_3 + x_4 = 4 \end{cases}$

26. $\begin{cases} x_1 - x_2 + x_4 = 3 \\ x_2 + x_3 = 3 \\ 2x_1 - x_2 + x_3 + 2x_4 = 9 \\ 2x_1 - x_2 + x_3 + x_4 = 5 \end{cases}$

27. $\begin{cases} x_1 + 2x_3 + x_4 = 3 \\ x_1 + x_2 + 3x_3 + x_4 = 2 \\ 3x_1 + 3x_2 + 9x_3 + 4x_4 = 7 \\ 2x_1 + 4x_3 + 2x_4 = 7 \end{cases}$

28. $\begin{cases} x_1 + x_2 + x_3 + x_4 = 2 \\ x_1 + x_2 + x_3 + x_4 = 1 \\ x_1 + x_2 + x_3 + x_4 = 0 \\ x_1 + x_2 + x_3 + x_4 = 3 \end{cases}$

Formulate each situation as a system of linear equations in an appropriate number of variables and be sure to state clearly the meaning of each. Solve the system of equations by row-reducing the corresponding augmented matrix using the Gauss–Jordan method. State your final answer in terms of the original question.

⊞ If permitted by your instructor, you may use a graphing calculator.

29. Nursery Management   The Nyack Nursery starts plants from seeds and sells potted plants to garden stores for resale. Each dahlia costs 16¢ to start and needs a 10¢ flowerpot and 5 ounces of soil; each chrysanthemum costs 11¢ to start and needs a 12¢ flowerpot and 6 ounces of soil; and each daisy costs 13¢ to start and needs an 8¢ flowerpot and 5 ounces of soil. If the nursery has $50 to spend on starting the plants, $48 to spend on flowerpots, and 153 pounds of potting soil, how many of each plant can it raise using all of the available resources?

30. Family Entertainment   The Family Fun Center in Asheville offers the package specials listed in the table for an afternoon of fun at its go-kart track, miniature golf course, house of funny mirrors, and snack stand serving hot dogs and sodas. How much does one hot dog cost? How much does a go-kart ride cost if the mirror house is $1.50 and sodas are $1?

Go-kart	Golf	Mirror House	Hot Dog	Soda	Package Price
2	1	1	1	1	$15
3	2	1	2	1	$23
4	2	2	3	2	$31

3.4   Matrix Arithmetic

Find each matrix product.

31. $\begin{pmatrix} 1 & 2 & 3 & 4 \\ 4 & 3 & 2 & 1 \end{pmatrix} \begin{pmatrix} 1 & 0 \\ 0 & -1 \\ 1 & 0 \\ 0 & -1 \end{pmatrix}$

32. $\begin{pmatrix} 1 & 0 \\ 0 & -1 \\ 1 & 0 \\ 0 & -1 \end{pmatrix} \begin{pmatrix} 1 & 2 & 3 & 4 \\ 4 & 3 & 2 & 1 \end{pmatrix}$

Use the given matrices to find each matrix expression.

$$A = \begin{pmatrix} -1 & 2 & -1 \\ 2 & -1 & 2 \end{pmatrix} \qquad B = \begin{pmatrix} 3 & 6 & 8 \\ 7 & 5 & 4 \end{pmatrix}$$

$$C = \begin{pmatrix} 2 & 1 & 2 \\ 1 & -2 & 1 \\ 2 & 1 & 2 \end{pmatrix} \qquad D = \begin{pmatrix} 5 & 8 \\ 7 & 6 \end{pmatrix}$$

33. $3D - A \cdot B^t + I$ **34.** $3I + A^t \cdot B - C$

Rewrite each system of linear equations as a matrix equation $A \cdot X = B$.

35. $\begin{cases} x_1 + 4x_2 + x_3 = 15 \\ 2x_1 + 8x_2 + 3x_3 = 26 \\ x_1 + 5x_2 + 2x_3 = 17 \end{cases}$

36. $\begin{cases} 2x_1 + 3x_2 - x_3 + x_4 = 20 \\ 5x_1 + 4x_2 + x_3 + 2x_4 = 35 \\ 2x_1 + x_2 + x_3 + x_4 = 12 \end{cases}$

For each row operation (or sequence of row operations), find a 3 × 3 matrix R so that the matrix product $R \cdot A$ is the same as the result of carrying out the row operation(s) on the matrix

$$A = \begin{pmatrix} 2 & 5 & 23 \\ 1 & 3 & 13 \\ 2 & 4 & 20 \end{pmatrix}$$

37. $R'_1 = R_1 - R_2$

38. $R'_1 = R_2$ and $R'_2 = R_1$, $R'_2 = R_2 - R_3$, and $R'_3 = R_3 - 2R_1$

Formulate each situation in matrix form. Be sure to indicate the meaning of your rows and columns. Find the requested quantities using the appropriate matrix arithmetic.

39. Growing Grandchildren   A proud grandmother's record book shows that her grandson Thomas is now 61 inches tall and weighs 90 pounds while last year he was 58 inches tall and weighed 80 pounds; her grandson Richard is now 54 inches tall and weighs 75 pounds while last year he was 52 inches tall and weighed 70 pounds; and her granddaughter Harriet is now 47 inches tall and weighs 60 pounds while last year she was 46 inches tall and weighed 55 pounds. Represent these facts as a "this year" matrix and a "last year" matrix. Use these matrices to find how much each grandchild grew.

40. Spring Fashions   The buyer for the ladies sportswear division of a large department store needs 200 jackets, 300 blouses, 250 skirts, and 175 pairs of slacks. The spring lines shown by both an East Coast designer and an Italian team are acceptable to her, but the East Coast designer wants $195 for each jacket, $85 for each blouse, $145 for each skirt, and $130 for each pair of slacks, while the Italian company wants $190 for each jacket, $90 for each blouse, $150 for each skirt, and $125 for each pair of slacks. Represent her needs as a column matrix and the prices as a price matrix. Use these matrices to find the cost of her order from each source.

3.5   Inverse Matrices and Systems of Linear Equations

Find each matrix product. Identify each pair of matrices as "a matrix and its inverse" or "not a matrix and its inverse."

41. $\begin{pmatrix} 1 & 2 & 3 \\ 1 & 1 & 1 \\ 0 & 1 & 3 \end{pmatrix}$ and $\begin{pmatrix} -2 & 3 & 1 \\ 3 & -3 & -2 \\ -1 & 1 & 1 \end{pmatrix}$

42. $\begin{pmatrix} -3 & 0 & 1 \\ 1 & 3 & 1 \\ -3 & 2 & 2 \end{pmatrix}$ and $\begin{pmatrix} -4 & -2 & 3 \\ 5 & 3 & -4 \\ -11 & -6 & 8 \end{pmatrix}$

Row-reduce $(A \mid I)$ for each matrix A to find the inverse matrix A^{-1} or to identify A as a singular matrix.

43. $\begin{pmatrix} 1 & 1 & 0 \\ 0 & -3 & 1 \\ 2 & 3 & 0 \end{pmatrix}$ **44.** $\begin{pmatrix} 1 & 0 & 1 \\ 1 & 1 & 0 \\ 2 & 1 & 1 \end{pmatrix}$

Rewrite each system of equations as a matrix equation $A \cdot X = B$ and use the inverse of A to find the solution. Be sure to check the solution in the original system of equations.

45. $\begin{cases} 13x_1 + 4x_2 = 33 \\ 3x_1 + x_2 = 8 \end{cases}$ **46.** $\begin{cases} 3x_1 + 8x_2 = 25 \\ 2x_1 + 5x_2 = 16 \end{cases}$

47. $\begin{cases} 5x_1 + x_2 + 2x_3 = 11 \\ 2x_1 + 2x_2 + x_3 = 7 \\ 2x_1 + x_2 + x_3 = 5 \end{cases}$

48. $\begin{cases} 3x_1 + 2x_2 + x_3 + 2x_4 = 7 \\ 2x_1 + 5x_2 + 2x_3 + 2x_4 = 10 \\ x_1 + 2x_2 + x_3 + x_4 = 4 \\ 2x_1 + 2x_2 + x_3 + 2x_4 = 6 \end{cases}$

Formulate each situation as a collection of systems of linear equations. Be sure to state clearly the meaning of each variable. Solve each collection of systems of equations by finding the inverse of the common coefficient matrix and using matrix multiplications of this inverse times the various constant-term matrices. State your final answer in terms of the original question.

If permitted by your instructor, you may use a graphing calculator.

49. Retail Displays   A chain of furniture stores sells living room and bedroom suites and displays them in the store windows and inside on the showroom floor. A living room suite window display requires 3 square yards of window space and then the same furniture group is allowed 7 square yards of showroom space, while a bedroom suite window display requires 4 square yards of window space and then the same furniture group is allowed 9 square yards of showroom space. For furniture styles not given window space but shown only on the showroom floor, each living room suite is allowed 8 square yards and each bedroom suite is allowed 10 square yards. The setup costs per square yard are $50 for furniture shown in the window and on the showroom floor, while for furniture shown only on the showroom floor, $60 for living room suites and $70 for bedroom suites. The table lists the available space (in square yards) at the chain's stores in Kingman, Prescott, and Holbrook, together with the approved setup budgets and the amount of the total showroom space that must be given to furniture not displayed in the window. How many different living room and bedroom suites can be shown at each location?

	Window Space	Floor Space for Showroom Floor	Nonwindow Suites	Setup Budget
Kingman	18	131	90	$ 8,850
Prescott	21	182	134	$12,190
Holbrook	17	187	148	$12,680

50. Income Taxes Find the federal, state, and city income taxes on the taxable incomes of the individuals listed in the table if the federal income tax is 20% of the taxable income after first deducting the state and city taxes; the state income tax is 10% of the taxable income after first deducting the federal and city taxes; and the city income tax is 5% of the taxable income after first deducting the federal and state taxes.

Taxpayer	Mr. Dahlman	Mrs. Farrell	Ms. Mazlin	Mr. Seidner
Taxable income	$96,700	$48,350	$77,360	$145,050

3.6 Two Applications

Leontief "Open" Input–Output Models

Find the technology matrix for each economy diagram.

51.

25¢
15¢   30¢   20¢

52.

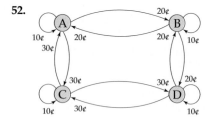

20¢
10¢   20¢   20¢   10¢
30¢
30¢   30¢   20¢
10¢   30¢   10¢

Draw an economy diagram for each technology matrix.

53.
$$\begin{pmatrix} 0.10 & 0 & 0.20 \\ 0.10 & 0.15 & 0 \\ 0 & 0.15 & 0.20 \end{pmatrix}$$

54.
$$\begin{pmatrix} 0.15 & 0.10 & 0 & 0.15 \\ 0 & 0.10 & 0.20 & 0 \\ 0.15 & 0 & 0.10 & 0 \\ 0 & 0.10 & 0 & 0.15 \end{pmatrix}$$

Find the excess production Y of each economy with technology matrix A and economic activity level X.

55. $A = \begin{pmatrix} 0.10 & 0.05 & 0.15 \\ 0.15 & 0.10 & 0.05 \\ 0.10 & 0.05 & 0.10 \end{pmatrix}, X = \begin{pmatrix} 540 \\ 620 \\ 560 \end{pmatrix}$

56. $A = \begin{pmatrix} 0.15 & 0.15 & 0.10 & 0.05 \\ 0.10 & 0.10 & 0.05 & 0.10 \\ 0.10 & 0.05 & 0.10 & 0.15 \\ 0.05 & 0.10 & 0.15 & 0.10 \end{pmatrix}, X = \begin{pmatrix} 280 \\ 200 \\ 260 \\ 240 \end{pmatrix}$

Find, for each economy with technology matrix A, the economic activity level X necessary to generate excess production Y.

57. $A = \begin{pmatrix} 0.10 & 0.35 \\ 0.40 & 0.15 \end{pmatrix}, Y = \begin{pmatrix} 175 \\ 225 \end{pmatrix}$

58. $A = \begin{pmatrix} 0.10 & 0.10 & 0.20 \\ 0.30 & 0.20 & 0.10 \\ 0.10 & 0.30 & 0.20 \end{pmatrix}, Y = \begin{pmatrix} 245 \\ 294 \\ 196 \end{pmatrix}$

Represent each situation as a Leontief "open" input–output model by constructing an economy diagram and the corresponding technology matrix. Find the required excess production or level of economic activity, and be sure to state your final answer in terms of the original question.

59. A 5&10 Problem Each of the four divisions of the Woolworth Corporation depends on the others in the following way: Each dollar of production from each division requires 10¢ of production from that division and 5¢ of production from each of the others. How much must each division produce to yield an excess production of $3 million from each?

60. Energy Dependence As a result of treaty obligations, the domestic and foreign oil consumption of an industrialized nation is linked to the protection provided by its military forces. Each dollar of domestic oil produced requires 5¢ of domestic oil and 10¢ of military protection; each dollar of foreign oil requires 20¢ of military protection; and each dollar of military protection consumes 10¢ of domestic oil, 10¢ of foreign oil, and 5¢ of military protection. If the domestic oil production level is $940 million, the foreign production is $2000 million, and the military protection budget is $520 million, how much of each can be used elsewhere in the nation's economy?

Least Squares

Solve each system of equations by row-reducing the corresponding augmented matrix. Rewrite the system of equations in the matrix form $A \cdot X = B$, calculate $X = (A^tA)^{-1} \cdot (A^tB)$, and verify that this is the same solution.

61. $\begin{cases} x - y = 1 \\ 3x + y = 15 \\ x + 2y = 10 \end{cases}$

62. $\begin{cases} 4x + 3y = 12 \\ x + y = 5 \\ -2x + y = 14 \end{cases}$

 Row-reduce the corresponding augmented matrix for each system of equations to verify that the system is inconsistent. Rewrite the system of equations in the matrix form $A \cdot X = B$, calculate $X = (A^tA)^{-1} \cdot (A^tB)$, and verify that this compromise solution "almost" satisfies the original equations.

63. $\begin{cases} x - 3y = -9 \\ 2x - y = 12 \\ x + 2y = 6 \end{cases}$

64. $\begin{cases} x + 3y = 3 \\ 3x - y = 9 \\ x + y = -9 \end{cases}$

 Find the least squares best approximation line $y = mx + b$ for each collection of x and y data pairs.

65.

x	2	3	4
y	22	32	36

66.

x	3	5	6
y	238	364	462

67.

x	2	3	5	6
y	40	50	80	90

68.

x	10	11	12	13	14
y	140	150	180	200	210

 Use the least squares best approximation line to make each prediction. Be sure to state your final answer in terms of the original question.

69. Store Hours   The owners of a "Mom and Pop" corner store have had their store open various numbers of hours on recent Mondays, the usual day off. The table shows the number of hours the store was open and the sales receipts for the day. How much could they expect to sell next Monday if they keep their store open for 12 hours?

Hours open	6	8	10	14
Total sales	$2230	$3035	$3770	$5065

70. Production Accidents   The supervisor of a factory assembly line gathered the data in the table over the last year to compare the number of minor accidents each month with the number of extra five-minute mini-breaks she allows during the day. How many minor accidents could she expect next month if she allows four mini-breaks during the day?

Mini-breaks	0	2	6	8
Minor accidents	21	16	8	3

Projects and Essays

The following projects and essays are based on Chapter 3. There are no right or wrong answers—the results depend only on your imagination and resourcefulness.

1. Go to a library (especially a business or technical library) and browse through the last several issues of the journal *Management Science* or *Operations Research*. Write a one-page report on the use of matrices to analyze a real-world problem.

2. Read the essay "Gauss, the Prince of Mathematicians" by Eric Temple Bell in Volume One of *The World of Mathematics* by James R. Newman (New York: Simon and Schuster, 1956), pages 295–339, and write a one-page report on what you learn.

3. Read the essay "Invariant Twins, Cayley and Sylvester" by Eric Temple Bell in Volume One of *The World of Mathematics* by James R. Newman (New York: Simon and Schuster, 1956), pages 341–365, and write a one-page report on what you learn.

4. Read the article "The Structure of the U.S. Economy" by Wassily W. Leontief on pages 25–35 of the April 1965 issue of *Scientific American* (Volume 212, Number 4), and write a one-page report on what you find.

5. Read the article "The World Economy of the Year 2000" by Wassily W. Leontief on pages 206–231 of the September 1980 issue of *Scientific American* (Volume 243, Number 3), and write a one-page report on the progress made since this article was written on reducing the gap between the rich and poor of the world.

6. Find examples of matrices representing "independent" and "dependent" systems of equations and representing "consistent" and "inconsistent" systems of equations that are different from those given on page 191.

7. Make up your own example of a Leontief "open" input–output economic model and explore how each one-dollar change in the production level of the first product changes the total excess production.

8. Find an example of an inconsistent system of equations $A \cdot X = B$ such that the square matrix $A^t \cdot A$ is not invertible.

9. Show by example that if the x and y data for a least squares best approximation line contains two pairs with the same x-value, then the matrix $A^t \cdot A$ is not invertible.

10. If the statistical commands for your graphing calculator will automatically find the least squares best approximation line, compare the calculator answer to your matrix solution of Exercise 36 on page 250.

4

Linear Programming

An assembly line can work only if enough parts and workers are always ready at the correct times and places. Such management problems can be solved by the method of linear programming, as explained in this chapter.

Linear Inequalities

Northwest Airlines and Crew Schedules

Northwest Airlines spends over $1 billion each year on salaries, benefits, overtime, and other expenses for its pilots and flight attendants. If the available personnel could be used more effectively, a minor improvement of just 1% would represent a $10 million savings that year. For this reason, Northwest Airlines employs mathematicians and computer analysts in its schedule development department and provides them with the latest supercomputers. Because the airline's flight schedule changes monthly, the scheduling department is never out of work. Its objective is to minimize the crew costs while adequately staffing the scheduled flights and satisfying Federal Aviation Agency regulations and union contract requirements. These constraints may be expressed as inequalities: Each flight must have at least a pilot and copilot in the cockpit, each crew member can fly for no more than a given number of hours and must have at least a corresponding number of rest hours each day, each crew member's flight plans must form a round trip back to the starting airport, and so on. The solution of each month's problem is found in two stages: finding a "feasible solution" that provides a crew for each flight while satisfying all of the constraints, and then searching for ways to alter this initial solution to lower the costs while still respecting all the restrictions. Although the large size of Northwest Airline's scheduling problem makes it impossible to find the "best" solution even using today's computers, the scheduling department's solutions (found using the methods described in this chapter) are sufficiently close to the best to provide the company with significant savings.

Introduction

This chapter describes and solves a large class of problems known as *linear programming problems*. These problems and their solutions by the *simplex method* are the cornerstone of the science of *operations research*. Many problems in modern business may be addressed by these methods. Although the simplex method is an algebraic procedure, it is based on the geometry of linear inequalities, where we now begin.

Inequalities

On page 4 we discussed inequalities such as $x < y$ ("x is less than y") and $x \geq y$ ("x is greater than or equal to y" or, equivalently, "x is at least y"). Many everyday relationships are easily expressed using inequalities.

EXAMPLE 1   Writing an Inequality in Algebraic Notation

Express the statement "Joe is richer than Fred" as an inequality.

Solution

This means that "Joe has more money than Fred," so using J for Joe's money and F for Fred's money, we could express this as $J > F$. The statement could also be rephrased "Fred has less money than Joe" and then written as $F < J$. Because a number minus a smaller number is positive, we might even write that $J - F > 0$ or, looking at it from the other viewpoint, $F - J < 0$. From this we see that every inequality may be written in several different ways.   ∎

Because inequalities with two variables can be understood in the same manner as one-variable inequalities, we start with the simpler situation.

Inequalities in One Variable

To *sketch* an inequality means to graph the points that satisfy the inequality. The *boundary* of an inequality is the corresponding *equality*. For example, the boundary of the inequality $x \geq 2$ is the equality $x = 2$. On the number line, $x = 2$ represents just a single point, and the numbers satisfying $x \geq 2$ are all on the same side of $x = 2$. That is, we sketch $x \geq 2$ on the number line in three steps:

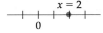

1. Draw the boundary

2. $x > 2$ is on the *right* side of the boundary

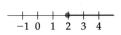

3. Shade the correct side of the boundary

You can also find the correct side of the boundary by checking the inequality at a "test point," substituting a number for the variable. For example, given the inequality $x \geq 2$, substituting 5 for x gives $5 \geq 2$, which is *true*, so the point $x = 5$ is on the *correct* side of $x \geq 2$. On the other hand, substituting -3 for x gives $-3 \geq 2$, which is *false*, so $x = -3$ is on the *wrong* side of $x \geq 2$. Any one point on either side will do; finding the "wrong" side just means that the other side is "right."

The set of all the points satisfying the inequality is called the *feasible region* of the inequality. To graph the feasible region, sketch the inequality using the following three steps.

How to Sketch a Linear Inequality

> **1.** Draw the boundary.
> **2.** Choose the side of the boundary corresponding to the inequality (and verify it with a "test point").
> **3.** Shade the correct side to show the feasible region.

A *system* of linear inequalities is two or more linear inequalities joined by a brace, {, meaning that *each* of the inequalities holds. The feasible region of a system is the set of points where *each* of the inequalities holds, and it can be graphed using the same three steps. For example, the numbers x on the real number line that satisfy the system
$$\begin{cases} -1 \leq x \\ x \leq 2 \end{cases} \text{ (which means } -1 \leq x \text{ and } x \leq 2\text{) are found as follows:}$$

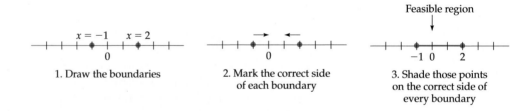

Because there *are* points between $x = -1$ and $x = 2$, the system of linear inequalities $-1 \leq x$ and $x \leq 2$ is said to be *feasible*. If there are *no* feasible points, then the system is *infeasible*. For example, there are no points such that $x \geq 3$ and $x \leq 1$:

Two-Variable Inequalities

Inequalities in two variables can be sketched by a similar method, but now in the *plane* rather than on the number line. A test point now has two coordinates.

EXAMPLE 2 **Sketching a Linear Inequality in the Plane**

Sketch the linear inequality $2x + 3y \geq 12$.

Solution

The boundary is the line $2x + 3y = 12$. Such a line can be sketched by plotting the *x*- and *y*-intercepts. Since the *x*-intercept has $y = 0$, the equation becomes $2x + 0 = 12$, so $x = 12/2 = 6$ and the *x*-intercept is $(6, 0)$. Similarly, since the *y*-intercept has $x = 0$, the equation becomes $0 + 3y = 12$, so $y = 12/3 = 4$ and the *y*-intercept is $(0, 4)$. Continuing as before:

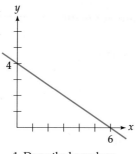

1. Draw the boundary

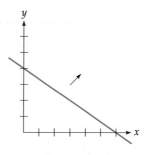

2. Using $(0, 0)$ as a test point,
$2 \cdot 0 + 3 \cdot 0$ is not ≥ 12.
The origin $(0, 0)$ is *not* on
the correct side of the boundary

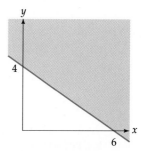

3. Shade the correct
side of the boundary

In general:

Boundary of a Linear Inequality

> The boundary of $ax + by \leq c$ (or $\geq c$) is the line $ax + by = c$ with *x*-intercept at $\dfrac{c}{a}$ and *y*-intercept at $\dfrac{c}{b}$.

Notice that if the ratio c/a or c/b does not exist, then neither does the corresponding intercept. For example, the boundary of the inequality $1x + 0y \geq 2$ is the line $x = 2$. This vertical line does not have a

y-intercept, and the ratio "2/0" does not exist (since division by zero is not defined).

> If $c = 0$, then the boundary line $ax + by = 0$ passes through the origin $(0, 0)$ and the points $(b, -a)$ and $(-b, a)$.

For example, the boundary of $2x - 3y \geq 0$ is the line $2x - 3y = 0$, which passes through the origin $(0, 0)$ and the points $(-3, -2)$ and $(3, 2)$. Any two of these points will suffice to sketch the boundary line.

Graphing Calculator Exploration

You can view any nonvertical boundary line $ax + by = c$ by entering it as $y = (c - ax)/b$. For example, to see the boundary line $2x + 3y = 12$ from Example 2:

a. Enter the boundary line as $y_1 = (12 - 2x)/3$ and graph it on the window $[-5, 10]$ by $[-5, 10]$.

b. Use TRACE or EVALUATE to check the x- and y-intercepts.

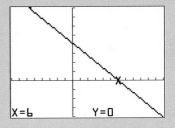

X=6   Y=0

PRACTICE PROBLEM 1

Sketch the linear inequality $3x - 5y \leq 60$.   *Solution at the back of the book*

EXAMPLE 3 **Sketching a System of Linear Inequalities**

a. Sketch the system $\begin{cases} x + y \geq 3 \\ x - y \geq 3. \\ x \leq 6 \end{cases}$

b. Is this system feasible?

c. What happens if the last inequality ($x \leq 6$) is replaced by $x \leq 2$?

Solution

a. The boundaries are the lines $x + y = 3$ [with intercepts $(3, 0)$ and $(0, 3)$], $x - y = 3$ [with intercepts $(3, 0)$ and $(0, -3)$], and $x = 6$ [a vertical line with x-intercept $(6, 0)$ and no y-intercept]. The origin $(0, 0)$ does not satisfy either $x + y \geq 3$ or $x - y \geq 3$, but it does satisfy $x \leq 6$.

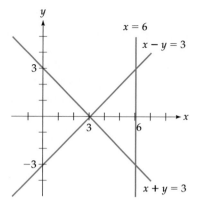

1. Draw the boundary lines

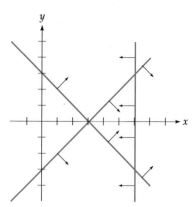

2. Mark the correct sides

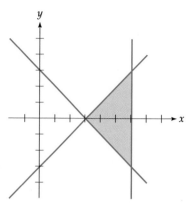

3. Shade the feasible region

b. Since there are points that lie on the correct sides of *all* the boundaries, this system of linear inequalities *is* feasible.

c. If we replace the third inequality ($x \leq 6$) by $x \leq 2$, the vertical boundary line $x = 6$ is replaced by the vertical boundary line $x = 2$:

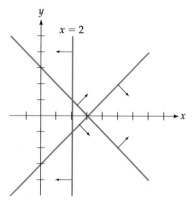

This new system of linear inequalities is *infeasible* because there are no points that lie on the correct sides of all the boundaries at once (as you can see from the sketch).

———————————————————————————————————— ■

PRACTICE PROBLEM 2

Sketch the system $\begin{cases} x + 2y \leq 20 \\ x + y \geq 10 \\ x \leq 10 \end{cases}$. Is this system feasible?

Solution at the back of the book

Convex Regions and Vertices

A region is *convex* if it contains the line segment joining any two points in the region. Because any linear inequality represents just one side of the boundary line, the feasible region of any linear inequality is convex. Thus *the feasible region of any system of linear inequalities is convex* because, given any two points in the region, those two points and the line segment between them are on the correct side of each of the boundary lines, and so the line segment is also in the region.

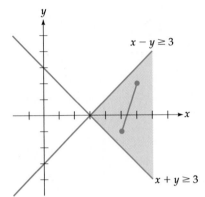

The feasible region for the system $\begin{cases} x - y \geq 3 \\ x + y \geq 3 \end{cases}$ is convex.

On the other hand, the following region is *not* convex because the line segment between (1, 1) and (5, 1) does *not* lie completely inside the region. (You cannot get a region such as the one below for the feasible region of a system of linear inequalities, because it contains points on *both* sides of a boundary line.)

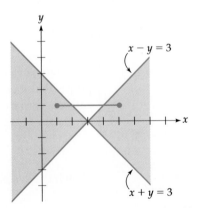

This region is *not* convex and so *cannot* be a feasible region.

The "corner points" of the feasible region are called *vertices*. More precisely:

> A *vertex* of a system of linear inequalities is an intersection point of two (or more) of the boundaries that satisfies all of the inequalities.

Each vertex may be found by solving one of the boundaries for x or y in terms of the other variable, substituting this expression into the other boundary, solving for the value of the remaining variable, and then finding the value of the first variable. If this point satisfies all of the inequalities, then it is a vertex of the region.

EXAMPLE 4 **Finding the Vertices of a Feasible Region**

Sketch the system $\begin{cases} x + 2y \geq 12 \\ x - y \geq 0 \\ x \leq 8 \end{cases}$ and find the vertices of the feasible region.

Solution

The boundaries are the lines $x + 2y = 12$ [with intercepts (12, 0) and (0, 6)], $x - y = 0$ [with intercept (0, 0) and passing through (1, 1)], and $x = 8$ [a vertical line with x-intercept (8, 0) and no y-intercept]. The test point (0, 3) does not satisfy $x + 2y \geq 12$, does not satisfy $x - y \geq 0$, but does satisfy $x \leq 8$.

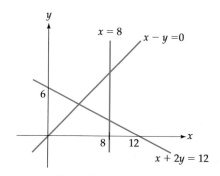

1. Draw the boundary lines

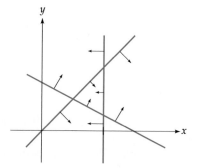

2. Mark the correct sides

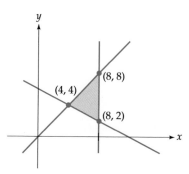

3. Shade the feasible region

This region has three vertices. Two of them lie on the vertical line $x = 8$. Substituting $x = 8$ into $x + 2y = 12$ gives $(8) + 2y = 12$, so $y = 2$ and the intersection point is $(8, 2)$. Substituting $x = 8$ into $x - y = 0$ gives $(8) - y = 0$, so $y = 8$ and the intersection point is $(8, 8)$. The third vertex is the intersection of $x + 2y = 12$ and $x - y = 0$:

$$x = 12 - 2y \qquad \text{Solving } x + 2y = 12 \text{ for } x$$

$$(12 - 2y) - y = 0 \qquad \text{Substituting into } x - y = 0$$

$$12 - 3y = 0 \text{ so } 12 = 3y \text{ and } y = 12/3 = 4 \quad \text{Simplifying and solving for } y$$

$$x = 12 - 2(4) = 12 - 8 = 4 \qquad x = 12 - 2y \text{ with } y = 4$$

$$(4, 4) \qquad \text{Vertex at } x = 4, y = 4$$

The vertices of this region are the points $(8, 2)$, $(8, 8)$, and $(4, 4)$, as shown in the preceding drawing.

A region is *bounded* if it can be completely contained inside a rectangular region of the form $L \leq x \leq R$ and $B \leq y \leq T$ for some values of L, R, B, and T (for "left," "right," "bottom," and "top," respectively). A region is *unbounded* if it is not bounded. Of course, for a bounded region there are many possible choices for the values of L, R, B, and T. The region in Example 4 is bounded because all of the x-values are between 2 and 10 and all of the y-values are between 1 and 10.

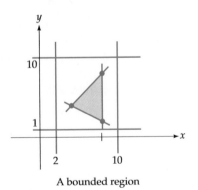

A bounded region

PRACTICE PROBLEM 3

Find the vertices of the feasible region for the system $\begin{cases} x + 2y \leq 20 \\ x + y \geq 10 \\ x \leq 10 \end{cases}$

from Practice Problem 2. Is this region bounded?

Solution at the back of the book

Two Typical Regions

In many applications, the variables represent quantities of materials and thus may not take negative values. The conditions $x \geq 0$ and $y \geq 0$ are *nonnegativity constraints* and force the feasible region to be in the first quadrant. The next two examples display typical feasible regions.

EXAMPLE 5   Finding the Feasible Region for a Word Problem

A small jewelry company prepares and mounts semiprecious stones. There are 10 lapidaries (who cut and polish the stones) and 12 jewelers (who mount the stones in gold settings). Each employee works 7 hours each day. Each tray of agates requires 5 hours of cutting and polishing and 4 hours of mounting, while each tray of onyxes requires 2 hours of cutting and polishing and 3 hours of mounting. How many trays of each stone can be processed each day?

Formulate this situation as a system of linear inequalities, sketch the feasible region, and find the vertices.

Solution

Clearly, there are many different quantities of stones that this company could process each day: It could choose to do nothing and give the employees the day off, it could process just one kind of stone, or it could process some combination. But it could not exceed the amount of time the lapidaries could work (10 workers at 7 hours each is 70 work-hours) or the amount of time the jewelers could work (12 workers at 7 hours each is 84 work-hours). Nor could it process a negative number of trays.

Let

$$x = \begin{pmatrix} \text{Number of trays} \\ \text{of agates} \end{pmatrix} \quad \text{and} \quad y = \begin{pmatrix} \text{Number of trays} \\ \text{of onyxes} \end{pmatrix}$$

For the lapidaries,

$$5x \quad + \quad 2y \quad \leq \quad 70$$

| x trays @ 5 hours each | y trays @ 2 hours each | No more than | 10 workers @ 7 hours each |

$$\begin{pmatrix} \text{Time} \\ \text{for} \\ \text{agates} \end{pmatrix} + \begin{pmatrix} \text{Time} \\ \text{for} \\ \text{onyxes} \end{pmatrix} \leq \begin{pmatrix} \text{Total} \\ \text{time for} \\ \text{lapidaries} \end{pmatrix}$$

For the jewelers,

$$4x \quad + \quad 3y \quad \leq \quad 84$$

| x trays @ 4 hours each | y trays @ 3 hours each | No more than | 12 workers @ 7 hours each |

$$\begin{pmatrix} \text{Time} \\ \text{for} \\ \text{agates} \end{pmatrix} + \begin{pmatrix} \text{Time} \\ \text{for} \\ \text{onyxes} \end{pmatrix} \leq \begin{pmatrix} \text{Total} \\ \text{time for} \\ \text{jewelers} \end{pmatrix}$$

Combining these constraints with the nonnegativity conditions $x \geq 0$ and $y \geq 0$, we can represent the problem by the system of linear inequalities

$$\begin{cases} 5x + 2y \leq 70 \\ 4x + 3y \leq 84 \\ x \geq 0 \\ y \geq 0 \end{cases}$$

We can now proceed as usual. The intercepts of the boundary line $5x + 2y = 70$ are $(70/5, 0) = (14, 0)$ and $(0, 70/2) = (0, 35)$. The intercepts of the boundary line $4x + 3y = 84$ are $(84/4, 0) = (21, 0)$ and $(0, 84/3) = (0, 28)$. The nonnegativity conditions, $x \geq 0$ and $y \geq 0$, place the feasible region in the first quadrant. Since the test point $(0, 0)$ satisfies the constraints, this region has the origin for one of its vertices.

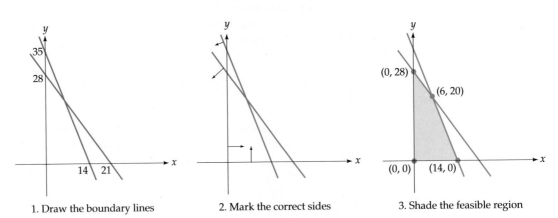

1. Draw the boundary lines   2. Mark the correct sides   3. Shade the feasible region

Three of the four vertices of this bounded region are already known from the x- and y-intercepts of the boundary lines. The fourth is the intersection of $5x + 2y = 70$ and $4x + 3y = 84$:

$$5x = 70 - 2y \quad \text{so} \quad x = 14 - \tfrac{2}{5}y \qquad \text{Solving } 5x + 2y = 70 \text{ for } x$$

$$4\left(14 - \tfrac{2}{5}y\right) + 3y = 84 \qquad \text{Substituting into } 4x + 3y = 84$$

$$56 - \tfrac{8}{5}y + 3y = 84 \quad \text{so} \quad \tfrac{7}{5}y = 28 \quad \text{and} \quad y = \tfrac{5}{7} \cdot 28 = 20 \qquad \text{Simplifying and solving for } y$$

$$x = 14 - \tfrac{2}{5}(20) = 14 - 8 = 6 \qquad x = 14 - \tfrac{2}{5}y \text{ with } y = 20$$

$$(6, 20) \qquad \text{Vertex at } x = 6, y = 20$$

The four vertices of this region are $(0, 0)$, $(14, 0)$, $(6, 20)$, and $(0, 28)$, as shown above.

Graphing Calculator Exploration

You can sketch the boundary lines from Example 5 on your graphing calculator by entering them as $y_1 = (70 - 5x)/2$ and $y_2 = (84 - 4x)/3$.

a. Set the window to $[-10, 60]$ by $[-10, 40]$ and graph your lines. Remember that the non-negativity conditions $x \geq 0$ and $y \geq 0$ put this region in the first quadrant.

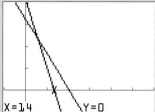

X=14   Y=0

b. Use TRACE or EVALUATE to check the *x*- and *y*-intercepts.

c. You can use TRACE or INTERSECT to find the intersection point of the lines y_1 and y_2.

EXAMPLE 6   **An Unbounded Feasible Region**

The Marshall County trash incinerator in Norton burns 10 tons of trash per hour and co-generates 6 kilowatts (kW) of electricity, while the Wiseburg incinerator burns 5 tons per hour and co-generates 4 kilowatts. If the county needs to burn at least 70 tons of trash and co-generate at least 48 kilowatts of electricity each day, how many hours should each plant operate?

Formulate this situation as a system of linear inequalities, sketch the feasible region, and find the vertices.

Solution

Again, there are many different schedules of operating times that could burn all the trash and co-generate enough electricity. But there must be at least enough time to get the job done, and neither plant can operate a negative number of hours. Let

$$x = \binom{\text{Number of hours}}{\text{Norton operates}} \quad \text{and} \quad y = \binom{\text{Number of hours}}{\text{Wiseburg operates}}$$

For the amount of trash to be burned, we have the inequality

$$\underbrace{10x}_{\substack{x \text{ hours} \\ @ \text{ 10 tons} \\ \text{each}}} + \underbrace{5y}_{\substack{y \text{ hours} \\ @ \text{ 5 tons} \\ \text{each}}} \underbrace{\geq}_{\substack{\text{at} \\ \text{least}}} \underbrace{70}_{\substack{\text{Tons of} \\ \text{trash} \\ \text{to burn}}} \qquad \binom{\text{Tons}}{\substack{\text{burned} \\ \text{at Norton}}} + \binom{\text{Tons}}{\substack{\text{burned at} \\ \text{Wiseburg}}} \geq \binom{\text{Tons}}{\substack{\text{to} \\ \text{burn}}}$$

For the electricity to be produced,

$$\underbrace{6x}_{\substack{x\ \text{hours} \\ @\ 6\ \text{kWs} \\ \text{each}}} + \underbrace{4y}_{\substack{y\ \text{hours} \\ @\ 4\ \text{kWs} \\ \text{each}}} \underbrace{\geq}_{\substack{\text{at} \\ \text{least}}} \underbrace{48}_{\substack{\text{kWs} \\ \text{needed}}} \qquad \left(\begin{array}{c}\text{Norton} \\ \text{electricity}\end{array}\right) + \left(\begin{array}{c}\text{Wiseburg} \\ \text{electricity}\end{array}\right) \geq \left(\begin{array}{c}\text{Electricity} \\ \text{needed}\end{array}\right)$$

Combining these constraints with the nonnegativity conditions $x \geq 0$ and $y \geq 0$, we can represent the problem by the system of linear inequalities

$$\begin{cases} 10x + 5y \geq 70 \\ 6x + 4y \geq 48 \\ x \geq 0 \\ y \geq 0 \end{cases}$$

The intercepts of the boundary line $10x + 5y = 70$ are $(70/10, 0) = (7, 0)$ and $(0, 70/5) = (0, 14)$. The intercepts of the boundary line $6x + 4y = 48$ are $(48/6, 0) = (8, 0)$ and $(0, 48/4) = (0, 12)$. The nonnegativity conditions place the feasible region in the first quadrant. Since the origin does not satisfy either of the first two inequalities, this region does *not* have the origin for one of its vertices.

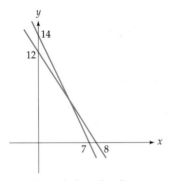

1. Draw the boundary lines

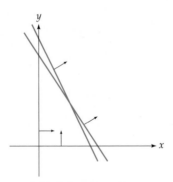

2. Mark the correct sides

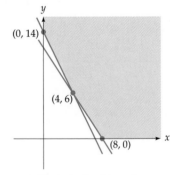

3. Shade the feasible region

Two of the three vertices of this unbounded region are already known from the x- and y-intercepts of the boundary lines. The third is the intersection of $10x + 5y = 70$ and $6x + 4y = 48$:

$$5y = 70 - 10x \quad \text{so} \quad y = 14 - 2x \qquad \text{Solving } 10x + 5y = 70 \text{ for } y$$

$$6x + 4(14 - 2x) = 48 \qquad \text{Substituting into } 6x + 4y = 48$$

$$6x + 56 - 8x = 48 \quad \text{means} \quad -2x = -8 \quad \text{so} \quad x = 4 \qquad \text{Simplifying and solving for } x$$

$$y = 14 - 2 \cdot 4 = 14 - 8 = 6 \qquad y = 14 - 2x \text{ with } x = 4$$

$$(4, 6) \qquad \text{Vertex at } x = 4, y = 6$$

The three vertices of this region are (8, 0), (4, 6), and (0, 14), as shown in the preceding diagram.

∎

Section Summary

The *feasible region* of a *system of linear inequalities* $\begin{cases} ax + by \le c \\ \quad \cdot \cdot \end{cases}$ consists of all the (x, y) points on the correct sides of the *boundary lines* $\begin{cases} ax + by = c \\ \quad \cdot \cdot \end{cases}$. Each boundary line $ax + by = c$ has x-intercept $\left(\frac{c}{a}, 0\right)$ and y-intercept $\left(0, \frac{c}{b}\right)$ [if $c = 0$, the boundary passes through the points $(0, 0)$, $(-b, a)$, and $(b, -a)$]. The correct side of the boundary can be found by trying a "test point" in the inequality [the origin $(0, 0)$ is usually the easiest to use].

The *vertices* are the "corner points" of the region and are the intersections of two (or more) boundary lines that satisfy the inequalities. These intersections can be found by solving the equations representing the lines.

A feasible region is *bounded* if it is contained within a rectangular box, and *unbounded* if it is not. The feasible region of any system of linear inequalities is *convex* because it contains the line segment joining any two points in the region.

Two special inequalities are the *nonnegativity conditions* $x \ge 0$ and $y \ge 0$, which place the feasible region in the first quadrant and frequently appear in word problems when the variables represent amounts of materials or other real objects.

EXERCISES 4.1

For each region, select the linear inequality it represents.

1. a. $5x + 8y \ge 40$
   b. $8x + 5y \ge 40$
   c. $5x + 8y \le 40$
   d. $8x + 5y \le 40$
   e. No such inequality because the region is not convex

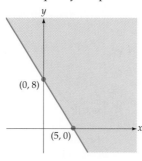

2. a. $7x + 6y \ge 42$
   b. $6x + 7y \ge 42$
   c. $7x + 6y \le 42$
   d. $6x + 7y \le 42$
   e. No such inequality because the region is not convex

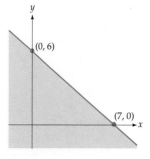

3. a. $-12x + 5y \geq -60$
   b. $5x - 12y \geq -60$
   c. $5x - 12y \leq 60$
   d. $10x + 24y \leq 60$
   e. No such inequality because the region is not convex

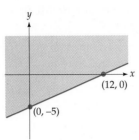

4. a. $5x - 3y \leq -15$
   b. $5x - 3y \geq -15$
   c. $-3x + 5y \leq 15$
   d. $-5x + 3y \leq 15$
   e. No such inequality because the region is not convex

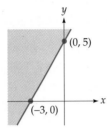

For each region, select the system of linear inequalities it represents.

5. a. $\begin{cases} 5x + 2y \leq 20 \\ y \geq 0 \end{cases}$
   b. $\begin{cases} 4x + 10y \leq 40 \\ x \leq 0 \end{cases}$
   c. $\begin{cases} 10x - 4y \leq 40 \\ y \leq 0 \end{cases}$
   d. $\begin{cases} -2x + 5y \geq -20 \\ y \geq 0 \end{cases}$
   e. No such inequalities because the region is not convex

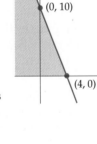

6. a. $\begin{cases} 5x - 3y \leq -30 \\ x \geq 0 \end{cases}$
   b. $\begin{cases} -6x + 10y \leq -60 \\ x \geq 0 \end{cases}$
   c. $\begin{cases} -5x + 3y \leq 30 \\ x \leq 0 \end{cases}$
   d. $\begin{cases} 5x + 3y \leq -30 \\ x \geq 0 \end{cases}$
   e. No such inequalities because the region is not convex

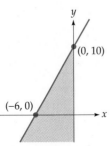

7. a. $\begin{cases} -x + 2y \leq 2 \\ 2x - y \leq 2 \end{cases}$
   b. $\begin{cases} x - 2y \leq 2 \\ x - 2y \leq -2 \end{cases}$
   c. $\begin{cases} x + 2y \leq 2 \\ x + 2y \geq -2 \end{cases}$
   d. $\begin{cases} x - 2y \leq 2 \\ -x + 2y \leq 2 \end{cases}$
   e. No such inequalities because the region is not convex

8. a. $\begin{cases} x + y \leq 8 \\ x - y \leq 0 \end{cases}$
   b. $\begin{cases} x + y \geq 8 \\ x + y \geq 0 \end{cases}$
   c. $\begin{cases} x + 8y \leq 8 \\ x - y \geq 0 \end{cases}$
   d. $\begin{cases} x - y \leq -8 \\ -x + y \leq 0 \end{cases}$
   e. No such inequalities because the region is not convex

9. a. $\begin{cases} x - 3y \leq -60 \\ x \geq 0 \\ y \geq 0 \end{cases}$
   b. $\begin{cases} 2x - 6y \geq -120 \\ x \geq 0 \\ y \geq 0 \end{cases}$
   c. $\begin{cases} -60x + 20y \leq 240 \\ y \geq 0 \end{cases}$
   d. $\begin{cases} 3x - y \geq -60 \\ x \leq 0 \\ y \geq 0 \end{cases}$
   e. No such inequalities because the region is not convex

10. a. $\begin{cases} 2x + y \geq 16 \\ x \geq 0 \end{cases}$
   b. $\begin{cases} 8x + 16y \leq 32 \\ y \leq 0 \\ x \geq 0 \end{cases}$
   c. $\begin{cases} x + 2y \geq 16 \\ x \geq 0 \end{cases}$
   d. $\begin{cases} 2x + y \geq -16 \\ y \geq 0 \end{cases}$
   e. No such inequalities because the region is not convex

Sketch each system of linear inequalities. List all vertices and identify the region as "bounded" or "unbounded."

11. $\begin{cases} x + 2y \le 40 \\ x \ge 0, y \ge 0 \end{cases}$

12. $\begin{cases} 3x + y \le 90 \\ x \ge 0, y \ge 0 \end{cases}$

13. $\begin{cases} -2x + y \le 10 \\ x \le 10 \\ x \ge 0, y \ge 0 \end{cases}$

14. $\begin{cases} 2x - y \le 20 \\ y \le 40 \\ x \ge 0, y \ge 0 \end{cases}$

15. $\begin{cases} x + 2y \le 8 \\ x + y \le 6 \\ x \ge 0, y \ge 0 \end{cases}$

16. $\begin{cases} 2x + y \le 10 \\ x + y \le 8 \\ x \ge 0, y \ge 0 \end{cases}$

17. $\begin{cases} 5x + 2y \ge 20 \\ x \ge 0, y \ge 0 \end{cases}$

18. $\begin{cases} 4x + 5y \ge 20 \\ x \ge 0, y \ge 0 \end{cases}$

19. $\begin{cases} 4x + 3y \ge 24 \\ y \ge 4 \\ x \ge 0, y \ge 0 \end{cases}$

20. $\begin{cases} 3x + 4y \ge 12 \\ x \le 8 \\ x \ge 0, y \ge 0 \end{cases}$

21. $\begin{cases} 3x + y \ge 12 \\ x + y \ge 8 \\ x \ge 0, y \ge 0 \end{cases}$

22. $\begin{cases} x + 3y \ge 15 \\ x + y \ge 9 \\ x \ge 0, y \ge 0 \end{cases}$

23. $\begin{cases} x + y \le 20 \\ x \le 15 \\ y \le 10 \\ x \ge 0, y \ge 0 \end{cases}$

24. $\begin{cases} 2x + y \le 20 \\ x \le 8 \\ y \le 10 \\ x \ge 0, y \ge 0 \end{cases}$

25. $\begin{cases} 2x + y \le 80 \\ x + 3y \ge 30 \\ x \ge 0, y \ge 0 \end{cases}$

26. $\begin{cases} 3x + y \le 90 \\ x + 2y \ge 20 \\ x \ge 0, y \ge 0 \end{cases}$

27. $\begin{cases} x - 3y \le 15 \\ 2x + y \ge 30 \\ x \ge 0, y \ge 0 \end{cases}$

28. $\begin{cases} 3x + y \ge 30 \\ x - 2y \ge -60 \\ x \ge 0, y \ge 0 \end{cases}$

29. $\begin{cases} 2x + y \le 18 \\ x + y \le 10 \\ x + 3y \le 24 \\ x \ge 0, y \ge 0 \end{cases}$

30. $\begin{cases} 3x + y \ge 24 \\ 3x + 2y \ge 42 \\ x + 2y \ge 18 \\ x \ge 0, y \ge 0 \end{cases}$

APPLIED EXERCISES

Formulate each situation as a system of linear inequalities, sketch the feasible region, and find the vertices. Be sure to state clearly the meaning of your x- and y-variables.

31. **Livestock Management** A rancher raises goats and llamas on his 400-acre ranch. Each goat needs 2 acres of land and requires $100 of veterinary care per year, while each llama needs 5 acres of land and requires $80 of veterinary care per year. If the rancher can afford no more than $13,200 for veterinary care this year, how many of each animal can he raise?

32. **Agriculture Management** A farmer grows wheat and barley on her 500-acre farm. Each acre of wheat requires 3 days of labor to plant, tend, and harvest, while each acre of barley requires 2 days of labor. If the farmer and her hired field hands can provide no more than 1200 days of labor this year, how many acres of each crop can she grow?

33. **Production Planning** A boat company manufactures aluminum dinghies and rowboats. The amounts of metal work and painting needed for each are shown in the table, together with the number of hours of skilled labor available for each task.

How many of each kind of boat can the company manufacture?

	Dinghy	Rowboat	Labor Available
Metal work	2 hours	3 hours	120 hours
Painting	2 hours	2 hours	100 hours

34. **Resource Allocation** A sailboat company manufactures fiberglass prams and yawls. The amounts of molding, painting, and finishing needed for each are shown in the table, together with the number of hours of skilled labor available for each task. How many of each kind of sailboat can the company manufacture?

	Pram	Yawl	Labor Available
Molding	3 hours	6 hours	150 hours
Painting	3 hours	2 hours	114 hours
Finishing	2 hours	6 hours	132 hours

35. Nutrition Joshua loves "junk food" but wants to stay within the recommended daily limits of 80 grams of fat and 2250 calories. If each serving of SugarSnaks contains 5 grams of fat and 125 calories and each bag of Gobbl'Ems contains 8 grams of fat and 250 calories, how much of each food may he eat today?

36. Diet Planning Justin plays a lot of sports and wants to make sure he gets at least 70 grams of protein and 20 milligrams of iron each day. If each serving of Pro-Team Power Bars has 10 grams of protein and 2 milligrams of iron and each glass of Bulk-Up-Delight has 5 grams of protein and 2 milligrams of iron, how much of each food should he eat each day?

37. Pollution Control A smelting company refines metals at two factories located in Ohio and in Pennsylvania. The smokestacks release both sulfur dioxide (which combines with water vapor to form "acid rain") and particulates (solid matter such as soot that can cause respiratory problems) at the rates shown in the table. The EPA has obtained an injunction against the company preventing it from releasing more than 64 pounds of sulfur dioxide and 60 pounds of particulates into the atmosphere each day. How many hours can the company operate these factories each day?

(Pounds Per Hour)	Sulfur Dioxide	Particulates
Ohio factory	4	5
Pennsylvania factory	4	3

38. Pollution Control A chemical company manufactures batteries at two factories located in Connecticut and in Alabama. The factories discharge both heavy metals (such as mercury and cadmium, which are very toxic) and nitric acid into the local river systems at the rates shown in the table. An environmental organization has obtained an injunction against the company preventing it from discharging more than 54 pounds of heavy metals and 60 pounds of nitric acid each day. How many hours can the company operate these factories each day?

(Pounds Per Hour)	Heavy Metals	Nitric Acid
Connecticut factory	6	4
Alabama factory	3	4

39. Investment Strategy An investment portfolio manager has $8 million to invest in stock and bond funds. If the amount invested in stocks can be no more than the amount invested in bonds, how much can be invested in each type of fund?

40. Financial Planning A retired couple want to invest their $20,000 life savings in bank certificates of deposit and Treasury bonds. If they want at least $5000 in each type of investment, how much can they invest in each?

4.2 Two-Variable Linear Programming Problems

Managing an Investment Portfolio

The adage "never put all your eggs in one basket" particularly applies to investing money. There is no such thing as a low-risk yet high-return opportunity, so each investor must reconcile the desire for high returns with the possibility that even the principal might be lost. An "investment strategy" balances greed against safety and thus determines both the types of investments made and the amounts of each. Given the many different opportunities in today's global economy, the actual solution of a particular instance of this portfolio management problem can be very difficult.

TOT   AME   RD   HI SBE DNB
50s$4\frac{1}{2}$ $2\frac{1}{2}$ $1\frac{1}{2}$ 5   4   8

Let us imagine a simplified situation in which there are only two possible investments: one with low risk and low return and the other with high risk and corresponding high return. Suppose you have $10,000 to invest and wish to protect yourself by investing at least as much in the low-risk investment as in the high-risk investment. As in the previous section, we could express these restrictions as inequalities and sketch the feasible region of possible investments. The portfolio management problem is to find the point that represents the investment choice with the greatest total return.

It should be immediately clear that in this simple situation, the greatest return is gotten by investing as much as possible in the high-return investment and subject to at least half being in the low-return investment. In this section, we shall see that the solution of any problem of this general type is found by "pushing" the values to an extreme point of the feasible region.

Introduction

In this section we will explain what a linear programming problem is and then use the geometry of feasible regions from the previous section to show that the solution of a linear programming problem occurs at a vertex of the region.

Linear Programming Problems

A *linear programming problem* asks for the greatest or smallest value of a linear *objective function* subject to *constraints* in the form of a system of linear inequalities.

EXAMPLE 1   A Linear Programming Problem

A farmer grows corn and soybeans on his 200-acre farm. To maintain soil fertility, the farmer rotates the crops and always plants at least as many acres of soybeans as acres of corn. If each acre of corn yields a profit of $150 and each acre of soybeans yields a profit of $100, how many acres of each crop should the farmer plant to obtain the greatest possible profit?

 Formulate this situation as a linear programming problem by identifying the variables, the objective function, and the constraints.

Solution

Since the question asks "how many acres of each crop," we let

$$x = \begin{pmatrix} \text{Number of} \\ \text{acres of corn} \end{pmatrix} \quad \text{and} \quad y = \begin{pmatrix} \text{Number of} \\ \text{acres of soybeans} \end{pmatrix}$$

The objective is to maximize the farmer's profit, and this profit is $P = 150x + 100y$ because the profits per acre are $150 for corn and $100 for soybeans. The 200-acre size of the farm leads to the constraint $x + y \leq 200$, while the crop rotation requirement that $y \geq x$ can be written as $x - y \leq 0$. Since x and y cannot be negative, we also have the nonnegativity constraints $x \geq 0$ and $y \geq 0$. This maximum linear programming problem may be written as

Objective function

$$\text{Maximize } P = 150x + 100y \qquad \text{Corn and soybean profits}$$

$$\text{Subject to } \begin{cases} x + y \leq 200 & \text{Size of farm} \\ x - y \leq 0 & \text{Crop rotation} \\ x \geq 0 \text{ and } y \geq 0 & \text{Nonnegativity} \end{cases}$$

Constraints

Since $\geq$ can be changed to $\leq$ by multiplying by -1, every linear programming problem in two variables may be written in one of the following forms:

Maximize $P = Mx + Ny$

Subject to $\begin{cases} ax + by \leq c \\ \ddots \end{cases}$

or

Minimize $C = Mx + Ny$

Subject to $\begin{cases} ax + by \geq c \\ \ddots \end{cases}$

We have named the objective function P for profit in the maximum problem and C for cost in the minimum problem, but problems with different goals are perfectly acceptable. Nonnegativity conditions should be included only when appropriate.

Fundamental Theorem of Linear Programming

A *solution* of a *maximum* linear programming problem is a feasible point for the system of linear inequalities that gives the *largest* possible value of the objective function. Similarly, a solution of a *minimum* linear programming problem is a feasible point for the system of linear inequalities that gives the *smallest* possible value of the objective function. With either problem, if the constraints are infeasible, then the problem has no solution.

How can the solution be found? As with systems of linear inequalities, we begin with a simpler problem in just one variable. Omitting the y-variables from the maximum linear programming problem above, we are left with the task of finding the largest value of Mx subject to constraints of the form $x \leq R$. It is immediate that the solution of this simple problem will occur at one of the extreme values allowed for x. (For instance, the largest value of $5x$ where $x \leq 4$ is 20 when $x = 4$.)

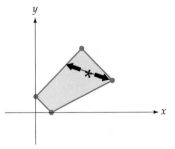

Turning back to two-variable problems, let us first suppose that the feasible region is bounded. For any interior point and any vertex (see the sketch on the left), the line segment from the vertex to the interior point can be extended to meet the boundary (because the region is bounded) and this line segment lies completely within the feasible region (because it is convex). On just this line segment, the largest (or smallest) value must occur at one end or the other of the segment, just as in the one-variable case.

Thus the solution must occur on the boundary of the region. The boundary consists of line segments joining pairs of vertices, so the same reasoning shows that the solution must occur at an endpoint of these segments. That is, *the solution of the linear programming problem with a bounded region must occur at a vertex of the region.*

On an unbounded region, it is possible that a solution does not exist because the line segment from a vertex to an interior point may extend "forever," and we may wish to move in this unbounded direction to improve the value of the objective function. However, if improving the value of the objective function always means moving toward a boundary, then the solution will exist and it must occur at a vertex.

Combining these observations, we have established the following result about the solution of a linear programming problem.

The Fundamental Theorem of Linear Programming

> If a linear programming problem has a solution, then it occurs at a *vertex* of the region determined by the constraints.

Graphing Calculator Exploration

You can use the program* ViewLP to explore the values of an objective function on a feasible region. The following demonstration uses the problem from Example 1.

a. Enter the boundary line $x + y = 200$ as $y_1 = (200 - 1x)/1$ (or just $y_1 = 200 - x$). Then enter the boundary line $x - y = 0$ as $y_2 = (0 - 1x)/- 1$ (or just $y_2 = x$).

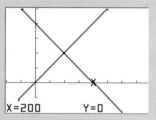

b. Graph them on the window $[-80, 390]$ by $[-80, 230]$ and check that you see the expected region.

c. Run the program ViewLP and enter the values for the objective function: 150 for M and 100 for N (press ENTER after each value).

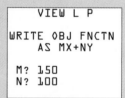

d. Use the arrow buttons to move the dot into the region. At each (x, y) point, the bottom half of the screen shows the value of the objective function MX + NY at that X and Y.

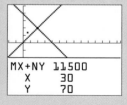

* See Preface for information on how to obtain this and other graphing calculator programs.

e.   Move around in the region and watch how the values of the objective function change. Be careful not to move outside the region. Move to each of the vertices and note the value of the objective function. Does the greatest value occur at a vertex?

f.   When you are finished exploring the region, press ENTER to reset the viewing screen and exit the program.

ViewLP can be used in this way to explore and verify the solution of any linear programming problem in two variables, provided that you can find an appropriate viewing window.

The Fundamental Theorem of Linear Programming gives us the following procedure to find a solution.

How to Solve a Linear Programming Problem

On a bounded region, list the vertices, calculate the value of the objective function at each, and select the vertex that gives the largest (or smallest) value.

On an unbounded region, first check whether the objective function "improves" in an unbounded direction. If it does, there is *no solution*; if it does not, list the vertices, calculate the value of the objective function at each, and select the vertex that gives the largest (or smallest) value.

EXAMPLE 2   **Solution of Example 1**

Solve the linear programming problem from Example 1 on page 280.

$$\text{Maximize } P = 150x + 100y \qquad \text{Objective function}$$

$$\text{Subject to } \begin{cases} x + y \leq 200 \\ x - y \leq 0 \\ x \geq 0 \text{ and } y \geq 0 \end{cases} \qquad \text{Constraints}$$

Solution

We begin by sketching the region determined by the constraints. The nonnegativity conditions $x \geq 0$ and $y \geq 0$ place the region in the first quadrant. The boundary line $x + y = 200$ has intercepts (200, 0) and (0, 200), whereas the boundary line $x - y = 0$ passes through the origin (0, 0) and the point (1, 1). The origin satisfies the inequality $x + y \leq 200$, and the point (0, 200) satisfies the inequality $x - y \leq 0$.

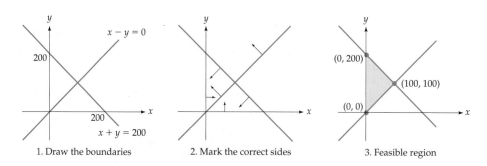

1. Draw the boundaries   2. Mark the correct sides   3. Feasible region

The region has three vertices, and two are already known from the x- and y-intercepts of the boundary lines. The third is the intersection of $x - y = 0$ and $x + y = 200$:

$x - y = 0$ so $x = y$	Solving $x - y = 0$ for x
$(y) + y = 200$	Substituting into $x + y = 200$
$2y = 200$ so $y = 200/2 = 100$	Simplifying and solving for y
$x = 100$	$x = y$ with $y = 100$
$(100, 100)$	Vertex at $x = 100, y = 100$

The vertices are $(0, 0)$, $(0, 200)$, and $(100, 100)$, as shown above.

Since the region is bounded, this problem *does* have a solution, and it occurs at a vertex. Evaluating the objective function at the vertices, we find the following values:

Vertex	Value of $P = 150x + 100y$	
$(0, 0)$	0	$0 = 150 \cdot 0 + 100 \cdot 0$
$(0, 200)$	20,000	$20,000 = 150 \cdot 0 + 100 \cdot 200$
$(100, 100)$	25,000 ← Largest	$25,000 = 150 \cdot 100 + 100 \cdot 100$

Since the largest value of the objective function occurs at the vertex $(100, 100)$, the solution of this problem is: The maximum value of P is 25,000 at the vertex $(100, 100)$. In terms of the original word problem in Example 1, the maximum profit is $25,000 when the farmer plants 100 acres of corn and 100 acres of soybeans.

■

PRACTICE PROBLEM 1

Solve the linear programming problem:

$$\text{Maximize } P = 5x + 2y$$

$$\text{Subject to } \begin{cases} 3x + y \le 60 \\ y \le 36 \\ x \ge 0 \text{ and } y \ge 0 \end{cases}$$

Solution at the back of the book

You may also want to use the program ViewLP (page 282) to explore this problem on your graphing calculator.

EXAMPLE 3   **A Minimum Problem on an Unbounded Region**

Solve the linear programming problem:

$$\text{Minimize } C = 3x + 5y$$

$$\text{Subject to } \begin{cases} 2x + y \geq 12 \\ x + y \geq 8 \\ x \geq 0 \text{ and } y \geq 0 \end{cases}$$

Solution

We begin by sketching the region determined by the constraints. The nonnegativity conditions $x \geq 0$ and $y \geq 0$ place the region in the first quadrant. The boundary line $2x + y = 12$ has intercepts $(12/2, 0) = (6, 0)$ and $(0, 12)$, while the boundary line $x + y = 8$ has intercepts $(8, 0)$ and $(0, 8)$. Since the test point $(0, 0)$ does not satisfy $2x + y \geq 12$ and $x + y \geq 8$, the region is on the sides of these boundaries *away* from the origin.

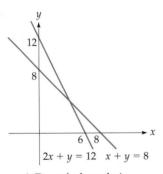

1. Draw the boundaries

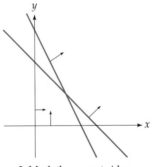

2. Mark the correct sides

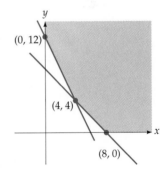

3. Feasible region

This unbounded region has three vertices, and two are already known from the *x*- and *y*-intercepts of the boundary lines. The third is the intersection of $2x + y = 12$ and $x + y = 8$. Solving the second as $x = 8 - y$ and substituting into the first equation, $2(8 - y) + y = 12$ means $16 - 2y + y = 12$ so $y = 4$ and then $x = 8 - 4 = 4$. The vertices are $(8, 0)$, $(4, 4)$, and $(12, 0)$, as shown above.

The region is unbounded, so before looking at the vertices, we must find whether a solution exits. The value of the objective function C at the interior point $(10, 10)$, for example, is $C = 3 \cdot 10 + 5 \cdot 10 = 80$.

If we shift this point upward to (10, 15) or right to (15, 10), the larger values for x or y will *increase* the value of $C = 3x + 5y$, which is the *opposite* of what we should do to minimize C. (You may want to use the program ViewLP (page 282) to explore this further.) Thus the objective function does not "improve" in an unbounded direction and the solution *does* exist. Evaluating the objective function at the vertices, we find the following values.

Vertex	$C = 3x + 5y$		
(8, 0)	24	← Smallest	$24 = 3 \cdot 8 + 5 \cdot 0$
(4, 4)	32		$32 = 3 \cdot 4 + 5 \cdot 4$
(0, 12)	60		$60 = 3 \cdot 0 + 5 \cdot 12$

Since the smallest value of the objective function occurs at the vertex (8, 0), the solution of this problem is: The minimum value of C is 24 at the vertex (8, 0). ▪

PRACTICE PROBLEM 2

Solve the linear programming problem:

$$\text{Minimize } C = 5x + 11y$$

$$\text{Subject to } \begin{cases} x + 3y \geq 60 \\ x + 2y \geq 50 \\ x \geq 0 \text{ and } y \geq 0 \end{cases}$$

Solution at the back of the book

You may also want to use the program ViewLP (page 282) to explore this problem on your graphing calculator.

EXAMPLE 4   **A Manufacturing Problem**

A fully automated plastics factory produces two toys, a racing car and a jet airplane, in three stages: molding, painting, and packaging. After allowing for routine maintenance, the equipment for each stage can operate no more than 150 hours per week. Each batch of racing cars requires 6 hours of molding, 2.5 hours of painting, and 5 hours of packaging, while each batch of jet airplanes requires 3 hours of molding, 7.5 hours of painting, and 5 hours of packaging. If the profit per batch of toys is $120 for cars and $100 for airplanes, how many batches of each toy should be produced each week to obtain the greatest possible profit?

Solution

Let

$$x = \begin{pmatrix} \text{Number of batches} \\ \text{of racing car toys} \end{pmatrix} \quad \text{and} \quad y = \begin{pmatrix} \text{Number of batches} \\ \text{of jet airplane toys} \end{pmatrix}$$

made during the week. The profit is $P = 120x + 100y$ and the problem is to maximize P subject to the constraints that the molding, painting, and packaging processes can each take no more than 150 hours:

$$6x + 3y \leq 150 \qquad \text{Molding time}$$

$$2.5x + 7.5y \leq 150 \qquad \text{Painting time}$$

$$5x + 5y \leq 150 \qquad \text{Packaging time}$$

and the nonnegativity conditions $x \geq 0$ and $y \geq 0$. Simplifying the constraints by removing common factors (that is, by dividing the molding constraint by 3, the painting constraint by 2.5, and the packaging constraint by 5), we can rewrite this problem as the linear programming problem

$$\text{Maximize } P = 120x + 100y \qquad \text{Objective function}$$

$$\text{Subject to } \begin{cases} 2x + y \leq 50 \\ x + 3y \leq 60 \\ x + y \leq 30 \\ x \geq 0 \text{ and } y \geq 0 \end{cases} \qquad \text{Constraints}$$

The nonnegativity conditions $x \geq 0$ and $y \geq 0$ place the region in the first quadrant. The origin satisfies all the constraints, the region is bounded, and one vertex is the origin. For the boundary line intercepts, $2x + y = 50$ has intercepts $(50/2, 0) = (25, 0)$ and $(0, 50)$, $x + 3y = 60$ has intercepts $(60, 0)$ and $(0, 60/3) = (0, 20)$, and $x + y = 30$ has intercepts $(30, 0)$ and $(0, 30)$.

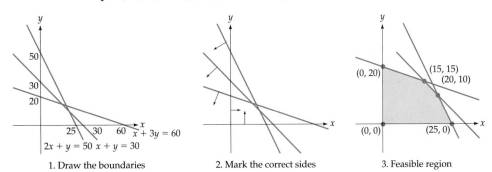

1. Draw the boundaries   2. Mark the correct sides   3. Feasible region

The region has five vertices, and three are already known from the x- and y-intercepts of the boundary lines. The other two vertices are intersections of the lines $2x + y = 50$, $x + 3y = 60$, and $x + y = 30$. The intersection of $2x + y = 50$ and $x + y = 30$ is $(20, 10)$, and this point *is* a vertex of the region because it also satisfies the inequality $x + 3y \le 60$. The intersection of $x + 3y = 60$ and $x + y = 30$ is $(15, 15)$, and this point *is* a vertex because it also satisfies the inequality $2x + y \le 50$. [The intersection of $2x + y = 50$ and $x + y = 60$ is the point $(18, 14)$, but this is *not* a vertex of the region because it does not also satisfy $x + y \le 30$: $18 + 14$ is not ≤ 30.] The vertices are $(0, 0)$, $(25, 0)$, $(20, 10)$, $(15, 15)$, and $(0, 20)$ as shown in the preceding diagram.

The region is bounded, so this problem *has* a solution, and it occurs at a vertex. Evaluating the objective function at the vertices, we find the following values:

Vertex	$P = 120x + 100y$
$(0, 0)$	0
$(25, 0)$	3000
$(20, 10)$	3400 $\leftarrow$ Largest
$(15, 15)$	3300
$(0, 20)$	2000

Since the largest value of the objective function occurs at the vertex $(20, 10)$, the solution is: The maximum value is 3400 at the vertex $(20, 10)$. In terms of the original question, the maximum profit is \$3400 when the factory produces 20 batches of racing car toys and 10 batches of jet airplane toys each week.

_____ ∎

Be careful! Except in very simple problems, some boundary line intersections are *not* vertices of the feasible region because they violate at least one of the other constraints. Particularly when sketching regions by hand, it is best to find all the intersection points anyway and then verify which are feasible and which are not.

Extensions to Larger Problems

There are two difficulties with our geometric solution of linear programming problems. First, as the number of constraints increases, the number of intersection points rapidly increases, with relatively fewer and fewer being vertices. For a problem with many constraints, this will cause an enormous amount of wasted effort in calculating and discarding these irrelevant points. Second, having just two variables is not realistic: a manufacturing problem such as Example 4 should really be concerned with *many* different toys instead of just two. But three or more variables make the feasible regions difficult if not impossible to sketch.

One solution to the first difficulty would be to find only the vertices needed for the solution of the problem and omit the calculation of all the other intersection points. In a region such as in Example 4, it is clear that the origin is a vertex (the company could choose to manufacture nothing). It might then be possible to start at this first vertex (which we have found with no effort) and move to the solution by finding a "path of vertices" giving larger and larger profits until there is nowhere to go that could make the profit greater. For the region of Example 4, this might result in the path shown below using only the vertices $(0, 0)$, $(25, 0)$, and $(20, 10)$ to arrive at the solution.

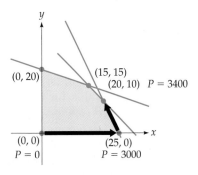

A solution method based on this idea is the subject of the next section.

The second difficulty of how to deal with more variables suggests that we must leave the geometric method and search for an *algebraic* method of calculating the solution. One possibility would be to use the constraints to make upper estimates on the size of the objective function. If these upper estimates could be made small enough, then perhaps the solution might be found. For the problem from Example 4, if we multiply the first constraint by 20 and add it to 80 times the third, we find

Maximize $P = 120x + 100y$

Subject to $\begin{cases} 2x + y \le 50 \\ x + 3y \le 60 \\ x + y \le 30 \\ x \ge 0 \text{ and } y \ge 0 \end{cases}$

$\times$ 20 is $\qquad 40x + 20y \le 1000$

$\times$ 80 is $\qquad \dfrac{80x + 80y \le 2400}{}$

and add to get $\quad 120x + 100y \le 3400$

The resulting inequality, $120x + 100y \le 3400$, not only contains the exact objective function but also gives an estimate that is the best possible because there actually is a vertex where the objective function takes the value 3400. Although this may seem too good to be true in general, in Section 4.4 we will see that this idea leads to a powerful method for solving linear programming problems.

Section Summary

A *linear programming problem* consists of a linear *objective function* to be *maximized* or *minimized* subject to *constraints*, written as a system of linear inequalities. Every such problem can be written in the form

Maximize $P = Mx + Ny$ or   Minimize $C = Mx + Ny$

Subject to $\begin{cases} ax + by \leq c \\ \quad \vdots \end{cases}$ Subject to $\begin{cases} ax + by \geq c \\ \quad \vdots \end{cases}$

where the objective function is labeled above, and Constraints is labeled below.

A *solution* is a feasible point that gives the largest (or smallest) possible value for the objective function. The *fundamental theorem of linear programming* says that if there is a solution, it occurs at a *vertex* of the feasible region. Thus the problem can be solved by sketching the feasible region of the constraints, evaluating the objective function at the vertices, and selecting the vertex with the optimal value. If the region is unbounded, you must check that a solution exists before selecting the "best" vertex.

EXERCISES 4.2

For Exercises 1–5: Find each maximum or minimum value for the given region. If such a value does not exist, explain why not.

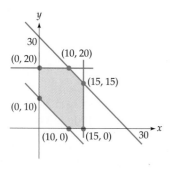

1. Maximum of $P = 2x + y$

2. Maximum of $P = x + 2y$

3. Minimum of $C = 3x + 4y$

4. Minimum of $C = 4x + 3y$

5. Maximum of $P = x - y$

For Exercises 6–10: Find each maximum or minimum value for the given region. If such a value does not exist, explain why not.

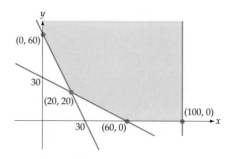

6. Minimum of $C = x + y$

7. Minimum of $C = 3x + y$

8. Maximum of $P = x + 3y$

9. Maximum of $P = 5x + 8y$

10. Maximum of $P = -x - y$.

Solve each linear programming problem by sketching the region and labeling the vertices, deciding whether a solution exists, and then finding it if it does exist.

11. Maximize $P = 30x + 40y$

Subject to $\begin{cases} 2x + y \le 16 \\ x + y \le 10 \\ x \ge 0, y \ge 0 \end{cases}$

12. Maximize $P = 80x + 70y$

Subject to $\begin{cases} x + 2y \le 18 \\ x + y \le 10 \\ x \ge 0, y \ge 0 \end{cases}$

13. Minimize $C = 15x + 45y$

Subject to $\begin{cases} 2x + 5y \ge 20 \\ x \ge 0, y \ge 0 \end{cases}$

14. Minimize $C = 35x + 25y$

Subject to $\begin{cases} 5x + 3y \ge 60 \\ x \ge 0, y \ge 0 \end{cases}$

15. Maximize $P = 4x + 5y$

Subject to $\begin{cases} 2x + y \le 50 \\ x + 3y \le 75 \\ x \ge 0, y \ge 0 \end{cases}$

16. Maximize $P = 7x + 8y$

Subject to $\begin{cases} 3x + y \le 90 \\ x + 2y \le 60 \\ x \ge 0, y \ge 0 \end{cases}$

17. Minimize $C = 12x + 10y$

Subject to $\begin{cases} 4x + y \ge 40 \\ 2x + 3y \ge 60 \\ x \ge 0, y \ge 0 \end{cases}$

18. Minimize $C = 20x + 30y$

Subject to $\begin{cases} 3x + 2y \ge 120 \\ x + 4y \ge 80 \\ x \ge 0, y \ge 0 \end{cases}$

19. Maximize $P = 5x + 7y$

Subject to $\begin{cases} 4x + 3y \le 60 \\ x - y \le 8 \\ x \ge 0, y \ge 0 \end{cases}$

20. Maximize $P = 2x + y$

Subject to $\begin{cases} 3x - 4y \le 60 \\ x + y \le 48 \\ x \ge 0, y \ge 0 \end{cases}$

21. Minimize $C = 6x + 10y$

Subject to $\begin{cases} -x + 4y \le 60 \\ 2x + y \ge 60 \\ x \ge 0, y \ge 0 \end{cases}$

22. Minimize $C = 5x - 10y$

Subject to $\begin{cases} -3x + y \le 60 \\ y \le 120 \\ x \ge 0, y \ge 0 \end{cases}$

23. Maximize $P = 5x + 3y$

Subject to $\begin{cases} 2x + y \le 90 \\ x + y \le 50 \\ x + 2y \le 90 \\ x \ge 0, y \ge 0 \end{cases}$

24. Maximize $P = 6x + 5y$

Subject to $\begin{cases} x + 2y \le 96 \\ x + y \le 54 \\ 2x + y \le 96 \\ x \ge 0, y \ge 0 \end{cases}$

25. Maximize $P = 10x + 12y$

Subject to $\begin{cases} 3x + 2y \le 180 \\ 4x + y \le 120 \\ 3x + y \le 105 \\ x \ge 0, y \ge 0 \end{cases}$

26. Maximize $P = 20x + 15y$

Subject to $\begin{cases} 2x + 3y \le 60 \\ x + 4y \le 40 \\ x + 3y \le 33 \\ x \ge 0, y \ge 0 \end{cases}$

27. Minimize $C = 20x + 25y$

$$\text{Subject to } \begin{cases} 3x + y \geq 60 \\ x + y \geq 42 \\ x + 3y \geq 60 \\ x \geq 0, y \geq 0 \end{cases}$$

28. Minimize $C = 50x + 35y$

$$\text{Subject to } \begin{cases} x + 3y \geq 72 \\ x + y \geq 48 \\ 3x + y \geq 72 \\ x \geq 0, y \geq 0 \end{cases}$$

29. Minimize $C = 5x + 6y$

$$\text{Subject to } \begin{cases} 4x + 3y \geq 840 \\ 2x + 5y \geq 700 \\ 4x + 5y \geq 1280 \\ x \geq 0, y \geq 0 \end{cases}$$

30. Minimize $C = 15x + 10y$

$$\text{Subject to } \begin{cases} 3x + 4y \geq 336 \\ 5x + 2y \geq 280 \\ 5x + 4y \geq 520 \\ x \geq 0, y \geq 0 \end{cases}$$

APPLIED EXERCISES

Formulate each situation as a linear programming problem by identifying the variables, the objective function, and the constraints. Be sure to state clearly the meaning of each variable. Determine whether a solution exists, and if it does, find it. State your final answer in terms of the original question.

31. Livestock Management A rancher raises goats and llamas on his 400-acre ranch. Each goat needs 2 acres of land and requires $100 of veterinary care per year, while each llama needs 5 acres of land and requires $80 of veterinary care per year. The rancher can afford no more than $13,200 for veterinary care this year. If the expected profit is $60 for each goat and $90 for each llama, how many of each animal should he raise to obtain the greatest possible profit?

32. Agriculture Management A farmer grows wheat and barley on her 500-acre farm. Each acre of wheat requires 3 days of labor to plant, tend, and harvest, while each acre of barley requires 2 days of labor. The farmer and her hired field hands can provide no more than 1200 days of labor this year. If the expected profit is $50 for each acre of wheat and $40 for each acre of barley, how many acres of each crop should she grow to obtain the greatest possible profit?

33. Resource Allocation A sailboat company manufactures fiberglass prams and yawls. The amount of molding, painting, and finishing needed for each is shown in the table, together with the number of hours of skilled labor available for each task. If the expected profit is $150 for each pram and $180 for

each yawl, how many of each kind of sailboat should the company manufacture to obtain the greatest possible profit?

	Pram	Yawl	Labor Available
Molding	3 hours	6 hours	150 hours
Painting	3 hours	2 hours	114 hours
Finishing	2 hours	6 hours	132 hours

34. Production Planning A small jewelry company prepares and mounts semiprecious stones. There are 10 lapidaries (who cut and polish the stones) and 12 jewelers (who mount the stones in gold settings). Each employee works 7 hours each day. Each tray of agates requires 5 hours of cutting and polishing and 4 hours of mounting, while each tray of onyxes requires 2 hours of cutting and polishing and 3 hours of mounting. If the profit is $15 for each tray of agates and $10 for each tray of onyxes, how many trays of each stone should be processed each day to obtain the greatest possible profit?

35. Waste Management The Marshall County trash incinerator in Norton burns 10 tons of trash per hour and co-generates 6 kilowatts of electricity, while the Wiseburg incinerator burns 5 tons per hour and co-generates 4 kilowatts. The county needs to burn at least 70 tons of trash and co-generate at least 48 kilowatts of electricity every day. If the Norton incinerator costs $80 per hour to operate and the Wiseburg incinerator costs $50, how many hours should each incinerator operate each day with the least cost to the county?

36. Disaster Relief An international relief agency has been asked to provide medical support to a Caribbean island devastated by a recent hurricane. The agency estimates that it must be able to perform at least 50 major surgeries, 78 minor surgeries, and 130 outpatient services each day. The daily capacities of its portable field hospitals and clinics are given in the table, along with the daily operating costs. How many field hospitals and clinics should the agency airlift to the island to provide the needed help at the least daily cost?

min

	Major Surgeries	Minor Surgeries	Outpatient Services	Daily Cost
Field hospitals	5	3	2	$2000
Clinics	0	2	10	$500

37. Nutrition A pet store owner raises baby bunnies and feeds them leftover salad greens from a nearby restaurant. The weekly nutritional needs of each bunny are given in the table, as well as the nutrition provided by the salad greens and nutritional supplement drops. If the salad greens cost 2¢ per handful and the nutritional supplement costs 1¢ per drop, how many handfuls of greens and drops of supplement should each bunny receive weekly to minimize the owner's costs?

	Salad Greens (Per Handful)	Nutritional Supplement (Per Drop)	Minimum Weekly Requirement
Fiber (grams)	3	0	90
Vitamins (mg)	2	5	140
Minerals (mg)	1	6	84

38. Diet Planning The director of a school district's hot lunch program estimates that each lunch should contain at least 1000 calories and 15 grams of protein and no more than 30 grams of fat. The nutritional contents per ounce of the cafeteria's famous mystery foods X and Y are given in the table. If each ounce of X costs 5¢ and each ounce of Y costs 6¢, how much of each food should be served to meet the director's nutritional goals at the least cost?

	Calories	Protein	Fat
Mystery food X	200	1	3
Mystery food Y	100	3	3

39. Land Reclamation A state government included $1.2 million in its current appropriations bill to reclaim some of the land at a 3000-acre strip mine site. It will cost $800 per acre to return the land to productive grassland suitable for livestock and $500 per acre to return the land to forest suitable for commercial timber production. The appropriations bill also requires that at least 1800 acres be reclaimed immediately and then the income from leasing it to ranchers and/or paper companies be used to reclaim the rest of the site in the coming years. If long-term-use agreements yield $200 per acre of grassland and $150 per acre of forest, how many acres of each should be reclaimed this year to raise the greatest amount for next year's reclamation efforts?

40. Urban Renewal A nonprofit urban development corporation has agreed to rebuild at least 24 city blocks in the south side of Megatropolis in which at least 15 blocks will be semidetached single family homes and at least 3 but not more than 10 blocks will be commercial buildings (retail stores and/or light industry). If it will cost $6 million to rebuild 1 block with homes and $7 million to rebuild 1 block for commercial use, how many blocks of each should be rebuilt to meet the goals at the least cost?

4.3 The Simplex Method for Standard Maximum Problems

APPLICATION PREVIEW

George Dantzig and the Origins of the Simplex Method*

The military term "program" refers to a training, supply, and troop deployment schedule. During World War II, such programs became critically important and took enormous effort to develop. George Dantzig became an expert at solving such problems while serving as the mathematical advisor to the U.S. Air Force Comptroller. There were no computers, so finding a solution meant managing large rooms of clerks carrying out their assigned computations on mechanical calculators (with numbers handed them on slips of paper), writing their results on more pieces of paper, and passing these to other clerks for further calculations.

After the war, Dantzig sought a way to compute solutions in less time. He was familiar with the matrix input–output model of the American economy proposed by Wassily Leontief in 1932, and he saw the need to generalize it to include alternative activities and make it more easily computable. He wrote years later that "initially there was no objective function: explicit goals did not exist because practical planners had no way to implement such a concept," but by mid-1947 he "decided that the objective had to be made explicit." He formulated a "linear programming problem" as the optimization of "a linear form subject to linear equations and inequalities." Assuming that economists were familiar with this type of problem, he visited T. C. Koopmans (who had worked on transportation models for the Allied Shipping Board during the war) at the University of Chicago. Koopmans immediately saw the implications of Dantzig's work for general economic planning but could not offer a method of solution. Dantzig at first rejected on intuitive grounds the obvious idea of moving along the edges from vertex to vertex as being too inefficient. However, when he tried it, "by good luck it worked!" This "simplex" method was named for the geometric objects from which the convex feasible region can be built, just as in two variables the region can be viewed as a collection of adjacent triangles whose vertices form the corners of the region.

In a 1981 address to the American Mathematical Society, Dantzig concluded by remarking that "The ability to state general

* George B. Dantzig, "Reminiscences about the Origins of Linear Programming," *Essays in the History of Mathematics*, Memoirs of the American Mathematical Society, Number 298, March 1984, pp. 1–11.

objectives and then find optimal policy solutions to practical decision problems of great complexity is a revolutionary development. . . . [But in] areas such as modeling the dynamics of growing populations of the world against a diminishing resource base, its potential for raising the standard of living has scarcely been realized."

Introduction

Using the geometric method from the previous section as a guide, we shall explain and demonstrate the *simplex method* for solving linear programming problems. This method finds a "path of vertices" leading from an initial vertex to a solution without finding all the intersection points of the boundary lines or even all the vertices of the feasible region.

Standard Maximum Problems

A linear programming problem is a *standard maximum problem* if the objective function is to be maximized, the constraints include nonnegativity conditions for all the variables, and the origin is a vertex of the feasible region. Writing $x_1, x_2, x_3, \ldots, x_n$ for the variables instead of just x and y, we can express a standard maximum problem in the following form:

Standard Maximum Problem

Maximize $P = c_1 x_1 + c_2 x_2 + \cdots + c_n x_n$

Subject to
$$\begin{cases} a_{1,1} x_1 + a_{1,2} x_2 + \cdots + a_{1,n} x_n \leq b_1 \\ a_{2,1} x_1 + a_{2,2} x_2 + \cdots + a_{2,n} x_n \leq b_2 \\ \qquad \ddots \\ a_{m,1} x_1 + a_{m,2} x_2 + \cdots + a_{m,n} x_n \leq b_m \\ x_1 \geq 0, x_2 \geq 0, \ldots, x_n \geq 0 \end{cases}$$

A "larger" version of
Maximize $P = Mx + Ny$
Subject to $\begin{cases} ax + by \leq c \\ \qquad \ddots \end{cases}$

where $b_1 \geq 0, b_2 \geq 0, \ldots, b_m \geq 0$

Here n is the number of variables and m is the number of constraints other than the nonnegativity conditions $x_1 \geq 0, x_2 \geq 0, \ldots, x_n \geq 0$. Because the numbers $b_1, b_2, \ldots, b_m$ after the "at most" inequalities are all nonnegative, $x_1 = 0, x_2 = 0, \ldots, x_n = 0$ satisfies all the constraints, and so the origin *is* a vertex of the region.

A *matrix inequality* is satisfied if every pair of corresponding entries satisfies the inequality; for instance, $\begin{pmatrix} 2 \\ 4 \end{pmatrix} \le \begin{pmatrix} 6 \\ 4 \end{pmatrix}$ because $2 \le 6$ and $4 \le 4$. Define the matrices c^t, A, X, and b to be

$$c^t = \underbrace{(c_1 \ c_2 \ \dots \ c_n)}_{\substack{\text{From the objective} \\ \text{function}}}, \quad A = \underbrace{\begin{pmatrix} a_{1,1} & a_{1,2} & \cdots & a_{1,n} \\ \vdots & \vdots & \ddots & \vdots \\ a_{m,1} & a_{m,2} & \cdots & a_{m,n} \end{pmatrix}}_{\text{From constraints before } \le}$$

$$X = \underbrace{\begin{pmatrix} x_1 \\ \vdots \\ x_n \end{pmatrix}}_{\text{Variables}}, \quad \text{and} \quad b = \underbrace{\begin{pmatrix} b_1 \\ \vdots \\ b_m \end{pmatrix}}_{\substack{\text{From constraints} \\ \text{after } \le}}$$

We may use matrices to write this standard maximum problem in the following compact form:

Standard Maximum Problem

$$\text{Maximize } P = c^t X$$

$$\text{Subject to } \begin{cases} AX \le b \\ X \ge 0 \end{cases}$$

$$\text{where } b \ge 0$$

The Initial Simplex Tableau

The constraints in the standard maximum problem are written as inequalities. Equations are easier to solve than inequalities, so we first simplify the problem by changing the inequalities into equations by adding to each a *slack variable* that represents the "amount not used." For example, the two sides of the inequality $4 \le 7$ differ by 3, so adding 3 to the left side gives an *equation* $4 + 3 = 7$. Similarly, the inequality $x \le 7$ can be rewritten as the equation $x + s = 7$ by adding a slack variable $s \ge 0$ to the left side to make up the difference between the sides. We do this for each inequality, using a different slack variable $s_1, s_2, s_3, \dots, s_m$ for each inequality. Using zeros to indicate the slacks that do not appear in a constraint, we can rewrite the constraints as *equations*:

$$a_{1,1}x_1 + a_{1,2}x_2 + \cdots + a_{1,n}x_n + 1s_1 + 0s_2 + \cdots + 0s_m = b_1$$

$$a_{2,1}x_1 + a_{2,2}x_2 + \cdots + a_{2,n}x_n + 0s_1 + 1s_2 + \cdots + 0s_m = b_2$$

$$a_{m,1}x_1 + a_{m,2}x_2 + \cdots + a_{m,n}x_n + 0s_1 + 0s_2 + \cdots + 1s_m = b_m$$

Constraints with variables arranged in columns

To express the objective function $P = c_1x_1 + c_2x_2 + \cdots + c_nx_n$ in the same form, we write the equation with all of the x-variables on the left side, and then include the slacks by adding $0s_1 + 0s_2 + \cdots + 0s_m$ to obtain

$$P - c_1x_1 - c_2x_2 - \cdots - c_nx_n + 0s_1 + 0s_2 + \cdots + 0s_m = 0$$

Because both the objective function and the constraints are now written as equations with the variables appearing in the same order, this standard maximum problem can be represented by just a rectangular table of numbers. Dantzig named this representation the "simplex tableau" (using the French word for table, whose plural is tableaux).

Initial Simplex Tableau

The *initial simplex tableau* of the standard maximum linear programming problem is the rectangular table of variables and coefficients

	x_1	x_2	$\cdots$	x_n	s_1	s_2	$\cdots$		s_m	
s_1	$a_{1,1}$	$a_{1,2}$	$\cdots$	$a_{1,n}$	1	0	0	$\cdots$	0 0	b_1
s_2	$a_{2,1}$	$a_{2,2}$	$\cdots$	$a_{2,n}$	0	1	0	$\cdots$	0 0	b_2
$\vdots$	$\vdots$	$\vdots$	$\ddots$	$\vdots$	$\vdots$	$\vdots$	$\ddots$		$\vdots$	$\vdots$
s_m	$a_{m,1}$	$a_{m,2}$	$\cdots$	$a_{m,n}$	0	0	0	$\cdots$	0 1	b_m
P	$-c_1$	$-c_2$	$\cdots$	$-c_m$	0	0	0	$\cdots$	0 0	0

Column variables

Constraints

Objective function

Represents an equals sign

When we use the matrices defined on the previous page, the initial simplex tableau takes the following form:

Initial Simplex Tableau

	X	S	
S	A	I	b
P	$-c^t$	0	0

$$I = \begin{pmatrix} 1 & 0 & \cdots & 0 \\ 0 & 1 & \cdots & 0 \\ \vdots & \vdots & \ddots & \vdots \\ 0 & 0 & \cdots & 1 \end{pmatrix}$$

The matrix I is the "m by m identity matrix" consisting of all zeros except for 1's along the diagonal from top left to bottom right.

EXAMPLE 1 Constructing an Initial Simplex Tableau

Rewrite the following problem in matrix form, determine whether it is a standard maximum problem, and if it is, construct the initial simplex tableau.

$$\text{Maximize } P = 9x_1 - 2x_2 + 8x_3$$

$$\text{Subject to } \begin{cases} -16x_1 + 4x_2 + 11x_3 \leq 176 \\ 15x_1 + 3x_2 + 9x_3 \leq 135 \\ x_1 \geq 0, x_2 \geq 0, \text{ and } x_3 \geq 0 \end{cases}$$

Solution

The matrix form of the problem is

$$\text{Maximize } P = \overset{c^t}{(9 \quad -2 \quad 8)} \overset{X}{\begin{pmatrix} x_1 \\ x_2 \\ x_3 \end{pmatrix}} \qquad \text{Maximize } P = c^t X$$

$$\text{Subject to } \begin{cases} \overset{A}{\begin{pmatrix} -16 & 4 & 11 \\ 15 & 3 & 9 \end{pmatrix}} \begin{pmatrix} x_1 \\ x_2 \\ x_3 \end{pmatrix} \leq \overset{b}{\begin{pmatrix} 176 \\ 135 \end{pmatrix}} \\[2em] \text{and } \begin{pmatrix} x_1 \\ x_2 \\ x_3 \end{pmatrix} \geq 0 \end{cases} \qquad \text{Subject to } \begin{cases} A\,X \leq b \\ X \geq 0 \end{cases}$$

None of the numbers in the matrix $b = \begin{pmatrix} 176 \\ 135 \end{pmatrix}$ is negative, so this *is* a standard maximum problem. Because there are two constraint inequalities besides the nonnegativity conditions, there will be two slack variables. The initial simplex tableau is

	x_1	x_2	x_3	s_1	s_2	
s_1	−16	4	11	1	0	176
s_2	15	3	9	0	1	135
P	−9	2	−8	0	0	0

	X	S	
S	A	I	b
P	$-c^t$	0	0

where $I = \begin{pmatrix} 1 & 0 \\ 0 & 1 \end{pmatrix}$

PRACTICE PROBLEM 1

Rewrite the following problem in matrix form, determine whether it is a standard maximum problem, and if it is, construct the initial simplex tableau.

$$\text{Maximize } P = 9x_1 - x_2 + 10x_3 + 12x_4$$

$$\text{Subject to } \begin{cases} x_1 + x_2 + 2x_3 + 2x_4 \le 16 \\ 2x_1 + 2x_2 + x_3 + x_4 \le 20 \\ x_1 + 2x_2 + 2x_3 + x_4 \le 18 \\ x_1 \ge 0, x_2 \ge 0, x_3 \ge 0, \text{ and } x_4 \ge 0 \end{cases}$$

Solution at the back of the book

Basic and Nonbasic Variables

In an initial simplex tableau such as

	x_1	x_2	x_3	s_1	s_2	
s_1	-16	4	11	1	0	176
s_2	15	3	9	0	1	135
P	-9	2	-8	0	0	0

the slack variables s_1 and s_2 have two special properties:

1. Each of the columns for these variables contains all zeros except for exactly one 1.
2. Each of the rows above the bottom row has exactly one of these special 1s.

These special variables determine a feasible point in the region without any further calculation: Just let each take the value at the right end of its row and set all the remaining variables equal to zero. For the tableau above, this means

	x_1	x_2	x_3	s_1	s_2	
s_1	-16	4	11	1	0	176
s_2	15	3	9	0	1	135
P	-9	2	-8	0	0	0

$\leftarrow s_1 = 176$
$\leftarrow s_2 = 135$
and $x_1 = x_2 = x_3 = 0$ Other variables are zero

In general, any m variables in a simplex tableau that satisfy properties (1) and (2) above are called *basic* variables and form a *basis*. The other variables are the *nonbasic* variables. Because of their importance, the basic variables are listed on the left side of the simplex tableau, each next to the row containing its special 1. The tableau is to be interpreted as specifying a *basic feasible point* (a vertex) of the region: Each basic variable takes the value at the right end of its row and the nonbasic variables take the value zero. Moreover, the tableau also displays the value of the objective function at this point in the bottom right corner. For example, the simplex tableau below has basic variables $s_1, s_2, \ldots, s_m$, has nonbasic variables $x_1, x_2, \ldots, x_n$, and displays the basic feasible point shown on the right.

		x_1	x_2	$\cdots$	x_n	s_1	s_2	$\cdots$	s_m		
Basic variables	s_1	$a_{1,1}$	$a_{1,2}$	$\cdots$	$a_{1,n}$	1	0	0	0	b_1	Values for the basic variables
	$\vdots$	$\vdots$	$\vdots$	$\ddots$	$\vdots$	$\vdots$	$\ddots$	$\ddots$	$\vdots$	$\vdots$	
	s_m	$a_{m,1}$	$a_{m,2}$	$\cdots$	$a_{m,n}$	0	0	0	1	b_m	
	P	$-c_1$	$-c_2$	$\cdots$	$-c_m$	0	0	0	0	0	

$$s_1 = b_1$$
$$\vdots$$
$$s_m = b_m$$

$$x_1 = 0, \ldots, x_n = 0$$

↑—— Values of P at this vertex

The simplex method begins by taking the slack variables as the basic variables and then finding successively better basic variables (ones that increase the value of the objective function) until the solution is found. This procedure mimics the geometric procedure that we used earlier (see page 283) for two-variable problems: calculating ratios to find intercepts (an operation now called *finding the pivot element*) and then using the substitution method to find intersection points (now called *performing the pivot operation*).

The Pivot Element

To find the pivot element, we first find its column and then its row.

The best variable to use to increase the objective function is the variable with the largest positive coefficient, because increasing it can have the greatest effect. Because the bottom row of the tableau contains the *negatives* of these coefficients and each column corresponds to a variable, this means choosing the column with the *smallest negative entry in the bottom row.*

Pivot Column

	x_1	x_2	s_1	s_2	
s_1	3	4	1	0	24
s_2	4	2	0	1	10
p	−3	−6	0	0	0

↑
(smallest negative)

Some students choose the smallest negative entry by looking for the "most negative" number. The smallest negative number among −5, 10, −4 is −5.

Pivot Column

The *pivot column* is the column with the smallest negative entry in the bottom row (omitting the right column). If there is a tie, the pivot column is the leftmost such column.

Having chosen the variable, how much can we increase it without violating a constraint? Recall from the previous section that in a constraint like $3x + 4y \leq 24$, the x cannot be increased further than the intercept $\frac{24}{3} = 8$ (the ratio of the rightmost number divided by the coefficient of x). Analogously, the chosen variable cannot exceed the ratio of the rightmost number of any constraint divided by its pivot column entry. Not exceeding *each* ratio means not exceeding the *smallest* of them. Of course, ratios involving division by zero or by negative numbers should not even be considered: they correspond to boundary lines parallel to an axis or intercepts not in the first quadrant. Therefore:

Pivot Row

| Pivot Row | | | | | | |

	x_1	x_2	s_1	s_2		
s_1	3	4	1	0	24	$\frac{24}{4} = 6$
s_2	4	2	0	1	10	$\frac{10}{2} = 5 \leftarrow$
						(smallest ratio)
P	-3	-6	0	0	0	

↑

Pivot Row

> For each row except the bottom row, divide the rightmost entry by the pivot column entry (omitting any row with a zero or negative pivot column entry). The *pivot row* is the row with the *smallest* (nonnegative) ratio. If there is a tie, the pivot row is the uppermost such row.

Pivot Element

	x_1	x_2	s_1	s_2	
s_1	3	4	1	0	24
s_2	4	2	0	1	10	←
P	-3	-6	0	0	0

↑

Pivot Element

> The *pivot element* is the entry in the pivot column and the pivot row.

If the tableau does not have a pivot column, then the objective function cannot be increased, so the solution has been found: The maximum value of the objective function appears in the bottom right corner of the tableau, and it occurs at the basic feasible point given by the tableau (see pages 299–300).

If the tableau has a pivot column but no pivot row, then there is *no solution* because there is a direction that will increase the objective function but there is no boundary to prevent the variable from increasing in this direction without bound.

The Pivot Operation

Once we have found the pivot element, we want to increase the objective function as much as possible, stopping when we reach the boundary of the feasible region. We accomplish this by the *pivot operation,* which is just the tableau version of the elimination method that we used for solving equations on pages 165–166.

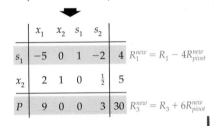

	x_1	x_2	s_1	s_2	
s_1	3	4	1	0	24
s_2	4	2	0	1	10
P	-3	-6	0	0	0

	x_1	x_2	s_1	s_2		
s_1	3	4	1	0	24	$(PCE) = 4$
x_2	2	1	0	$\frac{1}{2}$	5	$R_{pivot}^{new} = \frac{1}{2}R_{pivot}$
P	-3	-6	0	0	0	$(PCE) = -6$

	x_1	x_2	s_1	s_2		
s_1	-5	0	1	-2	4	$R_1^{new} = R_1 - 4R_{pivot}^{new}$
x_2	2	1	0	$\frac{1}{2}$	5	
P	9	0	0	3	30	$R_3^{new} = R_3 + 6R_{pivot}^{new}$

The Pivot Operation

1. Divide every entry in the pivot row by the pivot element to obtain a new pivot row, which we denote R_{pivot}^{new}. Replace the variable written at the left end of this row by the variable corresponding to the pivot column.

2. Subtract multiples of the new pivot row from all *other* rows of the tableau to get zeros in the pivot column. Specifically, if R_{other} is any *other* row of the tableau, and (PCE) is the element of that row in the pivot column, perform the row operation

$$R_{other}^{new} = R_{other} - (PCE)\, R_{pivot}^{new}$$

EXAMPLE 2 **The Pivot Element and the Pivot Operation**

Find the pivot element in the following simplex tableau and carry out the pivot operation.

	x_1	x_2	x_3	x_4	s_1	s_2	s_3	
s_1	1	1	-2	1	1	0	0	8
s_2	6	-2	2	4	0	1	0	10
s_3	5	3	1	3	0	0	1	6
P	-8	12	-10	-6	0	0	0	0

Solution

The smallest negative entry in the bottom row is -10, so the pivot column is column 3 (as shown on the following page).

To find the pivot row, we divide the last entry of each row by the pivot column entry for that row (except the bottom row and any row having a zero or negative pivot column entry) and choose the row with the smallest nonnegative ratio. The calculations on the right show that the pivot row is row 2.

The pivot element is the 2 at the intersection of the pivot column and the pivot row.

Pivot element

	x_1	x_2	x_3	x_4	s_1	s_2	s_3	
s_1	1	1	-2	1	1	0	0	8
s_2	6	-2	2	4	0	1	0	10
s_3	5	3	1	3	0	0	1	6
P	-8	12	-10	-6	0	0	0	0

(Can't use because -2 is < 0)
$\frac{10}{2} = 5$ ← Smallest ratio ← Pivot row
$\frac{6}{1} = 6$

↑
Smallest
negative

Pivot column ↗

Having found the pivot element, we carry out the pivot operation. The first step is to divide the pivot row by the pivot element, so we divide the second row by 2. We also update the basis on the left (the pivot column variable replaces the pivot row variable).

Pivot element

x_3 replaces s_2 in basis →

	x_1	x_2	x_3	x_4	s_1	s_2	s_3	
s_1	1	1	-2	1	1	0	0	8
x_3	3	-1	1	2	0	1/2	0	5
s_3	5	3	1	3	0	0	1	6
P	-8	12	-10	-6	0	0	0	0

$R_{pivot}^{new} = \frac{1}{2}R_{pivot}$

Other rows remain the same

The second step of the pivot operation is to subtract multiples of the *new* pivot row from the other rows to "zero out" the rest of the pivot column. For the first row, the pivot column entry (*PCE*) is -2. Thus the formula $R_{other}^{new} = R_{other} - (PCE)\,R_{pivot}^{new}$ says to multiply the pivot row by -2 and subtract it from the first row: $R_1^{new} = R_1 - (-2)R_{pivot}^{new}$. Since this is the same as $R_1^{new} = R_1 + (2)R_{pivot}^{new}$, we multiply the pivot row by 2 and add it to the first row:

This is a "scratch work" calculation to find the first row of the next tableau.

$$
\begin{array}{rrrrrrrrll}
  1 & 1 & -2 & 1 & 1 & 0 & 0 & 8 & R_1 & \text{Row 1 (above)} \\
+\ 6 & -2 & 2 & 4 & 0 & 1 & 0 & 10 & (2)R_{pivot}^{new} & \text{Twice pivot row} \\
\hline
  7 & -1 & 0 & 5 & 1 & 1 & 0 & 18 & R_1^{new} = R_1 + (2)R_{pivot}^{new} & \text{New row 1}
\end{array}
$$

For the third row, the pivot column entry (PCE) is 1, so the formula on page 302 gives $R_3^{new} = R_3 - (1)R_{pivot}^{new}$ or simply $R_3^{new} = R_3 - R_{pivot}^{new}$. For the bottom row, the PCE is -10, so the formula gives $R_3^{new} = R_3 - (-10)R_{pivot}^{new}$ or simply $R_3^{new} = R_3 + 10R_{pivot}^{new}$. Carrying out these two calculations on scratch paper (as we did above) and writing each of the four new rows in its proper place gives the new simplex tableau:

		x_1	x_2	x_3	x_4	s_1	s_2	s_3		
	s_1	7	-1	0	5	1	1	0	18	$R_1^{new} = R_1 + (2)R_{pivot}^{new}$
	x_3	3	-1	1	2	0	1/2	0	5	$R_{pivot}^{new} = R_{pivot}/2$
	s_3	2	4	0	1	0	$-1/2$	1	1	$R_3^{new} = R_3 - R_{pivot}^{new}$
	P	22	2	0	14	0	5	0	50	$R_4^{new} = R_4 + (10)R_{pivot}^{new}$

The new basis takes the values at the ends of the rows

← Do pivot row first

All of these steps—finding the pivot element, updating the basic variables on the left of the tableau, and changing each row according to the formulas in the Pivot Operation box on page 302 (always beginning with the pivot row)—constitute *one* pivot operation. You will have to practice with scrap paper before you fully master this process. Be sure to begin by finding the new pivot row, writing the new basic varible on the left, and then working on the other rows from the top down. From now on we will just state the resulting tableau, omitting the separate calculations.

Graphing Calculator Exploration

The program* PIVOT carries out the pivot operation after you specify the pivot column and pivot row. To carry out the pivot operation on your calculator with the simplex tableau from Example 2 on page 302, proceed as follows:

a. Enter the simplex tableau as one large matrix [A]. The tableau from Example 2 has four rows of numbers and eight columns. Enter the numbers one by one until the tableau is complete. Because this tableau is too large to fit on the calculator screen, it will scroll from left to right and back again as you enter your numbers.

```
MATRIX[A]  4 X8
[1         1    -2    -
[6        -2     2    -
[5         3     1    -
[-8       12   -10    -

4,3=-10
```

* See Preface for information on how to obtain this and other graphing calculator programs.

b. Run the program PIVOT. It will display the current tableau in matrix [A], and you can use the arrows to scroll the screen to see the rest of it. Press ENTER to enter the Pivot Column and then again to enter the Pivot Row.

```
prgmPIVOT
[[1   1  -2   1
 [6  -2   2   4  ...
 [5   3   1   3  ...
 [-8 12 -10  -6 ...
P C? 3
P R? 2
```

c. After you ENTER the pivot row, the calculator will perform the pivot operation and display the new tableau. Check this result for Example 2 with the new simplex tableau on page 304. (Of course, if you select a zero for your pivot element, the program will respond with an error message.)

Press ENTER to exit the program.

```
P C? 3
P R? 2
[[7  -1   0   5   1 ...
 [3  -1   1   2   0 ...
 [2   4   0   1   0 ...
 [22  2   0  14   0 ...
              Done
```

If your problem requires several pivot operations, you can rerun the program with the new tableau by pressing the ENTER key again.

PRACTICE PROBLEM 2

In each of the following simplex tableaux, find the pivot element and carry out the pivot operation.

a.

	x_1	x_2	x_3	x_4	s_1	s_2	
s_1	-1	2	1	1	1	0	4
s_2	3	6	3	2	0	1	9
P	-5	-3	-6	-2	0	0	0

b.

	x_1	x_2	x_3	x_4	s_1	s_2	
s_1	3	6	-1	2	1	0	18
s_2	2	-8	0	3	0	1	24
P	-3	-1	-4	-2	0	0	0

Solution at the back of the book

The Simplex Method

The *simplex method* solves a standard maximum problem by repeating the pivot operation until the tableau does not have a pivot element, and this final tableau solves the problem.

The Simplex Method

To solve a standard maximum problem by the simplex method:

1. Construct the initial simplex tableau (page 297).

2. Locate the pivot element (pages 300–301) and go to step 3. If the tableau does not have a pivot element, go to step 4.

3. Perform the pivot operation (page 302) using the pivot element and return to step 2.

4. If the final tableau does not have a pivot *column* (see page 300), then the solution occurs at the vertex given by the basic variables (whose values appear at the right ends of their rows, with the nonbasic variables equal to zero), and the maximum value of the objective function appears in the bottom right corner of the tableau. If the final tableau *does* have a pivot *column* but no pivot *row* (see page 301), then there is *no solution* to the problem.

The simplex method uses the pivot operation to make the bottom row "less negative" while keeping the right column nonnegative. That is, *the simplex method makes the tableau "more nearly optimal" while keeping it feasible*. When the bottom row contains no negative entries, the tableau is *optimal* because the largest value for the objective function has been found. For example, the tableau that we obtained at the end of Example 2 on page 304 *is* a final tableau because it has no pivot column (no negative numbers in the bottom row).

	x_1	x_2	x_3	x_4	s_1	s_2	s_3	
s_1	7	−1	0	5	1	1	0	18
x_3	3	−1	1	2	0	1/2	0	5
s_3	2	4	0	1	0	−1/2	1	1
P	22	2	0	14	0	5	0	50

We may then interpret this tableau to give the solution to the original linear programming problem represented by the tableau on page 302 as follows: The maximum value of P is in the bottom right-hand corner (50), with the variables listed in the extreme left column (s_1, x_3, and s_3) taking the values in the extreme right column (18, 5, and 1), with the other variables (x_1, x_2, and s_2) being zero. We will usually state the solution by giving just the values of P and the x's:

The maximum value of P is 50, when $x_1 = 0$, $x_2 = 0$, and $x_3 = 5$

EXAMPLE 3   **The Simplex Method**

Solve the following linear programming problem by the simplex method.

$$\text{Maximize } P = 3x_1 + 5x_2$$

$$\text{Subject to } \begin{cases} 2x_1 + x_2 \le 8 \\ x_1 - x_2 \le 1 \\ x_2 \le 4 \\ x_1 \ge 0 \text{ and } x_2 \ge 0 \end{cases}$$

Solution

This problem in matrix form is

$$\text{Maximize } P = \begin{pmatrix} 3 & 5 \end{pmatrix} \begin{pmatrix} x_1 \\ x_2 \end{pmatrix}$$

Maximize $P = c^t X$

$$\text{Subject to } \begin{cases} \begin{pmatrix} 2 & 1 \\ 1 & -1 \\ 0 & 1 \end{pmatrix} \begin{pmatrix} x_1 \\ x_2 \end{pmatrix} \le \begin{pmatrix} 8 \\ 1 \\ 4 \end{pmatrix} \\ \\ \text{and } \begin{pmatrix} x_1 \\ x_2 \end{pmatrix} \ge 0 \end{cases}$$

Subject to $\begin{cases} A\ X \le b \\ X \ge 0 \end{cases}$

None of the numbers in the matrix $b = \begin{pmatrix} 8 \\ 1 \\ 4 \end{pmatrix}$ is negative, so this *is* a standard maximum problem. Besides the nonnegativity conditions, there are three constraint inequalities, so there will be three slack variables. In the initial simplex tableau, these slacks will be the basic variables:

		x_1	x_2	s_1	s_2	s_3	
Basic variables	s_1	2	1	1	0	0	8
	s_2	1	-1	0	1	0	1
	s_3	0	1	0	0	1	4
	P	-3	-5	0	0	0	0

	X	S	
S	A	I	b
P	$-c^t$	0	0

where $I = \begin{pmatrix} 1 & 0 & 0 \\ 0 & 1 & 0 \\ 0 & 0 & 1 \end{pmatrix}$

The pivot column is column 2 (with the smallest negative number in the bottom row). The pivot row is row 3 (with the smallest ratio of the rightmost number divided by the pivot column entry). The pivot element is the 1 in column 2 and row 3:

Pivot element

	x_1	x_2	s_1	s_2	s_3		
s_1	2	1	1	0	0	8	$\frac{8}{1} = 8$
s_2	1	−1	0	1	0	1	(Omit since −1 < 0)
s_3	0	−1	0	0	1	4	$\frac{4}{1} = 4 \leftarrow$ Pivot row
P	−3	−5	0	0	0	0	

↑
Smallest
negative
↗
Pivot column ⟶

Carrying out the pivot operation (applying the formulas on page 302 to each row, showing the row operations on the right) gives the following tableau.

	x_1	x_2	s_1	s_2	s_3			
s_1	2	0	1	0	−1	4	$R_1^{new} = R_1 - (1)R_{pivot}^{new}$	
s_2	1	0	0	1	1	5	$R_2^{new} = R_2 - (-1)R_{pivot}^{new}$	Or just $R_2 + R_{pivot}^{new}$
x_2	0	1	0	0	1	4	$R_{pivot}^{new} = R_{pivot}$	Unchanged since divided by 1
P	−3	0	0	0	5	20	$R_4^{new} = R_4 - (-5)R_{pivot}^{new}$	Or just $R_4 + 5R_{pivot}^{new}$

x_2 replaces s_3
in basis →

The −3 at the bottom of the first column means that there is yet another pivot column. The smallest ratio from the rows is $\frac{4}{2} = 2$ for the first row, so row 1 is the pivot row. Carrying out the pivot operation on the 2 in the first column and first row gives the tableau shown below (as you should verify).

x_1 replaces s_1
in basis →

	x_1	x_2	s_1	s_2	s_3			
x_1	1	0	1/2	0	−1/2	2	$R_{pivot}^{new} = R_{pivot}/2$	← Do this first
s_2	0	0	−1/2	1	3/2	3	$R_2^{new} = R_2 - (1)R_{pivot}^{new}$	
x_2	0	1	0	0	1	4	$R_3^{new} = R_3 - (0)R_{pivot}^{new}$	Unchanged
P	0	0	3/2	0	7/2	26	$R_4^{new} = R_4 - (-3)R_{pivot}^{new}$	Or just $R_4 + 3R_{pivot}^{new}$

Because the bottom row of this tableau does not contain any negative entries, *this is the final tableau.* The basic variables take the values at the ends of their rows, so $x_1 = 2$, $s_2 = 3$, and $x_2 = 4$; the nonbasic variables are zero, so $s_1 = 0$ and $s_3 = 0$; and the objective function takes the value at the end of the bottom row, so $P = 26$. (You should check that these values satisfy the original constraints and give the claimed value in the objective function.)

Answer: The maximum value of P is 26 at the vertex $x_1 = 2$, $x_2 = 4$.

■

We can compare this solution with the graphical method of the previous section by graphing the feasible region with x_1 on the horizontal axis and x_2 on the vertical axis, as follows. The five vertices and the values of the objective function are given in the table, and the "path of vertices" visited by the simplex method is marked in bold.

	Vertex (x_1, x_2)	$P = 3x_1 + 5x_2$
Initial tableau →	$(0, 0)$	0
	$(1, 0)$	3
Second tableau →	$(0, 4)$	20
	$(3, 2)$	19
Final tableau →	$(2, 4)$	26

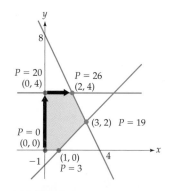

Graphing Calculator Exploration

Use the program PIVOT with the initial simplex tableau from Example 3 on page 307 to carry out the pivot operation calculations on the pivot elements in [column 2, row 3] and in [column 1, row 1], thus verifying that the final tableau given on page 308 is correct.

PRACTICE PROBLEM 3

Find the solution of the following linear programming problem from its final tableau shown on the right. [*Hint:* No calculations are necessary.]

Maximize $P = 5x_1 + 7x_2$

Subject to $\begin{cases} x_1 + 2x_2 \le 16 \\ x_1 - x_2 \le 4 \\ 2x_1 + x_2 \le 14 \\ x_1 \ge 0 \text{ and } x_2 \ge 0 \end{cases}$

$\xrightarrow[\text{leads to}]{\text{Simplex method}}$

	x_1	x_2	s_1	s_2	s_3	
x_2	0	1	3/2	0	−1/3	6
s_2	0	0	1	1	−1	6
x_1	1	0	−1/3	0	2/3	4
P	0	0	3	0	1	62

Solution at the back of the book

EXAMPLE 4 **A Manufacturing Problem with Many Variables**

A pottery shop manufactures dinnerware in four different patterns by shaping the clay, decorating it, and then kiln-firing it. The numbers of hours required per place setting for the four designs are given in the following table, together with the expected profits. The shop employs two skilled workers to do the initial shaping and three artists to do the decorating, none of whom will work more than 40 hours each week. The kiln can be used no more than 55 hours per week. How many place settings of each pattern should be made this week to obtain the greatest possible profit?

	Classic	Modern	Art Deco	Floral
Shaping	2 hours	1 hour	4 hours	2 hours
Decorating	3 hours	1 hour	6 hours	4 hours
Kiln-firing	1 hour	1 hour	1 hour	1 hour
Expected profit	$10	$6	$9	$8

Solution

Let

$$x_1 = \left(\begin{array}{c}\text{Number of Classic}\\ \text{place settings}\end{array}\right), \quad x_2 = \left(\begin{array}{c}\text{Number of Modern}\\ \text{place settings}\end{array}\right),$$

$$x_3 = \left(\begin{array}{c}\text{Number of Art Deco}\\ \text{place settings}\end{array}\right), \quad \text{and} \quad x_4 = \left(\begin{array}{c}\text{Number of Floral}\\ \text{place settings}\end{array}\right)$$

made during the week, so $x_1 \geq 0$, $x_2 \geq 0$, $x_3 \geq 0$, and $x_4 \geq 0$. From the last line in the table, the profit to be maximized is $P = 10x_1 + 6x_2 + 9x_3 + 8x_4$. The constraints on the time spent shaping, decorating, and firing come from the other numbers in the table, along with the time available for each:

$$2x_1 + x_2 + 4x_3 + 2x_4 \leq 80 \qquad \text{2 shapers @ 40 hours each}$$

$$3x_1 + x_2 + 6x_3 + 4x_4 \leq 120 \qquad \text{3 decorators @ 40 hours each}$$

$$x + x_2 + x_3 + x_4 \leq 55 \qquad \text{Time kiln can be used}$$

In matrix form this problem is

$$\text{Maximize } P = (10 \quad 6 \quad 9 \quad 8) \begin{pmatrix} x_1 \\ x_2 \\ x_3 \\ x_4 \end{pmatrix}$$

Subject to $\left\{ \begin{array}{l} \begin{pmatrix} 2 & 1 & 4 & 2 \\ 3 & 1 & 6 & 4 \\ 1 & 1 & 1 & 1 \end{pmatrix} \begin{pmatrix} x_1 \\ x_2 \\ x_3 \\ x_4 \end{pmatrix} \le \begin{pmatrix} 80 \\ 120 \\ 55 \end{pmatrix} \\ \\ \text{and } \begin{pmatrix} x_1 \\ x_2 \\ x_3 \\ x_4 \end{pmatrix} \ge 0 \end{array} \right.$

Standard since

$$b = \begin{pmatrix} 80 \\ 120 \\ 55 \end{pmatrix} \ge 0$$

The initial simplex tableau is

Pivot element

	x_1	x_2	x_3	x_4	s_1	s_2	s_3	
s_1	(2)	1	4	2	1	0	0	80
s_2	3	1	6	4	0	1	0	120
s_3	1	1	1	1	0	0	1	55
P	−10	−6	−9	−8	0	0	0	0

$\frac{80}{2} = 40 \leftarrow$ Smallest ratio $\leftarrow$ Pivot row
$\frac{120}{3} = 40$ (Tie for smallest ratio, so take top row)
$\frac{55}{1} = 55$

↑
Smallest
negative

Pivot column ↗

| When we pivot (on the 2 in column 1, row 1), the tableau becomes

Pivot element

		x_1	x_2	x_3	x_4	s_1	s_2	s_3	
x_1 replaces → x_1 s_1 in basis	x_1	1	1/2	2	1	1/2	0	0	40
	s_2	0	−1/2	0	1	−3/2	1	0	0
	s_3	0	(1/2)	−1	0	−1/2	0	1	15
	P	0	−1	11	2	5	0	0	400

$\frac{40}{1/2} = 80$
(Cannot use since $-\frac{1}{2}$ is < 0)
$\frac{15}{1/2} = 30 \leftarrow$
Smallest
ratio

↑
Smallest
negative

Pivot column ↗

Pivot row ↗

Pivoting again (on the 1/2 in column 2, row 3), we reach the final tableau:

x_2 replaces s_3 in the basis →

	x_1	x_2	x_3	x_4	s_1	s_2	s_3	
x_1	1	0	3	1	1	0	−1	25
s_2	0	0	−1	1	−2	1	1	15
x_2	0	1	−2	0	−1	0	2	30
P	0	0	9	2	4	0	2	430

No negatives in bottom row
so final tableau

The basic variables take the values at the right ends of their rows, the nonbasic variables are zero, and the objective function value appears in the bottom right corner. Therefore, the maximum value of P is 430 when $x_1 = 25$, $x_2 = 30$, $x_3 = 0$, and $x_4 = 0$. The fact that $s_2 = 15$ means that 15 hours of the available decorating time will not be needed.

Answer: The pottery shop should manufacture 25 place settings of Classic, 30 of Modern, and none of Art Deco and Floral.

How Good Is the Simplex Method?

Will the simplex method always find the solution? Even if it does find the solution, will it get to the answer after a reasonable number of tableaux?

In 1955, E. M. I. Beale published a linear programming problem with the awful property that the simplex method found the same sequence of tableaux over and over again. This repetition of tableaux is known as "cycling" because the tableaux repeat in a cyclic pattern. If a computer were solving this problem, it would never stop. Several examples of cycling and changes to the simplex method that prevent cycling are given in Exercises 56–60.

In 1972, V. Klee and G. J. Minty described a class of problems that take many more tableaux than their small size would suggest: With each new variable, the number of tableaux doubles. This means that a Klee–Minty problem with 2 variables takes 4 tableaux to solve, one with 3 variables takes 8 tableaux, one with 4 variables takes 16 tableaux, and so on. The 10-variable problem in this sequence will take 1024 tableaux with larger problems becoming even more unmanageable. Examples of these kinds of problems are given in Exercises 61–65.

Most "real world" computer programs for solving linear programming problems include subroutines to protect against cycling and Klee–Minty situations. S. Smale has shown that ". . . the number of pivots required to solve a linear programming problem grows in proportion to the number of variables on average." Several other methods (by L. Khachiyan and N. Karmarkar) to solve linear programming problems guarantee that the number of steps required does not grow too fast for every problem. However, the simplex method continues to provide the fastest solutions to commercial problems.

Section Summary

The maximum linear programming problem

$$\text{Maximize } P = c^t X$$

$$\text{Subject to } \begin{cases} AX \leq b \\ X \geq 0 \end{cases}$$

is a *standard maximum problem* if $b \geq 0$. The *initial simplex tableau* represents this linear system in the form

	X	S	
S	A	I	b
P	$-c^t$	0	0

The variables listed on the far left form a *basis*, and setting them equal to the values on the far right of their rows and setting the nonbasic variables equal to zero gives a *basic feasible point* that is a vertex of the region. The pivot column is the column with the smallest negative entry in the bottom row (omitting the right column), and the pivot row is the row with the smallest ratio between the last entry and the pivot column entry (omitting any row with a zero or negative pivot column entry, and the bottom row). The *pivot element* is the entry in the pivot column and pivot row.

The *pivot operation* divides the pivot row by the pivot element and then subtracts from each of the other rows the pivot column entry of that row times the new pivot row.

$$R_{pivot}^{new} = R_{pivot}/(\text{Pivot Element})$$

$$R_{other}^{new} = R_{other} - \begin{pmatrix} \text{Pivot} \\ \text{Column} \\ \text{Entry} \end{pmatrix} R_{pivot}^{new}$$

The pivot column variable enters the basis by replacing the variable of the pivot row. The pivot operation rewrites the tableau as an equivalent linear system. The *simplex method* solves a standard maximum problem by pivoting until the tableau does not have a pivot element. If there is a pivot *column* but no pivot *row*, then there is no solution. If there is no pivot *column*, then the solution occurs at the vertex given by the basic variables, and the maximum value of the objective function appears in the bottom right corner.

EXERCISES 4.3

Construct the initial simplex tableau for each standard maximum problem.

1. Maximize $P = 8x_1 + 9x_2 + 7x_3$

Subject to $\begin{cases} 3x_1 + 2x_2 + 4x_3 \le 12 \\ 6x_1 + x_2 + 5x_3 \le 15 \\ x_1 \ge 0, x_2 \ge 0, x_3 \ge 0 \end{cases}$

2. Maximize $P = 10x_1 + 15x_2 + 12x_3$

Subject to $\begin{cases} 5x_1 + 2x_2 + 6x_3 \le 30 \\ 3x_1 + 4x_2 + x_3 \le 36 \\ x_1 \ge 0, x_2 \ge 0, x_3 \ge 0 \end{cases}$

3. Maximize $P = 13x_1 + 7x_2$

Subject to $\begin{cases} 4x_1 + 3x_2 \le 12 \\ 5x_1 + 2x_2 \le 20 \\ x_1 + 6x_2 \le 12 \\ x_1 \ge 0, x_2 \ge 0 \end{cases}$

4. Maximize $P = 8x_1 + 33x_2$

Subject to $\begin{cases} 3x_1 + 2x_2 \le 30 \\ x_1 + 4x_2 \le 20 \\ 5x_1 + 6x_2 \le 60 \\ x_1 \ge 0, x_2 \ge 0 \end{cases}$

5. Maximize $P = 5x_1 - 2x_2 + 10x_3 - 5x_4$

Subject to $\begin{cases} 2x_1 + x_2 + x_3 + 3x_4 \le 6 \\ x_1 + 4x_2 - 2x_3 + x_4 \le 8 \\ x_1 \ge 0, x_2 \ge 0, x_3 \ge 0, x_4 \ge 0 \end{cases}$

6. Maximize $P = 5x_1 + 10x_2 - 30x_3 + 5x_4$

Subject to $\begin{cases} 3x_1 + x_2 - 2x_3 + 4x_4 \le 12 \\ 2x_1 - x_2 + 3x_3 + x_4 \le 12 \\ x_1 \ge 0, x_2 \ge 0, x_3 \ge 0, x_4 \ge 0 \end{cases}$

7. Maximize $P = 10x_1 + 20x_2 + 15x_3$

Subject to $\begin{cases} 8x_1 + x_2 + 4x_3 \le 32 \\ 3x_1 + 5x_2 + 7x_3 \le 30 \\ 6x_1 + 2x_2 + 9x_3 \le 28 \\ x_1 \ge 0, x_2 \ge 0, x_3 \ge 0 \end{cases}$

8. Maximize $P = 10x_1 + 15x_2 + 5x_3$

Subject to $\begin{cases} 7x_1 + 3x_2 + x_3 \le 21 \\ 6x_1 + 5x_2 + 4x_3 \le 20 \\ 8x_1 + 9x_2 + 2x_3 \le 45 \\ x_1 \ge 0, x_2 \ge 0, x_3 \ge 0 \end{cases}$

9. Maximize $P = 90x_1 + 80x_2 + 100x_3$

Subject to $\begin{cases} x_1 + 2x_2 + 3x_3 \le 45 \\ 6x_1 + 5x_2 + 4x_3 \le 40 \\ 7x_1 + 8x_2 + 9x_3 \le 63 \\ 12x_1 + 11x_2 + 10x_3 \le 60 \\ x_1 \ge 0, x_2 \ge 0, x_3 \ge 0 \end{cases}$

10. Maximize $P = 10x_1 + 20x_2 + 5x_3$

Subject to $\begin{cases} 9x_1 + x_2 + 8x_3 \le 20 \\ 10x_1 - 2x_2 + 7x_3 \le 40 \\ -11x_1 + 3x_2 - 6x_3 \le 80 \\ 12x_1 - 4x_2 + 5x_3 \le 60 \\ x_1 \ge 0, x_2 \ge 0, x_3 \ge 0 \end{cases}$

For each simplex tableau, find the pivot element and carry out one complete pivot operation. If there is no pivot element, explain what the tableau shows about the solution of the original standard maximum problem.

11.

	x_1	x_2	x_3	s_1	s_2	s_3	
s_1	2	0	1	1	0	0	4
s_2	1	1	1	0	1	0	5
s_3	1	-1	3	0	0	1	6
P	-7	-8	-9	0	0	0	0

12.

	x_1	x_2	x_3	s_1	s_2	s_3	
s_1	1	-2	1	1	0	0	5
s_2	3	1	2	0	1	0	6
s_3	2	0	3	0	0	1	7
P	-4	-5	-3	0	0	0	0

13.

	x_1	x_2	x_3	s_1	s_2	s_3	
s_1	1	0	1	1	0	0	3
s_2	0	1	1	0	1	0	4
s_3	1	1	0	0	0	1	5
P	-6	-7	-8	0	0	0	0

14.

	x_1	x_2	x_3	s_1	s_2	s_3	
s_1	0	1	1	1	0	0	5
s_2	1	0	1	0	1	0	3
s_3	1	1	0	0	0	1	4
P	-7	-8	-6	0	0	0	0

15.

	x_1	x_2	x_3	x_4	s_1	s_2	
s_1	4	2	6	2	1	0	12
s_2	3	1	2	1	0	1	8
P	-4	-5	6	-3	0	0	0

16.

	x_1	x_2	x_3	x_4	s_1	s_2	
s_1	5	3	1	3	1	0	30
s_2	6	2	2	4	0	1	16
P	-4	-6	-5	7	0	0	0

17.

	x_1	x_2	x_3	x_4	s_1	s_2	s_3	
s_1	0	0	2	1	1	1	-1	10
x_2	0	1	1	1	0	1	0	15
x_1	1	0	-2	0	0	-1	1	10
P	0	0	0	2	0	1	3	90

18.

	x_1	x_2	x_3	x_4	s_1	s_2	s_3	
x_2	0	1	-1	0	1	-1	0	20
x_4	1	0	1	1	0	1	0	40
s_3	1	0	2	0	-1	1	1	30
P	2	0	8	0	3	3	0	300

19.

	x_1	x_2	s_1	s_2	s_3	s_4	
s_1	0	0	1	-1	0	0	10
x_1	1	0	0	1	-1	0	10
x_2	0	1	0	1	0	0	20
s_4	0	0	0	-1	0	1	5
P	0	0	0	25	-10	0	400

20.

	x_1	x_2	s_1	s_2	s_3	s_4	
s_1	0	0	1	0	-1	0	5
x_2	0	1	0	1	-1	0	10
x_1	1	0	0	1	0	0	20
s_4	0	0	0	-1	0	1	5
P	0	0	0	35	-15	0	550

Solve each problem by the simplex method. (Exercises 21, 22, 31, and 32 can also be solved by the graphical method.)

You may use the PIVOT program if permitted by your instructor.

21. Maximize $P = x_1 + 2x_2$

Subject to $\begin{cases} 3x_1 + x_2 \le 24 \\ x_1 + x_2 \le 14 \\ x_1 \ge 0, x_2 \ge 0 \end{cases}$

22. Maximize $P = x_1 + 3x_2$

Subject to $\begin{cases} 2x_1 + x_2 \le 24 \\ x_1 + x_2 \le 15 \\ x_1 \ge 0, x_2 \ge 0 \end{cases}$

23. Maximize $P = 3x_1 + 4x_2 + 5x_3 + 2x_4$

Subject to $\begin{cases} x_1 + 2x_2 + x_3 + x_4 \le 8 \\ x_1 + x_2 + 2x_4 \le 7 \\ 2x_1 + x_2 + x_3 + x_4 \le 6 \\ x_1 \ge 0, x_2 \ge 0, x_3 \ge 0, x_4 \ge 0 \end{cases}$

24. Maximize $P = 3x_1 + 6x_2 + 5x_3 + 4x_4$

Subject to $\begin{cases} 3x_1 + x_2 + x_3 + 2x_4 \le 10 \\ x_1 + 2x_3 + x_4 \le 2 \\ x_1 + x_2 + 3x_3 + x_4 \le 8 \\ x_1 \ge 0, x_2 \ge 0, x_3 \ge 0, x_4 \ge 0 \end{cases}$

25. Maximize $P = 30x_1 + 40x_2 + 15x_3$

Subject to $\begin{cases} 2x_1 + x_2 + 3x_3 \le 150 \\ 3x_1 + 2x_2 + x_3 \le 100 \\ x_1 \ge 0, x_2 \ge 0, x_3 \ge 0 \end{cases}$

26. Maximize $P = 30x_1 + 60x_2 + 15x_3$

Subject to $\begin{cases} 2x_1 + 3x_2 + x_3 \le 75 \\ 3x_1 + x_2 + 2x_3 \le 50 \\ x_1 \ge 0, x_2 \ge 0, x_3 \ge 0 \end{cases}$

27. Maximize $P = 6x_1 + 4x_2 + 5x_3$

Subject to $\begin{cases} x_1 - x_2 + x_3 \le 20 \\ x_1 + 2x_3 \le 10 \\ x_1 \ge 0, x_2 \ge 0, x_3 \ge 0 \end{cases}$

28. Maximize $P = 6x_1 + 5x_2 + 8x_3$

Subject to $\begin{cases} 2x_1 + x_3 \le 40 \\ x_1 - x_2 + x_3 \le 30 \\ x_1 \ge 0, x_2 \ge 0, x_3 \ge 0 \end{cases}$

29. Maximize $P = 4x_1 + 3x_2 - 5x_3 + 6x_4$

Subject to $\begin{cases} x_1 + x_2 + x_4 \le 60 \\ x_1 + x_3 + x_4 \le 40 \\ x_1 + x_2 + x_3 \le 50 \\ x_1 \ge 0, x_2 \ge 0, x_3 \ge 0, x_4 \ge 0 \end{cases}$

30. Maximize $P = 2x_1 + 3x_2 + 4x_3 + x_4$

Subject to $\begin{cases} x_1 + x_2 + x_4 \le 20 \\ x_1 + x_3 + x_4 \le 15 \\ x_1 + x_2 + x_3 \le 25 \\ x_1 \ge 0, x_2 \ge 0, x_3 \ge 0, x_4 \ge 0 \end{cases}$

31. Maximize $P = 4x_1 + 2x_2$

Subject to $\begin{cases} x_1 - x_2 \le 1 \\ -x_1 + x_2 \le 3 \\ x_1 + x_2 \le 5 \\ x_1 \ge 0, x_2 \ge 0 \end{cases}$

32. Maximize $P = 12x_1 + 6x_2$

Subject to $\begin{cases} x_1 + x_2 \le 8 \\ x_1 - x_2 \le 2 \\ -x_1 + 2x_2 \le 10 \\ x_1 \ge 0, x_2 \ge 0 \end{cases}$

33. Maximize $P = 40x_1 + 60x_2 + 50x_3$

Subject to $\begin{cases} 2x_1 + x_2 + 4x_3 \le 400 \\ 4x_1 + 3x_2 + 2x_3 \le 600 \\ x_1 \ge 0, x_2 \ge 0, x_3 \ge 0 \end{cases}$

34. Maximize $P = 140x_1 + 80x_2 + 100x_3$

Subject to $\begin{cases} 4x_1 + 2x_2 + x_3 \le 100 \\ 2x_1 + 3x_2 + 3x_3 \le 150 \\ x_1 \ge 0, x_2 \ge 0, x_3 \ge 0 \end{cases}$

35. Maximize $P = x_1 + 2x_2 + 3x_3$

Subject to $\begin{cases} x_1 + x_2 + x_3 \le 15 \\ x_2 + x_3 \le 10 \\ x_3 \le 5 \\ x_1 \ge 0, x_2 \ge 0, x_3 \ge 0 \end{cases}$

36. Maximize $P = 15x_1 + 10x_2 + 5x_3$

Subject to $\begin{cases} x_1 + x_2 + x_3 \le 3 \\ x_1 + x_2 \le 2 \\ x_1 \le 1 \\ x_1 \ge 0, x_2 \ge 0, x_3 \ge 0 \end{cases}$

37. Maximize $P = 50x_1 + 30x_2 + 40x_3$

Subject to $\begin{cases} 4x_1 + 2x_2 + x_3 \le 80 \\ 2x_1 + x_2 + 3x_3 \le 120 \\ x_1 \ge 0, x_2 \ge 0, x_3 \ge 0 \end{cases}$

38. Maximize $P = 64x_1 + 56x_2 + 80x_3$

Subject to $\begin{cases} 3x_1 + x_2 + 2x_3 \le 240 \\ x_1 + 3x_2 + 4x_3 \le 160 \\ x_1 \ge 0, x_2 \ge 0, x_3 \ge 0 \end{cases}$

39. Maximize $P = 14x_1 + 5x_2 + 10x_3$

Subject to $\begin{cases} 4x_1 + x_2 + 3x_3 \le 130 \\ 3x_1 + x_2 + 2x_3 \le 95 \\ 4x_1 + 2x_2 + 3x_3 \le 140 \\ x_1 \ge 0, x_2 \ge 0, x_3 \ge 0 \end{cases}$

40. Maximize $P = 8x_1 + 12x_2 + 7x_3$

Subject to $\begin{cases} 4x_1 + 4x_2 + 3x_3 \le 440 \\ x_1 + 2x_2 + x_3 \le 150 \\ 3x_1 + 3x_2 + 2x_3 \le 320 \\ x_1 \ge 0, x_2 \ge 0, x_3 \ge 0 \end{cases}$

APPLIED EXERCISES

Formulate each situation as a linear programming problem by identifying the variables, the objective function, and the constraints. Be sure to state clearly the meaning of each variable. Check that the problem is a standard maximum problem and then solve it by the simplex method. State your final answer in terms of the original question.

You may use the PIVOT program if permitted by your instructor.

41. **Production Planning** An automotive parts shop rebuilds carburetors, fuel pumps, and alternators. The numbers of hours to rebuild and then inspect and pack each part are shown in the following table, together with the number of hours of skilled labor available for each task. If the profit is $12 for each carburetor, $14 for each fuel pump, and $10 for each alternator, how many of each should the shop rebuild to obtain the greatest possible profit?

	Carburetor	Fuel Pump
Rebuilding	5 hours	4 hours
Inspection and packaging	1 hour	1 hour

	Alternator	Labor Available
Rebuilding	3 hours	200 hours
Inspection and packaging	0.5 hour	45 hours

42. **Pollution Control** An empty storage yard at a coal-burning electric power plant can hold no more than 100,000 tons of coal. Two grades of coal are available: low-sulfur (1%) with an energy content of 20 million British thermal units (BTU) per ton and high-sulfur (2%) with an energy content of 30 million BTU per ton. If the next coal purchase may contain no more than 1400 tons of sulfur, how many tons of each type of coal should be purchased to obtain the most energy?

43. **Recycling Management** A volunteer recycle center accepts both used paper and empty glass bottles, which are then sorted and sold to a reprocessing company. The center has room to accept 800 crates of paper and glass each week and 50 hours of volunteer help to do the sorting. Each crate of paper products takes 5 minutes to sort and sells for 8¢, while each crate of bottles takes 3 minutes to sort and sells for 7¢. How many crates of each should the center accept each week to raise the most money for its ecology scholarship fund?

44. **Production Planning** A small jewelry company prepares and mounts semiprecious stones. There are 20 lapidaries (who cut and polish the stones) and 24 jewelers (who mount the stones in gold settings). Each employee works 7 hours each day. Each tray of agates requires 5 hours of cutting and polishing and 4 hours to mount, each tray of onyxes requires 2 hours of cutting and polishing and 3 hours to mount, and each tray of garnets requires 6 hours of cutting and polishing and 3 hours to mount. If the profit is $15 for each tray of agates, $10 for each tray of onyxes, and $12 for each tray of garnets, how many trays of each stone should be processed each day to obtain the greatest possible profit?

45. **Production Planning** Repeat the previous problem but with a profit of $13 per tray of garnets.

46. **Resource Allocation** A furniture shop manufactures wooden desks, tables, and chairs. The numbers of hours to assemble and finish each piece are shown in the table, together with the numbers of hours of skilled labor available for each task. If the profit is $80 for each desk, $84 for each table, and $68 for each chair, how many of each should the shop manufacture to obtain the greatest possible profit?

	Desk	Table
Assembly	2 hours	1 hour
Finishing	2 hours	3 hours

	Chair	Labor Available
Assembly	2 hours	200 hours
Finishing	1 hour	150 hours

47. *Advertising*   In the last few days before the election, a politician can afford to spend no more than $27,000 on TV advertisements and can arrange for no more than 10 ads. Each day-time ad costs $2000 and reaches 4000 viewers, each prime-time ad costs $3000 and reaches 5000 viewers, and each late-night ad costs $1000 and reaches 2000 viewers. Ignoring repeated viewings by the same person and assuming every viewer can vote, how many ads in each of the time periods will reach the most voters?

48. *Advertising*   The manager of a new mall may spend up to $18,000 on grand opening announcements in newspapers, on radio, and on TV. Each newspaper ad costs $300 and reaches 6000 readers, each one-minute radio commercial costs $800 and is heard by 10,000 listeners, and each 15-second TV spot costs $900 and is seen by 11,000 viewers. If there is time to arrange for no more than 5 newspaper ads and no more than 20 radio commercials and TV spots combined, how many of each should be placed to reach the largest number of potential customers? (Ignore multiple exposures to the same consumer.)

49. *Agriculture*   A farmer grows corn, peanuts, and soybeans on his 240-acre farm. To maintain soil fertility, the farmer rotates the crops and always plants at least as many acres of soybeans as the total acres of the other crops. Each acre of corn requires 2 days of labor and yields a profit of $150, each acre of peanuts requires 5 days of labor and yields a profit of $300, and each acre of soybeans requires 1 day of labor and yields a profit of $100. If the farmer and his family can put in at most 630 days of labor, how many acres of each crop should the farmer plant to obtain the greatest possible profit?

50. *Agriculture*   A farmer grows wheat, barley, and oats on her 500-acre farm. Each acre of wheat requires 3 days of labor (to plant, tend, and harvest) and costs $21 (for seed, fertilizer, and pesticides), each acre of barley requires 2 days of labor and costs $27, and each acre of oats requires 3 days of labor and costs $24. The farmer and her hired field hands can provide no more than 1200 days of labor this year, and she can afford to spend no more than $15,120. If the expected profit is $50 for each acre of wheat, $40 for each acre of barley, and $45 for each acre of oats, how many acres of each crop should she grow to obtain the greatest possible profit?

Explorations and Excursions

The following problems extend and augment the material presented in the text.

 The PIVOT program may be helpful (if permitted by your instructor).

More About the Pivot Operation

51. a. Find the intersection of the lines $5x + 2y = 70$ and $4x + 3y = 84$ by solving the first equation for x and substituting this expression for x into the second equation. Then solve for y and find the value of x. (See Section 4.1, Example 5 on pages 271–272.)

b. Carry out one complete pivot operation using the 5 for the pivot element in the mini-tableau

	x	y	
	5	2	70
	4	3	84

How do the rows of your new mini-tableau compare with the two equations you created for the first sentence of part (a)?

52. Show that pivoting on a tableau obeys the diagram

$$
\begin{array}{cc}
\vdots \ \vdots & \vdots \ \vdots \\
\cdots \ c \ q \ \cdots & \cdots \ 0 \ q-\dfrac{rc}{p} \ \cdots \\
\longrightarrow & \\
\cdots \ p \ r \ \cdots & \cdots \ 1 \ r/p \ \cdots \\
\vdots \ \vdots & \vdots \ \vdots
\end{array}
$$

where p is the pivot element, c is any other entry in the pivot column, r is any other entry in the pivot row, and q is any entry in the tableau not in the pivot row or pivot column. In words, this diagram defines the pivot operation in four steps: "(1) the pivot element becomes 1, (2) the rest of the pivot column becomes 0, (3) the rest of the pivot row is divided by the pivot element, and (4) every entry not in the pivot row or pivot column is decreased by the product of the pivot column entry in the same row with the pivot row entry in the same column divided by the pivot element." This form of the pivot operation makes it easy to check any particular number in a new tableau without repeating the row operations.

53. In Example 2 (pages 302–304), we carried out the pivot operation on the 2 in column 3 and row 2:

	x_1	x_2	x_3	x_4	s_1	s_2	s_3	
s_1	1	1	−2	1	1	0	0	8
s_2	6	−2	2	4	0	1	0	10
s_3	5	3	1	3	0	0	1	6
P	−8	12	−10	−6	0	0	0	0

	x_1	x_2	x_3	x_4	s_1	s_2	s_3	
s_1	7	−1	0	5	1	1	0	18
→ x_3	3	−1	1	2	0	1/2	0	5
s_3	2	4	0	1	0	−1/2	1	1
P	22	2	0	14	0	5	0	50

To undo this pivot operation, pivot on the $\frac{1}{2}$ in column 6 and row 2 of the final tableau. Does this return you to the initial simplex tableau? (This shows how you can undo a mistaken pivot operation and return to your previous tableau.)

54. The final tableaux in Exercises 17 and 18 both came from initial simplex tableaux having the form

	x_1	x_2	x_3	x_4	s_1	s_2	s_3	
s_1	?	?	?	?	1	0	0	?
s_2	?	?	?	?	0	1	0	?
s_3	?	?	?	?	0	0	1	?
P	?	?	?	?	0	0	0	0

To find the initial simplex tableaux for these problems, begin with the actual final tableaux and pivot where necessary to return the columns of the slack variables to their initial form. Reconstruct the original problems from these initial tableaux and check that the answers you found in Exercises 17 and 18 satisfy these original constraints and objective functions.

55. The final tableaux in Exercises 19 and 20 both came from initial simplex tableaux having the form

	x_1	x_2	s_1	s_2	s_3	s_4	
s_1	?	?	1	0	0	0	?
s_2	?	?	0	1	0	0	?
s_3	?	?	0	0	1	0	?
s_4	?	?	0	0	0	1	?
P	?	?	0	0	0	0	0

To find the initial simplex tableaux for these problems, begin with the actual final tableaux and pivot where necessary to return the columns of the slack variables to their initial form. Reconstruct the original problems from these initial tableaux and sketch the regions determined by their constraints. Do these regions confirm your answers to Exercises 19 and 20?

More About Cycling

56. Beale's Cycling Example Attempt to solve the following problem by the simplex method:

$$\text{Maximize } P = (3/4 \quad -20 \quad 1/2 \quad -6) \begin{pmatrix} x_1 \\ x_2 \\ x_3 \\ x_4 \end{pmatrix}$$

Subject to
$$\begin{pmatrix} 1/4 & -8 & -1 & 9 \\ 1/2 & -12 & -1/2 & 3 \\ 0 & 0 & 1 & 0 \end{pmatrix} \begin{pmatrix} x_1 \\ x_2 \\ x_3 \\ x_4 \end{pmatrix} \leq \begin{pmatrix} 0 \\ 0 \\ 1 \end{pmatrix}$$

and $\begin{pmatrix} x_1 \\ x_2 \\ x_3 \\ x_4 \end{pmatrix} \geq 0$

You should return to the initial simplex tableau after pivoting at [column 1, row 1], [column 2, row 2], [column 3, row 1], [column 4, row 2], [column 5, row 1], and [column 6, row 2].

57. Rescaling Changing the size of the numbers in a problem without altering the feasible region can prevent cycling.

a. Multiply the first constraint in Beale's cycling example (Exercise 56) by 4 to get the equivalent problem

$$\text{Maximize } P = (3/4 \quad -20 \quad 1/2 \quad -6)\begin{pmatrix} x_1 \\ x_2 \\ x_3 \\ x_4 \end{pmatrix}$$

$$\text{Subject to } \begin{cases} \begin{pmatrix} 1 & -32 & -4 & 36 \\ 1/2 & -12 & -1/2 & 3 \\ 0 & 0 & 1 & 0 \end{pmatrix}\begin{pmatrix} x_1 \\ x_2 \\ x_3 \\ x_4 \end{pmatrix} \le \begin{pmatrix} 0 \\ 0 \\ 1 \end{pmatrix} \\ \text{and } \begin{pmatrix} x_1 \\ x_2 \\ x_3 \\ x_4 \end{pmatrix} \ge 0 \end{cases}$$

You should reach the final simplex tableau after pivoting at [column 1, row 1], [column 2, row 2], [column 3, row 1], [column 4, row 2], [column 1, row 3], and [column 5, row 2].

b. Divide every coefficient of x_2 in Beale's cycling example (Exercise 56) by 4 to get the equivalent problem

$$\text{Maximize } P = (3/4 \quad -5 \quad 1/2 \quad -6)\begin{pmatrix} x_1 \\ x_2 \\ x_3 \\ x_4 \end{pmatrix}$$

$$\text{Subject to } \begin{cases} \begin{pmatrix} 1/4 & -2 & -1 & 9 \\ 1/2 & -3 & -1/2 & 3 \\ 0 & 0 & 1 & 0 \end{pmatrix}\begin{pmatrix} x_1 \\ x_2 \\ x_3 \\ x_4 \end{pmatrix} \le \begin{pmatrix} 0 \\ 0 \\ 1 \end{pmatrix} \\ \text{and } \begin{pmatrix} x_1 \\ x_2 \\ x_3 \\ x_4 \end{pmatrix} \ge 0 \end{cases}$$

You should reach the final simplex tableau after pivoting at [column 1, row 1], [column 3, row 2], [column 4, row 3] and [column 5, row 3].

58. Bland's Rule Eliminate cycling by changing the choice of the pivot column to be *the leftmost column with a negative entry in the bottom row* (that is, ignore the size of the numbers and just look for the first negative). Use Bland's rule to solve Beale's cycling example (Exercise 56). You should reach the final simplex tableau after pivoting at [column 1, row 1], [column 2, row 2], [column 3, row 1], [column 4, row 2], [column 1, row 3], and [column 5, row 2].

59. A Nondeterministic Simplex Algorithm The simplex method (page 306) is a *deterministic* procedure to solve standard maximum problems because once the process is started, the rules *always* select the *same* sequence of pivot elements. If we allow *chance* to play a role in the selection of the pivot column or pivot row, the resulting solution method is nondeterministic in that the sequence of pivot elements in the solution is no longer determined but rather may change from one solution attempt to another. Let us change the rule on page 301 for the selection of the pivot row to state that *if there is a tie for the smallest ratio between two of the rows, the pivot row is to be selected by a coin flip.* Solve Beale's cycling example (Exercise 56) several times using this modified procedure. Do your solutions all use the same number of pivot operations?

60. More Cycling Examples A large class of cyclic examples was found by K. T. Marshall and J. W. Suurballe in 1969.

a. Try to solve this typical Marshall–Suurballe problem by the simplex method.

$$\text{Maximize } P = (1 \quad -28 \quad -2 \quad -2)\begin{pmatrix} x_1 \\ x_2 \\ x_3 \\ x_4 \end{pmatrix}$$

$$\text{Subject to } \begin{cases} \begin{pmatrix} 1/2 & -18 & -2 & 4 \\ 1/4 & -5 & -1/2 & 1/2 \end{pmatrix}\begin{pmatrix} x_1 \\ x_2 \\ x_3 \\ x_4 \end{pmatrix} \le \begin{pmatrix} 0 \\ 0 \end{pmatrix} \\ \text{and } \begin{pmatrix} x_1 \\ x_2 \\ x_3 \\ x_4 \end{pmatrix} \ge 0 \end{cases}$$

You should return to the initial simplex tableau after pivoting at [column 1, row 1], [column 2, row 2], [column 3, row 1], [column 4, row 2], [column 5, row 1], and [column 6, row 2].

b. Rescale the constraints of the problem in part (a) by multiplying the first 2 and the second by 4. How many pivots does it now take to solve this rescaled problem?

c. Solve the problem in part (a) using Bland's rule (see Exercise 58).

d. Solve the problem in part (a) using the nondeterministic simplex algorithm given in Exercise 59.

Klee–Minty Problems

61. Solve the following two-variable problem by the simplex method. (Your fourth tableau should be the final tableau.)

$$\text{Maximize } P = \begin{pmatrix} 1 & 10 \end{pmatrix}\begin{pmatrix} x_1 \\ x_2 \end{pmatrix}$$

$$\text{Subject to } \begin{cases} \begin{pmatrix} 0 & 1 \\ 1 & 20 \end{pmatrix}\begin{pmatrix} x_1 \\ x_2 \end{pmatrix} \le \begin{pmatrix} 1 \\ 100 \end{pmatrix} \\ \text{and } \begin{pmatrix} x_1 \\ x_2 \end{pmatrix} \ge 0 \end{cases}$$

62. Solve the following three-variable problem by the simplex method. (Your eighth tableau should be the final tableau.)

$$\text{Maximize } P = \begin{pmatrix} 1 & 10 & 100 \end{pmatrix}\begin{pmatrix} x_1 \\ x_2 \\ x_3 \end{pmatrix}$$

$$\text{Subject to } \begin{cases} \begin{pmatrix} 0 & 0 & 1 \\ 0 & 1 & 20 \\ 1 & 20 & 200 \end{pmatrix}\begin{pmatrix} x_1 \\ x_2 \\ x_3 \end{pmatrix} \le \begin{pmatrix} 1 \\ 100 \\ 10{,}000 \end{pmatrix} \\ \text{and } \begin{pmatrix} x_1 \\ x_2 \\ x_3 \end{pmatrix} \ge 0 \end{cases}$$

63. Following the form of Exercises 61 and 62, write down a problem using four variables that will continue the pattern and require sixteen tableaux to solve.

64. Following the form of Exercises 61–63, write down a problem using five variables that will continue the pattern and require thirty-two tableaux to solve.

65. Using the problems from Exercises 61–64, check that each of this family of Klee–Minty problems can be solved in one pivot by using Bland's rule to choose the pivot column.

4.4 Duality and Standard Minimum Problems

APPLICATION PREVIEW

Diet Problems and "The Cost of Subsistence"

In 1945 the University of Minnesota economist George J. Stigler* published "The Cost of Subsistence," in which he pointed out that although "elaborate investigations have been made of the adequacy of diets at various income levels, . . . no one has determined the minimum cost of obtaining the amounts of calories, protein, minerals, and vitamins which these studies accept as adequate or optimum. This will be done in the present paper, not only for its

* George J. Stigler, "The Cost of Subsistence," *Journal of Farm Economics* **27**:303–314, 1945.

own interest but because it sheds much light on the meaning of conventional 'low-cost' diets." He examined 77 foods whose nutritive values and 1939 prices were known. This preliminary food list was first reduced down to 15 foods by excluding those whose nutritive values per dollar of expenditure were less than some others. "For example, white bread has less than half the nutrients (per dollar) of white flour, except for calcium, for which neither is an economical source." Stigler then remarked that "... *there does not appear to be any direct method of finding the minimum of a linear function subject to linear conditions.* . . . There is no reason to believe that the cheapest combination was found, for only a handful of the 510 possible combinations . . . were examined."

STIGLER'S MINIMUM-COST ($39.93) ANNUAL DIET (1939 PRICES)

Commodity	Quantity	Cost
Wheat flour	370 pounds	$13.33
Evaporated milk	57 cans	3.84
Cabbage	111 pounds	4.11
Spinach	23 pounds	1.85
Dried navy beans	285 pounds	16.80

In the fall of 1947, Laderman of the Mathematical Tables Project in New York computed the optimal solution of Stigler's problem in a test of Dantzig's new simplex method. Interestingly enough, this minimum solution gave a yearly cost of $36.64, while Stigler's trial-and-error solution (using the intuition of an economist) cost just $3.29 more. Comparing his minimal diet to those recommended by professional dieticians, Stigler found that they cost two to three times as much as his. He ended his paper with the following remarks.

Why do these conventional diets cost so much? The answer is evident from their composition. The dieticians take account of the palatability of foods, variety of diet, prestige of various foods, and other cultural facets of consumption. . . . No one can now say with any certainty what the cultural requirements of a particular person may be. . . . If the dieticians persist in presenting minimum diets, they should at least report separately the physical and cultural components. . . .

Costs between 1939 and 2000 increased by a factor of 12.1, so Stigler's $39.93 annual subsistence diet would now cost about $485.

Introduction

The simplex method for standard maximum problems begins at the origin, with an objective function value of zero, and moves along a path of vertices to arrive at a solution vertex, where the objective function has the largest possible value. On page 289 we remarked that it might also be possible to find the solution by making smaller and smaller upper estimates on the objective function using multiples of the constraints. In this section, we shall see that this idea leads to a pairing between maximum and minimum problems in such a way that the solution of one gives the solution of the other. We can already solve maximum problems using the simplex method, so this observation will allow us to solve minimum problems by solving the corresponding maximum problems.

The Dual of a Standard Maximum Problem

Although we could solve the standard maximum problem

$$\text{Maximize } P = 24x_1 + 16x_2 + 42x_3$$

$$\text{Subject to } \begin{cases} 3x_1 + x_2 + 3x_3 \le 9 \\ x_1 + x_2 + 2x_3 \le 8 \\ x_1 \ge 0, x_2 \ge 0, \text{ and } x_3 \ge 0 \end{cases}$$

by the simplex method, let us instead attempt to estimate P in terms of the constraints. If we multiply the first constraint by 11 and add it to 5 times the second, we have that P can be no more than 139:

$$\text{Maximize } P = 24x_1 + 16x_2 + 42x_3$$

$$\text{Subject to } \begin{cases} 3x_1 + x_2 + 3x_3 \le 9 & \times 11 \text{ is} & 33x_1 + 11x_2 + 33x_3 \le 99 \\ x_1 + x_2 + 2x_3 \le 8 & \times 5 \text{ is} & \underline{5x_1 + 5x_2 + 10x_3 \le 40} \\ x_1 \ge 0, x_2 \ge 0, x_3 \ge 0 & \text{Add to give} & 38x_1 + 16x_2 + 43x_3 \le 139 \end{cases}$$

$$\text{so } P = 24x_1 + 16x_2 + 42x_3 \text{ is} \le 139$$

We want the best possible estimate, so we should make more estimates using other multipliers. But we should (1) never multiply an inequality by a negative number (a negative multiple would become a lower estimate), (2) always get at least 24 of the x_1's, at least 16 of the x_2's, and at least 42 of the x_3's, and (3) make the upper estimate for P as small as possible. If we multiply the first constraint by a number y_1 and add it to y_2 times the second constraint, we get

$3x_1 + x_2 + 3x_3 \le 9$	$\times y_1$ is	$3y_1x_1$	$+$	y_1x_2	$+$	$3y_1x_3 \le 9y_1$
$x_1 + x_2 + 2x_3 \le 8$	$\times y_2$ is	y_2x_1	$+$	y_2x_2	$+$	$2y_2x_3 \le 8y_2$
	Add to give	$(3y_1 + y_2)x_1$	$+$	$(y_1 + y_2)x_2$	$+$	$(3y_1 + 2y_2)x_3 \le 9y_1 + 8y_2$

Rule (1) now means that $y_1 \geq 0$ and $y_2 \geq 0$, rule (2) becomes $3y_1 + y_2 \geq 24$, $y_1 + y_2 \geq 16$, and $3y_1 + 2y_2 \geq 42$, and rule (3) means we want to minimize $9y_1 + 8y_2$. This new linear programming problem:

$$\text{Minimize } C = 9y_1 + 8y_2$$

$$\text{Subject to } \begin{cases} 3y_1 + y_2 \geq 24 \\ y_1 + y_2 \geq 16 \\ 3y_1 + 2y_2 \geq 42 \\ y_1 \geq 0 \text{ and } y_2 \geq 0 \end{cases}$$

is called the *dual* of the original problem. These two problems have the property that $P \leq C$ for any x_1, x_2, x_3 and y_1, y_2 values in the feasible regions. Furthermore, if we could find feasible points such that $P = C$, then we would have the solutions to both problems because the value of P could be made no larger and the value of C could be made no smaller. Fortunately, as we will see on page 327, the simplex method developed in the previous section finds feasible points for both problems.

The relation between our maximum problem and its dual minimum problem becomes clearer when we write both problems in matrix form.

$$\text{Maximize } P = (24 \quad 16 \quad 42) \begin{pmatrix} x_1 \\ x_2 \\ x_3 \end{pmatrix}$$

and

$$\text{Minimize } C = (9 \quad 8) \begin{pmatrix} y_1 \\ y_2 \end{pmatrix}$$

$$\text{Subject to } \begin{cases} \begin{pmatrix} 3 & 1 & 3 \\ 1 & 1 & 2 \end{pmatrix} \begin{pmatrix} x_1 \\ x_2 \\ x_3 \end{pmatrix} \leq \begin{pmatrix} 9 \\ 8 \end{pmatrix} \\ \\ \text{and } \begin{pmatrix} x_1 \\ x_2 \\ x_3 \end{pmatrix} \geq 0 \end{cases}$$

$$\text{Subject to } \begin{cases} \begin{pmatrix} 3 & 1 \\ 1 & 1 \\ 3 & 2 \end{pmatrix} \begin{pmatrix} y_1 \\ y_2 \end{pmatrix} \geq \begin{pmatrix} 24 \\ 16 \\ 42 \end{pmatrix} \\ \\ \text{and } \begin{pmatrix} y_1 \\ y_2 \end{pmatrix} \geq 0 \end{cases}$$

Notice that rows of numbers in one problem become columns of the same numbers in the other problem.

Two matrices with the property that the columns of one are the rows of the other in the same order are *transposes* of each other. The *transpose* of a matrix M is written M^t and is found by turning each row into a column. For example, the transpose of

$$M = \begin{pmatrix} 1 & 2 & 3 & 4 \\ 5 & 6 & 7 & 8 \\ 9 & 10 & 11 & 12 \end{pmatrix} \text{ is } M^t = \begin{pmatrix} 1 & 5 & 9 \\ 2 & 6 & 10 \\ 3 & 7 & 11 \\ 4 & 8 & 12 \end{pmatrix}$$

It makes no difference whether you change rows into columns or columns into rows. Furthermore, the transpose of a transpose is the original matrix again: $(M^t)^t = M$.

Dual Problems and the Duality Theorem

Giving the matrices in the maximum problem the usual names, we have the following definition:

Dual Problems

Each of the following linear programming problems is the *dual* of the other.

$$\text{Maximize } P = c^t X \qquad \text{and} \qquad \text{Minimize } C = b^t Y$$

$$\text{Subject to } \begin{cases} AX \le b \\ X \ge 0 \end{cases} \qquad \qquad \text{Subject to } \begin{cases} A^t Y \ge c \\ Y \ge 0 \end{cases}$$

The dual of a maximum problem is a minimum problem, and the dual of a minimum problem is a maximum problem. The numbers in the objective function of one problem are the numbers after the inequalities in the constraints of the other. The maximum problem has $\le$ constraints while the minimum problem has $\ge$ constraints. There are as many slack variables (one for each constraint) in the *maximum* problem as variables $(y_1, y_2, \ldots, y_m)$ in the *minimum* problem. There are as many slack variables (one for each constraint) in the *minimum* problem as variables $(x_1, x_2, \ldots, x_n)$ in the *maximum* problem. We shall write t_1, $t_2, \ldots, t_n$ for the slack variables in the minimum problem constraints $A^t Y \ge c$, and $t_1 \ge 0, t_2 \ge 0, \ldots, t_n \ge 0$ because we *subtract* them to lower $A^t Y$ down to equal c.

EXAMPLE 1   The Dual of a Minimum Problem

Find the dual of the following minimum problem and then construct the initial simplex tableau for this dual maximum problem.

$$\text{Minimize } C = 440y_1 + 300y_2 + 200y_3$$

$$\text{Subject to } \begin{cases} 3y_1 + 2y_2 + 2y_3 \ge 156 \\ 2y_1 + y_2 + 4y_3 \ge 120 \\ 2y_1 + 2y_2 + 3y_3 \ge 132 \\ 4y_1 + 3y_2 - y_3 \ge 180 \\ y_1 \ge 0, y_2 \ge 0, \text{ and } y_3 \ge 0 \end{cases}$$

Solution

In matrix form, this problem and its dual maximum problem are

$$\text{Minimize } C = (\begin{array}{ccc} 440 & 300 & 200 \end{array}) \begin{pmatrix} y_1 \\ y_2 \\ y_3 \end{pmatrix}$$

$$\text{Maximize } P = (\begin{array}{cccc} 156 & 120 & 132 & 180 \end{array}) \begin{pmatrix} x_1 \\ x_2 \\ x_3 \\ x_4 \end{pmatrix}$$

and

$$\text{Subject to } \begin{cases} \begin{pmatrix} 3 & 2 & 2 \\ 2 & 1 & 4 \\ 2 & 2 & 3 \\ 4 & 3 & -1 \end{pmatrix} \begin{pmatrix} y_1 \\ y_2 \\ y_3 \end{pmatrix} \geq \begin{pmatrix} 156 \\ 120 \\ 132 \\ 180 \end{pmatrix} \\ \text{and } \begin{pmatrix} y_1 \\ y_2 \\ y_3 \end{pmatrix} \geq 0 \end{cases}$$

$$\text{Subject to } \begin{cases} \begin{pmatrix} 3 & 2 & 2 & 4 \\ 2 & 1 & 2 & 3 \\ 2 & 4 & 3 & -1 \end{pmatrix} \begin{pmatrix} x_1 \\ x_2 \\ x_3 \\ x_4 \end{pmatrix} \leq \begin{pmatrix} 440 \\ 300 \\ 200 \end{pmatrix} \\ \text{and } \begin{pmatrix} x_1 \\ x_2 \\ x_3 \\ x_4 \end{pmatrix} \geq 0 \end{cases}$$

The minimum problem has 3 variables y_1, y_2, y_3 and 4 slacks t_1, t_2, t_3, t_4, while the dual maximum problem has 4 variables x_1, x_2, x_3, x_4 and 3 slacks s_1, s_2, s_3. The initial simplex tableau for this dual maximum problem is

	x_1	x_2	x_3	x_4	s_1	s_2	s_3	
s_1	3	2	2	4	1	0	0	440
s_2	2	1	2	3	0	1	0	300
s_3	2	4	3	-1	0	0	1	200
P	-156	-120	-132	-180	0	0	0	0

PRACTICE PROBLEM 1

Find the dual of the following minimum problem and then construct the initial simplex tableau for this dual maximum problem.

$$\text{Minimize } C = 11y_1 + 9y_2 + 7y_3$$

$$\text{Subject to } \begin{cases} 2y_1 + y_2 + y_3 \geq 4 \\ y_1 + y_2 \geq 7 \\ y_2 + y_3 \geq 5 \\ y_1 + y_2 + 3y_3 \geq 6 \\ y_1 \geq 0, y_2 \geq 0, \text{ and } y_3 \geq 0 \end{cases}$$

Solution at the back of the book

How can we solve a minimum problem? The following theorem explains that any tableau for a maximum problem displays *at the same time* information about the dual minimum problem.

Duality Theorem for the Simplex Method

The bottom row of any simplex tableau for a maximum linear programming problem displays values for the dual minimum problem's slack variables, variables, and objective function:

These values for the minimum problem satisfy its constraints and yield the given value in its objective function even if some are negative.

Because dual problems are related by transposition, it is not surprising to find that the solution of the *minimum* problem appears in the bottom *row,* just as the solution of the *maximum* problem appears in the last *column.* Because the choice of the pivot element makes the pivot operation keep the right column of the tableau nonnegative while making the bottom row "less negative," *the simplex method makes the maximum problem optimal while keeping it feasible and makes the dual minimum problem feasible while keeping it optimal.* If the tableau represents a feasible point for both the maximum and the dual minimum problems, the tableau is the final tableau for both problems.

Solving Standard Minimum Problems

Using the duality theorem, we can solve a minimum problem by the simplex method, provided its dual maximum problem is a *standard* problem.

Standard Minimum Problem

Minimize $C = b^t Y$

Subject to $\begin{cases} A^t Y \geq c \\ Y \geq 0 \end{cases}$

where $b^t \geq 0$

The solution of a standard minimum problem can be found by solving the dual maximum problem by the simplex method and interpreting the final tableau as a statement about the original minimum problem.

Solution of a Standard Minimum Problem

To solve a standard minimum problem:

1. Construct the dual maximum problem (page 325).

2. Solve the dual maximum problem by the simplex method (page 306).

3. If there is a solution to the dual maximum problem, then there is a solution to the minimum problem and the values of the slacks, the variables, and the objective function appear (in that order) in the bottom row of the final tableau of the dual maximum problem. If there is no solution to the dual maximum problem, then the minimum problem has no solution.

EXAMPLE 2 **Solution of a Standard Minimum Problem**

Solve the following standard minimum problem by finding the dual maximum problem and using the simplex method.

$$\text{Minimize } C = 50y_1 + 60y_2$$

$$\text{Subject to } \begin{cases} y_1 + 4y_2 \geq 20 \\ 2y_1 + 3y_2 \geq 30 \\ y_1 - y_2 \geq 5 \\ y_1 \geq 0 \text{ and } y_2 \geq 0 \end{cases}$$

Solution

In matrix form, this problem and its dual maximum problem are

$$\text{Minimize } C = (50 \quad 60)\begin{pmatrix} y_1 \\ y_2 \end{pmatrix}$$

and

$$\text{Subject to } \begin{cases} \begin{pmatrix} 1 & 4 \\ 2 & 3 \\ 1 & -1 \end{pmatrix}\begin{pmatrix} y_1 \\ y_2 \end{pmatrix} \geq \begin{pmatrix} 20 \\ 30 \\ 5 \end{pmatrix} \\ \text{and } \begin{pmatrix} y_1 \\ y_2 \end{pmatrix} \geq 0 \end{cases}$$

$$\text{Maximize } P = (20 \quad 30 \quad 5)\begin{pmatrix} x_1 \\ x_2 \\ x_3 \end{pmatrix}$$

$$\text{Subject to } \begin{cases} \begin{pmatrix} 1 & 2 & 1 \\ 4 & 3 & -1 \end{pmatrix}\begin{pmatrix} x_1 \\ x_2 \\ x_3 \end{pmatrix} \leq \begin{pmatrix} 50 \\ 60 \end{pmatrix} \\ \text{and } \begin{pmatrix} x_1 \\ x_2 \\ x_3 \end{pmatrix} \geq 0 \end{cases}$$

The initial simplex tableau for the dual maximum problem is

	x_1	x_2	x_3	s_1	s_2	
s_1	1	2	1	1	0	50
s_2	4	3	−1	0	1	60
P	−20	−30	−5	0	0	0

Bottom row states that
$t_1 = -20, t_2 = -30, t_3 = -5,$
$y_1 = 0, y_2 = 0,$ and $C = 0$

To solve the dual maximum problem, we first pivot on the 3 in column 2 and row 2 to find the tableau

	x_1	x_2	x_3	s_1	s_2	
s_1	−5/3	0	5/3	1	−2/3	10
x_2	4/3	1	−1/3	0	1/3	20
P	20	0	−15	0	10	600

Bottom row states that
$t_1 = 20, t_2 = 0, t_3 = -15,$
$y_1 = 0, y_2 = 10,$
and $C = 600$

Then we pivot again on the 5/3 in column 3 and row 1 to reach the final tableau:

	x_1	x_2	x_3	s_1	s_2	
x_3	−1	0	1	3/5	−2/5	6
x_2	1	1	0	1/5	1/5	22
P	5	0	0	9	4	690

Bottom row states that
$t_1 = 5, t_2 = 0, t_3 = 0,$
$y_1 = 9, y_2 = 4,$ and
$C = 690$

Answer: The minimum value is 690 when $y_1 = 9$ and $y_2 = 4$.

To compare this solution with the graphical method, we graph the feasible region of the minimum problem with y_1 on the horizontal axis and y_2 on the vertical axis. The three vertices and the values of the objective function are given in the table. The path of nonfeasible points visited by the simplex tableaux of the dual maximum problem and leading to a feasible vertex is marked in bold.

Vertex (y_1, y_2)	$C = 50y_1 + 60y_2$
(20, 0)	1000
(12, 2)	720
(9, 4)	690

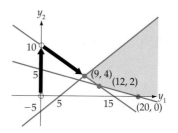

The dual variables are multipliers for the constraints of the other problem, so let us check this with both the values for y_1, y_2 and for x_1, x_2, x_3. Using $y_1 = 9$, $y_2 = 4$ as multipliers for the dual maximum problem, we find the estimate $P \leq 690$:

Maximize $P = 20x_1 + 30x_2 + 5x_3$

Subject to $\begin{cases} x_1 + 2x_2 + x_3 \leq 50 \\ 4x_1 + 3x_2 - x_3 \leq 60 \\ x_1 \geq 0, x_2 \geq 0, x_3 \geq 0 \end{cases}$ $\begin{matrix} \times 9 \text{ is} \\ \times 4 \text{ is} \\ \text{Add to give} \end{matrix}$ $\begin{matrix} 9x_1 + 18x_2 + 9x_3 \leq 450 \\ 16x_1 + 12x_2 - 4x_3 \leq 240 \\ \overline{25x_1 + 30x_2 + 5x_3 \leq 690} \end{matrix}$

so $P = 20x_1 + 30x_2 + 5x_3$ is ≤ 690

If P is the profit from goods manufactured by processes limited by the constraints, the multipliers $y_1, y_2, \ldots, y_m$ are called the *marginal values* for these processes because they show how much value each contributes to the final profit. For example, if the above problem represented the profit of a pottery shop as in Example 4 on page 310, but now with just two constraints ("shaping" and "decorating"), $y_1 = 9$ and $y_2 = 4$ show that the shaping process contributes \$9 per hour to the profit and the decorating process contributes \$4 per hour.

Using $x_1 = 0$, $x_2 = 22$, and $x_3 = 6$ as multipliers for the minimum problem, we similarly find the estimate $C \geq 690$:

Minimize $C = 50y_1 + 60y_2$

Subject to $\begin{cases} y_1 + 4y_2 \geq 20 \\ 2y_1 + 3y_2 \geq 30 \\ y_1 - y_2 \geq 5 \\ y_1 \geq 0, y_2 \geq 0 \end{cases}$ $\begin{matrix} \times 0 \text{ is} \\ \times 22 \text{ is} \\ \times 6 \text{ is} \\ \text{Add to give} \end{matrix}$ $\begin{matrix} 0y_1 + 0y_2 \geq 0 \\ 44y_1 + 66y_2 \geq 660 \\ 6y_1 - 6y_2 \geq 30 \\ \overline{50y_1 + 60y_2 \geq 690} \end{matrix}$

so $C = 50y_1 + 60y_2$ is ≥ 690

If C is the cost of meeting the requirements represented by the constraints, the multipliers $x_1, x_2, \ldots, x_n$ are called the *shadow prices* of these requirements because they show how much each contributes to the total cost. For example, if the above problem represented the cost of meeting nutritional needs as in Exercise 38 on page 293 with the constraints representing "calories," "protein," and "fat," then $x_1 = 0$, $x_2 = 22$, and $x_3 = 6$ show that the calorie requirement does not contribute to the cost, but the price of the protein requirement is 22¢ per gram and the price of the fat requirement is 6¢ per gram.

PRACTICE PROBLEM 2

Find the solution of the linear programming problem

Minimize $C = 255y_1 + 435y_2 + 300y_3 + 465y_4$

Subject to $\begin{cases} 2y_1 + 3y_2 + 2y_3 + 4y_4 \geq 25 \\ y_1 + 3y_2 + 2y_3 + 2y_4 \geq 18 \\ 3y_1 + 4y_2 + 3y_3 + 2y_4 \geq 36 \\ y_1 \geq 0, y_2 \geq 0, y_3 \geq 0, \text{ and } y_4 \geq 0 \end{cases}$

from the fact that the final tableau of its dual maximum problem is

	x_1	x_2	x_3	s_1	s_2	s_3	s_4	
x_3	0	0	1	0	−2	3	0	30
x_1	1	0	0	1	3	−5	0	60
x_2	0	1	0	−1	0	1	0	45
s_4	0	0	0	−2	−8	12	1	75
P	0	0	0	7	3	1	0	3390

Solution at the back of the book

Graphing Calculator Exploration

The initial simplex tableau of the dual maximum problem in Practice Problem 2 is

	x_1	x_2	x_3	s_1	s_2	s_3	s_4	
s_1	2	1	3	1	0	0	0	255
s_2	3	3	4	0	1	0	0	435
s_3	2	2	3	0	0	1	0	300
s_4	4	2	2	0	0	0	1	465
P	−25	−18	−36	0	0	0	0	0

Use the program PIVOT to pivot at [column 3, row 1], [column 2, row 3], and [column 1, row 2] and thus verify that the final tableau is as given.

Mixed Constraints: A Transportation Problem

Although we have been careful to write our standard minimum problem constraints as $A^t Y \geq c$, we do not mean to exclude the possibility of *mixed constraints*, some $\leq$ and some $\geq$, from the initial statement of the problem. However, we must write them all as $\geq$ inequalities (by multiplying by −1 as necessary) before forming the matrix inequality $A^t Y \geq c$. (The requirement $b \geq 0$ refers to the objective function of the minimum problem and places no restrictions on the inequalities.) The following example belongs to a general type of minimization problem with mixed constraints having the nice property that the pivot elements will all be 1s.

EXAMPLE 3   **A Transportation Problem**

A retail store chain has cartons of goods stored at warehouses in Maryland and Washington that must be distributed to its stores in

Ohio and Louisiana. The cost to ship each carton from Maryland to Ohio is $6 and from Maryland to Louisiana is $7, while the cost to ship each carton from Washington to Ohio is $8 and from Washington to Louisiana is $9. There are 300 cartons at the Maryland warehouse and 300 at the warehouse in Washington. If the Ohio stores need 200 cartons and the Louisiana stores need 300 cartons, how many cartons should be shipped from each warehouse to each state to incur the smallest shipping costs?

Solution

This is a minimization problem that requires four variables: one for the amount shipped from each warehouse to each state. Let

$$y_1 = \begin{pmatrix} \text{Cartons shipped} \\ \text{from MD to OH} \end{pmatrix}, \quad y_2 = \begin{pmatrix} \text{Cartons shipped} \\ \text{from MD to LA} \end{pmatrix},$$

$$y_3 = \begin{pmatrix} \text{Cartons shipped} \\ \text{from WA to OH} \end{pmatrix}, \quad \text{and} \quad y_4 = \begin{pmatrix} \text{Cartons shipped} \\ \text{from WA to LA} \end{pmatrix}$$

Because each warehouse can ship no more than the number of cartons stored there, and each state must receive at least the required number of cartons, this problem is the linear programming problem

Minimize $C = 6y_1 + 7y_2 + 8y_3 + 9y_4$ Total shipping costs

$$\text{Subject to} \begin{cases} y_1 + y_2 & \leq 300 & \text{Have 300 in MD} \\ y_3 + y_4 & \leq 300 & \text{Have 300 in WA} \\ y_1 + y_3 & \geq 200 & \text{Need 200 in OH} \\ y_2 + y_4 & \geq 300 & \text{Need 300 in LA} \\ y_1 \geq 0, y_2 \geq 0, y_3 \geq 0, \text{and } y_4 \geq 0 & & \text{Nonnegativity} \end{cases}$$

When we multiply the first two constraints by -1 to put them into $\geq$ form, this problem becomes

$$\text{Minimize } C = \begin{pmatrix} 6 & 7 & 8 & 9 \end{pmatrix} \begin{pmatrix} y_1 \\ y_2 \\ y_3 \\ y_4 \end{pmatrix}$$

$$\text{Subject to} \begin{cases} \begin{pmatrix} -1 & -1 & 0 & 0 \\ 0 & 0 & -1 & -1 \\ 1 & 0 & 1 & 0 \\ 0 & 1 & 0 & 1 \end{pmatrix} \begin{pmatrix} y_1 \\ y_2 \\ y_3 \\ y_4 \end{pmatrix} \geq \begin{pmatrix} -300 \\ -300 \\ 200 \\ 300 \end{pmatrix} \\ \\ \text{and} \begin{pmatrix} y_1 \\ y_2 \\ y_3 \\ y_4 \end{pmatrix} \geq 0 \end{cases}$$

The dual maximum problem is

$$\text{Maximize } P = (-300 \quad -300 \quad 200 \quad 300) \begin{pmatrix} x_1 \\ x_2 \\ x_3 \\ x_4 \end{pmatrix}$$

$$\text{Subject to } \begin{cases} \begin{pmatrix} -1 & 0 & 1 & 0 \\ -1 & 0 & 0 & 1 \\ 0 & -1 & 1 & 0 \\ 0 & -1 & 0 & 1 \end{pmatrix} \begin{pmatrix} x_1 \\ x_2 \\ x_3 \\ x_4 \end{pmatrix} \le \begin{pmatrix} 6 \\ 7 \\ 8 \\ 9 \end{pmatrix} \\ \\ \text{and } \begin{pmatrix} x_1 \\ x_2 \\ x_3 \\ x_4 \end{pmatrix} \ge 0 \end{cases}$$

The initial simplex tableau is

	x_1	x_2	x_3	x_4	s_1	s_2	s_3	s_4	
s_1	-1	0	1	0	1	0	0	0	6
s_2	-1	0	0	1	0	1	0	0	7
s_3	0	-1	1	0	0	0	1	0	8
s_4	0	-1	0	1	0	0	0	1	9
P	300	300	-200	-300	0	0	0	0	0

We pivot at [column 4, row 2], [column 3, row 1], and [column 1, row 3] to reach the final tableau:

	x_1	x_2	x_3	x_4	s_1	s_2	s_3	s_4	
x_3	0	-1	1	0	0	0	1	0	8
x_4	0	-1	0	1	-1	1	1	0	9
x_1	1	-1	0	0	-1	0	1	0	2
s_4	0	0	0	0	1	-1	-1	1	0
P	0	100	0	0	0	300	200	0	3700
					y_1	y_2	y_3	y_4	C

Bottom row states that $t_1 = 0, t_2 = 100,$ $t_3 = 0, t_4 = 0,$ $y_1 = 0, y_2 = 300,$ $y_3 = 200, y_4 = 0,$ and $C = 3700$

The Maryland warehouse should send nothing to Ohio and 300 cartons to Louisiana, while the Washington warehouse should send 200 cartons to Ohio and nothing to Louisiana to achieve the least shipping cost of $3700. The second slack shows that there will be 100 cartons unused in the Washington warehouse.

■

SPREADSHEET EXPLORATION

The spreadsheet below shows the solution of Example 3 using the `Solver` command. The entries in column E combine the coefficients and the current values of the variables to find the values that the `Solver` compares to the bounds and the goal of obtaining the minimum of the objective function.

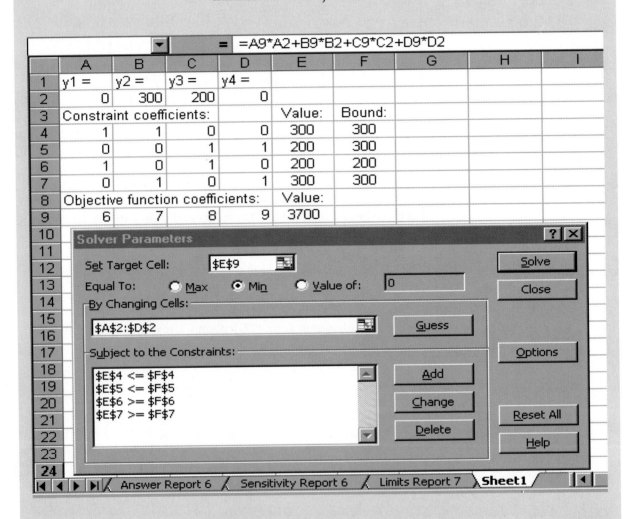

The answer, sensitivity, and limits reports generated by the `Solver` provide additional information about the solution. How do these reports compare to the values in the bottom row and right column of the final tableau in the solution of Example 3 on page 333?

Section Summary

The dual of the minimum problem

$$\text{Minimize } C = b^t Y$$

$$\text{Subject to } \begin{cases} A^t Y \geq c \\ Y \geq 0 \end{cases}$$

is the maximum problem

$$\text{Maximize } P = c^t X$$

$$\text{Subject to } \begin{cases} AX \leq b \\ X \geq 0 \end{cases}$$

Both problems are standard problems if $b \geq 0$. The rows of one problem form the columns of the other. The minimum constraints are all written as $\geq$ inequalities, and the maximum constraints are all written as $\leq$ inequalities. The maximum of P is the same as the minimum of C. The simplex method solution of the maximum problem displays in the bottom row the solution of the dual minimum problem (slack variables, variables, and objective function). If one problem has no solution, then neither does its dual.

EXERCISES 4.4

Write each of the following standard minimum problems in the matrix form "minimize $C = b^t Y$ subject to $A^t Y \geq c$ and $Y \geq 0$ where $b^t \geq 0$," and construct the dual maximum problem and its initial simplex tableau.

1. Minimize $C = 60y_1 + 100y_2 + 300y_3$

Subject to $\begin{cases} y_1 + 2y_2 + 3y_3 \geq 180 \\ 4y_1 + 5y_2 + 6y_3 \geq 120 \\ y_1 \geq 0, y_2 \geq 0, y_3 \geq 0 \end{cases}$

2. Minimize $C = 100y_1 + 60y_2 + 280y_3$

Subject to $\begin{cases} 5y_1 + 9y_2 + 7y_3 \geq 315 \\ 2y_1 + 6y_2 + 4y_3 \geq 480 \\ y_1 \geq 0, y_2 \geq 0, y_3 \geq 0 \end{cases}$

3. Minimize $C = 3y_1 + 20y_2$

Subject to $\begin{cases} 3y_1 + 2y_2 \geq 150 \\ y_1 + 4y_2 \geq 100 \\ 3y_1 + 4y_2 \geq 228 \\ y_1 \geq 0, y_2 \geq 0 \end{cases}$

4. Minimize $C = 60y_1 + 16y_2$

Subject to $\begin{cases} y_1 + 4y_2 \geq 20 \\ 3y_1 + 2y_2 \geq 30 \\ 3y_1 + 4y_2 \geq 48 \\ y_1 \geq 0, y_2 \geq 0 \end{cases}$

5. Minimize $C = 84y_1 + 21y_2$

Subject to $\begin{cases} 3y_1 + y_2 \geq 21 \\ 4y_1 - y_2 \geq 0 \\ y_1 \geq 0, y_2 \geq 0 \end{cases}$

6. Minimize $C = 7y_1 + 7y_2$

Subject to $\begin{cases} 2y_1 + 3y_2 \geq 42 \\ 3y_1 + y_2 \geq 21 \\ y_1 \geq 0, y_2 \geq 0 \end{cases}$

7. Minimize $C = 15y_1 + 20y_2 + 5y_3$

Subject to $\begin{cases} y_1 - y_2 - 2y_3 \leq 30 \\ y_1 + 2y_2 + y_3 \geq 30 \\ y_1 \geq 0, y_2 \geq 0, y_3 \geq 0 \end{cases}$

8. Minimize $C = 30y_1 + 40y_2 + 80y_3$

Subject to $\begin{cases} -3y_1 + y_2 - y_3 \leq 60 \\ y_1 + y_2 + 2y_3 \geq 60 \\ y_1 \geq 0, y_2 \geq 0, y_3 \geq 0 \end{cases}$

9. Minimize $C = 105y_1 + 40y_2$

Subject to $\begin{cases} 7y_1 + 5y_2 \geq 70 \\ 3y_1 + y_2 \geq 45 \\ 5y_1 + 2y_2 \geq 80 \\ y_1 \geq 0, y_2 \geq 0 \end{cases}$

10. Minimize $C = 40y_1 + 105y_2$

Subject to $\begin{cases} 3y_1 + 5y_2 \geq 75 \\ 2y_1 + 5y_2 \geq 80 \\ y_1 + 3y_2 \geq 45 \\ y_1 \geq 0, y_2 \geq 0 \end{cases}$

Solve each of the following standard minimum problems by finding the dual maximum problem and using the simplex method. (Exercises 13–22 can also be solved by the graphical method.)

You may use the PIVOT program if permitted by your instructor.

11. Minimize $C = 10y_1 + 20y_2 + 10y_3$

Subject to $\begin{cases} -y_1 + y_2 + y_3 \geq 50 \\ y_1 + y_2 - y_3 \geq 30 \\ y_1 \geq 0, y_2 \geq 0, y_3 \geq 0 \end{cases}$

12. Minimize $C = 30y_1 + 50y_2 + 30y_3$

Subject to $\begin{cases} -y_1 + y_2 + y_3 \geq 10 \\ y_1 + y_2 - y_3 \geq 20 \\ y_1 \geq 0, y_2 \geq 0, y_3 \geq 0 \end{cases}$

13. Minimize $C = 4y_1 + 5y_2$

Subject to $\begin{cases} y_1 + y_2 \geq 10 \\ y_1 \geq 2 \\ y_2 \geq 3 \\ y_1 \geq 0, y_2 \geq 0 \end{cases}$

14. Minimize $C = 3y_1 + 2y_2$

Subject to $\begin{cases} y_1 + y_2 \geq 20 \\ y_1 \geq 4 \\ y_2 \geq 5 \\ y_1 \geq 0, y_2 \geq 0 \end{cases}$

15. Minimize $C = 15y_1 + 10y_2$

Subject to $\begin{cases} y_1 + 2y_2 \geq 20 \\ 3y_1 - y_2 \geq 60 \\ y_1 \geq 0, y_2 \geq 0 \end{cases}$

16. Minimize $C = 40y_1 + y_2$

Subject to $\begin{cases} 5y_1 - y_2 \geq 5 \\ 4y_1 + y_2 \geq 4 \\ y_1 \geq 0, y_2 \geq 0 \end{cases}$

17. Minimize $C = 4y_1 + 3y_2$

Subject to $\begin{cases} y_1 + y_2 \geq 15 \\ 3y_1 + y_2 \leq 60 \\ y_1 \geq 0, y_2 \geq 0 \end{cases}$

18. Minimize $C = 4y_1 + 5y_2$

Subject to $\begin{cases} y_1 + y_2 \geq 20 \\ 2y_1 + y_2 \leq 50 \\ y_1 \geq 0, y_2 \geq 0 \end{cases}$

19. Minimize $C = 2y_1 + y_2$

Subject to $\begin{cases} -y_1 + y_2 \leq 20 \\ y_1 - y_2 \leq 20 \\ y_1 + y_2 \geq 10 \\ y_1 \geq 0, y_2 \geq 0 \end{cases}$

20. Minimize $C = 2y_1 + 3y_2$

Subject to $\begin{cases} -y_1 + y_2 \leq 30 \\ y_1 - y_2 \leq 30 \\ y_1 + y_2 \geq 10 \\ y_1 \geq 0, y_2 \geq 0 \end{cases}$

21. Minimize $C = 30y_1 + 90y_2$

Subject to $\begin{cases} -y_1 + 2y_2 \geq 60 \\ 3y_1 - y_2 \geq 45 \\ y_1 \geq 0, y_2 \geq 0 \end{cases}$

22. Minimize $C = 60y_1 + 45y_2$

Subject to $\begin{cases} 2y_1 - y_2 \ge 90 \\ -y_1 + 3y_2 \ge 30 \\ y_1 \ge 0, y_2 \ge 0 \end{cases}$

23. Minimize $C = 30y_1 + 19y_2 + 30y_3$

Subject to $\begin{cases} 2y_1 + y_2 + y_3 \ge 6 \\ y_1 + y_2 + 2y_3 \ge 4 \\ y_1 \ge 0, y_2 \ge 0, y_3 \ge 0 \end{cases}$

24. Minimize $C = 60y_1 + 39y_2 + 60y_3$

Subject to $\begin{cases} 2y_1 + y_2 + y_3 \ge 6 \\ y_1 + y_2 + 2y_3 \ge 8 \\ y_1 \ge 0, y_2 \ge 0, y_3 \ge 0 \end{cases}$

25. Minimize $C = 20y_1 + 30y_2 + 40y_3$

Subject to $\begin{cases} y_1 - y_2 + y_3 \ge 15 \\ y_1 + y_2 + y_3 \ge 20 \\ y_1 - y_2 + y_3 \le 10 \\ y_1 \ge 0, y_2 \ge 0, y_3 \ge 0 \end{cases}$

26. Minimize $C = 20y_1 + 50y_2 + 30y_3$

Subject to $\begin{cases} 2y_1 - y_2 + y_3 \le 10 \\ y_1 + y_2 + y_3 \ge 30 \\ 2y_1 - y_2 + y_3 \ge 20 \\ y_1 \ge 0, y_2 \ge 0, y_3 \ge 0 \end{cases}$

27. Minimize $C = 5y_1 + 3y_2 + 2y_3 + 4y_4$

Subject to $\begin{cases} y_1 + y_2 \le 20 \\ y_3 + y_4 \le 30 \\ y_1 + y_2 + y_3 + y_4 \ge 40 \\ y_1 \ge 0, y_2 \ge 0, y_3 \ge 0, y_4 \ge 0 \end{cases}$

28. Minimize $C = 7y_1 + 6y_2 + 9y_3 + 8y_4$

Subject to $\begin{cases} y_1 + y_3 \le 40 \\ y_2 + y_4 \le 50 \\ y_1 + y_2 + y_3 + y_4 \ge 60 \\ y_1 \ge 0, y_2 \ge 0, y_3 \ge 0, y_4 \ge 0 \end{cases}$

29. Minimize $C = 132y_1 + 102y_2 + 60y_3$

Subject to $\begin{cases} 3y_1 + 2y_2 + y_3 \ge 48 \\ 4y_1 + 3y_2 + 2y_3 \ge 72 \\ 2y_1 + 2y_2 + y_3 \ge 42 \\ y_1 \ge 0, y_2 \ge 0, y_3 \ge 0 \end{cases}$

30. Minimize $C = 96y_1 + 144y_2 + 84y_3$

Subject to $\begin{cases} 2y_1 + 3y_2 + 2y_3 \ge 102 \\ 3y_1 + 4y_2 + 2y_3 \ge 132 \\ y_1 + 2y_2 + y_3 \ge 60 \\ y_1 \ge 0, y_2 \ge 0, y_3 \ge 0 \end{cases}$

APPLIED EXERCISES

Formulate each situation as a standard minimum linear programming problem by identifying the variables, the objective function, and the constraints. Be sure to state clearly the meaning of each variable. Solve it by finding the dual maximum problem and using the simplex method. State your final answer in terms of the original question.

 You may use the PIVOT program if permitted by your instructor.

31. **Nutrition**   An athlete's training diet needs at least 44 more grams of carbohydrates, 12 more grams of fat, and 16 more grams of protein each day. A

dietician recommends two food supplements, Bulk-Up Bars (costing 48¢ each) and Power Drink (costing 45¢ per can), with nutritional contents (in grams) as given in the table. How much of each food supplement will provide the extra needed nutrition at the least cost?

	Bulk-Up Bar	Power Drink
Carbohydrates	4	3
Fat	1	1
Protein	2	1

32. Diet Planning   The residents of an elder care facility need at least 156 more milligrams of calcium, 180 more micrograms of folate, and 66 more grams of protein in their weekly diets. The staff chef has created a new fish entree and a salad that the residents should find appealing, and the nutritional contents of the new menu items are given in the accompanying table. If the fish entree costs 21¢ per ounce and the salad costs 15¢ per ounce, how many ounces of each should be added to every resident's weekly menu to provide the additional nutrition at the least cost?

	Fish Entree (per Ounce)	Salad (per Ounce)
Calcium (milligrams)	6	4
Folate (micrograms)	6	5
Protein (grams)	2	2

33. Purchasing   The office manager of a large accounting firm needs at least 315 more boxes of pens and 120 more boxes of pencils but has only $1500 left in his budget. Jack's Office Supplies has packages of 5 boxes of pens with 2 boxes of pencils on sale for $20, while John's Discount offers packages of 3 boxes of pens and 1 box of pencils for $11. How many packages from each store should he buy to restock the store room at the least cost?

34. Highway Construction   The project engineer for a highway construction company needs to cut through a small hill to make way for a new road. She estimates that at least 3500 cubic yards of dirt and at least 2400 cubic yards of crushed rock will have to be hauled away. A heavy-duty dump truck can haul either 10 cubic yards of dirt or 6 cubic yards of crushed rock, while a regular dump truck can haul either 5 cubic yards of dirt or 4 cubic yards of crushed rock. The dirt and crushed rock cannot be mixed, because they go to different dumping sites. The union contract demands that this job use at least 400 loads carried in heavy-duty dump trucks. If each heavy-duty dump truck load costs $90 and each regular dump truck load costs $50, how many truckloads with each type of truck and cargo will be needed to complete the job at the least cost?

35. Highway Construction   Repeat Exercise 34 with the additional requirement that there can be no more than 800 truckloads hauled from the job site in order to get the project finished on time.

36. Purchasing   Repeat Problem 33 with the additional requirement that the office manager feels he must purchase at least $1200 worth of pens and pencils from Jack's Office Supplies because it gave him such a good deal on some filing cabinets last month.

37. Agriculture   A soil analysis of a farmer's field showed that he needs to apply at least 3000 pounds of nitrogen, 2400 pounds of phosphoric acid, and 2100 pounds of potash. Plant fertilizer is labeled with three numbers giving the percentages of nitrogen, phosphoric acid, and potash. The local farm supply store sells 15-30-15 Miracle Mix for 15¢ per pound and a 10-5-5 store brand for 8¢ per pound. How many pounds of each fertilizer should the farmer buy to meet the needs of the field at the least cost?

38. Gardening   A weekend gardener's vegetable patch needs at least 10.2 pounds of nitrogen, 7.8 pounds of phosphoric acid, and 6.6 pounds of potash. Plant fertilizer is labeled with three numbers giving the percentages of nitrogen, phosphoric acid, and potash. The local garden center sells 15-30-15 Miracle Mix for 16¢ per pound, 15-10-10 Grow Great for 13¢ per pound, and a 10-5-5 store brand for 8¢ per pound. How many pounds of each fertilizer should the gardener buy to meet the needs of the vegetable patch at the least cost?

39. Transportation   A retail store chain has cartons of goods stored at warehouses in Kentucky and Utah that must be distributed to its stores in Kansas, Texas, and Oregon. Each carton shipped from the Utah warehouse costs $2 whether it goes to Kansas, Texas, or Oregon. However, the cost to ship one carton from the Kentucky warehouse to Kansas is $2, to Texas is $4, and to Oregon is $5. There are 200 cartons at the Utah warehouse and 400 at the warehouse in Kentucky. If the Kansas stores need 200 cartons, the Texas stores need 300 cartons, and the Oregon stores need 100 cartons, how many cartons should be shipped from each warehouse to each state to incur the smallest shipping costs?

40. Transportation A soda distributor has warehouses in Seaford and Centerville and needs to supply stores in Huntington and Towson. The costs of shipping one case of sodas are 8¢ from Seaford to Huntington, 5¢ from Seaford to Towson, 6¢ from Centerville to Huntington, and 4¢ from Centerville to Towson. There are 800 cases in the Seaford warehouse and 1200 in the Centerville warehouse. If the Huntington stores need at least 1000 cases and the Towson stores need at least 600 cases, how many cases should be shipped from each warehouse to each city to incur the smallest shipping costs?

Explorations and Excursions

The following problems extend and augment the material presented in the text.

41. Why $P \leq C$ for Dual Problems (*Requires matrix algebra*) For the pair of dual problems

Maximize $P = c^t X$

and

Minimize $C = b^t Y$

Subject to $\begin{cases} AX \leq b \\ X \geq 0 \end{cases}$

Subject to $\begin{cases} A^t Y \geq c \\ Y \geq 0 \end{cases}$

show that $P \leq C$ for any feasible X and Y by justifying each = and ≤ in the chain of statements

$$P = c^t \cdot X = X^t \cdot c \leq X^t \cdot A^t Y$$

$$= (X^t A^t) \cdot Y = (AX)^t \cdot Y \leq b^t \cdot Y = C$$

42. (*Requires matrix algebra*) Check the chain of matrix statements in Exercise 41 for the matrices

$$A = \begin{pmatrix} 1 & 2 & 1 \\ 4 & 3 & -1 \end{pmatrix}, \quad b = \begin{pmatrix} 50 \\ 60 \end{pmatrix}, \quad c = \begin{pmatrix} 20 \\ 30 \\ 5 \end{pmatrix}, \quad X = \begin{pmatrix} x_1 \\ x_2 \\ x_3 \end{pmatrix},$$

and $Y = \begin{pmatrix} y_1 \\ y_2 \end{pmatrix}$ from Example 2 on page 328.

43. A Proof of the Duality Theorem (*Requires matrix algebra*) The following sequence of statements provides a proof of the Duality Theorem on page 327. Justify each statement to verify this proof.

a. If $mx + b$ is the same as $0x + v$ for every value of x, then $m = 0$ and $b = v$.

b. The dual problems in Exercise 41 may be rewritten as

Maximize $P = c^t X$

and

Minimize $C = b^t Y$

Subject to $\begin{cases} AX + S = b \\ X, S \geq 0 \end{cases}$

Subject to $\begin{cases} A^t Y - T = c \\ Y, T \geq 0 \end{cases}$

to clearly show both the variables (X and Y) and the slack variables (S and T).

c. The simplex tableau for the maximum problem (both the initial tableau and after *any* sequence of pivot operations) has numbers in the bottom row:

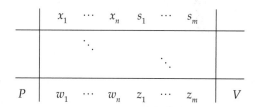

To prove the Duality Theorem, we must show that these values can be used for the values of Y and T in the minimum problem. That is, if we set $y_1 = z_1, \ldots, y_m = z_m$, and $t_1 = w_1, \ldots, t_n = w_n$, then the number V is the same as the value $C = b^t Y$ and the constraint $A^t Y - T = c$ is satisfied.

d. If we use the notations $W = \begin{pmatrix} w_1 \\ \vdots \\ w_n \end{pmatrix}$ and $Z = \begin{pmatrix} z_1 \\ \vdots \\ z_m \end{pmatrix}$,

the bottom row of the tableau is the same as the matrix equation $P + W^t X + Z^t S = V$, and this equation holds for *any* values of X and S such that $AX + S = b$.

e. Since $P = c^t X$ and $S = b - AX$, we may rewrite the matrix equation $P + W^t X + Z^t S = V$ as

$$c^t X + W^t X + Z^t (b - AX) = V.$$

f. That is, $X^t c + X^t W + b^t Z - (AX)^t Z = V$ and so

$$X^t \cdot (c + W - A^t Z) + b^t Z = V.$$

Thus $(c + W - A^t Z) = 0$ and $b^t Z = V$. So the number V is the value of the minimum problem objective function $C = b^t Y$ when Y takes the value Z and this value for Y satisfies the constraint $A^t Y - T = c$ when the slack T takes the value W.

44. Complementary Slackness Use the Duality Theorem for the simplex method (see page 327) to show that *at least* one of *every* pair of variables x_1 and t_1, x_2 and t_2, . . . , x_n and t_n, y_1 and s_1, y_2 and s_2, . . . , y_m and s_m is zero for the solution points of a linear programming problem and its dual. [*Hint:* What can you say about the bottoms of the basic variable columns in the simplex tableau of a maximum problem?]

45. Verify the Complementary Slackness Theorem (Exercise 44) for the final tableau given in Practice Problem 2 (see pages 330–331).

4.5 Nonstandard Problems: The Dual Pivot Element and the Two-Stage Method

This section and the next present alternative methods for solving nonstandard linear programming problems. The choice of which to cover is left to the instructor. The examples and exercises duplicate those in Section 4.6.

APPLICATION PREVIEW

Kantorovich and Production Planning

The earliest discussion and solution of what we now call "linear programming problems" was published in 1939 by L. V. Kantorovich, a young professor at Leningrad State University in the USSR. In his introduction, he pointed out that "[t]here are two ways of increasing the efficiency of the work of a shop, an enterprise, or a whole branch of industry. One way is by various improvements in technology. . . . The other way—thus far much less used—is improvement in the organization of planning and production. Here are included, for instance, such questions as the distribution of work among individual machines . . . , the correct distribution of orders among enterprises, the correct distribution of different kinds of raw materials, fuel, and other factors." He then discussed many examples and described a method of solution.

Kantorovich was careful to note the differences between a capitalist society and the Soviet system of economic planning and management. In particular, he included requirements that the machines be in use for all the available time and that the workers be fully occupied. He thus considered maximum linear programming problems that included "at least" inequalities as well as the familiar "at most" restrictions. For some reason, probably based on an ideological assertion that an abstract subject like mathematics could have no possible use in the economic problems of the Soviet Union, Kantorovich was ignored. Dantzig discovered his article while collecting references to linear programming problems and arranged for its translation and publication* in 1960. In 1975, Kantorovich and Koopmans shared the Nobel Prize in Economics "for their contributions to the theory of optimum allocation of resources."

Introduction

In this section we extend the simplex method to maximum problems that include constraints with $\geq$ inequalities as well as $\leq$ inequalities. Because a $\geq$ inequality can be changed into a $\leq$ inequality by multiplying through by -1, this extension is equivalent to dropping the requirement that $b \geq 0$ in the constraint $AX \leq b$.

The Dual Pivot Element

The origin is not a vertex for a nonstandard maximum problem because an inequality like $x_1 \geq 3$ separates the feasible region from the origin. The initial tableau begins at the origin so it is not feasible as well as not optimal. The question becomes: How can we move to a feasible point so that we can then work on making it optimal? On page 327 we saw that the pivot operations for a standard maximum problem make the dual minimum problem feasible. This gives the clue to the answer to our question: Pivoting on "dual pivot elements" would make the problem feasible, and then pivoting on (regular) pivot elements would make it optimal.

Since dual problems interchange the roles of rows and columns, the following definition is just the definition of the pivot element with the roles of the rows and columns reversed, and we must take the *largest* ratio instead of the smallest because we are now dividing by negative numbers.

Dual Pivot Element

> The *dual pivot row* is the row with the smallest negative entry in the rightmost column of the tableau (omitting the bottom row). If there is a tie, the dual pivot row is the uppermost such row.
>
> The *dual pivot column* is the column with the largest ratio found by dividing the bottom entry by the dual pivot row entry, omitting any column with a zero or positive dual pivot row entry and omitting the rightmost column. If there is a tie, the dual pivot column is the leftmost such column.
>
> The *dual pivot element* is the entry in the dual pivot row and the dual pivot column.

---

* L. V. Kantorovich, "Mathematical Methods of Organizing and Planning Production," *Management Science* 6:366–422, 1960.

EXAMPLE 1 **The Dual Pivot Element**

Find the dual pivot element in the following simplex tableau or explain why the tableau does not have a dual pivot element.

	x_1	x_2	x_3	x_4	s_1	s_2	s_3	
s_1	2	1	1	2	1	0	0	30
s_2	−1	−2	−1	−1	0	1	0	−5
s_3	−1	−1	−1	−2	0	0	1	−10
P	−8	12	−10	−14	0	0	0	0

Solution

The dual pivot row is row 3 because −10 is the smallest negative entry on the right (omitting the bottom row). The dual pivot column is column 3 because the ratio $\frac{-10}{-1} = 10$ is greater than the others (the remaining columns are not considered because their dual pivot row entries are zero or positive and the rightmost column is never considered).

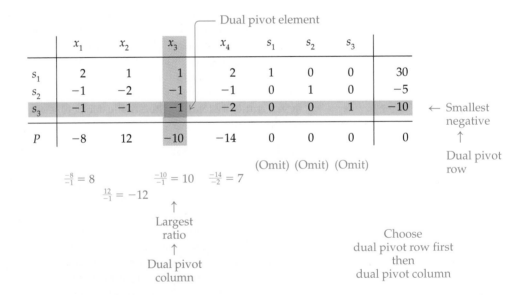

The dual pivot element is the −1 in row 3 and column 3.

PRACTICE PROBLEM 1

For each of the following simplex tableaux, find the dual pivot element or explain why the tableau does not have a dual pivot element.

a.

	x_1	x_2	x_3	x_4	s_1	s_2	s_3	
s_1	-1	-1	2	-2	1	0	0	-10
s_2	1	1	-1	-1	0	1	0	20
s_3	0	-1	3	2	0	0	1	-5
P	-5	-6	14	8	0	0	0	0

b.

	x_1	x_2	x_3	x_4	s_1	s_2	s_3	s_4	
s_1	1	1	5	1	1	0	0	0	25
s_2	-1	-1	-2	-1	0	1	0	0	-15
s_3	3	1	1	2	0	0	1	0	30
s_4	2	1	1	4	0	0	0	1	-20
P	-6	-8	10	-10	0	0	0	0	0

Solutions at the back of the book

The Two-Stage Simplex Method

Pivoting on dual pivot elements will make the tableau feasible. After the tableau becomes feasible, pivoting on *regular* pivot elements will lead to the solution of the problem.

Two-Stage Simplex Method

To solve any maximum problem by the simplex method:

1. Write the constraints in the form $AX \leq b$ with $X \geq 0$, and construct the initial simplex tableau.

2. If the tableau is not feasible (at least one basic variable has a negative value in the rightmost column), go to step 3. If the tableau is feasible (all the basic variables have nonnegative values in the rightmost column), go to step 4.

3. Locate the dual pivot element (page 341), perform the pivot operation (page 302), and return to step 2. If the tableau has a dual pivot row but no dual pivot column, then there is *no solution* to the problem (the constraints are infeasible).

(continued)

4. Locate the pivot element (pages 300–301), perform the pivot operation (page 302), and return to step 2. If the tableau does not have a pivot column, then the solution occurs at the vertex given by the basic variables, and the maximum value of the objective function appears in the bottom right corner of the tableau. If the tableau has a pivot column but no pivot row, then there is *no solution* to the problem (the region is unbounded, in a direction for which the objective function increases).

EXAMPLE 2   The Two-Stage Simplex Method

Solve the following nonstandard linear programming problem by the two-stage simplex method.

$$\text{Maximize } P = 5x_1 + 7x_2$$

$$\text{Subject to } \begin{cases} x_1 + x_2 \leq 8 \\ x_1 + x_2 \geq 4 \\ 2x_1 + x_2 \geq 6 \\ x_1 \geq 0 \text{ and } x_2 \geq 0 \end{cases}$$

"Mixed constraints" because they have $\leq$ and $\geq$ inequalities

Solution

We first change the inequalities from $\geq$ to $\leq$ by multiplying by -1 as needed: The second inequality becomes $-x_1 - x_2 \leq -4$ and the third becomes $-2x_1 - x_2 \leq -6$. This problem in matrix form is

$$\text{Maximize } P = (5 \quad 7)\begin{pmatrix} x_1 \\ x_2 \end{pmatrix}$$

$$\text{Subject to } \begin{cases} \begin{pmatrix} 1 & 1 \\ -1 & -1 \\ -2 & -1 \end{pmatrix}\begin{pmatrix} x_1 \\ x_2 \end{pmatrix} \leq \begin{pmatrix} 8 \\ -4 \\ -6 \end{pmatrix} \\ \text{and } \begin{pmatrix} x_1 \\ x_2 \end{pmatrix} \geq 0 \end{cases}$$

The initial simplex tableau is

	x_1	x_2	s_1	s_2	s_3	
s_1	1	1	1	0	0	8
s_2	-1	-1	0	1	0	-4
s_3	-2	-1	0	0	1	-6
P	-5	-7	0	0	0	0

$\left.\begin{array}{l} s_1 = 8 \\ s_2 = -4 \\ s_3 = -6 \end{array}\right\}$ Not feasible because some basic variables are negative

The tableau is not feasible because s_2 and s_3 are negative. The dual pivot row is row 3 (-6 is the smallest negative entry in the right column, omitting the bottom row). The dual pivot column is column 2 (the ratio $\frac{-7}{-1} = 7$ is greater than $\frac{-5}{-2} = 2.5$, and none of the other columns may be considered). The dual pivot element is the -1 in row 3 and column 2. Pivoting on this dual pivot element, the tableau becomes feasible:

	x_1	x_2	s_1	s_2	s_3	
s_1	-1	0	1	0	1	2
s_2	1	0	0	1	-1	2
x_2	2	1	0	0	-1	6
P	9	0	0	0	-7	42

$\left.\begin{array}{l} s_1 = 2 \\ s_2 = 2 \\ x_2 = 6 \end{array}\right\}$ Feasible because all basic variables are nonnegative

Not optimal because bottom row has a negative entry

If it were still not feasible, we would look for another dual pivot element. The tableau is now feasible but not optimal, so we look for a (regular) pivot element. The pivot column is column 5 (-7 is the only negative entry in the bottom row). The pivot row is row 1 (the other rows cannot be considered because their pivot column entries are negative). The pivot element is the 1 in column 5 and row 1. Pivoting on this pivot element, the tableau becomes optimal:

	x_1	x_2	s_1	s_2	s_3	
s_3	-1	0	1	0	1	2
s_2	0	0	1	1	0	4
x_2	1	1	1	0	0	8
P	2	0	7	0	0	56

$\left.\begin{array}{l} \\ \\ \end{array}\right\}$ Feasible: all are nonnegative

Optimal: all are nonnegative

This is the final tableau because it is both feasible and optimal. The maximum is $P = 56$ when $x_1 = 0$ and $x_2 = 8$.

∎

To compare this solution with the graphical method, we graph the feasible region with x_1 on the horizontal axis and x_2 on the vertical axis.

The five vertices and the values of the objective function are given in the table. The path of vertices visited by the two-stage simplex method is marked in bold.

Vertex (x_1, x_2)	$P = 5x_1 + 7x_2$
$(8, 0)$	40
$(4, 0)$	20
$(2, 2)$	24
$(0, 6)$	42
$(0, 8)$	56

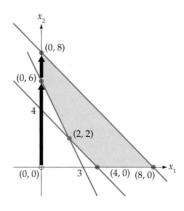

PRACTICE PROBLEM 2

Solve the following nonstandard linear programming problem by the two-stage simplex method.

$$\text{Maximize } P = 4x_1 + x_2 + 3x_3$$

$$\text{Subject to } \begin{cases} x_1 + 2x_2 + x_3 \leq 50 \\ 2x_1 + x_2 + 2x_3 \geq 10 \\ x_1 \geq 0, x_2 \geq 0, \text{ and } x_3 \geq 0 \end{cases}$$

Solution at the back of the book

EXAMPLE 3 **Managing an Investment Portfolio**

The manager of an $80 million mutual fund has three investment possibilities: secure government bonds yielding 4%, a blue chip growth stock paying 6%, and a biotechnology start-up company returning 12%. If the manager wants to invest at least $50 million in a combination of safe government bonds and blue chip stock, and at least as much in the bonds as the total in blue chip and biotechnology stocks, how much should be invested in each to obtain the greatest possible return?

Solution

Let

$$x_1 = \begin{pmatrix} \text{Amount invested in} \\ \text{government bonds} \end{pmatrix}$$

$$x_2 = \begin{pmatrix} \text{Amount invested in} \\ \text{the blue chip stock} \end{pmatrix}$$

$$x_3 = \begin{pmatrix} \text{Amount invested in the} \\ \text{biotechnology company} \end{pmatrix}$$

be given in millions of dollars. The objective is to maximize the return

$$P = 0.04x_1 + 0.06x_2 + 0.12x_3$$

Rate × amount
for each investment

subject to the restrictions

$$x_1 + x_2 + x_3 \le 80$$ $80 million available to invest

$$x_1 + x_2 \ge 50$$ $50 million in safe investments

$$x_1 \ge x_2 + x_3$$ Bonds ≥ total in stocks

$$x_1 \ge 0, x_2 \ge 0, x_3 \ge 0$$ Nonnegativity

Multiplying the second constraint by -1 and rewriting the third as $-x_1 + x_2 + x_3 \le 0$, we find that this problem in matrix form is

$$\text{Maximize } P = (0.04 \quad 0.06 \quad 0.12) \begin{pmatrix} x_1 \\ x_2 \\ x_3 \end{pmatrix}$$

$$\text{Subject to} \begin{cases} \begin{pmatrix} 1 & 1 & 1 \\ -1 & -1 & 0 \\ -1 & 1 & 1 \end{pmatrix} \begin{pmatrix} x_1 \\ x_2 \\ x_3 \end{pmatrix} \le \begin{pmatrix} 80 \\ -50 \\ 0 \end{pmatrix} \\ \\ \text{and } \begin{pmatrix} x_1 \\ x_2 \\ x_3 \end{pmatrix} \ge 0 \end{cases}$$

The initial simplex tableau is not feasible:

	x_1	x_2	x_3	s_1	s_2	s_3	
s_1	1	1	1	1	0	0	80
s_2	−1	−1	0	0	1	0	−50 $s_2 < 0$ ← Dual pivot row
s_3	−1	1	1	0	0	1	0
P	−0.04	−0.06	−0.12	0	0	0	0

We look for a dual pivot element. The dual pivot row is row 2 and the dual pivot column is column 2 (because the ratio $\frac{-0.06}{-1} = 0.06$ is greater than $\frac{-0.04}{-1} = 0.04$ and none of the other columns may be considered). Pivoting on the -1 in row 2 and column 2, the tableau is still not feasible:

	x_1	x_2	x_3	s_1	s_2	s_3	
s_1	0	0	1	1	1	0	30
x_2	1	1	0	0	−1	0	50
s_3	−2	0	1	0	1	1	−50 $s_3 < 0$ ← Dual pivot row
P	0.02	0	−0.12	0	−0.06	0	3

Pivoting on the dual pivot element in row 3 and column 1, the tableau becomes feasible but not optimal:

	x_1	x_2	x_3	s_1	s_2	s_3	
s_1	0	0	1	1	1	0	30
x_2	0	1	1/2	0	−1/2	1/2	25
x_1	1	0	−1/2	0	−1/2	−1/2	25
P	0	0	−0.11	0	−0.05	0.01	2.5

Feasible: all are ≥ 0

Not optimal: some are < 0

Pivoting on the (regular) pivot element in column 3 and row 1, the tableau becomes both feasible and optimal:

	x_1	x_2	x_3	s_1	s_2	s_3	
x_3	0	0	1	1	1	0	30
x_2	0	1	0	−1/2	−1	1/2	10
x_1	1	0	0	1/2	0	−1/2	40
P	0	0	0	0.11	0.06	0.01	5.8

Feasible: all are ≥ 0

Optimal: all are ≥ 0

This is the final tableau because it is both feasible and optimal. The maximum is $P = 5.8$ when $x_1 = 40$, $x_2 = 10$, and $x_3 = 30$. In terms of the original situation, the manager should invest $40 million in the government bonds, $10 million in the blue chip stock, and $30 million in the biotechnology start-up company for a maximum return of $5.8 million.

Graphing Calculator Exploration

Formulate the following situation as a linear programming problem. Use the program PIVOT to solve it by the two-stage simplex method.

The mutual fund has grown, and now the manager has $100 million to invest among three possibilities: secure government bonds still yielding 4%, a blue chip growth stock now paying 10% (and looking not so secure), and a biotechnology start-up company returning 14%. If the manager now decides to invest at least $60 million in safe government bonds and at least as much in the bonds as the total in blue chip and biotechnology stocks, how much should be invested in each to obtain the greatest possible return?

(*Answer:* The manager should invest $60 million in the government bonds, nothing in the blue chip stock, and $40 million in the biotechnology start-up company for a maximum return of $8 million.)

Can Any Linear Programming Problem Now Be Solved?

The two-stage simplex method allows us to solve *any* maximum linear programming problem, no matter what mixture of ≤ and ≥ inequalities are present in the constraints, provided that the variables are nonnegative. A direct way to change a minimum problem into a maximum problem is described in Exercises 41–42. Should the initial formulation of a problem include equality constraints, there are several ways to change the problem into a form that we can solve. Exercises 43–45 explore equality constraints by discussing several solutions of an example of Kantorovich, and additional equality constraint problems are given in Exercises 46–47. Exercises 48–50 explain how to reformulate a problem so that all the variables are nonnegative. Taken together, these techniques allow us to write *any* linear programming problem as a maximum problem with inequality constraints and nonnegative variables. Thus we may conclude that we are now able to solve *any* linear programming problem.

Section Summary

A nonstandard maximum problem has both ≥ and ≤ constraints. Change the ≥ inequalities into ≤ form by multiplying by −1, and then construct the initial simplex tableau. In general, this tableau will not be feasible (it will have negative numbers in the rightmost column) as well as not optimal (it will have negative numbers in the bottom row).

The dual pivot *row* is the row with the smallest negative entry in the rightmost column of the tableau (omitting the bottom row). The dual pivot *column* is the column with the largest ratio of the bottom entry divided by the dual pivot row entry, omitting any column with a zero or positive dual pivot row entry and omitting the rightmost column.

The two-stage simplex method pivots on dual pivot elements until the tableau is feasible and then continues to pivot on (regular) pivot elements until the tableau is optimal. The problem has no solution if the dual pivot column or the (regular) pivot row does not exist.

EXERCISES 4.5

Write each of the following nonstandard linear programming problems in the matrix form "maximize $P = c^t X$ subject to $AX \leq b$ and $X \geq 0$," construct the initial simplex tableau, and locate the dual pivot element.

1. Maximize $P = 15x_1 + 20x_2 + 18x_3$

Subject to $\begin{cases} 3x_1 + 2x_2 + 8x_3 \leq 96 \\ 5x_1 + x_2 + 6x_3 \geq 30 \\ x_1 \geq 0, x_2 \geq 0, x_3 \geq 0 \end{cases}$

2. Maximize $P = 60x_1 + 90x_2 + 30x_3$

Subject to $\begin{cases} 4x_1 + 5x_2 + x_3 \geq 40 \\ 10x_1 + 18x_2 + 3x_3 \leq 150 \\ x_1 \geq 0, x_2 \geq 0, x_3 \geq 0 \end{cases}$

3. Maximize $P = 6x_1 + 4x_2 + 6x_3$

Subject to $\begin{cases} 2x_1 + x_2 + 3x_3 \geq 30 \\ x_1 + x_2 + 2x_3 \geq 20 \\ x_1 \geq 0, x_2 \geq 0, x_3 \geq 0 \end{cases}$

4. Maximize $P = 9x_1 + 5x_2 + 4x_3$

Subject to $\begin{cases} 3x_1 + x_2 + x_3 \geq 40 \\ 2x_1 + 2x_2 + x_3 \geq 30 \\ x_1 \geq 0, x_2 \geq 0, x_3 \geq 0 \end{cases}$

5. Maximize $P = 20x_1 + 30x_2 + 10x_3$

Subject to $\begin{cases} x_1 + x_2 + x_3 \geq 8 \\ x_1 + 2x_2 + 3x_3 \leq 30 \\ x_1 + 2x_2 + x_3 \leq 18 \\ x_1 \geq 0, x_2 \geq 0, x_3 \geq 0 \end{cases}$

6. Maximize $P = 24x_1 + 18x_2 + 30x_3$

Subject to $\begin{cases} x_1 + x_2 + x_3 \geq 4 \\ x_1 + 2x_2 + 3x_3 \leq 15 \\ x_1 + 2x_2 + x_3 \leq 9 \\ x_1 \geq 0, x_2 \geq 0, x_3 \geq 0 \end{cases}$

7. Maximize $P = 3x_1 + 2x_2 + 5x_3 + 4x_4$

Subject to $\begin{cases} x_1 + x_2 + x_3 + x_4 \geq 30 \\ 2x_1 + 3x_2 + 2x_3 + x_4 \geq 20 \\ 4x_1 + 2x_2 + x_3 + 2x_4 \leq 80 \\ x_1 \geq 0, x_2 \geq 0, x_3 \geq 0, x_4 \geq 0 \end{cases}$

8. Maximize $P = 6x_1 + 8x_2 + 3x_3 + 5x_4$

Subject to $\begin{cases} x_1 + x_2 + x_3 + x_4 \geq 70 \\ 3x_1 + 2x_2 + 4x_3 + 3x_4 \geq 60 \\ 4x_1 + x_2 + 2x_3 + 5x_4 \leq 100 \\ x_1 \geq 0, x_2 \geq 0, x_3 \geq 0, x_4 \geq 0 \end{cases}$

9. Maximize $P = 2x_1 + 2x_2 + x_3$

Subject to $\begin{cases} x_1 + 2x_2 + x_3 \geq 40 \\ 2x_1 + x_2 + x_3 \geq 50 \\ 5x_1 + 3x_2 + 2x_3 \leq 120 \\ 3x_1 + x_2 + 2x_3 \leq 150 \\ x_1 \geq 0, x_2 \geq 0, x_3 \geq 0 \end{cases}$

10. Maximize $P = x_1 + 2x_2 + 3x_3$

Subject to $\begin{cases} x_1 + x_2 + 2x_3 \geq 30 \\ x_1 + 2x_2 + x_3 \geq 60 \\ 2x_1 + 5x_2 + 3x_3 \leq 150 \\ 2x_1 + 3x_2 + x_3 \leq 120 \\ x_1 \geq 0, x_2 \geq 0, x_3 \geq 0 \end{cases}$

Solve each of the following nonstandard linear programming problems by the two-stage simplex method. (Exercises 11, 12, 15, 16, 19, and 20 can also be solved by the graphical method.)

⊠ You may use the PIVOT program if permitted by your instructor.

11. Maximize $P = 4x_1 + x_2$

Subject to $\begin{cases} 3x_1 + 2x_2 \leq 120 \\ x_1 + x_2 \geq 50 \\ 2x_1 + x_2 \leq 60 \\ x_1 \geq 0, x_2 \geq 0 \end{cases}$

12. Maximize $P = x_1 + 4x_2$

Subject to $\begin{cases} 3x_1 + 2x_2 \leq 120 \\ x_1 + x_2 \leq 50 \\ 2x_1 + x_2 \geq 60 \\ x_1 \geq 0, x_2 \geq 0 \end{cases}$

13. Maximize $P = 30x_1 + 20x_2 + 28x_3$

Subject to $\begin{cases} 3x_1 + x_2 + 2x_3 \geq 30 \\ 4x_1 + x_2 + 3x_3 \leq 60 \\ x_1 \geq 0, x_2 \geq 0, x_3 \geq 0 \end{cases}$

14. Maximize $P = 18x_1 + 10x_2 + 20x_3$

Subject to $\begin{cases} 3x_1 + x_2 + 4x_3 \geq 30 \\ 2x_1 + x_2 + 5x_3 \leq 50 \\ x_1 \geq 0, x_2 \geq 0, x_3 \geq 0 \end{cases}$

15. Maximize $P = 4x_1 + 5x_2$

Subject to $\begin{cases} x_1 + 2x_2 \leq 12 \\ x_1 + x_2 \geq 15 \\ 2x_1 + x_2 \leq 12 \\ x_1 \geq 0, x_2 \geq 0 \end{cases}$

16. Maximize $P = 5x_1 + 4x_2$

Subject to $\begin{cases} 2x_1 + 3x_2 \leq 12 \\ x_1 + x_2 \geq 10 \\ 3x_1 + 2x_2 \leq 12 \\ x_1 \geq 0, x_2 \geq 0 \end{cases}$

17. Maximize $P = 15x_1 + 12x_2 + 18x_3$

Subject to $\begin{cases} 5x_1 + x_2 + 2x_3 \geq 30 \\ 2x_1 + x_2 + 3x_3 \leq 24 \\ x_1 \geq 0, x_2 \geq 0, x_3 \geq 0 \end{cases}$

18. Maximize $P = 4x_1 + 5x_2 + 6x_3$

Subject to $\begin{cases} 5x_1 + x_2 + 3x_3 \leq 30 \\ 2x_1 + x_2 + 2x_3 \geq 24 \\ x_1 \geq 0, x_2 \geq 0, x_3 \geq 0 \end{cases}$

19. Maximize $P = x_1 + x_2$

Subject to $\begin{cases} x_1 \leq 8 \\ x_2 \leq 5 \\ x_1 + 2x_2 \geq 6 \\ x_1 \geq 0, x_2 \geq 0 \end{cases}$

20. Maximize $P = x_1 + x_2$

Subject to $\begin{cases} x_1 \leq 6 \\ x_2 \leq 9 \\ 3x_1 + x_2 \geq 6 \\ x_1 \geq 0, x_2 \geq 0 \end{cases}$

21. Maximize $P = 2x_1 + 3x_2 + 2x_3$

Subject to $\begin{cases} 2x_1 + x_2 + x_3 \geq 20 \\ x_1 - x_2 + x_3 \leq 10 \\ x_1 \geq 0, x_2 \geq 0, x_3 \geq 0 \end{cases}$

22. Maximize $P = 6x_1 + 5x_2 + 3x_3$

Subject to $\begin{cases} 2x_1 + x_2 - x_3 \leq 20 \\ 3x_1 + 2x_2 + x_3 \geq 30 \\ x_1 \geq 0, x_2 \geq 0, x_3 \geq 0 \end{cases}$

23. Maximize $P = 4x_1 + 6x_2 + 12x_3 + 10x_4$

Subject to $\begin{cases} x_1 + x_2 + x_3 + x_4 \leq 60 \\ 2x_1 + x_2 + x_3 + 2x_4 \geq 10 \\ 2x_1 + 2x_2 + x_3 + x_4 \leq 100 \\ x_1 \geq 0, x_2 \geq 0, x_3 \geq 0, x_4 \geq 0 \end{cases}$

24. Maximize $P = 8x_1 - x_2 + 5x_3 + 14x_4$

Subject to $\begin{cases} x_1 + 2x_2 + 3x_4 \leq 55 \\ x_1 + x_2 + x_3 + 2x_4 \geq 25 \\ x_1 + 3x_3 + 3x_4 \leq 45 \\ x_1 \geq 0, x_2 \geq 0, x_3 \geq 0, x_4 \geq 0 \end{cases}$

25. Maximize $P = 5x_1 + 32x_2 + 3x_3 + 15x_4$

Subject to $\begin{cases} 5x_1 + 8x_2 + x_3 + 3x_4 \geq 18 \\ 2x_1 - x_2 + x_3 + x_4 \leq 10 \\ 3x_1 - 3x_2 + x_3 + 2x_4 \geq 6 \\ x_1 \geq 0, x_2 \geq 0, x_3 \geq 0, x_4 \geq 0 \end{cases}$

26. Maximize $P = 70x_1 + 12x_2 + 60x_3 + 20x_4$

Subject to $\begin{cases} 2x_1 + x_2 + x_3 + x_4 \leq 14 \\ 3x_1 + 3x_2 + x_3 + 2x_4 \leq 24 \\ 5x_1 + 3x_2 + 4x_3 + 2x_4 \geq 44 \\ x_1 \geq 0, x_2 \geq 0, x_3 \geq 0, x_4 \geq 0 \end{cases}$

27. Maximize $P = -x_1 + x_2 - 2x_3$

Subject to $\begin{cases} x_1 + x_2 + x_3 \leq 50 \\ x_1 + x_3 \geq 10 \\ x_2 + x_3 \geq 20 \\ x_1 \geq 0, x_2 \geq 0, x_3 \geq 0 \end{cases}$

28. Maximize $P = x_1 + 2x_2 + x_3$

Subject to $\begin{cases} x_1 + x_2 + x_3 \leq 30 \\ x_1 + x_2 \leq 25 \\ x_2 + x_3 \geq 15 \\ x_1 \geq 0, x_2 \geq 0, x_3 \geq 0 \end{cases}$

29. Maximize $P = 12x_1 + 2x_2 + 14x_3 + 8x_4$

Subject to $\begin{cases} x_1 + x_2 + x_3 + x_4 \geq 20 \\ 2x_1 + x_2 + 2x_3 + x_4 \leq 30 \\ x_3 + x_4 \geq 10 \\ x_1 \geq 0, x_2 \geq 0, x_3 \geq 0, x_4 \geq 0 \end{cases}$

30. Maximize $P = 8x_1 + 10x_2 + 6x_3 + 8x_4$

Subject to $\begin{cases} x_1 + x_2 + x_4 \leq 30 \\ 2x_1 + 2x_2 + x_3 + x_4 \leq 80 \\ x_1 + x_2 + x_3 + x_4 \geq 20 \\ x_1 \geq 0, x_2 \geq 0, x_3 \geq 0, x_4 \geq 0 \end{cases}$

APPLIED EXERCISES

Formulate each situation as a linear programming problem by identifying the variables, the objective function, and the constraints. Be sure to state clearly the meaning of each variable. Check that the problem is a nonstandard maximum problem, and then solve it by the two-stage simplex method. State your final answer in terms of the original question.

 You may use the PIVOT program if permitted by your instructor.

31. Financial Planning A retired couple want to invest their $20,000 life savings in bank certificates of deposit yielding 6% and Treasury bonds yielding 5%. If they want at least $5000 in each type of investment, how much should they invest in each to receive the greatest possible income?

32. Recycling Management A volunteer recycling center accepts both used paper and empty glass bottles, which it then sorts and sells to a reprocessing company. The center has room to accept a total of 800 crates of paper and glass each week and has 50 hours of volunteer help to do the sorting. Each crate of paper products takes 5 minutes to sort and sells for 8¢, while each crate of bottles takes 3 minutes to sort and sells for 7¢. To support the city's "grab that glass" recycle theme this week, the center wants to accept at least 600 crates of glass. How many crates of each should the center accept this week to raise the most money for its ecology scholarship fund?

33. Advertising The manager of a new mall may spend up to $18,000 on "grand opening" announcements in newspapers, on radio, and on TV. Each newspaper ad costs $300 and reaches 5000 readers, each one-minute radio commercial costs $800 and is heard by 13,000 listeners, and each 15-second TV spot costs $900 and is seen by 15,000 viewers. If the manager wants at least 5 newspaper ads and at least 20 radio commercials and TV spots combined, how many of each should be placed to reach the largest number of potential customers? (Ignore multiple exposures to the same customer.)

34. Advertising In the last few days before the election, a politician can afford to spend no more than $27,000 on TV advertisements and can arrange for no more than 10 ads. Each daytime ad costs $2000 and reaches 4000 viewers, each prime time ad costs $3000 and reaches 5000 viewers, and each late night ad costs $1000 and reaches 2000 viewers. To be sure to reach the widest variety of voters, the politician's advisors insist on at least 5 ads scheduled during daytime and late night combined. Ignoring repeated viewings by the same person and assuming every viewer can vote, how many ads in each of the time periods will reach the most voters?

35. Resource Allocation A furniture shop manufactures wooden desks, tables, and chairs. The numbers of hours to assemble and finish each piece are shown in the table, together with the number of hours of skilled labor available for each task. To meet expected demand, a total of at least 30 desks and tables combined must be made. If the profit is $75 for each desk, $84 for each table, and $66 for each chair, how many of each should the company manufacture to obtain the greatest possible profit?

	Desk	Table	Chair	Labor Available
Assembly	2 hours	1 hour	2 hours	210 hours
Finishing	2 hours	3 hours	1 hour	150 hours

36. Production Planning An automotive parts shop rebuilds carburetors, fuel pumps, and alternators. The numbers of hours to rebuild and then inspect and pack each part are shown in the table, together with the number of hours of skilled labor available for each task. At least 12 carburetors must be rebuilt. If the profit is $12 for each carburetor, $14 for each fuel pump, and $10 for each alternator, how many of each should the shop rebuild to obtain the greatest possible profit?

	Carburetor	Fuel Pump
Rebuilding	5 hours	4 hours
Inspection and packaging	1 hour	1 hour

	Alternator	Labor Available
Rebuilding	3 hours	200 hours
Inspection and packaging	0.5 hour	45 hours

37. Agriculture A farmer grows wheat, barley, and oats on her 500-acre farm. Each acre of wheat requires 3 days of labor (to plant, tend, and harvest) and costs $21 (for seed, fertilizer, and pesticides), each acre of barley requires 2 days of labor and costs $27, and each acre of oats requires 3 days of labor and costs $24. The farmer and her hired field hands can provide no more than 1200 days of labor this year. She wants to grow at least 100 acres of oats and can afford to spend no more than $15,120. If the profit is $50 for each acre of wheat, $40 for each acre of barley, and $45 for each acre of oats, how many acres of each crop should she grow to obtain the greatest possible profit?

38. Agriculture A farmer grows corn, peanuts, and soybeans on his 240-acre farm. To maintain soil fertility, the farmer rotates the crops and always plants at least as many acres of soybeans as the total acres of the other crops. Because he has promised to sell some of his corn to a neighbor who raises cattle, he must plant at least 42 acres of corn. Each acre of corn requires 2 days of labor and yields a profit of $150, each acre of peanuts requires 5 days of labor and yields a profit of $300, and each acre of soybeans requires 1 day of labor and yields a profit of $100. If the farmer and his children can put in at most 630 days of labor, how many acres of each crop should the farmer plant to obtain the greatest possible profit?

39. Pollution Control An empty storage yard at a coal burning electric power plant can hold no more than 100,000 tons of coal. Two grades of coal are available: low sulfur (1%) with an energy content of 20 million BTU per ton and high sulfur (2%) with an energy content of 30 million BTU per ton. If existing contracts with the high-sulfur coal mine operator require that at least 50,000 tons of high sulfur coal be purchased, and if the next coal purchase may contain no more than 1400 tons of sulfur, how many tons of each type of coal should be purchased to obtain the most energy?

40. Production Planning A small jewelry company prepares and mounts semiprecious stones. There are 20 lapidaries (who cut and polish the stones) and 24 jewelers (who mount the stones in gold settings). Each employee works 7 hours each day, 5 days each week. Each tray of agates requires 5 hours of cutting and polishing and 4 hours to mount, each tray of onyxes requires 2 hours of cutting and polishing and 3 hours to mount, and each tray of garnets requires 6 hours of cutting and polishing and 3 hours to mount. Furthermore, the company's owner has decided that the company will process at least 12 trays of agates each week. If the profit is $15 for each tray of agates, $10 for each tray of onyxes, and $13 for each tray of garnets, how many trays of each stone should be processed each week to obtain the greatest possible profit?

Explorations and Excursions

The following problems extend and augment the material presented in the text.

 The PIVOT program may be helpful (if permitted by your instructor).

More About the Dual Pivot Element

41. a. Solve both of the following problems by the graphical method from Section 4.2. Is it true that the minimum of C is the same as -1 times the maximum of $P = -C$?

$$\text{Minimize } C = 3x + 4y$$
$$\text{Subject to } \begin{cases} x + y \le 10 \\ 2x + y \ge 8 \\ x \ge 0 \text{ and } y \ge 0 \end{cases}$$

$$\text{Maximize } P = -3x - 4y$$
$$\text{Subject to } \begin{cases} x + y \le 10 \\ 2x + y \ge 8 \\ x \ge 0 \text{ and } y \ge 0 \end{cases}$$

b. Solve the following problem by finding the dual maximum problem and using the (regular) simplex method as we did in Section 4.4.

$$\text{Minimize } C = 3y_1 + 4y_2$$
$$\text{Subject to } \begin{cases} -y_1 - y_2 \ge -10 \\ 2y_1 + y_2 \ge 8 \\ y_1 \ge 0 \text{ and } y_2 \ge 0 \end{cases}$$

c. Solve the following nonstandard problem by the two-stage simplex method.

$$\text{Maximize } P = -3x_1 - 4x_2$$

$$\text{Subject to } \begin{cases} x_1 + x_2 \le 10 \\ -2x_1 - x_2 \le -8 \\ x_1 \ge 0 \text{ and } x_2 \ge 0 \end{cases}$$

d. Compare the tableaux and pivot elements in parts (b) and (c). Are the pivot elements the same numbers? Is the arithmetic to choose each pair of pivot elements the same? Is the "dual pivot element" just the "dual" of the (regular) pivot element?

42. Repeat the process of Exercise 41 to find "dual" solutions to the following problems.

$$\text{Minimize } C = 20y_1 + 18y_2$$

$$\text{Subject to } \begin{cases} 2y_1 + y_2 \ge 18 \\ y_1 + y_2 \ge 14 \\ 4y_1 + 3y_2 \ge 48 \\ y_1 \ge 0 \text{ and } y_2 \ge 0 \end{cases}$$

$$\text{Maximize } P = -20x_1 - 18x_2$$

$$\text{Subject to } \begin{cases} -2x_1 - x_2 \le -18 \\ -x_1 - x_2 \le -14 \\ -4x_1 - 3x_2 \le -48 \\ x_1 \ge 0 \text{ and } x_2 \ge 0 \end{cases}$$

Kantorovich's First Example

The first example from the 1939 paper "Mathematical Methods for Organizing and Planning Production" by L. V. Kantorovich may be phrased as follows:

A machine shop has three types of equipment that can produce either or both of two kinds of parts (see the table). These parts are used to make finished items containing one of each, so the shop must produce the same number of both parts. Furthermore, the equipment must never be idle. How should the production be divided so that the greatest number of finished items are produced?

Productivity of the Machines for Two Parts

Type of Machine	Number of Machines	Output per Machine	
		First Part	Second Part
Milling machines	3	10	20
Turret lathes	3	20	30
Automatic lathe	1	30	80

Type of Machine	Number of Machines	Total Output	
		First Part	Second Part
Milling machines	3	30	60
Turret lathes	3	60	90
Automatic lathe	1	30	80

Let x_1 represent the fraction of the work day that the milling machines make the first part, so that they make the second part for the remaining $1 - x_1$ fraction of the work day. Similarly, let x_2 and x_3 be the corresponding fractions for the turret lathes and the automatic lathe. We must have that $0 \le x_1 \le 1$, $0 \le x_2 \le 1$, and $0 \le x_3 \le 1$ for these fractions. The number of first parts produced is then $30x_1 + 60x_2 + 30x_3$, and since the remaining time is spent making the second part, the number of second parts produced is $60(1 - x_1) + 90(1 - x_2) + 80(1 - x_3)$. In order to make the finished items without leftover parts, these must be equal:

$$30x_1 + 60x_2 + 30x_3 =$$
$$60(1 - x_1) + 90(1 - x_2) + 80(1 - x_3)$$

That is,

$$9x_1 + 15x_2 + 11x_3 = 23 \qquad \text{Divide by 10 and combine terms}$$

Since the number of finished parts will be $30x_1 + 60x_2 + 30x_3$, we have the linear programming problem

$$\text{Maximize } P = 30x_1 + 60x_2 + 30x_3$$

$$\text{Subject to } \begin{cases} x_1 \le 1 \\ x_2 \le 1 \\ x_3 \le 1 \\ 9x_1 + 15x_2 + 11x_3 = 23 \quad \leftarrow \text{Equality constraint} \\ x_1 \ge 0, x_2 \ge 0, \text{ and } x_3 \ge 0 \end{cases}$$

Exercises 43–45 explore three possible ways to solve this problem with an equality constraint.

43. One way to solve Kantorovich's First Example is to just go ahead and write down the initial simplex tableau for the constraints just as they are:

	x_1	x_2	x_3	s_1	s_2	s_3	
s_1	1	0	0	1	0	0	1
s_2	0	1	0	0	1	0	1
s_3	0	0	1	0	0	1	1
??	9	15	11	0	0	0	23
P	−30	−60	−30	0	0	0	0

a. This tableau does not display a basis, so we must select one of the nonbasic variables to replace the "??" in the basis listed on the left of the tableau. The "best" variable to enter the basis is the one with the smallest negative value in the bottom row. Show that pivoting on the 15 in column 2 and row 4 of the tableau will enter x_2 into the basis and result in an optimal but not feasible tableau.

b. Find the dual pivot element in the tableau from part (a) and pivot to show that the maximum occurs when $x_1 = 8/9$, $x_2 = 1$, and $x_3 = 0$. (In terms of Kantorovich's original problem, the milling machines should produce the first part $\frac{8}{9}$ of the time and the second part the remaining $\frac{1}{9}$, the turret lathes should produce the first part all of the time, and the automatic lathe should produce the second part all of the time.)

44. Since the linear equation $9x_1 + 15x_2 + 11x_3 = 23$ is the same as the two linear inequalities $9x_1 + 15x_2 + 11x_3 \leq 23$ and $9x_1 + 15x_2 + 11x_3 \geq 23$, another way of finding the answer to Kantorovich's First Example is to use the two-stage simplex method to solve the nonstandard problem

$$\text{Maximize } P = 30x_1 + 60x_2 + 30x_3$$

$$\text{Subject to } \begin{cases} x_1 \leq 1 \\ x_2 \leq 1 \\ x_3 \leq 1 \\ 9x_1 + 15x_2 + 11x_3 \leq 23 \\ 9x_1 + 15x_2 + 11x_3 \geq 23 \\ x_1 \geq 0, x_2 \geq 0, \text{ and } x_3 \geq 0 \end{cases}$$

Carry out the solution using the two-stage simplex method and check that it agrees with the solution found in Exercise 43.

45. Another approach to Kantorovich's First Example is to solve the equation $9x_1 + 15x_2 + 11x_3 = 23$ for x_3 and use this information to rewrite the problem in just x_1 and x_2.

a. Show that $x_3 = \frac{23}{11} - \frac{9}{11}x_1 - \frac{15}{11}x_2$ and that Kantorovich's First Example may be rewritten (remember that $x_3 \geq 0$) as

$$\text{Maximize } P = \tfrac{1}{11}(60x_1 + 210x_2 + 690)$$

$$\text{Subject to } \begin{cases} x_1 \leq 1 \\ x_2 \leq 1 \\ -9x_1 - 15x_2 \leq -12 \\ 9x_1 + 15x_2 \leq 23 \\ x_1 \geq 0 \text{ and } x_2 \geq 0 \end{cases}$$

b. Explain how the solution to the problem in part (a) can be found from the solution of the problem

$$\text{Maximize } P = 2x_1 + 7x_2$$

$$\text{Subject to } \begin{cases} x_1 \leq 1 \\ x_2 \leq 1 \\ -3x_1 - 5x_2 \leq -4 \\ 9x_1 + 15x_2 \leq 23 \\ x_1 \geq 0 \text{ and } x_2 \geq 0 \end{cases}$$

c. Solve the problem in part (b) by the two-stage simplex method.

d. Use the solution found in part (c) to find the solution of the problem in part (a) and then deduce the solution to Kantorovich's First Example. Does this solution agree with the solutions found in Exercises 43 and 44?

46. Use the methods discussed in Exercises 43–45 to solve the following equality constraint linear programming problem. Check your answer by solving this problem by the graphical method from Section 4.2.

$$\text{Maximize } P = 5x_1 + 3x_2$$

$$\text{Subject to } \begin{cases} 2x_1 + x_2 = 6 \\ x_1 + x_2 \leq 4 \\ x_1 \geq 0 \text{ and } x_2 \geq 0 \end{cases}$$

47. Use the methods discussed in Exercises 43–45 to solve the following equality constraint linear programming problem.

Maximize $P = 5x_1 + 4x_2 + 6x_3 + 3x_4$

Subject to $\begin{cases} 2x_1 + x_2 + x_3 \le 40 \\ 2x_1 + 2x_3 + x_4 \le 60 \\ x_1 + x_2 = 20 \\ x_3 + x_4 = 30 \\ x_1 \ge 0, x_2 \ge 0, x_3 \ge 0, \text{ and } x_4 \ge 0 \end{cases}$

More About Nonnegativity Conditions The following problems show how to reformulate linear programming problems to have nonnegativity conditions on *all* the variables when the original versions do not.

48. Show that the first problem may be rewritten as the second by replacing the variable x_1 by a new variable u_1 where $u_1 = x_1 + 8$ and $u_1 \ge 0$. Solve the first problem by the graphical method from Section 4.2, solve the second by the simplex method from Section 4.3, and check that your answers agree.

Maximize $P = 5x_1 + 3x_2$

Subject to $\begin{cases} 2x_1 + x_2 \le 6 \\ x_1 + x_2 \le 4 \\ x_1 \ge -8 \text{ and } x_2 \ge 0 \end{cases}$

Maximize $P = 5u_1 + 3x_2 - 40$

Subject to $\begin{cases} 2u_1 + x_2 \le 22 \\ u_1 + x_2 \le 12 \\ u_1 \ge 0 \text{ and } x_2 \ge 0 \end{cases}$

49. Show that the first problem may be rewritten as the second by replacing the variable x_1 by a new variable u_1 where $u_1 = -x_1$ and $u_1 \ge 0$. Solve the first

problem by the graphical method from Section 4.2, solve the second by the simplex method from Section 4.3, and check that your answers agree.

Maximize $P = 5x_1 + 3x_2$

Subject to $\begin{cases} 2x_1 + x_2 \le 10 \\ -x_1 + x_2 \le 4 \\ x_1 \le 0 \text{ and } x_2 \ge 0 \end{cases}$

Maximize $P = -5u_1 + 3x_2$

Subject to $\begin{cases} -2u_1 + x_2 \le 10 \\ u_1 + x_2 \le 4 \\ u_1 \ge 0 \text{ and } x_2 \ge 0 \end{cases}$

50. A variable is *unrestricted* if it may take *any* value (positive, negative, or zero). Show that the first problem may be rewritten as the second by replacing the variable x_1 by two new variables u_1 and u_2 where $u_1 - u_2 = x_1$ and $u_1, u_2 \ge 0$. Solve the first problem by the graphical method from Section 4.2, solve the second by the simplex method from Section 4.3, and check that your answers agree.

Maximize $P = 3x_1 + 5x_2$

Subject to $\begin{cases} 2x_1 + x_2 \le 6 \\ x_2 \le 4 \\ x_1 \text{ unrestricted and } x_2 \ge 0 \end{cases}$

Maximize $P = 3u_1 - 3u_2 + 5x_2$

Subject to $\begin{cases} 2u_1 - 2u_2 + x_2 \le 6 \\ x_2 \le 4 \\ u_1 \ge 0, u_2 \ge 0, \text{ and } x_2 \ge 0 \end{cases}$

[*Hint:* Can any real number be written as the difference of two nonnegative numbers?]

Nonstandard Problems: Artificial Variables and the Big-M Method

This and the previous section present alternative methods for solving non-standard linear programming problems. The choice of which to cover is left to the instructor. The examples and exercises duplicate those in Section 4.5.

Kantorovich and Production Planning

The earliest discussion and solution of what we now call "linear programming problems" was published in 1939 by L.V. Kantorovich, a young professor at Leningrad State University in the USSR. In his introduction, he pointed out that "[t]here are two ways of increasing the efficiency of the work of a shop, an enterprise, or a whole branch of industry. One way is by various improvements in technology. . . . The other way—thus far much less used—is improvement in the organization of planning and production. Here are included, for instance, such questions as the distribution of work among individual machines . . . , the correct distribution of orders among enterprises, the correct distribution of different kinds of raw materials, fuel and other factors." He then discussed many examples and described a method of solution.

Kantorovich was careful to note the differences between a capitalist society and the Soviet system of economic planning and management. In particular, he included requirements that the machines be in use for all the available time and that the workers be fully occupied. He thus considered maximum linear programming problems that included "at least" inequalities as well as the familiar "at most" restrictions. For some reason, probably based on an ideological assertion that an abstract subject like mathematics could have no possible use in the economic problems of the Soviet Union, Kantorovich was ignored. Dantzig discovered his article while collecting references to linear programming problems and arranged for its translation and publication* in 1960. In 1975, Kantorovich and Koopmans shared the Nobel Prize in economics "for their contributions to the theory of optimum allocation of resources."

* L.V. Kantorovich, "Mathematical Methods of Organizing and Planning Production," *Management Science* **6**: 366–422, 1960.

Introduction

In this section we extend the simplex method to maximum problems with $\geq$ and $=$ as well as the usual $\leq$ constraints. Because a basis for such a problem is not necessarily feasible or even immediately apparent, we will modify the problem to obtain a basic feasible point in the initial tableau and to force the simplex method to find a basic feasible point for the original problem if at all possible.

Artificial Variables

Since multiplying an inequality by -1 reverses the sense of the inequality ($5 \geq -3$ becomes $-5 \leq 3$), each constraint of a linear programming problem may be written with a nonnegative constant term. Expanding our notion of constraint to include equalities, we can write any nonstandard maximum linear programming problem in the following form:

Maximum Problem in Normal Form

Maximize $P = c_1 x_1 + c_2 x_2 + \cdots + c_n x_n$

Subject to
$$\begin{cases} a_{1,1}x_1 + a_{1,2}x_2 + \cdots + a_{1,n}x_n \leq b_1 \\ \quad\quad\quad\ddots \\ a_{k,1}x_1 + a_{k,2}x_2 + \cdots + a_{k,n}x_n \geq b_k \\ \quad\quad\quad\ddots \\ a_{m,1}x_1 + a_{m,2}x_2 + \cdots + a_{m,n}x_n = b_m \\ x_1 \geq 0, x_2 \geq 0, \dots, \text{and } x_n \geq 0 \end{cases}$$

$\leq$, $\geq$, and $=$ constraints allowed, but constant terms must be nonnegative

where $b_1 \geq 0$, $b_2 \geq 0$, ..., and $b_m \geq 0$

Although each $\leq$ constraint leads naturally to a slack variable that can be used in the basis for a basic feasible point, a $\geq$ constraint does not: $x_1 + x_2 \geq 3$ leads to $x_1 + x_2 - s = 3$, and $s = -3$ does not satisfy $s \geq 0$. Even worse, each $=$ constraint needs no slack variable and thus makes no natural contribution to a feasible basis.

To obtain a basis representing a basic feasible point in the initial tableau, we expand the original problem by introducing *artificial variables*, one for each of the $\geq$ and $=$ constraints. These new variables are "artificial" because they have no interpretation in terms of the original problem and are introduced only to obtain an initial basic feasible point. We will write a_k for the artificial variable inserted into the kth constraint. We could simply introduce artificial variables into every constraint, but we find it easier for hand calculations if we only insert artificial variables into the $\geq$ and $=$ constraints (and then build a feasible basis from them and the slack variables coming from the $\leq$ constraints). To ensure that the simplex method removes these new variables from the basis and never returns to them, we subtract a large multiple of each from the objective function. Writing M for this multiplier, we can write the expanded problem as follows:

$$\text{Maximize } P = c_1 x_1 + c_2 x_2 + \cdots + c_n x_n - M a_k - \cdots - M a_m$$

Subtract M times each artificial variable from the objective function

$$\text{Subject to } \begin{cases} a_{1,1} x_1 + a_{1,2} x_2 + \cdots + a_{1,n} x_n + s_1 = b_1 \\ \qquad\qquad \ddots \\ a_{k,1} x_1 + a_{k,2} x_2 + \cdots + a_{k,n} x_n - s_k + a_k = b_k \\ \qquad\qquad \ddots \\ a_{m,1} x_1 + a_{m,2} x_2 + \cdots + a_{m,n} x_n + a_m = b_m \\[4pt] x_1 \geq 0, x_2 \geq 0, \dots, \text{ and } x_n \geq 0 \\ s_1 \geq 0, \dots, s_k \geq 0, \dots, \text{ and} \\ a_k \geq 0, \dots, a_m \geq 0 \end{cases}$$

Change each $\leq$ to $=$ by adding a slack variable

Change each $\geq$ to $=$ by subtracting a slack variable and adding an artificial variable

Change each $=$ by adding an artificial variable

Every variable is nonnegative

where $b_1 \geq 0, b_2 \geq 0, \dots,$ and $b_m \geq 0$ and M is a positive number larger than the absolute value of any value in every possible simplex tableau for this problem.

It is an immediate and fair question to ask whether such a number M exists and, if so, how can it be estimated. Because there is a finite number of variables and constraints, there are a finite number of basic feasible points, and so M may be any number that is larger than the absolute values of a finite list of numbers. If the solution is to be calculated on a computer, a number near the largest possible value the computer can represent can be used for M (if this doesn't suffice, the solution can't be calculated on the computer anyway). This large "machine number" value for M explains the name given to this use of artificial variables: "big-M method."

Initial Tableau

The tableau for the expanded version of the maximum problem is the following *preliminary* tableau.

	x_1	$\cdots$	x_n	s_1	$\cdots$	s_k	$\cdots$	a_k	$\cdots$	a_m		
s_1	$a_{1,1}$	$\cdots$	$a_{1,n}$	1	$\cdots$	0	$\cdots$	0	$\cdots$	0	b_1	Each $\le$ constraint has $+1$ slack variable
$\vdots$	$\vdots$	$\ddots$	$\vdots$	$\vdots$	$\ddots$	$\vdots$	$\ddots$	$\vdots$	$\ddots$	$\vdots$	$\vdots$	
a_k	$a_{k,1}$	$\cdots$	$a_{k,n}$	0	$\cdots$	-1	$\cdots$	1	$\cdots$	0	b_k	Each $\ge$ constraint has -1 slack variable and $+1$ artificial variable
$\vdots$	$\vdots$	$\ddots$	$\vdots$	$\vdots$	$\ddots$	$\vdots$	$\ddots$	$\vdots$	$\ddots$	$\vdots$	$\vdots$	
a_m	$a_{m,1}$	$\cdots$	$a_{m,n}$	0	$\cdots$	0	$\cdots$	0	$\cdots$	1	b_m	Each $=$ constraint has $+1$ artificial variable
P	$-c_1$	$\cdots$	$-c_n$	0	$\cdots$	0	$\cdots$	M	$\cdots$	M	0	

Before we can use $s_1, \ldots, a_k, \ldots, a_m$ as the basis for a basic feasible point, the M's at the bottom of the artificial-variable columns must be removed. We can bring the artificial variables into the basis by pivoting on their 1's, but this requires only that we subtract M times those rows from the bottom row.

Initial Tableau

The initial simplex tableau for a maximum problem in normal form is

	x_1	$\cdots$	x_n	s_1	$\cdots$	s_k	$\cdots$	a_k	$\cdots$	a_m	
s_1	$a_{1,1}$	$\cdots$	$a_{1,n}$	1	$\cdots$	0	$\cdots$	0	$\cdots$	0	b_1
$\vdots$	$\vdots$	$\ddots$	$\vdots$	$\vdots$	$\ddots$	$\vdots$	$\ddots$	$\vdots$	$\ddots$	$\vdots$	$\vdots$
a_k	$a_{k,1}$	$\cdots$	$a_{k,n}$	0	$\cdots$	-1	$\cdots$	1	$\cdots$	0	b_k
$\vdots$	$\vdots$	$\ddots$	$\vdots$	$\vdots$	$\ddots$	$\vdots$	$\ddots$	$\vdots$	$\ddots$	$\vdots$	$\vdots$
a_m	$a_{m,1}$	$\cdots$	$a_{m,n}$	0	$\cdots$	0	$\cdots$	0	$\cdots$	1	b_m
P	$-v_1 M - c_1$	$\cdots$	$-v_n M - c_n$	0	$\cdots$	M	$\cdots$	0	$\cdots$	0	$-v_b M$

where each value v_i is the sum of the coefficients in the column above it that appear in a row with an artificial variable.

EXAMPLE 1   **Initial Tableau Using Artificial Variables**

Construct the preliminary and initial simplex tableaux for the non-standard problem

$$\text{Maximize } P = 8x_1 + 12x_2 - 10x_3 + 14x_4$$

$$\text{Subject to } \begin{cases} 2x_1 + x_2 + x_3 + 2x_4 \leq 30 \\ x_1 + 2x_2 + x_3 + x_4 \geq 5 \\ x_1 + x_2 + x_3 + 2x_4 = 10 \\ x_1 \geq 0, x_2 \geq 0, x_3 \geq 0, \text{ and } x_4 \geq 0 \end{cases}$$

Solution

Because the 30, 5, and 10 are all nonnegative, this maximum problem is in normal form. The slack variable s_1 for the first constraint can be used in the initial basis, but the second and third constraints require the introduction of artificial variables a_2 and a_3. The preliminary tableau is

	x_1	x_2	x_3	x_4	s_1	s_2	a_2	a_3	
s_1	2	1	1	2	1	0	0	0	30
a_2	1	2	1	1	0	-1	1	0	5
a_3	1	1	1	2	0	0	0	1	10
P	-8	-12	10	-14	0	0	M	M	0

Subtracting M times the sum of the column entries in each row that has an artificial variable from the objective function gives the initial tableau

	x_1	x_2	x_3	x_4	s_1	s_2	a_2	a_3	
s_1	2	1	1	2	1	0	0	0	30
a_2	1	2	1	1	0	-1	1	0	5
a_3	1	1	1	2	0	0	0	1	10
P	$-2M - 8$	$-3M - 12$	$-2M + 10$	$-3M - 14$	0	M	0	0	$-15M$

Notice that the names of the artificial variables correspond to their rows. Of course, it is not necessary to place the constraints in $\leq$, $\geq$, $=$ order before constructing the tableau, and with a little practice you can directly write down an initial tableau from the problem without first writing out the preliminary tableau.

PRACTICE PROBLEM 1

Construct the initial simplex tableau for the nonstandard problem

$$\text{Maximize } P = 5x_1 + 6x_2 - 14x_3 - 8x_4$$

$$\text{Subject to } \begin{cases} x_1 + x_2 - 2x_3 + 2x_4 \geq 10 \\ x_1 + x_2 - x_3 - x_4 \leq 20 \\ x_2 - 3x_3 - 2x_4 \geq 5 \\ x_1 \geq 0,\, x_2 \geq 0,\, x_3 \geq 0, \text{ and } x_4 \geq 0 \end{cases}$$

Solution at the back of the book

The Big-M Method

The big-M method of solving a nonstandard maximum linear programming problem begins by introducing the appropriate artificial variables to form an initial simplex tableau that displays a basic feasible point. If the solution is carried out on a computer, a large value is used in place of the symbol M, and then the extended problem is solved by the usual simplex method. If the solution is carried out by hand, the usual simplex method is applied to solve the extended problem, but the calculations for the bottom row are done *symbolically* and expressed in terms of the large but otherwise unspecified value M. The results of solving this extended problem give the solution of the original problem.

The Big-M Method

To solve any maximum linear programming problem by the big-M method:

1. Write the problem in normal form (page 358).

2. Construct the initial simplex tableau for the extended problem using the appropriate artificial variables (page 360).

3. Solve the extended problem by the simplex method (page 306).

4. If there is a solution of the extended problem *and* every artificial variable is *zero*, then the values of the original variables give the solution of the original problem. Otherwise, there is no solution to the original problem.

EXAMPLE 2 **The Big-M Method**

Solve the following nonstandard linear programming problem by the big-M method.

$$\text{Maximize } P = 5x_1 + 7x_2$$

$$\text{Subject to } \begin{cases} x_1 + x_2 \le 8 \\ x_1 + x_2 \ge 4 \\ 2x_1 + x_2 \ge 6 \\ x_1 \ge 0 \text{ and } x_2 \ge 0 \end{cases}$$

Solution

Because the constant terms 8, 4, and 6 in the constraints are nonnegative, we leave the inequalities as they are given. The first slack s_1 can be used in the initial basic feasible point, but the second and third cannot, so we need two artificial variables, a_2 and a_3:

	x_1	x_2	s_1	s_2	s_3	a_2	a_3	
s_1	1	1	1	0	0	0	0	8
a_2	1	1	0	-1	0	1	0	4
a_3	2	1	0	0	-1	0	1	6
P	$-3M - 5$	$-2M - 7$	0	M	M	0	0	$-10M$

The pivot column is column 1 (because $-3M - 5$ is smaller than $-2M - 7$ for large positive values of M), and the pivot row is row 3 (because $\frac{6}{2} = 3$ is smaller than $\frac{8}{1} = 8$ and $\frac{4}{1} = 4$). Pivoting on the 2 in the first column and third row, we obtain

	x_1	x_2	s_1	s_2	s_3	a_2	a_3	
s_1	0	1/2	1	0	1/2	0	$-1/2$	5
a_2	0	1/2	0	-1	1/2	1	$-1/2$	1
x_1	1	1/2	0	0	$-1/2$	0	1/2	3
P	0	$-\frac{1}{2}M - \frac{9}{2}$	0	M	$-\frac{1}{2}M - \frac{5}{2}$	0	$\frac{3}{2}M + \frac{5}{2}$	$-M + 15$

The bottom row of this new tableau was found by multiplying the new pivot row by $3M + 5$ and adding it to the old bottom row:

$$
\begin{array}{cccccccccc}
-3M + 5 & -2M - 7 & 0 & M & M & 0 & 0 & -10M & & R_4 \\
+ \;\; 1 \cdot (3M + 5) & \frac{1}{2} \cdot (3M + 5) & 0 & 0 & -\frac{1}{2} \cdot (3M + 5) & 0 & \frac{1}{2} \cdot (3M + 5) & 3 \cdot (3M + 5) & & (3M + 5) \cdot R_{pivot}^{new} \\
\hline
0 & -\frac{1}{2}M - \frac{9}{2} & 0 & M & -\frac{1}{2}M - \frac{5}{2} & 0 & \frac{3}{2}M + \frac{5}{2} & -M + 15 & & R_4^{new} = R_4 + (3M + 5)R_{pivot}^{new}
\end{array}
$$

The pivot column is column 2 $\left(\text{because } -\frac{1}{2}M - \frac{9}{2} \text{ is smaller than} -\frac{1}{2}M - \frac{5}{2}\right)$, and the pivot row is row 2 $\left(\text{because } \frac{1}{1/2} = 2 \text{ is smaller than} \frac{5}{1/2} = 10 \text{ and } \frac{3}{1/2} = 6\right)$. Pivoting on the $1/2$ in the second column and second row, we obtain

	x_1	x_2	s_1	s_2	s_3	a_2	a_3	
s_1	0	0	1	1	0	-1	0	4
x_2	0	1	0	-2	1	2	-1	2
x_1	1	0	0	1	-1	-1	1	2
P	0	0	0	-9	2	$M + 9$	$M - 2$	24

Notice that the artificial variables have left the basis and they can never re-enter it: All further pivot operations will add (and subtract) only numbers but not multiples of M to (and from) the $M + 9$ and $M - 2$ at the bottom of the a_2 and a_3 columns. Pivoting on the 1 in column 4 and row 3 gives

	x_1	x_2	s_1	s_2	s_3	a_2	a_3	
s_1	-1	0	1	0	1	0	-1	2
x_2	2	1	0	0	-1	0	1	6
s_2	1	0	0	1	-1	-1	1	2
P	9	0	0	0	-7	M	$M + 7$	42

Pivoting on the 1 in column 5 and row 1, we reach the final tableau for the extended problem:

	x_1	x_2	s_1	s_2	s_3	a_2	a_3	
s_3	-1	0	1	0	1	0	-1	2
x_2	1	1	1	0	0	0	0	8
s_2	0	0	1	1	0	-1	0	4
P	2	0	7	0	0	M	M	56

The solution of the extended problem is that the maximum of P is 56 when $x_1 = 0$ and $x_2 = 8$ ($s_3 = 2$, and the nonbasic variables s_1, a_2, and a_3 are all zero). Because the extended problem has a solution and the artificial variables are all zero, there is a solution to the original problem and it is the same: The maximum of P is 56 when $x_1 = 0$ and $x_2 = 8$.

To compare this solution with the graphical method, we graph the feasible region with x_1 on the horizontal axis and x_2 on the vertical axis. The five vertices and the values of the objective function are given in the table. The path of vertices visited by the big-M method is marked in bold.

Vertex (x_1, x_2)	$P = 5x_1 + 7x_2$
$(8, 0)$	40
$(4, 0)$	20
$(2, 2)$	24
$(0, 6)$	42
$(0, 8)$	56

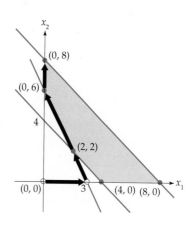

PRACTICE PROBLEM 2

Solve the following nonstandard linear programming problem by the big-M method.

$$\text{Maximize } P = 4x_1 + x_2 + 3x_3$$

$$\text{Subject to } \begin{cases} x_1 + 2x_2 + x_3 \le 50 \\ 2x_1 + x_2 + 2x_3 \ge 10 \\ x_1 \ge 0,\ x_2 \ge 0 \text{ and } x_3 \ge 0 \end{cases}$$

Solution at the back of the book

Graphing Calculator Exploration

Since many graphing calculators perform only numeric calculations, the big-M method can be carried out only by using a particular large value for M. The following screens use the program PIVOT (see pages 304–305) to start the solution of Example 2 using the value $M = 100$.

```
[[1    1    1 0...
 [1    1    0 -...
 [2    1    0 0...
 [-305 -207 0 1...
P C? 1
P R? 3
```

```
... 1  0    0   0 0...
... 0 -1    0   1 0...
... 0  0   -1   0 1...
... 0 100 100   0 0...
```

```
... 0    0 0  8     ]
... 0    1 0  4     ]
... -1   0 1  6     ]
... 100  0 0 -100]]
```

```
[-305 -207 0   1...
P C? 1
P R? 3
[[0  1/2     1 0 ...
 [0  1/2     0 -1...
 [1  1/2     0 0 ...
 [0 -109/2   0 10...
```

```
[-305 -207 0   1...
P C? 1
P R? 3
... 0    1/2    0 ...
... -1   1/2    1 ...
... 0   -1/2    0 ...
... 100 -105/2  0 ...
```

```
...100  0 0 -100]]
P C?  1
P R?  3
... 0  -1/2   5    ]
... 1  -1/2   1    ]
... 0   1/2   3    ]
..2  0  305/2 -85]]
```

Do you see how these values correspond to our symbolic solution on page 363 with M instead of the particular value of 100?

Some advanced graphing calculators can perform symbolic as well as numeric calculations. The following TI-92 screens use the program PIVOT to show the last two tableaux of the solution to Example 2 on page 364.

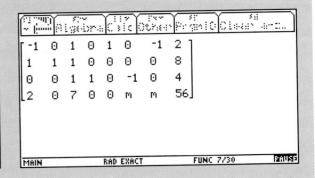

EXAMPLE 3   **Managing an Investment Portfolio**

The manager of an $80 million mutual fund has three investment possibilities: secure government bonds yielding 4%, a blue chip growth stock paying 6%, and a biotechnology start-up company returning 12%. If the manager wants to invest at least $50 million in a combination of safe government bonds and blue chip stock, and at least as much in the bonds as the total in blue chip and biotechnology stocks, how much should be invested in each to obtain the greatest possible return?

Solution

Let

$$x_1 = \begin{pmatrix} \text{Amount invested in} \\ \text{government bonds} \end{pmatrix}$$

$$x_2 = \begin{pmatrix} \text{Amount invested in} \\ \text{the blue chip stock} \end{pmatrix}$$

$$x_3 = \begin{pmatrix} \text{Amount invested in the} \\ \text{biotechnology company} \end{pmatrix}$$

be given in millions of dollars. The objective is to maximize the return

$$P = 0.04x_1 + 0.06x_2 + 0.12x_3 \qquad \text{Rate} \times \text{amount for each investment}$$

subject to the restrictions

$$x_1 + x_2 + x_3 \le 80 \qquad \text{\$80 million available to invest}$$
$$x_1 + x_2 \ge 50 \qquad \text{\$50 million in safe investments}$$
$$x_1 \ge x_2 + x_3 \qquad \text{Bonds} \ge \text{total in stocks}$$
$$x_1 \ge 0, x_2 \ge 0, x_3 \ge 0 \qquad \text{Nonnegativity}$$

Rewriting the third constraint as $x_1 - x_2 - x_3 \ge 0$, we find that the second and third constraints require artificial variables:

	x_1	x_2	x_3	s_1	s_2	s_3	a_2	a_3	
s_1	1	1	1	1	0	0	0	0	80
a_2	1	1	0	0	−1	0	1	0	50
a_3	1	−1	−1	0	0	−1	0	1	0
P	$-2M - 0.04$	-0.06	$M - 0.12$	0	M	M	0	0	$-50M$

| Pivoting on the 1 in column 1 and row 3 yields

	x_1	x_2	x_3	s_1	s_2	s_3	a_2	a_3	
s_1	0	2	2	1	0	1	0	−1	80
a_2	0	2	1	0	−1	1	1	−1	50
x_1	1	−1	−1	0	0	−1	0	1	0
P	0	−2M − 0.10	−M − 0.16	0	M	−M − 0.04	0	2M + 0.04	−50M

| Pivoting on the 2 in column 2 and row 2, there are now no artificial variables in the basis:

	x_1	x_2	x_3	s_1	s_2	s_3	a_2	a_3	
s_1	0	0	1	1	1	0	−1	0	30
x_2	0	1	1/2	0	−1/2	1/2	1/2	−1/2	25
x_1	1	0	−1/2	0	−1/2	−1/2	1/2	1/2	25
P	0	0	−0.11	0	−0.05	0.01	M + 0.05	M − 0.01	2.5

| Pivoting on the 1 in column 3 and row 1 brings us to the final tableau:

	x_1	x_2	x_3	s_1	s_2	s_3	a_2	a_3	
x_3	0	0	1	1	1	0	−1	0	30
x_2	0	1	0	−1/2	−1	1/2	1	−1/2	10
x_1	1	0	0	1/2	0	−1/2	0	1/2	40
P	0	0	0	0.11	0.06	0.01	M − 0.06	M − 0.01	5.8

| Because the extended problem has a solution with the artificial variables all equal to zero, we have the solution of the original problem: The maximum is $P = 5.8$ when $x_1 = 40$, $x_2 = 10$, and $x_3 = 30$. In terms of the original situation, the manager should invest $40 million in the government bonds, $10 million in the blue chip stock, and $30 million in the biotechnology start-up company for a maximum return of $5.8 million. ∎

COMPUTER EXPLORATION

Besides spreadsheet solutions of linear programming problems (see page 334), specialized computer programs are available to solve optimization problems. One of the best known, both in education and in industry, is LINDO (for **l**inear, **i**nteractive, and **d**iscrete **o**ptimizer).* The following screen shows the LINDO 6.1 solution of Example 3. Note the similarity between the problem input format and our notation for linear programming problems.

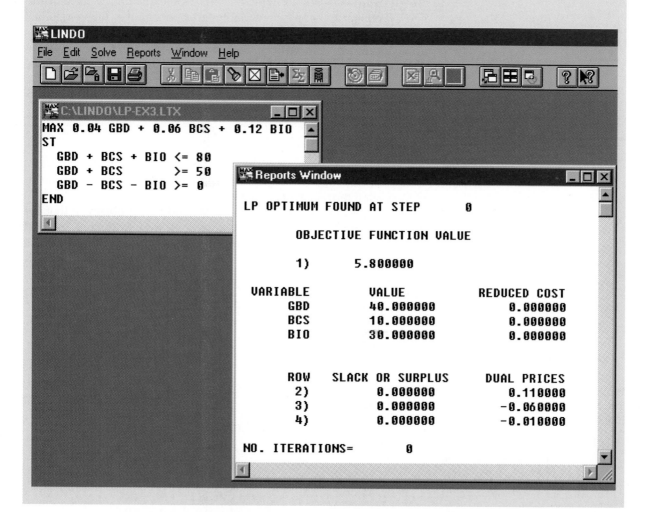

* Further information about LINDO can be found at the Internet site *http://www.lindo.com* or obtained from LINDO Systems, Inc., Chicago, Illinois.

Can Any Linear Programming Problem Now Be Solved?

The big-M method allows us to solve *any* maximum linear programming problem, no matter what mixture of ≤ and ≥ inequalities or = equalities are present in the constraints, provided that the variables are nonnegative. A direct way to change a minimum problem into a maximum problem is described in Exercises 41–42. Exercises 43–45 further explore equality constraints by discussing several solutions of an example of Kantorovich, and additional equality constraint problems are given in Exercises 46–47. Exercises 48–50 explain how to reformulate a problem so that all the variables are nonnegative. Taken together, these techniques allow us to write *any* linear programming problem as a maximum problem with nonnegative variables. Thus we may conclude that we are now able to solve *any* linear programming problem.

Section Summary

A nonstandard maximum problem can have any combination of ≥, ≤, and = constraints, and can be written in *normal form* with a nonnegative constant term in every constraint (see page 358).

The big-M method (see page 362) extends the problem so that the initial tableau displays a basic feasible point by introducing artificial variables into the ≥ and = constraints and subtracting M times these artificial variables from the objective function (see page 360). If this extended problem has a solution using the simplex method in which every artificial variable is zero, then the original problem has the same solution (otherwise, the original problem does not have a solution).

EXERCISES 4.6

Write each of the following nonstandard linear programming problems in normal form and construct the initial simplex tableau.

1. Maximize $P = 15x_1 + 20x_2 + 18x_3$

   Subject to $\begin{cases} 3x_1 + 2x_2 + 8x_3 \leq 96 \\ 5x_1 + x_2 + 6x_3 \geq 30 \\ x_1 \geq 0,\, x_2 \geq 0,\, x_3 \geq 0 \end{cases}$

2. Maximize $P = 60x_1 + 90x_2 + 30x_3$

   Subject to $\begin{cases} 4x_1 + 5x_2 + x_3 \geq 40 \\ 10x_1 + 18x_2 + 3x_3 \leq 150 \\ x_1 \geq 0,\, x_2 \geq 0,\, x_3 \geq 0 \end{cases}$

3. Maximize $P = 6x_1 + 4x_2 + 6x_3$

   Subject to $\begin{cases} 2x_1 + x_2 + 3x_3 \geq 30 \\ x_1 + x_2 + 2x_3 \geq 20 \\ x_1 \geq 0,\, x_2 \geq 0,\, x_3 \geq 0 \end{cases}$

4. Maximize $P = 9x_1 + 5x_2 + 4x_3$

   Subject to $\begin{cases} 3x_1 + x_2 + x_3 \geq 40 \\ 2x_1 + 2x_2 + x_3 \geq 30 \\ x_1 \geq 0,\, x_2 \geq 0,\, x_3 \geq 0 \end{cases}$

5. Maximize $P = 20x_1 + 30x_2 + 10x_3$

Subject to $\begin{cases} x_1 + x_2 + x_3 \geq 8 \\ x_1 + 2x_2 + 3x_3 \leq 30 \\ x_1 + 2x_2 + x_3 \leq 18 \\ x_1 \geq 0, x_2 \geq 0, x_3 \geq 0 \end{cases}$

6. Maximize $P = 24x_1 + 18x_2 + 30x_3$

Subject to $\begin{cases} x_1 + x_2 + x_3 \geq 4 \\ x_1 + 2x_2 + 3x_3 \leq 15 \\ x_1 + 2x_2 + x_3 \leq 9 \\ x_1 \geq 0, x_2 \geq 0, x_3 \geq 0 \end{cases}$

7. Maximize $P = 3x_1 + 2x_2 + 5x_3 + 4x_4$

Subject to $\begin{cases} x_1 + x_2 + x_3 + x_4 \geq 30 \\ 2x_1 + 3x_2 + 2x_3 + x_4 \geq 20 \\ 4x_1 + 2x_2 + x_3 + 2x_4 = 80 \\ x_1 \geq 0, x_2 \geq 0, x_3 \geq 0, x_4 \geq 0 \end{cases}$

8. Maximize $P = 6x_1 + 8x_2 + 3x_3 + 5x_4$

Subject to $\begin{cases} x_1 + x_2 + x_3 + x_4 \geq 70 \\ 3x_1 + 2x_2 + 4x_3 + 3x_4 \geq 60 \\ 4x_1 + x_2 + 2x_3 + 5x_4 = 100 \\ x_1 \geq 0, x_2 \geq 0, x_3 \geq 0, x_4 \geq 0 \end{cases}$

9. Maximize $P = 2x_1 + 2x_2 + x_3$

Subject to $\begin{cases} x_1 + 2x_2 + x_3 \geq 40 \\ 2x_1 + x_2 + x_3 \geq 50 \\ 5x_1 + 3x_2 + 2x_3 \leq 120 \\ 3x_1 + x_2 + 2x_3 \leq 150 \\ x_1 \geq 0, x_2 \geq 0, x_3 \geq 0 \end{cases}$

10. Maximize $P = x_1 + 2x_2 + 3x_3$

Subject to $\begin{cases} x_1 + x_2 + 2x_3 \geq 30 \\ x_1 + 2x_2 + x_3 \geq 60 \\ 2x_1 + 5x_2 + 3x_3 \leq 150 \\ 2x_1 + 3x_2 + x_3 \leq 120 \\ x_1 \geq 0, x_2 \geq 0, x_3 \geq 0 \end{cases}$

Solve each of the following nonstandard linear programming problems by the big-M method. (Problems 11, 12, 15, 16, 19, and 20 can also be solved by the graphical method.)

You may use the PIVOT program if permitted by your instructor.

11. Maximize $P = 4x_1 + x_2$

Subject to $\begin{cases} 3x_1 + 2x_2 \leq 120 \\ x_1 + x_2 \geq 50 \\ 2x_1 + x_2 \leq 60 \\ x_1 \geq 0, x_2 \geq 0 \end{cases}$

12. Maximize $P = x_1 + 4x_2$

Subject to $\begin{cases} 3x_1 + 2x_2 \leq 120 \\ x_1 + x_2 \leq 50 \\ 2x_1 + x_2 \geq 60 \\ x_1 \geq 0, x_2 \geq 0 \end{cases}$

13. Maximize $P = 30x_1 + 20x_2 + 28x_3$

Subject to $\begin{cases} 3x_1 + x_2 + 2x_3 \geq 30 \\ 4x_1 + x_2 + 3x_3 \leq 60 \\ x_1 \geq 0, x_2 \geq 0, x_3 \geq 0 \end{cases}$

14. Maximize $P = 18x_1 + 10x_2 + 20x_3$

Subject to $\begin{cases} 3x_1 + x_2 + 4x_3 \geq 30 \\ 2x_1 + x_2 + 5x_3 \leq 50 \\ x_1 \geq 0, x_2 \geq 0, x_3 \geq 0 \end{cases}$

15. Maximize $P = 4x_1 + 5x_2$

Subject to $\begin{cases} x_1 + 2x_2 \leq 12 \\ x_1 + x_2 \geq 15 \\ 2x_1 + x_2 \leq 12 \\ x_1 \geq 0, x_2 \geq 0 \end{cases}$

16. Maximize $P = 5x_1 + 4x_2$

Subject to $\begin{cases} 2x_1 + 3x_2 \leq 12 \\ x_1 + x_2 \geq 10 \\ 3x_1 + 2x_2 \leq 12 \\ x_1 \geq 0, x_2 \geq 0 \end{cases}$

17. Maximize $P = 15x_1 + 12x_2 + 18x_3$

Subject to $\begin{cases} 5x_1 + x_2 + 2x_3 \geq 30 \\ 2x_1 + x_2 + 3x_3 \leq 24 \\ x_1 \geq 0, x_2 \geq 0, x_3 \geq 0 \end{cases}$

18. Maximize $P = 4x_1 + 5x_2 + 6x_3$

Subject to $\begin{cases} 5x_1 + x_2 + 3x_3 \leq 30 \\ 2x_1 + x_2 + 2x_3 \geq 24 \\ x_1 \geq 0, x_2 \geq 0, x_3 \geq 0 \end{cases}$

19. Maximize $P = x_1 + x_2$

Subject to $\begin{cases} x_1 \leq 8 \\ x_2 \leq 5 \\ x_1 + 2x_2 \geq 6 \\ x_1 \geq 0, x_2 \geq 0 \end{cases}$

20. Maximize $P = x_1 + x_2$

Subject to $\begin{cases} x_1 \leq 6 \\ x_2 \leq 9 \\ 3x_1 + x_2 \geq 6 \\ x_1 \geq 0, x_2 \geq 0 \end{cases}$

21. Maximize $P = 2x_1 + 3x_2 + 2x_3$

Subject to $\begin{cases} 2x_1 + x_2 + x_3 \geq 20 \\ x_1 - x_2 + x_3 \leq 10 \\ x_1 \geq 0, x_2 \geq 0, x_3 \geq 0 \end{cases}$

22. Maximize $P = 6x_1 + 5x_2 + 3x_3$

Subject to $\begin{cases} 2x_1 + x_2 - x_3 \leq 20 \\ 3x_1 + 2x_2 + x_3 \geq 30 \\ x_1 \geq 0, x_2 \geq 0, x_3 \geq 0 \end{cases}$

23. Maximize $P = 4x_1 + 6x_2 + 12x_3 + 10x_4$

Subject to $\begin{cases} x_1 + x_2 + x_3 + x_4 \leq 60 \\ 2x_1 + x_2 + x_3 + 2x_4 \geq 10 \\ 2x_1 + 2x_2 + x_3 + x_4 = 100 \\ x_1 \geq 0, x_2 \geq 0, x_3 \geq 0, x_4 \geq 0 \end{cases}$

24. Maximize $P = 8x_1 - x_2 + 5x_3 + 14x_4$

Subject to $\begin{cases} x_1 + 2x_2 + 3x_4 \leq 55 \\ x_1 + x_2 + x_3 + 2x_4 \geq 25 \\ x_1 + 3x_3 + 3x_4 = 45 \\ x_1 \geq 0, x_2 \geq 0, x_3 \geq 0, x_4 \geq 0 \end{cases}$

25. Maximize $P = 5x_1 + 32x_2 + 3x_3 + 15x_4$

Subject to $\begin{cases} 5x_1 + 8x_2 + x_3 + 3x_4 \geq 18 \\ 2x_1 - x_2 + x_3 + x_4 \leq 10 \\ 3x_1 - 3x_2 + x_3 + 2x_4 \geq 6 \\ x_1 \geq 0, x_2 \geq 0, x_3 \geq 0, x_4 \geq 0 \end{cases}$

26. Maximize $P = 70x_1 + 12x_2 + 60x_3 + 20x_4$

Subject to $\begin{cases} 2x_1 + x_2 + x_3 + x_4 \leq 14 \\ 3x_1 + 3x_2 + x_3 + 2x_4 \leq 24 \\ 5x_1 + 3x_2 + 4x_3 + 2x_4 \geq 44 \\ x_1 \geq 0, x_2 \geq 0, x_3 \geq 0, x_4 \geq 0 \end{cases}$

27. Maximize $P = -x_1 + x_2 - 2x_3$

Subject to $\begin{cases} x_1 + x_2 + x_3 \leq 50 \\ x_1 + x_3 \geq 10 \\ x_2 + x_3 \geq 20 \\ x_1 \geq 0, x_2 \geq 0, x_3 \geq 0 \end{cases}$

28. Maximize $P = x_1 + 2x_2 + x_3$

Subject to $\begin{cases} x_1 + x_2 + x_3 \leq 30 \\ x_1 + x_2 \leq 25 \\ x_2 + x_3 \geq 15 \\ x_1 \geq 0, x_2 \geq 0, x_3 \geq 0 \end{cases}$

29. Maximize $P = 12x_1 + 2x_2 + 14x_3 + 8x_4$

Subject to $\begin{cases} x_1 + x_2 + x_3 + x_4 \geq 20 \\ 2x_1 + x_2 + 2x_3 + x_4 \leq 30 \\ x_3 + x_4 \geq 10 \\ x_1 \geq 0, x_2 \geq 0, x_3 \geq 0, x_4 \geq 0 \end{cases}$

30. Maximize $P = 8x_1 + 10x_2 + 6x_3 + 8x_4$

Subject to $\begin{cases} x_1 + x_2 + x_4 \leq 30 \\ 2x_1 + 2x_2 + x_3 + x_4 \leq 80 \\ x_1 + x_2 + x_3 + x_4 \geq 20 \\ x_1 \geq 0, x_2 \geq 0, x_3 \geq 0, x_4 \geq 0 \end{cases}$

APPLIED EXERCISES

Formulate each situation as a linear programming problem by identifying the variables, the objective function, and the constraints. Be sure to state clearly the meaning of each variable. Write the problem as a non-standard maximum problem in normal form, and then solve it by the big-M method. State your final answer in terms of the original question.

 You may use the PIVOT program if permitted by your instructor.

31. Financial Planning A retired couple want to invest their $20,000 life savings in bank certificates of deposits yielding 6% and Treasury bonds yielding 5%. If they want at least $5000 in each type of investment, how much should they invest in each to receive the greatest possible income?

32. Recycling Management A volunteer recycling center accepts both used paper and empty glass bottles which it then sorts and sells to a reprocessing company. The center has room to accept a total of 800 crates of paper and glass each week and has 50 hours of volunteer help to do the sorting. Each crate of paper products takes 5 minutes to sort and sells for 8¢, while each crate of bottles takes 3 minutes to sort and sells for 7¢. To support their city's "grab that glass" recycle theme this week, the center wants to accept at least 600 crates of glass. How many crates of each should the center accept this week to raise the most money for their ecology scholarship fund?

33. Advertising The manager of a new mall may spend up to $18,000 on "grand opening" announcements in newspapers, on radio and on TV. Each newspaper ad costs $300 and reaches 5000 readers, each one-minute radio commercial costs $800 and is heard by 13,000 listeners, and each 15-second TV spot costs $900 and is seen by 15,000 viewers. If the manager wants at least 5 newspaper ads and at least 20 radio commercials and TV spots combined, how many of each should be placed to reach the largest number of potential customers? (Ignore multiple exposures to the same customer.)

34. Advertising   In the last few days before the election, a politician can afford to spend no more than $27,000 on TV advertisements and can arrange for no more than 10 ads. Each day time ad costs $2000 and reaches 4000 viewers, each prime time ad costs $3000 and reaches 5000 viewers, and each late night ad costs $1000 and reaches 2000 viewers. To be sure to reach the widest variety of voters, the politician's advisors insist on at least 5 ads scheduled during day time and late night combined. Ignoring repeated viewings by the same person and assuming every viewer can vote, how many ads in each of the time periods will reach the most voters?

35. Resource Allocation   A furniture shop manufactures wooden desks, tables and chairs. The numbers of hours to assemble and finish each piece are shown in the table together with the number of hours of skilled labor available for each task. To meet expected demand, a total of at least 30 desks and tables combined must be made. If the profit is $75 for each desk, $84 for each table, and $66 for each chair, how many of each should the company manufacture to obtain the greatest possible profit?

	Desk	Table	Chair	Labor Available
Assembly	2 hours	1 hour	2 hours	210 hours
Finishing	2 hours	3 hours	1 hour	150 hours

36. Production Planning   An automotive parts shop rebuilds carburetors, fuel pumps, and alternators. The numbers of hours to rebuild and then inspect and pack each part are shown in the table, together with the number of hours of skilled labor available for each task. At least 12 carburetors must be rebuilt. If the profit is $12 for each carburetor, $14 for each fuel pump, and $10 for each alternator, how many of each should the shop rebuild to obtain the greatest possible profit?

	Carburetor	Fuel Pump	Alternator	Labor Available
Rebuilding	5 hours	4 hours	3 hours	200 hours
Inspection and packaging	1 hour	1 hour	0.5 hour	45 hours

37. Agriculture   A farmer grows wheat, barley, and oats on her 500 acre farm. Each acre of wheat requires 3 days of labor (to plant, tend, and harvest) and costs $21 (for seed, fertilizer, and pesticides), each acre of barley requires 2 days of labor and costs $27, and each acre of oats requires 3 days of labor and costs $24. The farmer and her hired field hands can provide no more than 1200 days of labor this year. She wants to grow at least 100 acres of oats and can afford to spend no more than $15,120. If the profit is $50 for each acre of wheat, $40 for each acre of barley, and $45 for each acre of oats, how many acres of each crop should she grow to obtain the greatest possible profit?

38. Agriculture   A farmer grows corn, peanuts, and soybeans on his 240-acre farm. To maintain soil fertility, the farmer rotates the crops and always plants at least as many acres of soybeans as the total acres of the other crops. Because he has promised to sell some of his corn to a neighbor who raises cattle, he must plant at least 42 acres of corn. Each acre of corn requires 2 days of labor and yields a profit of $150, each acre of peanuts requires 5 days of labor and yields a profit of $300, and each acre of soybeans requires 1 day of labor and yields a profit of $100. If the farmer and his children can put in at most 630 days of labor, how many acres of each crop should the farmer plant to obtain the greatest possible profit?

39. Pollution Control   An empty storage yard at a coal burning electric power plant can hold no more than 100,000 tons of coal. Two grades of coal are available: low sulfur (1%) with an energy content of 20 million BTU per ton and high sulfur (2%) with an energy content of 30 million BTU per ton. If existing contracts with the high-sulfur coal mine operator require that at least 50,000 tons of high sulfur coal be purchased, and if the next coal purchase may contain no more than 1400 tons of sulfur, how many tons of each type of coal should be purchased to obtain the most energy?

40. Production Planning A small jewelry company prepares and mounts semiprecious stones. There are 20 lapidaries (who cut and polish the stones) and 24 jewelers (who mount the stones in gold settings). Each employee works 7 hours each day, 5 days each week. Each tray of agates requires 5 hours of cutting and polishing and 4 hours to mount, each tray of onyxes requires 2 hours of cutting and polishing and 3 hours to mount, and each tray of garnets requires 6 hours of cutting and polishing and 3 hours to mount. Furthermore, the company's owner has decided that they will process at least 12 trays of agates each week. If the profit is $15 for each tray of agates, $10 for each tray of onyxes, and $13 for each tray of garnets, how many trays of each stone should be processed each week to obtain the greatest possible profit?

Explorations and Excursions

The following problems extend and augment the material presented in the text.

 The PIVOT program may be helpful (if permitted by your instructor).

Minimum Problems and the Big-M Method

41. a. Solve both of the following problems by the graphical method from Section 4.2. Is it true that the minimum of C is the same as -1 times the maximum of $P = -C$?

$$\text{Minimize } C = 3x + 4y$$

$$\text{Subject to } \begin{cases} x + y \leq 10 \\ 2x + y \geq 8 \\ x \geq 0 \text{ and } y \geq 0 \end{cases}$$

$$\text{Maximize } P = -3x - 4y$$

$$\text{Subject to } \begin{cases} x + y \leq 10 \\ 2x + y \geq 8 \\ x \geq 0 \text{ and } y \geq 0 \end{cases}$$

b. Solve the following problem by finding the dual maximum problem and using the (regular) simplex method as we did in Section 4.4.

$$\text{Minimize } C = 3y_1 + 4y_2$$

$$\text{Subject to } \begin{cases} -y_1 - y_2 \geq -10 \\ 2y_1 + y_2 \geq 8 \\ y_1 \geq 0 \text{ and } y_2 \geq 0 \end{cases}$$

c. Solve the following nonstandard problem by the big-M method.

$$\text{Maximize } P = -3x_1 - 4x_2$$

$$\text{Subject to } \begin{cases} x_1 + x_2 \leq 10 \\ 2x_1 + x_2^{7,8} \\ x_1 \geq 0 \text{ and } x_2 \geq 0 \end{cases}$$

d. For each sequence of tableaux from parts (b) and (c), trace the basic points on the graphs from part (a). Which solution method is more efficient?

42. Repeat the process of Exercise 41 to compare the "dual" solution to the following minimum problem with the big-M solution of the corresponding maximum problem.

$$\text{Minimize } C = 20y_1 + 18y_2$$

$$\text{Subject to } \begin{cases} 2y_1 + y_2 \geq 18 \\ y_1 + y_2 \geq 14 \\ 4y_1 + 3y_2 \geq 48 \\ y_1 \geq 0 \text{ and } y_2 \geq 0 \end{cases}$$

$$\text{Maximize } P = -20x_1 - 18x_2$$

$$\text{Subject to } \begin{cases} 2x_1 + x_2 \geq 18 \\ x_1 + x_2 \geq 14 \\ 4x_1 + 3x_2 \geq 48 \\ x_1 \geq 0 \text{ and } x_2 \geq 0 \end{cases}$$

Kantorovich's First Example The first example from the 1939 paper "Mathematical Methods for Organizing and Planning Production" by L. V. Kantorovich may be phrased as follows:

A machine shop has three types of equipment that can produce either or both of two kinds of parts (see the table). These parts are used to make finished items containing one of each, so the shop must produce the same number of both parts. Furthermore, the equipment must never be idle. How should the production be divided so that the greatest number of finished items are produced?

Productivity of the Machines for Two Parts

Type of Machine	Number of Machines	Output per Machine	
		First Part	**Second Part**
Milling machines	3	10	20
Turret lathes	3	20	30
Automatic lathe	1	30	80

Type of Machine	Number of Machines	Total Output	
		First Part	**Second Part**
Milling machines	3	30	60
Turret lathes	3	60	90
Automatic lathe	1	30	80

Let x_1 represent the fraction of the work day that the milling machines make the first part, so that they make the second part for the remaining $1 - x_1$ fraction of the work day. Similarly, let x_2 and x_3 be the corresponding fractions for the turret lathes and the automatic lathe. We must have that $0 \leq x_1 \leq 1, 0 \leq x_2 \leq 1$, and $0 \leq x_3 \leq 1$ for these fractions. The number of first parts produced is then $30x_1 + 60x_2 + 30x_3$, and since the remaining time is spent making the second part, the number of second parts produced is $60(1 - x_1) + 90(1 - x_2) + 80(1 - x_3)$. In order to make the finished items without leftover parts, these must be equal:

$$30x_1 + 60x_2 + 30x_3 =$$
$$60(1 - x_1) + 90(1 - x_2) + 80(1 - x_3)$$

That is,

$$9x_1 + 15x_2 + 11x_3 = 23 \qquad \begin{array}{l}\text{Divide by 10 and}\\ \text{combine terms}\end{array}$$

Since the number of finished parts will be $30x_1 + 60x_2 + 30x_3$, we have the linear programming problem

Maximize $P = 30x_1 + 60x_2 + 30x_3$

$$\text{Subject to} \begin{cases} x_1 \leq 1 \\ x_2 \leq 1 \\ x_3 \leq 1 \\ 9x_1 + 15x_2 + 11x_3 = 23 \quad \leftarrow \begin{array}{l}\text{Equality}\\ \text{constraint}\end{array} \\ x_1 \geq 0, x_2 \geq 0, \text{ and } x_3 \geq 0 \end{cases}$$

Exercises 43–45 explore three possible ways to solve this problem with an equality constraint.

43. One way to solve Kantorovich's First Example is to use the big-M method. Show that the maximum occurs when $x_1 = 8/9$, $x_2 = 1$, and $x_3 = 0$. (In terms of Kantorovich's original problem, the milling machines should produce the first part $\frac{8}{9}$ of the time and the second part the remaining $\frac{1}{9}$, the turret lathes should produce the first part all of the time, and the automatic lathe should produce the second part all of the time.)

44. Since the linear equation $9x_1 + 15x_2 + 11x_3 = 23$ is the same as the two linear inequalities $9x_1 + 15x_2 + 11x_3 \leq 23$ and $9x_1 + 15x_2 + 11x_3 \geq 23$, another way of finding the answer to Kantorovich's First Example is to use the big-M method to solve the nonstandard problem

Maximize $P = 30x_1 + 60x_2 + 30x_3$

$$\text{Subject to} \begin{cases} x_1 \leq 1 \\ x_2 \leq 1 \\ x_3 \leq 1 \\ 9x_1 + 15x_2 + 11x_3 \leq 23 \\ 9x_1 + 15x_2 + 11x_3 \geq 23 \\ x_1 \geq 0, x_2 \geq 0, \text{ and } x_3 \geq 0 \end{cases}$$

Carry out the solution using the big-M method and check that it agrees with the solution found in Exercise 43.

45. Another approach to Kantorovich's First Example is to solve the equation $9x_1 + 15x_2 + 11x_3 = 23$ for x_3 and use this information to rewrite the problem in just x_1 and x_2.

a. Show that $x_3 = \frac{23}{11} - \frac{9}{11}x_1 - \frac{15}{11}x_2$ and that Kantorovich's First Example may be rewritten (remember that $x_3 \geq 0$) as

Maximize $P = \frac{1}{11}(60x_1 + 210x_2 + 690)$

$$\text{Subject to} \begin{cases} x_1 \leq 1 \\ x_2 \leq 1 \\ 9x_1 + 15x_2 \geq 12 \\ 9x_1 + 15x_2 \leq 23 \\ x_1 \geq 0 \text{ and } x_2 \geq 0 \end{cases}$$

b. Explain how the solution to the problem in part (a) can be found from the solution of the problem

$$\text{Maximize } P = 2x_1 + 7x_2$$

$$\text{Subject to } \begin{cases} x_1 \le 1 \\ x_2 \le 1 \\ 3x_1 + 5x_2 \ge 4 \\ 9x_1 + 15x_2 \le 23 \\ x_1 \ge 0 \text{ and } x_2 \ge 0 \end{cases}$$

c. Solve the problem in part (b) by the big-M method.

d. Use the solution found in part (c) to find the solution of the problem in part (a) and then deduce the solution to Kantorovich's First Example. Does this solution agree with the solutions found in Exercises 43 and 44?

46. Use the methods discussed in Exercises 43–45 to solve the following equality constraint linear programming problem. Check your answer by solving this problem by the graphical method from Section 4.2.

$$\text{Maximize } P = 5x_1 + 3x_2$$

$$\text{Subject to } \begin{cases} 2x_1 + x_2 = 6 \\ x_1 + x_2 \le 4 \\ x_1 \ge 0 \text{ and } x_2 \ge 0 \end{cases}$$

47. Use the methods discussed in Exercises 43–45 to solve the following equality constraint linear programming problem.

$$\text{Maximize } P = 5x_1 + 4x_2 + 6x_3 + 3x_4$$

$$\text{Subject to } \begin{cases} 2x_1 + x_2 + x_3 \le 40 \\ 2x_1 + 2x_3 + x_4 \le 60 \\ x_1 + x_2 = 20 \\ x_3 + x_4 = 30 \\ x_1 \ge 0, x_2 \ge 0, x_3 \ge 0, \text{ and } x_4 \ge 0 \end{cases}$$

More About Nonnegativity Conditions The following problems show how to reformulate linear programming problems to have nonnegativity conditions on *all* the variables when the original versions do not.

48. Show that the first problem may be rewritten as the second by replacing the variable x_1 by a new variable u_1 where $u_1 = x_1 + 8$ and $u_1 \ge 0$. Solve the first problem by the graphical method from Section 4.2, solve the second by the simplex method from Section 4.3, and check that your answers agree.

$$\text{Maximize } P = 5x_1 + 3x_2$$

$$\text{Subject to } \begin{cases} 2x_1 + x_2 \le 6 \\ x_1 + x_2 \le 4 \\ x_1 \ge -8 \text{ and } x_2 \ge 0 \end{cases}$$

$$\text{Maximize } P = 5u_1 + 3x_2 - 40$$

$$\text{Subject to } \begin{cases} 2u_1 + x_2 \le 22 \\ u_1 + x_2 \le 12 \\ u_1 \ge 0 \text{ and } x_2 \ge 0 \end{cases}$$

49. Show that the first problem may be rewritten as the second by replacing the variable x_1 by a new variable u_1 where $u_1 = -x_1$ and $u_1 \ge 0$. Solve the first problem by the graphical method from Section 4.2, solve the second by the simplex method from Section 4.3, and check that your answers agree.

$$\text{Maximize } P = 5x_1 + 3x_2$$

$$\text{Subject to } \begin{cases} 2x_1 + x_2 \le 10 \\ -x_1 + x_2 \le 4 \\ x_1 \le 0 \text{ and } x_2 \ge 0 \end{cases}$$

$$\text{Maximize } P = -5u_1 + 3x_2$$

$$\text{Subject to } \begin{cases} -2u_1 + x_2 \le 10 \\ u_1 + x_2 \le 4 \\ u_1 \ge 0 \text{ and } x_2 \ge 0 \end{cases}$$

50. A variable is *unrestricted* if it may take *any* value (positive, negative, or zero). Show that the first problem may be rewritten as the second by replacing the variable x_1 by two new variables u_1 and u_2 where $u_1 - u_2 = x_1$ and $u_1, u_2 \ge 0$. Solve the first problem by the graphical method from Section 4.2, solve the second by the simplex method from Section 4.3, and check that your answers agree.

$$\text{Maximize } P = 3x_1 + 5x_2$$

$$\text{Subject to } \begin{cases} 2x_1 + x_2 \le 6 \\ x_2 \le 4 \\ x_1 \text{ unrestricted and } x_2 \ge 0 \end{cases}$$

$$\text{Maximize } P = 3u_1 - 3u_2 + 5x_2$$

$$\text{Subject to } \begin{cases} 2u_1 - 2u_2 + x_2 \le 6 \\ x_2 \le 4 \\ u_1 \ge 0, u_2 \ge 0, \text{ and } x_2 \ge 0 \end{cases}$$

[*Hint:* Can any real number be written as the difference of two nonnegative numbers?]

Chapter Summary with Hints and Suggestions

Reading the text and doing the exercises in this chapter have helped you to master the following skills, which are listed by section (in case you need to review them) and are keyed to particular Review Exercises. Answers for all Review Exercises are given at the back of the book, and full solutions can be found in the Student Solutions Manual.

4.1   LINEAR INEQUALITIES

- Sketch a linear inequality $ax + by \leq c$ by drawing the boundary line $ax + by = c$ from its intercepts, using a test point to choose the correct side, and shading in the feasible region. *(Review Exercises 1–2.)*

$$\left(\frac{c}{a}, 0\right) \quad \text{and} \quad \left(0, \frac{c}{b}\right)$$

- Sketch a system of linear inequalities $\begin{cases} ax + by \leq c \\ \ddots \end{cases}$

  by drawing the boundary lines, finding the correct sides of the boundaries, and shading in the feasible region. List the vertices or "corners" of the region by finding the intersections of the boundary lines that are feasible. Identify the region as bounded or unbounded. *(Review Exercises 3–8.)*

- Formulate an applied situation as a system of linear inequalities and sketch the feasible region with the vertices. Many situations include nonnegativity conditions for the variables because of what they represent. *(Review Exercises 9–10.)*

4.2   TWO-VARIABLE LINEAR PROGRAMMING PROBLEMS

- Find the maximum or minimum of a linear function on a given region or explain why such a value does not exist. *(Review Exercises 11–14.)*

- Solve a two-variable linear programming problem by sketching the feasible region, determining whether a solution exists, and then finding it from the values of the objective function at the vertices. *(Review Exercises 15–18.)*

- Formulate an application as a linear programming problem by identifying the variables, the objective function, and the constraints, and then solve it using the vertices of the feasible region. *(Review Exercises 19–20.)*

4.3   THE SIMPLEX METHOD FOR STANDARD MAXIMUM PROBLEMS

- Rewrite a maximum linear programming problem in matrix form, check that it is a standard problem, and construct the initial simplex tableau. *(Review Exercises 21–22.)*

$$\text{Maximize } P = c^t X$$

$$\text{Subject to } \begin{cases} AX \leq b \\ X \geq 0 \end{cases}$$

where $b \geq 0$

	X	S	
S	A	I	b
P	$-c^t$	0	0

- Find the pivot element in a given simplex tableau and carry out one complete pivot operation. If there is no pivot element, explain what the tableau shows about the original standard maximum problem. *(Review Exercises 23–24.)*

- Solve a standard maximum problem by the simplex method (construct the initial simplex tableau, pivot until the tableau does not have a pivot element, and interpret this final tableau). *(Review Exercises 25–28.)*

- Formulate an application as a linear programming problem and check that the problem is a standard maximum problem. Solve it by the simplex method. *(Review Exercises 29–30.)*

4.4 DUALITY AND STANDARD MINIMUM PROBLEMS

- Rewrite a minimum linear programming problem in matrix form, check that it is a standard problem, and construct the dual maximum problem and the initial simplex tableau. *(Review Exercises 31–34.)*

$$\text{Minimize } C = b^t Y$$
$$\text{Subject to } \begin{cases} A^t Y \geq c \\ Y \geq 0 \end{cases}$$
$$\text{where } b^t \geq 0$$

$$\text{Maximize } P = c^t X$$
$$\text{Subject to } \begin{cases} AX \leq b \\ X \geq 0 \end{cases}$$
$$\text{where } b \geq 0$$

- Solve a standard minimum problem by finding the dual maximum problem and using the simplex method. *(Review Exercises 35–38.)*

- Formulate an application as a standard minimum problem and solve it by finding the dual maximum problem and using the simplex method. *(Review Exercises 39–40.)*

4.5 NONSTANDARD PROBLEMS: THE DUAL PIVOT ELEMENT AND THE TWO-STAGE METHOD

- Rewrite a nonstandard maximum problem in matrix form, construct the initial simplex tableau, and locate the dual pivot element. *(Review Exercises 41–42.)*

- Solve a nonstandard maximum problem by the two-stage simplex method. *(Review Exercises 43–48.)*

- Formulate an application as a nonstandard maximum problem and solve it by the two-stage simplex method. *(Review Exercises 49–50.)*

4.6 NONSTANDARD PROBLEMS: ARTIFICIAL VARIABLES AND THE BIG-M METHOD

- Write a nonstandard maximum problem in normal form and construct the initial simplex tableau. *(Review Exercises 51–52.)*

- Solve a nonstandard maximum problem by the big-M method. *(Review Exercises 53–58.)*

- Formulate an application as a nonstandard maximum problem and solve it by the big-M method. *(Review Exercises 59–60.)*

HINTS AND SUGGESTIONS

- *(Overview)* A linear programming problem asks for the maximum or minimum of a linear objective function subject to constraints in the form of linear inequalities. If there is a solution to the problem, it occurs at a *vertex* of the feasible region determined by the constraints. A problem with more than two variables must be solved algebraically using the simplex method, duality (for a standard minimum problem), the two-stage simplex or the big-M method (for a nonstandard problem).

- While a problem with a *bounded* region always has a solution, a problem with an *unbounded* region may or may not have a solution (if it *does,* then the solution occurs at a vertex).

- When setting up a word problem, look first for the question "how many" or "how much" to help identify the variables. Be sure that finding values for your variables will answer the question stated in the problem. Use the rest of the given information to find the objective function and constraints. Include nonnegativity conditions for your variables as appropriate (for example, a farmer *cannot* raise a negative number of cows).

- A simplex tableau is *feasible* if it has no negative numbers in the rightmost *column* and is *optimal* if it has no negative numbers in the bottom *row.* If a simplex tableau is both feasible and optimal, it displays the solution of the problem.

- The pivot operation is not completed until *all* the rows of the tableau have been recalculated.

- If you are solving a standard maximum problem by the simplex method and you have a negative number in the rightmost column after pivoting, you've made an error in your choice of the pivot element and/or your arithmetic.

- Many simplex tableaux require several pivots to solve, so don't give up: Keep pivoting until the tableau does not have a pivot element.

- Practice for test: Review Exercises 1, 7, 9, 11, 13, 16, 17, 20, 25, 29, 35, 39, 41, 45, 50, 51, 55, 60.

Review Exercises for Chapter 4 *Practice test exercise numbers are in blue.*

4.1 Linear Inequalities

Sketch each linear inequality.

1. $3x - 2y \le 24$

2. $x + y \ge -6$

Sketch each system of linear inequalities. List all vertices and identify the region as "bounded" or "unbounded."

3. $\begin{cases} x + y \le 20 \\ x \ge 0, y \ge 0 \end{cases}$

4. $\begin{cases} x + 2y \ge 10 \\ x \ge 0, y \ge 0 \end{cases}$

5. $\begin{cases} 2x + y \le 20 \\ x + y \le 15 \\ x \ge 0, y \ge 0 \end{cases}$

6. $\begin{cases} 2x + y \ge 12 \\ x + y \ge 8 \\ x \ge 0, y \ge 0 \end{cases}$

7. $\begin{cases} 3x + 2y \le 24 \\ x + y \le 10 \\ x \ge 2 \\ x \ge 0, y \ge 0 \end{cases}$

8. $\begin{cases} x + y \le 10 \\ -x + y \le 4 \\ x \le 9 \\ x \ge 0, y \ge 0 \end{cases}$

Formulate each application as a system of linear inequalities, sketch the feasible region, and find the vertices.

9. **Dog Training** A dog trainer works with Irish setters and Labrador retrievers for no more than 6 hours each day. Each training session with a "setter" takes half an hour and requires 8 dog treats, while each session with a "lab" takes three-quarters of an hour and also requires 8 dog treats. If the trainer has only 80 dog treats left, how many of each breed can he train today?

10. **Blending Bird Seed** A pet store owner blends custom bird seed by mixing SongBird and MeadowMix brands together. Each pound of Song-Bird contains 2 ounces of sunflower hearts and 3 ounces of crushed peanuts (along with a lot of cheap filler seed), while each pound of Meadow-Mix contains 4 ounces of sunflower hearts and 2 ounces of crushed peanuts. If each bag of the custom blend is labeled "contains at least 104 ounces of sunflower hearts and 84 ounces of crushed peanuts," how many pounds of each brand must be put in each bag?

4.2 Two-Variable Linear Programming Problems

Find each maximum or minimum value for the given region. If such a value does not exist, explain why not.

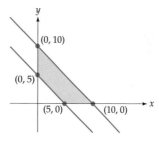

11. Maximum of $P = 3x + 4y$

12. Minimum of $C = 3x + 2y$

Find each maximum or minimum value for the given region. If such a value does not exist, explain why not.

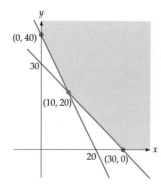

13. Maximum of $P = 4x + 3y$

14. Minimum of $C = 2x + 5y$

Solve each linear programming problem by sketching the region, labeling the vertices, checking whether a solution exists, and then finding it if it does.

15. Maximize $P = 20x + 15y$

Subject to $\begin{cases} x + 3y \le 18 \\ x + y \le 12 \\ x \ge 0, y \ge 0 \end{cases}$

16. Minimize $C = 30x + 40y$

Subject to $\begin{cases} 2x + y \geq 120 \\ x + 3y \geq 120 \\ x \geq 0, y \geq 0 \end{cases}$

17. Maximize $P = 20x + 30y$

Subject to $\begin{cases} 2x + y \leq 24 \\ x + 4y \leq 40 \\ x + y \leq 13 \\ x \geq 0, y \geq 0 \end{cases}$

18. Minimize $C = 7x + 11y$

Subject to $\begin{cases} x + y \geq 20 \\ x - y \geq -10 \\ x \geq 0, y \geq 0 \end{cases}$

Formulate each application as a linear programming problem by identifying the variables, the objective function, and the constraints. Determine whether a solution exists, and if it does, find it using the vertices of the feasible region. State your final answer in terms of the original question.

19. **Production Planning** A wood-working shop makes antique reproduction cases for wall and mantel clocks. The amount of cutting and finishing needed for each is shown in the table, together with the number of hours of skilled labor available for each task. If the profit for each case is $250 for wall clocks and $200 for mantel clocks, how many of each kind of clock case should the shop make to obtain the greatest possible profit?

	Wall Clock Case	Mantel Clock Case	Labor Available
Cutting	5 hours	3 hours	165 hours
Finishing	3 hours	4 hours	132 hours

20. **Nutrition** A large and active cat needs at least 70 grams of fat and 134.4 grams of protein each week. Each ounce of canned cat food contains 2 grams of fat and 3.2 grams of protein, while each ounce of dry food contains 2.5 grams of fat and 9.6 grams of protein. If canned cat food costs 5¢ per ounce and dry food costs 8¢ per ounce, how many ounces of each type of food will meet the cat's nutritional needs for the week at the least cost?

4.3 The Simplex Method for Standard Maximum Problems

Construct the initial simplex tableau for each standard maximum problem.

21. Maximize $P = 4x_1 + 10x_2 + 9x_3$

Subject to $\begin{cases} 5x_1 + 2x_2 + 3x_3 \leq 30 \\ 2x_1 + 3x_2 + 4x_3 \leq 24 \\ x_1 \geq 0, x_2 \geq 0, x_3 \geq 0 \end{cases}$

22. Maximize $P = 5x_1 + 7x_2$

Subject to $\begin{cases} x_1 + x_2 \leq 10 \\ 2x_1 + x_2 \leq 14 \\ x_1 - x_2 \leq 4 \\ x_1 \geq 0, x_2 \geq 0 \end{cases}$

For each simplex tableau, find the pivot element and carry out one complete pivot operation. If there is no pivot element, explain what the tableau shows about the solution of the original standard maximum problem.

23.

	x_1	x_2	x_3	s_1	s_2	s_3	
s_1	1	1	1	1	0	0	6
s_2	3	1	2	0	1	0	8
s_3	1	1	-3	0	0	1	6
P	-6	4	-8	0	0	0	0

24.

	x_1	x_2	s_1	s_2	s_3	
s_1	2	0	1	0	-2	8
s_2	6	0	0	1	3	42
x_2	1	1	0	0	1	10
P	5	0	0	0	15	150

Solve each problem by the simplex method.

25. Maximize $P = 7x_1 + 8x_2 + 7x_3$

Subject to $\begin{cases} 2x_1 + x_2 + 4x_3 \leq 16 \\ x_1 + 2x_2 - x_3 \leq 30 \\ x_1 + x_2 + 3x_3 \leq 12 \\ x_1 \geq 0, x_2 \geq 0, x_3 \geq 0 \end{cases}$

26. Maximize $P = 8x_1 + 10x_2 + 9x_3$

Subject to $\begin{cases} 3x_1 + 2x_2 + 3x_3 \leq 18 \\ 2x_1 + x_2 + 4x_3 \leq 14 \\ x_1 \geq 0, x_2 \geq 0, x_3 \geq 0 \end{cases}$

27. Maximize $P = 8x_1 + 7x_2 + 5x_3 + 7x_4$

Subject to $\begin{cases} 3x_1 + 2x_2 + 4x_3 + 3x_4 \le 12 \\ x_1 + x_2 + x_3 + x_4 \le 5 \\ 3x_1 + x_2 + 5x_3 + 3x_4 \le 15 \\ x_1 \ge 0, x_2 \ge 0, x_3 \ge 0, x_4 \ge 0 \end{cases}$

28. Maximize $P = 5x_1 + 4x_2 + 3x_3$

Subject to $\begin{cases} x_1 + x_2 \le 8 \\ x_2 + x_3 \le 7 \\ x_1 + x_2 - x_3 \le 6 \\ x_1 \ge 0, x_2 \ge 0, x_3 \ge 0 \end{cases}$

Formulate each application as a linear programming problem by identifying the variables, the objective function, and the constraints. Check that the problem is a standard maximum problem and then solve it by the simplex method. State your final answer in terms of the original question.

29. Book Publishing A book publisher needs more copies of its latest best seller. By rescheduling several other projects over the next two weeks, it can arrange up to 300 hours of printing and 250 hours of binding. For each thousand copies, the trade edition requires 3 hours of printing and 2 hours of binding, the book club edition requires 2 hours of printing and 2 hours of binding, and the paperback edition requires 1 hour of printing and 1 hour of binding. If the profit per book is $7 for the trade edition, $6 for the book club edition, and $4 for the paperback, how many of each edition should be printed to obtain the greatest possible profit?

30. Production Planning A fully automated plastics factory produces three toys—a racing car, a jet airplane, and a speed boat—in three stages: molding, painting, and packaging. After allowing for routine maintenance, the equipment can operate no more than 150 hours per week. Each batch of racing cars requires 6 hours of molding, 2.5 hours of painting, and 5 hours of packaging; each batch of jet airplanes requires 3 hours of molding, 7.5 hours of painting, and 5 hours of packaging; and each batch of speed boats requires 3 hours of molding, 5 hours of painting, and 5 hours of packaging. If the profits per batch of toys are $120 for cars, $100 for airplanes, and $110 for boats, how many batches of each toy should be produced each week to obtain the greatest possible profit?

4.4 Duality and Standard Minimum Problems

Rewrite each minimum linear programming problem in matrix form, check that it is a standard problem, and then construct the dual maximum problem and the initial simplex tableau. (Do not solve the problem any further.)

31. Minimize $C = 10y_1 + 15y_2 + 20y_3$

Subject to $\begin{cases} 2y_1 - y_2 + y_3 \ge 40 \\ y_1 + y_2 + 3y_3 \ge 30 \\ y_1 \ge 0, y_2 \ge 0, y_3 \ge 0 \end{cases}$

32. Minimize $C = 30y_1 + 20y_2$

Subject to $\begin{cases} 5y_1 + 2y_2 \ge 210 \\ 3y_1 + 4y_2 \ge 252 \\ 5y_1 + 4y_2 \le 380 \\ y_1 \ge 0, y_2 \ge 0 \end{cases}$

33. Minimize $C = 42y_1 + 36y_2$

Subject to $\begin{cases} 7y_1 + 4y_2 \ge 84 \\ y_1 + y_2 \ge 18 \\ y_1 \ge 0, y_2 \ge 0 \end{cases}$

34. Minimize $C = 130y_1 + 40y_2 + 98y_3$

Subject to $\begin{cases} 3y_1 + y_2 + 2y_3 \ge 51 \\ 4y_1 + y_2 + 2y_3 \ge 60 \\ 3y_1 + y_2 + 3y_3 \ge 57 \\ y_1 \ge 0, y_2 \ge 0, y_3 \ge 0 \end{cases}$

Solve each standard minimum problem by finding the dual maximum problem and using the simplex method.

35. Minimize $C = 9y_1 + 24y_2 + 6y_3$

Subject to $\begin{cases} y_1 + 2y_2 \ge 18 \\ 3y_2 + y_3 \ge 15 \\ y_1 \ge 0, y_2 \ge 0, y_3 \ge 0 \end{cases}$

36. Minimize $C = 30y_1 + 36y_2$

Subject to $\begin{cases} 5y_1 + 3y_2 \ge 60 \\ y_1 + y_2 \ge 16 \\ y_2 \ge 5 \\ y_1 \ge 0, y_2 \ge 0 \end{cases}$

37. Minimize $C = 3y_1 + 4y_2 + 10y_3 + 8y_4$

Subject to $\begin{cases} -y_1 + y_2 + y_3 - 2y_4 \ge 10 \\ y_1 - y_2 - 2y_3 + y_4 \ge 20 \\ y_1 \ge 0, y_2 \ge 0, y_3 \ge 0, y_4 \ge 0 \end{cases}$

38. Minimize $C = 12y_1 + 5y_2 + 6y_3$

$$\text{Subject to } \begin{cases} 3y_1 + y_2 + 3y_3 \geq 9 \\ 2y_1 + y_2 - 3y_3 \geq 12 \\ 4y_1 + 2y_2 + 2y_3 \geq 8 \\ y_1 \geq 0, y_2 \geq 0, y_3 \geq 0 \end{cases}$$

Formulate each application as a standard minimum problem and then solve it by finding the dual maximum problem and using the simplex method. State your final answer in terms of the original question.

39. Pens and Ink   A student needs at least 3 new pens and 2 dozen ink cartridges. The bookstore sells single pens for 99¢, packages of 6 ink cartridges for 89¢, and a $1.19 "writer's combo" package containing 1 pen and 2 ink cartridges. How many of each will meet the student's needs at the least cost?

40. Production Planning   A pasta company is expanding its linguini production facility. Two machines are available: a small-capacity machine costing $5000 that produces 20 pounds per minute and needs 1 operator, and a large-capacity machine costing $6000 that produces 30 pounds per minute and needs 2 operators. The company wants to hire no more than 34 additional employees yet increase production by at least 600 pounds per minute. How many of each machine should the company buy to expand its production at the least cost?

4.5   Nonstandard Problems: The Dual Pivot Element and the Two-Stage Method

Rewrite each nonstandard maximum problem in matrix form, construct the initial simplex tableau, and locate the dual pivot element. (Do not carry out the pivot operation.)

41. Maximize $P = 80x_1 + 30x_2$

$$\text{Subject to } \begin{cases} 4x_1 + x_2 \leq 40 \\ 2x_1 + 3x_2 \leq 60 \\ x_1 + x_2 \geq 10 \\ x_1 \geq 0, x_2 \geq 0 \end{cases}$$

42. Maximize $P = 4x_1 + 8x_2 + 6x_3 + 10x_4$

$$\text{Subject to } \begin{cases} 2x_1 + 3x_2 - 3x_3 + x_4 \leq 30 \\ 7x_1 - 5x_2 + x_3 + 5x_4 \geq 35 \\ x_1 \geq 0, x_2 \geq 0, x_3 \geq 0, x_4 \geq 0 \end{cases}$$

Solve each nonstandard maximum problem by the two-stage simplex method.

43. Maximize $P = 3x_1 + 4x_2$

$$\text{Subject to } \begin{cases} x_1 + x_2 \leq 10 \\ x_1 + x_2 \geq 5 \\ x_1 \geq 0, x_2 \geq 0 \end{cases}$$

44. Maximize $P = 8x_1 - 20x_2 - 18x_3$

$$\text{Subject to } \begin{cases} x_1 + 3x_2 + 2x_3 \leq 55 \\ x_1 + x_2 + x_3 \geq 25 \\ 2x_1 + 3x_2 + 3x_3 \leq 70 \\ x_1 \geq 0, x_2 \geq 0, x_3 \geq 0 \end{cases}$$

45. Maximize $P = 6x_1 - 9x_2$

$$\text{Subject to } \begin{cases} 3x_1 + x_2 \leq 24 \\ x_1 + 3x_2 \leq 24 \\ x_1 + x_2 \geq 6 \\ x_1 \geq 0, x_2 \geq 0 \end{cases}$$

46. Maximize $P = 12x_1 - 21x_2 + 2x_3$

$$\text{Subject to } \begin{cases} 2x_1 - 3x_2 + x_3 \geq 20 \\ x_1 - x_2 + x_3 \leq 25 \\ x_1 - 3x_2 \geq 5 \\ x_1 \geq 0, x_2 \geq 0, x_3 \geq 0 \end{cases}$$

47. Maximize $P = 6x_1 + 7x_2$

$$\text{Subject to } \begin{cases} x_1 \geq 2 \\ x_2 \geq 3 \\ x_1 \leq 4 \\ x_2 \leq 5 \\ x_1 \geq 0, x_2 \geq 0 \end{cases}$$

48. Maximize $P = 8x_1 + 2x_2 + 10x_3$

$$\text{Subject to } \begin{cases} 2x_1 + x_2 + x_3 \leq 30 \\ x_1 + 2x_2 + x_3 \geq 5 \\ x_1 - x_2 + x_3 \geq 10 \\ x_1 \geq 0, x_2 \geq 0, x_3 \geq 0 \end{cases}$$

Formulate each application as a nonstandard maximum problem and solve it by the two-stage simplex method. State your final answer in terms of the original question.

49. Lumber Production   A saw mill rough-cuts lumber and planes some of it to make finished-grade boards. The saw cuts 1 thousand board-feet per hour and can operate up to 10 hours each day, while the plane finishes 1 thousand board-feet per hour and can operate up to 8 hours each day. The profit per thousand board-feet is $120 for rough-cut lumber and $100 for finished-grade boards. If at least 4 thousand board-feet of finished-grade boards must be produced each day, how many board-feet of each type of lumber should be produced daily to obtain the greatest possible profit?

50. **Financial Planning** An international money manager wishes to invest no more than $150,000 in United States and Canadian stocks and bonds, with at least twice as much in bonds as in stocks. The current market yields are 10% on U.S. stocks, 5% on U.S. bonds, 11% on Canadian stocks, and 4% on Canadian bonds. If at least $30,000 must be invested in Canadian stocks, how much should be invested in each to obtain the greatest possible return?

4.6   Nonstandard Problems: Artificial Variables and the Big-M Method

Write each nonstandard maximum problem in normal form and construct the initial simplex tableau.

51. Maximize $P = 80x_1 + 30x_2$

Subject to
$$\begin{cases} 4x_1 + x_2 \le 40 \\ 2x_1 + 3x_2 \le 60 \\ x_1 + x_2 \ge 10 \\ x_1 \ge 0, x_2 \ge 0 \end{cases}$$

52. Maximize $P = 4x_1 + 8x_2 + 6x_3 + 10x_4$

Subject to
$$\begin{cases} 7x_1 - 5x_2 + x_3 + 5x_4 \ge 35 \\ 2x_1 + 3x_2 - 3x_3 + x_4 \le 30 \\ x_1 \ge 0, x_2 \ge 0, x_3 \ge 0, x_4 \ge 0 \end{cases}$$

Solve each nonstandard maximum problem by the big-M method.

53. Maximize $P = 3x_1 + 4x_2$

Subject to
$$\begin{cases} x_1 + x_2 \le 10 \\ x_1 + x_2 \ge 5 \\ x_1 \ge 0, x_2 \ge 0 \end{cases}$$

54. Maximize $P = 8x_1 - 20x_2 - 18x_3$

Subject to
$$\begin{cases} x_1 + 3x_2 + 2x_3 \le 55 \\ x_1 + x_2 + x_3 \ge 25 \\ 2x_1 + 3x_2 + 3x_3 \le 70 \\ x_1 \ge 0, x_2 \ge 0, x_3 \ge 0 \end{cases}$$

55. Maximize $P = 6x_1 - 9x_2$

Subject to
$$\begin{cases} 3x_1 + x_2 \le 24 \\ x_1 + 3x_2 \le 24 \\ x_1 + x_2 \ge 6 \\ x_1 \ge 0, x_2 \ge 0 \end{cases}$$

56. Maximize $P = 12x_1 - 21x_2 + 2x_3$

Subject to
$$\begin{cases} 2x_1 - 3x_2 + x_3 \ge 20 \\ x_1 - x_2 + x_3 \le 25 \\ x_1 - 3x_2 \ge 5 \\ x_1 \ge 0, x_2 \ge 0, x_3 \ge 0 \end{cases}$$

57. Maximize $P = 6x_1 + 7x_2$

Subject to
$$\begin{cases} x_1 \ge 2 \\ x_2 \ge 3 \\ x_1 \le 4 \\ x_2 \le 5 \\ x_1 \ge 0, x_2 \ge 0 \end{cases}$$

58. Maximize $P = 8x_1 + 2x_2 + 10x_3$

Subject to
$$\begin{cases} 2x_1 + x_2 + x_3 \le 30 \\ x_1 + 2x_2 + x_3 \ge 5 \\ x_1 - x_2 + x_3 \ge 10 \\ x_1 \ge 0, x_2 \ge 0, x_3 \ge 0 \end{cases}$$

Formulate each application as a nonstandard maximum problem in normal form and solve it by the big-M method. State your final answer in terms of the original question.

59. **Lumber Production** A saw mill rough-cuts lumber and planes some of it to make finished-grade boards. The saw cuts 1 thousand board-feet per hour and can operate up to 10 hours each day, while the plane finishes 1 thousand board-feet per hour and can operate up to 8 hours each day. The profit per thousand board-feet is $120 for rough-cut lumber and $100 for finished-grade boards. If at least 4 thousand board-feet of finished-grade boards must be produced each day, how many board-feet of each type of lumber should be produced daily to obtain the greatest possible profit?

60. **Financial Planning** An international money manager wishes to invest no more than $150,000 in United States and Canadian stocks and bonds, with at least twice as much in bonds as in stocks. The current market yields are 10% on U.S. stocks, 5% on U.S. bonds, 11% on Canadian stocks, and 4% on Canadian bonds. If at least $30,000 must be invested in Canadian stocks, how much should be invested in each to obtain the greatest possible return?

Projects and Essays

The following projects and essays are based on Chapter 4. There are no right or wrong answers—the results depend only on your imagination and resourcefulness.

1. Visit your business school's library and browse through the last several issues of the journals *Management Science* or *Operations Research*. Write a one-page report on the application of linear programming to a "real world" problem.

2. Use an internet connection to visit the Web site of the Operations Research Department of Stanford University. Write a report about that department, linear programming problems, and George Dantzig.

3. Improve the program PIVOT so that it will find the pivot column and row for you. Be sure your new program checks that the tableau is feasible before looking for a pivot column. What should the program do if the tableau is not feasible?

4. Another method of solving linear programming problems is the "ellipsoid method" or "Khachiyan's algorithm." Look up the articles "Some References for the Ellipsoid Algorithm" by Philip Wolfe in *Management Science* **26**(8):747–749, August 1980, and "Khachiyan's Algorithm" by G. C. Berresford, A. M. Rockett, and J. C. Stevenson in *Byte* **5**(8):198–208, August 1980, and *Byte* **5**(9):242–255, September 1980. Write a one-page report on what you find.

5. Another method of solving linear programming problems is N. Karmarkar's "interior point" method. Look up the article "Karmarkar's Algorithm" by A. M. Rockett and J. C. Stevenson in *Byte* **12**(10):146–160, September 1987, and write a one-page report on what you find.

5

Probability

Probability, first developed in the seventeenth century to answer questions arising from gambling, has now become an essential tool for dealing with the randomness and uncertainty that are a part of our world.

5.1 Sets, Counting, and Venn Diagrams

The Chevalier de Méré

Probability theory as a mathematical discipline began with the analysis by Blaise Pascal (1623–1662) and Pierre de Fermat (1601–1665) of gambling problems posed by a French nobleman, Antoine Gombaude, Chevalier de Méré. The Chevalier's concerns were practical: After winning so often that he had no more takers for his wager that he could throw at least one six in 4 rolls of a die, he had changed his bet to be that he could throw at least one double six in 24 rolls of two dice, and he was now losing his bets. Since 4 rolls with chance $\frac{1}{6}$ on each gives the same ratio as 24 rolls with chance $\frac{1}{36}$, the Chevalier was at a loss to understand his change of luck. Rather than give up on this paradox, the Chevalier had the good sense to ask Blaise Pascal, one of the foremost mathematicians of his era. Pascal's analysis of the Chevalier's problems and his correspondence with Pierre de Fermat (the other great French mathematician of this period) showed that reasonable explanations followed from careful counting of all the possible throws of the dice. We will analyze the Chevalier's bets on pages 415–416, 418–419, and 431.

Introduction

Although the word "probability" has various meanings in everyday speech, it has a precise meaning in mathematics. Before defining it mathematically, we must clarify some basic counting principles on which the definition will be based. We begin by reviewing sets and how to count their members.

Sets and Set Operations

A *set* is any well-defined collection of objects (also called *elements* or *members*). For example, we may speak of *the set of American citizens*, since there are specified conditions for being an American citizen. On the other hand, we cannot speak of *the set of all thin people*, since the word "thin" is not precisely defined. There are several *operations* on sets. We will illustrate these operations by *Venn diagrams**, in which sets are represented as regions enclosed by ellipses.

*After the English logician John Venn (1843–1923), author of *Logic of Chance*.

The *intersection* of sets A and B, denoted $A \cap B$, is the set of all elements that are in *both A and B.*

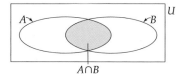

$A \cap B$

Two sets A and B are *disjoint* if their intersection is empty (that is, if they have no elements in common).

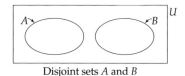

Disjoint sets A and B

The *union* of sets A and B, denoted $A \cup B$, is the set of all elements that are in *either A or B* (or both).

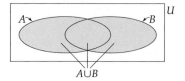

$A \cup B$

The *complement* of a set A, denoted A^c, is the set of all elements *not in A.*

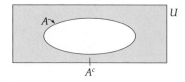

A^c

In general, the words "and," "or," and "not" translate into "intersection," "union," and "complement." We will use the word "or" in the *inclusive* sense, meaning one possibility or the other *or both*. We always draw sets inside a larger rectangle, labeled U for *universal* set, which represents the set of all possible elements being considered. The complement of the universal set is the *empty* set, also called the *null* set, denoted $\varnothing$, which has no members. A *subset* of a set is a set (possibly empty) of elements from the original set. We use the symbol $n(A)$ to mean the number of elements in a set A:

$$n(A) = \begin{pmatrix} \text{Number of} \\ \text{elements in } A \end{pmatrix}$$

For example, $n(\varnothing) = 0$ because the null set has no elements. From the previous Venn diagram, it is clear that the elements of the set A^c are precisely the elements of the universal set U that are *not* in A. Therefore:

Complementary Principle of Counting

$n(A^c) = n(U) - n(A)$	The number of elements in the *complement* of a set is the number of elements in the universal set minus the number of elements in the original set

We will find this principle useful when it is easier to count the elements in the complement of a set rather than the elements in the set itself.

Addition Principle for Counting

Notice that in the following Venn diagram there are $n(A) = 4 + 3 = 7$ elements (dots) in A and $n(B) = 3 + 2 = 5$ elements in B.

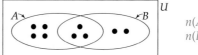

$n(A) = 7$
$n(B) = 5$

Adding $n(A) = 7$ and $n(B) = 5$ gives 12, but this is *not* the number of elements in $A \cup B$, since $n(A \cup B) = 4 + 3 + 2 = 9$. Why doesn't adding $n(A)$ and $n(B)$ give $n(A \cup B)$? The problem is that the elements in the middle of the diagram (that is, in $A \cap B$) get counted *twice*, once in A and once in B. To get the correct number of elements in $A \cup B$ we must add $n(A)$ to $n(B)$ but then *subtract* the number of elements in $A \cap B$ (to correct for the double-counting). This gives the general rule:

Addition Principle of Counting

$n(A \cup B) = n(A) + n(B) - n(A \cap B)$	The number of elements in the *union* of two sets is the number of elements in one plus the number of elements in the other minus the number of elements in both

Of course, if A and B are disjoint, then $A \cap B$ is empty, so $n(A \cap B) = 0$, giving a simpler addition principle for disjoint sets:

$$n(A \cup B) = n(A) + n(B) \qquad \text{for } A, B \text{ disjoint}$$

EXAMPLE 1   Counting Cars in a Parking Lot

A mall parking lot contains 150 convertibles, 200 cars with sound systems, and 90 convertibles with sound systems. How many cars in the lot are convertibles or have sound systems?

Solution

Let C be the set of convertibles, and let S be the set of cars with sound systems. Since "or" means "union," we want $n(C \cup S)$. We are given that $n(C) = 150$, $n(S) = 200$, and $n(C \cap S) = 90$, and using the addition principle, we find that

$$n(C \cup S) = n(C) + n(S) - n(C \cap S) = \underbrace{150}_{150} + \underbrace{200}_{200} - \underbrace{90}_{90} = 260$$

There are 260 cars that are convertibles or have sound systems.

■

PRACTICE PROBLEM 1

A survey of insurance coverage in 300 metropolitan businesses revealed that 150 offer their employees dental insurance, 150 offer vision coverage, and 100 offer both dental and vision coverage. How many of these businesses offer their employees dental or vision insurance?

Solution at the back of the book

We may identify all parts of a Venn diagram in terms of sets and complements.

$A \cap B$	in A and B
$A \cap B^c$	in A but outside B
$A^c \cap B$	outside A but in B
$A^c \cap B^c$	outside A and outside B

EXAMPLE 2   Counting Customers

A regional telephone company with 200,000 customers provides 60,000 with call waiting and 70,000 with call forwarding. If 20,000 customers have both, how many have neither?

Solution

We could solve this problem by the addition principle, as before, but instead we use Venn diagrams. Let W be the set of customers with call waiting, and let F be the set of customers with call forwarding. We want to find $n(W^c \cap F^c)$, and we are given that $n(U) = 200{,}000$, $n(W) = 60{,}000$, $n(F) = 70{,}000$, and $n(W \cap F) = 20{,}000$. We fill in the "pieces" of a Venn diagram in stages, beginning with the only piece that we are given, $n(W \cap F)$. For simplicity, we enter the numbers in thousands.

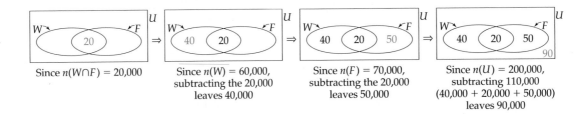

Since $n(W \cap F) = 20{,}000$

Since $n(W) = 60{,}000$, subtracting the 20,000 leaves 40,000

Since $n(F) = 70{,}000$, subtracting the 20,000 leaves 50,000

Since $n(U) = 200{,}000$, subtracting 110,000 $(40{,}000 + 20{,}000 + 50{,}000)$ leaves 90,000

This last number, $n(W^c \cap F^c) = 90{,}000$, is the answer we wanted: 90,000 people have neither call waiting nor call forwarding.

∎

Notice that in this example we used the *complementary principle* of counting, finding the "outside" by finding the inside parts and subtracting.

The Multiplication Principle for Counting

Suppose that you are choosing an outfit to wear, and may choose any one of 3 shirts, S_1, S_2, or S_3, and either of 2 pairs of pants, P_1 or P_2. Since the 3 shirts can be combined freely with the 2 pairs of pants, there are $3 \times 2 = 6$ different possible outfits, namely S_1P_1, S_1P_2, S_2P_1, S_2P_2, S_3P_1, and S_3P_2. More generally, we have the following *multiplication principle* that will be very useful throughout this chapter.

Multiplication Principle for Counting

If two choices are to be made, and there are m possibilities for the first choice and n possibilities for the second choice, and if any first choice can be combined with any second choice, then the *combination* of the two choices can be made in $m \cdot n$ ways.

The multiplication principle can be proved by enumerating all of the possibilities, as in the following example.

EXAMPLE 3   **Counting Different Products**

A toy company makes red, green, blue, and yellow plastic cars, trucks, and planes. How many different kinds of toys do they make?

Solution

Let the set of colors be C = {red, green, blue, yellow}, and let the set of shapes be S = {car, truck, plane}. Each possible toy is described by its color and shape:

	Red	**Green**	**Blue**	**Yellow**
Car	(red, car)	(green, car)	(blue, car)	(yellow, car)
Truck	(red, truck)	(green, truck)	(blue, truck)	(yellow, truck)
Plane	(red, plane)	(green, plane)	(blue, plane)	(yellow, plane)

This table contains all possible combination of colors and shapes, and since the table is a rectangle, the number of boxes is found by multiplying length times width: There are $4 \cdot 3 = 12$ possible toys that can be made.

■

We can also show all of the possibilities of a multi-stage choice by means of a *tree diagram*. The following tree diagram for the toy company first lists all of the possible *colors,* and then it "branches" from each color to a list of all the possible shapes:

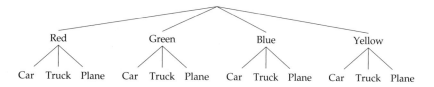

By counting all the "leaves" at the ends of the branches, we can find the number of pairs. Again we see that there are $12 = 4 \cdot 3$ possible toys. This tree has two layers of branches, corresponding to the two choices (color, shape) to be made. If there were three choices (such as color, shape, and material), then the tree diagram would have *three* layers of branches. Tree diagrams with many layers (or many items in each layer) quickly become too cumbersome to draw.

A convenient way of using the multiplication principle is to imagine making up a typical combination of the two choices. For instance, in the preceding example there were 4 choices for the color of the toy and 3 choices for its shape, giving

$$4 \cdot 3 = 12 \text{ possible toys.}$$

(_____ , _____)
  ↑   ↑
 Color   Shape
(4 choices) (3 choices)

The Multiplication Principle for Counting generalizes to *more* than two choices.

Generalized Multiplication Principle for Counting

> If k choices are to be made, and there are m_1 possibilities for the first choice, m_2 possibilities for the second choice, m_3 possibilities for the third choice, and so on down to m_k possibilities for the kth choice, and if the choices can be combined in any way, then the *combination* of the k choices can be made in $m_1 \cdot m_2 \cdot \cdots \cdot m_k$ ways.

EXAMPLE 4   Counting Parking Permits

A parking permit displays an identification code consisting of a letter (A to Z) followed by two digits (0 to 9). How many different permits can be issued?

Solution

Since each identification code consists of three symbols, we need to fill three blanks:

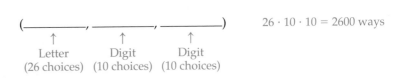

$$26 \cdot 10 \cdot 10 = 2600 \text{ ways}$$

(_____ , _____ , _____)
   ↑   ↑   ↑
 Letter   Digit   Digit
(26 choices) (10 choices) (10 choices)

Therefore, 2600 different permits can be issued.   ∎

PRACTICE PROBLEM 2

A computer network password consists of four letters (A to Z) followed by four digits (0 to 9). How many different passwords are there?   *Solution at the back of the book*

Section Summary

A prerequisite to studying probability is mastery of several counting techniques for accurately enumerating large collections of possibilities without having to count "one by one." Three of the most important techniques are summarized below.

$$n(A^c) = n(U) - n(A)$$ Complementary principle

$$n(A \cup B) = n(A) + n(B) - n(A \cap B)$$ Addition principle

$$\left(\begin{array}{c}\text{Number of pairs } (a, b) \\ \text{with } a \text{ in } A \text{ and } b \text{ in } B\end{array}\right) = n(A) \cdot n(B)$$ Multiplication principle

EXERCISES 5.1

Find each quantity using the following Venn diagram.

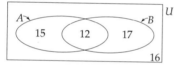

1. $n(A) = 27$
2. $n(B)$
3. $n(U) = 60$
4. $n(A \cap B)$
5. $n(A \cup B) = 44$
6. $n(A^c)$
7. $n(B^c) = 31$
8. $n(A \cap B^c)$
9. $n(A^c \cap B) = 17$
10. $n(A \cup B^c)$

Identify each numbered part in the following Venn diagram in terms of intersections of the sets A, B, and C and their complements.

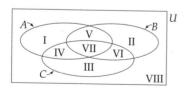

11. Part I
12. Part II
13. Part III
14. Part IV
15. Part V
16. Part VI
17. Part VII
18. Part VIII

For exercises 19–22, draw a tree diagram for each situation, and count the number of leaves to find the number of possible pairs.

19. A parking permit sticker displays an identification code consisting of a letter (from A, B, and C) followed by a digit (from 1, 2, 3, and 4). How many different permits can be issued?

20. A computer network password consists of two letters, the first being X, Y, or Z and the second being A, B, C, D, or E. How many different passwords are there?

21. A clothing store sells windbreakers, ski jackets, and overcoats in your choice of red or blue. How many different kinds of coats do they sell?

22. Ted, Bob, Fred, Jim, Bill, and Sam share a student dorm suite. How many ways can they choose who takes out the garbage and who sweeps the floor if they can choose the same person to do both? What if different men must do the chores?

23. Is it practical to solve Example 4 (page 392) using a tree diagram? Sketch at least part of the diagram to see how large it might be.

24. Is it practical to solve Practice Problem 2 (page 392) using a tree diagram? Sketch at least part of the diagram to see how large it might be.

APPLIED EXERCISES (▦ or ▦ may be helpful.) _____

25. Lacrosse The sophomore lacrosse team has 24 players, of whom 10 played defense last year, 12 played offense, and 5 played both defense and offense, while the rest of the players did not play last year. How many members of the team played last year?

26. Political Fund Raising If 12,300 individuals contributed to the governor's first election campaign, 15,200 contributed to her second, and 7800 contributed to both, how many individuals made contributions?

27. Current Events A survey of 1200 residents of New Orleans revealed that 760 get information on international events by watching television news programs, 530 from listening to radio news shows, and 290 from both television and radio, while the rest have no interest in international events. How many have no interest in international events?

28. International Marketing Of the 30 freshman majoring in international marketing, 12 speak French, 8 speak German, and 5 speak both languages. How many do not speak either French or German?

29. Parking Permits Look back at Example 4 on page 392 and find the number of parking permits that can be issued if the letter O and the number 0 are omitted to avoid confusion.

30. Computer Security A computer network password is made up of letters A to Z and may be either four or five letters long. How many such passwords are there? Find the answer in two different ways:

   a. Find the number of four-letter passwords and add it to the number of five-letter passwords.

   b. Think of a four-letter password as really having length 5, where in the fifth position is a "blank" that can be thought of as one more member of an extended "alphabet" for the last position.

31. Computer Security How many eight-symbol computer passwords can be formed using the letters A to J and the digits 2 to 6?

32. Bagels A bagel shop offers 12 different kinds of bagels and 18 flavors of cream cheese. How many different orders for a bagel and cream cheese can customers place?

33. Combination Locks A suitcase combination lock has four wheels, each labeled with the digits 0 to 9. How many lock combinations are possible?

34. Awards A small community consists of 25 families, each consisting of two parents and three children. If one parent and one of his or her children are to be honored as "parent and child of the year," in how many different ways can the award be made?

35. Committees A college student governance committee is made up of 3 freshman, 4 sophomores, 4 juniors, and 5 seniors. A subcommittee of 4 consisting of one person from each class is to go to a national convention. How many different subcommittees are there?

36. Mix & Match Fashions A designer has created a completely mixable line of five shirts, nine ties, and four pairs of slacks and advertises that "the possibilities are endless." Exactly how "endless" are the possible different outfits consisting of a shirt, a tie, and a pair of slacks?

Explorations and Excursions

The following problems extend and augment the material presented in the text.

De Morgan's Laws explain the relationship between the complement of a set and the operations of union and intersection. (A symbolic-logic version of De Morgan's Laws will be discussed on pages 570 and 597.)

37. Show the first of De Morgan's Laws, $(A \cup B)^c = A^c \cap B^c$, by carefully sketching Venn diagrams for $(A \cup B)^c$ and for $A^c \cap B^c$.

38. Show the second of De Morgan's Laws, $(A \cap B)^c = A^c \cup B^c$, by carefully sketching Venn diagrams for $(A \cap B)^c$ and for $A^c \cup B^c$.

39. Verify that $n((A \cup B)^c) = n(A^c \cap B^c)$ for the sets $A = W$ and $B = F$ discussed in Example 2 (pages 389–390).

40. Verify that $n((A \cap B)^c) = n(A^c \cup B^c)$ for the sets $A = W$ and $B = F$ discussed in Example 2 (pages 389–390).

Permutations, Combinations, and the Binomial Theorem

Starting Lineups

A sports broadcaster, irritated that a baseball manager kept changing his lineup, once said that the manager should "choose his most talented 9 players, try them in all different batting orders, and then stick with the best one." Does this advice make sense? How many possible orderings for nine players are there? In this section we will see that there are $9 \cdot 8 \cdot 7 \cdot 6 \cdot 5 \cdot 4 \cdot 3 \cdot 2 \cdot 1 = 362,880$ different orderings. How long would it take to try them out? If the manager were to try each one for just 15 minutes and hold tryouts 8 hours a day every day of the year, it would take more than 30 years to try them all! The techniques we develop in this section would have helped the broadcaster avoid making such an impractical suggestion.

Introduction

In this section we will develop two very useful formulas for counting various types of choices, known as *permutations* and *combinations*, and then apply these formulas to a wide variety of problems. We begin by describing *factorial notation*.

Factorials

Products of successive integers from a number down to 1, such as $5 \cdot 4 \cdot 3 \cdot 2 \cdot 1$, are called *factorials*. We denote factorials by exclamation points, so the preceding product would be written 5!. Formally:

Factorials

For any positive integer n,	
$n! = n(n-1) \cdot \cdots \cdot 1$	n factorial is the product of the integers from n down to 1
$0! = 1$	Zero factorial is 1

Exercise 23 shows why zero factorial is defined to be 1.

EXAMPLE 1 **Calculating Factorials**

Find: **a.** $4!$ **b.** $\dfrac{7!}{6!}$ **c.** $\dfrac{6!}{3!}$ **d.** $\dfrac{100!}{99!}$

Solution

a. $4! = 4 \cdot 3 \cdot 2 \cdot 1 = 24$

b. $\dfrac{7!}{6!} = \dfrac{7 \cdot 6 \cdot 5 \cdot 4 \cdot 3 \cdot 2 \cdot 1}{6 \cdot 5 \cdot 4 \cdot 3 \cdot 2 \cdot 1} = \dfrac{7 \cdot \cancel{6 \cdot 5 \cdot 4 \cdot 3 \cdot 2 \cdot 1}}{\cancel{6 \cdot 5 \cdot 4 \cdot 3 \cdot 2 \cdot 1}} = 7$

When finding quotients of factorials, look for cancellation

c. $\dfrac{6!}{3!} = \dfrac{6 \cdot 5 \cdot 4 \cdot 3 \cdot 2 \cdot 1}{3 \cdot 2 \cdot 1} = \dfrac{6 \cdot 5 \cdot 4 \cdot \cancel{3 \cdot 2 \cdot 1}}{\cancel{3 \cdot 2 \cdot 1}} = 6 \cdot 5 \cdot 4 = 120$

d. $\dfrac{100!}{99!} = \dfrac{100 \cdot 99 \cdot \,\cdots\, \cdot 1}{99 \cdot \,\cdots\, \cdot 1} = 100$

Canceling $99 \cdot \,\cdots\, \cdot 1$

We will use factorials to count *orderings* of objects.

Permutations

How many different orderings are there for the letters *a*, *b*, and *c*? We may list the orderings as *abc*, *acb*, *bac*, *bca*, *cab*, and *cba*, so there are 6. Each of these orderings is called a *permutation*. Instead of listing them all, we could observe that there are 3 ways of choosing the first letter, 2 ways of choosing the second (because one letter was "used up" in the first choice), and 1 way of choosing the last (whichever is left), so by the multiplication principle there are $3 \cdot 2 \cdot 1 = 6$ possible orderings, just as we found before. How many permutations are there of *n* distinct objects? By the same reasoning, the answer is $n(n - 1) \cdot \,\cdots\, \cdot 1 = n!$.

EXAMPLE 2 **Counting Batting Orders**

How many different batting orders are there for a 9-player baseball team?

Solution

There are $9! = 9 \cdot 8 \cdot 7 \cdot 6 \cdot 5 \cdot 4 \cdot 3 \cdot 2 \cdot 1 = 362{,}880$ possible batting orders.

PRACTICE PROBLEM 1

You have six different tasks to do today. In how many different orders can they be done?

Solution at the back of the book

EXAMPLE 3 **Counting Orderings of Books**

You have 3 math books, 4 history books, and 5 English books. In how many ways can these 12 books be arranged on the shelf if all books of the same subject are together?

Solution

There are 3! ways to arrange the math books among themselves, 4! ways to arrange the history books, and 5! ways to arrange the English books, so by the multiplication principle there are $3! \cdot 4! \cdot 5!$ ways if the order of the subjects is math then history then English. However, these three subjects can *themselves* be arranged in any of $3 \cdot 2 \cdot 1 = 3!$ orderings. For each of the 3! orderings of the subjects the books can be arranged in $3! \cdot 4! \cdot 5!$ ways, so the total number of such arrangements is

$$3! \cdot 3! \cdot 4! \cdot 5! = 6 \cdot 6 \cdot 24 \cdot 120 = 103{,}680$$

∎

To return to our baseball example, how many different orderings are there for just *the first 3 batters* on the 9-player team? Clearly any one of the 9 players can bat first, then any one of the remaining 8, then any one of the remaining 7, for a total of $9 \cdot 8 \cdot 7$ orderings (by the multiplication principle). In general, if we have n distinct objects and want to count all possible orderings of any r of them, there are n choices for the first, $n - 1$ choices for the second, $n - 2$ choices for the third, down to $n - r + 1$ choices for the rth. (Notice that taking r of them means leaving $n - r$, so the last one kept is the preceding one, $n - r + 1$.) Such orderings are called *permutations* and can be counted using the following formula.

Permutations*

The number of permutations (ordered arrangements) of n distinct objects taken r at a time is

$$\overset{r \text{ factors}}{\overbrace{{}_nP_r = n \cdot (n - 1) \cdot \cdots \cdot (n - r + 1)}}$$

Product of r numbers from n down

We define ${}_nP_0 = 1$ since there is exactly one way of taking zero objects from n—namely, taking nothing.

* Alternative expressions for ${}_nP_r$ are $P_{n,r}$, P_r^n, and $P(n,r)$.

EXAMPLE 4 **Counting Nonsense Words**

How many five letter "nonsense" words (that is, strings of letters without regard to meaning) can be made from the letters A to Z with no letter repeated?

Solution

Words are ordered arrangements of letters, so we use the permutation formula with $n = 26$ and $r = 5$.

$$_{26}P_5 = 26 \cdot 25 \cdot 24 \cdot 23 \cdot 22 = 7{,}893{,}600$$

Product of 5 numbers from 26 down

There are 7,893,600 five-letter nonsense words with distinct letters.

Instead of using the permutation formula, we may think of building each word by filling in 5 blanks with distinct letters and multiplying the numbers of choices:

$$(\underline{\qquad}, \underline{\qquad}, \underline{\qquad}, \underline{\qquad}, \underline{\qquad})$$

↑	↑	↑	↑	↑
26	25	24	23	22
choices	choices	choices	choices	choices

This gives $26 \cdot 25 \cdot 24 \cdot 23 \cdot 22 = 7{,}893{,}600$ nonsense words, just as before.

Graphing Calculator Exploration

Values of $_nP_r$ are easy to find on some graphing calculators. The following displays show $_nP_r$ for $n = 26$ and several values of r, showing that $_nP_r$ gets large quickly as r increases.

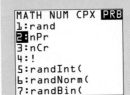

```
MATH NUM CPX PRB
1:rand
2:nPr
3:nCr
4:!
5:randInt(
6:randNorm(
7:randBin(
```

```
26 nPr 1
                26
26 nPr 2
               650
26 nPr 3
             15600
```

```
26 nPr 4
            358800
26 nPr 5
           7893600
26 nPr 6
         165765600
```

EXAMPLE 5 **Counting License Plates**

If a car license plate consists of three distinct letters followed by three distinct digits, how many different license plates are possible?

Solution

The number of three-letter patterns with no repeats is $_{26}P_3 = 26 \cdot 25 \cdot 24$, and the number of three-digit patterns with no repeats is $_{10}P_3 = 10 \cdot 9 \cdot 8$. By the multiplication principle, the number of possible license plates is

$$\underbrace{26 \cdot 25 \cdot 24}_{_{26}P_3} \cdot \underbrace{10 \cdot 9 \cdot 8}_{_{10}P_3} = 11{,}232{,}000$$

∎

PRACTICE PROBLEM 2

Jurors at an art exhibition must select the first-, second-, and third-place winners from an exhibition of 35 paintings. In how many different ways can the winning paintings be chosen?   *Solution at the back of the book*

Combinations

Sometimes we want to count choices *where order does not matter;* such choices are called *combinations.* For example, how many different pairs of letters can be made from the letters *a, b, c,* and *d* without repetition and *without regard to order*? We may write out the pairs as *ab, ac, ad, bc, bd,* and *cd,* so there are 6. Notice that combinations are counted *without regard to order,* so *ba* is considered to be the same object as *ab* (as distinct from *permutations,* in which they would be counted as *different* objects). If we *were* considering order, then there would clearly be $_4P_2 = 4 \cdot 3 = 12$ orderings, but *not* considering order means dividing this number by 2 (since each pair can be reordered the other way) to get 6, just as we found by listing them. This observation leads to a way of counting combinations: Count the number of *permutations* and then divide by the number of possible *reorderings* of each permutation to eliminate duplicates.

In general, let $_nC_r$ be the number of combinations of n things taken r at a time. To find $_nC_r$ we simply take the number of permutations, $_nP_r = n \cdot \cdots \cdot (n - r + 1)$, and then divide by $r!$ to eliminate multiple countings (reorderings). That is,

Combinations*

The number of combinations (*unordered* arrangements) of n distinct objects taken r at a time is

$$_nC_r = \frac{n \cdot (n - 1) \cdot \cdots \cdot (n - r + 1)}{r \cdot (r - 1) \cdot \cdots \cdot (1)} \qquad \longleftarrow r \text{ numbers beginning with } n$$
$$\longleftarrow r!$$

$$= \frac{n!}{r!(n - r)!} \qquad\qquad \text{In factorial form}$$

* Alternative expressions for $_nC_r$ are $C_{n,r}$, C_r^n, $C(n,r)$, and $\binom{n}{r}$.

We get the second formula from the first by multiplying the numerator and denominator by $(n - r) \cdot \cdots \cdot 1 = (n - r)!$, which "completes" the factorial in the numerator. We define $_nC_0 = 1$ since there is clearly one way to choose zero objects from n—namely, taking nothing.

EXAMPLE 6   Counting Committees

How many committees of 2 can be formed from a group of 5 people? How many committees of 3?

Solution

For a committee, the *order* in which the members are chosen does not matter, only who is eventually *on* the committee, so we want *combinations*. The number of committees of 2 is

$$_5C_2 = \frac{5 \cdot 4}{2 \cdot 1} = 10 \qquad \begin{matrix} \longleftarrow \text{Two numbers beginning with 5} \\ \longleftarrow 2! \end{matrix}$$

The factorial formula $_5C_2 = \dfrac{5!}{2!\,3!}$ would give the same answer, 10

The number of committees of 3 is

$$_5C_3 = \frac{5 \cdot 4 \cdot 3}{3 \cdot 2 \cdot 1} = 10$$

Or, using factorials, $_5C_3 = \dfrac{5!}{3!\,2!} = 10$

Notice that with 5 people there are exactly as many committees of 2 as committees of 3, since $_5C_2$ and $_5C_3$ were each 10. There is a simple reason for this. Each time you choose a committee of 2 from a group of 5, you are, in a sense, selecting *another* committee of 3 people, those *not* chosen. Any new choice of 2 also selects a new *non*choice of 3, so there are as many committees of 2 as committees of 3 that can be formed from 5 people. This is also obvious from the factorial formulas: $_5C_2 = \frac{5!}{2!\,3!}$ and $_5C_3 = \frac{5!}{3!\,2!}$ are clearly equal. For the same reason, given n people, there are as many committees of r people as committees of $n - r$ people, so we have in general that $_nC_r = {_nC_{n-r}}$.

EXAMPLE 7   Counting Permutations and Combinations

A student club has 15 members. (a) How many ways can a president, vice president, and treasurer be chosen? (b) How many ways can a committee of three members be chosen?

Solution

Each question involves choosing three members from the club, but for the officers we want *ordered* arrangements (the order determines the offices: the president is listed first, the vice president second, and the treasurer third), while for the committee we want *unordered* arrangements. Thus part (a) asks for permutations, part (b) for combinations.

a. For the officers: $_{15}P_3 = 15 \cdot 14 \cdot 13 = 2730$
3 numbers from 15 down

b. For the committee: $_{15}C_3 = \dfrac{15 \cdot 14 \cdot 13}{3 \cdot 2 \cdot 1} = 455$
Permutations divided by 3!

There are 2730 different ways of choosing the president, vice president, and treasurer, and 455 ways of choosing the committee.

■

PRACTICE PROBLEM 3

A college business major can also minor in computer science by taking any 6 courses from an approved list of 10 courses. How many different collections of courses will satisfy the requirements for the computer science minor?   *Solution at the back of the book*

Graphing Calculator Exploration

Values of $_nC_r$ are easy to find if your calculator includes this command. The following screens show several values of $_nC_r$ for $n = 15$. Notice that $_nC_r$ gets larger and then smaller as r increases.

```
MATH NUM CPX PRB
1:rand
2:nPr
3:nCr
4:!
5:randInt(
6:randNorm(
7:randBin(
```

```
15 nCr 2
              105
15 nCr 3
              455
15 nCr 4
             1365
```

```
15 nCr 12
              455
15 nCr 13
              105
15 nCr 14
               15
```

Observe also that $_{15}C_2$ and $_{15}C_{13}$ are both 105 and that $_{15}C_3$ and $_{15}C_{12}$ are both 455. These are two more examples of the relationship $_nC_r = {}_nC_{n-r}$.

A *standard deck* of 52 playing cards contains four *suits* (spades, hearts, diamonds, and clubs), each of which contains three *face cards* (jack, queen, and king), cards numbered 2 through 10, and an *ace*. A *hand* of cards is a selection of cards from the deck.

EXAMPLE 8   **Counting 5-Card Hands**

How many different 5-card hands are there?

Solution

Since the order in which the cards are dealt makes no difference to the final hand, we want the number of *combinations* of 52 cards taken 5 at a time.

$$_{52}C_5 = \frac{52 \cdot 51 \cdot 50 \cdot 49 \cdot 48}{5 \cdot 4 \cdot 3 \cdot 2 \cdot 1} = 2{,}598{,}960$$

There are 2,598,960 possible 5-card hands.

∎

EXAMPLE 9   **Counting Specified Hands**

How many 5-card hands will have two 8's and three jacks?

Solution

We can choose the two 8's (from the four 8's) in $_4C_2$ different ways, and we can choose the three jacks (from the four jacks) in $_4C_3$ different ways. Then, by the multiplication principle, the number of ways in which the combined choice can be made is

$$_4C_2 \cdot {_4C_3} = \frac{4 \cdot 3}{2 \cdot 1} \cdot \frac{4 \cdot 3 \cdot 2}{3 \cdot 2 \cdot 1} = 6 \cdot 4 = 24$$

There are 24 possible hands with two 8's and three jacks.

∎

Binomial Theorem

Recall that

$$(x + y)^2 = x^2 + 2xy + y^2$$

See page 26 or $(x + y)(x + y) =$
$x^2 + xy + yx + y^2 = x^2 + 2xy + y^2$

To find a formula for $(x + y)^n$ for *any* power n, we multiply n copies of $(x + y)$ together:

$$(x + y)^n = \underbrace{(x + y)} \cdot \underbrace{(x + y)} \cdot \cdots \cdot \underbrace{(x + y)}$$

1st
factor

2nd
factor

nth
factor

Multiplying these n factors means finding all products consisting of one term (either x or y) from each factor and then combining similar terms. We start by taking the x from each of the n factors, giving $x \cdot \cdots \cdot x = x^n$, which may be written $_nC_0x^n$ (since $_nC_0 = 1$). Next, we take the x from *all factors except one*, and the y from the remaining factor, obtaining $x^{n-1}y$. There are n of these products since there are n factors from which we may choose the single y, giving $nx^{n-1}y$, which may be written $_nC_1x^{n-1}y$ (since $_nC_1 = n$). Next we choose the x from *all but two* factors, obtaining $x^{n-2}y^2$. There are $_nC_2$ of these products since there are $_nC_2$ ways of choosing from the n factors the 2 that provide the y's, giving $_nC_2x^{n-2}y^2$. Continuing in this way, taking the x from *all but three* factors gives $_nC_3x^{n-3}y^3$, and so on, ending with $_nC_ny^n$. Adding all of these terms gives a formula for the expansion of the "binomial" $x + y$ to a power.

Binomial Theorem*

$$(x + y)^n = {_nC_0}x^n + {_nC_1}x^{n-1}y + {_nC_2}x^{n-2}y^2 + \cdots + {_nC_n}y^n$$

For any positive integer n

The coefficients $_nC_k$ are called "binomial coefficients" since they appear in the binomial theorem. The first and last coefficients, $_nC_0$ and $_nC_n$, may be omitted because they are both equal to 1.

EXAMPLE 10   **Using the Binomial Theorem**

Expand $(x + y)^4$.

Solution

$$(x + y)^4 = {_4C_0}x^4 + {_4C_1}x^{4-1}y + {_4C_2}x^{4-2}y^2 + {_4C_3}x^{4-3}y^3 + {_4C_4}y^4$$

Binomial theorem with $n = 4$

$$= x^4 + 4x^3y + \tfrac{4 \cdot 3}{2 \cdot 1}x^2y^2 + \tfrac{4 \cdot 3 \cdot 2}{3 \cdot 2 \cdot 1}x^1y^3 + y^4$$

Expanding the combinations

$$= x^4 + 4x^3y + 6x^2y^2 + 4xy^3 + y^4$$

Simplifying

Notice the pattern in successive terms:

1. The powers of x and y always add to 4 [or for $(x + y)^n$ they add to n].

2. The powers of x begin with 4 and decrease to 0, while the powers of y begin with 0 and increase to 4 (or, more generally to n).

3. The coefficients are symmetric, being the same when read "inward" from either end.

* The binomial theorem may also be written
$$(x + y)^n = \binom{n}{0}x^n + \binom{n}{1}x^{n-1}y + \binom{n}{2}x^{n-2}y^2 + \cdots + \binom{n}{n}y^n.$$

SPREADSHEET EXPLORATION

The spreadsheet below verifies the binomial theorem with $n = 9$ for the x- and y-values entered in cells B2 and C2. The value of $(x + y)^n$ is calculated in cell E2, while the value of $_9C_0x^9 + {}_9C_1x^8y^1 + {}_9C_2x^7y^2 + \cdots + {}_9C_8x^1y^8 + {}_9C_9y^9$ shown in cell D16 is the sum of the $_nC_rx^{n-r}y^r$ values in cells D5 to D14. The COMBIN(..., ...) function was used to find the values of $_nC_r$ in B5:B14.

	A	B	C	D	E
1	n = 9	x =	y =		(x+y)^n =
2		7	-5		512
3					
4	r =	nCr =		nCr x^(n-r) y^r =	
5	0	1		40353607	
6	1	9		-259416045	
7	2	36		741188700	
8	3	84		-1235314500	
9	4	126		1323551250	
10	5	126		-945393750	
11	6	84		450187500	
12	7	36		-137812500	
13	8	9		24609375	
14	9	1		-1953125	
15					
16	Binomial Sum =			512	

Changing the x in cell B2 to 8 and the y in cell C2 to −6 will certainly change the values in cells D5 to D14. But will the values in cells E2 and D16 change?

EXAMPLE 11   **Using the Binomial Theorem**

Expand $(z - 2)^3$

Solution

$(z - 2)^3 = {}_3C_0z^3 + {}_3C_1z^2(-2) + {}_3C_2z(-2)^2 + {}_3C_3(-2)^3$ Binomial theorem with $n = 3$, $x = z$, $y = -2$

$= z^3 - 6z^2 + 12z - 8$ $_3C_0 = 1, {}_3C_1 = 3,$ $_3C_2 = 3, {}_3C_3 = 1$

∎

EXAMPLE 12   **Interpreting the Binomial Theorem**

Evaluate the binomial theorem with $x = 1$ and $y = 1$, and interpret the result in terms of numbers of committees of n people.

Solution

The binomial theorem with $x = 1$ and $y = 1$ gives

$$2^n = {}_nC_0 + {}_nC_1 + {}_nC_2 + \cdots + {}_nC_n$$

Since $(1 + 1)^n = 2^n$ and 1 to any power is 1

The left-hand side gives the number of committees of *any* size that can be formed from n people since there are two possibilities for the first person (*include* or *exclude* that person), two possibilities for the second person (include or exclude), and so on for each of the n people, giving (by the multiplication principle) 2^n possible committees. The right-hand side gives the numbers of committees of size 0, of size 1, of size 2, and so on up to size n, all of which add up to the *total* number of committees.

$$2^n \quad = \quad {}_nC_0 \quad + \quad {}_nC_1 \quad + \quad {}_nC_2 \quad + \cdots + \quad {}_nC_n$$

Number of committees of *any* size	Number of committees of size 0	Number of committees of size 1	Number of committees of size 2	Number of committees of size n

We spoke of n people and the committees made from them only to make the ideas specific. More generally, we may speak of a *set* of n elements and the possible *subsets* of this set. Therefore, we have the following result:

Number of Subsets of a Set

> A set with n elements has 2^n subsets.

EXAMPLE 13   **Finding the Subsets of a Set**

List all subsets of the set $\{a, b, c\}$ and verify that there are $2^3 = 8$ of them.

Solution

The subsets are

$$\{a\}, \{b\}, \{c\}, \{a, b\}, \{a, c\}, \{b, c\}, \{a, b, c\}, \text{ and } \varnothing$$

Indeed there are 8 subsets (including, of course, the empty set $\varnothing$).

EXAMPLE 14 **Counting Subsets**

A restaurant offers pizza with mushrooms, peppers, onions, pepperoni, and sausage. How many different types of pizza can be ordered? How many with exactly two toppings?

Solution

Because the set of toppings has 5 members, there are $2^5 = 32$ possible subsets and so 32 possible pizzas. (Do you see which subset corresponds to "plain" pizza?) There are $_5C_2 = \frac{5 \cdot 4}{2 \cdot 1} = 10$ pizzas that have exactly two toppings.

∎

Section Summary

Counting *permutations* means counting *orderings;* counting *combinations* means counting collections *without regard to order.* That is, in *permutations* a different order means a *different object* (*abc* and *bca* are different), but in combinations changing the order does *not* represent a new object (*abc* and *bca* are counted as the same). If order matters (as with letters in words or listings of people for president, vice president, and treasurer), use *permutations;* if order does not matter (as with committees or hands of cards), use *combinations.*

$$_nP_r = n \cdot (n - 1) \cdot \cdots \cdot (n - r + 1)$$

Permutations
(orderings)

$$_nC_r = \frac{n \cdot (n - 1) \cdot \cdots \cdot (n - r + 1)}{r \cdot (r - 1) \cdot \cdots \cdot 1} = \frac{n!}{r!(n - r)!}$$

Combinations
(collections)

The *binomial theorem* gives the expansion of $x + y$ to *any* positive integer power:

$$(x + y)^n = {_nC_0}x^n + {_nC_1}x^{n-1}y + {_nC_2}x^{n-2}y^2 + \ldots + {_nC_n}y^n$$

EXERCISES 5.2

Find each factorial or quotient of factorials.

1. a. $2!$ **b.** $6!$ **2. a.** $1!$ **b.** $5!$

3. $\dfrac{10!}{7!}$ **4.** $\dfrac{12!}{10!}$

5. $\dfrac{200!}{198!}$ **6.** $\dfrac{1000!}{998!}$

Find each number of permutations.

7. $_6P_3$ **8.** $_6P_2$

9. $_8P_1$ **10.** $_9P_8$

 11. a. $_{13}P_4$ **b.** $_{13}P_5$ **c.** $_{13}P_6$ **12. a.** $_{11}P_5$ **b.** $_{11}P_6$ **c.** $_{11}P_7$

 13. a. $_{10}P_1$ **b.** $_{10}P_5$ **c.** $_{10}P_{10}$ **14. a.** $_{12}P_1$ **b.** $_{12}P_6$ **c.** $_{12}P_{12}$

Find each number of combinations.

15. $_6C_2$

16. $_5C_3$

17. $_7C_3$

18. $_8C_1$

  19. a. $_{11}C_4$ **b.** $_{11}C_5$ **c.** $_{11}C_6$ **d.** $_{11}C_7$

20. a. $_{13}C_5$ **b.** $_{13}C_6$ **c.** $_{13}C_7$ **d.** $_{13}C_8$

21. a. $_{12}C_1$ **b.** $_{12}C_2$ **c.** $_{12}C_6$ **d.** $_{12}C_{11}$ **e.** $_{12}C_{12}$

22. a. $_{10}C_1$ **b.** $_{10}C_2$ **c.** $_{10}C_5$ **d.** $_{10}C_9$ **e.** $_{10}C_{10}$

23. a. Show that if $n > 1$ is an integer, then $\frac{n!}{n} = (n-1)!$

  b. Show that if the equation in part (a) is to hold for $n = 1$, then we must define $0! = 1$.

24. a. Explain why the following calculator screens show that 70! is larger than 10^{100} and that this particular calculator cannot evaluate 70!.

  b. Explain why the following calculator screens show that this calculator does not use the factorial formulas to evaluate permutations and combinations.

Use the binomial theorem to expand each expression.

25. $(x + y)^3$

26. $(x + y)^5$

27. $(a + 2)^5$

28. $(a + 3)^3$

29. $(w - 3)^4$

30. $(p - 2)^4$

31. $(x + 2y)^5$

32. $(x - 3y)^4$

33. $(2x + 3y)^4$

34. $(3x - 2y)^4$

35–36. *Pascal's triangle* is the following arrangement of positive integers with 1's down each side and with each interior value being the sum of the two numbers diagonally above it.

$$
\begin{array}{ccccccccc}
 & & & & 1 & & & & \\
 & & & 1 & & 1 & & & \\
 & & 1 & & 2 & & 1 & & \\
 & 1 & & 3 & & 3 & & 1 & \\
1 & & 4 & & 6 & & 4 & & 1
\end{array}
$$

35. Calculate the next two rows.

36. Compare the numbers in a row with the binomial coefficients [that is, the coefficients in the expansion of $(x + y)^n$] for $n = 0, 1, 2, 3$, and 4.

37. Find the first three terms in the binomial expansion of $(x + y)^{20}$.

38. Find the first three terms in the binomial expansion of $(x + y)^{12}$.

39. How many subsets does a set consisting of 8 elements have?

40. How many subsets does a set consisting of 10 elements have?

41. Show that $_nC_0 - {_nC_1} + {_nC_2} - {_nC_3} + \cdots \pm {_nC_n} = 0$ (the signs alternate). [*Hint:* Find appropriate x- and y-values to use in the binomial theorem.]

42. Show that $_nC_k + {_nC_{k-1}} = {_{n+1}C_k}$ in two ways:

  a. By algebra.

  b. By interpreting the two sides in terms of committees. [*Hint:* Think of forming committees of k people from n people plus yourself, and interpret the left-hand side as the number of committees excluding you plus the number of committees including you.]

  Do you see how this equation relates to Pascal's triangle (Exercises 35–36) where the sum of two adjacent numbers gives the number below them?

43. Evaluate both sides of the formula $2^n = {_nC_0} + {_nC_1} + {_nC_2} + \cdots + {_nC_n}$ for $n = 4$ and interpret the result in terms of numbers of committees.

44. Evaluate both sides of the formula $2^n = {_nC_0} + {_nC_1} + {_nC_2} + \cdots + {_nC_n}$ for $n = 5$ and interpret the result in terms of numbers of committees.

APPLIED EXERCISES

45. **Election Ballots**   How many different ways can the eight candidates for the school board election be listed on the ballot?

46. **Employee Training**   Six candidates in an employee training program are evaluated on the time it takes them to perform a task. Assuming that all of the times are different, how many different rankings are possible?

47. **Commercial Art**   A bank lobby wall has space to display 4 paintings by local artists. If 12 paintings are available, how many different ways can they be displayed on the wall?

48. **Postal Codes**   How many five-digit ZIP codes have no repeated digits?

49. **Telephone Numbers**   A telephone number consists of seven digits and the first digit cannot be a zero or a one. How many telephone numbers have no repeated digits?

50. **Horse Racing**   How many different ways can the 14 horses in a race finish first, second, and third?

51. **Computer Passwords**   A computer password is to consist of 4 alphanumeric characters with no repeats. (An alphanumeric character is a letter from A to Z or a digit from 0 to 9.) How may such passwords are there? How many are there if the letter O and the digit 0 are excluded to avoid confusion?

52. **Social Security Numbers**   How many nine-digit Social Security numbers start with three odd digits, then have two even digits, and end with four distinct digits from 0 to 9?

53. **Sports**   A league has 6 college teams, and each team must host each of the others once in a season. How many games are played per season?

54. **Books**   You have 2 science books, 3 language books, and 4 philosophy books. In how many ways can they be arranged on the shelf if all books of the same subject are to be together?

55. **Basketball Teams**   A junior high girls' basketball team is to consist of 5 players. How many different teams can the manager select from a roster of 12 girls?

56. **Novels**   In an English class you are to read 3 novels from a list of 21. How many choices are possible?

57. **Playing Cards**   How many five-card hands from a standard deck will contain only spades? How many will contain only cards of the same suit?

58. **Playing Cards**   How many five-card hands from a standard deck will contain only face cards? How many will contain no face cards?

59. **Labor Negotiations**   The union leadership and the management strike team have agreed to each select 4 representatives to enter into a "closed-room, around-the-clock" marathon in a final attempt to reach a settlement. If there are 10 members of the union leadership and 12 members of the management team, how many ways can the marathon negotiators be chosen?

60. **Lines and Circles**   Ten distinct points have been marked on the circumference of a circle. How many lines can be drawn through pairs of these points?

61. **Test Taking**   A test consists of 12 questions, from which each student chooses 10 to answer. How many choices are possible?

62. **Homework Grading**   A sociology professor assigned 5 problems from Unit One, 10 problems from Unit Two, and 8 problems from Unit Three for next Wednesday and announced that she will grade only 2 of the assigned problems from Unit One, 4 of those from Unit Two, and 3 from Unit Three. How many different ways can she choose the problems that she will grade?

63. **Committees**   The U.S. Senate consists of 100 members, two from each state. A committee of 5 members is to be chosen. How many such committees are possible? How many if the committee cannot have more than one senator from the same state?

64. **Juries**   In how many ways can a civil jury of 6 members split 4 to 2 on a decision?

65. **Cable TV Channels**   A cable TV company offers regular service plus 6 premium channels that must be ordered separately. How many different options are there? How many options including exactly two premium channels are there?

66. Book Clubs You join a book club that each month offers five books, of which any number (or none) can be ordered. How many possibilities are there? What if at least one book must be ordered?

67. Classes There are 10 students who have the prerequisites to take a math course. How many different possible classes are there? (Assume that a class must have at least one student.)

68. Making Change You have a penny, a nickel, a dime, a quarter, a half-dollar, and a dollar in your pocket. How many different amounts of money can you make?

Explorations and Excursions

The following problems extend and augment the material presented in the text.

Stirling's Approximation The formula below provides a good approximation to $n!$ in the sense that for large values of n, the ratio of the two quantities is very close to 1.

$$n! \approx \sqrt{2\pi n}\ n^n e^{-n}$$

69. a. Make a table of the values of $n!$ and $\sqrt{2\pi n}\ n^n e^{-n}$ for $n = 1, 2, 5, 10$, and 50.

b. For each of these values of n, calculate the ratio

$$\frac{n!}{\sqrt{2\pi n}\ n^n e^{-n}}$$

c. For each of these values of n, calculate the *percentage error*

$$\left(\frac{n!}{\sqrt{2\pi n}\ n^n e^{-n}} - 1 \right) \times 100$$

70. a. Use Stirling's approximation to show that

$$\frac{10!}{5!\ 5!} \approx \frac{2^{10}}{\sqrt{5\pi}}$$

b. For a group of 10 people, interpret the relationship in part (a) in terms of the number of committees of size 5 compared to the total number of committees of any size.

71. a. Use Stirling's approximation to show that

$$\frac{(2n)!}{n!\ n!} \approx \frac{2^{2n}}{\sqrt{n\pi}}$$

b. For a group of $2n$ people, interpret the relationship in part (a) in terms of the number of committees of size n compared to the total number of committees of any size.

Probability Spaces

APPLICATION PREVIEW

Coincidences

We often hear of "amazing" coincidences. For example, a woman named Evelyn Marie Adams won the New Jersey Lottery twice, in 1985 and in 1986, an event that was widely reported to have a probability of 1 in 17 trillion.

Actually, the figure 1 in 17 trillion is misleading: such events are not all that unlikely. In fact, given enough tries, the most outrageous things are virtually certain to happen. For example, if a coincidence is defined as an event with a one-in-a-million chance of happening to you today, then in the United States, with over

250 million people, we should expect more than 250 coincidences each day, and almost 100,000 in a year.

Returning to the supposed 1-in-17-trillion double lottery winning, that figure is the right answer to the wrong question: What is the probability that a *preselected* person who buys just *two* tickets for separate lotteries will win on both? The more relevant question is: What is the probability that *some* person, among the many millions who buy lottery tickets (most buying multiple tickets), will win twice in a lifetime? It has been calculated* that such a double winning is likely to occur once in 7 years, with the likelihood approaching certainty for longer periods.

Some knowledge of probability is necessary for an understanding of the world, if only to cast doubt on the misleading statements that one often hears. For further information on probability, see the two books listed below.†

Introduction

Some experiments always have the same outcome, while others involve "chance" or "random" effects that produce a variety of outcomes. In this section we will use such "chance" experiments as models for probability spaces, which provide the mathematical framework for our discussion of probability.

Random Experiments and Sample Spaces

A *random* experiment is one that, when repeated under identical conditions, may produce different outcomes. Each repetition of the experiment is a called a *trial,* and each result is an *outcome.* The set of all possible outcomes is called the *sample space* for the experiment. The outcomes of an experiment can often be described in several different ways, giving several possible sample spaces, depending on the interest of the observer. The only requirement is that exactly *one* of the possible outcomes occurs whenever the experiment is performed. Some simple random experiments and sample spaces.

* See Persi Diaconis and Frederick Mosteller, "Methods for Studying Coincidences," *Journal of the American Statistical Association,* **84**(408):853–861, December 1989.

† For a readable introduction to probability, see Warren Weaver, *Lady Luck* (New York: Dover Publications, 1982). For a more complete exposition, see William Feller, *An Introduction to Probability Theory and Its Applications,* Vol. 1, 3rd ed. (New York: Wiley, 1957).

Random Experiment	Sample Space
a. Flip a coin to obtain *heads* (H) or *tails* (T).	{H,T}
b. Flip a coin twice to obtain H or T followed by another H or T.	{(H,H), (H,T), (T,H), (T,T)}
c. Flip a coin twice to see whether the outcomes match (M) or differ (D).	{M,D}
d. Flip a coin twice and count the number of heads.	{0, 1, 2}
e. Flip a coin to obtain H or T and roll a die to obtain a number 1, 2, 3, 4, 5, or 6.	{(H,1), (H,2), (H,3), (H,4), (H,5), (H,6), (T,1), (T,2), (T,3), (T,4), (T,5), (T,6)}

If we denote the possible outcomes by e_1, e_2, $\cdots$, e_n, then the sample space S is the set of these outcomes,

$$S = \{e_1, e_2, \cdots, e_n\}$$

n is the number of possible outcomes

For example, flipping a coin twice [experiment (b) above] involves $n = 4$ possible outcomes, which may be listed $e_1 = (H,H)$, $e_2 = (H,T)$, $e_3 = (T,H)$, and $e_4 = (T,T)$.

Events

An *event* is a *subset* of the sample space. An event *occurs* if the outcome of the experiment is an element of the event. We will usually pick a sample space consisting of the *most basic* possible outcomes of the experiment, since more complicated events can then be described in terms of them. For instance, when tossing a coin twice, we will generally use the sample space $S = \{(H,H),(H,T),(T,H),(T,T)\}$ [experiment (b) above], since it can be used for questions like the "match or differ" experiment (c) by defining the events M and D to be

$$M = \{(H,H),(T,T)\} \qquad \text{Coins match}$$

$$D = \{(H,T),(T,H)\} \qquad \text{Coins differ}$$

PRACTICE PROBLEM 1

In the experiment of tossing a coin twice and recording the number of heads [experiment (d) above], represent the events 0, 1, and 2 using events from the sample space $S = \{(H,H),(H,T),(T,H),(T,T)\}$.

Solution at the back of the book

Among all events there is one *certain* event, the entire sample space S, which consists of *all* possible events, and therefore *must* occur. There is also one *impossible* event, the *empty* set (or *null* set) $\varnothing$, which contains *no* possible outcomes and therefore *cannot* occur.

Assigning Probabilities to Possible Outcomes

Having identified the possible outcomes of an experiment, we now assign to each of them a probability from 0 (impossible) to 1 (certain). How do we assign probabilities? This depends on our knowledge of the experiment. For example, if there are three possible outcomes, and each seems equally likely, then we assign probability $\frac{1}{3}$ to each; if there are 10 outcomes that seem equally likely, then we assign probability $\frac{1}{10}$ to each. In general,

Equally Likely Outcomes

> If each of the n possible outcomes in the sample space S is equally likely to occur, then we assign probability $\frac{1}{n}$ to each.

EXAMPLE 1   **Assigning Equal Probabilities**

a. A coin is said to be *fair* if "heads" and "tails" are equally likely. Therefore, for a fair coin we assign probabilities $P(H) = \frac{1}{2}$ and $P(T) = \frac{1}{2}$.

b. A die is said to be *fair* if each of its six faces is equally likely. Therefore, for a fair die we assign $P(1) = P(2) = P(3) = P(4) = P(5) = P(6) = \frac{1}{6}$.

c. A card randomly drawn from a standard deck is equally likely to be any one of the 52 cards. Therefore, the probability of drawing any particular card (such as the queen of hearts) is $\frac{1}{52}$.

From now on when we speak of coins or dice, we will assume that they are fair, unless stated otherwise.

EXAMPLE 2   **Assigning Equal Probability Using Combinations**

A student club has 15 members. If a 3-member fund-raising committee is selected at random, what is the probability that the committee will consist of Bob, Sue, and Tim?

Solution

Since there are $_{15}C_3 = \frac{15 \cdot 14 \cdot 13}{3 \cdot 2 \cdot 1} = 455$ different 3-member committees that can be selected from the 15 club members, and each committee is equally likely, the probability that any one particular committee is selected is $\frac{1}{455}$.

Not all outcomes are equally likely. For instance, when we toss a coin twice and count the number of heads [experiment (d) on page 411], there are three possible outcomes (0, 1, and 2), but assigning probability $\frac{1}{3}$ to each would be unrealistic since there is only one way to get 0 (namely *T,T*) but there are two ways to get 1 (*H,T* or *T,H*). Probabilities must be assigned on the basis of experience and knowledge of the basic experiment, but they are subject to two conditions: Each probability must be between 0 and 1 (inclusive), and they must add to 1.

EXAMPLE 3   Assigning Unequal Probabilities

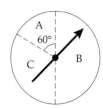

The arrow of the spinner on the right can point to any one of three regions labeled A, B, and C. If the probability of pointing to a region is proportional to its area, find the probability of pointing to each of the areas A, B, and C.

Solution

Let the sample space be $S = \{A,B,C\}$, representing the events that the spinner lands in regions A, B, or C. Clearly, region A is $\frac{60°}{360°} = \frac{1}{6}$ of the circle, region B is $\frac{1}{2}$ of the circle, and region C is the remaining $\frac{1}{3}$ of the circle. Therefore

$$P(A) = \tfrac{1}{6} \qquad P(B) = \tfrac{1}{2} \qquad P(C) = \tfrac{1}{3}$$

■

PRACTICE PROBLEM 2

Suppose that the spinner on the right is spun. What are the probabilities of the three outcomes?   *Solution at the back of the book*

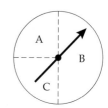

Probabilities of Events

We have assigned probabilities to events that consist of exactly one possible outcome. How do we find the probabilities of other events? First, two extreme cases. Since the sample space S includes all possible outcomes and one of them *must* occur, the probability of the event S is 1.

$$P(S) = 1 \qquad\qquad \text{Something } must \text{ happen}$$

Similarly, since the null set $\emptyset$ is empty, it contains *no* possible events and so cannot occur. Therefore, the probability of the event $\emptyset$ is zero.

$$P(\emptyset) = 0 \qquad\qquad \text{"Nothing" cannot happen}$$

Between these two extremes, an event consisting of several possible outcomes has probability equal to the *sum* of the probabilities of the possible outcomes that it contains:

Probability Summation Formula

$$P(E) = \sum_{\text{All } e_i \text{ in } E} P(e_i)$$	The probability of an event is the sum of the probabilities of the possible outcomes in it

Σ is the Greek capital letter sigma, which is equivalent to our capital S, and stands for "sum."

EXAMPLE 4 Finding the Probability of an Event

A fair die is rolled. What is the probability of rolling an even number?

Solution

Each of the six outcomes in the sample space $S = \{1, 2, 3, 4, 5, 6\}$ has probability $\frac{1}{6}$. The event "rolling an even number" is $E = \{2, 4, 6\}$. Applying the summation formula, we add the probabilities of these three possible outcomes:

$$P(E) = \sum_{\substack{\text{All } e_i \text{ in} \\ \{2,4,6\}}} P(e_i) = P(2) + P(4) + P(6) = \frac{1}{6} + \frac{1}{6} + \frac{1}{6} = \frac{3}{6} = \frac{1}{2}$$

The probability of rolling an even number is $\frac{1}{2}$.

In the preceding example, we could instead have taken the number of "favorable" outcomes (the 3 even faces) divided by the *total* number of possible outcomes (6 faces) to get $\frac{3}{6} = \frac{1}{2}$, the same answer as before. Although this method is sometimes stated as a basic rule of probability, it is important to remember that it holds only when the outcomes are *equally likely.*

$$P(E) = \frac{\left(\begin{array}{c}\text{Number of outcomes} \\ \textit{favorable} \text{ to } E\end{array}\right)}{\left(\begin{array}{c}\textit{Total} \text{ number} \\ \text{of outcomes}\end{array}\right)}$$	For equally likely outcomes

Probability That an Event Does *Not* Occur

An event E is a subset of the sample space, so the event that E does *not* occur consists of those outcomes that are *not* in the subset E, which is the complement, E^c. Since E and E^c are disjoint and between them contain all outcomes, we must have $P(E) + P(E^c) = 1$. Solving this equation for $P(E^c)$ gives

Complementary Probability

$$P(E^c) = 1 - P(E)$$	The probability that an event does *not* occur is 1 minus the probability that it *does* occur

EXAMPLE 5   **Using Complementary Probabilities**

For a student club consisting of 15 members, find the probability that a randomly chosen committee of 3 does *not* consist of Bob, Sue, and Tim.

Solution

Let E be the event that the committee *does* consist of Bob, Sue, and Tim. In Example 2 on page 412 we found that $P(E) = \frac{1}{455}$. The probability that the committee does *not* consist of Bob, Sue, and Tim is the probability of the complement E^c. The complementary probability formula gives

$$P(E^c) = 1 - P(E) = 1 - \tfrac{1}{455} = \tfrac{454}{455}$$

Therefore, the probability that the committee does *not* consist of Bob, Sue, and Tim is $\frac{454}{455}$, or about 99.8%.

■

Be careful! To say that the committee does not consist of Bob, Sue, and Tim does not mean that *none* of them is on it. For example, a committee consisting of Bob, Sue, and Zachary *would* count in the complementary event; only the single committee {Bob, Sue, Tim} is excluded.

EXAMPLE 6   **Why the Chevalier Won His First Bet**

Recall from the Application Preview on page 386 that the Chevalier de Méré's first bet was that he could roll at least one six in four rolls of a die. Find the probability that he won his bet.

Solution

In one roll of a die there are 6 possible outcomes, so by the multiplication principle, for *four* rolls there are $6 \cdot 6 \cdot 6 \cdot 6 = 6^4 = 1296$ possible outcomes, each being equally likely and each with probability $\frac{1}{1296}$. For these four rolls, let W (for "win") be the event that he rolls at least one six, so that the complement W^c is the event of *no* sixes. How many ways are there to roll *no* sixes in four rolls? Each roll must come up 1 through 5, and by the same reasoning as before, the four rolls can be done in $5 \cdot 5 \cdot 5 \cdot 5 = 5^4 = 625$ ways. According to the summation formula, we add up the probability of these 625 events, each of which is $\frac{1}{1296}$:

$$P(W^c) = \sum_{\text{All } e_i \text{ in } W^c} P(e_i) = \underbrace{\tfrac{1}{1296} + \cdots + \tfrac{1}{1296}}_{\text{625 times}} = \tfrac{625}{1296} \approx 0.482 \qquad \text{Probability of no sixes}$$

Using complementary probability principle yields

$$P(W) = 1 - P(W^c) \approx 1 - 0.482 = 0.518 \qquad \begin{array}{l}\text{Probability of} \\ \text{at least one six}\end{array}$$

Therefore, the Chevalier's guess was correct: Betting on rolling at least one six in four rolls will result in a win about 52% of the time. ∎

PRACTICE PROBLEM 3

What if the Chevalier choose *three* rolls? That is, find the probability of rolling at least one six in *three* rolls of a fair die.

Solution at the back of the book

Probability Space

A sample space and an assignment of probabilities to the possible outcomes make up a *probability space.* To summarize:

Probability Space

A *probability space* is a sample space $S = \{e_1, e_2, \ldots, e_n\}$ of possible outcomes together with probabilities $P(e_1), P(e_2), \ldots, P(e_n)$ satisfying two conditions:

1. $0 \leq P(e_i) \leq 1$ for each outcome e_i in S Probabilities are between 0 and 1

2. $P(e_1) + P(e_2) + \ldots + P(e_n) = 1$ Probabilities sum to 1

An *event* E is a subset of S and has probability

$$P(E) = \sum_{\text{All } e_i \text{ in } E} P(e_i) \qquad \begin{array}{l}\text{Sum of the probabilities} \\ \text{of the possible outcomes}\end{array}$$

In particular, $P(S) = 1$, $P(\varnothing) = 0$, and $P(E^c) = 1 - P(E)$.

Addition Rule for Probability

On page 388 we found that the number of elements in the union of two sets is given by the formula $n(A \cup B) = n(A) + n(B) - n(A \cap B)$, with the subtraction at the end done to avoid double-counting the elements that are in both A and B. For exactly the same reason, to find the probability of a union $A \cup B$, we add the probabilities of events A and B but then must subtract the probability of the intersection (to avoid double-counting outcomes in both events):

Addition Rule for Probability

$$P(A \cup B) = P(A) + P(B) - P(A \cap B)$$	The probability of a *union* is the sum of the probabilities minus the probability of the intersection

For disjoint events (also called *mutually exclusive events*), the intersection is empty and so has probability zero, leading to a simpler addition rule:

Addition Rule for Disjoint Events

$$P(A \cup B) = P(A) + P(B)$$	For *disjoint* events, the probability of the *union* is the sum of the probabilities

EXAMPLE 7 **Probability of a Union**

If you flip a coin and roll a die, what is the probability that the coin comes up heads or the die comes up four?

Solution

Let A be the event that the coin comes up heads, and let B be the event that the die comes up four. Then the sample space can be diagrammed as follows [see experiment (e) on page 411]:

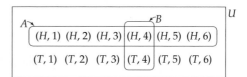

Since each of the 12 possible outcomes is equally likely, each has probability $\frac{1}{12}$. To find the probabilities of an event, we add up $\frac{1}{12}$ as many times as there are elements in it.

$$P(A) = \frac{6}{12}$$ *A has 6 elements*

$$P(B) = \frac{2}{12}$$ *B has 2 elements*

$$P(A \cap B) = \frac{1}{12}$$ *A and B intersect at one element, (H, 4)*

The addition rule gives

$$P(A \cup B) = P(A) + P(B) - P(A \cap B) = \frac{6}{12} + \frac{2}{12} - \frac{1}{12} = \frac{7}{12}$$

The probability of tossing a head or rolling a four is $\frac{7}{12}$.

EXAMPLE 8   **Finding the Probability of a Disjoint Union**

In the same probability space as in Example 7 (flipping a coin and rolling a die) what is the probability of rolling a three or a five?

Solution

We want the union of the events A and B as shown below.

		A		B		u
(H, 1)	(H, 2)	(H, 3)	(H, 4)	(H, 5)	(H, 6)	
(T, 1)	(T, 2)	(T, 3)	(T, 4)	(T, 5)	(T, 6)	

These events are disjoint, so we use the simpler version of the addition rule with $P(A) = \frac{2}{12}$ and $P(B) = \frac{2}{12}$.

$$P(A \cup B) = P(A) + P(B) = \frac{2}{12} + \frac{2}{12} = \frac{4}{12} = \frac{1}{3}$$

The probability of rolling three or five is $\frac{1}{3}$.

EXAMPLE 9   **Why the Chevalier Lost his Second Bet**

The Chevalier de Méré's second bet was that he could roll at least one *double* six in 24 rolls of two dice. Find the probability that he won this bet.

Solution

Rolling two dice has $6 \cdot 6 = 6^2 = 36$ possible outcomes. Repeating this 24 times gives 36^{24} possible outcomes, each being equally likely, so each has probability $\frac{1}{36^{24}}$. For these 24 rolls, let D be the event of rolling at least one double six, so that the complement D^c represents 24 rolls with *no* double sixes. *One* roll of two dice gives 36 possible outcomes, of which one is a double six and 35 are not. Repeating this 24 times gives (by the multiplication principle) 35^{24} ways of *not* rolling a double six. Therefore, D^c contains 35^{24} elements, each having probability $\frac{1}{36^{24}}$, and so has probability

$$P(D^c) = \frac{35^{24}}{36^{24}} = \left(\frac{35}{36}\right)^{24} \approx 0.509 \qquad \text{Using a calculator}$$

Therefore,

$$P(D) = 1 - P(D^c) \approx 1 - 0.509 = 0.491 \qquad \text{Complementary probability principle}$$

The probability of throwing at least one double six in 24 rolls of a pair of dice is (approximately) 0.49.

∎

Although both of the Chevalier's bets had probability close to $\frac{1}{2}$, his observation that in the long run the first was a winning bet and the second was a losing bet is borne out by these calculations.

Section Summary

For a random experiment, the *sample space* is the set of possible outcomes $S = \{e_1, e_2, \ldots, e_n\}$. We assign to each possible outcome a probability between 0 and 1 such that the sum of the probabilities is 1. If the n possible events are equally likely, we assign each probability $\frac{1}{n}$. A *probability space* is a sample space S together with an assignment of probabilities. An *event E* is a subset of S, and the probability of the event E is

$$P(E) = \sum_{\text{All } e_i \text{ in } E} P(e_i) \qquad \text{Adding the probabilities for each outcome in the event}$$

In particular, $P(S) = 1$, $P(\varnothing) = 0$, and $P(E^c) = 1 - P(E)$. The probability of a *union $A \cup B$* of two events is found by the *addition rule:*

$$P(A \cup B) = P(A) + P(B) - P(A \cap B)$$

If events A and B are *mutually exclusive* (that is, $A \cap B = \varnothing$), this rule simplifies to $P(A \cup B) = P(A) + P(B)$.

EXERCISES 5.3

1. Committees Find the sample space for a committee of two chosen from Alice, Bill, Carol, and Dan. Then find the sample space if both sexes must be represented. (*Use initials.*)

2. Committees Find the sample space for a committee of two formed from Alice, Bill, Carol, Dan, and Edgar. Then find the sample space if the committee must include at least one woman. (*Use initials.*)

3. Wardrobes Find the sample space for choosing an outfit consisting of one of four shirts (S_1, S_2, S_3, or S_4) and one of two jackets (J_1 or J_2).

4. Wardrobes Find the sample space for choosing an outfit consisting of one of two shirts (S_1 or S_2), one of two jackets (J_1 or J_2), and one of two pairs of pants (P_1 or P_2).

5. Marbles A box contains three marbles, one red, one green, and one blue. A first marble is chosen, its color recorded, and then it is replaced in the box and a second marble is chosen, and its color is recorded. Find the sample space. Then find the sample space if the first marble is *not* replaced before the second is chosen.

6. Coin Tossing You toss a coin until you get the first head or until you have tossed it five times, whichever comes first. Find the sample space.

Dice A die is rolled, and $A = \{1, 3, 5\}$ (rolling an odd number), and $B = \{3, 6\}$ (rolling a three or a six). Specify each event as a subset of the sample space $S = \{1, 2, 3, 4, 5, 6\}$.

7. a. $A \cap B$
 b. $A^c \cup B$

8. a. $A \cup B$
 b. $A \cap B^c$

Dice Two dice are rolled. E is the event that the sum is even, F is the event of rolling at least one six, and G is the event that the sum is eight. List the outcomes for the following events.

9. a. $E \cap F$
 b. $E^c \cap G$

10. a. $F \cap G$
 b. $E^c \cap F \cap G$

Probabilities For the sample space $S = \{e_1, e_2, e_3, e_4, e_5\}$ with probabilities $P(e_1) = 0.10$, $P(e_2) = 0.30$, $P(e_3) = 0.40$,

$P(e_4) = 0.05$, and $P(e_5) = 0.15$, and events $A = \{e_1, e_2, e_3\}$, $B = \{e_1, e_3, e_5\}$, and $C = \{e_2, e_4\}$, find each probability.

11. a. $P(A)$
 b. $P(B \cup C)$

12. a. $P(B)$
 b. $P(A \cup C)$

13. a. $P(A^c)$
 b. $P(A^c \cap B)$

14. a. $P(B^c)$
 b. $P(B^c \cap C)$

15. Committees If a committee of 3 is to be chosen at random from a class of 12 students, what is the probability of any particular committee being selected? What if the committee is to consist of a president, a vice president, and a treasurer?

16. Committees A committee of 3 is to be chosen at random from a class of 20 students. What is the probability that a particular committee will be selected? What if the 3 are to be a president, a vice president, and a treasurer?

17. Marbles One marble is selected at random from a box containing 6 red and 4 blue marbles, and its color noted, so the sample space is $\{R, B\}$. What probability should be assigned to each outcome?

Spinners Find the probability of the arrow landing in each numbered region. Assume that the probability of any region is proportional to its area and that areas that look the same size are the same size.

18.

19.

20. Dartboards A circular dartboard has a radius of 12 inches with a circular bull's-eye of radius 2 inches at the center. You throw a dart and hit the dartboard. Assume that the probability of hitting any region is proportional to the area of the region.

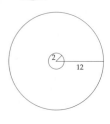

 a. What is the probability of hitting the bull's-eye?
 b. What is the probability of hitting the rest of the dartboard?

21. Baseball A baseball hits the 12-foot-by-10-foot wall of your house. Assume that the probability of its hitting any region is proportional to the area of the region.

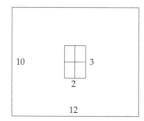

a. What is the probability that it hits the 3-foot-by-2-foot window?

b. What is the probability that it misses the window?

22. Choosing Colors One ball is to be chosen from a box containing red, blue, and green balls. If the probability that the chosen ball is red is $\frac{2}{7}$ and the probability that it is blue is $\frac{2}{7}$, what is the probability that it is green?

23. Political Contributions In a town, 38% of the citizens contributed to the Republicans, 42% contributed to the Democrats, and 12% contributed to both. What percentage contributed to neither party?

24. Credit Cards A store accepts both Visa cards and Mastercards. If 61% of its customers carry Visa cards, 52% carry Mastercards, and 28% carry both, what proportion carry a card that the store will accept?

25. Surveys A college survey claimed that 63% of students took English composition, 48% took calculus, 15% took both, and 10% took neither. Show that these figures cannot be correct.

26. Dice One die is rolled. Find the probability of:
a. rolling at least 3.
b. rolling an odd number.

27. Marbles A box contains 4 red and 8 green marbles. You reach in and remove 3 marbles all at once. Find the probability that these 3 marbles:
a. are all red. $\frac{4}{12}$
b. are all of the same color. $\frac{3}{4}$ $\frac{3}{8}$

28. Lottery In a lottery you choose 5 numbers out of 40. Then 6 numbers are announced, and you win something if you have 5 of the 6 numbers. What is the probability that you win something?

29. Guessing Numbers Someone picks a number between 1 and 10 (inclusive) and you have three guesses. What is the probability that you will get it?

30. Words Among all 5-letter nonsense words (that is, without regard to meaning), what is the probability that a word has:
a. no vowels?
b. at least one vowel?

31. Cards If you are dealt 5 cards at random from an ordinary deck, what is the probability that your hand contains all four aces?

32. Cards If you are dealt 5 cards at random from an ordinary deck, what is the probability of being dealt a flush (all 5 cards of the same suit)?

33. Elevator Stops An elevator has 5 people and makes 7 stops. What is the probability that no two people get off on the same floor?

34. Defective Products A carton of 24 CD players includes 4 that are missing a part. If you choose 4 at random, what is the probability that you get those 4?

35. Defective Products A box of 100 screws contains 10 that are defective. If you choose 10 at random, what is the probability that none are defective?

36. Committees A committee of 12 is to be formed from your class of 100 students. What is the probability that you and your best friend will be on the committee?

37. Senate The United States Senate consists of 100 members, 2 from each state. A committee of 8 senators is formed. What is the probability that it contains at least one senator from your state?

38. Light Bulbs Your house uses fifteen light bulbs, five in each of three different wattages, and you keep one spare bulb of each wattage. Two bulbs burn out. What is the probability that you have spares for both?

39. Keys You carry six keys in your pocket, two of which are for the two locks on your front door. You lose one key. What is the probability that you can get into your house through the front door?

40. Shoes You are rushing to leave on a trip, and you randomly grab four shoes from the five pairs in your closet. What is the probability that you take at least one pair?

Explorations and Excursions

The following problems extend and augment the material presented in the text.

Odds If E is an event with probability $P(E)$ such that $0 < P(E) < 1$, then "the *odds for* the event E" means the ratio $P(E):P(E^c)$ [read: "$P(E)$ to $P(E^c)$"] while "the *odds against* the event E" means the ratio $P(E^c):P(E)$ [read: "$P(E^c)$ to $P(E)$"]. Since odds are ratios, we multiply to clear the fractions, so that, for example, $\frac{2}{3}:\frac{1}{3}$ becomes 2:1. For equally likely outcomes, the odds for an event can be interpreted as the ratio of the number of *favorable* outcomes to the number of *unfavorable* outcomes.

41. Show that the odds for heads in a flip of a coin are 1:1.

42. Show that the odds for a two on a roll of a six-sided die are 1:5.

43. Show that if $P(E) = \frac{n}{m}$ then the odds for E are $n:(m - n)$.

44. Show that if $P(E) = \frac{n}{m}$ then the odds against E are $(m - n):n$.

45. Show that if the odds for E are $n:m$, then $P(E) = \frac{n}{n + m}$.

46. Show that if the odds against E are $n:m$, then $P(E) = \frac{m}{n + m}$.

47. Baseball   A radio announcer says the odds that the Yankees will beat the Orioles are 7:4. What does the announcer believe is the probability that the Yankees will beat the Orioles?

48. Horse Racing   A bookie has changed his odds for Win-By-A-Neck from 3:14 to 5:24. Does he think the probability that this three-year-old will win has gone up or down?

49. Stocks   A stock exchange announces that gainers beat losers by 7 to 2. What does this mean about the probability that a stock gained value?

5.4 Conditional Probability and Independence

Independence and the Law

In 1968 in Los Angeles, a woman was assaulted from behind. She told the police that the assailant was a young woman with blond hair. Other witnesses said that they saw a Caucasian woman with blond hair and a ponytail run from the scene and get into a yellow car driven by a black man with a mustache and beard. A few days later the police arrested a couple matching this description. At the trial, the prosecution presented the following probabilities for these characteristics:

$$P(\text{yellow car}) = \tfrac{1}{10}$$
$$P(\text{man with mustache}) = \tfrac{1}{4}$$
$$P(\text{woman with ponytail}) = \tfrac{1}{10}$$
$$P(\text{woman with blond hair}) = \tfrac{1}{3}$$
$$P(\text{black man with beard}) = \tfrac{1}{10}$$
$$P(\text{interracial couple in car}) = \tfrac{1}{1000}$$

Multiplying these probabilities together gives $\frac{1}{12,000,000}$, which suggests that only 1 in 12 million couples have these characteristics. Los Angeles had roughly 1.5 million couples at that time, and largely on this evidence, the jury convicted the couple. However, the Supreme Court of California overturned the conviction, stating that the characteristics were not "independent," so the multiplication of the probabilities was not justified. In this section we will discuss the very important concept of independence which explains when probabilities may be multiplied together.

Introduction

We often ask about the probability of one event *given another*. For example, a player being dealt cards may want to know the probability that a third card will be an ace given that the first two were aces. A teenager might ask about the probability of developing cancer given that one continues to smoke. Such questions involve *conditional probability*, which is the subject of this section. Conditional probability will then lead to a discussion of independence.

Conditional Probability

When two dice (always fair unless stated otherwise) are rolled, there are $6 \cdot 6 = 36$ possible outcomes in the sample space:

$$S = \{(1, 1), (1, 2), \ldots, (2, 1), (2, 2), \ldots, (6, 6)\}$$

EXAMPLE 1 **Finding a Conditional Probability**

Suppose that you roll two dice. What is the probability that the sum is 5, given the additional information that the first die comes up 3?

Solution

Let B be the event that the first die comes up 3, so $B = \{(3, 1), (3, 2), (3, 3), (3, 4), (3, 5), (3, 6)\}$. B contains six possible outcomes, each equally likely, and only one, $(3, 2)$, has the sum 5. Therefore, the probability that event A (the sum is 5) occurs *given that B occurred* is 1 out of 6, which we write as

$$P(A \text{ given } B) = \tfrac{1}{6}$$

In the preceding example, we restricted ourselves to a new, *smaller* sample space corresponding to the "given" event B. The conditional probability was then the probability of A in B—that is, $P(A$ and $B)$ *relative* to this new sample space B. This leads to the following general definition.

Conditional Probability*

> Let A and B be events with $P(B) > 0$. Then the *conditional probability* that A occurs given that B occurs is
>
> $$P(A \text{ given } B) = \frac{P(A \text{ and } B)}{P(B)}$$
>
> The probability of *A given B* is the probability of *A and B* divided by the probability of *B*

We assume $P(B) > 0$ to avoid a zero denominator. If we represent events by a Venn diagram and probability by area, then the conditional probability is just the ratio of two areas.

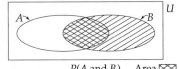

$$P(A \text{ given } B) = \frac{P(A \text{ and } B)}{P(B)} = \frac{\text{Area} \boxtimes}{\text{Area} \boxtimes}$$ Area of the intersection over the area of B

We may check the result of Example 1 using this formula.

$$P(A \text{ and } B) = \tfrac{1}{36}$$ First roll 3 and the sum is 5 means only (3, 2)

$$P(B) = \tfrac{6}{36} = \tfrac{1}{6}$$ The six outcomes were listed in Example 1

Therefore,

$$P(A \text{ given } B) = \frac{P(A \text{ and } B)}{P(B)} = \frac{\tfrac{1}{36}}{\tfrac{1}{6}} = \tfrac{6}{36} = \tfrac{1}{6}$$ Same answer as before

The *unconditional* probability of rolling a sum of 5 is $\tfrac{1}{9}$ (happening on 4 outcomes, (1, 4), (2, 3), (3, 2), and (4, 1), out of the 36 possibilities), so *conditioning* on the first roll being 3 changed the probability from $\tfrac{1}{9}$ to $\tfrac{1}{6}$.

It is important to remember that when we speak of "event A given event B," we do not mean to imply that event B occurred first: The random experiment has been performed, the outcome has been observed, and we are just discussing how the information that the outcome is part of event B influences the chance that it is also part of event A.

---

* For *events* we prefer to say A *and* B rather than $A \cap B$, although they are equivalent. The conditional probability $P(A$ given $B)$ can also be written $P(A|B)$.

EXAMPLE 2   **Finding Probabilities of Hands of Cards**

You are playing cards with a friend, and each of you has been dealt 5 cards at random from a standard deck. If you have no face cards, what is the probability that your friend doesn't either?

Solution

Let A be the event that your friend has no face cards, and let B be the event that you have no face cards. The event A *and* B means that *neither* of you has face cards, and since the deck has 12 face cards, this means that the first 10 cards dealt were from the 40 nonface cards. Therefore,

$$P(A \text{ and } B) = \frac{_{40}C_{10}}{_{52}C_{10}} \approx 0.0536$$

⟵ Ways of choosing 10 from the 40 nonface cards
⟵ Ways of choosing 10 from 52

Using a calculator

$$P(B) = \frac{_{40}C_5}{_{52}C_5} \approx 0.2532$$

⟵ Ways of choosing 5 from the 40 nonface cards
⟵ Ways of choosing 5 from 52

Therefore,

$$P(A \text{ given } B) = \frac{P(A \text{ and } B)}{P(B)} \approx \frac{0.0536}{0.2532} \approx 0.2117$$

The probability that your friend has no face cards given that you don't have any is about 21%.

We could also solve this problem by looking directly at the restricted sample space: Given that you have 5 nonface cards, your friend's cards come from the remaining 47 cards, 12 of which are face cards and 35 of which are not. Therefore,

$$P(A \text{ given } B) = \frac{_{35}C_5}{_{47}C_5} \approx 0.2117$$

⟵ 5 from the remaining 35 nonface cards
⟵ 5 from the remaining 47

Same answer as before

Solving the conditional probability formula for $P(A$ and $B)$ gives the following formula, which is very useful when a conditional probability is given.

Probability Conditioning Formula

$$P(A \text{ and } B) = P(A \text{ given } B) \cdot P(B)$$

The probability of A and B is the probability of A *given* B times the probability of B

We speak of this formula as finding the probability of *A and B* by *conditioning* on the event *B*.

EXAMPLE 3 **Using Conditioning**

You need to be somewhere within 30 minutes, and your parents are out with the car. If they come back soon (you give this a 50–50 chance), the probability that you will get there in time is 90%. Otherwise you will walk, with only a 60% chance of arriving on time. What is the probability that you will arrive on time and in your parents' car?

Solution

Let *A* be the event that you arrive on time (by whatever means), and let *H* be the event that your parents come home soon. You want to find $P(A \text{ and } H)$.

$$P(A \text{ and } H) = P(A \text{ given } H) \cdot P(H) \qquad \text{Conditioning on } H$$

$$= (0.90) \cdot (0.50) = 0.45 \qquad \text{Using the given information}$$

The probability that you will arrive on time and in your parents' car is 0.45, or 45%. ∎

PRACTICE PROBLEM 1

Using the information in the above example, what is the probability that you will arrive on time and on foot?   *Solution at the back of the book*

Partitions and Total Probability

A set *E* such that $0 < P(E) < 1$ and its complement E^c are said to form a *partition* of a sample space *S* because they are disjoint and their union is the whole sample space: $E \cap E^c = \varnothing$ and $E \cup E^c = S$. More generally, a *partition* of a sample space *S* is any collection of events $S_1, S_2, \ldots, S_m$ with positive probabilities that are (pairwise) disjoint and whose union is the whole sample space: $S = S_1 \cup S_2 \cup \cdots \cup S_m$. (Such events are said to be *mutually exclusive and collectively exhaustive*.) A partition can be illustrated in a Venn diagram by dividing the sample space into "strips."

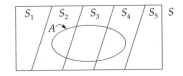

$S_1, S_2, S_3, S_4,$ and S_5 form a *partition* of *S*.
The partition divides the event *A* into parts.

Notice that the partition divides the event A into disjoint parts, so we have

$$P(A) = \sum_{\substack{\text{All } S_i \text{ in} \\ \text{the partition}}} P(A \text{ and } S_i)$$

The probability of A is the sum of the probabilities of its parts in $S_1, \ldots, S_m$

If we now use the probability conditioning formula (page 425) to replace each $P(A \text{ and } S_i)$ ˎ by $P(A \text{ given } S_i) \cdot P(S_i)$,   we obtain

Total Probability Formula

$$P(A) = \sum_{\substack{\text{All } S_i \text{ in} \\ \text{the partition}}} P(A \text{ given } S_i) \cdot P(S_i)$$

Total probability can be found by conditioning on a partition and adding

This formula is so named because it, like the previous formula, gives the probability of an event by separating it into parts corresponding to the partition and then adding up to find the total. It can be thought of as a *weighted average* of conditional probabilities formed by *conditioning* on the partition. The total probability formula may be shown as a tree diagram, where the probability $P(A)$ is found by multiplying along each branch and adding the results.

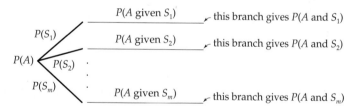

For instance, in Example 3 we found that the probability of your arriving on time and by car was

$$P(A \text{ and } H) = P(A \text{ given } H) \cdot P(H) = (0.90) \cdot (0.50) = 0.45$$

In Practice Problem 1, you found that the probability of arriving on time and on foot was

$$P(A \text{ and } H^c) = P(A \text{ given } H^c) \cdot P(H^c) = (0.60) \cdot (0.50) = 0.30$$

Adding these two together gives $P(A)$, the *total probability* of arriving on time:

$$P(A) = P(A \text{ given } H) \cdot P(H) + P(A \text{ given } H^c) \cdot P(H^c) = 0.45 + 0.30 = 0.75$$

Therefore, your probability of arriving on time (by *any* means) is 75%, which we found by *conditioning* on whether your parents arrive home soon or not.

EXAMPLE 4   **Sampling Without Replacement**

A box contains three blue balls and two green balls. A ball is randomly chosen (and not replaced), and then a second ball is chosen. What is the probability that the second ball is green?

Solution

To find a probability concerning the second ball, we will *condition* on the color of the first ball. We write B_1 for the event that the first ball is blue, G_1 for a green first ball and, similarly, B_2 and G_2 for the second ball. Since the first ball must be blue or green, B_1 and G_1 form a *partition* of the sample space. Their probabilities are $P(B_1) = \frac{3}{5}$ and $P(G_1) = \frac{2}{5}$ (since of the five balls, three are blue and two are green). We also have $P(G_2 \text{ given } B_1) = \frac{2}{4}$ (since given B_1 there are four balls left, of which two are green), and similarly $P(G_2 \text{ given } G_1) = \frac{1}{4}$. From the total probability formula with $A = G_2$, $S_1 = B_1$ and $S_2 = G_1$, we obtain

$$P(G_2) = P(G_2 \text{ given } B_1) \cdot P(B_1) + P(G_2 \text{ given } G_1) \cdot P(G_1)$$

$$= \frac{2}{4} \cdot \frac{3}{5} + \frac{1}{4} \cdot \frac{2}{5} = \frac{3}{10} + \frac{1}{10} = \frac{2}{5}$$

The probability that the second ball is green is $\frac{2}{5}$, or 40%.

Independent Events

Roughly speaking, two events are said to be *independent* if one has nothing to do with the other, so that the occurrence of one has no bearing on the probability of the other. In terms of conditional probability, this means that for independent events A and B, we have $P(A \text{ given } B) = P(A)$. Using the definition of conditional probability, this equation becomes

$$\frac{P(A \text{ and } B)}{P(B)} = P(A) \qquad \text{for } P(B) \neq 0$$

Multiplying each side by $P(B)$ gives the following equivalent condition, which we take as the *definition* of independence.

Independent Events

Events A and B are *independent* if

$$P(A \text{ and } B) = P(A) \cdot P(B)$$ The probability of A *and* B is the *product* of the probabilities

Events that are not independent are *dependent*.

Be careful! Independent does not mean the same thing as *disjoint*, or *mutually exclusive*. If events A and B have positive probabilities and are disjoint, then $P(A \text{ and } B) = 0$ but $P(A) \cdot P(B)$ is positive, so they cannot be independent. Intuitively, disjointness is a very strong kind of *dependence:* The occurrence of one of two disjoint events guarantees the *non*occurrence of the other, so they cannot be independent.

The definition above gives a simple test for the independence of events A and B: Find $P(A \text{ and } B)$ and find the product $P(A) \cdot P(B)$. If the results are *equal*, then A and B are *independent;* otherwise, A and B are *dependent*.

EXAMPLE 5   **Independent Coin Tosses**

A coin is tossed twice. Let A be the event that the first toss is heads, and let B be the event that the second toss is heads. Are the events A and B independent?

Solution

The sample space is

$$S = \{(H,H), (H,T), (T,H), (T,T)\}$$

and

$$A = \{(H,H), (H,T)\}, \quad B = \{(H,H), (T,H)\}, \quad \text{and} \quad A \cap B = \{(H,H)\}$$

Since the outcomes are equally likely,

$$P(A) = \frac{2}{4} = \frac{1}{2}, \qquad P(B) = \frac{2}{4} = \frac{1}{2}, \qquad \text{and} \qquad P(A \text{ and } B) = \frac{1}{4}$$

Since $P(A \text{ and } B) = \frac{1}{4}$ and $P(A) \cdot P(B) = \frac{1}{2} \cdot \frac{1}{2} = \frac{1}{4}$ are the same value, we have shown that $P(A \text{ and } B) = P(A) \cdot P(B)$, and the events A and B *are* independent.

■

If $P(A \text{ and } B)$ and $P(A) \cdot P(B)$ had *not* been equal, the events would have been *dependent*. The fact that successive coin tosses are independent is sometimes expressed by saying that the coin has "no memory" and that each toss is a new experiment with "no influence from the past."

EXAMPLE 6 **Assessing Independence**

A coin is tossed three times. Let A be the event that at most one head occurs, and let B be the event that the tosses include both heads and tails. Are the events A and B independent?

Solution

The sample space is $S = \begin{Bmatrix} (H,H,H), & (H,H,T), & (H,T,H), & (H,T,T) \\ (T,H,H), & (T,H,T), & (T,T,H), & (T,T,T) \end{Bmatrix}$.

We then have

$A = \{(H,T,T), \quad (T,H,T), \quad (T,T,H), \quad (T,T,T)\}$ At most 1 head

$B = \begin{Bmatrix} (H,H,T), & (H,T,H), & (H,T,T) \\ (T,H,H), & (T,H,T), & (T,T,H) \end{Bmatrix}$ Both heads and tails

$A \cap B = \{(H,T,T) \quad (T,H,T) \quad (T,T,H)\}$ Intersection

Since the outcomes are equally likely,

$$P(A) = \frac{4}{8} = \frac{1}{2}, \qquad P(B) = \frac{6}{8} = \frac{3}{4}, \qquad \text{and} \qquad P(A \text{ and } B) = \frac{3}{8}$$

Since $P(A) \cdot P(B) = \frac{1}{2} \cdot \frac{3}{4} = \frac{3}{8} = P(A \text{ and } B)$, the events A and B *are* independent. ∎

The concept of independence can be extended to more than two events.

Many Independent Events

A collection $E_1, E_2, \cdots, E_m$ of events is *independent* if any subcollection of them satisfies the multiplication formula:

$$P(E_i \text{ and } E_j \text{ and } \ldots \text{ and } E_k) = P(E_i) \cdot P(E_j) \cdot \cdots \cdot P(E_k)$$

The probability of E_i and . . . and E_k is the *product* of the probabilities

In practice, the most important applications of independence do not involve proving it but *assuming* it. That is, many applications involve events that clearly have nothing to do with each other, and so we will *assume* that they are independent, which enables us to find probabilities simply by multiplying. With the concept of independence, we can solve the Chevalier's second problem (see pages 418–419) more easily than before.

EXAMPLE 7   **The Chevalier de Méré, Again**

What is the probability of at least one double six in 24 rolls of two fair dice?

Solution

Clearly, different rolls of dice are independent. On any one roll of two dice, the probability of *not* getting a double six is $\frac{35}{36}$, so the probability of not getting a double six repeatedly on 24 rolls is

$$\underbrace{\frac{35}{36} \cdot \frac{35}{36} \cdot \ \cdots \ \cdot \frac{35}{36}}_{24 \text{ rolls}} = \left(\frac{35}{36}\right)^{24} \approx 0.509$$

Therefore, the probability of getting *at least one* double six is $1 - \left(\frac{35}{36}\right)^{24} \approx 0.491$, which agrees with our calculation in Example 9 on pages 418–419.

EXAMPLE 8   **Telemarketing**

Each of five salespeople at a telemarketing company can sell a magazine subscription to a customer within ten minutes of calling them 80% of the time. If all five salespeople have started talking with new customers, what is the probability that all will sell subscriptions within the next ten minutes?

Solution

Since the salespeople work independently, the probability that all will make sales is the product of the probabilities that each does:

$$\underbrace{0.80 \cdot 0.80 \cdot 0.80 \cdot 0.80 \cdot 0.80}_{5 \text{ salespeople}} = 0.80^5 \approx 0.328 \qquad \text{Multiplying the probabilities}$$

The probability that all five will make sales is about 33%.

PRACTICE PROBLEM 2

For the situation described in Example 8, find the probability that none of the salespeople will make sales within the next ten minutes.

Solution at the back of the book

Section Summary

For events A and B with $P(B) > 0$, the *conditional probability* that A occurs given that B occurs is

$$P(A \text{ given } B) = \frac{P(A \text{ and } B)}{P(B)} \qquad \text{for } P(B) > 0$$

This definition may be rewritten to give the probability of *A and B:*

$$P(A \text{ and } B) = P(A \text{ given } B) \cdot P(B) \qquad \text{Conditioning on } B$$

A *partition* of a sample space S is any collection of events $S_1, S_2, \ldots, S_m$ with positive probabilities that are *mutually exclusive* (any two are disjoint) and *collectively exhaustive* ($S = S_1 \cup S_2 \cup \ldots \cup S_m$). For any event A,

$$P(A) = \sum_{\substack{\text{All } S_i \text{ in} \\ \text{the partition}}} P(A \text{ given } S_i) \cdot P(S_i) \qquad \begin{array}{l}\text{Total probability} \\ \text{formula}\end{array}$$

Two events A and B are *independent* if

$$P(A \text{ and } B) = P(A) \cdot P(B) \qquad \begin{array}{l}\text{The probabilities} \\ \text{multiply}\end{array}$$

A collection $E_1, E_2, \ldots, E_m$ of events is independent if every subcollection of the events obeys a similar multiplication property.

EXERCISES 5.4

1. If $P(A) = 0.6$, $P(B) = 0.4$, and $P(A \cap B) = 0.2$, find:

   a. $P(A \text{ given } B)$ **b.** $P(B \text{ given } A)$.

2. If $P(A) = 0.5$, $P(B) = 0.3$, and $P(A \cap B) = 0.1$, find:

   a. $P(A \text{ given } B)$ **b.** $P(B \text{ given } A)$.

3. If $P(A) = 0.4$, $P(B) = 0.5$, and $P(A \cup B) = 0.6$, find:

   a. $P(A \text{ given } B)$ **b.** $P(B \text{ given } A)$.

4. If $P(A) = 0.6$, $P(B) = 0.5$, and $P(A \cup B) = 0.8$, find:

   a. $P(A \text{ given } B)$ **b.** $P(B \text{ given } A)$.

5. Marbles A box contains 4 white, 2 red, and 4 black marbles. One marble is chosen at random, and it is not black. Find the probability that it is white.

6. Cards You select two cards at random from an ordinary deck. If the first card is a spade, what is the probability that the second card is a spade?

7. Gender Your friend has two children, and you know that at least one is a girl. What is the probability that both are girls? (Assume that girls and boys are equally likely.)

8. Eye Color Suppose that each of two children in a family has probability $\frac{1}{5}$ of having blue eyes, independently of each other. If at least one child has blue eyes, what is the probability that both have blue eyes?

9. Cards A deck contains three cards: One is red on both sides, one is blue on both sides, and the third is red on one side and blue on the other. One card is chosen at random from the deck, and the color on one side is observed. If this side is blue, what is the probability that the other side is blue?*

* Many people mistakenly believe that if one side is blue, then the probability that the other side is blue is $\frac{1}{2}$ since it can only be the blue–blue card or the blue–red card, and these are

10. Credit Cards Looking back at Exercise 24 on page 421, if a customer has at least one credit card, what is the probability that the customer has a Visa card?

11. Political Contributions Looking back at Exercise 23 on page 421, if a person is a contributor, what is the probability that the person contributes to the Republican party?

12. Cards In the game of bridge, each of four players is dealt 13 cards. If a certain player has no aces, find the probability that that person's partner has:

a. no aces   **b.** at least two aces

13. Choosing Courses You will take either a basket weaving course or a philosophy course, depending on what your advisor decides. You estimate that the probability of your getting an A in basket weaving is 0.95, while in philosophy it is 0.70. However, the chances of your advisor choosing the basket weaving course is only 20%, while there is an 80% chance of his putting you in the philosophy course. What is the probability of your ending up with an A?

14. Multiple-Choice Tests On a multiple-choice test you know the answers to 70% of the questions (and so get them right), and for the remaining 30% you choose randomly among the 5 answers. What percent of the answers should you expect to get right?

15. Driving Suppose that 70% of drivers are "careful" and 30% are "reckless." Suppose further that a careful driver has a 0.1 probability of being in an accident in a given year, while for a reckless driver the probability is 0.3. What is the probability that a randomly selected driver will have an accident within a year?

16. Quality Control A computer manufacturer has assembly plants in three states. The Delaware plant produces 25% of the company's computers, the Michigan plant produces 35%, and the California plant produces the other 40%. The probabilities that a computer will pass inspection are 93% for the Delaware plant, 89% for the Michigan plant, and 94% for the California plant. What is the probability that a randomly selected computer from this company will pass inspection?

Independence For the experiment of tossing a coin twice, find whether events A and B are independent or dependent.

17. A: heads on the first toss
B: different results on the two tosses

18. A: heads on the second toss
B: the same results on both tosses

Independence For the experiment of rolling two dice, find whether events A and B are independent or dependent.

19. A: odd number on the first roll
B: sum of the numbers is 4

20. A: even number on the first roll
B: sum of the numbers is 10

21. Dice A pair of dice is rolled three times in succession. Find the probability that each of the rolls has a sum of seven.

22. Brand Loyalty Suppose that each time that you buy a car, you choose between Ford and General Motors. Suppose that each time after the first, you stay with the same company with probability $\frac{2}{3}$ and switch with probability $\frac{1}{3}$. If you are equally likely to choose either company for your first car, what is the probability that your first and second choices will be Ford cars and your third and fourth choices will be General Motors cars?

23. Class Attendance Two students are registered for the same class and attend independently of each other, student A 90% of the time and student B 70% of the time. The teacher remembers that on a given day, at least one of them is in class. What is the probability that student A was in class?

24. Class Attendance For the students in Exercise 23, what is the probability that on a given day:

a. both will be in class?
b. neither will be in class?
c. at least one will be in class?

equally likely. The error comes from not realizing that the blue side is equally likely to be any one of the *three* blue sides. More than one probability student has made money by knowing the correct conditional probabilities and offering bets on the outcome.

25. Dice   Three dice are rolled. Find the probability of getting:

   a. all sixes

   b. all the same outcomes

   c. all different outcomes

Explorations and Excursions

The following problems extend and augment the material presented in the text.

26. If event A is such that $P(A) = 0$, and B is any other event, show that events A and B are independent.

27. If events A and B are independent, show that events A and B^c are independent.

28. If events A and B are independent, show that events A^c and B^c are independent. [*Hint:* Use the result of the previous exercise.]

29. In the experiment of tossing two coins, define events as follows:

$$A: \text{heads on first toss}$$

$$B: \text{heads on second coin}$$

$$C: \text{outcomes agree}$$

Show that A, B, and C are *pairwise* independent but do *not* satisfy $P(A \text{ and } B \text{ and } C) = P(A) \cdot P(B) \cdot P(C)$ and so are not independent.

30. To see why the events in the preceding exercise are not independent, calculate $P(C)$ and $P(C$ given $(A \cap B))$.

5.5   Bayes' Theorem

Airline Hijacking

In 1980 the Federal Aviation Administration developed a "profile" of airplane hijackers that was claimed to fit 90% of hijackers and only 0.05% of legitimate travelers. Is the FAA justified in using such a profile to detain and search anyone who fits the characteristics? In terms of conditional probability, the question comes down to this: Granted that there is a high probability that you fit the profile given that you are a hijacker, what is the probability that you are a hijacker given that you fit the profile? (Notice that the second question *reverses* the conditioning in the first.) In Exercise 7 we will see that the probability that a randomly selected person who fits the profile is actually a hijacker is less than 0.02%, the other 99.98% being legitimate travelers. Nonetheless, the federal courts (*U.S.* v. *Lopez*) upheld the use of hijacker profiles as sufficient for probable cause to detain a passenger, although the court did refer to "disquieting possibilities." In this section we will see how to calculate such probabilities.

Introduction

Sometimes we will want to *reverse the conditioning* in conditional probability. For example, it would be useful to find not only the probability that one develops cancer given that one smokes, but also the probability that one smoked given that one developed cancer. In this section we develop a formula to answer such questions.

Bayes' Formula

Let S_1 be an event in a partition $S_1, S_2, \ldots, S_m$, and let A be an event with positive probability. Then

$$P(S_1 \text{ given } A) = \frac{P(S_1 \text{ and } A)}{P(A)}$$

Definition of conditional probability

$$= \frac{P(A \text{ and } S_1)}{P(A)}$$

Reversing the order

$$= \frac{P(A \text{ given } S_1) \cdot P(S_1)}{P(A)}$$

Using the probability conditioning formula (page 425)

$$= \frac{P(A \text{ given } S_1) \cdot P(S_1)}{\displaystyle\sum_{\substack{\text{All } S_i \text{ in} \\ \text{the partition}}} P(A \text{ given } S_i) \cdot P(S_i)}$$

Using the total probability formula (page 427)

This result is known as Bayes' formula.*

Bayes' Formula

Let $S_1, S_2, \ldots, S_m$ be a partition of the sample space S, and let A be any event with positive probability. Then

$$P(S_1 \text{ given } A) = \frac{P(A \text{ given } S_1) \cdot P(S_1)}{\displaystyle\sum_{\substack{\text{All } S_i \text{ in} \\ \text{the partition}}} P(A \text{ given } S_i) \cdot P(S_i)}$$

Top conditions ← on just S_1

← Bottom conditions on *all* S_i (including S_1)

Notice the reversal of the conditioning: On the left is S_1 *given A*, while on the right is *A given* S_i. Do not, however, think of Bayes' Formula as *reversing the arrow of time*: $P(S_1$ given $A)$ simply means the probability that the outcome is in S_1 given that it is in A. We may represent Bayes' formula using the same tree diagram as on page 427 by noticing that

*After Thomas Bayes (1702–1761), an English theologian and mathematician.

the numerator is the product from the uppermost branch, while the denominator is the sum of the products from *all* of the branches.

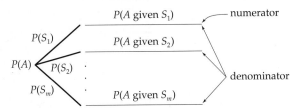

EXAMPLE 1   Assessing Voting Patterns

Registered voters in Marlin County are 45% Democrats, 30% Republicans, and 25% Independents. In the last election for county supervisor, 70% of the Democrats voted, 80% of the Republicans voted, and 90% of the Independents voted. What is the probability that a randomly selected voter in this last election was a Democrat?

Solution

Democrats, Republicans, and Independents form a *partition* of the voters, which we represent by D, R, and I. We have that

$$P(D) = 0.45, \qquad P(R) = 0.30, \qquad \text{and} \qquad P(I) = 0.25$$

Let V be the event that a randomly selected voter voted in the last election. We are given

$$P(V \text{ given } D) = 0.70, \quad P(V \text{ given } R) = 0.80, \quad \text{and} \quad P(V \text{ given } I) = 0.90$$

We are asked to find the probability that a randomly selected voter was a Democrat, which means the probability of being a Democrat *given* that one voted, which is $P(D \text{ given } V)$. Bayes' Formula yields

$$P(D \text{ given } V)$$

$$= \frac{P(V \text{ given } D) \cdot P(D)}{P(V \text{ given } D) \cdot P(D) + P(V \text{ given } R) \cdot P(R) + P(V \text{ given } I) \cdot P(I)}$$

$$= \frac{(0.70) \cdot (0.45)}{(0.70) \cdot (0.45) + (0.80) \cdot (0.30) + (0.90) \cdot (0.25)} \approx 0.404$$

The probability that a voter in the last election was a Democrat is about 40%.

■

How did we know that we should use Bayes' formula in this example? Because we were asked for the probability of being a Democrat *given that one voted*, and we were told the probability of voting *given that one is a Democrat* (or a Republican or an Independent). Such a reversal of conditioning calls for Bayes' formula.

PRACTICE PROBLEM

Using the information in the previous example, find the probability that a randomly selected voter was an Independent.

Solution at the back of the book

EXAMPLE 2 Management Training Programs

A corporation is reviewing the success rate of its management training program. It finds that 80% of those entering the program are college graduates and that 90% of them successfully complete it, while just 60% of those who are not college graduates successfully complete it. What is the probability that an employee who successfully completes the training program is not a college graduate?

Solution

Let C be the event that the employee entering the program is a college graduate. We are given that $P(C) = 0.80$. Let N be the event that the employee entering the program is *not* a college graduate, so that $P(N) = 1 - P(C) = 0.20$. Then C and $N = C^c$ form a *partition* of those entering the program. Let S be the event that the employee entering the program completes it successfully. We are told that $P(S$ given $C) = 0.90$ and that $P(S$ given $N) = 0.60$. The probability that an employee who successfully completes the program is *not* a college graduate is $P(N$ given $S)$, which we find using Bayes' formula:

$$P(N \text{ given } S) = \frac{P(S \text{ given } N) \cdot P(N)}{P(S \text{ given } N) \cdot P(N) + P(S \text{ given } C) \cdot P(C)}$$

$$= \frac{(0.60) \cdot (0.20)}{(0.60) \cdot (0.20) + (0.90) \cdot (0.80)} \approx 0.143$$

The probability that an employee who successfully completes the training program is not a college graduate is about 14%. ∎

Section Summary

$$P(A|B) = \frac{P(A) \cdot P(B/A)}{P(A) \cdot P(B/A) + P(A') \cdot P(B/A')}$$

To reverse conditioning, we use *Bayes' Formula:*

$$P(S_1 \text{ given } A) = \frac{P(A \text{ given } S_1) \cdot P(S_1)}{\sum_{\substack{\text{All } S_i \text{ in} \\ \text{the partition}}} P(A \text{ given } S_i) \cdot P(S_i)} \qquad \begin{array}{l} \text{For a partition} \\ S_1, S_2, \cdots, S_m \end{array}$$

EXERCISES 5.5 (▦ will be helpful.)

1. **Voting** In a town, 60% of the citizens are Republicans and 40% are Democrats. In the last election 55% of the Republicans voted and 65% of the Democrats voted. If a voter is randomly selected, what is the probability that the person is a Republican?

2. **Colorblindness** An estimated 8% of men and 0.5% of women are colorblind. If a colorblind person is selected at random, what is the probability that the person is a man? (Assume that men and women occur in equal numbers.)

3. **Medical Testing** A new test is developed to test for a certain disease, giving "positive" or "negative" results to indicate that the person does or does not have the disease. For a person who actually *has* the disease, the test will give a positive result with probability 0.95 and a negative result with probability 0.05 (a so-called "false negative"). For a person who does *not* have the disease, the test will give a positive result with probability 0.05 (a "false positive") and a negative result with probability 0.95. Furthermore, only one person in 1000 actually has this disease. If a randomly selected person is given the test and tests positive, what is the probability that the person actually has the disease?

4. **Home Pregnancy Testing** Suppose that 400 pregnant women take a home pregnancy test, and 397 of them test "positive" and the other 3 test "negative." Suppose also that 200 nonpregnant women take the test, and 184 of them test "negative" and the remaining 16 test "positive." What is the probability that a woman who tests positive is actually pregnant?

5. **Manufacturing Defects** A computer chip factory has three machines, A, B, and C, for producing the memory chips. Machine A produces 50% of the factory's chips, machine B produces 30%, and machine C produces 20%. It is known that 3% of the chips produced by machine A are defective, as are 2% of chips produced by machine B and 1% of the chips from machine C. If a randomly selected chip from the factory's output is found to be defective, what is the probability that it was produced by machine B?

6. **Manufacturing Defects** For the information in Exercise 5, if a randomly selected chip is found *not* to be defective, what is the probability that it came from machine B?

7. **Airline Hijacking** The "hijacker profile" developed by the Federal Aviation Administration fits 90% of hijackers and only 0.05% of legitimate passengers (see the Application Preview on page 434). Based on historical data, assume that only 30 of 300 million passengers are hijackers. What is the probability that a person who fits the profile is actually a hijacker?

8. **Polygraph Tests** Although seldom admissible in courts, polygraphs (sometimes called "lie detectors") are used by many businesses and branches of government. The Office of Technology Assessment has estimated that polygraphs have a "false negative" rate of 0.11 and a "false positive" rate of 0.20. Suppose that you are trying to find a single thief in your company of 100 employees. What is the probability that a person who fails a polygraph test is *not* the thief?

9. **Random Drug Testing** In 1986 the Reagan administration issued an executive order allowing agency heads to subject all employees to urine tests for drugs. Suppose that the test is 95% accurate both in identifying drug users and in clearing nonusers. Suppose also that 1% of employees use drugs. What is the probability that a person who tests positive is *not* a drug user?

10. **Grades** A probability class consists of 5 math majors, 3 science majors, and 2 whose major is "undecided." If the probability of earning an A is 90% for math majors, 85% for science majors, and 70% for "undecideds," what is the probability that a randomly selected A student is a math major?

11. **Grades** For the information in Exercise 10, what is the probability that a randomly selected A student is "undecided"?

12. **Olympic Athletes** Suppose that 10% of olympic athletes use steroids. A blood test is 95% correct in identifying drug users and has a 2% rate of false positives (that is, of *incorrectly* indicating steroid use). A champion cycler tests positive. What is the probability that he uses steroids?

13. **Mammograms** Many doctors recommend that women in their forties have annual mammograms to detect breast cancer. Approximately 2% of women in their forties will develop breast cancer

during that decade, and the early mammograms (no longer used) for that age group had a 30% rate of false positives and a 25% rate of false negatives. What is the probability that a woman who tested positive on this test had breast cancer?

14. **Unemployment**   A city has five districts, and 30% of its citizens live in district I, 25% in district II, 20% in district III, 15% in district IV, and 10% in district

V. The unemployment rate in district I is 3%, while in districts II, III, IV, and V it is 4%, 4%, 5%, and 6%, respectively. The local newspaper interviews a randomly selected unemployed resident. What is the probability that this resident lives in district I?

15. **Unemployment**   For the information in Exercise 14, what is the probability that this resident lives in district V?

5.6   Random Variables and Distributions

Home Run Records

In 1998, a baseball record that had stood for 37 years, the maximum number of home runs in a season (66), was broken by Mark McGuire (with 70 home runs), and tied by Sammy Sosa (with 66). Was this a fluke? Should we expect the new record to be broken soon? Suppose that we let X stand for the number of home runs that McGuire hits in a season, and Y for the number that Sosa hits (these are examples of *random variables*, to be discussed in this section). How can we calculate probabilities such as $P(X > 70)$ or $P(Y > 70)$? The number of home runs can be modeled by the *binomial* random variables discussed in this section, and Exercise 44 involves finding the probability $P(X > 70)$ that McGuire will break his own record in a given season. The answer is about 5%, which shows that 70 home runs really was a remarkable record. However, the probability is not so small as to suggest that this record will last for another 37 years.

Introduction

Many times a random experiment results in a *number*, such as the sum on the faces of two dice or your winnings in a lottery. Such numeric quantities are called *random variables* and are the central objects of probability and statistics. In this section (and also in the following chapter) we will use random variables to answer questions about chance events and to introduce some of the random variables that have proved most useful in applications.

Random Variables

A random variable is an assignment of a number to each outcome in the sample space. We will use capital letters such as X and Y for random variables.

EXAMPLE 1 Defining a Random Variable

Let X be the number of heads in four tosses of a coin. Find the possible values for X and the outcomes corresponding to each of its possible values.

Solution

The number of heads in four tosses can be 0, 1, 2, 3, or 4, so these are the possible values for X. The sample space S consists of the following 16 sequences of H's and T's.

$$
S = \begin{cases}
(H,H,H,H), & (H,H,H,T), & (H,H,T,H), & (H,T,H,H), \\
(T,H,H,H), & (H,H,T,T), & (H,T,H,T), & (H,T,T,H), \\
(T,H,H,T), & (T,H,T,H), & (T,T,H,H), & (H,T,T,T), \\
(T,H,T,T), & (T,T,H,T), & (T,T,T,H), & (T,T,T,T)
\end{cases}
$$

Since X is the number of heads, for any particular outcome we can find the value of X by counting H's. For example, (H,T,H,H) gives $X = 3$. The following table lists the possible values of X and the outcomes for which it takes those values (as you should check by "counting heads").

Values of X	Outcomes
$X = 0$	(T,T,T,T)
$X = 1$	$(H,T,T,T),(T,H,T,T),(T,T,H,T),(T,T,T,H)$
$X = 2$	$(H,H,T,T),(H,T,H,T),(H,T,T,H),(T,H,H,T),(T,H,T,H),(T,T,H,H)$
$X = 3$	$(H,H,H,T),(H,H,T,H),(H,T,H,H),(T,H,H,H)$
$X = 4$	(H,H,H,H)

A random variable taking a particular value, such as $X = 2$ above, corresponds to a *subset* of the sample space and so is an *event*. The probability of such an event is just the sum of the probabilities of the outcomes it contains. The *probability distribution* of a random variable is the collection of these probabilities for its various values.

Random Variable

A *random variable* X is an assignment of a number to each element in the sample space. The *probability distribution* of the random variable X is the collection of all probabilities $P(X = x)$ for each possible value x.

EXAMPLE 2   **Finding a Probability Distribution**

Find and graph the probability distribution for

$$X = \begin{pmatrix} \text{Number of heads in} \\ \text{four tosses of a coin} \end{pmatrix}.$$

Solution

Using the table in Example 1, $X = 0$ occurs only for the outcome (T,T,T,T), which is one out of 16 equally likely outcomes, so $P(X = 0) = \frac{1}{16}$. The event $X = 1$ corresponds to 4 outcomes in the table, so $P(X = 1) = \frac{4}{16} = \frac{1}{4}$. The other probabilities $P(X = 2)$, $P(X = 3)$, and $P(X = 4)$ are similarly found by counting outcomes and dividing by 16 (which you should check), giving the probabilities in the following table. These probabilities are graphed on the right, the height of each bar being the probability that X takes the value at the bottom of the bar.

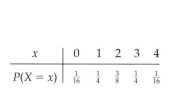

x	0	1	2	3	4
$P(X = x)$	$\frac{1}{16}$	$\frac{1}{4}$	$\frac{3}{8}$	$\frac{1}{4}$	$\frac{1}{16}$

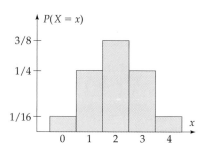

Probability distribution for the number of heads in four tosses of a coin

---

Observe that the *most likely* number of heads in 4 tosses is 2 and that the least likely are the extreme values 0 and 4, just as you might expect. The probabilities add to 1: $\frac{1}{16} + \frac{1}{4} + \frac{3}{8} + \frac{1}{4} + \frac{1}{16} = \frac{16}{16} = 1$, so the events $P(X = x)$ for the possible values x form a *partition* of the sample space. The sum of the probabilities being 1 means that *the area under the graph is 1* (since each bar has width 1).

EXAMPLE 3 **Sampling Without Replacement**

A box contains 4 red balls and 3 green balls. Two balls are selected at random and removed from the box. Find and graph the probability distribution of $X = \begin{pmatrix} \text{Number of red} \\ \text{balls removed} \end{pmatrix}$.

Solution

Let R_1 be the event that the first ball is red, R_2 the event that the second ball is red, G_1 the event that the first ball is green, and G_2 the event that the second ball is green. Then $P(R_1) = \frac{4}{7}$ (since initially there are 4 red balls out of 7) and $P(R_2 \text{ given } R_1) = \frac{3}{6}$ (since after a red is chosen, there are only 3 reds left out of 6). Finding the other probabilities similarly and using the conditioning formula (page 425), we obtain

$P(R_1 \text{ and } R_2) = P(R_2 \text{ given } R_1) \cdot P(R_1) = \frac{3}{6} \cdot \frac{4}{7} = \frac{2}{7}$ 2 red balls removed

$P(R_1 \text{ and } G_2) = P(G_2 \text{ given } R_1) \cdot P(R_1) = \frac{3}{6} \cdot \frac{4}{7} = \frac{2}{7}$

$P(G_1 \text{ and } R_2) = P(R_2 \text{ given } G_1) \cdot P(G_1) = \frac{4}{6} \cdot \frac{3}{7} = \frac{2}{7}$ } 1 red ball removed

$P(G_1 \text{ and } G_2) = P(G_2 \text{ given } G_1) \cdot P(G_1) = \frac{2}{6} \cdot \frac{3}{7} = \frac{1}{7}$ 0 red balls removed

These probabilities, in view of the notations on the right, give the probabilities for X, the number of red balls removed. The probabilities are listed in the table and graphed below on the right.

x	0	1	2
$P(X = x)$	$\frac{1}{7}$	$\frac{4}{7}$	$\frac{2}{7}$

From $\frac{2}{7} + \frac{2}{7}$

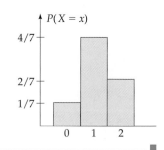

As is always the case for probability distributions, the area under the graph is 1.

EXAMPLE 4 **Finding Gambling Probabilities**

A die is rolled and you win $5 if the die rolls evens, you win $1 if the die rolls one or three, and you loose $10 if the die comes up five. Find and graph the probability distribution of your winnings.

Solution

Let X represent your winnings, and let the sample space be $S = \{1, 2, 3, 4, 5, 6\}$. The possible values for X are 5, 1, and -10 (indicating a loss). Since each face of the die has probability $\frac{1}{6}$, we obtain the probability distribution given in the table below and graphed on the right.

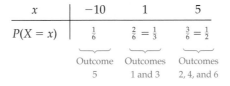

x	-10	1	5
$P(X = x)$	$\frac{1}{6}$	$\frac{2}{6} = \frac{1}{3}$	$\frac{3}{6} = \frac{1}{2}$
	Outcome 5	Outcomes 1 and 3	Outcomes 2, 4, and 6

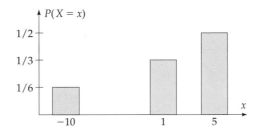

Expected Values

If you were to play the game in Example 4 many times, in the long run would you expect to win money, loose money, or break even? Since you would expect to win \$5 half the time $\left(\text{its probability is } \frac{1}{2}\right)$, win \$1 a third of the time $\left(\text{its probability is } \frac{1}{3}\right)$, and loose \$10 a sixth of the time, you might reasonably expect your *long-run average* winnings to be

$$5 \cdot \tfrac{1}{2} + 1 \cdot \tfrac{1}{3} - 10 \cdot \tfrac{1}{6} = \tfrac{15 + 2 - 10}{6} = \tfrac{7}{6}$$

Each value times its probability

The result is positive, so the game is *favorable* to you. If the result were zero, the game would be *fair*, while if it were negative, it would be *unfavorable* to you. For any random variable, we can similarly calculate an *expected* or *mean value*:

Expected Value

The *expected value* of a random variable X is

$$E(X) = \sum_{\substack{\text{All} \\ \text{possible } x}} x \cdot P(X = x)$$

Sum of the possible values times their probabilities

The expected value $E(X)$ is also called the *expectation* or the *mean* of the random variable (or of the probability distribution) and is often denoted μ (the Greek letter mu). The expected value is a *weighted average* of the possible values, each value weighted by the probability that the random variable takes that value. If the possible values are equally likely, then the expected value is just their average.

EXAMPLE 5   Expected Value of a Roll of a Die

Find the expected value of one roll of a die.

Solution

The possible values are 1, 2, 3, 4, 5, and 6, each with probability $\frac{1}{6}$, so

$$E(X) = 1 \cdot \tfrac{1}{6} + 2 \cdot \tfrac{1}{6} + 3 \cdot \tfrac{1}{6} + 4 \cdot \tfrac{1}{6} + 5 \cdot \tfrac{1}{6} + 6 \cdot \tfrac{1}{6} = \tfrac{21}{6} = 3.5$$

Notice that 3.5 is *not* one of the possible values for a die, so the expected value of a random variable need not be one of the possible values. If a random variable represents your *winnings* in a game, then the mean or expected value may be interpreted as the *fair price* for the game. That is, if you were to roll a die and win in dollars whatever the die showed, then a fair price for you to pay to play that game would be $3.50, because that would make your average winnings zero.

EXAMPLE 6   Expected Value of a Raffle Ticket

Four hundred raffle tickets are sold for $100 each to raise money for a local children's hospital. First prize is a $3000 Florida vacation for two, second prize is a $1000 credit at the town's supermarket, and the five third prizes are $100 "dinner for two" gift certificates at a local restaurant. What is the expected value of a raffle ticket?

Solution

Let X represent winnings of a ticket. Then $P(X = 3000) = \frac{1}{400}$, $P(X = 1000) = \frac{1}{400}$, $P(X = 100) = \frac{5}{400}$, (since there are five third prizes), and $P(X = 0) = \frac{393}{400}$ (since the rest of the tickets win nothing). The expected value is

$$E(X) = 3000 \cdot \tfrac{1}{400} + 1000 \cdot \tfrac{1}{400} + 100 \cdot \tfrac{5}{400} + 0 \cdot \tfrac{393}{400} = 11.25$$

Each winning times its probability

The expected value of the raffle ticket is $11.25.

This figure represents your average winnings, so for each $100 ticket that you buy, you are, on average, losing $88.75. On a more positive note, this latter figure represents your true generosity to the hospital.

PRACTICE PROBLEM 1

Suppose that the raffle prizes in Example 6 are changed to one $3000 first prize, two $1000 second prizes, and ten $100 third prizes. What is the expected value of a raffle ticket now?   *Solution at the back of the book*

Standard Deviation

A random variable may have hundreds of possible values, and it is sometimes useful to try to *summarize* it by one or two numbers to make it easier to grasp. To represent its "average" or "typical" value, we use its mean or expected value (although other choices are possible). However, its values may be tightly grouped near this mean, or they may be widely distributed around it. For example, a random variable that is equally likely to be $+1$ or -1 is much more closely spaced around its mean of zero than a random variable that is equally likely to be $+100$ or -100, whose mean is also zero.

Mean 0, closely spaced   Mean 0, widely spaced

To measure the *dispersion* or *spread* of a random variable, we calculate how far it is, on average, from its mean. To avoid negative deviations canceling positive ones, we square the deviations. This leads to the following definitions.

Variance and Standard Deviation

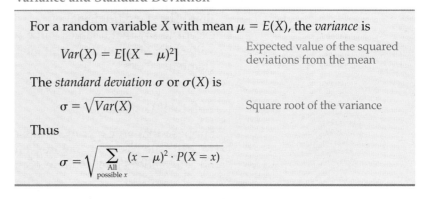

For a random variable X with mean $\mu = E(X)$, the *variance* is

$$Var(X) = E[(X - \mu)^2]$$

Expected value of the squared deviations from the mean

The *standard deviation* σ or $\sigma(X)$ is

$$\sigma = \sqrt{Var(X)}$$

Square root of the variance

Thus

$$\sigma = \sqrt{\sum_{\substack{\text{All} \\ \text{possible } x}} (x - \mu)^2 \cdot P(X = x)}$$

The Greek letter σ, pronounced "sigma," is the lowercase version of Σ. The variance is sometimes denoted σ^2. Finding the variance or standard deviation involves first finding the mean μ. The variance and standard deviation both measure the *spread* of the random variable away from its mean, but the standard deviation has the advantage of being in the same units as the original random variable (such as feet or seconds) and so is easier to interpret.

EXAMPLE 7   Calculating a Standard Deviation

Find the standard deviation of the random variable X from Example 3 .

Solution

This X is defined by the table of values and probabilities on page 442:

x	0	1	2
$P(X = x)$	$\frac{1}{7}$	$\frac{4}{7}$	$\frac{2}{7}$

First we find the mean.

$$\mu = E(X) = \sum_{\substack{\text{All} \\ \text{possible } x}} x \cdot P(X = x) = 0 \cdot \tfrac{1}{7} + 1 \cdot \tfrac{4}{7} + 2 \cdot \tfrac{2}{7} = \tfrac{8}{7}$$

Values times probabilities

Then

$$\sigma = \sqrt{\sum_{\substack{\text{All} \\ \text{possible } x}} (x - \mu)^2 \cdot P(X = x)}$$

$$= \sqrt{\left(0 - \tfrac{8}{7}\right)^2 \cdot \tfrac{1}{7} + \left(1 - \tfrac{8}{7}\right)^2 \cdot \tfrac{4}{7} + \left(2 - \tfrac{8}{7}\right)^2 \cdot \tfrac{2}{7}} \approx 0.639$$

The mean $\mu = \tfrac{8}{7}$ means that, on average, $1\tfrac{1}{7}$ red balls are removed each time. The standard deviation $\sigma = 0.639$ is harder to interpret by itself, except to say that a larger value would mean more spread, a smaller value less. More on this point later in this section and in the following chapter.

Binomial Distribution

Suppose that an experiment has only two possible outcomes, which we call "success" and "failure," and that success happens with probability p, and therefore failure occurs with probability $1 - p$. We may represent this situation by a random variable that takes only the values 0 and 1 as follows:

$$X = \begin{cases} 1 & \text{with probability } p \\ 0 & \text{with probability } 1 - p \end{cases}$$

$X = 1$ means "success"
$X = 0$ means "failure"

Such a random variable is called a *Bernoulli* random variable.* Note that the term "success" can mean whatever we choose, such as *tossing heads* or *catching a cold*. If such success-or-failure experiments are repeated, with successive repetitions being independent, the repetitions are called *Bernoulli trials*. The number of successes in n Bernoulli trials is called a *binomial* random variable.

Binomial Distribution

Let X be the number of successes in n Bernoulli trials, with probability p of success on each trial. Then X is called a *binomial random variable with parameters n and p*, and its distribution, called the *binomial distribution*, is

$$P(X = x) = {}_nC_x\, p^x(1 - p)^{n-x}$$

n = number of trials
p = probability of success
x = number of successes

for $x = 0, 1, \ldots, n$. The mean is $\mu = np$ and the standard deviation is $\sigma = \sqrt{np(1 - p)}$.

To verify the formula for the probability distribution, recall that any sequence of n Bernoulli trials with x successes (and so $n - x$ failures) will, by independence, have probability $p^x(1 - p)^{n-x}$. How many such sequences are there? The x successes may occur in any x of the n trials, the other trials then being failures. But choosing the x trials for the successes out of a total of n can be done in exactly ${}_nC_x$ ways (see page 399). Adding together these ${}_nC_x$ probabilities, each of which is $p^x(1 - p)^{n-x}$, gives the binomial distribution formula in the box above. The formulas for the mean and standard deviation are proved in the Explorations and Excursions on pages 454–455.

* After the Swiss mathematician James Bernoulli (1654–1705), who first recognized the importance of such random variables. His main work was *Ars Conjectandi (Art of Conjecturing)*.

EXAMPLE 8   **A Binomial Distribution**

Let X be the number of heads in four tosses of a coin. Find the probability distribution for X.

Solution

Coin tossing is merely Bernoulli trials, so we want the probability distribution of a binomial random variable with $n = 4$ and $p = \frac{1}{2}$.

$$P(X = 0) = 1 \cdot \left(\tfrac{1}{2}\right)^0 \cdot \left(\tfrac{1}{2}\right)^4 = \tfrac{1}{16} \qquad\qquad {}_4C_0\, p^0(1 - p)^4$$

$$P(X = 1) = 4 \cdot \left(\tfrac{1}{2}\right)^1 \cdot \left(\tfrac{1}{2}\right)^3 = \tfrac{4}{16} = \tfrac{1}{4} \qquad\qquad {}_4C_1\, p^1(1 - p)^3$$

$$P(X = 2) = 6 \cdot \left(\tfrac{1}{2}\right)^2 \cdot \left(\tfrac{1}{2}\right)^2 = \tfrac{6}{16} = \tfrac{3}{8} \qquad\qquad {}_4C_2\, p^2(1 - p)^2$$

$$P(X = 3) = 4 \cdot \left(\tfrac{1}{2}\right)^3 \cdot \left(\tfrac{1}{2}\right)^1 = \tfrac{4}{16} = \tfrac{1}{4} \qquad\qquad {}_4C_3\, p^3(1 - p)^1$$

$$P(X = 4) = 1 \cdot \left(\tfrac{1}{2}\right)^4 \cdot \left(\tfrac{1}{2}\right)^0 = \tfrac{1}{16} \qquad\qquad {}_4C_4\, p^4(1 - p)^0$$

These five probabilities make up the probability distribution for X, the number of heads in four tosses. They could also have been found using a graphing calculator. For the mean μ and standard deviation σ we use the formulas in the preceding box:

```
binompdf(4,.5)
(.0625 .25 .375...
Ans▶Frac
(1/16 1/4 3/8 1...
```

$$\mu = np = 4 \cdot \tfrac{1}{2} = 2 \qquad\qquad \sigma = \sqrt{npq} = \sqrt{4 \cdot \tfrac{1}{2} \cdot \tfrac{1}{2}} = 1$$

The fact that the expected number of heads in four tosses of a fair coin is two should come as no surprise. We found exactly this distribution in Example 2 (page 441). Here we found it using the binomial probability formula, and there we found it from basic principles. The results agree.

PRACTICE PROBLEM 2

Let X be the number of heads in six tosses of a coin. Find $P(X = 3)$ and the mean and standard deviation of X.

Solution at the back of the book

EXAMPLE 9   **Employee Retention**

A restaurant manager estimates the probability that a newly hired waiter will still be working at the restaurant six months later is only

60%. For the five new waiters just hired, what is the probability that at least four of them will still be working at the restaurant in six months?

Solution

Assuming that the waiters decide independently of each other, they make up five Bernoulli trials. Counting a waiter who stays as a "success," the number X of waiters who stay is a binomial random variable with $n = 5$ and $p = 0.6$. We want $P(X \geq 4)$, which means $P(X = 4 \text{ or } 5) = P(X = 4) + P(X = 5)$.

$$P(X = 4) + P(X = 5) = {}_5C_4(0.6)^4(0.4)^1 + {}_5C_5(0.6)^5(0.4)^0 \approx 0.337$$

$$\underbrace{\qquad}_{5} \qquad \underbrace{\qquad}_{1}$$

The probability that at least four of the new waiters will stay for six months is only about 34%.

```
binompdf(5,.6,4)
+binompdf(5,.6,5
)
          .33696
```

EXAMPLE 10   **Money-Back Guarantees**

A manufacturer of computer diskettes sells them in packs of 10 with a "double your money back" guarantee if more than one diskette is defective. If each diskette is defective with a 1% probability independently of the others, what proportion of packages will require refunds?

Solution

If X is the number defective diskettes in a package, then X is a binomial random variable with $n = 10$ and $p = 0.01$. Rather than calculating the probability of a refund (2 or more defectives), we find the complementary probability:

$$P(X = 0) + P(X = 1) = {}_{10}C_0(0.01)^0(0.99)^{10} + {}_{10}C_1(0.01)^1(0.99)^9 \approx 0.996$$

$$\underbrace{\qquad}_{1} \qquad \underbrace{\qquad}_{10}$$

```
binompdf(10,.01,
0)+binompdf(10,.
01,1)
       .9957337998
1-Ans
        .0042662002
```

Subtracting this answer from 1 (since it is the complementary probability) gives 0.004. Therefore, the company will have to replace only about 0.4% of its packs.

Graphing Calculator Exploration

The program* BINOMIAL graphs the binomial distribution for given values of n and p. For example, the following screens show the graph of the binomial probability distribution with $n = 10$ but with different values of p.

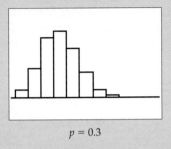

$p = 0.3$

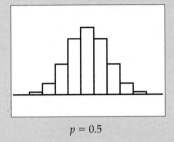

$p = 0.5$

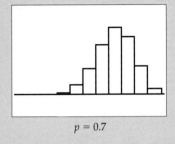

$p = 0.7$

Notice that the second graph is symmetric, and that values of p away from 0.5 skew the graph to one side or the other. This program also gives numeric values for the probabilities and the mean and standard deviation, and it shows graphically where the mean and standard deviation fall on the graph.

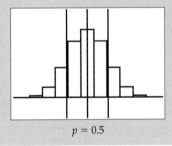

$p = 0.5$

Chebyshev's Inequality

The standard deviation measures the *spread* or *dispersion* of a random variable away from its mean, but is there any general result that shows exactly how much spread a given standard deviation allows? The answer is yes, and the result is known as Chebyshev's inequality.[†]

Chebyshev's Inequality

The probability that the value of a random variable is within k standard deviations on either side of its mean is at least $1 - \frac{1}{k^2}$.

---

* See the Preface for information on how to obtain this and other graphing calculator programs.
† After the Russian mathematician Pafnuti Lvovich Chebychev (1821–1894), who first discovered and proved it.

(A proof of this inequality is given in the Explorations and Excursions on page 455.) Chebyshev's inequality says that for *any* random variable, the probability of being within k standard deviations of the mean is at least $1 - \frac{1}{k^2}$. For example, for $k = 2, 3$, and 4, the values of $1 - \frac{1}{k^2}$ are 0.75, 0.89, and 0.94, respectively. Therefore, Chebyshev's inequality for these values of k says that a random variable will be

within 2 standard deviations of its mean with probability at least 75%.

within 3 standard deviations of its mean with probability at least 89%.

within 4 standard deviations of its mean with probability at least 94%.

Chebyshev's inequality gives precision to the intuitive notion that any random variable is unlikely to be far (measured in standard deviation units) from its mean.

Section Summary

A *random variable* X is an assignment of a number to each outcome in the sample space. The *expected value* (or *expectation* or *mean*), *variance*, and *standard deviation* of a random variable X are

$$\mu = E(X) = \sum_{\substack{\text{All} \\ \text{possible } x}} x \cdot P(X = x)$$

$$Var(X) = \sum_{\substack{\text{All} \\ \text{possible } x}} (x - \mu)^2 \cdot P(X = x)$$

$$\sigma = \sqrt{\sum_{\substack{\text{All} \\ \text{possible } x}} (x - \mu)^2 \cdot P(X = x)}$$

Bernoulli trials are repeated independent experiments that result in *success* (with probability p) or *failure* (with probability $1 - p$). The number X of successes in n Bernoulli trials is a *binomial* random variable with parameters n and p, and the *binomial probability distribution* is

$$P(X = x) = {}_nC_x \, p^x (1 - p)^{n-x} \quad \text{for } x = 0, 1, \ldots, n$$

The mean is $\mu = np$ and the standard deviation is $\sigma = \sqrt{np(1 - p)}$.

    Chebyshev's inequality states that the probability that the value of a random variable is within k standard deviations of its mean is at least $1 - \frac{1}{k^2}$.

EXERCISES 5.6 (▦ or ▦ will be helpful.)

1. **Coins**   A coin is tossed three times, and X is the number of heads. Find and graph the probability distribution of X.

2. **Dice**   Two dice are rolled, and X is the sum of the faces. Find and graph the probability distribution of X.

3. **Coins**   You toss three coins and win $11 if they all agree (all heads or all tails), and otherwise you lose $1. Find and graph the probability distribution of your winnings.

4. **Marbles**   Two marbles are chosen at random (without replacement) from a box containing 3 red and 5 green marbles. Let X be the number of green marbles chosen. Find and graph the probability distribution of X.

5. **Dice**   A die is rolled, and you win $2 if it comes up odds, you lose $12 if it comes up 2, and you win $3 if it comes up 4 or 6. Find and graph the probability distribution of your winnings.

6. **Dice**   A die is rolled and you win $8 if it comes up odds, you lose $15 if it comes up 2 or 4, and you win $6 if it comes up 6. Find and graph the probability distribution of your winnings.

7. **Dice**   Two dice are rolled, and X is the larger of the two numbers that come up. Find and graph the probability distribution of X.

8. **Dice**   Two dice are rolled, and X is the smaller of the two numbers that come up. Find and graph the probability distribution of X.

9. **Mean**   Find the mean of the random variable in Exercise 1.

10. **Mean**   Find the mean of the random variable in Exercise 2.

11. **Mean and Standard Deviation**   Find the mean and standard deviation of the random variable in Exercise 3.

12. **Mean**   Find the mean of the random variable in Exercise 4.

13. **Mean and Standard Deviation**   Find the mean and standard deviation of the random variable in Exercise 5.

14. **Mean and Standard Deviation**   Find the mean and standard deviation of the random variable in Exercise 6.

15. **Mean**   Find the mean of the random variable in Exercise 7.

16. **Mean**   Find the mean of the random variable in Exercise 8.

17. **Raffle Tickets**   One thousand raffle tickets are sold, and there is one first prize worth $2000, one second prize worth $250, and 20 third prizes worth $50 each. Find the expected value of a ticket.

18. **Standard Deviations**   Find the standard deviation of a random variable that is equally likely to be 1 or -1. Then find the standard deviation if the values are changed to 50 and -50 (still equally likely).

19. **Standard Deviations**   Find the standard deviation of a random variable that is equally likely to be 1 or -1. Then find the standard deviation if the values are changed to 100 and -100 (still equally likely).

20. **Binomial Distribution**   Find and graph the probability distribution of a binomial random variable with parameters $n = 4$ and $p = \frac{1}{5}$. Use the formulas on page 447 to find its mean and standard deviation.

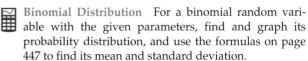

Binomial Distribution   For a binomial random variable with the given parameters, find and graph its probability distribution, and use the formulas on page 447 to find its mean and standard deviation.

21. $n = 20$ and $p = \frac{1}{2}$

22. $n = 25$ and $p = \frac{1}{2}$

23. $n = 20$ and $p = 0.8$

24. $n = 25$ and $p = 0.64$

25. **Coins**   For 6 tosses of a coin, find the probability that the number of heads is between 2 and 4 (inclusive).

26. **Coins**   What is the probability of getting exactly 5 heads in 10 tosses of a coin?

27. **Coins**   What is the probability of getting exactly 4 heads in 8 tosses of a coin?

28. **Coins**   For an unfair coin whose probability of heads is 0.7, what is the most likely number of

heads in 10 tosses, and what is its probability? What is the *least* likely number of heads, and what is its probability?

 29. Coins For an unfair coin whose probability of heads is $\frac{1}{3}$, what is the most likely number of heads in 6 tosses, and what is its probability? What is the *least* likely number of heads, and what is its probability?

 30. Dice In rolling a die 10 times, what is the most likely number of sixes, and what is its probability? What is the mean number of sixes?

 31. Dice In rolling a die 8 times, what is the most likely number of sixes, and what is its probability? What is the mean number of sixes?

32. Testing You know that one of three batteries is dead. If you test them one at a time until you find the defective one, what is the expected number of tests?

33. Sales An automobile salesperson predicts that a customer will buy a $30,000 car with probability $\frac{1}{10}$, a $25,000 car with probability $\frac{1}{5}$, a $20,000 car with probability $\frac{3}{10}$, and otherwise buy nothing. What is the expected value of the sale?

34. Insurance An insurance company estimates that on a typical policy it will have to pay out $10,000 with probability 0.05, $5000 with probability 0.1, and $1000 with probability 0.2. If the company wants to charge $200 more than the expected payout of the policy, what should it charge?

35. Product Quality A company manufactures products of which 2% have hidden defects. If you buy 10 of them, then the number of defective ones in your purchase is a binomial random variable. Find its mean and standard deviation.

36. Multiple-Choice Testing A multiple-choice test has 5 possible answers for each of 10 questions. For a student who guesses randomly:
  a. What are the mean and standard deviation of the number of right answers?
  b. What is the probability that this student gets 4 or more correct answers?

37. ESP A person claims to have ESP (extrasensory perception) and calls 7 out of 10 tosses of a fair coin correctly. What is the probability of doing at least that well by guessing randomly?

 38. Juries Suppose that on a jury of 12 people it takes at least 9 votes to convict. If each person decides correctly with probability 0.9, what is the probability of that a guilty person is convicted? What is the probability that an innocent person is found innocent? What assumptions are you making about independence?

39. Communications A one-digit message (0 or 1) is to be transmitted over a communications line that has a $\frac{1}{10}$ probability of changing the digit. Because of this, the message is to be transmitted in triplicate, 000 or 111, with decoding by *majority,* which means that the digit that occurs two or three times will be taken to be the message. What is the probability that the message is received correctly? How does this compare with the probability of correct reception when only *one* digit is transmitted?

40. Family Distribution Assuming that boys and girls are equally likely, what is the probability that a family of 6 children consists of 3 boys and 3 girls?

 41. Cards A bridge hand consists of 13 cards. What is the probability that you get no aces in three consecutive hands?

42. Target Practice If the probability of hitting a target is $\frac{1}{4}$ and 5 shots are fired, what is the probability of hitting the target at least once?

43. Sports Suppose that in a sports contest with no ties, the stronger team has probability $\frac{3}{5}$ of winning any particular game. Find the probability that the stronger team wins:
  a. A two-out-of-three series. [*Hint:* For ease of calculation, assume that all three games are played. What assumptions about independence are you making?]
  b. A three-out-of-five series.
  c. A four-out-of-seven series.

 44. Home Runs Reread the Application Preview on page 439. On the basis of records from 1996 to 1998, the probability that Mark McGuire will hit a home run on any particular at-bat is $p_M = 0.110$. Model the number X of home runs by McGuire in a season by a binomial random variable with parameters $n = 530$ (a typical number of at-bats in a season) and p_M. Find $P(X > 50)$, $P(X > 60)$, and $P(X > 70)$. Then do the same for Sammy Sosa, using his home-run-per-at-bat probability $p_S = 0.095$.

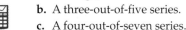

Explorations and Excursions

The following exercises extend and augment the material presented in the text.

The mean and standard deviation of the binomial distribution   The following exercises prove the formulas $\mu = np$ and $\sigma = \sqrt{np(1-p)}$ for binomial random variables with parameters n (number of trials) and p (probability of success on each trial). We will use the fact that adding together n binomial random variables, each with parameters $n = 1$ and p, gives a binomial random variable with parameters n and p (since adding up the number of "successes" in each of n Bernoulli trials gives the total number of successes in all n trials).

45. If X is binomial with $n = 1$, show that $E(X) = 1 \cdot p + 0 \cdot (1 - p) = p$.

46. Justify each numbered equals sign. For X and Y independent,

$$E(X + Y) \overset{1}{=} \sum_{\text{All } x,y} (x + y) \cdot P(X = x \text{ and } Y = y)$$

$$\overset{2}{=} \sum_{\text{All } x,y} (x + y) \cdot P(X = x) \cdot P(Y = y)$$

$$\overset{3}{=} \sum_{\text{All } x,y} x \cdot P(X = x) \cdot P(Y = y) + \sum_{\text{All } x,y} y \cdot P(X = x) \cdot P(Y = y)$$

$$\overset{4}{=} \left(\sum_{\text{All } x} x \cdot P(X = x) \right) \cdot \left(\sum_{\text{All } y} P(Y = y) \right) + \left(\sum_{\text{All } y} y \cdot P(Y = y) \right) \cdot \left(\sum_{\text{All } x} P(X = x) \right)$$

$$\overset{5}{=} \left(\sum_{\text{All } x} x \cdot P(X = x) \right) \cdot 1 + \left(\sum_{\text{All } y} y \cdot P(Y = y) \right) \cdot 1 = E(X) + E(Y)$$

That is, for independent random variables, $E(X + Y) = E(X) + E(Y)$ (in words, the expectation of the sum is the sum of the expectations). This result immediately generalizes to any number of independent random variables:
$$E(X_1 + X_2 + \cdots + X_n) = E(X_1) + E(X_2) + \cdots + E(X_n).$$

47. If X is a binomial random variable with parameters n and p, we may write $X = X_1 + X_2 + \cdots + X_n$, where $X_1, X_2, \ldots, X_n$ are independent binomial random variables each with parameters $n = 1$ and p. Justify:

$$E(X) = E(X_1 + X_2 + \cdots + X_n) = E(X_1) + E(X_2) + \cdots + E(X_n) = \underbrace{p + p + \cdots + p}_{n \text{ terms}} = np$$

Therefore, a binomial random variable with parameters n and p has mean $\mu = np$.

48. We now prove the formula for the standard deviation. If X is a binomial random variable with parameters $n = 1$ and p, show that

$$Var(X) = (1 - p)^2 \cdot p + (0 - p)^2 \cdot (1 - p) = p(1 - p)(1 - p + p) = p(1 - p)$$

49. Justify each numbered equals sign. For X and Y independent,

$$Var(X + Y) \overset{1}{=} \sum_{\text{All } x,y} ((x + y) - E(X + Y))^2 \cdot P(X = x \text{ and } Y = y)$$

$$\overset{2}{=} \sum_{\text{All } x,y} ((x - E(X)) + (y - E(Y)))^2 \cdot P(X = x) \cdot P(Y = y)$$

$$\overset{3}{=} \sum_{\text{All } x,y} ((x - E(X))^2 + 2(x - E(X)) \cdot (y - E(Y)) + (y - E(Y))^2) \cdot P(X = x) \cdot P(Y = y)$$

$$\overset{4}{=} \sum_{\text{All } x,y} (x - E(X))^2\, P(X = x) \cdot P(Y = y) + \sum_{\text{All } x,y} 2(x - E(X)) \cdot (y - E(Y)) \cdot P(X = x) \cdot P(Y = y)$$

$$+ \sum_{\text{All } x,y} (y - E(Y))^2 \cdot P(X = x) \cdot P(Y = y)$$

$$\overset{5}{=} \sum_{\text{All } x} (x - E(X))^2\, P(X = x) \cdot \left(\sum_{\text{All } y} P(Y = y) \right) + 2 \cdot \left(\sum_{\text{All } x} (x - E(X)) \cdot P(X = x) \right) \cdot \left(\sum_{\text{All } y} (y - E(Y)) \cdot P(Y = y) \right)$$

$$+ \sum_{\text{All } y} (y - E(Y))^2\, P(Y = y) \cdot \left(\sum_{\text{All } x} P(X = x) \right)$$

$$\overset{6}{=} Var(X) \cdot 1 + 2 \cdot E(X - E(X)) \cdot E(Y - E(Y)) + Var(Y) \cdot 1$$

$$\overset{7}{=} Var(X) + Var(Y)$$

That is, for independent random variables, $Var(X + Y) = Var(X) + Var(Y)$ (in words, the variance of the sum is the sum of the variances). This result immediately generalizes to any number of independent random variables: $Var(X_1 + X_2 + \cdots + X_n) = Var(X_1) + Var(X_2) + \cdots + Var(X_n)$

50. If X is a binomial random variable with parameters n and p, we may write $X = X_1 + X_2 + \cdots + X_n$, where $X_1, X_2, \cdots, X_n$ are independent binomial random variables each with parameters $n = 1$ and p. Then

$$Var(X) = Var(X_1 + X_2 + \cdots + X_n) = Var(X_1) + Var(X_2) + \cdots + Var(X_n)$$
$$= \underbrace{p(1 - p) + p(1 - p) + \cdots + p(1 - p)}_{n \text{ terms}} = np(1 - p)$$

Therefore, a binomial random variable with parameters n and p has variance $np(1 - p)$. Its standard deviation is then the square root of the variance, $\sigma = \sqrt{Var(X)} = \sqrt{np(1 - p)}$.

Chebyshev's inequality   Exercises 51–55 prove Chebyshev's inequality. Justify each numbered equality or inequality in Exercise 51 and then the statements in the remaining exercises.

51. $\sigma^2 \overset{1}{=} \sum_{\text{All } x} (x - \mu)^2 P(X = x)$

$$\overset{2}{=} \sum_{\substack{\text{Those } x \text{ with} \\ (x-\mu)^2 \ge (k\sigma)^2}} (x - \mu)^2 P(X = x) + \sum_{\substack{\text{Those } x \text{ with} \\ (x-\mu)^2 < (k\sigma)^2}} (x - \mu)^2 P(X = x)$$

$$\overset{3}{\ge} \sum_{\substack{\text{Those } x \text{ with} \\ (x-\mu)^2 \ge (k\sigma)^2}} (X - \mu)^2 P(X = x) \overset{4}{\ge} (k\sigma)^2 \sum_{\substack{\text{Those } x \text{ with} \\ (x-\mu)^2 \ge (k\sigma)^2}} P(X = x)$$

52. $\sigma^2 \ge (k\sigma)^2\, P\big((X - \mu)^2 \ge (k\sigma)^2\big)$

53. $P\big(|X - \mu| \ge k\sigma\big) \le \dfrac{1}{k^2}$

54. $P\big(|X - \mu| > k\sigma\big) < 1 - \dfrac{1}{k^2}$

55. Explain why the statement of Exercise 54 is Chebychev's inequality.

56. Explain why Chebyshev's inequality is clearly true for $k = 1$.

57. Use Chebyshev's inequality to support the following claim: "Almost all of *any* probability distribution is within 5 standard deviations of the mean."

Chapter Summary with Hints and Suggestions

Reading the text and doing the exercises in this chapter have helped you to master the following skills, which are listed by section (in case you need to review them) and are keyed to particular Review Exercises. Answers for all Review Exercises are given at the back of the book, and full solutions can be found in the Student Solutions Manual.

5.1 SETS, COUNTING, AND VENN DIAGRAMS

- Read and interpret a Venn diagram. *(Review Exercise 1.)*

- Use the complementary or the addition principles of counting or Venn diagrams to solve an applied problem. *(Review Exercise 2.)*

- Use the multiplication principle of counting to solve an applied problem. *(Review Exercise 3.)*

5.2 PERMUTATIONS, COMBINATIONS, AND THE BINOMIAL THEOREM

- Calculate numbers of permutations and combinations. *(Review Exercise 4.)*

- Use the permutation and combination formulas (possibly using ▦) to solve an applied problem. *(Review Exercise 5–6.)*

- Use the binomial theorem to expand an expression. *(Review Exercise 7.)*

5.3 PROBABILITY SPACES

- Find an appropriate sample space for a random experiment. *(Review Exercises 8–9.)*

- Describe an event in terms of outcomes from a sample space. *(Review Exercise 10.)*

- Assign probabilities to outcomes in a sample space, and find probabilities of events. *(Review Exercises 11–15.)*

- Use the techniques of the section to find a probability in an applied problem. *(Review Exercises 16–24.)*

5.4 CONDITIONAL PROBABILITY AND INDEPENDENCE

- Find a conditional probability using either the definition or a restricted sample space. *(Review Exercises 25–26.)*

- Use conditioning and the total probability formula to solve an applied problem. *(Review Exercises 27–32.)*

- Determine whether two events are independent or dependent. *(Review Exercise 33.)*

- Use independence to find a probability in an applied problem. *(Review Exercises 34–35.)*

5.5 BAYES' THEOREM

- Use Bayes' formula to solve an applied problem. *(Review Exercises 36–38.)*

5.6 RANDOM VARIABLES AND DISTRIBUTIONS

- Determine the outcomes that correspond to the values of a random variable. *(Review Exercise 39.)*

- Find and graph the probability distribution of a random variable, and find the mean and standard deviation. *(Review Exercises 40–44.)*

- For a binomial random variable, find and graph the probability distribution, and find its mean and standard deviation. *(Review Exercise 45.)*

- Use the binomial probability distribution to solve an applied problem. *(Review Exercises 46–51.)*

HINTS AND SUGGESTIONS

- Much of probability depends on counting, and there are several principles to simplify counting large numbers of objects. Roughly speaking, the *complementary* principle says that you can count the *opposite* set and then subtract; the *addition* principle says that you can add numbers from two sets but you must then subtract what was double-counted; and the *multiplication* principle says that the number of two-part choices is the product of the number of first-part choices times the number of second-part choices.

- *Permutations* are arrangements where a different order means a different object (such as letters in a word or rankings of people). *Combinations* are arrangements where reordering does *not* make a new object (such as hands of cards or committees of people).

- The *binomial theorem* is a formula for expanding $(x + y)^n$ for any positive integer n. The formula involves combinations because multiplying $(x + y)$ by itself a total of n times involves counting the number of ways that x can be chosen from some factors and y from others.

- A set of n objects has 2^n subsets (counting the "empty" subset and the set itself).

- Probabilities are assigned to possible outcomes of a random experiment so that each probability is between 0 and 1 (inclusive) and they add to 1. Equally likely outcomes should be assigned equal probabilities. Probabilities of more complicated events are found by adding up the probabilities of the outcomes in the event.

- Conditional probability is the relative probability of the intersection of the events compared to the probability of the *given* event. Conditional probabilities are sometimes more easily found from the restricted sample space than from the definition.

- A complicated probability can sometimes be found by *conditioning* on the occurrence of some underlying events (a *partition*).

- Independent events "have nothing to do with each other," and the probability of both occurring is the *product* of their probabilities. This is not the same as *disjoint* events, where one *precludes* the other, and which therefore *cannot* be independent.

- Bayes' formula is useful for finding events of the form S_1 *given* A in terms of events of the form A *given* S_i where $S_1, \cdots, S_m$ is a partition.

- The mean of a random variable gives a *representative* or *typical* value, and the standard deviation measures the *spread* of its values about the mean.

- Bernoulli trials are repeated independent experiments with only two outcomes, *success* and *failure*. The number of successes in several Bernoulli trials is a *binomial* random variable.

- A graphing calculator that finds permutations, combinations, and binomial and other probabilities is very useful.

- Practice for test: Review Exercises 1, 2, 3, 5, 7, 9, 10, 11, 16, 17, 18, 22, 23, 26, 30, 32, 35, 37, 40, 42, 43, 44, 47, 49, 51.

Review Exercises for Chapter 5   *Practice test exercise numbers are in blue.*

5.1   Sets, Counting, and Venn Diagrams

1. **Venn Diagrams**   For the Venn diagram below, find

    a. $n(A)$ **b.** $n(A \cup B)$
    c. $n(B^c)$ **d.** $n(A^c \cap B)$

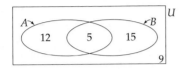

2. **Purchases**   A survey of 1000 homeowners found that during the last year, 230 bought an automobile, 340 bought a major appliance, and 540 bought neither. How many homeowners had bought both an automobile and a major appliance?

3. **Initials**   Find how many three-letter monograms there are if:

    a. Repeated letters are allowed.
    b. Repeated letters are not allowed.

5.2   Permutations, Combinations, and the Binomial Theorem

4. **Permutations and Combinations**   Find:

    a. $_6P_3$ **b.** $_6C_3$

5. **Committees**   How many 4-member committees can be formed from a club consisting of 20 students? What if the committee is to consist of a president, vice president, secretary, and treasurer?

6. **Cards**   How many 5-card hands are there that contain only spades?

7. **Binomial Theorem**   Use the binomial theorem to:

    a. Write out the first three terms of $(x + y)^{10}$.
    b. Expand $(2a - 1)^4$.

5.3   Probability Spaces

8. **Wardrobes**   Find the sample space for choosing an outfit consisting of one of two coats (C_1 or C_2), one of two scarves (S_1 or S_2), and one of two hats (H_1 or H_2).

9. **Marbles**   A box contains three marbles: one blue, one yellow, and one red. A first marble is chosen, then it is replaced, and a second marble is chosen. Find the sample space. Then find the sample space if the first marble is *not* replaced before the second is chosen.

10. **Events**   For the experiment of tossing a coin twice, describe each event as a subset of the sample space $\{(H,H), (H,T), (T,H), (T,T)\}$.

    a. At least one head
    b. At most one head
    c. Different faces

11. **Probabilities of Events**   Consider the sample space in Exercise 10.

    a. What probability should be assigned to each outcome?
    b. What is the probability of getting at least one head?
    c. What is the probability of getting different faces?

12. **Committees**   A committee of 2 is to be chosen at random from a class of 15 students. What probability should be assigned to any particular committee? What if one of the 2 members is to be designated committee spokesperson?

Spinners   Find the probability of the arrow landing in each numbered region. Assume that the probability of any region is proportional to its area, and assume that areas that look the same size are the same size.

13.   **14.**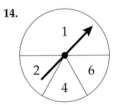

15. **Marbles**   A marble is drawn at random from a box containing 3 red, 6 green, and 9 black marbles, and the color is noted. What is the probability of each outcome in the sample space $\{R, G, B\}$?

16. **Smoking and Weight** For a randomly selected person, the probability of being a smoker is 0.35, the probability of being overweight is 0.40, and the probability of being both a smoker and overweight is 0.20. What is the probability of being *neither* a smoker *nor* overweight?

17. **Dice** If you roll one die, find the probability of:
    a. Rolling at most 2.
    b. Rolling an even number.

18. **Lottery** In a lottery you choose 4 numbers out of 40. Then 5 numbers are announced, and you win something if you have 4 of the 5 numbers. What is the probability that you win something?

19. **Cards** You are dealt 5 cards at random from an ordinary deck. Find the probability of being dealt:
    a. All face cards.
    b. No face cards.

20. **Cards** You are dealt 13 cards at random from an ordinary deck. Find the probability of being dealt:
    a. No spades.
    b. No face cards.

21. **Defective Products** You have 30 computer diskettes, and 2 of them contain computer viruses. If you lend 5 to a friend, what is the probability that they are virus-free?

22. **Defective Products** A store shelf has 50 light bulbs, of which 2 are defective. If you buy 4, what is the probability that you get both defective bulbs? Neither defective bulb?

23. **Committees** A committee of 3 is to be formed from your class of 30. What is the probability that both you and your best friend will be on it?

24. **Memory** You remember the names of 6 of the 10 people you just met. If you run into 2 of the 10 on the street, what is the probability that you will remember their names?

5.4 Conditional Probability and Independence

25. If $P(A) = 0.5$, $P(B) = 0.4$, and $P(A \cap B) = 0.3$, find
    a. $P(A$ given $B)$
    b. $P(B$ given $A)$

26. **Dice** If you roll two dice, what is the probability of at least one six given that the sum of the numbers is seven?

27. **Gender** If you know that a family of three children has at least one boy, what is the probability that all three are boys? Assume that boys and girls are equally likely.

28. **Cards** You draw three cards at random from a standard deck. If the first is an ace, what is the probability that the other two are also aces?

29. **Cards** If you are being dealt five cards and the first four are hearts, what is the probability that the fifth is also a heart?

30. **Market Share** An automobile salesperson estimates that with a male customer she can make a sale with probability 0.3 and that with a female customer the probability is 0.4. If 80% of her customers are male and 20% female, what is the probability that for a randomly selected customer she can make a sale?

31. **Searches** An airplane is missing, and you estimate that with probability 0.6 it crashed in the mountains and with probability 0.4 it crashed in the valley. If it is in the mountains, a search there will find it with probability 0.8, whereas if it is in the valley, a search there will find it with probability 0.9. What is the probability that the plane will be found? (The complementary probabilities to the numbers 0.8 and 0.9 are called the *overlook probabilities* because they give the probability that it will *not* be found in a region if it is there.)

32. **Births** Twelve percent of all births are by cesarean (surgical) section, and of these births 96% survive. Overall, 98% of all babies survive delivery. For a randomly chosen mother who did not have a cesarean, what is the probability that her baby survives?

33. **Coins** For the experiment of tossing a coin twice, with sample space {(H,H), (H,T), (T,H), (T,T)}, are the following events independent or dependent?
    a. *At least one head* and *at most one head*
    b. *Heads on first toss* and *same face on both tosses*

34. Computer Malfunctions An airline reservations computer breaks down with probability 0.01, at which time a second computer takes over, but it also has probability 0.01 of failing. Find the probability that the airlines will be able to take reservations.

35. Coins Five coins are tossed. Find the probability of getting

a. All heads.
b. All the same outcome.
c. Alternating outcomes.

5.5 Bayes' Theorem

36. Coins You have two coins in your pocket, one with two heads, and one with a head and a tail. You choose one at random and toss it, and it comes up heads. What is the probability that it is the two-headed coin?

37. Manufacturing Defects A soft drink bottler has three bottling machines: A, B, and C. Machine A bottles 25% of the company's output, machine B bottles 35%, and machine C bottles 40%. It is known that 4% of the bottles produced by machine A are defective, as are 3% of bottles produced by machine B and 2% of the bottles from machine C. If a randomly selected bottle from the factory's output is found to be defective, what is the probability that it was produced by machine A?

38. Manufacturing Defects For the information in Exercise 37, if a randomly selected bottle is *not* defective, what is the probability that it came from machine A?

5.6 Random Variables and Distributions

39. Coins Five coins are tossed, and X is the number of heads. List the outcomes (from the usual sample space) corresponding to each of the events $X = 5$, $X = 1$, and $X = 0$.

40. Coins You toss two coins. You win $34 if you get double heads, and otherwise you lose $2. Find and graph the probability distribution of your winnings.

41. Mean and Standard Deviation Find the mean and standard deviation of the random variable in Exercise 40.

42. Marbles Two marbles are chosen at random and without replacement from a box containing 5 red and 4 green marbles. If X is the number of green marbles chosen, find and graph the probability distribution of X.

43. Mean Find the mean of the random variable in Exercise 42.

44. Raffle Tickets Five hundred raffle tickets are sold for the following prizes: one first prize worth $1000, two second prizes worth $150, and 10 third prizes worth $25. Find the expected value of a ticket.

45. Binomial Distribution Find and graph the distribution of a binomial random variable with parameters $n = 4$ and $p = \frac{4}{5}$. Find its mean and standard deviation using the formulas developed for a binomial random variable.

46. Coins For 8 tosses of a coin, find the probability that the number of heads is between 3 and 5 inclusive.

47. Sales A television salesperson estimates that a customer will buy a $900 TV with probability $\frac{1}{10}$, a $500 TV with probability $\frac{1}{2}$, a $200 TV with probability $\frac{1}{10}$, and otherwise buy nothing. What is the expected value of the sale?

48. Testing You know that one of three keys will work in a lock. If you try them one at a time until you find the right one, what is the expected number of tries?

49. Product Quality Of the products a company produces, 1% have hidden defects. If you buy a dozen of them, what are the mean and standard deviation of the number of defective ones in your purchase?

50. Product Quality A company makes screws, and 98% of them are usable. If you buy a box of 100 screws, what is the probability of at least 97 being usable?

51. Target Practice You shoot at a target 5 times. If the probability of a hit on each shot is 75%, what is the probability that you hit it at least 4 times?

Projects and Essays

The following projects and essays are based on the material in this chapter. There are no right or wrong answers—the results depend only on your imagination and resourcefulness.

1. Research and write a page about the early history of probability, including the contributions of Pierre de Fermat, Blaise Pascal, and the Chevalier de Méré (for more, include Gerolamo Cardano and Abraham de Moivre). Include a discussion of the Chavalier's dice questions (see pages 386, 415–416, 418–419, and 431), along with the results of your own experiments with his dice games. (What is a chevalier anyway, and why did he have so much time to play dice?)

2. Reread the Application Preview about coincidences (pages 409–410) and then try to imagine some "one in a million" coincidences that have happened to you, like running into an old friend in a city, seeing something happen that you had dreamed of, or anything else. Then try to estimate the number of times that the coincidence did *not* occur before it actually *did* come true, and so on. Write about whether coincidences are as unlikely as they seem.

3. Imagine that you have an *unfair* coin, whose probability of heads is 0.6 and whose probability of tails is therefore 0.4. For the experiment of tossing it once, define X to be 1 for heads and 0 for tails. Find the mean and standard deviation of X. Is the standard deviation more or less than that for a fair coin? Can you give an intuitive reason for this? What if the coin were even *more* unfair—how would the standard deviation change? What value of p makes the standard deviation the greatest (and the least), and why, both mathematically and intuitively? Write a report discussing all of this.

4. Do Exercise 3 on page 438 about medical tests and "false positives." Then try to find out the accuracy rate (including false positives and false negatives) for acquired immune deficiency syndrome (AIDS) tests, especially for people in low-risk groups. Analyze the results using conditional probability. Discuss the consequences of false positives and false negatives on people's lives. Much information can

be found in libraries, on the Internet, or from AIDS information sources such as the National AIDS Clearing House, many of which have 800 telephone numbers.

5. The "wallet paradox" can be described as follows: Al and Betty wonder whose wallet contains more money. They agree that if they have different amounts of money in their wallets, then each of them has the same chance of being the one with the larger amount. They decide to take out their wallets and look, and the person with more money gets to keep *both* amounts. Both agree that if they exchange money, they have an equal chance of losing what they have or of getting *more* than they have. Therefore, both feel that the game is favorable to them. However, such a game cannot be favorable to both. Write about ways to analyze, understand, and resolve this paradox.

6. There is a joke that goes as follows: A man is frightened that when he flies there will be a bomb on his plane. Therefore, whenever he flies he carries his own bomb in his suitcase, because he figures that the chances of there being *two* crazy people with bombs on one plane is infinitesimal. Discuss in terms of probability and conditional probability.

7. If your state (or some other entity) has a lottery, find out about the probability of a ticket being a winner. Find the expected value of a ticket. Find the probability that, of ten tickets, at least one is a winner. Find the probability that, of n tickets, at least one is a winner, for any value of n. Discuss two types of lotteries: one with a *fixed* number of winning tickets and the other with a *variable* number of winning tickets determined by how many chose the winning numbers. Discuss, for both types, whether different tickets have independent chances of winning. For each type, find a formula for the probability that, of n tickets, at least one is a winner. How do the two formulas compare for small and large values of n?

6

Statistics

Surveys of just a few people can provide estimates of the result of asking everyone, and they can do so in less time and at lower cost. This chapter discusses the statistics of random samples.

6.1 Random Samples and Data Organization

Roosevelt and Landon in 1936

1936 was an election year and Governor Alf Landon of Kansas was trying to unseat the incumbent President Franklin Delano Roosevelt. A highly respected magazine, *Literary Digest*, conducted a poll by sending 10 million questionnaires to names taken from telephone directories and automobile registration forms. Only a small percentage of the questionnaires were returned, but the results were overwhelming: Alf Landon was the favorite, and the magazine confidently predicted a Landon victory. The election results were equally overwhelming, but the other way: Roosevelt won, with 28 million votes to Landon's 17 million.

How could such a large poll have been so wrong? Besides the small return rate, basing the poll on the opinions of people who had telephones or owned automobiles during the depths of the Great Depression yielded a highly *unrepresentative* sample, one decidedly skewed toward the wealthy. Today we understand more about the need to use *random samples*, which is one of the subjects of this section.

Introduction

In life you have to make decisions based on incomplete information. For example, to find out how many Americans watch a television program, you can't ask everyone, and to find out how long a light bulb lasts, a company can't test all of its bulbs, particularly if it wants to have any left to sell. These and many other "real world" problems depend on taking samples and analyzing the data. Statistics is the branch of mathematics concerned with the collection, organization, analysis, and interpretation of numeric data.

Random Samples

A *statistical population* is the entire collection of whatever data you are studying, such as the ages of citizens of the United States, the size of bank accounts in Illinois, or the brands of automobiles in California. Although it is possible in principle to measure the entire population, the time, cost, or intrusiveness of such an undertaking usually dictates that we measure only part of the population. For that purpose we use a *random sample* to represent the population.

Random Sample

> A (simple) *random sample* of size n is a selection of n members of the population satisfying two requirements:
>
> 1. Every member of the population is equally likely to be included in the sample, and
> 2. Every possible sample of size n from the population is equally likely to be chosen.

A sample that does not meet these criteria is not representative and cannot be used to infer characteristics of the entire population. For example, a telephone survey to homes at 10 A.M. on a weekday cannot produce valid information about work skills. Random samples can be chosen using random number tables or by other methods, and what we say in this chapter is based on the assumption that the data we are analyzing is from a random sample.

Levels of Measurement

We shall consider only numerical data, which we classify into four types: *nominal, ordinal, interval,* and *ratio* data.

Nominal data (*nominal* means "in name only") means numbers used only to *identify* objects. For example, the numbers on the backs of football jerseys are nominal data, since they do not mean that "number 24" can run twice as fast or throw twice as far as "number 12."

Ordinal data (the word comes from "order") means numbers that can be arranged in order but differences between them are not meaningful. For example, you might rate the teachers you have had in college as "best" (number 1), "second best" (number 2), and so on. However, the difference between "1" and "2" may not be the same as the difference between "7" and "8."

Interval data means numbers that can be put in order *and* differences between them can be compared, but ratios cannot. For example, it makes sense to talk about the *difference* between two temperatures, like 40° and 80°, but it makes no sense to say that 80° is *twice* as hot as 40°, since the zero point in the Fahrenheit (or Celsius) system is arbitrary and does not mean *zero heat.*

Ratio data means numbers that can be put in order, whose differences can be compared, *and* whose *ratios* have meaning. Examples of ratio data are amounts of money and weights, since 80 of either really *is* twice 40. In general, with ratio data, the zero level *does* mean "none" of something.

When we refer to the *type* of a data collection, we mean the highest of these levels that applies, from nominal (lowest) to ratio (highest).

EXAMPLE 1 **Judging Levels of Data Measurement**

Identify the level of measurement of each data collection.

a. Ten vehicles passing through a turnpike toll booth were classified as 1 for a car, 2 for a bus, and 3 for a truck, giving data {1, 3, 2, 3, 1, 1, 1, 1, 3, 1}.

b. A rock concert enthusiast rates the last eight performances she attended on a scale from 1 (awful) to 10 (awesome) as {8, 5, 7, 9, 7, 10, 6, 9}.

c. The temperatures of five students at the college infirmary are {99.3, 102.1, 101.8, 101.5, 100.9} degrees Fahrenheit.

d. The weights of twelve members of the cross-country team are {124, 151, 132, 153, 142, 147, 120, 127, 154, 119, 118, 116} pounds.

Solution

a. *Nominal data:* The numbers are used only to identify the type of vehicle. The number 2 does not mean more or less than the number 3, so differences between these numbers have no meaning.

b. *Ordinal data:* The rankings put the performances in relative order, but no meaning can be given to differences between the rankings. The difference between ratings 5 and 6 may be more or less than the difference between ratings 9 and 10.

c. *Interval data:* The temperatures can be put in order from "lowest" to "highest," and differences between them can be compared. However, this is not ratio data because the zero point in temperature is arbitrary—101° is not 1% hotter than 100°.

d. *Ratio data:* The weights can be put in order, "10 pounds heavier" has meaning, *and* 200 pounds really is twice as heavy as 100 pounds.

PRACTICE PROBLEM 1

Before they left the final examination, a random sample of students was asked to complete a course evaluation form requesting the following information: (a) gender: 1—male, 2—female: (b) course rating: 1—poor, 2—acceptable, 3—good, 4—outstanding: and (c) current grade point average. Identify the level of measurement of the data gathered from each question.　　*Solution at the back of the book*

Bar Chart

A *bar chart* provides a visual summary of data that contains just a few different values. The number of times each value appears in the data corresponds to the length of the bar for that value. The bars may be drawn vertically or horizontally, but they have spaces between them. A bar chart is appropriate for *any* level of data measurement.

EXAMPLE 2　**Constructing a Bar Chart**

The colors of a random sample of thirty cars in the Country Corners Mall parking lot were recorded using: 1—white, 2—blue, 3—green, 4—yellow, 5—brown, and 6—black. Construct a bar chart for the data $\{3, 5, 2, 4, 1, 5, 1, 1, 1, 2, 4, 1, 6, 1, 6, 5, 6, 3, 1, 1, 1, 1, 1, 6, 3, 1, 2, 2, 4, 2\}$.

Solution

We first tally the frequency of each color and then sketch the bar chart.

Color	Tally	Frequency
1	ⅠⅡⅠⅡⅠⅡ ‖	12
2	ⅠⅡⅠⅡ	5
3	‖‖	3
4	‖‖	3
5	‖‖	3
6	‖‖‖	4

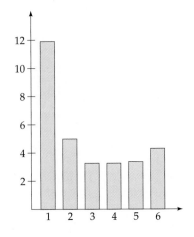

SPREADSHEET EXPLORATION

The spreadsheet below shows the car colors and frequencies from Example 2, together with some of the many graphs that can be constructed from such data. Though it is easily drawn by a computer, the *pie chart* at the lower right is difficult to draw accurately by hand because each "slice" from the circular "pie" must be the proper fraction of the whole.

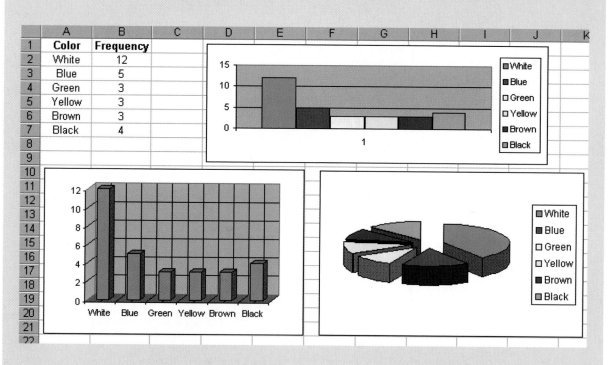

Can you modify this spreadsheet to calculate the frequencies from a column list of the original data given in Example 2.

PRACTICE PROBLEM 2

Construct a bar chart for the data given in Example 1(a) (page 465).

Solution at the back of the book

Stem-and-Leaf Display

A *stem-and-leaf display* is a quick way of visualizing data when the values consist of at least two digits. For each number, the "stem" is the leading digit and the "leaf" consists of the remaining digits. Thus

34 would have a stem of 3 and a leaf of 4, while in a different collection of data, 1.5 would have a stem of 1 and a leaf of 5. The numbers are then listed by row, with a different stem beginning each row. No attempt is made to put the leaves in order—they are left in the order they came in. A stem-and-leaf display is appropriate for levels of measurement higher than the nominal level.

EXAMPLE 3 **Constructing a Stem-and-Leaf Display**

Construct a stem-and-leaf display of the data {34, 10, 53, 50, 80, 38, 39, 31, 52, 41, 46, 46, 41, 69, 73, 57, 40, 52, 47, 68, 22, 33, 51, 65, 23, 47, 64, 45, 26, 74}.

Solution

Since the values lie between 0 and 100, we choose the tens digit for the stem and the units digit for the leaf. Filling in the display in the order of the values, we obtain

Stem	Leaf	
0		
1	0	
2	2, 3, 6	
3	4, 8, 9, 1, 3	← 34 is the first number entered
4	1, 6, 6, 1, 0, 7, 7, 5	
5	3, 0, 2, 7, 2, 1	← 53 is the third number entered
6	9, 8, 5, 4	
7	3, 4	← 74 is the last number entered
8	0	
9		

This stem-and-leaf display shows that the most frequent values are those in the forties, with the values ranging from the tens to the eighties.

PRACTICE PROBLEM 3

Construct a stem-and-leaf display for the data {7.8, 6.2, 5.4, 6.0, 6.4, 4.6, 6.2, 5.6, 3.7, 5.7, 2.5, 6.7, 4.4, 7.6, 8.1}.   *Solution at the back of the book*

Histogram

A *histogram* is similar to a bar chart and to a stem-and-leaf display, but now the *width* of the rectangles has meaning. The data is divided into *classes*, usually numbering from 5 to 15, one for each bar, and each class has the same *width*:

$$\begin{pmatrix}\text{Class} \\ \text{width}\end{pmatrix} \approx \frac{\begin{pmatrix}\text{Largest} \\ \text{data value}\end{pmatrix} - \begin{pmatrix}\text{Smallest} \\ \text{data value}\end{pmatrix}}{\begin{pmatrix}\text{Number} \\ \text{of classes}\end{pmatrix}}$$

Round the class width *up* so that all data will be covered

We then make a tally of the number of data values that fall into each class. Some histograms use the convention that any value on the boundary between two classes belongs to the upper class, while others are designed so that no data value falls on a class boundary. The histogram is then drawn in the same way as a bar chart but with the sides of the rectangles meeting at their common class boundary. It is often helpful to make a quick stem-and-leaf display before constructing a histogram. A histogram is appropriate for both interval and ratio levels of data measurement.

EXAMPLE 4   Constructing a Histogram

Construct a histogram for the data values from Example 3 (page 468).

Solution

Since we know from the stem-and-leaf display of this data that the smallest value is 10 and the largest is 80, if we choose 6 classes, we obtain a class width of $\frac{80-10}{6} \approx 11.7$, which we round *up* to 12. Since we rounded up, we may choose our first class to start just below the lowest data point of 10, say at 9.5. We then add successively the class width of 12 to get the other class boundaries.

Class boundaries	Tally	Frequency
9.5–21.5	|	1
21.5–33.5	ЖН	5
33.5–45.5	ЖН ||	7
45.5–57.5	ЖН ЖН	10
57.5–69.5	||||	4
69.5–81.5	|||	3

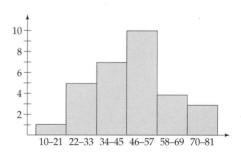

The base of each rectangle may be labeled with either the range of data values in that class or the class boundaries.

■

The histogram will vary depending on the number of bars chosen and the particular class boundaries.

Graphing Calculator Exploration

Histograms can be drawn on graphing calculators using the STAT PLOT command. To explore several histograms of the data from Example 3 (page 468), proceed as follows:

a. Enter the data values as a list and store it in the list L_1.

{34,10,53,50,80,
38,39,31,52,41,4
6,46,41,69,73,57
,40,52,47,68,22,
33,51,65,23,47,6
4,45,26,74}→L_1

b. Turn off the axes in the window FORMAT menu, turn on STAT PLOT 1, select the histogram icon with Xlist L_1, and set the WINDOW parameters Xmin = 9.5, Xmax = 81.5, Xscl = 12 to match the start, finish, and class width we used in Example 4 on the previous page.

c. GRAPH and then TRACE to explore your histogram.

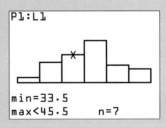

Of course, different choices of the class width, number of classes, and starting value will result in slightly different histograms. The following are several possibilities:

 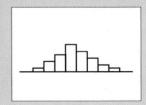

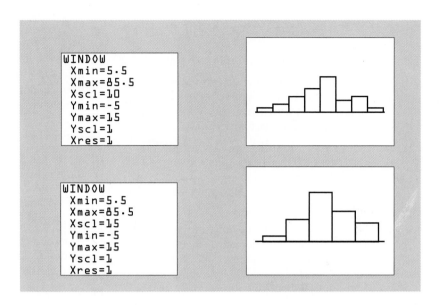

Section Summary

A *random sample* of a statistical population requires that every member of the population is equally likely to be included and that every possible sample of the same size from the population is equally likely to be chosen.

There are four levels of data measurement:

Level of Measurement	Suitable Calculation
Nominal	Classify by name
Ordinal	Compare rank
Interval	Compare rank and differences
Ratio	Compare rank, differences, and ratios

Data of all levels of measurement may be displayed in a *bar chart*, with the length of a bar indicating the frequency of the corresponding data value. A *stem-and-leaf display* shows the number of data values having the same stem and is appropriate for data of at least the ordinal level. A *histogram* groups data values into classes and shows the frequency of values in each class, and is appropriate for data of at least the interval level.

EXERCISES 6.1

Identify the level of data measurement in each situation.

1. **Discount Brokers** A survey of ten randomly selected traders at a discount broker service records the numbers of transactions involving stocks, bonds, and mutual funds last Wednesday.

2. **Travel Agencies** An airline surveys twenty randomly selected travel agencies in and around San Francisco and records the numbers of tickets sold at each last month to Hawaii, Alaska, Mexico, and New York.

3. **Mortgage Banking** The home office of the First & National Bank sends questionnaires to two hundred randomly selected families who applied for mortgages last summer, asking for ratings (from "poor" to "outstanding") of the courtesy of the mortgage staff at the branch offices.

4. **Shoe Sales** The national personnel manager for the Feets-R-Us shoe store franchise reviews the supervisor reports on thirty randomly selected sales employees and counts the number of "below average," "average," and "above average" performance evaluations.

5. **SAT Scores** The test results on the verbal SAT administered by the Educational Testing Service are numeric scores from 200 to 800. A survey records the verbal SAT scores of twelve randomly selected seniors at J. T. Marshall High School.

6. **Midwest Aviation** A survey records the years of birth of fifty randomly selected commercial pilots residing in Missouri.

7. **IQ Scores** An educational research team working on a grant from the National Institutes of Health records the IQ scores of eighty randomly selected fifth graders in the Baltimore County Public School system.

8. **Quality Control** A major cereal manufacturer randomly selects thirty boxes of Corn Chex Crunches from the shipping warehouse and records the actual net weights.

9. **Disposable Income** The Chamber of Commerce tabulates the disposable family incomes of ninety randomly selected households in Thomsonville.

10. **Financial Trends** The *Get Rich Slowly* investment newsletter reports the price-to-earnings ratios of sixty randomly selected technology stocks traded last month on the American Stock Exchange.

Construct a bar chart of the data in each situation.

11. **Marital Status** A random sample of shoppers in the Kingston Valley Mall counted 39 "never married," 97 "married," 16 "widowed," and 8 "divorced" individuals.

12. **Place of Birth** A random sample of skiers in Grand Junction last February counted 12 from Colorado, 27 from California, 19 from New York, 8 from Florida, and 6 from other states or nations.

13. **State Legislators** A random sample of candidates running for election to the Nevada legislature counted 4 business executives, 7 real estate developers, 18 lawyers, 1 doctor, 3 certified public accountants, and 2 retirees.

14. **Voter Concerns** A random sample of voters leaving the polls identified the most important issue in the Higginsville school board election as follows: 8 for teacher tenure, 54 for budget, 11 for after-school day care, 19 for the bond proposal for a new football stand at the high school, and 3 making "no comment."

Construct a stem-and-leaf display of the data in each situation.

15. **Carry-on Baggage** At the Logan County Airport, a random sample of thirty passengers passing through the security checkpoint found the following total weights (in pounds) of their carry-on baggage:

22	29	23	34	27	0	21	38	9	35
48	29	18	32	32	42	29	50	25	19
43	12	0	27	25	21	21	25	24	0

16. **Textbook Prices** A random sample of eighteen college textbooks at the Student Book Exchange found the following prices (in dollars):

67	80	91	35	51	67	49	56	86
52	56	29	51	73	60	52	52	68

17. Starting Salaries A random sample of twenty-four recent graduates from the Gainsburg College School of Business found the following starting salaries (in thousands of dollars):

```
31  48  22  41  34  28  26  26
42  33  37  24  38  46  34  54
33  36  34  35  32  37  35  37
```

18. Life Expectancy At the Peaceful Valley Memorial Gardens cemetery, a random sample of forty headstones found the following ages at death (in years):

```
53  25  47  54  49  48  71  47  31  51
66  76  71  52  52  35  32  56  61  72
51   9  71  38  64  48  55  35  53  47
73  50  44  67  57  54  38  95  24  58
```

19. Car Speeds A random sample of thirty-two cars passing through a radar trap on I-95 in southern Georgia found the following speeds (in miles per hour):

```
52  59  68  76  74  46  66  65
40  70  49  52  42  76  53  55
56  74  60  56  78  66  59  77
61  54  52  71  58  63  75  59
```

20. Gasoline Consumption A random sample of twenty customers at Jack's Pump & Go Express Service found the following weekly gasoline usages (in gallons) for commuters to Evansville:

```
36  26  50  46  33  29  37  31  37  35
20  28  19  17  39  45  49  36  35  28
```

Construct a histogram of the data in each situation, using an appropriate class width and starting class lower boundary.

You may use a graphing calculator, if your instructor permits it.

21. Shipping Weights A random sample of twenty packages shipped from the Holiday Values Corporation mail order center found the following weights (in ounces). (Use six classes of width 8 starting at 12.5.)

```
29  21  60  23  39  24  33  49  37  57
13  27  56  21  30  43  34  24  58  53
```

22. Blue Book Values A random sample of thirty-two cars parked at Emmett's Field last June found the following "blue book" values (in thousands of dollars). (Use six classes of width 4 starting at 2.5.)

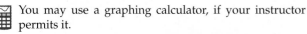
```
15   5   8  17  18  17   8   9
 5  13   6   9   5  19   8  13
20   9   7  13  25  16   5  25
25  17  24  10  23  14  15   4
```

23. Web Sites A random sample of twenty-eight student accounts at the campus Academic Computing Center showed the following number of web site "hits" between 9 and 12 P.M. last Wednesday. (Use eight classes of width 10 starting at 10.5.)

```
38  29  16  35  69  63  13
58  11  58  34  73  45  57
70  16  45  37  39  88  69
47  30  40  12  47  20  61
```

24. Football Players A random sample of twenty-four defensive linemen in the East Coast Division II college conference found the following weights (in pounds). (Use eight classes of width 15 starting at 140.5.)

```
173  171  211  169  258  213  143  236
191  163  192  235  196  208  157  221
174  200  195  198  228  152  188  201
```

25. Airline Tickets A random sample of forty passengers traveling economy class from New York to Los Angeles last Thanksgiving weekend found the following one-way ticket costs (in dollars):

```
425  236  481  230  258  473  423  480
249  451  244  440  240  565  445  569
255  515  412  377  453  425  490  301
379  439  229  422  465  252  219  237
565  477  440  227  518  232  230  437
```

26. Insurance Claims A random sample of thirty accident repair claims received in December at the western division office of the Some States Insurance Company were for the following dollar amounts:

```
5197  1018  1859  6207  3497  1097
4785  1253  3652  6279  3639  4553
6249  1250  6601  1193  5961  4625
1308  4289  2948  1272  1329  5483
 968  5297  1423  5159  5527  4543
```

27. Law Practice A random sample of twenty-one attorneys hired within the past year at the U.S. Department of Justice found the following numbers of hours worked last week:

$$
\begin{array}{ccccccc}
48 & 44 & 50 & 49 & 44 & 45 & 51 \\
50 & 57 & 36 & 53 & 52 & 59 & 54 \\
60 & 52 & 52 & 39 & 47 & 49 & 57
\end{array}
$$

28. Light Bulbs A random sample of thirty-six 75-watt light bulbs found the following lifetimes (number of hours until burnout):

$$
\begin{array}{cccccc}
1070 & 1210 & 1280 & 1230 & 1340 & 1280 \\
1200 & 1120 & 1310 & 1100 & 1170 & 1000 \\
1290 & 1420 & 1190 & 1280 & 1160 & 1290 \\
1250 & 1130 & 1180 & 1150 & 1110 & 1200 \\
1230 & 1160 & 1240 & 1250 & 1230 & 1240 \\
1000 & 1280 & 1180 & 1280 & 1270 & 1210
\end{array}
$$

29. Medical Insurance A random sample of twenty private practice physicians in the Lehigh Valley found the following numbers of insurance plans accepted by each:

$$
\begin{array}{cccccccccc}
16 & 21 & 9 & 23 & 19 & 23 & 20 & 23 & 18 & 19 \\
19 & 29 & 20 & 21 & 20 & 23 & 20 & 14 & 11 & 19
\end{array}
$$

30. Dining Out A random sample of forty dinner checks at the Lotus East Chinese Restaurant last Saturday evening showed the following total costs (rounded to the nearest dollar):

$$
\begin{array}{cccccccc}
43 & 104 & 45 & 60 & 25 & 103 & 33 & 96 \\
108 & 41 & 67 & 53 & 43 & 48 & 99 & 45 \\
35 & 40 & 66 & 103 & 27 & 110 & 60 & 33 \\
46 & 126 & 112 & 84 & 42 & 90 & 74 & 46 \\
102 & 41 & 48 & 41 & 54 & 44 & 82 & 33
\end{array}
$$

6.2 Measures of Central Tendency

APPLICATION PREVIEW

Household Incomes

The following histogram shows the 1998 percentages of U.S. households with annual incomes from $0 to $100,000 grouped in blocks of $5000 (the remaining 10.5% of the households have incomes above $100,000).

Source: U.S. Census

What is the "most typical" household income? The national "average" of $51,855 is much higher than that of most households, since half have incomes below $38,885 and the other half have more. But the $10,000 to $14,999 income class contains a greater percentage of households than any other income block. This section explores several ways of describing the "most typical" member of a data collection and discusses the relative merits of each.

Introduction

We often hear of averages, from grade point averages to batting averages. Given a collection of data, we often use the *average* as a representative or typical value. In this section we will see that not all types of data are appropriately summarized by an average, and sometimes another kind of "typical" value is more representative. To avoid confusion with casual meanings of "average," these different kinds of typical values are called *measures of central tendency.*

Mode

The *mode* of a collection of data is the *most frequently occurring* value. If all of the values in a data set occur the same number of times, there is no mode, while if only a few of the values occur the same maximal number of times, there are several modes. The mode is an appropriate measure of central tendency for every level of data measurement.

EXAMPLE 1   **Finding Modes**

Find the mode of each collection of values.

a. {2, 2, 2, 3, 3, 3, 4, 4, 4}
b. {1, 2, 3, 3, 3, 3, 4, 5, 6, 7, 7, 7, 7, 8, 9, 9, 10}
c. {1, 2, 3, 3, 4, 4, 4, 5, 5, 6, 6, 6, 6, 7, 8, 8, 9, 50}

Solution

a. Since each value occurs the same number of times, there is no mode.
b. Since both 3 and 7 occur four times and every other value occurs fewer times, the modes are 3 and 7. A collection of values with two modes is said to be *bimodal.*
c. Since 6 occurs more often than any other value, the mode is 6. A collection of values with just one mode is said to be *unimodal.* Notice that the presence of one value that is very different from the others, in this case the 50, has no effect on the value of the mode. If the 50 were replaced by a 100, a 1000, or even a 1,000,000, the mode would still remain unchanged.

■

PRACTICE PROBLEM 1

Find the mode of the values {17, 20, 20, 13, 10, 17, 10, 20}.

Solution at the back of the book

Graphing Calculator Exploration

When finding the mode of a long list of values, it is often easier to first sort the list so that all the same values appear grouped within the list. For instance, to find the mode of {62, 56, 52, 53, 58, 64, 68, 67, 60, 61, 57, 58, 57, 63, 55}, proceed as follows:

a. Enter the data values as a list and STORE it in the list L_1 (or enter the numbers into L_1 directly using STAT and EDIT).

```
{62,56,52,53,58,
64,68,67,60,61,5
7,58,57,63,55}→L
1
```

b. From the STAT or LIST menu, select SortA (to sort in *ascending* order). (You could also use SortD for descending order.)

```
EDIT CALC TESTS
1:Edit...
2:SortA(
3:SortD(
4:ClrList
5:SetUpEditor
```

c. Complete the ascending sort command by entering the name of the list of values, L_1.

```
{62,56,52,53,58,
64,68,67,60,61,5
7,58,57,63,55}→L
1
{62 56 52 53 58…

SortA(L1)
```

d. Examine the sorted list L_1 to identify the value(s) appearing most often. The modes of this collection of values are 57 and 58 because they appear twice and the others appear only once each.

```
SortA(L1)
            Done
L1
{52 53 55 56 57…
…57 58 58 60 61…
…62 63 64 67 68}
```

Median

The *median* is the *middle* value of a list of data values when sorted in ascending or descending order. If there is an even number of data values, the median is the number halfway between the two middle values. Since the median depends on arranging the values in order, the median is not appropriate for nominal data. For instance, it would be meaningless to sort the data for car colors from Example 2 (page 466) from 1—white to 6—black, and then find a "median color," since the assignment of numbers to colors was arbitrary. The median is an appropriate measure of central tendency for every level of data measurement except the nominal level.

EXAMPLE 2   **Finding Medians**

Find the median of each data set. Assume that each represents at least ordinal data.

a. {1, 3, 4, 7, 11, 18, 39}

b. {1, 3, 4, 7, 11, 18, 1000}

c. {26, 16, 10, 6, 4, 2}

d. {62, 56, 52, 53, 58, 64, 68, 67, 60, 61, 57, 58, 57, 63, 55}

Solution

a. Since these seven numbers are arranged in ascending order, the median is the fourth number, 7, because three numbers are smaller and three are larger.

b. Just as in part (a), the median is 7. Notice that replacing the largest data value in the previous data set with a much larger value did not change the median.

c. Since these six values are arranged in descending order, the median is 8, the number halfway between the two middle values of 10 and 6. Thus the median need not be one of the data values.

d. Since these fifteen values are not in ascending or descending order, we first order them (as in the Graphing Calculator Exploration on the previous page):

{52, 53, 55, 56, 57, 57, 58, 58, 60, 61, 62, 63, 64, 67, 68}

The median of these is the middle value, 58, since there are seven data values no larger and seven values no smaller.

PRACTICE PROBLEM 2

Find the median of the following values representing nominal data: {4, 5, 5, 6, 7, 8}.   *Solution at the back of the book*

Graphing Calculator Exploration

The median of an unsorted list of values can be found with a single calculator command. To find the median of {62, 56, 52, 53, 58, 64, 68, 67, 60, 61, 57, 58, 57, 63, 55}, the unordered values from Example 2(d) on the previous page, proceed as follows:

Enter the data values as a list and store it in the list L_1. From the LIST MATH menu, select the median command and apply it to the list L_1.

```
{62,56,52,53,58,
64,68,67,60,61,5
7,58,57,63,55}→L
1
```

```
NAMES OPS MATH
1:min(
2:max(
3:mean(
4:median(
5:sum(
6:prod(
7↓stdDev(
```

```
{62,56,52,53,58,
64,68,67,60,61,5
7,58,57,63,55)→L
1
...57 58 57 63 55}
median(L1)
                58
```

Mean

The *mean* is the usual arithmetic average found by summing the values and dividing by the number of them:

Mean

The *mean* $\bar{x}$ of the n values $x_1, x_2, \ldots, x_n$ is

$$\bar{x} = \frac{1}{n}(x_1 + x_2 + \cdots + x_n) = \frac{1}{n} \sum_{\text{All values}} x_k$$

The mean of a collection of numbers is equivalent to the mean of a random variable that is equally likely to be any of the numbers. The mean of a data set is sometimes called the *sample mean* to distinguish it from the mean μ of a probability distribution. Calculating the mean assumes that the distance between values is meaningful, so the mean is not an appropriate measure for ordinal or nominal data. For instance, it would be meaningless to find the mean of the concert ratings in Example 1(b) (page 465), from 1—awful to 10—awesome, because we do not know if a 6 and an 8 should average to a 7, since the intervals between them may not be the same. The mean is an appropriate measure of central tendency only for interval and ratio data.

EXAMPLE 3   **Finding Means**

Find the mean of each data set. Assume that each represents at least interval data.

a. {1, 3, 4, 7, 11, 18, 39}

b. {1, 3, 4, 7, 11, 18, 1000}

c. {62, 56, 52, 53, 58, 64, 68, 67, 60, 61, 57, 58, 57, 63, 55}

Solution

a. $\bar{x} = \frac{1}{7}(1 + 3 + 4 + 7 + 11 + 18 + 39) = \frac{1}{7}(83) = 11\frac{6}{7}$.

b. $\bar{x} = \frac{1}{7}(1 + 3 + 4 + 7 + 11 + 18 + 1000) = \frac{1}{7}(1044) = 149\frac{1}{7}$.
Notice that the values are the same as in part (a) except that the last number changed from 39 to 1000, with the result that the mean increased substantially. Thus, unlike the median and the mode, the mean is sensitive to changes in just one of the values.

c. There is no need to arrange the values in any particular order when finding the mean:

$$\bar{x} = \frac{1}{15}(62 + 56 + 52 + 53 + 58 + 64 + 68 + 67 + 60 + 61 + 57$$
$$+ 58 + 57 + 63 + 55) = \frac{1}{15}(891) = 59\frac{2}{5}$$

Notice that the mean is usually not one of the original data values.

EXAMPLE 4   **Baseball Salaries**

In the 1995 baseball strike it was disclosed that the mean salary of major league baseball players was $1.2 million, while the median salary was only $500,000. What does this say about the distribution of baseball salaries?

Solution

If the *middle* salary is $500,000, then for the mean to be higher, there must be a few very large salaries. That is, a few baseball players must be receiving enormously high salaries to skew the mean that much. In such cases, the median is usually considered a more representative value for most players than the mean.

PRACTICE PROBLEM 3

Find the mode, median, and mean of the values {1, 1, 2, 3, 8} representing ratio data.
                                Solution at the back of the book

Graphing Calculator Exploration

Like the median, the mean of a list of values can be found with a single calculator command. To find the mean of {62, 56, 52, 53, 58, 64, 68, 67, 60, 61, 57, 58, 57, 63, 55}, the values from Examples 2(d) and 3(c) on the previous pages, enter the data values as a list and store it in the list L_1. From the LIST MATH menu, select the mean command and apply it to the list L_1.

```
{62,56,52,53,58,
64,68,67,60,61,5
7,58,57,63,55}→L
1
```

```
NAMES OPS MATH
1:min(
2:max(
3:mean(
4:median(
5:sum(
6:prod(
7↓stdDev(
```

```
{62,56,52,53,58,
64,68,67,60,61,5
7,58,57,63,55}→L
1
...57 58 57 63 55}
mean(L1)
         59.4
```

Section Summary

Although there are three measures of central tendency, not all of them can be used with every type of data. For nominal data only the mode is appropriate, for ordinal data either the mode or the median is appropriate, and for interval or ratio data all three are appropriate.

Measure of Central Tendency	Definition	Appropriate Data Levels
Mode	Most frequent	All levels
Median	Middle value	Ordinal, interval, and ratio
Mean	$\bar{x} = \dfrac{1}{n} \displaystyle\sum_{\text{All values}} x_k$	Interval and ratio

The mean is sensitive to changes in the extremes, but the median and mode are not. The mean of a collection of numbers is equivalent to the mean of a random variable that is equally likely to be any of the numbers. The mean of a data set is sometimes called the *sample mean* to distinguish it from the mean of a probability distribution.

[handwritten notes at top:] $\bar{x} = AVG$ $\sigma x = SIGMA (STD. DEV)$ $n = \#\ of\ NUMBERS$

EXERCISES 6.2

(📱 will be helpful throughout.)

Find the mode, median, and mean of each (ratio) data set.

1. {6, 17, 12, 10, 15, 16, 5, 8, 9, 14, 9}

2. {10, 7, 9, 13, 11, 14, 15, 14, 8, 9}

3. {15, 14, 5, 7, 5, 14, 9, 7, 7, 14, 7, 14, 12}

4. {11, 19, 20, 11, 12, 18, 11, 20, 16, 18, 17, 10, 13, 11, 15}

5. {19, 11, 10, 19, 18, 12, 19, 10, 14, 12, 13, 17, 20, 19, 15}

6. {12, 5, 10, 6, 7, 6, 15, 7, 6, 5, 6, 5, 13, 10, 7, 13, 8, 11, 12}

7. {8, 15, 10, 14, 13, 14, 5, 10, 7, 9, 13, 14, 15, 14, 15, 19, 13, 12, 8}

8. {11, 8, 23, 21, 14, 16, 15, 11, 9, 22, 15, 11, 12, 11, 10, 15, 10, 25, 18, 6, 11}

9. {19, 9, 15, 8, 18, 15, 12, 22, 16, 19, 15, 19, 20, 24, 7, 14, 21, 19, 20, 5, 19}

10. {10, 13, 8, 15, 8, 12, 12, 12, 14, 8, 12, 11, 14, 9, 8, 14, 14, 14, 10, 10}

APPLIED EXERCISES

Find the mode, median, and mean for the data in each situation.

11. **Emergency Services** A random sample from the records of the Farmington Volunteer Ambulance Corps found the following twenty response times (in minutes) to 911 calls received during February:

| 14 | 9 | 25 | 14 | 12 | 16 | 5 | 20 | 20 | 25 |
| 20 | 24 | 12 | 7 | 19 | 12 | 6 | 13 | 24 | 18 |

12. **Car Sales** A random sample of fifteen General Motors salesmen at Tri-State Dealers found the following earnings (in dollars) for the first week of September:

1200	1200	1100	1300	900
600	800	700	1200	500
1000	1200	600	900	900

13. **Volunteer Recycling** A random sample of thirty volunteers helping sort newspapers, aluminum, and glass at the Parkerville Regional Recovery Center found the following hours contributed last week:

10	7	13	11	7	11	10	5	9	14
14	12	12	5	10	12	19	10	7	10
9	21	13	10	7	18	10	15	10	12

14. **Commuting Times** A random sample of twenty-four employees at the Grover-Smith Forge found the following morning commute times (in minutes) for last Monday:

75	60	50	55	90	53	65	71
33	81	60	74	71	73	83	53
66	75	47	36	77	54	67	65

15. **Unemployment** A random sample of nineteen newly unemployed workers in Wilmington last August reported the following times (in weeks) until they found new jobs:

| 12 | 6 | 9 | 9 | 11 | 14 | 8 | 15 | 12 | 13 |
| 9 | 8 | 9 | 7 | 9 | 5 | 6 | 13 | 15 | |

Explorations and Excursions.

The following problems extend and augment the material presented in the text.

Mean of Grouped Data Sometimes the original data are not available and only the class intervals and frequencies are known. The mean of these grouped data may be estimated by summing each midpoint value at the center of each class interval as many times as is indicated by the frequency and then dividing that total by the sum of the frequencies.

Verify the calculation of each mean of grouped data.

16. For the grouped data

Class interval	14.5–19.5	19.5–24.5	24.5–29.5
Frequency	7	9	4

the mean is

$$\bar{x} = \frac{17 \cdot 7 + 22 \cdot 9 + 27 \cdot 4}{7 + 9 + 4}$$

$$= \frac{119 + 198 + 108}{20} = \frac{425}{20} = 21\tfrac{1}{4}$$

17. For the grouped data

Class interval	6–12	13–19	20–26	27–33
Frequency	5	10	4	6

the mean is

$$\bar{x} = \frac{9 \cdot 5 + 16 \cdot 10 + 23 \cdot 4 + 30 \cdot 6}{5 + 10 + 4 + 6}$$

$$= \frac{477}{25} = 19.08$$

Use the midpoints of the class intervals to estimate the mean of each set of grouped data.

18. *Truck Repairs* The Wink-Quick package service maintains its own statewide fleet of delivery trucks. A random sample of the repair records found the following numbers of days last year that fifty trucks were not available for use because their repairs could not be completed overnight:

Number of days	0–2	3–5	6–8	9–11
Frequency	30	7	12	1

19. *Overtime Pay* A random sample of payroll records for one hundred and ten craftspeople at the Old Fashion Vermont Country Furniture factory found the following amounts of overtime pay (in dollars) for the last week in November:

Overtime pay	0–19	20–39	40–59	60–79	80–99
Frequency	29	11	42	12	16

20. *Charity Donations* Donors at the Elktonburg Hospital Charity Ball are designated as "friends" for donations from $51 to $100, "benefactors" for $101 to $150, and "founders" for $151 to $200. A random sample of thirty hospital supporters found the following designations:

Designation	Friends	Benefactors	Founders
Frequency	9	6	15

6.3   Measures of Variation

Pain, Suffering, and Statistics

In 1994, three secretaries sued the Digital Equipment Corporation, the manufacturer of the equipment on which they had been typing, claiming that the keyboards had caused numbness in their hands and pain in their arms and backs. In 1996, all three won large monetary judgments from the company, with separate amounts for "damages" (medical expenses and lost earnings) and for "pain and suffering" (noneconomic loss). In 1997, federal judge Jack B. Weinstein ruled that one of the pain and suffering awards was within reasonable limits, but another was unreasonably large, and that the case should be retried (for other reasons as well). (The third award was eliminated because the case had been filed too late.) On what reasoning did the judge base his conclusion about the size of a "reasonable" award? He employed a standard statistical measure called the *sample standard deviation,* which will be defined in this section.

The judge asked the lawyers to find similar court cases from the past, and from these he selected a group of 22 cases as being applicable to one of the cases (that of Jeanette Rotolo), and another group of 27 as being comparable to the other case (that of Patricia Geressy), by calculating the means and standard deviations of the awards for each of these groups. He then decided that a reasonable award should be within two standard deviations of the mean. For the group similar to the Rotolo case the mean was $404,214 and the standard deviation was $465,489, which meant that her award should not exceed 404,214 + 2 · 465,489 = $1,335,192. The award of $100,000 to Rotolo was far below this and so was deemed not excessive. For the Geressy case the mean was $747,372 and the standard deviation was $606,873, so her award should not exceed 747,372 + 2 · 606,873 = $1,961,118. The award of $3,490,000 to Geressy was far above this amount, so the judge ruled it to be excessive. Judge Weinstein's decision received much favorable attention as an innovative use of standard deviation to bring some rationality to the enormous disparity in pain and suffering awards.*

Although this particular use of statistics was new, probability has a long history in the legal profession. In fact, Pierre de Fermat, one of the founders of probability (see page 386), was a lawyer and would be intrigued with this judicial application of the subject that he founded.

Introduction

It is helpful to summarize a given data set by a few numbers, as we did with random variables in the previous chapter. For a *representative* or *typical* value, we use one of the measures of central tendency (mode, median, or mean, whichever is most appropriate for the level of data measurement). But how do we measure whether these values are grouped tightly or spread widely about this central value? In this section we discuss three ways of describing the spread of the values about the center. Since all three use differences among data values, we must now restrict our discussion to data values at the interval or ratio level of measurement.

---

* U.S. District Court, Eastern District of New York; Geressy, Rotolo, *et al.* v. Digital Equipment Corp.; Amended Judgment, Memorandum and Decision on post-trial motions; 94-CV-1427; Jack B. Weinstein, Senior District Judge.

Range

The simplest measure of the spread of a data set is the *range,* which is the difference between the smallest and largest data values.

EXAMPLE 1   **Finding Ranges**

Find the range of each data set.

a. {1, 5, 6, 7, 7, 7, 8, 15}

b. {1, 2, 10, 10, 12, 13, 15}

Solution

a. The largest value is 15 and the smallest is 1, so the range is 15 − 1 = 14.

b. Again, the range is 15 − 1 = 14. Notice that the range is *not* sensitive to the distribution of values between the two extremes.

■

PRACTICE PROBLEM 1

Find the range of the data values {15, 17, 27, 12, 30, 15, 10, 27, 25, 29}.

Solution at the back of the book

Box-and-Whisker Plot

A *box-and-whisker plot* provides a quick way to visualize how data values are distributed between the largest and smallest by graphically displaying a *five-point summary* of the data. The *minimum* is the smallest value, the *maximum* is the largest, and we have already discussed the *median.* The *first quartile* is the median of the data values below the median (but not including the median), and the *third quartile* is the median of the data values above the median (but not including the median). The median may then be called the *second quartile.* The box extends from the first quartile to the third quartile with a vertical bar at the median. The whiskers extend from the box out to the minimum and maximum data values.

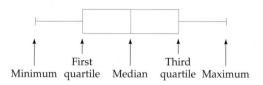

Notice that the distance between the extreme ends of the whiskers is the *range* of the data and that 50% of the data is contained in the box. The length of the box, from the first quartile to the third quartile, is sometimes called the *interquartile range.* In general, the quartiles will not be evenly spaced.

EXAMPLE 2 **Constructing a Box-and-Whisker Plot**

Make a five-point summary and draw the box-and-whisker plot for the data values {20, 37, 65, 77, 78, 79, 81, 82, 83, 85, 87, 90}.

Solution

Since these values are already in order, the minimum is 20 and the maximum is 90. Since there are twelve values, the median is halfway between the sixth value (79) and the seventh (81), so the median is 80. The values below the median are the six values {20, 37, 65, 77, 78, 79}, and the first quartile is the median of these values (that is, the value halfway between 65 and 77), so the first quartile is 71. Similarly, the values above the median are the six values {81, 82, 83, 85, 87, 90}, and the third quartile is the median 84 because it is the value halfway between 83 and 85. The five-point summary of this data set and the corresponding box-and-whisker plot are shown below.

Minimum = 20

First quartile = 71

Median = 80

Third quartile = 84

Maximum = 90

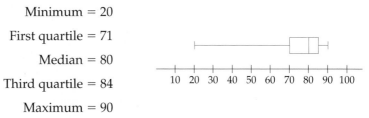

The box-and-whisker plot shows clearly that the bottom quarter of the values are spread rather thinly from 20 to 71, whereas the entire top half is concentrated between 80 and 90 (with a quarter of the values in the narrow range between 80 and 83). Such clustering is much easier to see from the box-and-whisker plot than from simply looking at the data values.

PRACTICE PROBLEM 2

Make a five-point summary and draw the box-and-whisker plot for the data values {8, 10, 13, 14, 15, 16, 17, 19, 23, 24, 29}. From your plot, which quarter is the most tightly clustered?

Solution at the back of the book

Graphing Calculator Exploration

Box-and-whisker plots can be drawn on graphing calculators by using the STAT PLOT command. To explore several box-and-whisker plots, including one of the data from Example 2 on the previous page, proceed as follows.

a. Turn off the axes in the window FORMAT menu and set the WINDOW parameters to $Xmin = 0$ and $Xmax = 100$.

b. Enter the data values {20, 37, 65, 77, 78, 79, 81, 82, 83, 85, 87, 90} as the list L_1, the values {20, 31, 39, 41, 63, 80, 80, 80, 81, 87, 88, 90} as the list L_2, and {20, 55, 58, 60, 60, 61, 63, 64, 65, 79, 81, 90} as the list L_3.

c. Turn on STAT PLOT 1 and select the box-and-whisker plot icon with Xlist L_1, and do the same for STAT PLOT 2 with Xlist L_2 and STAT PLOT 3 with Xlist L_3.

d. GRAPH and then TRACE to explore your box-and-whisker plots. The screen shows that the first quartile of the second list of values is 40.

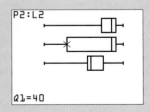

Sample Standard Deviation

The third way of measuring variation is the *sample standard deviation,* which estimates the typical variation of the data values from the (sample) mean of the data in much the same way that in probability theory the standard deviation σ measures the variation from the expectation μ of a random variable.

Sample Standard Deviation

The *sample standard deviation* s of the data values $x_1, x_2, \cdots, x_n$ from a random sample of size n is

$$s = \sqrt{\frac{(x_1 - \bar{x})^2 + \cdots + (x_n - \bar{x})^2}{n - 1}}$$

$$= \sqrt{\frac{1}{n - 1} \sum_{\text{All } x_k \text{values}} (x_k - \bar{x})^2}$$

$\bar{x} = \text{mean}$

Unlike the formula for the sample mean (page 478), for the sample standard deviation we divide by $n - 1$ instead of n. This is because the n data values are related by the fact that their sum divided by n is the mean $\bar{x}$, so that any one value can be determined by the others together with the mean, leaving only $n - 1$ independent values in the sum.

EXAMPLE 3   Calculating a Sample Standard Deviation

Find the sample standard deviation of the data values $\{9, 13, 16, 18, 19\}$.

Solution

Since there are five values, $n = 5$. The mean is

$$\bar{x} = \tfrac{1}{5}(9 + 13 + 16 + 18 + 19) = \tfrac{1}{5}(75) = 15$$

Then

$$s = \sqrt{\frac{(9 - 15)^2 + (13 - 15)^2 + (16 - 15)^2 + (18 - 15)^2 + (19 - 15)^2}{5 - 1}}$$

$$= \sqrt{\frac{36 + 4 + 1 + 9 + 16}{4}} = \sqrt{\frac{66}{4}} \approx 4.06 \qquad s = \sqrt{\frac{(x_1 - \bar{x})^2 + \cdots + (x_n - \bar{x})^2}{n - 1}}$$

The sample standard deviation s is (approximately) 4.06.

■

PRACTICE PROBLEM 3

Find the sample standard deviation of the data values $\{2, 6, 7, 8, 9, 15, 16, 17\}$.

Solution at the back of the book

Graphing Calculator Exploration

The sample standard deviation of a list of values can be found with a single calculator command. To find the sample standard deviation of $\{9, 13, 16, 18, 19\}$, the values from Example 3 on the previous page, enter the data values as a list, and store it in the list L_1. From the LIST MATH menu, select the standard deviation command and apply it to the list L_1.

```
{9,13,16,18,19}→
L₁
```

```
NAMES OPS MATH
1:min(
2:max(
3:mean(
4:median(
5:sum(
6:prod(
7↓stdDev(
```

```
{9,13,16,18,19}→
L₁
  {9 13 16 18 19}
stdDev(L₁)
        4.062019202
```

Section Summary

There are three ways to measure the spread of a data set. The simplest is the *range*, which is the largest data value minus the smallest. The range gives no indication of the typical variation away from the mean, just the difference between the extremes.

A *box-and-whisker plot* graphically shows the range of each quarter of the data and can be drawn easily using a graphing calculator.

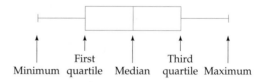

The *sample standard deviation* measures the typical spread of the data around the mean $\bar{x}$ as a single number and can be found easily using a calculator:

$$s = \sqrt{\frac{(x_1 - \bar{x})^2 + \cdots + (x_n - \bar{x})^2}{n - 1}} = \sqrt{\frac{1}{n - 1} \sum_{\text{All } x_k \text{values}} (x_k - \bar{x})^2}$$

EXERCISES 6.3

(will be helpful throughout.)

Find the range of each (ratio) data set.

1. {21, 44, 48, 52, 83}

2. {26, 35, 47, 59, 86, 92, 116}

3. {5, 25, 6, 9, 7, 7, 21, 19, 23, 16}

4. {2.0, 3.8, 5.2, 1.6, 4.3, 2.3, 5.4, 3.5, 2.5, 3.1}

5. {121, 115, 147, 163, 171, 116, 167, 147, 169, 131}

Make a five-point summary and draw the box-and-whisker plot for each (ratio) data set.

6. {5, 8, 11, 13, 16, 20, 26}

7. {3, 10, 12, 14, 17, 21, 23}

8. {2, 3, 5, 9, 10, 12, 15, 17, 20}

9. {8, 9, 11, 12, 13, 14, 15, 19, 26}

10. {12, 8, 19, 17, 20, 8, 13, 14, 23, 5, 6, 9, 23, 12, 9}

11. {21, 13, 18, 14, 24, 8, 24, 24, 9, 12, 25, 20, 16, 17, 12}

12. {14, 17, 17, 13, 11, 6, 10, 21, 20, 18, 14, 11, 19, 14, 14}

13. {18, 7, 8, 18, 6, 15, 12, 15, 18, 24, 6, 18, 16, 17, 24}

14. {16, 9, 14, 12, 18, 15, 21, 19, 14, 18, 8, 4, 17, 15, 21, 14, 23, 13, 20, 17}

15. {19, 18, 7, 16, 5, 8, 9, 16, 18, 11, 20, 16, 17, 17, 8, 10, 15, 10, 8, 20}

Find the sample standard deviation of each (ratio) data set.

16. {8, 13, 14, 17, 13}

17. {9, 2, 17, 13, 14}

18. {18, 11, 18, 1, 5, 19}

19. {22, 19, 11, 2, 20, 4}

20. {14, 11, 11, 9, 2, 13}

APPLIED EXERCISES

Analyze the data in each situation by finding the range, the five-point summary, and the sample standard deviation and then drawing the box-and-whisker plot.

21. **Stock Prices** Last Thursday was a very active trading day for shares of the new DiNextron technology stock. A random sample of twenty purchase transactions found the following share selling prices (in dollars):

16	12	17	11	12	7	12	17	11	20
10	11	14	17	16	14	25	9	17	5

22. **Kitchen Cabinets** A random sample of fifteen electric bills at Rick's Always Cooking cabinet factory found the following monthly electricity consumptions (in thousands of kilowatt hours):

10	12	6	15	13
15	12	9	14	6
5	8	9	12	9

23. **Airline Cancellations** A random sample of thirty airplane departures from Dallas International found the following numbers of "no shows" at the boarding gates:

2	6	8	4	10	5	10	8	7	5
10	10	9	6	11	9	6	6	10	8
11	6	7	13	12	8	9	5	9	10

24. **Malpractice Settlements** A random sample of twenty-four malpractice claims against a midwestern health maintenance organization (HMO) that were settled out of court found the following settlement amounts (in tens of thousands of dollars):

16	22	16	18	29	22	30	18
19	28	18	37	17	32	24	31
19	16	16	6	14	15	8	18

25. **Trial Schedules** A random sample from the clerk's files at the Marksburg District Court found the following jail waiting times (in weeks) between arraignment and trial (or plea bargain) for twenty individuals charged with felonies:

11	35	15	33	20	21	2	21	23	24
18	16	9	25	12	20	20	25	21	33

Explorations and Excursions

The following problems extend and augment the material presented in the text.

Calculating Standard Deviations   Although the formula on page 487 defines the sample standard deviation, there is a simpler way to calculate its value that is used in most computer programs and calculators.

26. Verify each step in the following transformation of the definition formula for the sample standard deviation

$$s = \sqrt{\frac{1}{n-1} \sum (x_k - \bar{x})^2}$$

The sum inside the square root can be written as follows:

$$\sum (x_k - \bar{x})^2$$
$$= \sum (x_k^2 - 2x_k\bar{x} + \bar{x}^2)$$
$$= \left(\sum x_k^2\right) - 2\bar{x} \cdot \left(\sum x_k\right) + \bar{x}^2 \cdot \left(\sum 1\right)$$
$$= \left(\sum x_k^2\right) - 2 \cdot \left(\frac{1}{n} \cdot \sum x_k\right) \cdot \left(\sum x_k\right) + \left(\frac{1}{n} \cdot \sum x_k\right)^2 \cdot n$$
$$= \left(\sum x_k^2\right) - \frac{1}{n} \cdot \left(\sum x_k\right)^2$$

Then

$$s = \sqrt{\frac{\left(\sum x_k^2\right) - \frac{1}{n} \cdot \left(\sum x_k\right)^2}{n-1}}$$

That is, s can be found from the sum of the data values and the sum of their squares without calculating the mean $\bar{x}$.

27. Verify the following calculation of the sample standard deviation of the data $\{19, 18, 27, 5, 11\}$.

$$\sum x_k = 19 + 18 + 27 + 5 + 11 = 80$$

and

$$\sum x_k^2 = 19^2 + 18^2 + 27^2 + 5^2 + 11^2 = 1560$$

so

$$s = \sqrt{\frac{1560 - \frac{1}{5}(80)^2}{5 - 1}}$$
$$= \sqrt{\frac{1560 - 1280}{4}} = \sqrt{70} \approx 8.37$$

28. Verify the following calculation of the sample standard deviation of the data $\{11, 18, 29, 14, 24, 6\}$.

$$\sum x_k = 11 + 18 + 29 + 14 + 24 + 6 = 102$$

and

$$\sum x_k^2 = 11^2 + 18^2 + 29^2 + 14^2 + 24^2 + 6^2$$
$$= 2094$$

so

$$s = \sqrt{\frac{2094 - \frac{1}{6}(102)^2}{6 - 1}} = \sqrt{\frac{2094 - 1734}{5}}$$
$$= \sqrt{72} \approx 8.49$$

29. Use the final formula from Exercise 26 to find the sample standard deviation of the data $\{9, 2, 17, 13, 14\}$ and then compare your value with your answer to Exercise 17.

30. Use the final formula from Exercise 26 to find the sample standard deviation of the data $\{22, 19, 11, 2, 20, 4\}$ and then compare your value with your answer to Exercise 19.

6.4   Normal Distributions and Binomial Approximation

APPLICATION PREVIEW

The Bell-Shaped Curve

In the machine pictured on the following page, 30,000 steel balls fall, one at a time, from the center of the top, each ball making its way downward through the rows of pins, turning right or left at each pin, and finally falling into one of the slots at the bottom. A few balls will turn right at each pin, ending up in the extreme right slot; a few will turn left at each pin, ending up in the extreme left slot; but most will make a combination of the right and left turns, ending up somewhere in the middle. As the picture shows, the balls fill out what is known as a *bell-shaped* or *normal curve*, highest in the middle and symmetrically lower farther away on either side. This machine illustrates the principle that the cumulative effect of many small changes (right and left detours around each pin) often leads to the normal curve. In this section we will define the normal curve and use it to predict such quantities as people's heights and weights, lengths of hospital stays, and orders received by a company.

Introduction

The normal distribution, with its famous bell-shaped graph, is perhaps the most important of all probability distributions. It is used to predict everything from sizes of newborn babies to stock market fluctuations to retirement benefits. After we examine some of its most important features, we will see how it is used in applications and examine its relation to the binomial distribution.

Normal Distribution

The normal distribution depends on two parameters: the mean μ and the standard deviation σ. The curve is given by the following formula, although we will not use it explicitly in what follows.

Normal Probability Distribution

> The normal probability distribution with mean μ and standard deviation σ is given by the curve
>
> $$f(x) = \frac{1}{\sigma\sqrt{2\pi}} e^{-\frac{1}{2}\left(\frac{x-\mu}{\sigma}\right)^2} \qquad \text{for } -\infty < x < \infty$$

The curve has a central peak at μ and then falls back symmetrically on either side to approach the x-axis. Different values of μ change the location of the peak, as is shown below by three normal distributions with different means.

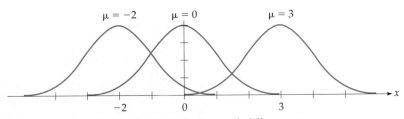

Three normal distributions with different means

The standard deviation σ measures how the values spread away from the mean. The distribution will be higher and more peaked if σ is small, and lower and more rounded if σ is large.

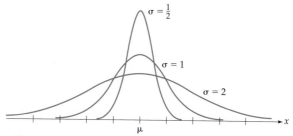

Three normal distributions with different standard deviations

As the distribution becomes "shorter" it also becomes "wider," and the area under the curve always stays at 1, although we will not prove this fact. The peaked shape of the normal distribution concentrates most of the probability (or area) near the center at μ. About 68% of the area under the normal curve is within one standard deviation of the mean, about 95% is within two standard deviations, and more than 99% is within three standard deviations. Notice also that the curve rises or falls most steeply at one standard deviation from the mean.

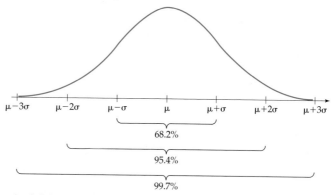

Area (probability) under the curve within 1, 2, and 3 standard deviations of the mean

The probability that a normal random variable has its value between two x-values corresponds to the *area* under the normal curve between those x-values. The probability is most easily found using a graphing calculator, but it can also be found using tables of values of the normal distribution.

Readers who have graphing calculators that find normal probabilities should continue reading this section. Readers who do not have such calculators should now turn to page 607 to read the appendix near the end of this book.

Graphing Calculator Exploration

Areas under normal distribution curves can be found easily on a graphing calculator. To find the area under the normal distribution curve with mean $\mu = 20$ and standard deviation $\sigma = 5$ from 10 (which is $\mu - 2\sigma$) to 30 (which is $\mu + 2\sigma$), proceed as follows:

a. From the DISTRIBUTION menu, select the normalcdf command to find the cumulative values of the normal distribution (that is, the area under the curve) between two given values.

```
DISTR DRAW
1:normalpdf(
2:normalcdf(
3:invNorm(
4:tpdf(
5:tcdf(
6:X²pdf(
7↓X²cdf(
```

b. Enter the (left) starting value and the (right) ending value as well as the mean and standard deviation. This area is the same as the probability that the normal random variable is between the two given values.

```
normalcdf(10,30,
20,5)
        .954499876
```

c. To *see* this area shaded under the normal curve, use the ShadeNorm command from the DISTRIBUTION DRAW menu with an appropriate window.

```
WINDOW
 Xmin=0
 Xmax=40
 Xscl=5
 Ymin=-.05
 Ymax=.1
 Yscl=1
 Xres=1
```

```
DISTR DRAW
1:ShadeNorm(
2:Shade_t(
3:ShadeX²(
4:ShadeF(
```

```
ShadeNorm(10,30,
20,5)
```

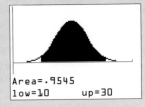

```
Area=.9545
low=10      up=30
```

EXAMPLE 1   **Finding a Probability for a Normal Random Variable**

The heights of American men are approximately normally distributed with mean $\mu = 68.1$ inches and standard deviation $\sigma = 2.7$ inches.* Find the proportion of men who are between 5 feet 9 inches and 6 feet tall.

* These and other data in this section are from *Handbook of Human Factors* by Gavriel Salvendy (New York: John Wiley-Interscience, 1987).

Solution

Converting to inches, we want the probability of a man's height being between 69 inches and 72 inches. From the graphing calculator screen below, about 30% of American men are between 5 feet 9 inches and 6 feet tall.

```
normalcdf(69,72,
68.1,2.7)
        .2951343633
```

■

PRACTICE PROBLEM 1

The weights of American women are approximately normally distributed with mean 134.7 pounds and standard deviation 30.4 pounds. Find the proportion of women who weigh between 130 and 150 pounds.

Solution at the back of the book

z-Scores

The *standard normal distribution* is the normal distribution with mean $\mu = 0$ and standard deviation $\sigma = 1$ (the word "standard" indicates mean zero and standard deviation one). The letter z is traditionally used to denote the variable for this special distribution. Since $\mu = 0$, the highest point on the curve is at $z = 0$, and the curve is symmetric about this value.

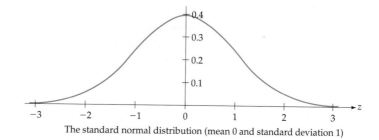

The standard normal distribution (mean 0 and standard deviation 1)

To change any x-value for a normal distribution with mean μ and standard deviation σ into the corresponding z-score for the *standard normal distribution*, subtract the mean and then divide by the standard deviation.

z-Score

$$z = \frac{x - \mu}{\sigma}$$

Standardizing an x-value

The z-score shows how many standard deviations the value is away from its mean.

EXAMPLE 2   Finding a z-Score

Convert each x-value into a z-score using the given values for μ and σ.

a. $x = 8$ with $\mu = 4$ and $\sigma = 2$ **b.** $x = 8$ with $\mu = 12$ and $\sigma = 4$

Solution

a. $z = \dfrac{8 - 4}{2} = 2$ **b.** $z = \dfrac{8 - 12}{4} = -1$

In part (a) the z-score $z = 2$ means that the original 8 was *two* standard deviations to the right of its mean. In part (b) the z-score $z = -1$ means that the 8 in that distribution was one standard deviation to the *left* of *its* mean.   ∎

PRACTICE PROBLEM 2

Convert $x = 20$ with $\mu = 15$ and $\sigma = 1$ into a z-score. Is this value far from the mean?   *Solution at the back of the book*

The *central limit theorems* developed by Carl F. Gauss (1777–1855) and other mathematicians during the nineteenth century proved that the errors in observed values, the means of random samples, and many other statistical quantities were normally distributed. Extensive tables of the probabilities of collections of z-scores were laboriously calculated, and the results were worth the effort because they could be used for any normally distributed x-values by first converting them to z-scores. Although graphing calculators can easily compute values for a normal distribution with any given mean and standard deviation, z-scores remain useful as a quick tool to place a particular x-value in its proper position relative to μ and σ.

The Normal and Binomial Distributions

As we saw on page 450, the binomial distribution too has a kind of "bell" shape, with a peak at the expected value $\mu = np$ and falling on both sides to very small probabilities several standard deviations $\sigma = \sqrt{np(1 - p)}$ away.

$n = 20, p = 0.4$ $n = 15, p = 0.5$ $n = 25, p = 0.7$

Several binomial distributions drawn using the program BINOMIAL

In the eighteenth century, Abraham de Moivre (1667–1754) and Pierre-Simon Laplace (1749–1827) discovered and proved that for any choice of p between 0 and 1, the binomial distribution *approaches* the normal distribution as n becomes large. This fundamental fact is known as the de Moivre–Laplace theorem.

Graphing Calculator Exploration

The program* SEENORML draws binomial distributions for $p = \frac{1}{2}$ with n from 2 to 30. Run this program to watch the "box-like" binomial distributions change into "bell-shaped" curves as n increases. The window for each graph is different so that each fills as much as possible of the viewing area. When n is a multiple of 5, the program also draws the corresponding normal distribution approximation.

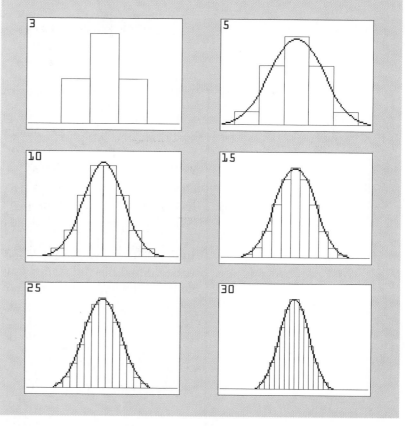

We may use the de Moivre–Laplace theorem to approximate binomial distributions by the normal distribution.

* See the Preface for information on how to obtain this and other graphing calculator programs.

Normal Approximation to the Binomial

> Let X be a binomial random variable with parameters n and p. If $np > 5$ and $n(1 - p) > 5$, then the distribution of X is approximately normal with mean $\mu = np$ and standard deviation $\sigma = \sqrt{np(1 - p)}$.

To have the same width for a "slice" of the normal distribution as for a slice of the binomial, we adopt the convention that the binomial probability $P(X = x)$ corresponds to the area under the normal distribution curve from $x - \frac{1}{2}$ to $x + \frac{1}{2}$. This is called the *continuous correction*.

how did they get those #'s

explain

EXAMPLE 3   Normal Approximation of a Binomial Probability

Estimate $P(X = 12)$ for the binomial random variable X with $n = 25$ and $p = 0.6$ using the corresponding normal distribution.

Solution

Since $np = (25)(0.6) = 15$ and $n(1 - p) = (25)(0.4) = 10$ are both greater than 5, we may use the normal distribution with $\mu = np = (25)(0.6) = 15$ and $\sigma = \sqrt{np(1 - p)} = \sqrt{(25)(0.6)(0.4)} = \sqrt{6}$. By the continuous correction, we interpret the event $X = 12$ as $11\frac{1}{2} \le X \le 12\frac{1}{2}$ to include numbers that would round to 12. Using a graphing calculator, we could find this probability as we did in Example 1 (pages 494–495), but the following calculator commands also show the picture.

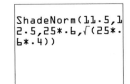

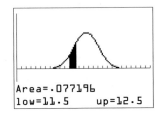

The required probability is (about) 0.077. This is a good approximation to the actual value of $_{25}C_{12}(0.6)^{12}(1 - 0.6)^{13} \approx 0.0759667$.

np

$n(1-p)$

$\overline{\sqrt{np(1-p)}}$

explain this

EXAMPLE 4   **Management MBA's**

At a major Los Angeles accounting firm, 73% of the managers have MBA degrees. In a random sample of 40 of these managers, what is the probability that between 27 and 32 will have MBA's?

$P = .73$

Solution $n = 40$ between 27 & 32

Because each manager either has or does not have an MBA, presumably independently of each other, the question asks for the probability that a binomial random variable X with $n = 40$ and $p = 0.73$ satisfies $27 \le X \le 32$. Since $np = (40)(0.73) = 29.2$ and $n(1 - p) = (40)(0.27) = 10.8$ are both greater than 5, this probability can be approximated as the area under the normal distribution curve with mean $\mu = np = (40)(0.73) = 29.2$ and standard deviation $\sigma = \sqrt{np(1 - p)} = \sqrt{(40)(0.73)(0.27)}$ from $26\frac{1}{2}$ to $32\frac{1}{2}$ (again using the continuous correction to include values that would round to between 27 and 32):

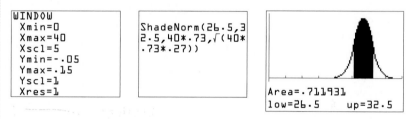

The probability is (about) 0.712, or about 71%. (The exact answer to this problem, from summing the binomial distribution from 27 to 32, is 71.5%, so the normal approximation is very accurate.)

PRACTICE PROBLEM 3

A brand of imported VCR is known to have defective tape rewind mechanisms in 8% of the units imported last April. If Jerry's Discount Electronics received a shipment of 80 of these VCRs, what is the probability that 10 or more are defective?   *Solution at the back of the book*

Section Summary

The graph of the normal distribution with mean μ and standard deviation σ is a bell-shaped curve that peaks at μ in the center and then falls back symmetrically on either side to approach the x-axis. The area under this curve between any two x-values is the probability that the normal random variable lies between these values. These areas and probabilities can be found easily with a graphing calculator.

Any value x of a normal random variable with mean μ and standard deviation σ can be converted into a z-score using the formula

$$z = \frac{x - \mu}{\sigma}$$

This z-score indicates the number of standard deviations the x-value is away from the mean.

The normal distribution with mean $\mu = np$ and standard deviation $\sigma = \sqrt{np(1 - p)}$ is a good approximation to the binomial distribution provided that both np and $n(1 - p)$ are greater than 5.

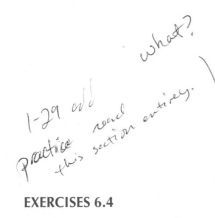

EXERCISES 6.4

1-29 odd
read this section entirely.
practice

(⊞ will be helpful throughout.)

Let X be the normal random variable with mean $\mu = 12$ and standard deviation $\sigma = 3$. Find each probability as an area under the normal curve.

$(x \times \mu, \sigma)$

1. $P(9 \le X \le 15)$

2. $P(6 \le X \le 18)$ $(6, 18, 12, 3)$

3. $P(3 \le X \le 21)$

4. $P(0 \le X \le 24)$

5. $P(12 \le X \le 15)$

6. $P(9 \le X \le 12)$

7. $P(15 \le X \le 16)$

8. $P(7 \le X \le 10)$ $(7, 10, 12, 3)$

9. $P\left(6\frac{1}{2} \le X \le 11\right)$

10. $P\left(10\frac{1}{2} \le X \le 14\right)$

Find the z-score of each x-value of the normal random variable with mean μ and standard deviation σ.

11. $x = 6$ with $\mu = 4$ and $\sigma = 2$ $\frac{6-4}{2}$ $\frac{2}{2} = 1$

12. $x = 20$ with $\mu = 15$ and $\sigma = 5$

$\frac{20-15}{5} = \frac{5}{5} = 1$

13. $x = 10$ with $\mu = 10$ and $\sigma = 3$

14. $x = 6$ with $\mu = 6$ and $\sigma = 5$ $\frac{6-6}{5}$ $\frac{0}{5} = 0$

15. $x = 15$ with $\mu = 20$ and $\sigma = 5$

16. $x = 6$ with $\mu = 8$ and $\sigma = 2$

17. $x = 15$ with $\mu = 20$ and $\sigma = 1$

18. $x = 6$ with $\mu = 8$ and $\sigma = 1$

19. $x = 130$ with $\mu = 100$ and $\sigma = 15$

20. $x = 90$ with $\mu = 100$ and $\sigma = 10$

Use the normal approximation to the binomial random variable X with $n = 20$ and $p = 0.7$ to find each probability as an area under the normal curve.

21. $P(X = 10)$

22. $P(X = 12)$ ✓

23. $P(11 \le X \le 12)$ $np75$

24. $P(8 \le X \le 10)$

25. $P(3 \le X \le 8)$

26. $P(13 \le X \le 19)$

27. $P(15 \le X \le 20)$

28. $P(2 \le X \le 7)$

29. $P(0 \le X \le 20)$

30. $P(14 \le X \le 20)$

APPLIED EXERCISES

Answer each question using an appropriate normal distribution probability.

31. **Weights** Weights of men are approximately normally distributed with mean 163 pounds and standard deviation 28 pounds. If the minimum and maximum weights in order to be a volunteer fireman in Martinsville are 128 and 254, what proportion of men meet this qualification?

$\mu = 163$

$\sigma = 28$ 89%

$128, 254$

32. **Advertising** The percentage of early afternoon television viewers who watch episodes of "As the World Spins" is a normal random variable with mean 22% and standard deviation 3%. The producers guarantee advertisers that the viewer percentage will be better than 20% or else the advertisers will receive free air time. What is the probability that the producers will have to give free air time after any particular episode? $np >$

$\mu = .22$

$\sigma = .03$

$v = 97 \qquad 1,100$

$\sigma = 6$

33. Airline Overbooking In order to avoid empty seats, airlines generally sell more tickets than there are seats. Suppose that the number of ticketed passengers who show up for a flight with 100 seats is a normal random variable with mean 97 and standard deviation 6. What is the probability that at least one person will have to be "bumped" from the flight?

34. SAT Scores The scores at Centerville High School on last year's mathematics SAT test were approximately normally distributed with mean 490 and standard deviation 140. What proportion of the scores were between 550 and 750?

35. Smoking In a large-scale study carried out in the 1960s of male smokers 35–45 years old, the number of cigarettes smoked daily was approximately normally distributed with mean 28 and standard deviation 10. What proportion of the smokers smoked between 30 and 40 (two packs!) each day?

36. Coin Tosses A fair coin is tossed fifty times. What is the probability that it comes up heads at least thirty times?

37. Actuarial Exams The probability of passing the first actuarial examination is 60%. If 940 college students take the exam in March, what is the probability that the number that pass is between 575 and 725?

38. Flu Shots Although getting a flu shot in October is a good way of planning ahead for the winter flu season, approximately 1% of those getting the shots develop side effects. If 800 students at Micheles College get flu shots, what is the probability that between 6 and 12 of them will develop side effects? $-\cap$

39. Real Estate Nationwide, 40% of the sales force at the Greener Pastures real estate franchises are college graduates. If 200 of these sales personnel are chosen at random, what is the probability that between 85 and 100 of them are college graduates?

40. Old Age If 3% of the population lives to age 90, what is the probability that between 10 and 15 of the 280 business majors at Henderson College will live to be that old?

Chapter Summary with Hints and Suggestions

Reading the text and doing the exercises in this chapter have helped you to master the following skills, which are listed by section (in case you need to review them) and are keyed to particular Review Exercises. Answers for all Review Exercises are given at the back of the book, and full solutions can be found in the Student Solutions Manual.

6.1 RANDOM SAMPLES AND DATA ORGANIZATION

- Identify levels of data measurement (nominal, ordinal, interval, and ratio). *(Review Exercises 1–4.)*
- Construct a bar chart, a stem-and-leaf display, or a histogram for a given data set. *(Review Exercises 5–10.)*

6.2 MEASURES OF CENTRAL TENDENCY

- Find the mode, median, and mean of a data set respecting the level of data measurement. *(Review Exercises 11–15.)*

Mode: most frequent

Median: middle

Mean: $\bar{x} = \dfrac{x_1 + \cdots + x_n}{n}$

6.3 MEASURES OF VARIATION

- Find the range of a data set. *(Review Exercise 16.)*

- Find the five-point summary of a data set and draw the box-and-whisker plot. *(Review Exercise 17.)*

- Find the sample standard deviation of a data set. *(Review Exercise 18.)*

$$s = \sqrt{\frac{(x_1 - \bar{x})^2 + \cdots + (x_n - \bar{x})^2}{n - 1}}$$

- Analyze the data in an application by finding the range, the five-point summary, and the sample standard deviation and then drawing the box-and-whisker plot. *(Review Exercises 19–20.)*

6.4 NORMAL DISTRIBUTIONS AND BINOMIAL APPROXIMATION

- Find the probability that the normal random variable with mean μ and standard deviation σ is between two given values. *(Review Exercises 21–22.)*

- Find the z-score of an x-value of the normal random variable with mean μ and standard deviation σ. *(Review Exercises 23–24.)*

$$z = \frac{x - \mu}{\sigma}$$

- Use the normal approximation to the binomial random variable X with n and p to find the probability that X takes given values. *(Review Exercises 25–26.)*

- Solve an applied problem involving probabilities using the normal distribution. *(Review Exercises 27–30.)*

HINTS AND SUGGESTIONS

- (*Overview*) Properties of "many" can be inferred from just "some" only if the "some" are a random sample from the population. There are four very different levels of data measurement: nominal, or-dinal, interval, and ratio. Data may be organized into bar charts, stem-and-leaf displays, and his-tograms. Measures of central tendency (mode, me-dian, and mean) summarize all the data as one typ-ical value, while measures of variation (range, box-and-whisker plot, and sample standard devia-tion) explain how closely the values cluster about the center. The bell-shaped curve of the normal distribution applies to many situations, and z-scores give the number of standard deviations the values are from the mean. The binomial distribu-tion is approximated by the normal distribution, provided both np and $n(1 - p)$ are sufficiently large.

- Bar charts are appropriate for any level of data measurement and the bars are separate. His-tograms can be drawn only for interval or ratio data and the bars touch each other on both sides.

- To find the mode or the median of data values, it is helpful to begin by sorting them into order.

- For measures of central tendency, with nominal ("in name only") data, only the mode is appropri-ate; with ordinal data, both the mode and the me-dian are appropriate; with interval or ratio data, all three (mode, median, and mean) are appropriate.

- All three measures of variation are appropriate only for interval or ratio data.

- The mean is sensitive to changes in a few extreme values, while the mode and median are not.

- Box-and-whisker plots graphically show how the data are distributed among the four quartiles.

- Areas under the normal distribution curve are probabilities for the normal random variable.

- When using the normal distribution to approxi-mate the binomial distribution, be sure to increase the largest value by $\frac{1}{2}$ and to decrease the smallest by $\frac{1}{2}$ (the continuous correction) to cover all the area represented by the boxes making up the bino-mial distribution.

- Practice for test: Review Exercises 1, 3, 5, 7, 9, 11, 19, 21, 23, 26, 28.

Review Exercises for Chapter 6   *Practice test exercise numbers are in blue.*

(will be helpful throughout.)

6.1   Random Samples and Data Organization

Identify the level of data measurement in each situation.

1. **Chicago Commuters**   A random sample of seventy Chicago area commuters counts the number traveling to work last Thursday by car, taxi, bus, and train.

2. **Gourmet Coffee**   At the House of Java Coffee Emporium, a random sample of eighteen customers rates the new Turkish Sultan flavor on a scale from 1 ("I'll try something else") to 5 ("Love at first sip").

3. **Fourth of July**   A random sample of five hundred suburban households records the backyard temperatures at 3 P.M. on Independence Day.

4. **Drinking Water**   A random sample of eighty small municipalities (less than 50,000 residents) lists the lead content (in parts per billion) of the tap water.

Construct a bar chart of the data in each situation.

5. **Veterans of Foreign Wars**   A random sample of forty members of the Hattiesburg V.F.W. identifies their branches of military service as 14 in the Army, 12 in the Navy, 6 in the Marines, and 8 in the Air Force.

6. **Homicides**   A random sample of twenty homicides investigated last year by the Macon County Coroner's Office lists the causes as 12 by handgun, 3 by firearms other than handguns, 2 by knives, 2 by blunt instruments, and 1 by use of hands.

Construct a stem-and-leaf display of the data in each situation.

7. **Five-Mile Run**   At the Norrisville Cardiac Fitness cross country five-mile run, the finish times for twenty randomly selected entrants were (in minutes):

59	36	59	38	55	47	56	43	57	45
33	50	29	58	60	87	40	54	46	55

8. **Study Times**   A random sample of thirty business majors interviewed during spring break in Daytona Beach found that they studied the following numbers of hours the previous week:

36	60	39	31	36	30	14	4	57	9
48	45	18	49	52	41	38	13	32	19
58	3	25	21	17	37	27	26	51	53

Construct a histogram of the data in each situation, using an appropriate class width and starting class lower boundary.

9. **Manufacturer's Rebates**   A random sample of twenty requests for the $5 rebate offered last summer on two pairs of Ultra-Cool Shades sunglasses found that the checks were received by the consumers after waits of the following numbers of days:

45	53	60	34	60	37	52	65	44	49
54	51	61	43	49	63	53	53	56	55

10. **Checking Accounts**   A random sample of thirty checking account balances on March 2 at the Lobsterman Trust Company branch in Bangor found the following amounts (rounded to the nearest dollar):

906	858	1168	1397	925	797
1036	1398	912	1059	931	815
698	787	711	1485	1048	937
1272	727	1339	1458	842	1264
1045	960	802	1204	1370	841

6.2   Measures of Central Tendency

Find the mode, median, and mean of each (ratio) data set.

11. {14, 11, 15, 8, 5, 4, 10, 3, 12, 11, 6}

12. {5, 6, 15, 13, 15, 6, 11, 13, 13, 6, 13, 6, 8}

13. {12, 13, 13, 5, 9, 14, 13, 8, 9, 14}

14. **Stock Earnings**   The third-quarter earnings per share for twelve randomly selected textile stocks on the New York Stock Exchange were (in dollars):

12	8	8	11	4	7
8	7	5	13	10	9

15. Personal Calls A random sample of the telephone records for fifteen office clerks at the southern regional office of the Marston Glassware Products Company counted the following numbers of personal calls last week:

$$
\begin{array}{ccccc}
4 & 15 & 23 & 26 & 2 \\
22 & 25 & 4 & 27 & 18 \\
30 & 28 & 11 & 20 & 6
\end{array}
$$

6.3 Measures of Variation

16. Find the range of the (ratio) data {38, 28, 32, 43, 41, 25, 35, 37, 43, 36}.

17. Find the five-point summary and draw the box-and-whisker plot for the (ratio) data {4, 5, 7, 9, 10, 13, 14, 15, 20}.

18. Find the sample standard deviation of the (ratio) data {5, 7, 9, 10, 13, 16}.

Analyze the data in each situation by finding the range, the five-point summary, and the sample standard deviation, and then drawing the box-and-whisker plot.

19. Bank Lines A random sample of fifteen customers at the Middletown branch of the Peoples & First National Bank found the following waiting times (in minutes):

$$
\begin{array}{ccccc}
5 & 6 & 6 & 5 & 3 \\
5 & 8 & 3 & 9 & 6 \\
7 & 9 & 5 & 9 & 6
\end{array}
$$

20. Mortgages A random sample of twenty mortgages filed last month at the Seaford Town Clerk office had the following values (in thousands of dollars):

$$
\begin{array}{ccccc}
132 & 141 & 154 & 143 & 115 \\
141 & 125 & 132 & 123 & 132 \\
119 & 126 & 123 & 114 & 126 \\
132 & 128 & 131 & 136 & 131
\end{array}
$$

6.4 Normal Distributions and Binomial Approximation

Let X be a normal random variable with mean $\mu = 50$ and standard deviation $\sigma = 10$. Find each probability as an area under the normal curve.

21. $P(30 \leq X \leq 60)$ **22.** $P(55 \leq X \leq 75)$

Find the z-score of each x value of a normal random variable with mean μ and standard deviation σ.

23. $x = 12$ with $\mu = 10$ and $\sigma = 1$

24. $x = 18$ with $\mu = 24$ and $\sigma = 3$

Use the normal approximation to the binomial random variable X with $n = 25$ and $p = 0.35$ to find each probability as an area under the normal curve.

25. $P(X = 14)$ **26.** $P(6 \leq X \leq 10)$

For exercises 27–30, answer each question using an appropriate normal distribution.

27. Heights Women's heights are approximately normally distributed with mean 63.2 inches and standard deviation 2.6 inches. Find the proportion of women with heights between 62 inches and 69 inches.

28. Tire Wear The life of Wear-Ever Super Tread automobile tires is normally distributed with mean 50,000 miles and standard deviation 5000 miles. What proportion of these tires last between 55,000 and 65,000 miles?

29. Lost Luggage During the Christmas rush, the chance that an airline will lose your suitcase increases to 1%. What is the probability that between 4 and 8 of the 690 suitcases checked in at Bellemeade Regional Airport this holiday season will be lost?

30. Heart Disease The leading cause of death in the United States is heart disease, which causes 32% of all deaths. In a random sample of 150 death certificates, what is the probability that between 50 and 60 list heart disease as the cause of death?

Projects and Essays

The following projects and essays are based on Chapter 6. There are no right or wrong answers—the results depend only on your imagination and resourcefulness.

1. Write a one-page report on an interesting situation discussed in the book *How to Lie with Statistics* by Darrell Huff (New York: Norton, 1954; paperback reissue 1993).

2. The bar chart was invented by William Playfair and first published by him in 1786. Write a one-page report on Playfair and his contributions to the graphical presentation of data after reading the articles "William Playfair: A Daring Worthless Fellow" and "Who Was Playfair?" by Ian Spence and Howard Wainer in the Winter 1997 issue of *Chance* magazine (Volume 10, Number 1), pages 31–34 and 35–37.

3. Stem-and-leaf displays and box-and-whisker plots are just two of many recent ideas for the presentation of numeric information. Write a one-page report on these and other methods after browsing through the book *Exploratory Data Analysis* by John Tukey (Reading, MA: Addison-Wesley, 1977).

4. A student at your college has taken three examinations in another section of this course, received grades of 70, 80, and 90, and has asked her professor to drop the lowest grade. Should her professor agree to her request? The class means and standard deviations on these tests are listed in the table below. Write a one-page essay in support of your answer.

Test	Her Score	Mean	Standard Deviation
No. 1	70	60	5
No. 2	80	75	10
No. 3	90	95	10

5. Look over your notes, homework, and this text, and write a one-page report about how your graphing calculator helped you in this chapter. Include examples of how it helped you explore concepts and how it helped to simplify your work. What was the most helpful or interesting use? What was the least helpful or interesting use? Are there problems that can be done on a graphing calculator but that are easier to do by hand?

+ 10 POINTS

DOUBLE SPACED

MIN 1 PAGE MAX 2

REPORT COVER + TITLE PAGE

DATE

7

Markov Chains

The sequence of deuce, advantage scores in a game between equally matched players, is an example of the Markov chains described in this chapter.

7.1 # States and Transitions

A P P L I C A T I O N   P R E V I E W

Weather in Sri Lanka*

The island nation of Sri Lanka, off the southeast coast of India, is a well-known producer of tea, coconuts, and rice. Only 16% of the land is arable and weather conditions vary from monsoon to drought, so accurate weather projections would do much to improve the country's economic output. Even if a month's total rainfall is adequate, crops can fail if the rain comes in a downpour after several dry weeks. A recent study modeled the weekly rainfall as a "Markov chain" (to be explained in this chapter), using 50 years of weather data to estimate the probabilities that a "wet" week will be followed by a "dry" week, a "wet" week will be followed by a "wet" week, and so on. The resulting probabilities were then used to simulate the rainfall in different areas over a year, with results such as the accompanying graph.

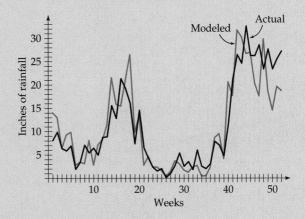

The model, which clearly tracks the actual rainfall quite well, can be used to predict the success of planting marginal areas with various crops whose growth rates and water requirements are known.

---

* Adapted in part from "On Development and Comparative Study of Two Markov Models of Rainfall in the Dry Zone of Sri Lanka" by B.V.R. Punyawardena and D. Kulasiri, Research Report 96/11, Centre for Computing and Biometrics, Lincoln University (Canterbury, New Zealand).

Introduction

When you read the morning newspaper or watch the evening TV news, much of the information is presented as the change since yesterday: The stock market is up, the Dodgers won again, and the weather is moderating. Sometimes the change is stated exactly ("The sun will rise two minutes earlier tomorrow") and sometimes it is given in terms of probabilities ("There is a 70% probability that the rain will end tomorrow"). In this chapter we will explore the behavior of repeated changes that depend on probabilities.

States and Transitions

The Russell 2000 Index is an average of the prices of 2000 stocks traded on the New York Stock Exchange. Each trading day this index *gains*, *loses*, or remains *unchanged* compared to the previous day, and we can denote these three possible *states* by the letters *G*, *L*, and *U*. Watching the market for several successive trading days might result in a sequence of states such as

$$U, G, G, L, G, L, L, L, U, G, \ldots$$

Each *transition* from one day's state to the next begins from one of the three states *G*, *L*, and *U* and ends in one of them, so there are 3 × 3 = 9 possible transitions. From past stock market records, it is possible to estimate the probabilities of these transitions, and we can specify them by a *state-transition diagram* or by a *transition matrix*.

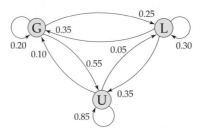

$$
\begin{array}{c}
\quad\quad G \quad\;\; L \quad\;\; U \\
\begin{array}{c} G \\ L \\ U \end{array}
\left(
\begin{array}{ccc}
0.20 & 0.25 & 0.55 \\
0.35 & 0.30 & 0.35 \\
0.10 & 0.05 & 0.85
\end{array}
\right)
\end{array}
$$

Transition Matrix

State-Transition Diagram

In the state-transition diagram, the number near the arrowhead from a starting state to an ending state gives the probability of the transition from that starting state to that ending state. For example, the diagram shows that the transition from state *L* to state *L* (which we denote as $G \to L$) has probability 0.25, while the transition $U \to U$

(that is, the market will remain unchanged) has probability 0.85. In the transition matrix on the right, the number in a particular row and column gives the probability of the transition from the *row* state to the *column* state. We always list the row states from top to bottom in the same order as the column states from left to right. The matrix shows that the transition $U \to L$ has probability 0.05 and that the transition $L \to L$ (the market will continue to lose) has probability 0.30.

Notice that the sum of the entries in each row is 1, since any given state must be followed by one of the states G, L, or U.

PRACTICE PROBLEM 1

What is the probability that the market will continue to gain?

Solution at the back of the book

Stochastic Process

In general, a (discrete) *stochastic process* (from the Greek word *stochastikos* meaning "skillful in guessing") is a collection of possible *states* together with the *transition probabilities* between them. These probabilities can be given in matrix form.

Stochastic Matrix

A square matrix is *stochastic* if the entries are nonnegative and the sum of each row is 1.

PRACTICE PROBLEM 2

Which of the following matrices are stochastic?

a. $\begin{pmatrix} 0.30 & 0.70 \\ 0.40 & 0.60 \end{pmatrix}$
b. $\begin{pmatrix} 0.80 & 0.10 \\ 0.60 & 0.40 \end{pmatrix}$
c. $\begin{pmatrix} 0.50 & 0.50 \\ 1.1 & -0.1 \end{pmatrix}$

Solution at the back of the book

If we omit the state names from a transition matrix, we will assume that the states are named S_1, S_2, . . . , S_n and that the rows (from top to bottom) and the columns (from left to right) correspond to these states in this order. An *observation* of the stochastic process is a list of observed states. The probability $P(S_i \to S_j)$ of passing from state S_i to state S_j is the entry in row i and column j of the transition matrix.

Markov Chain

One of the simplest and most widely used stochastic processes is the *Markov chain*, first studied by the Russian mathematician

A. A. Markov (1856–1922), in which the *same* transition probabilities are used for every transition. Thus the future states of a Markov chain depend only on the *present* state, not on any earlier states, and for this reason a Markov chain is sometimes said to "have no memory." Using the same probabilities for each transition greatly simplifies the mathematics but still leads to models that are surprisingly useful in applications.

Markov Chain

A (finite) *Markov chain* consists of states $S_1, S_2, \ldots, S_n$ together with an $n \times n$ stochastic matrix T such that the transition $S_i \to S_j$ has probability $t_{i,j}$ in row i and column j of T for *every* transition.

Types of Transition Matrices

What can we say about the effect of several successive transitions? Transition matrices may represent or contain several of the three basic types of transitions: *oscillating* (everything switching back and forth), *mixing* (moving among all possibilities), or *absorbing* (never leaving a state once it is reached).

EXAMPLE 1   Types of Transitions

Characterize the transition corresponding to each stochastic matrix as "oscillating," "mixing," or "absorbing."

a. $\begin{pmatrix} 0 & 1 \\ 1 & 0 \end{pmatrix}$
   b. $\begin{pmatrix} \frac{1}{3} & \frac{1}{3} & \frac{1}{3} \\ \frac{1}{3} & \frac{1}{3} & \frac{1}{3} \\ \frac{1}{3} & \frac{1}{3} & \frac{1}{3} \end{pmatrix}$
   c. $\begin{pmatrix} 1 & 0 & 0 \\ \frac{1}{2} & \frac{1}{2} & 0 \\ \frac{1}{2} & 0 & \frac{1}{2} \end{pmatrix}$

Solution

We make a state-transition diagram for each transition matrix.

$\begin{pmatrix} 0 & 1 \\ 1 & 0 \end{pmatrix}$

a. Since states S_1 and S_2 alternate, this is an *oscillating* transition. Repeating it several times would just flip back and forth between the two states.

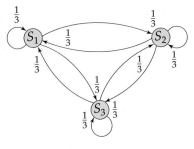

b. Since each state may lead to itself or to any of the other states, this is a *mixing* transition. Repeating it several times would give some combination of the states, and any mixture is possible.

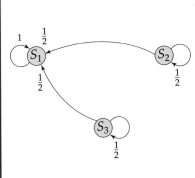

c. Since each of states S_2 and S_3 can reach state S_1 but then can never leave, this is an *absorbing* transition. Repeating it several times might lead to several occurrences of S_2 or S_3, but once the chain reached state S_1, it would remain there forever. ∎

Notice that the type of a transition *cannot* be inferred from a short observation of the chain. For example, $S_1S_2S_1S_2$ could be the result of either the oscillating transition of matrix (a) or the mixing transition of matrix (b) of Example 1, while $S_1S_1S_1S_1$ could be the result of either the mixing transition of matrix (b) or the absorbing transition of matrix (c).

PRACTICE PROBLEM 3

Describe the action of the transition matrix $\begin{pmatrix} 1 & 0 & 0 \\ 0 & 1 & 0 \\ 0 & 0 & 1 \end{pmatrix}$.

Solution at the back of the book

State Distribution Vectors

Given the transition matrix for a Markov chain, what can we say about the chain after a transition? Unless we have been following a particular observation, we do not know the current state, so let us apply the transition to an *estimate* of the current state and see how this estimate is changed. First consider an example.

EXAMPLE 2   **Calculating the Result of a Transition**

Returning to the stock market situation with states *G*, *L*, and *U* (for *gaining*, *losing*, and *unchanged*), suppose that the transition matrix is

$$
\begin{array}{c c}
 & \begin{array}{c c c} G & L & U \end{array} \\
\begin{array}{c} G \\ L \\ U \end{array} &
\begin{pmatrix} 0.20 & 0.25 & 0.55 \\ 0.35 & 0.30 & 0.35 \\ 0.10 & 0.05 & 0.85 \end{pmatrix}
\end{array}
$$

If we have a portfolio of 100 stocks of which 50 gained, 20 lost, and 30 remained unchanged yesterday, how many may we expect to gain, lose, and remain unchanged in today's market?

Solution

We are given the numbers of stocks in the states *G*, *L*, and *U*, and the first column of the transition matrix gives the probabilities of transitions from these states into the state *G*, so multiplying and adding will give the number of gaining stocks expected today:

$$(50) \cdot (0.20) + (20) \cdot (0.35) + (30) \cdot (0.10) = 20$$

Gained (*G*) $P(G \to G)$ Lost (*L*) $P(L \to G)$ Unchanged (*U*) $P(U \to G)$
yesterday   yesterday   yesterday

Expected value is the sum of all the $n \cdot p$

Thus we can expect 20 of the stocks to gain today. Similarly, multiplying the given numbers by the probabilities in the second column (the probabilities of transitions into the state *L*) gives today's expected number of losers:

$$(50) \cdot (0.25) + (20) \cdot (0.30) + (30) \cdot (0.05) = 20$$

Gained $P(G \to L)$ Lost $P(L \to L)$ Unchanged $P(U \to L)$
yesterday   yesterday   yesterday

Expected number of losing stocks

The remaining 60 stocks should remain unchanged today, and the corresponding calculation using the last column of the transition matrix verifies this count:

$$(50) \cdot (0.55) + (20) \cdot (0.35) + (30) \cdot (0.85) = 60$$

Gained $P(G \to U)$ Lost $P(L \to U)$ Unchanged $P(U \to U)$
yesterday   yesterday   yesterday

Expected number of unchanged stocks

Notice that these calculations are just the matrix product, using the usual row and column multiplication described on page 212:

$$(50 \quad 20 \quad 30) \cdot \begin{pmatrix} 0.20 & 0.25 & 0.55 \\ 0.35 & 0.30 & 0.35 \\ 0.10 & 0.05 & 0.85 \end{pmatrix} = (20 \quad 20 \quad 60)$$

Portfolio yesterday   Transition matrix   Portfolio today

Rephrasing our calculation in Example 2 in terms of probabilities for the 100 stocks in the portfolio, yesterday there were 50% gaining, 20% losing, and 30% unchanged, while after the transition to today, they became 20% gaining, 20% losing, and 60% unchanged. A row matrix consisting of such probabilities is called a *state distribution vector,* and the product of this vector with the transition matrix gives the state distribution vector for the *next* observation.

State Distribution Vector

A *state distribution vector* is a $1 \times n$ matrix $D = (d_1\ d_2\ \cdots\ d_n)$ of nonnegative numbers whose sum is 1. If D represents the current probability distribution for a Markov chain with transition matrix T, then the next probability distribution is the state distribution vector given by the matrix product $D \cdot T$.

Graphing Calculator Exploration

To verify that the transition matrix $T = \begin{pmatrix} 0.20 & 0.25 & 0.55 \\ 0.35 & 0.30 & 0.35 \\ 0.10 & 0.05 & 0.85 \end{pmatrix}$ from

Example 2 transforms the state distribution vector $D = (0.50\quad 0.20\quad 0.30)$ into $D \cdot T = (0.20\quad 0.20\quad 0.60)$, enter T and D into matrices [A] and [B] and find their product as follows.

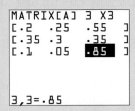

and then

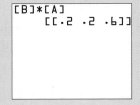

Repeat this calculation with other state distribution vectors of your choosing. What happens to the distribution vector $D = (0.140\quad 0.104\quad 0.756)$?

The *k*th State Distribution Vector

For a given Markov chain, what can we say about the future given some information about the present? If we have an *initial* state distribution vector D_0 and transition matrix T, then we have just seen that the state distribution vector after the first transition is

$$D_1 = D_0 \cdot T \qquad\qquad \text{\textit{T} gives one transition}$$

After the second transition the state distribution vector is

$$D_2 = D_1 \cdot T \qquad\qquad \text{Multiplying by \textit{T} again}$$
$$= (D_0 \cdot T) \cdot T = D_0 \cdot T^2 \qquad \text{\textit{T}}^2 \text{ gives two transitions}$$

Continuing in this way, we see that the *k*th state distribution vector D_k is the initial state distribution vector times the *k*th power of the transition matrix:

*k*th State Distribution Vector

$$D_k = D_0 \cdot T^k$$

To put this another way, T^k gives the probabilities for k successive transitions, so the entry in row i and column j of T^k gives the probability of going from state S_i to state S_j in exactly k transitions.

EXAMPLE 3   **Calculating State Distribution Vectors**

For the Markov chain with transition matrix $T = \begin{pmatrix} 1 & 0 & 0 \\ \frac{1}{2} & \frac{1}{2} & 0 \\ \frac{1}{2} & 0 & \frac{1}{2} \end{pmatrix}$ and

initial state distribution vector $D_0 = \begin{pmatrix} \frac{1}{5} & \frac{2}{5} & \frac{2}{5} \end{pmatrix}$, calculate the next two state distribution vectors D_1 and D_2.

Solution

We obtain the next state distribution vector by multiplying the previous one by the transition matrix:

$$D_1 = D_0 \cdot T = \begin{pmatrix} \frac{1}{5} & \frac{2}{5} & \frac{2}{5} \end{pmatrix} \cdot \begin{pmatrix} 1 & 0 & 0 \\ \frac{1}{2} & \frac{1}{2} & 0 \\ \frac{1}{2} & 0 & \frac{1}{2} \end{pmatrix}$$

$$= \begin{pmatrix} \frac{1}{5} \cdot 1 + \frac{2}{5} \cdot \frac{1}{2} + \frac{2}{5} \cdot \frac{1}{2} & 0 + \frac{2}{5} \cdot \frac{1}{2} + 0 & 0 + 0 + \frac{2}{5} \cdot \frac{1}{2} \end{pmatrix} \qquad \text{row} \times \text{columns}$$

$$= \begin{pmatrix} \frac{3}{5} & \frac{1}{5} & \frac{1}{5} \end{pmatrix}$$

and

$$D_2 = D_1 \cdot T = \begin{pmatrix} \frac{3}{5} & \frac{1}{5} & \frac{1}{5} \end{pmatrix} \cdot \begin{pmatrix} 1 & 0 & 0 \\ \frac{1}{2} & \frac{1}{2} & 0 \\ \frac{1}{2} & 0 & \frac{1}{2} \end{pmatrix} = \begin{pmatrix} \frac{4}{5} & \frac{1}{10} & \frac{1}{10} \end{pmatrix}$$

Omitting the details of the calculation

∎

Comparing $D_0 = \begin{pmatrix} \frac{1}{5} & \frac{2}{5} & \frac{2}{5} \end{pmatrix}$ with $D_2 = \begin{pmatrix} \frac{4}{5} & \frac{1}{10} & \frac{1}{10} \end{pmatrix}$, we see that the first entry has increased from $\frac{1}{5}$ to $\frac{4}{5}$. That is, in only two transitions, the probability of being in state S_1 has increased from 20% to 80%. Since we saw in Example 1c on pages 510–511 that the state S_1 is absorbing for this transition matrix, the probability of being in S_1 should increase with each transition.

Graphing Calculator Exploration

To check that the probability of being in state S_1 increases with each transition for the absorbing Markov chain in Example 3, enter T and D_0 into matrices [A] and [B] and calculate $D_3 = D_2 \cdot T$ and $D_4 = D_3 \cdot T$ as follows.

and then

```
[B]
        [[.2 .4 .4]]
Ans*[A]
        [[.6 .2 .2]]
        [[.8 .1 .1]]
      [[.9 .05 .05]]
[[.95 .025 .025...
```

Repeat this calculation with higher powers of T and with other initial state distribution vectors D_0. Are states S_2 and S_3 always absorbed into state S_1?

Duration in a Given State

Using state distribution vectors, we can watch the probability of being in a particular state change as the Markov chain continues. Suppose we observe that the Markov chain is now in state S_i. How long can we expect it to remain in state S_i before switching to another state?

The probability p that the chain stays S_i is the entry t_{ii} of the transition matrix T. If $p = 1$, then state S_i is an absorbing state and the chain will remain in state S_i forever. If $p = 0$, then the chain will leave state S_i immediately and will have been in this state just once. Between these extremes, if $0 < p < 1$, then the probability that it leaves for some other state is $1 - p$ (the probablility of the complementary event). The probability of its being in state S_i exactly n times and then switching to some other state is

$$\underbrace{p \cdot \cdots \cdot p}_{\substack{\text{Stay in } S_i \text{ for} \\ n - 1 \text{ transitions}}} \cdot \underbrace{(1 - p)}_{\substack{\text{and then} \\ \text{leave } S_i}} = p^{n-1}(1 - p)$$

Using the formula for "expected value" (page 443), the expected number of times in state S_i before changing to some other state is simply this probability multiplied by n and then summed over all values of n:

$$E = 1 \cdot (1 - p) + 2 \cdot p\,(1 - p) + 3 \cdot p^2\,(1 - p) + \cdots$$

$$= \underbrace{(1 - p)}_{+\, p} + \underbrace{(2\,p - 2\,p^2)}_{+\, p^2} + \underbrace{(3\,p^2 - 3\,p^3)}_{+\, p^3 + \cdots} + \cdots \qquad \text{Multiplying out}$$

$$= 1 + p^1 + p^2 + p^3 + \cdots \qquad \text{Collecting similar terms}$$

$$= \frac{1}{1 - p} \qquad \text{Using the result of Exercise 44 on pages 138–139}$$

We have established the following fact.

Expected Time in a Given State

Let p be the probability that a Markov chain stays in its present state for one transition. If $p = 1$, then the chain will remain in this state forever, while if $p < 1$, then the expected number of times it will be in this state before switching to some other state is

$$E = \frac{1}{1 - p}$$

EXAMPLE 4   **Finding an Expected Time**

The EasyDotCom Internet service provider classifies its residential customers by the number of connection hours used per week: H for high usage of more than 20 hours, M for moderate usage of between 5 and 20 hours, and I for infrequent usage of less than 5 hours. The company has found that in the San Diego metropolitan area, the usage behavior of their subscribers can be modeled as a Markov chain with transition matrix

$$
\begin{array}{c}
 \\
H \\
M \\
I
\end{array}
\begin{array}{ccc}
H & M & I \\
\left(\begin{array}{ccc}
0.75 & 0.20 & 0.05 \\
0.20 & 0.70 & 0.10 \\
0.25 & 0.20 & 0.55
\end{array}\right)
\end{array}
$$

If the Davidson household is presently a high user, how many weeks can it be expected to be a high user before changing to some other level of Internet use?

Solution

Since the probability that state H stays state H for one transition of this Markov chain is 0.75 (from the upper left entry in the transition matrix), the number of weeks (including the present week) that the Davidsons are expected to be heavy users before changing to some other level is

$$E = \frac{1}{1 - 0.75} = \frac{1}{0.25} = 4 \text{ weeks} \qquad E = \frac{1}{1 - p} \text{ with } p = 0.75$$

SPREADSHEET EXPLORATION

The spreadsheet below shows 500 observations of the Markov chain from Example 4 but labels the states 1 (yellow) for H, 2 (green) for M, and 3 (blue) for I. The transition matrix is stored in the upper left-hand corner, and the counters below it summarize the number of times each state occurred in this observation.

	A	B	C	D	E	F	G	H	I	J	K	L	M	N	O	P	Q	R	S	T	U	V	W	X	Y	Z	AA	AB
1	0.75	0.2	0.05																									
2	0.2	0.7	0.1																									
3	0.25	0.2	0.55																									
4				1	1	2	2	2	2	2	2	2	3	3	2	2	1	2	3	2	3	3	3	3	3	1	1	1
5	State #1 :			1	1	1	1	1	1	1	1	1	2	2	2	2	2	2	2	2	2	2	1	1	1	1	1
6	232	0.464		1	2	2	2	1	1	1	1	2	2	2	2	2	2	1	1	2	1	1	1	1	1	2	2	2
7				2	1	1	1	1	1	1	1	2	2	2	2	2	2	1	1	1	1	1	1	1	1	1	1
8	State #2 :			1	2	2	2	2	2	1	2	2	2	2	2	2	2	3	3	3	3	3	2	2	2	2	2	2
9	204	0.408		2	2	2	3	3	3	1	1	1	2	2	1	1	1	1	1	1	1	1	1	2	2	2	3	2
10				1	1	1	2	2	1	1	1	1	1	2	2	2	2	2	1	2	2	2	2	2	2	2	2
11	State #3 :			3	2	2	2	2	2	1	1	1	3	3	3	3	2	1	1	1	1	2	2	2	2	2	2	1
12	64	0.128		3	3	3	3	3	1	1	1	1	1	2	2	2	2	2	2	2	2	2	2	1	1	1	2
13				2	1	1	1	1	1	2	2	3	3	3	3	3	3	3	3	3	3	1	2	2	1	1	1	2
14				2	2	1	2	2	2	1	2	2	2	1	2	1	1	1	1	3	2	1	1	1	1	1	1	1
15				2	2	2	2	2	2	2	2	2	2	1	1	1	3	3	2	2	2	1	1	1	1	1
16				1	1	1	1	1	2	2	2	2	2	2	2	2	3	1	1	2	2	2	3	2	2	2	2	2
17				1	1	1	2	2	2	1	1	1	2	2	2	1	1	1	1	2	2	1	1	1	3	3	1	1
18				1	1	1	1	1	2	2	3	3	1	1	1	1	2	1	2	2	2	2	2	1	1	1	1	1
19				1	1	1	1	1	3	2	2	1	1	2	1	1	1	1	1	1	1	2	1	1	1	3	2
20				2	1	1	1	2	2	2	1	1	2	2	2	1	1	1	1	1	1	1	1	1	1	1	1	1
21				1	1	1	1	2	2	2	2	3	3	2	2	2	1	2	2	2	2	3	3	3	1	1	1	1
22				1	1	1	1	1	1	1	3	3	2	3	3	3	3	3	3	3	3	1	1	1	2	2	2	2
23				2	1	1	2	2	1	1	1	1	1	1	1	1	2	2	2	1	1	1	1	1	1	1	1	3

State H (yellow) sometimes appears just once and sometimes appears a dozen or more times in succession. Find the average length of these successions of H's, either by setting up your own modification of this spreadsheet or by counting from the screen image above. Is your answer consistent with our answer to Example 4?

Section Summary

A *stochastic matrix* is a square matrix of nonnegative entries with each row summing to 1. The entries are the transition probabilities between the states.

A *Markov chain* uses the same matrix for each transition. Transition matrices exhibit three basic transition behaviors:

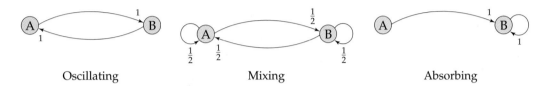

Oscillating   Mixing   Absorbing

An initial state distribution vector D_0 represents the probabilities of beginning in the various states, and the subsequent state distribution vectors $D_k = D_0 \cdot T^k$ are found by repeated multiplications by the transition matrix.

If the probability p that a Markov chain will stay in its present state for one transition is 1, then the chain will remain in that state forever. If it is less than 1, then the expected number of times it will be in this state before switching to some other state is $1/(1 - p)$.

EXERCISES 7.1

Construct a transition matrix T for the Markov chain represented by the state-transition diagram and identify the transition as "oscillating," "mixing," or "absorbing."

1.

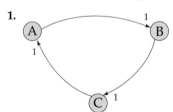

2.

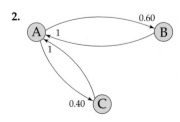

3.

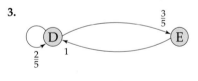

4.

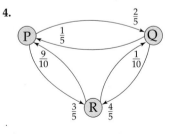

5.

6.

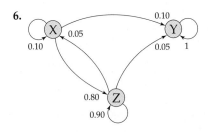

Construct a state-transition diagram for the Markov chain represented by the transition matrix and identify the transition as "oscillating," "mixing," or "absorbing."

7. $\begin{pmatrix} 0 & 1 & 0 \\ 0.20 & 0 & 0.80 \\ 0 & 1 & 0 \end{pmatrix}$

8. $\begin{pmatrix} 0 & 1 & 0 & 0 \\ 0 & 0 & 0.30 & 0.70 \\ 1 & 0 & 0 & 0 \\ 1 & 0 & 0 & 0 \end{pmatrix}$

9. $\begin{pmatrix} 0.10 & 0.90 \\ 0.35 & 0.65 \end{pmatrix}$

10. $\begin{pmatrix} 0 & 1 & 0 & 0 \\ 0 & 0 & 0.30 & 0.70 \\ 0 & 1 & 0 & 0 \\ 1 & 0 & 0 & 0 \end{pmatrix}$

11. $\begin{pmatrix} 1 & 0 & 0 \\ 0.10 & 0.20 & 0.70 \\ 0.10 & 0.20 & 0.70 \end{pmatrix}$

12. $\begin{pmatrix} 1 & 0 & 0 & 0 \\ 0 & 0 & 0.30 & 0.70 \\ 0 & 1 & 0 & 0 \\ 0 & 0 & 0 & 1 \end{pmatrix}$

For each Markov chain transition matrix T and initial state distribution vector D_0, calculate the next two state distribution vectors D_1 and D_2.

13. $T = \begin{pmatrix} \frac{1}{4} & \frac{3}{4} \\ \frac{1}{2} & \frac{1}{2} \end{pmatrix}$, $D_0 = \begin{pmatrix} \frac{3}{5} & \frac{2}{5} \end{pmatrix}$

14. $T = \begin{pmatrix} \frac{1}{2} & \frac{1}{2} \\ \frac{1}{3} & \frac{2}{3} \end{pmatrix}$, $D_0 = \begin{pmatrix} \frac{2}{5} & \frac{3}{5} \end{pmatrix}$

15. $T = \begin{pmatrix} 0 & 0.20 & 0.80 \\ 0.60 & 0.40 & 0 \\ 0.40 & 0 & 0.60 \end{pmatrix}$, $D_0 = (0.35 \quad 0.45 \quad 0.20)$

16. $T = \begin{pmatrix} 0.40 & 0.40 & 0.20 \\ 0.40 & 0 & 0.60 \\ 0.10 & 0.30 & 0.60 \end{pmatrix}$, $D_0 = (0.25 \quad 0.60 \quad 0.15)$

For each Markov chain transition matrix T and present state, find the expected number of times the chain will be in that state before moving to some other state.

17. $T = \begin{pmatrix} 0.60 & 0.20 & 0.20 \\ 0.10 & 0.80 & 0.10 \\ 0.60 & 0.20 & 0.20 \end{pmatrix}$, presently in state S_2

18. $T = \begin{pmatrix} 0.80 & 0.20 & 0 \\ 0.30 & 0.60 & 0.10 \\ 0.10 & 0 & 0.90 \end{pmatrix}$, presently in state S_3

19. $T = \begin{pmatrix} 0.80 & 0.10 & 0.10 & 0 \\ 0.20 & 0.50 & 0.10 & 0.20 \\ 0 & 0 & 1 & 0 \\ 0 & 0.30 & 0.10 & 0.60 \end{pmatrix}$, presently in state S_4

20. $T = \begin{pmatrix} 1 & 0 & 0 & 0 \\ 0 & 0.20 & 0.70 & 0.10 \\ 0.05 & 0.10 & 0.85 & 0 \\ 0 & 0 & 0 & 1 \end{pmatrix}$, presently in state S_3

Represent each situation as a Markov chain by constructing a state-transition diagram and the corresponding transition matrix. Be sure to state your final answer in terms of the original question.

21. Traffic Flow A traffic light on Main Street operates on a one-minute timer. If it is green (to the traffic), at the end of the minute it will change to red only if a pedestrian has pressed a button at the crosswalk. The probability of a pedestrian pressing the button during any one-minute interval is 0.25. If the light is red, it turns back to green at the end of the minute. If the light is green now for the traffic, find the expected number of minutes that the light is green.

22. Learning Theory The initial phase of a learning experiment is to establish that a rat will randomly wander in a maze prior to the introduction of food and other stimuli. The following maze has three connecting rooms, A, B, and C, and the location of the rat is observed at one-minute intervals. For both rooms A and B, the probability of remaining there is 50%, while the probability of remaining in room C is 75%. When the rat leaves a room, it is equally likely to choose either of the other two rooms. If the rat is observed in room B, find the expected number of minutes that the rat will remain there.

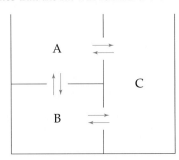

23. Breakfast Habits A survey of weekly breakfast eating habits found that although many people eat cereal, boredom and other reasons cause 5% to switch to something else (eggs, muffins, and so on) each week, while among those who are not eating cereal for breakfast, 15% will switch to cereal each week. If 70% of the residents of Cincinnati are eating cereal for breakfast this week, what percentage will be eating cereal next week? In two weeks?

24. Market Research A telephone survey of households in the Atlanta metropolitan area found that 85% of those using a DVD player planned to rent movies or otherwise use it next month, while 10% of those not using a DVD player this month planned on using one next month. If 2% of the households currently use a DVD player, what percentage will be using a DVD player after one year?

25. Retirement Sports A survey of retirees in Arizona who golf or fish every day found that 60% of those who played golf one day switched to fishing the next and that 90% of those who went fishing switched to golf the next day. If 30% of these retirees are golfing today, what percentage will be fishing tomorrow? Golfing the day after tomorrow?

26. Emergency Medical Care Authorization specialists at a national health maintenance organization spend twelve minutes to evaluate and process each request for emergency care, and their work loads are managed by a computerized distribution system. This system checks the work queue of each specialist every twelve minutes, transfers finished work to the data storage center, and adds new requests to the queue, but never allows any queue to contain more than four requests (one "active" and three "pending review"). The number of new requests the system has to add to any particular queue is variable: 15% of the time there are none, 55% there is one, 25% there are two, and 5% there are three new requests. If a particular specialist does not have any "pending review" requests in her queue, how many minutes can we expect this situation to continue? [*Hint:* Define four states by the number of unfinished requests in the queue at the end of each twelve-minute block of time, just before the system checks it to remove finished work and add new work.]

27. Portfolio Management An investment banker estimates the financial stability of mid-sized manufacturing companies as "secure," "doubtful," and "at risk." He has noticed that of the "secure" companies he follows, each year 5% decline to "doubtful" and the rest remain as they are; 10% of the "doubtful" companies improve to "secure," 5% decline to "at risk," and the rest remain as they are; and 5% of the "at risk" companies become bankrupt and never recover, 10% improve to "doubtful," and the rest remain as they are. If his current portfolio of investments is 80% "secure," 15% "doubtful," and 5% "at risk," what percentage will be "secure" in two years? (Assume that the present trends continue and that no changes are made to the portfolio.) In the long run, how many of the companies in the portfolio will become bankrupt?

28. Baseball Players A baseball fan tracks the careers of interesting players in the major and minor leagues and has found that from year to year, 80% of the major league players tracked stay in the majors, 10% drop down to the minor leagues, and 10% quit baseball or retire, while 20% of the minor league players tracked move up to the majors, 40% stay in the minors, and the rest quit or retire. If his current roster of interesting players is three-quarters major and one-quarter minor league players, what percentage of these players will have quit or retired in two years? What will happen to all the players in the long run?

29. Working Mothers A working mother of three teenagers expects them to help around the house and rates the laundry room each weekend as "empty," "manageable," or "you're all grounded!" She has noticed that if it is empty one weekend, then the following weekend with probability 0.40 it is empty, with probability 0.40 it is manageable, and with probability 0.20 it is so bad that "you're all grounded!" If it is manageable one weekend, then the following weekend with probability 0.50 it is empty and with probability 0.50 it is manageable. If it is so bad that she has to ground everybody, then the following weekend the laundry room is empty. If the laundry room is manageable this weekend, how many weekends will it be manageable before changing to either "you're all grounded!" or "empty"?

30. Jogging Routes Each morning a jogger runs either at the local high school track, around the neighborhood, or in the park. If she runs at the school one day, then the next day with probability 0.70 she runs there again and with probability 0.30 she switches to the park; if she runs around the neighborhood, then the next day with probability 0.90 she runs there again and with probability 0.10 she switches to the park; and if she runs in the park, then the next day she switches with probability 0.90 to the school and with probability 0.10 to around the neighborhood. If she runs around the neighborhood today, how many days can she be expected to run around the neighborhood before switching to the park?

7.2   Regular Markov Chains

Hidden Markov Models and Genetics

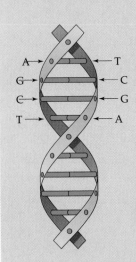

In 1953 Francis Crick and James Watson discovered the structure of the DNA (deoxyribonucleic acid) molecule that carries the genetic code of most living organisms. Every DNA molecule consists of two complementary strands wound together in a double helix, and each strand can be modeled as a sequence of the letters *A, G, C,* and *T* (for adenine, guanine, cytosine, and thymine). Although specific segments of this sequence contain identifiable genetic instructions, much of the rest is "junk" DNA, seemingly random jumbles of *A*'s, *G*'s, *C*'s, and *T*'s that are useless remnants of past evolutionary developments. The challenge today is to locate and decode the genetic information from the DNA of a particular cell in an effective and efficient manner.

A *hidden Markov model* for a sequence of observed states is a Markov chain that generates observations with the same random properties. Of course, a section of DNA encoding useful information is not a random jumble and thus can be detected by observing its deviation from its surroundings. Hidden Markov models have been successfully applied to find genes in the DNA of *Escherichia coli* bacteria by first using segments of a million nucleotides to estimate the transition probabilities and then testing the resulting model on other, shorter segments.* Other hidden Markov models are being developed and applied to problems in evolutionary DNA and protein structure. Hidden Markov models have many states and complicated transition probabilities, but they are based on the regular Markov chains developed in this section.

---

* See Anders Krogh, I. Saira Mian, and David Haussler, "A Hidden Markov Model That Finds Genes in *E. coli* DNA," *Nucleic Acids Research* 22 (1994), 4768–4778.

Introduction

Every coffee drinker knows how much cream and sugar to add to the cup to get just the right taste, and everyone knows that it makes no difference whether the cream or the sugar is added first, or on which side of the cup—a few random stirs mixes it uniformly. In this section we will explore the long-term behavior of *mixing* Markov chains. Unlike coffee, the resulting mixture is not, in general, a uniform distribution of the components but rather reflects the relative times spent in each state.

Ergodic Transitions

We consider first the simplest type of mixing stochastic transitions in which any state can be reached from every state in just one transition.

Ergodic* Matrix

> A stochastic matrix is *ergodic* if every entry is positive.

A Markov chain is ergodic if it has an ergodic transition matrix.

Just as with cream and sugar in coffee, the initial state distribution of a Markov chain with an ergodic transition matrix should not influence the long-term behavior of the state distribution, as shown in the following example.

EXAMPLE 1   Calculating Long Term Distributions

A rental truck company has branches in Springfield, Tulsa, and Little Rock, and each rents trucks by the week. The trucks may be returned to any of the branches. The return location probabilities are shown in the following state-transition diagram. Find the distribution of the company's 180 trucks among the three cities after one year if the trucks are initially distributed as follows:

a. 90 in Springfield and 45 each in Tulsa and in Little Rock;

b. 120 in Springfield, 20 in Tulsa, and 40 in Little Rock;

c. all 180 in Tulsa.

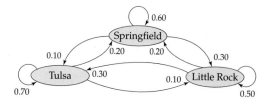

---

* The nineteenth-century physicist Ludwig Boltzmann (1844–1906) created the adjective "ergodic" from the Greek words *erg* (work) and *ode* (path) when formulating his "ergodic hypothesis" of gas dynamics to explain the "steady-state" uniformity of a gas in terms of the motion of the individual molecules.

Solution

We will describe what happens in terms of the *proportion* of trucks in each city, with S_1, S_2, and S_3 standing for Springfield, Tulsa, and Little Rock, respectively. From the state-transition diagram, the transition matrix is

$$T = \begin{pmatrix} 0.60 & 0.10 & 0.30 \\ 0.20 & 0.70 & 0.10 \\ 0.20 & 0.30 & 0.50 \end{pmatrix} \quad \begin{matrix} \text{Springfield} \\ \text{Tulsa} \\ \text{Little Rock} \end{matrix}$$

Since T is ergodic (there are no zero entries), the distribution of trucks among the rental offices over successive weeks should show that they "mix around" in some predictable fashion rather than "bunching up" in one city. For each initial state distribution D_0, the distribution after one year is the result of 52 weekly transitions:

$$D_{52} = D_0 \cdot T^{52}$$

a. For the numbers of trucks given in part (a), the initial state distribution vector is

$$D_0 = \left(\tfrac{90}{180} \quad \tfrac{45}{180} \quad \tfrac{45}{180} \right) = \left(\tfrac{1}{2} \quad \tfrac{1}{4} \quad \tfrac{1}{4} \right) \qquad \begin{matrix} \text{Each initial number} \\ \text{divided by 180 trucks} \end{matrix}$$

Then, after 52 weekly transitions, the distribution will be

$$D_{52} = \left(\tfrac{1}{2} \quad \tfrac{1}{4} \quad \tfrac{1}{4} \right) \cdot \begin{pmatrix} 0.60 & 0.10 & 0.30 \\ 0.20 & 0.70 & 0.10 \\ 0.20 & 0.30 & 0.50 \end{pmatrix}^{52} = \left(\tfrac{1}{3} \quad \tfrac{7}{18} \quad \tfrac{5}{18} \right) \qquad \begin{matrix} \text{Using a} \\ \text{calculator} \end{matrix}$$

After one year there will be $\tfrac{1}{3} \cdot 180 = 60$ trucks in Springfield, $\tfrac{7}{18} \cdot 180 = 70$ trucks in Tulsa, and $\tfrac{5}{18} \cdot 180 = 50$ trucks in Little Rock.

b. The initial numbers of trucks in part (b) give $D_0 = \left(\tfrac{120}{180} \quad \tfrac{20}{180} \quad \tfrac{40}{180} \right) = \left(\tfrac{2}{3} \quad \tfrac{1}{9} \quad \tfrac{2}{9} \right)$, so

$$D_{52} = \left(\tfrac{2}{3} \quad \tfrac{1}{9} \quad \tfrac{2}{9} \right) \cdot \begin{pmatrix} 0.60 & 0.10 & 0.30 \\ 0.20 & 0.70 & 0.10 \\ 0.20 & 0.30 & 0.50 \end{pmatrix}^{52} = \left(\tfrac{1}{3} \quad \tfrac{7}{18} \quad \tfrac{5}{18} \right) \qquad \begin{matrix} \text{Using a} \\ \text{calculator} \end{matrix}$$

This is the same distribution found in part (a), so again there will be 60 trucks in Springfield, 70 trucks in Tulsa, and 50 trucks in Little Rock

c. For the initial numbers in part (c), we have $D_0 = \left(0 \quad \tfrac{180}{180} \quad 0 \right) = (0 \quad 1 \quad 0)$, so

$$D_{52} = \left(0 \quad 1 \quad 0 \right) \cdot \begin{pmatrix} 0.60 & 0.10 & 0.30 \\ 0.20 & 0.70 & 0.10 \\ 0.20 & 0.30 & 0.50 \end{pmatrix}^{52} = \left(\tfrac{1}{3} \quad \tfrac{7}{18} \quad \tfrac{5}{18} \right) \qquad \begin{matrix} \text{Using a} \\ \text{calculator} \end{matrix}$$

Again we get the same distribution as in parts (a) and (b), so again there will be 60 trucks in Springfield, 70 trucks in Tulsa, and 50 trucks in Little Rock.

As we expected, the long-term distribution of the trucks did not depend on the initial distribution. Moreover, further mixing will not change the result:

$$\left(\tfrac{1}{3} \quad \tfrac{7}{18} \quad \tfrac{5}{18}\right) \cdot \begin{pmatrix} 0.60 & 0.10 & 0.30 \\ 0.20 & 0.70 & 0.10 \\ 0.20 & 0.30 & 0.50 \end{pmatrix} = \left(\tfrac{1}{3} \quad \tfrac{7}{18} \quad \tfrac{5}{18}\right)$$

This distribution remains unchanged by another transition

A distribution such as this that remains unchanged by a transition is called a *steady-state distribution*.

Steady-State Distribution

> A state distribution vector D is a *steady-state distribution* for the Markov chain with transition matrix T if $D \cdot T = D$.

PRACTICE PROBLEM 1

Is $D = \left(\tfrac{1}{2} \quad \tfrac{1}{2}\right)$ a steady-state distribution for the Markov chain with ergodic transition matrix $T = \begin{pmatrix} \tfrac{1}{4} & \tfrac{3}{4} \\ \tfrac{3}{4} & \tfrac{1}{4} \end{pmatrix}$?   *Solution at the back of the book*

If the ergodic Markov chain has a steady-state distribution, then the higher powers T^k of the transition matrix T must similarly "settle down" to represent the same mixing transformation.

Graphing Calculator Exploration

To see whether the higher powers of the transition matrix for the truck rental company in Example 1 have this property, we proceed as follows:

a. Enter the matrix $T = \begin{pmatrix} 0.60 & 0.10 & 0.30 \\ 0.20 & 0.70 & 0.10 \\ 0.20 & 0.30 & 0.50 \end{pmatrix}$ in matrix [A].

```
[A]
      [[.6 .1 .3]
       [.2 .7 .1]
       [.2 .3 .5]]
```

b. Find T^k for $k = 4, 6, 8,$ and 52. Display these values in fraction form if possible.

```
[A]^4►Frac
[[219/625 431/1…
 [203/625 527/1…
 [203/625 99/25…
```

```
[A]^6►Frac
[[.336064 .3791…
 [.331968 .3954…
 [.331968 .3913…
```

```
[A]^8►Frac
[[.33377024 .38…
 [.33311488 .39…
 [.33311488 .38…
```

```
[A]^52►Frac
[[1/3 7/18 5/18…
 [1/3 7/18 5/18…
 [1/3 7/18 5/18…
```

The last display shows T^{52}. Notice that each row is the steady-state distribution $\left(\tfrac{1}{3} \quad \tfrac{7}{18} \quad \tfrac{5}{18}\right)$ that we found on the previous page.

Why does each column "settle down" to a single value? Consider the product of one row of T with one column of T^*. If all the numbers in the column are the same, their product with the row will also be this same number because the row numbers add up to one. For instance, using the second row of the matrix from Example 1,

$$(0.2 \quad 0.7 \quad 0.1) \cdot \begin{pmatrix} t \\ t \\ t \end{pmatrix} = 0.2t + 0.7t + 0.1t = (0.2 + 0.7 + 0.1) \cdot t = t$$

Now suppose that the numbers in the column are different: Then there is a biggest value b and a smallest value s. Since some of the numbers in the column are smaller than b, the product of the row (with positive numbers adding to one) with the column must be smaller than the product of this same row with a column consisting of all b's:

$$(0.2 \quad 0.7 \quad 0.1) \cdot \begin{pmatrix} b \\ \vdots \\ s \end{pmatrix} < (0.2 \quad 0.7 \quad 0.1) \cdot \begin{pmatrix} b \\ b \\ b \end{pmatrix} = b$$

Similarly, the product of a row and column must be larger than the product of that row with a column consisting of all s's. Hence *every multiplication by T reduces the size of the largest entry in each column and increases the size of the smallest entry.* This means that as we raise T to higher and higher powers, the biggest and smallest values in each column will become closer and closer to each other, tending toward a single value, exactly as we saw in our calculator experiment. Thus, each of the resulting rows becomes the steady-state distribution, which is therefore unique. This reasoning would fail if any of the elements of the transition matrix were zero.

This fact is so important that it is called the *fundamental theorem* for ergodic Markov chains.

Fundamental Theorem of Ergodic Markov Chains

A Markov chain with an ergodic transition matrix T has exactly one steady-state distribution D solving $D \cdot T = D$. Higher powers T^* of the ergodic transition matrix approximate arbitrarily closely a matrix each of whose rows is equal to D.

How to Solve $D \cdot T = D$

How can we find the steady-state distribution D that solves $D \cdot T = D$ without having to calculate arbitrarily high powers of the transition matrix? We begin by writing the equation with zero on the right:

$$D \cdot T - D = 0 \qquad \text{From } D \cdot T = D$$

$$D \cdot (T - I) = 0 \qquad \text{Factoring (since } D \cdot I = D)$$

$$(T - I)^t \cdot D^t = 0 \qquad \text{Taking transposes since } (A \cdot B)^t = B^t \cdot A^t$$

$$(T^t - I) \cdot D^t = 0 \qquad \text{Since } (A - B)^t = A^t - B^t \text{ and } I^t = I$$

Since D is a state distribution vector, the sum of the entries of D must be 1, which we can write as the matrix equation

$$(1 \cdots 1) \cdot D^t = 1 \qquad \text{Entries of } D \text{ add up to 1}$$

Combining these last two matrix equations into one large matrix equation, we need to solve

$$\begin{pmatrix} T^t - I \\ 1 \cdots 1 \end{pmatrix} \cdot D^t = \begin{pmatrix} 0 \\ 1 \end{pmatrix} \qquad \begin{pmatrix} 0 \\ 1 \end{pmatrix} \text{ means } \begin{pmatrix} 0 \\ \vdots \\ 0 \\ 1 \end{pmatrix}$$

Since the coefficient matrix of this equation is *not* square, we must solve for D^t by row-reducing the augmented matrix $\left(\begin{array}{c|c} T^t - I & 0 \\ \hline 1 \cdots 1 & 1 \end{array} \right)$ to obtain $\left(\begin{array}{c|c} I & D^t \\ \hline 0 & 0 \end{array} \right)$, just as we solved equations by matrix row reduction in Section 3 of Chapter 3 (see page 201).

EXAMPLE 2 **Solving $D \cdot T = D$**

Find the steady-state distribution $D = (d_1 \ \ d_2 \ \ d_3)$ for the ergodic transition matrix $T = \begin{pmatrix} 0.6 & 0.1 & 0.3 \\ 0.2 & 0.7 & 0.1 \\ 0.2 & 0.3 & 0.5 \end{pmatrix}$ from the truck rental situation described in Example 1.

Solution

Following the procedure outlined above, we transpose T by changing rows into columns and then calculate $T^t - I$:

$$T^t - I = \begin{pmatrix} 0.6 & 0.2 & 0.2 \\ 0.1 & 0.7 & 0.3 \\ 0.3 & 0.1 & 0.5 \end{pmatrix} - \begin{pmatrix} 1 & 0 & 0 \\ 0 & 1 & 0 \\ 0 & 0 & 1 \end{pmatrix} = \begin{pmatrix} -0.4 & 0.2 & 0.2 \\ 0.1 & -0.3 & 0.3 \\ 0.3 & 0.1 & -0.5 \end{pmatrix}$$

We need to solve the matrix equation

$$\begin{pmatrix} -0.4 & 0.2 & 0.2 \\ 0.1 & -0.3 & 0.3 \\ 0.3 & 0.1 & -0.5 \\ 1 & 1 & 1 \end{pmatrix} \begin{pmatrix} d_1 \\ d_2 \\ d_3 \end{pmatrix} = \begin{pmatrix} 0 \\ 0 \\ 0 \\ 1 \end{pmatrix} \qquad \begin{pmatrix} T^t - I \\ 1 \cdots 1 \end{pmatrix} \cdot D^t = \begin{pmatrix} 0 \\ 1 \end{pmatrix}$$

We row-reduce the augmented matrix

$$\left(\begin{array}{ccc|c} -0.4 & 0.2 & 0.2 & 0 \\ 0.1 & -0.3 & 0.3 & 0 \\ 0.3 & 0.1 & -0.5 & 0 \\ 1 & 1 & 1 & 1 \end{array} \right) \quad \text{to obtain} \quad \left(\begin{array}{ccc|c} 1 & 0 & 0 & \frac{1}{3} \\ 0 & 1 & 0 & \frac{7}{18} \\ 0 & 0 & 1 & \frac{5}{18} \\ 0 & 0 & 0 & 0 \end{array} \right) \qquad \begin{array}{c} \left(\begin{array}{c|c} T^t - I & 0 \\ 1 \cdots 1 & 1 \end{array} \right) \\ \text{reduces to} \\ \left(\begin{array}{c|c} I & D^t \\ 0 & 0 \end{array} \right) \end{array}$$

(omitting the details). The last column (omitting the bottom zero) is the transpose D^t of the steady-state distribution

$$D = \begin{pmatrix} \frac{1}{3} & \frac{7}{18} & \frac{5}{18} \end{pmatrix}$$

This answer agrees with the fractions of the 180 trucks in the three cities that we found in Example 1 and also with the high powers of the transition matrix calculated in the Graphing Calculator Exploration on page 525.   ◼

Graphing Calculator Exploration

We may solve Example 2 using a graphing calculator. The program* MARKOV displays powers of a stochastic matrix T and the augmented matrix $\left(\begin{array}{c|c} T^t - I & 0 \\ 1 \cdots 1 & 1 \end{array} \right)$ together with its reduced row echelon form showing the steady-state distribution. To find the steady-state distribution in Example 2, we proceed as follows:

a.   Enter the stochastic matrix T in matrix [A].

```
MATRIX[A]  3 X3
[.6   .1   .3   ]
[.2   .7   .1   ]
[.2   .3   .5   ]



3,3=.5
```

b.   Run the program MARKOV and press ENTER several times to display T^k for $k = 1, 2, 3, 32,$ and $64,$ and then $\left(\begin{array}{c|c} T^t - I & 0 \\ 1 \cdots 1 & 1 \end{array} \right)$ followed by its reduced form.

*See the Preface for information on how to obtain this and other graphing calculator programs.

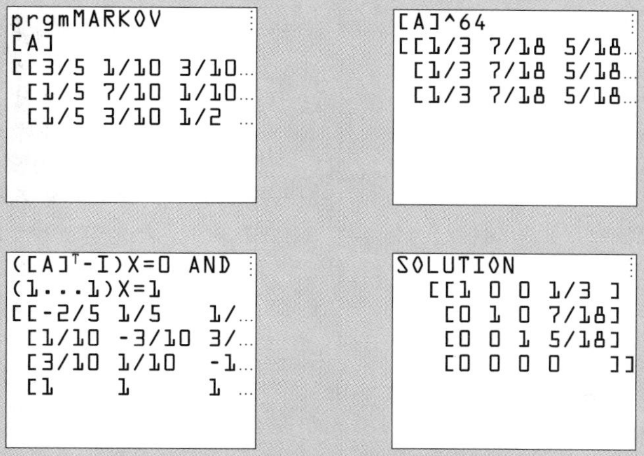

```
prgmMARKOV
[A]
[[3/5 1/10 3/10...
 [1/5 7/10 1/10...
 [1/5 3/10 1/2 ...
```

```
[A]^64
[[1/3 7/18 5/18...
 [1/3 7/18 5/18...
 [1/3 7/18 5/18...
```

```
([A]^T-I)X=0 AND
(1...1)X=1
[[-2/5 1/5    1/...
 [1/10 -3/10 3/...
 [3/10 1/10  -1...
 [1    1      1 ...
```

```
SOLUTION
[[1 0 0 1/3 ]
 [0 1 0 7/18]
 [0 0 1 5/18]
 [0 0 0 0   ]]
```

The steady state distribution $\left(\frac{1}{3} \quad \frac{7}{18} \quad \frac{5}{18}\right)$ shown in the right-hand column agrees with the answer found earlier.

Regular Transitions

Not all mixing transitions are ergodic. However, if it is possible to get from any state to every other state by *some* number of transitions, then a *power* of the transition matrix will be ergodic and the steady-state distribution will exist.

Regular Matrix

A stochastic matrix is *regular* if some power of it is ergodic.

A Markov chain is regular if it has a regular transition matrix.

EXAMPLE 3   Regular Matrices

Which of the following stochastic matrices are regular?

a. $A = \begin{pmatrix} \frac{1}{2} & \frac{1}{2} \\ 1 & 0 \end{pmatrix}$
     b. $B = \begin{pmatrix} \frac{1}{2} & \frac{1}{2} \\ 0 & 1 \end{pmatrix}$
     c. $C = \begin{pmatrix} 1 & 0 \\ 0 & 1 \end{pmatrix}$

Solution

We will first find the answer by calculating powers of the matrix and then check it by drawing the corresponding state-transition diagram.

a. Since all entries of A^2 (calculated below) are positive, the second power of A is ergodic, so A is regular.

$$A^2 = \begin{pmatrix} \frac{1}{2} & \frac{1}{2} \\ 1 & 0 \end{pmatrix} \cdot \begin{pmatrix} \frac{1}{2} & \frac{1}{2} \\ 1 & 0 \end{pmatrix} = \begin{pmatrix} \frac{3}{4} & \frac{1}{4} \\ \frac{1}{2} & \frac{1}{2} \end{pmatrix}$$

The corresponding state-transition diagram shows that this is a mixing transition. Any state can be reached from every state in two steps.

b. Since the entry in the second row and first column of B^k remains zero with each multiplication, B is not regular.

$$B = \begin{pmatrix} \frac{1}{2} & \frac{1}{2} \\ 0 & 1 \end{pmatrix}, B^2 = \begin{pmatrix} \frac{1}{4} & \frac{3}{4} \\ 0 & 1 \end{pmatrix}, B^3 = \begin{pmatrix} \frac{1}{8} & \frac{7}{8} \\ 0 & 1 \end{pmatrix}, \dots$$

Do you see why this entry will be 0 for *any* power of B?

The corresponding state-transition diagram shows that S_2 is an absorbing state for this transition, so S_1 is not reachable from S_2 in any number of steps.

c. Since every power of the identity matrix is still the identity matrix, $C = I$ is not regular. The state-transition diagram for C shows that both S_1 and S_2 are absorbing states.

The long-term behavior of a *regular* Markov chain can be calculated from the transition matrix in the same way as it is found for an ergodic transition.

The Steady-State Distribution for a Regular Markov Chain

The steady-state distribution D for a *regular* Markov chain with transition matrix T may be found by row-reducing the augmented matrix

$$\left(\begin{array}{c|c} T^t - I & 0 \\ \hline 1 \cdots 1 & 1 \end{array} \right) \qquad \text{to obtain} \qquad \left(\begin{array}{c|c} I & D^t \\ \hline 0 & 0 \end{array} \right)$$

and then transposing all but the bottom entry in the last column.

EXAMPLE 4 **Finding the Steady-State Distribution for a Regular Markov Chain**

Find the steady-state distribution D for the Markov chain with transition matrix $T = \begin{pmatrix} \frac{1}{3} & \frac{2}{3} \\ 1 & 0 \end{pmatrix}$.

Solution

$$T^t - I = \begin{pmatrix} \frac{1}{3} & 1 \\ \frac{2}{3} & 0 \end{pmatrix} - \begin{pmatrix} 1 & 0 \\ 0 & 1 \end{pmatrix} = \begin{pmatrix} -\frac{2}{3} & 1 \\ \frac{2}{3} & -1 \end{pmatrix} \text{ so } \left(\frac{T^t - I}{1 \cdots 1} \,\middle|\, \frac{0}{1} \right) = \left(\begin{array}{cc|c} -\frac{2}{3} & 1 & 0 \\ \frac{2}{3} & -1 & 0 \\ \hline 1 & 1 & 1 \end{array} \right)$$

There are many different sequences of row operations to reduce this augmented matrix, and all reach the same conclusion. One possible way is as follows:

$$\begin{pmatrix} -\frac{2}{3} & 1 & 0 \\ 0 & 0 & 0 \\ \frac{1}{3} & 2 & 1 \end{pmatrix} \begin{matrix} \\ R_2' = R_2 + R_1 \\ R_3' = R_3 + R_1 \end{matrix} \rightarrow \begin{pmatrix} 0 & 5 & 2 \\ 0 & 0 & 0 \\ \frac{1}{3} & 2 & 1 \end{pmatrix} R_1' = R_1 + 2R_3$$

$$\rightarrow \begin{pmatrix} \frac{1}{3} & 2 & 1 \\ 0 & 5 & 2 \\ 0 & 0 & 0 \end{pmatrix} \begin{matrix} R_1' = R_3 \\ R_2' = R_1 \\ R_3' = R_2 \end{matrix} \rightarrow \begin{pmatrix} 5 & 0 & 3 \\ 0 & 5 & 2 \\ 0 & 0 & 0 \end{pmatrix} R_1' = 15R_1 - 6R_2$$

$$\rightarrow \begin{pmatrix} 1 & 0 & \frac{3}{5} \\ 0 & 1 & \frac{2}{5} \\ 0 & 0 & 0 \end{pmatrix} \begin{matrix} R_1' = \frac{1}{5}R_1 \\ R_2' = \frac{1}{5}R_2 \end{matrix}$$

D^t is in the right hand column, above the 0

Transposing the upper entries in the right-hand column, we have found that the steady-state distribution for this regular Markov chain is $D = \begin{pmatrix} \frac{3}{5} & \frac{2}{5} \end{pmatrix}$.

∎

We could also have found this steady state solution by using the program MARKOV, as in the preceding graphing calculator exploration.

PRACTICE PROBLEM 2

Verify that the solution $D = \begin{pmatrix} \frac{3}{5} & \frac{2}{5} \end{pmatrix}$ to Example 4 satisfies the matrix equation $D \cdot \begin{pmatrix} \frac{1}{3} & \frac{2}{3} \\ 1 & 0 \end{pmatrix} = D$.

Solution at the back of the book

Section Summary

A stochastic matrix is *ergodic* if every entry is positive, and *regular* if some power of it is ergodic. The long-term behavior of an ergodic or regular Markov chain with transition matrix T is the solution D of the *steady-state distribution* equation

$$D \cdot T = D$$

D may be found by row-reducing the augmented matrix

$$\left(\begin{array}{c|c} T^t - I & 0 \\ \hline 1 \cdots 1 & 1 \end{array} \right) \quad \text{to obtain} \quad \left(\begin{array}{c|c} I & D^t \\ \hline 0 & 0 \end{array} \right)$$

and then transposing the last column (omitting the bottom zero).

EXERCISES 7.2

Identify each Markov transition matrix as "ergodic," "regular," or "neither."

1. $\begin{pmatrix} 0.15 & 0.85 \\ 0.40 & 0.60 \end{pmatrix}$

2. $\begin{pmatrix} 0 & 0 & 1 \\ 0 & 0 & 1 \\ 0.50 & 0.50 & 0 \end{pmatrix}$

3. $\begin{pmatrix} 0.75 & 0.25 \\ 1 & 0 \end{pmatrix}$

4. $\begin{pmatrix} 0 & 0.40 & 0.60 \\ 0 & 0.80 & 0.20 \\ 1 & 0 & 0 \end{pmatrix}$

5. $\begin{pmatrix} 0.75 & 0.25 \\ 0 & 1 \end{pmatrix}$

6. $\begin{pmatrix} 1 & 0 & 0 \\ 0 & 0.80 & 0.20 \\ 0 & 0.40 & 0.60 \end{pmatrix}$

7. $\begin{pmatrix} 0.70 & 0.30 & 0 \\ 0 & 0.30 & 0.70 \\ 0.70 & 0.30 & 0 \end{pmatrix}$

8. $\begin{pmatrix} 1 & 0 & 0 \\ 0 & 0 & 1 \\ 0 & 1 & 0 \end{pmatrix}$

For each Markov chain transition matrix T:

a. Find the steady-state distribution.

b. Calculate T^k for $k = 5, 10, 20,$ and 50 to verify that the rows of T^k approach the steady-state distribution.

9. $\begin{pmatrix} 0.40 & 0.60 \\ 0.60 & 0.40 \end{pmatrix}$

10. $\begin{pmatrix} 0.60 & 0.40 \\ 0.10 & 0.90 \end{pmatrix}$

11. $\begin{pmatrix} 0.55 & 0.45 \\ 0.30 & 0.70 \end{pmatrix}$

12. $\begin{pmatrix} 0.45 & 0.55 \\ 0.70 & 0.30 \end{pmatrix}$

13. $\begin{pmatrix} 0.10 & 0.80 & 0.10 \\ 0.20 & 0.70 & 0.10 \\ 0.40 & 0.50 & 0.10 \end{pmatrix}$

14. $\begin{pmatrix} 0.70 & 0.20 & 0.10 \\ 0.10 & 0.30 & 0.60 \\ 0.50 & 0.30 & 0.20 \end{pmatrix}$

15. $\begin{pmatrix} 0 & 0.60 & 0.40 \\ 0.10 & 0.90 & 0 \\ 0.30 & 0.50 & 0.20 \end{pmatrix}$

16. $\begin{pmatrix} 0 & 0.60 & 0.40 \\ 0.20 & 0.80 & 0 \\ 0 & 0.60 & 0.40 \end{pmatrix}$

For each regular Markov chain transition matrix T:

a. Construct a state-transition diagram and find the smallest positive number k of transitions needed to move from the given state S_i to state S_j.

b. Use this k to verify that T^k has a positive transition probability in row i and column j.

17. $\begin{pmatrix} 0 & 0 & 1 \\ 0.50 & 0.50 & 0 \\ 0 & 1 & 0 \end{pmatrix}, S_2 \to S_3$

18. $\begin{pmatrix} 0 & 0.50 & 0.50 \\ 1 & 0 & 0 \\ 0.50 & 0 & 0.50 \end{pmatrix}, S_2 \to S_2$

19. $\begin{pmatrix} 0 & 0 & 1 \\ 0.50 & 0 & 0.50 \\ 0 & 1 & 0 \end{pmatrix}, S_1 \to S_1$

20. $\begin{pmatrix} 0 & 0 & 0 & 1 \\ 0 & 0 & 1 & 0 \\ 0.50 & 0 & 0 & 0.50 \\ 0 & 1 & 0 & 0 \end{pmatrix}, S_4 \to S_1$

Represent each situation as a Markov chain by constructing a state-transition diagram and the corresponding transition matrix. Find the steady-state distribution and interpret it in terms of the original situation. Be sure to state your final answer in terms of the original question.

21. **Voting Patterns** The students in the political science summer program at Edson State College are studying voting patterns in Marston County. Half of the students reviewed voter records at the County Clerk's Office and found that a person who voted in an election has an 80% chance of voting in the next election, while someone who did not vote in an election has a 30% chance of voting in the next election. The other half of the students conducted surveys door to door and at shopping centers and found that voter perceptions were somewhat different: 90% of those who claimed to have voted in the last election said they would vote in the next, while 40% of those who said they hadn't voted in the last election said they would vote in the next. If these findings are valid for predicting long-term trends, which survey is consistent with the national average of about 61% of eligible voters actually voting? (*Source: Committee for the Study of the American Electorate*)

22. **Population Dynamics** Life is so good in Lucas, Marion, and Warren counties in upstate Maine that no one ever moves away from the tri-county region. However, each year 4% of the Lucas residents move to Marion and 2% move to Warren; 2% of the Marion residents move to Lucas and 2% move to Warren; and 2% of the Warren residents move to Lucas and 1% move to Marion. After many years, how many reside in each county if the combined population of these three counties is 11,200?

23. **Mass Transit** Commuters in the Pittsburgh metropolitan area either drive alone, join a car pool, or take the bus. Each month, of those who drive by themselves, 20% join a car pool, 30% switch to the bus, and the rest continue driving alone; of those who are in a car pool, 30% switch to driving alone, 20% switch to the bus, and the rest stay in a car pool; and of those who take the bus, 20% switch to driving alone, 30% join a car pool, and the rest continue taking the bus. After many months, how many of the three million commuters will be driving alone?

24. **Art Gallery Shows** The Harmon Gallery's showing of paintings by Clyberg, Stevensen, and Georgan attracted 1100 art dealers and collectors on its opening night. The security staff reported the following pattern of crowd movement every five minutes throughout the evening: Of the crowd in the Clyberg room, 10% stayed, 30% moved on to the Stevensen exhibit, and 60% went for more hors d'oeuvres. Of those in the Stevensen room, 30% stayed, 30% moved to the Clyberg exhibit, 20% moved to the Georgan exhibit, and 20% went for more hors d'oeuvres. Of those in the Georgan room, 20% stayed, 70% moved on to the Stevensen exhibit, and 10% went for more hors d'oeuvres. And of those in the central refreshments room, 40% stayed for more hors d'oeuvres, 10% went to the Clyberg exhibit, 20% went to the Stevensen exhibit, and 30% went to the Georgan exhibit. After several hours of milling about in this fashion, how many people were in each room at the gallery?

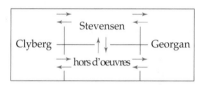

25. **Market Share** The Peerless Products Corporation has decided to market a new toothpaste, DentiMint, designed to compete successfully with the market leaders: TasteBest, SuperSmile, and MaxiWhite. Test marketing results from several cities indicate that each week, of those who used TasteBest the previous week, 40% will buy it again, 20% will switch to SuperSmile, 30% will switch to MaxiWhite, and 10% will switch to DentiMint. Of those who used SuperSmile the previous week, 40% will buy it again, 20% will switch to TasteBest, 10% will switch to MaxiWhite, and 30% will switch to DentiMint. Of those who used MaxiWhite the previous week, 40% will buy it again, 30% will switch to TasteBest, 10% will switch to SuperSmile, and 20% will switch to DentiMint. And of those who used DentiMint the previous week, 40% will buy it again, 10% will switch to TasteBest, 10% will switch to SuperSmile, and 40% will switch to MaxiWhite. If these buying patterns continue, what will the long-term market share be for DentiMint toothpaste?

26. **Apple Harvests** The apple harvest in the Shenandoah Valley is rated as excellent, average, or poor. Following an excellent harvest, the chances of having an excellent, average, or poor harvest are 0.50, 0.30, and 0.20, respectively. Following an average harvest, the chances of having an excellent, average, or poor harvest are 0.30, 0.50, and 0.20, respectively. And following a poor harvest, the chances of having an excellent, average, or poor harvest are 0.60, 0.20, and 0.20, respectively. Assuming these trends continue, find the long-term chance for an excellent harvest.

27. **Car Insurance** The records of an insurance broker in Boston classify the auto policy holders as preferred, satisfactory, poor, or in the assigned-risk pool. Each year 20% of the preferred policies are downgraded to satisfactory; 30% of the satisfactory policies are upgraded to preferred and another 20% are downgraded to poor; 60% of the poor policies are upgraded to satisfactory but 30% are placed in the assigned-risk pool; and just 20% of those in the assigned-risk pool are moved up to the poor classification. Assuming that these trends continue for many years, what percentage of the broker's auto policies are rated satisfactory or better?

28. **Rumor Accuracy** When repeating a rumor, gossips sometimes make mistakes and spread misinformation instead. Suppose the residents of a small town repeat a rumor that the mayor has been convicted of tax fraud at a trial held in the state capital, but that when they pass it on, 1% of the gossips reverse the result they were told. What is the long-term accuracy of the gossips' information?

29. **Renters Who Move** The rental records at a Denver apartment building show that each year 78% of those who lived in the building the entire previous year will remain for the next year, while only 44% of those who moved in during the year will remain for the next year. If these observations represent a long-term trend, are consistent with the U.S. Census statistic that about 34.3% of renters nationwide move each year?

30. **Homeowners Who Move** The property transfer records at a real estate agency in the Salt Lake City suburbs show that each year 96% of those who owned their own house the entire previous year will remain for the next year, while only 46% of those who moved into their own house during the year will remain for the next year. If these observations represent a long-term trend, are they consistent with the U.S. Census statistic that about 8.9% of homeowners nationwide move each year?

Explorations and Excursions

The following problems extend and augment the material presented in the text.

More About Steady-State Vectors

For each *oscillating* Markov chain transition matrix T:

a. Construct a state-transition diagram and find the *period* of the oscillation (that is, the smallest number k of transitions that return any state distribution vector to its initial form).

b. Verify that $T^j \neq T$ for $1 < j \leq k$ and that $T^{k+1} = T$, where k was found in part (a).

c. Find the "steady-state distribution" D by solving the matrix equation $D \cdot T = D$ using the row reduction method we used for regular Markov chains.

d. *Discuss:* Is D unique? Does D depend on the period of T? Is there a reasonable interpretation for D as a measure of a long-term observation of the oscillation?

31. $\begin{pmatrix} 0 & 1 & 0 \\ 0 & 0 & 1 \\ 1 & 0 & 0 \end{pmatrix}$

32. $\begin{pmatrix} 0 & 0 & 1 \\ 1 & 0 & 0 \\ 0 & 1 & 0 \end{pmatrix}$

33. $\begin{pmatrix} 0 & 1 & 0 & 0 \\ 0 & 0 & 1 & 0 \\ 0 & 0 & 0 & 1 \\ 1 & 0 & 0 & 0 \end{pmatrix}$

34. $\begin{pmatrix} 0 & 0 & 0 & 1 \\ 1 & 0 & 0 & 0 \\ 0 & 1 & 0 & 0 \\ 0 & 0 & 1 & 0 \end{pmatrix}$

35. a. $\begin{pmatrix} 0 & 0.40 & 0.60 \\ 1 & 0 & 0 \\ 1 & 0 & 0 \end{pmatrix}$

b. $\begin{pmatrix} 0 & t & 1-t \\ 1 & 0 & 0 \\ 1 & 0 & 0 \end{pmatrix}$ where $0 < t < 1$.

For each *absorbing* Markov chain transition matrix T:

a. Construct a state-transition diagram and identify the absorbing states.

b. Show that the state distribution vector D with a 1 in a position corresponding to an absorbing state and 0 everywhere else is a "steady-state distribution" for T because it satisfies $D \cdot T = D$.

c. Row-reduce $\left(\begin{array}{c|c} T^t - I & 0 \\ \hline 1 \cdots 1 & 1 \end{array} \right)$ and find all solutions of the matrix equation $D \cdot T = D$. [*Hint:* Review "dependent equations" and "parameterized solutions" in Example 1 from Section 3 of Chapter 3 on page 192.]

d. *Discuss:* What is the connection between the solutions in parts (b) and (c)? How many "steady-state distributions" does an absorbing Markov chain have?

36. $\begin{pmatrix} 1 & 0 & 0 \\ 0 & 1 & 0 \\ \frac{1}{4} & \frac{1}{4} & \frac{1}{2} \end{pmatrix}$

37. $\begin{pmatrix} 1 & 0 & 0 \\ \frac{1}{3} & \frac{1}{3} & \frac{1}{3} \\ 0 & 0 & 1 \end{pmatrix}$

38. $\begin{pmatrix} 1 & 0 & 0 & 0 \\ 0.25 & 0.50 & 0.25 & 0 \\ 0 & 0 & 1 & 0 \\ 0 & 0.50 & 0 & 0.50 \end{pmatrix}$

39. $\begin{pmatrix} 0 & 0 & 1 & 0 \\ 0 & 1 & 0 & 0 \\ 0 & 0.25 & 0.50 & 0.25 \\ 0 & 0 & 0 & 1 \end{pmatrix}$

40. $\begin{pmatrix} 0.60 & 0.15 & 0.05 & 0.20 \\ 0 & 1 & 0 & 0 \\ 0 & 0 & 1 & 0 \\ 0.10 & 0.05 & 0.15 & 0.70 \end{pmatrix}$

For each Markov chain transition matrix T, calculate T^2, T^3, and T^4. If T is ergodic, verify that as the exponent increases, the biggest entry in each column decreases and the smallest entry in each column increases. If T is neither ergodic nor regular, identify a column whose biggest entry stays the same and a column whose smallest entry stays the same as the exponent increases.

41. $\begin{pmatrix} 0.10 & 0.90 \\ 0.90 & 0.10 \end{pmatrix}$

42. $\begin{pmatrix} 0.40 & 0.60 \\ 0.20 & 0.80 \end{pmatrix}$

43. $\begin{pmatrix} 1 & 0 & 0 \\ 0.50 & 0.50 & 0 \\ 0 & 0.20 & 0.80 \end{pmatrix}$

44. $\begin{pmatrix} 0.90 & 0 & 0.10 \\ 0.20 & 0.80 & 0 \\ 0 & 0 & 1 \end{pmatrix}$

45. $\begin{pmatrix} 0.50 & 0.40 & 0.10 \\ 0.10 & 0.50 & 0.40 \\ 0.30 & 0.60 & 0.10 \end{pmatrix}$

7.3 **Absorbing Markov Chains**

Random Walks

Suppose you are standing midway between your mathematics classroom and the cafeteria, and with each flip of a coin, you take one step toward the cafeteria if it is heads or one step toward the classroom if it is tails. You continue this process until you arrive at either the classroom or the cafeteria, where you stay. Your position is then a *random walk* and might look like the graph shown below.

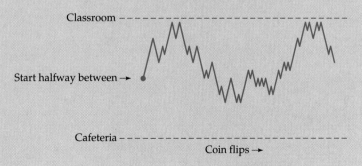

Random walks are Markov chains with "absorbing" states, which are the subject of this section and lead to many questions.* How many steps can you expect to take before getting to either destination? Which destination is more likely?

Because such random walk graphs resemble stock market fluctuations, they have been used to model economic behavior. For an entertaining and influential account of such models, see *A Random Walk Down Wall Street* by Burton G. Malkiel, first published in 1973 and now in its sixth edition.

Introduction

A state is *absorbing* if, once you enter it, you cannot leave it. If a Markov chain has several absorbing states, will it always end up in one of them? If so, which one and how soon? Unlike regular Markov chains, where the initial state distribution has no influence on the long-term outcome, the initial distribution *does* affect when and where the eventual absorption occurs. As in the regular case, we will see that much useful information is given by high powers of the transition matrix.

Absorbing Markov Chains

The transition probabilities of an *absorbing state* are 1 from that state to itself and 0 to every other state. A Markov chain is *absorbing* if it has at least one absorbing state *and* there is a positive integer k such that every nonabsorbing state has a positive probability of reaching an absorbing state after k transitions.

EXAMPLE 1   Finding Whether a Markov Chain is Absorbing

Determine whether each Markov chain is absorbing.

a.   **b.**

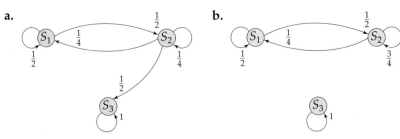

---

*Further information about the mathematical properties of random walks may be found in "Random Walks and Fluctuations," by Geoffrey C. Berresford, *Tools for Teaching, 1986* (Lexington, MA: COMAP).

Solution

a. This *is* an absorbing Markov chain since it has an absorbing state, S_3, which in $n = 2$ steps can be reached from S_1 $\left(\text{with probability } \frac{1}{2} \cdot \frac{1}{2} = \frac{1}{4}\right)$ and from S_2 $\left(\text{with probability } \frac{1}{4} \cdot \frac{1}{2} + \frac{1}{2} \cdot 1 = \frac{5}{8}\right)$.

b. This is *not* an absorbing Markov chain. Although S_3 is an absorbing state, it cannot be reached from either S_1 or S_2 (the nonabsorbing states).

EXAMPLE 2 **Wetlands Pollution**

The Portable Electric Company has filed an application with the Environmental Protection Agency to store heavy-metal waste (mostly cadmium and mercury) in a clay-lined storage pool at its manufacturing plant in Mayfield. An independent engineering consultant has certified that only 1% of the waste will leach out of the pool into the ground water each year. But a hydrologist working with the local chapter of People for a Cleaner Planet has pointed out that each year 2% of the contaminants in the ground water reach the Marlin Memorial Wetlands Conservation Area, where they remain indefinitely. Represent this situation by a transition diagram and matrix. Show that the wetlands are an absorbing state for the pollution.

Solution

The waste can be in one of three locations: at the factory (state *F*), in the ground water (state *G*), or in the wetlands ecosystem (state *W*). The given information is shown in the following state-transition diagram and matrix:

$$
\begin{array}{c}
 \begin{array}{ccc} F & G & W \end{array} \\
T = \begin{array}{c} F \\ G \\ W \end{array}\left(\begin{array}{ccc} 0.99 & 0.01 & 0 \\ 0 & 0.98 & 0.02 \\ 0 & 0 & 1 \end{array}\right)
\end{array}
$$

Since these changes occur each year, we have a Markov chain with transition matrix *T*. State *W* is an absorbing state because once the contamination reaches the Wetlands, it does not leave, and the other states *F* and *G* are nonabsorbing. The state-transition diagram shows that it is possible for some of the waste to move from the factory storage pool to the wetlands in just two transitions. In fact, squaring the transition matrix gives

$$T^2 = \begin{array}{c} \\ F \\ G \\ W \end{array} \begin{array}{ccc} F & G & W \\ \begin{pmatrix} 0.9801 & 0.0197 & 0.0002 \\ 0 & 0.9604 & 0.0396 \\ 0 & 0 & 1 \end{pmatrix} \end{array}$$

From the last column, every state has a positive probability of reaching the absorbing state in two transitions. Thus this is an absorbing Markov chain.

■

How much of the heavy-metal waste will reach the Wetlands, and how soon is it expected to get there? Will *all* the contamination ultimately reach the wetlands? In Example 4 on pages 541–542 we will answer these important questions.

Standard Form

The transition matrix of an absorbing Markov chain is in *standard form* if it is written so that the absorbing states are listed *before* the nonabsorbing states. Since the absorbing states themselves may be in any order (as may the nonabsorbing states), the standard form is not necessarily unique.

Standard Form of an Absorbing Transition Matrix

A $n \times n$ absorbing transition matrix T is in *standard form* if the absorbing states $A_1, \ldots, A_k$ appear before the nonabsorbing states $N_1, \ldots, N_{n-k}$, in which case T takes the form

$$T = \begin{array}{c} \\ A_1 \\ \vdots \\ A_k \\ \\ N_1 \\ \vdots \\ N_{n-k} \end{array} \begin{array}{cc} A_1 \ldots A_k & N_1 \ldots N_{n-k} \\ \left(\begin{array}{c|c} I & O \\ \hline R & Q \end{array} \right) \end{array}$$

I is an identity matrix, 0 is a zero matrix, *R* is the matrix of transition probabilities from the nonabsorbing states to the absorbing states, and *Q* is the matrix of transition probabilities from nonabsorbing states to nonabsorbing states.

Any absorbing transition matrix can be rewritten in standard form by reordering the rows to place the nonabsorbing states below the absorbing states and then reordering the columns to have the same order as the rows.

EXAMPLE 3   **Rewriting a Transition Matrix in Standard Form**

Rewrite the following absorbing transition matrix T in standard form and identify the matrices R and Q.

$$T = \begin{pmatrix} 0.80 & 0.03 & 0.10 & 0.02 & 0.05 \\ 0 & 1 & 0 & 0 & 0 \\ 0.10 & 0.07 & 0.20 & 0.03 & 0.60 \\ 0 & 0 & 0 & 1 & 0 \\ 0.05 & 0.01 & 0.10 & 0.04 & 0.80 \end{pmatrix}$$

Solution

We first identify the absorbing states. Since the row of an absorbing state consists of all zeros except one 1 in the *same* column as the row, the second and fourth rows represent absorbing states, which we will call A_1 and A_2. The other rows represent the nonabsorbing states, which we will call N_1, N_2 and N_3:

$$\begin{array}{c} N_1 \\ \text{Absorbing} \rightarrow A_1 \\ \text{rows (1's on} \\ \text{the main} \quad N_2 \\ \text{diagonal)} \rightarrow A_2 \\ N_3 \end{array} \begin{pmatrix} 0.80 & 0.03 & 0.10 & 0.02 & 0.05 \\ 0 & 1 & 0 & 0 & 0 \\ 0.10 & 0.07 & 0.20 & 0.03 & 0.60 \\ 0 & 0 & 0 & 1 & 0 \\ 0.05 & 0.01 & 0.10 & 0.04 & 0.80 \end{pmatrix}$$

Putting the rows in the order A_1, A_2, N_1, N_2, N_3 and *then* putting the columns in the same order, we have

$$\begin{array}{c} \\ A_1 \\ A_2 \\ N_1 \\ N_2 \\ N_3 \end{array} \begin{array}{ccccc} N_1 & A_1 & N_2 & A_2 & N_3 \\ \begin{pmatrix} 0 & 1 & 0 & 0 & 0 \\ 0 & 0 & 0 & 1 & 0 \\ 0.80 & 0.03 & 0.10 & 0.02 & 0.05 \\ 0.10 & 0.07 & 0.20 & 0.03 & 0.60 \\ 0.05 & 0.01 & 0.10 & 0.04 & 0.80 \end{pmatrix} \end{array} \quad \text{First fix the rows}$$

and then

$$\begin{array}{c} \\ A_1 \\ A_2 \\ N_1 \\ N_2 \\ N_3 \end{array} \begin{array}{ccccc} A_1 & A_2 & N_1 & N_2 & N_3 \\ \left(\begin{array}{cc|ccc} 1 & 0 & 0 & 0 & 0 \\ 0 & 1 & 0 & 0 & 0 \\ \hline 0.03 & 0.02 & 0.80 & 0.10 & 0.05 \\ 0.07 & 0.03 & 0.10 & 0.20 & 0.60 \\ 0.01 & 0.04 & 0.05 & 0.10 & 0.80 \end{array} \right) \end{array} \quad \text{Then fix the columns}$$

$$\underbrace{\qquad\qquad}_{R} \quad \underbrace{\qquad\qquad\qquad}_{Q}$$

The matrix R is the 3×2 matrix of transition probabilities from the nonabsorbing states to the absorbing states, while the matrix Q is the 3×3 matrix of transition probabilities from the nonabsorbing states to the nonabsorbing states:

$$R = \begin{array}{c} \\ N_1 \\ N_2 \\ N_3 \end{array} \begin{array}{cc} A_1 & A_2 \\ \begin{pmatrix} 0.03 & 0.02 \\ 0.07 & 0.03 \\ 0.01 & 0.04 \end{pmatrix} \end{array} \qquad Q = \begin{array}{c} \\ N_1 \\ N_2 \\ N_3 \end{array} \begin{array}{ccc} N_1 & N_2 & N_3 \\ \begin{pmatrix} 0.80 & 0.10 & 0.05 \\ 0.10 & 0.20 & 0.60 \\ 0.05 & 0.10 & 0.80 \end{pmatrix} \end{array}$$

∎

PRACTICE PROBLEM 1

Rewrite the transition matrix $\begin{array}{c} \\ F \\ G \\ W \end{array} \begin{array}{ccc} F & G & W \\ \begin{pmatrix} 0.99 & 0.01 & 0 \\ 0 & 0.98 & 0.02 \\ 0 & 0 & 1 \end{pmatrix} \end{array}$ from Example 2 in standard form.

Solution at the back of the book

Limiting Transition Matrix

Just as with regular Markov chains, higher powers of an absorbing transition matrix approach a "steady-state" matrix that gives important information about the ultimate behavior of the chain. The following results are established in Exercises 41–50.

Expected Times and Long-Term Absorption Probabilities

For an absorbing transition matrix in standard form,

$$T = \left(\begin{array}{c|c} I & 0 \\ \hline R & Q \end{array} \right)$$

Absorbing states listed before the nonabsorbing states

the powers T^k as k increases approach the matrix

$$T^* = \left(\begin{array}{c|c} I & 0 \\ \hline (I - Q)^{-1} \cdot R & 0 \end{array} \right)$$

T^* gives the "long-term" transition probabilities

The matrix $(I - Q)^{-1}$ is called the *fundamental matrix*, and it provides the following information.

1. The entry in row i and column j of the fundamental matrix $(I - Q)^{-1}$ is the expected number of times that the chain, if it begins in state N_i, will be in state N_j before being absorbed.

2. The sum of the row i entries of the fundamental matrix $(I - Q)^{-1}$ is the expected number of times the chain, if it begins in state N_i, will be in the nonabsorbing states before being absorbed.

3. The entry in row i and column j of the matrix $(I - Q)^{-1} \cdot R$ is the probability that the chain, if it begins in state N_i, will be absorbed into state A_j.

EXAMPLE 4   Finding Expected Time Until Absorption

Find the expected number of years until the heavy-metal waste in the factory storage pool (state F) will reach the Wetlands Conservation Area (state W) for the Markov chain described in Example 2 on pages 537–538.

Solution

Writing the transition matrix from Example 2 in standard form as

$$T = \begin{array}{c} \\ W \\ G \\ F \end{array} \begin{array}{c} W \quad G \quad F \\ \left(\begin{array}{c|cc} 1 & 0 & 0 \\ \hline 0.02 & 0.98 & 0 \\ 0 & 0.01 & 0.99 \end{array} \right) \end{array}$$

Listing first the absorbing state W (for "wetlands")

we have

$$Q = \begin{array}{c} G \\ F \end{array} \begin{array}{c} G \quad F \\ \left(\begin{array}{cc} 0.98 & 0 \\ 0.01 & 0.99 \end{array} \right) \end{array}$$

and

$$I - Q = \begin{pmatrix} 1 & 0 \\ 0 & 1 \end{pmatrix} - \begin{pmatrix} 0.98 & 0 \\ 0.01 & 0.99 \end{pmatrix} = \begin{pmatrix} 0.02 & 0 \\ -0.01 & 0.01 \end{pmatrix}$$

We calculate the fundamental matrix $(I - Q)^{-1}$ by row-reducing "$(A \mid I)$ to obtain $(I \mid A^{-1})$" (see pages 225–227 in Section 5 of Chapter 3).

$$\begin{pmatrix} 0.02 & 0 & | & 1 & 0 \\ -0.01 & 0.01 & | & 0 & 1 \end{pmatrix} \qquad (A \mid I)$$

$$\begin{pmatrix} 0.02 & 0 & | & 1 & 0 \\ 0 & 0.01 & | & \frac{1}{2} & 1 \end{pmatrix} R_2' = R_2 + \frac{1}{2}R_1$$

$$\begin{pmatrix} 1 & 0 & | & 50 & 0 \\ 0 & 1 & | & 50 & 100 \end{pmatrix} \begin{array}{l} R_1' = 50R_1 \\ R_2' = 100R_2 \end{array} \qquad (I \mid A^{-1})$$

The fundamental matrix is

$$(I - Q)^{-1} = \begin{array}{c} G \\ F \end{array} \begin{array}{c} G \quad F \\ \left(\begin{array}{cc} 50 & 0 \\ 50 & 100 \end{array} \right) \end{array}$$

By part (2) of the results on pages 540–541, the expected number of years until the heavy-metal waste in the factory storage pool (state F) will be absorbed into the Wetlands Conservation Area (state W) is simply the sum of the numbers in the F row of the fundamental matrix: $50 + 100 = 150$ years.

■

PRACTICE PROBLEM 2

How long is it expected that the heavy-metal waste will spend in the factory storage pool (state F) before absorption? How long in the ground water (state G)? [*Hint:* No work necessary. Just interpret the above fundamental matrix according to part (1) of the results on pages 540–541.]   *Solution at the back of the book*

Graphing Calculator Exploration

(*Continuation of Example 3*) To find the probability that the absorbing Markov chain in Example 3 (pages 539–540) will be absorbed into state A_2 given that it started in state N_3, proceed as follows.

a.   Enter the matrices R and Q from the solution of Example 3 into the calculator as [A] and [B], along with a 3×3 identity matrix in [I]:

```
MATRIXEA] 3 X3        MATRIXEB] 3 X2        MATRIXEI] 3 X3
[.8  .1  .05    ]     [.03 .02       ]      [1   0   0     ]
[.1  .2  .6     ]     [.07 .03       ]      [0   1   0     ]
[.05 .1  .8     ]     [.01 .04       ]      [0   0   1     ]

3,3=.8                3,2=.04               3,3=1
```

b.   Calculate $(I - Q)^{-1} \cdot R$ as $([I] - [A])^{-1} * [B]$:

```
([I]-[A])⁻¹*[B]
       [[.46  .54]
        [.43  .57]
        [.38  .62]]
```

By part (3) of the result on pages 540–541, the probability that the chain, beginning in state N_3, will be absorbed into state A_2 is the entry in row 3 and column 2 of $(I - Q)^{-1} \cdot R$, which is 0.62.

PRACTICE PROBLEM 3

From the last matrix in the preceding Graphing Calculator Exploration, what is the probability that the chain, if it begins in state N_1, will be absorbed into state A_2? *Solution at the back of the book*

Both the preceding Graphing Calculator Exploration and Practice Problem 3 found probabilities of being absorbed into state A_2, but beginning in different states. The fact that we obtained two different probabilities (0.62 and 0.54) shows that for an absorbing Markov chain, the initial state *does* affect the ultimate behavior. This is quite different from an *ergodic* or *regular* transition, where the initial state had no effect on the ultimate distribution.

EXAMPLE 5   Projection of Farm Size Distributions*

An analysis of Census data for farms in the Grand Forks metropolitan area from 1950 to 2000 found the following probabilities for transitions between small (state S) family-owned farms of 50 to 200 acres, medium (state M) family-owned farms of 201 to 600 acres, large (state L) family-owned farms of 601 to 1000 acres, suburban housing developments (state H), and corporate-owned agribusiness farms (state C). Notice that once a farm becomes a housing development or part of an agribusiness conglomerate, it never returns to family-owned farm status again.

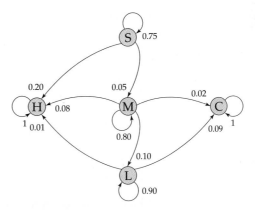

Assuming that these yearly transition patterns continue, what is the expected remaining lifetime of a medium family-owned farm (that is, the time until it becomes a housing development or owned by a corporation)? What is the probability that it will become a suburban housing development?

---

* This example is based on ideas in "Projection of Farm Numbers for North Dakota with Markov Chains" by Ronald D. Krenz, *Agricultural Economics Research* XVI, 3 (July 1964), pp. 77–83.

Solution

We represent the situation as a Markov chain. Since H and C are absorbing states, one standard form for the transition matrix is

$$
\begin{array}{c}
 \\
\begin{array}{c} H \\ C \\ S \\ M \\ L \end{array}
\end{array}
\begin{array}{ccccc}
H & C & S & M & L \\
\left(\begin{array}{cc|ccc}
1 & 0 & 0 & 0 & 0 \\
0 & 1 & 0 & 0 & 0 \\
\hline
0.20 & 0 & 0.75 & 0.05 & 0 \\
0.08 & 0.02 & 0 & 0.80 & 0.10 \\
0.01 & 0.09 & 0 & 0 & 0.90
\end{array}\right)
\end{array}
$$

so

$$
R = \begin{array}{c} S \\ M \\ L \end{array}
\begin{array}{c}
\begin{array}{cc} H & C \end{array} \\
\left(\begin{array}{cc}
0.20 & 0 \\
0.08 & 0.02 \\
0.01 & 0.09
\end{array}\right)
\end{array}
\quad \text{and} \quad
Q = \begin{array}{c} S \\ M \\ L \end{array}
\begin{array}{c}
\begin{array}{ccc} S & M & L \end{array} \\
\left(\begin{array}{ccc}
0.75 & 0.05 & 0 \\
0 & 0.80 & 0.10 \\
0 & 0 & 0.90
\end{array}\right)
\end{array}
$$

Using a calculator or pencil and paper, we find that the fundamental matrix $(I - Q)^{-1}$ is

$$
\left(\begin{pmatrix}
1 & 0 & 0 \\
0 & 1 & 0 \\
0 & 0 & 1
\end{pmatrix} -
\begin{pmatrix}
0.75 & 0.05 & 0 \\
0 & 0.80 & 0.10 \\
0 & 0 & 0.90
\end{pmatrix}\right)^{-1}
=
\begin{array}{c} S \\ M \\ L \end{array}
\begin{array}{c}
\begin{array}{ccc} S & M & L \end{array} \\
\begin{pmatrix}
4 & 1 & 1 \\
0 & 5 & 5 \\
0 & 0 & 10
\end{pmatrix}
\end{array}
\quad
\begin{array}{l}
\text{Expected} \\
\text{times} \\
(I - Q)^{-1}
\end{array}
$$

and $(I - Q)^{-1} \cdot R$ is

$$
\begin{array}{c} S \\ M \\ L \end{array}
\begin{pmatrix}
4 & 1 & 1 \\
0 & 5 & 5 \\
0 & 0 & 10
\end{pmatrix}
\cdot
\begin{array}{c}
\begin{array}{cc} H & C \end{array} \\
\begin{pmatrix}
0.20 & 0 \\
0.08 & 0.02 \\
0.01 & 0.09
\end{pmatrix}
\end{array}
=
\begin{array}{c} S \\ M \\ L \end{array}
\begin{array}{c}
\begin{array}{cc} H & C \end{array} \\
\begin{pmatrix}
0.89 & 0.11 \\
0.45 & 0.55 \\
0.10 & 0.90
\end{pmatrix}
\end{array}
\quad
\begin{array}{l}
\text{Probabilities} \\
(I - Q)^{-1} \cdot R
\end{array}
$$

The expected remaining lifetime of a medium family-owned farm (until it becomes housing or corporate-owned) is $0 + 5 + 5 = 10$ years, and the probability that it will become a suburban housing development is 0.45, or 45%.

PRACTICE PROBLEM 4

From these matrices, what is the expected remaining lifetime of a *small* family-owned farm (until it becomes housing or corporate-owned), and what is the probability that it will become part of an agribusiness conglomerate?
Solution at the back of the book

Section Summary

An *absorbing state* has transition probability 1 from itself to itself and 0 to every other state. An *absorbing Markov chain* has at least one absorbing state, and after enough transitions, every nonabsorbing state has a positive probability of reaching an absorbing state. The transition matrix T of an absorbing Markov chain can be written in *standard form* with the absorbing states first and then the nonabsorbing states:

$$T = \left(\begin{array}{c|c} I & 0 \\ \hline R & Q \end{array} \right)$$

In this form, R is the matrix of transition probabilities from the nonabsorbing states to the absorbing states and Q is the matrix of transition probabilities from the nonabsorbing states to the nonabsorbing states. The *fundamental matrix* of a standard form transition matrix for an absorbing Markov chain is the matrix

$$(I - Q)^{-1}$$

It provides the following information about the chain that starts in a nonabsorbing state N_i:

1. The expected number of times that the chain will be in state N_j before being absorbed is the entry in row i and column j of the fundamental matrix $(I - Q)^{-1}$.

2. The expected number of times that the chain will be in the nonabsorbing states before being absorbed is the sum of the row i entries of the fundamental matrix $(I - Q)^{-1}$.

3. The probability that the chain will be absorbed into state A_j is the entry in row i and column j of the matrix $(I - Q)^{-1} \cdot R$.

EXERCISES 7.3

For each absorbing Markov chain state-transition diagram, identify the absorbing and the nonabsorbing states. Find the smallest integer k such that after k transitions there is a positive probability that the given state will reach an absorbing state.

1.

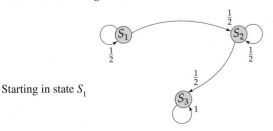

Starting in state S_1

2.

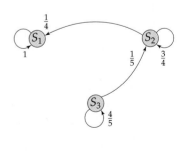

Starting in state S_2

3.

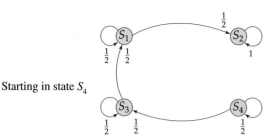

Starting in state S_4

4.

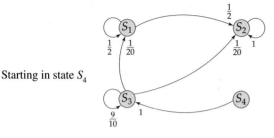

Starting in state S_4

For each absorbing Markov chain transition matrix T:

a. Draw a state-transition diagram, identify the absorbing and the nonabsorbing states, and find the smallest integer k such that after k transitions every nonabsorbing state has a positive probability of reaching an absorbing state.

 b. Verify your value for k by finding T^k and checking that every nonabsorbing state row has a positive probability in at least one absorbing state column.

5. $\begin{pmatrix} 1 & 0 & 0 \\ 0.20 & 0.55 & 0.25 \\ 0 & 0.25 & 0.75 \end{pmatrix}$

6. $\begin{pmatrix} 0.75 & 0 & 0 & 0.25 \\ 0 & 0.50 & 0.50 & 0 \\ 0 & 0 & 1 & 0 \\ 0 & 0.25 & 0.25 & 0.50 \end{pmatrix}$

7. $\begin{pmatrix} 0 & 1 & 0 & 0 \\ 0 & 1 & 0 & 0 \\ 0.20 & 0 & 0.60 & 0.20 \\ 0 & 0 & 0.20 & 0.80 \end{pmatrix}$

8. $\begin{pmatrix} 0.50 & 0.05 & 0.30 & 0.15 \\ 0 & 1 & 0 & 0 \\ 0.50 & 0 & 0.50 & 0 \\ 0 & 0 & 0 & 1 \end{pmatrix}$

Rewrite each absorbing Markov chain transition matrix T in standard form and identify the submatrices R and Q.

9. $\begin{pmatrix} 0.88 & 0.02 & 0.10 \\ 0 & 1 & 0 \\ 0.04 & 0.16 & 0.80 \end{pmatrix}$

10. $\begin{pmatrix} 1 & 0 & 0 & 0 \\ 0.10 & 0.85 & 0.05 & 0 \\ 0.05 & 0.10 & 0.80 & 0.05 \\ 0 & 0 & 0 & 1 \end{pmatrix}$

11. $\begin{pmatrix} 0.80 & 0 & 0.20 & 0 & 0 \\ 0 & 1 & 0 & 0 & 0 \\ 0 & 0.20 & 0.50 & 0.30 & 0 \\ 0 & 0 & 0 & 1 & 0 \\ 0.20 & 0 & 0.05 & 0 & 0.75 \end{pmatrix}$

12. $\begin{pmatrix} 1 & 0 & 0 & 0 & 0 \\ 0 & 0.65 & 0 & 0.30 & 0.05 \\ 0 & 0 & 1 & 0 & 0 \\ 0.05 & 0.30 & 0.05 & 0.60 & 0 \\ 0 & 0 & 0 & 0 & 1 \end{pmatrix}$

Find the fundamental matrix for each absorbing transition matrix in standard form. (You may use if permitted by your instructor.)

13. $\left(\begin{array}{c|cc} 1 & 0 & 0 \\ \hline 0.40 & 0.40 & 0.20 \\ 0.20 & 0.20 & 0.60 \end{array} \right)$ **14.** $\left(\begin{array}{c|cc} 1 & 0 & 0 \\ \hline 0.20 & 0.20 & 0.60 \\ 0.10 & 0.10 & 0.80 \end{array} \right)$

15. $\left(\begin{array}{cc|cc} 1 & 0 & 0 & 0 \\ 0 & 1 & 0 & 0 \\ \hline 0.10 & 0.10 & 0.20 & 0.60 \\ 0.05 & 0.15 & 0.20 & 0.60 \end{array} \right)$

16. $\left(\begin{array}{cc|cc} 1 & 0 & 0 & 0 \\ 0 & 1 & 0 & 0 \\ \hline 0.05 & 0 & 0.85 & 0.10 \\ 0.10 & 0.05 & 0.05 & 0.80 \end{array} \right)$

17. $\left(\begin{array}{c|ccc} 1 & 0 & 0 & 0 \\ \hline 0.10 & 0.20 & 0.10 & 0.60 \\ 0.10 & 0 & 0.90 & 0 \\ 0.20 & 0.20 & 0 & 0.60 \end{array} \right)$

18. $\left(\begin{array}{c|ccc} 1 & 0 & 0 & 0 \\ \hline 0.20 & 0.50 & 0.20 & 0.10 \\ 0.40 & 0 & 0.40 & 0.20 \\ 0.10 & 0 & 0.10 & 0.80 \end{array} \right)$

19. $\begin{pmatrix} 1 & 0 & 0 & 0 & 0 \\ 0 & 1 & 0 & 0 & 0 \\ 0.10 & 0.10 & 0.20 & 0.10 & 0.50 \\ 0.05 & 0.05 & 0.10 & 0.80 & 0 \\ 0.20 & 0 & 0.20 & 0.10 & 0.50 \end{pmatrix}$

20. $\begin{pmatrix} 1 & 0 & 0 & 0 & 0 \\ 0 & 1 & 0 & 0 & 0 \\ 0.04 & 0.16 & 0.60 & 0.10 & 0.10 \\ 0.06 & 0.04 & 0 & 0.90 & 0 \\ 0.08 & 0.02 & 0.20 & 0 & 0.70 \end{pmatrix}$

For each given initial state, find the expected number of times in the given state before absorption, and find the probability of absorption in the given final state using the indicated transition matrix.

	Initial State	Expected Number of Times in State	Probability of Absorption into State	Use Transition Matrix from Exercise
21.	N_1	N_1	A_1	13
22.	N_1	N_2	A_1	14
23.	N_2	N_1	A_2	15
24.	N_2	N_2	A_2	16
25.	N_2	N_2	A_1	17
26.	N_3	N_2	A_1	18
27.	N_1	N_3	A_1	19
28.	N_1	N_3	A_1	20
29.	N_2	any N_i	A_2	19
30.	N_2	any N_i	A_2	20

Represent each situation as an absorbing Markov chain by constructing a state-transition diagram and a corresponding transition matrix in standard form. Find the fundamental matrix of this transition. Be sure to state your final answers in terms of the original questions.

31. Smallpox To populations in the New World never before exposed to smallpox, its arrival with European explorers was devastating. Suppose the disease spread through a newly infected town in such a way that each month 20% of the uninfected population became sick, 15% of the sick recovered (and were therefore immune to further infections) and 35% of the sick perished.

a. How much of the town's original population survived?

b. For an infected person, what was the expected number of months for the disease to run its course?

32. Gambler's Ruin How long can two players gamble against each other before one wins all and the other is ruined? Suppose each player has $2 and they flip a coin with the loser paying $1 to the winner. Find the expected number of times the game will be played before it ends. [*Hint:* Assume the coin is fair, so the probability of winning or losing on each toss is $\frac{1}{2}$, and define five states for the game, each corresponding to one of the player's money: $0, $1, $2, $3, and $4. Which states are absorbing?]

33. Geriatric Care Each year at the Shady Oaks Assisted Living Facility, 10% of the independent living residents are reclassified as requiring assistance, 2% die, and 8% are transferred to a nursing home, while 10% of those needing assistance die and 15% are transferred to a nursing home. What is the probability that a resident requiring assistance will ultimately be transferred to a nursing home?

34. Inner-City Teachers A study of teacher careers in and around Chicago found that every five years 20% of the inner-city teachers planned to continue at their schools, 40% had decided to relocate to suburban schools, 20% were accepting administrative jobs, and the remaining 20% were either retiring or quitting; 10% of the suburban school teachers had decided to relocate to inner-city schools, 70% planned to continue at their schools, 10% were accepting administrative jobs, and the remaining 10% were either retiring or quitting; and 10% of the teachers working in administrative positions had decided to return to inner-city teaching, 20% had decided to relocate to suburban schools, 60% planned to continue as administrators, and the remaining 10% were either retiring or quitting. Assuming that these trends continue and that teachers who want to change their situations are able to do so, find the expected number of years an inner-city school teacher will teach at an inner-city school before retiring or quitting.

35. Term Insurance The term life insurance records of the All-County Insurance Company show that every five years those policies that were renewed are renewed again with probability 0.60, renewed with increased coverage with probability 0.20, closed with a death benefit payment with probability 0.07, and discontinued with probability 0.13, while those that had been renewed with increased coverage are renewed again with probability 0.30, renewed with increased coverage with probability 0.60, closed with a death benefit payment with probability 0.08, and discontinued with probability 0.02.

    a. What is the expected number of years that a policy that has just been renewed will be in force?

    b. What is the probability that a policy that is renewed with increased coverage will be closed with a death benefit payment?

36. Management Training The Big-Beige-Box Computer Company has an in-house program of classroom work and apprenticeships that gives technicians entry to management careers. Each year 25% of those in the classroom move on to the apprenticeship program, 25% drop out, and the remainder continue in the classroom, while 40% of those in the apprenticeship program are promoted to supervisor positions, 10% drop out, and the remainder continue as apprentices. What percentage of those in the classroom will become supervisors?

37. Dental Work A dentist's year-to-year records indicate that prior to extraction, a tooth needing no work remains that way for another year with probability 0.95, needs a filling with probability 0.03, needs a root canal with probability 0.01, and needs to be extracted with probability 0.01; a tooth with a filling one year needs no work the next with probability 0.90, needs a filling with probability 0.06, needs a root canal with probability 0.02, or needs to be extracted with probability 0.02; and (interestingly enough) a tooth with a root canal one year needs no work the next. Find the expected number of times a tooth needing a filling will need fillings before being extracted.

38. Genetics A rapidly mutating class of SMX viruses can be separated into distinct strains that are "successful," together with other "mutant" variations. After each year of replications, 90% of the successful strains are still successful, 5% have developed into mutants, and the remaining 5% have become extinct, while 20% of the previous mutants have further evolved to become successful strains, 40% are still mutants, and the remaining 40% have become extinct. What is the expected number of years until a mutant becomes extinct?

39. Third-World Economic Development The Henderholf Charitable Trust provides either business advice or low-interest loans to small entrepreneurs in developing economies around the world. Each year 9% of the companies they advise become successful and leave the program, 65% continue receiving advice, 25% enter the loan program, and 1% go bankrupt, while 7% of the companies receiving loans become successful and leave the program, 15% need no further loans and return to receiving advice, 75% continue receiving loans, and 3% go bankrupt. Is the Trust's claim that "more than three-quarters of the companies they help become successful" justified?

40. College Graduation Rates   As the students at Edson State College earn more credits, they progress from freshmen to sophomores to juniors to seniors and then graduate, unless they withdraw from college sometime along the way. If each year 64% of each level moves on to the next (including graduation), 20% remain at their current level, and 16% withdraw, what percentage of incoming freshmen will graduate?

Explorations and Excursions

The following problems extend and augment the material presented in the text.

Proof of the Long-Term Behavior of an Absorbing Markov Chain

The following exercises establish the results on pages 540–541 concerning the long-term behavior of an absorbing Markov chain with transition matrix $T = \left(\begin{array}{c|c} I & 0 \\ \hline R & Q \end{array}\right)$ in standard form.

41. Show that

a. $T^2 = \left(\begin{array}{c|c} I & 0 \\ \hline R & Q \end{array}\right) \cdot \left(\begin{array}{c|c} I & 0 \\ \hline R & Q \end{array}\right)$

$= \left(\begin{array}{c|c} I & 0 \\ \hline R + Q \cdot R & Q^2 \end{array}\right) = \left(\begin{array}{c|c} I & 0 \\ \hline (I + Q) \cdot R & Q^2 \end{array}\right)$

b. $T^3 = T^2 \cdot T = \left(\begin{array}{c|c} I & 0 \\ \hline R + Q \cdot R + Q^2 \cdot R & Q^3 \end{array}\right)$

$= \left(\begin{array}{c|c} I & 0 \\ \hline (I + Q + Q^2) \cdot R & Q^3 \end{array}\right)$

42. Verify Exercise 41 for

$$T = \left(\begin{array}{cc|cc} 1 & 0 & 0 & 0 \\ 0 & 1 & 0 & 0 \\ \hline 0.15 & 0.05 & 0.60 & 0.20 \\ 0.03 & 0.07 & 0.20 & 0.70 \end{array}\right).$$

43. Extend Exercise 41 to show that

$$T^m = \left(\begin{array}{c|c} I & 0 \\ \hline (I + Q + \cdots + Q^{m-1}) \cdot R & Q^m \end{array}\right).$$

44. Verify Exercise 43 for

$$T = \left(\begin{array}{cc|ccc} 1 & 0 & 0 & 0 & 0 \\ 0 & 1 & 0 & 0 & 0 \\ \hline 0.02 & 0.08 & 0.60 & 0.20 & 0.10 \\ 0.07 & 0.03 & 0.10 & 0.70 & 0.10 \\ 0.15 & 0.05 & 0.20 & 0.40 & 0.20 \end{array}\right) \quad \text{and} \quad m = 4.$$

45. Use the fact that T is absorbing to explain why every row of

$$(I + Q + \cdots + Q^{m-1}) \cdot R$$

has at least one positive entry for sufficiently large values of m. Then show that

$$I + Q + \cdots + Q^{m-1} \rightarrow (I - Q)^{-1}$$

as m increases. [*Hint:* Use Exercise 78 on page 238.]

46. Verify Exercise 45 for $Q = \begin{pmatrix} 0.11 & 0.02 \\ 0.05 & 0.10 \end{pmatrix}$ by calculating the *finite* sum $I + Q + Q^2 + Q^3 + Q^4$ and checking that rounding this value to four decimal places matches the value of

$$(I - Q)^{-1} = \begin{pmatrix} 1.1250 & 0.0250 \\ 0.0625 & 1.1125 \end{pmatrix}.$$

47. Verify Exercise 45 for $Q = \begin{pmatrix} 0.12 & 0.02 & 0.02 \\ 0.03 & 0.08 & 0.08 \\ 0.01 & 0.01 & 0.01 \end{pmatrix}$ by calculating the *finite* sum $I + Q + Q^2 + Q^3 + Q^4 + Q^5 + Q^6 + Q^7$ and checking that rounding this value to six decimal places matches the value of

$$(I - Q)^{-1} = \begin{pmatrix} 1.137500 & 0.025000 & 0.025000 \\ 0.038125 & 1.088750 & 0.088750 \\ 0.011875 & 0.011250 & 1.011250 \end{pmatrix}.$$

48. Adapt the discussion on page 516 of Section 1 to show that the entry in row i and column j of the fundamental matrix $(I - Q)^{-1}$ is the expected number of times that the chain, if it begins in state N_i, will be in state N_j before being absorbed.

49. Explain how the second result in the box on pages 540–541 follows from Exercise 48.

50. Explain how the third result in the box on pages 540–541 follows from Exercise 48 and the fact that the entry in row i and column j of the matrix R is the probability that state N_i moves to state A_j on one transition.

Chapter Summary with Hints and Suggestions

Reading the text and doing the exercises in this chapter have helped you to master the following skills, which are listed by section (in case you need to review them) and are keyed to particular Review Exercises. Answers for all Review Exercises are given at the back of the book, and full solutions can be found in the Student Solutions Manual.

7.1   STATES AND TRANSITIONS

- Construct a transition matrix for the Markov chain represented by a state-transition diagram and identify the transition as "oscillating," "mixing," or "absorbing."   *(Review Exercises 1–2.)*

- Construct a state-transition diagram for the Markov chain represented by a transition matrix and identify the transition as "oscillating," "mixing," or "absorbing."   *(Review Exercises 3–4.)*

- Calculate the kth state distribution vector for a Markov chain with transition matrix T and initial state distribution vector D_0.   *(Review Exercises 5–6.)*

$$D_k = D_0 \cdot T^k$$

- Find the expected number of times a Markov chain will be in a given state before moving to some other state.   *(Review Exercises 7–8.)*

$$E = \frac{1}{1 - p}$$

- Formulate an application as a Markov chain and answer the question by finding a state distribution vector or an expected time in a given state.   *(Review Exercises 9–10.)*

7.2   REGULAR MARKOV CHAINS

- Identify ergodic (all entries positive) and regular (some power is ergodic) Markov transition matrices.   *(Review Exercises 11–14.)*

- Find the steady state distribution $D \cdot T = D$ of an ergodic or regular Markov chain by row reduction, and then verify the result by calculating powers of the transition matrix (using 🖩).   *(Review Exercises 15–16.)*

$$\left(\begin{array}{c|c} T^t - I & 0 \\ \hline 1 \cdots 1 & 1 \end{array} \right) \rightarrow \left(\begin{array}{c|c} I & D^t \\ \hline 0 & 0 \end{array} \right)$$

- Find the smallest positive number k of transitions needed to move from state S_i to state S_j in a regular Markov chain, and then verify this value (using 🖩) by checking that T^k has a positive transition probability in row i and column j.   *(Review Exercises 17–18.)*

- Formulate an application as an ergodic or regular Markov chain and answer the question by finding the steady-state distribution.   *(Review Exercises 19–20.)*

7.3   ABSORBING MARKOV CHAINS

- Identify the absorbing and nonabsorbing states in a state-transition diagram, and find the smallest number of transitions needed to move from a given nonabsorbing state to an absorbing state.   *(Review Exercises 21–22.)*

- Identify absorbing and nonabsorbing states from a transition matrix, and find the smallest number of transitions needed to ensure that every nonabsorbing state has a positive probability of reaching an absorbing state.   *(Review Exercises 23–24.)*

- Rewrite an absorbing Markov chain transition matrix in standard form.   *(Review Exercises 25–26.)*

$$T = \left(\begin{array}{c|c} I & 0 \\ \hline R & Q \end{array} \right)$$

- Find the fundamental matrix of an absorbing Markov chain.   *(Review Exercises 27–30.)*

$$(I - Q)^{-1}$$

- Use the fundamental matrix to find the expected number of times in nonabsorbing states before absorption and the probability of absorption into a particular absorbing state.   *(Review Exercises 31–38.)*

- Formulate an application as an absorbing Markov chain and answer the question using the fundamental matrix.   *(Review Exercises 39–40.)*

HINTS AND SUGGESTIONS

- **Overview:** A Markov chain is a stochastic process with the same transition probabilities for each transition. Although oscillating transitions never "settle down," the long-term behavior of mixing and absorbing Markov chains can be expressed exactly as a limiting matrix approached by higher and higher powers of the transition matrix. An ergodic or regular Markov chain tends to a stable steady-state distribution that is independent of the initial state, while an absorbing Markov chain always tends to the absorbing states but the expected time before absorption and the probability of absorption into a given state depend on the initial state. Despite their simple mathematical properties, Markov chains have many important applications to science and business situations.

- When calculating the steady-state distribution by row-reducing $\left(\begin{array}{c|c} T^t - I & 0 \\ \hline 1 \cdots 1 & 1 \end{array}\right)$, be sure to transpose T, but don't transpose Q when finding the fundamental matrix $(I - Q)^{-1}$.

- A Markov chain with an absorbing state is *not* an absorbing Markov chain if at least one nonabsorbing state can't reach an absorbing state.

- When rewriting an absorbing transition matrix in standard form, be sure to keep the state names with the rows and columns as you switch the absorbing rows to the top positions and then put the columns in the same order.

- When setting up the transition matrix for an applied situation, always draw a state-transition diagram first and then double-check your work by verifying that the probabilities "leaving" each state total 1.

- Watch out for careless errors such as changing $\frac{1}{2}$ to 0.05 or 5% to 0.50.

- **Practice for Test:** Review Exercises 1, 3, 5, 7, 9, 13, 15, 20, 21, 25, 29, 35, 36, 39.

Review Exercises for Chapter 7 **Practice test exercise numbers are in blue.**

7.1 States and Transitions

Construct a transition matrix T for the Markov chain represented by the state-transition diagram and identify the transition as "oscillating," "mixing," or "absorbing."

1.

2.

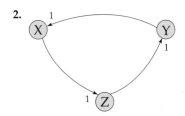

Construct a state-transition diagram for the Markov chain represented by the transition matrix and identify the transition as "oscillating," "mixing," or "absorbing."

3. $\begin{pmatrix} \frac{1}{3} & \frac{2}{3} \\ 0 & 1 \end{pmatrix}$

4. $\begin{pmatrix} 0 & \frac{2}{5} & \frac{3}{5} \\ 0 & \frac{4}{5} & \frac{1}{5} \\ \frac{1}{2} & \frac{1}{2} & 0 \end{pmatrix}$

For each Markov transition matrix T and initial state distribution vector D_0, calculate the next two state distribution vectors D_1 and D_2.

5. $T = \begin{pmatrix} 0.65 & 0.35 \\ 0.15 & 0.85 \end{pmatrix}$, $D_0 = (0.40 \quad 0.60)$

6. $T = \begin{pmatrix} 0 & 0.50 & 0.50 \\ 0.60 & 0.30 & 0.10 \\ 0.10 & 0.30 & 0.60 \end{pmatrix}$, $D_0 = (0 \quad 1 \quad 0)$

For each Markov transition matrix T and present state, find the expected number of times the chain will be in that state before moving to some other state.

7. $T = \begin{pmatrix} 0.95 & 0.05 \\ 0.45 & 0.55 \end{pmatrix}$, presently in state S_1

8. $T = \begin{pmatrix} 1 & 0 & 0 & 0 \\ 0.10 & 0.30 & 0 & 0.60 \\ 0.20 & 0 & 0.80 & 0 \\ 0.10 & 0.30 & 0 & 0.60 \end{pmatrix}$, presently in state S_3

Represent each situation as a Markov chain by constructing a state-transition diagram and the corresponding transition matrix. Be sure to state your final answer in terms of the original question.

9. Taxis Repairs The Yellow-Top Cab Company has a large fleet of aging taxi cabs. Each day 11% break down and are removed from service, while 99% of those in the repair shop from yesterday are returned to service. What is the expected number of days a working cab will remain in service before breaking down?

10. Two-Year Colleges Each year 48% of the first-year students at the Marsten Community College earn enough credits to become second-year students, 20% remain at the first-year level, and the remaining 32% withdraw. Also every year, 64% of the second-year students graduate, 20% remain at the second-year level, and the other 16% withdraw. If there are now 625 first-year and 250 second-year students, how many of these will be second-year students in two years?

7.2 Regular Markov Chains

Identify each Markov transition matrix as "ergodic," "regular," or "neither."

11. $\begin{pmatrix} 0.20 & 0.80 \\ 0.45 & 0.55 \end{pmatrix}$ **12.** $\begin{pmatrix} 0 & 1 \\ 1 & 0 \end{pmatrix}$

13. $\begin{pmatrix} 0 & 0.40 & 0.60 \\ 0 & 0.80 & 0.20 \\ 0.50 & 0 & 0.50 \end{pmatrix}$ **14.** $\begin{pmatrix} 0 & 0.20 & 0.80 \\ 0.10 & 0.20 & 0.70 \\ 0.30 & 0.20 & 0.50 \end{pmatrix}$

For each Markov transition matrix T:

a. Find the steady-state distribution.
b. Calculate T^k for $k = 5, 10, 20,$ and 50 to verify that the rows of T^k approach the steady-state distribution.

15. $\begin{pmatrix} 0.55 & 0.45 \\ 0.15 & 0.85 \end{pmatrix}$ **16.** $\begin{pmatrix} 0.30 & 0.60 & 0.10 \\ 0 & 0.40 & 0.60 \\ 0.30 & 0.60 & 0.10 \end{pmatrix}$

For each regular Markov transition matrix T:

a. Construct a state-transition diagram and find the smallest positive number k of transitions needed to move from the given state S_i to state S_j.
b. Use this k to verify that T^k has a positive transition probability in row i and column j.

17. $\begin{pmatrix} 0.40 & 0 & 0.60 \\ 0.30 & 0.70 & 0 \\ 0 & 0.30 & 0.70 \end{pmatrix}$, $S_1 \to S_2$

18. $\begin{pmatrix} 0.50 & 0.50 & 0 \\ 0.90 & 0 & 0.10 \\ 0.10 & 0 & 0.90 \end{pmatrix}$, $S_3 \to S_2$

Represent each situation as a Markov chain by constructing a state-transition diagram and the corresponding transition matrix. Find the steady-state distribution and interpret it in terms of the original situation. Be sure to state your final answer in terms of the original question.

19. Customer Satisfaction Surveys of customer satisfaction with the repair service departments of dealers for a major car manufacturer show that of those rated "below average" one year, the next year 10% will remain that way, 70% will improve to "satisfactory," and the remaining 20% will improve to "excellent;" of those rated "satisfactory" one year, the next year 50% will remain that way, 20% will improve to "excellent," and the remaining 30% will slip to "below average;" and of those rated "excellent" one year, the next year 70% will remain that way and the remaining 30% will slip to "satisfactory." Assuming that this pattern has repeated for many years, how many of the manufacturer's 2660 dealers nationwide have service departments rated "excellent" by their customers?

20. Weather   The old timers at Edna's Cafe in downtown Nora Springs will tell you the weather just keeps getting better and better for growing corn. In fact, if it was bad last year, there is a 20% chance it will now be terrific and a 60% chance it will be great this year; if last year was great, there is a 50% chance it will now be terrific and a 40% chance it will be great again this year; and if it was terrific last year, there is a 34% chance it will be that way again this year and a 54% chance it will be great this year. (Of course, they don't bother to mention that the weather might be bad this year because that could change their luck!) If they are right, how many of the next 25 years will have terrific weather for growing corn?

7.3   Absorbing Markov Chains

For each absorbing Markov chain state-transition diagram, identify the absorbing and the nonabsorbing states. Find the smallest integer k such that after k transitions there is a positive probability that the given state will reach an absorbing state.

21.

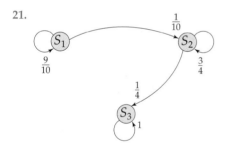

Starting in state S_1

22.

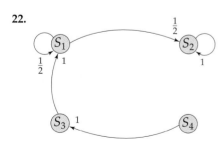

Starting in state S_4

For each absorbing Markov transition matrix T:

a. Draw a state-transition diagram, identify the absorbing and the nonabsorbing states, and find the smallest integer k such that after k transitions, every nonabsorbing state has a positive probability of reaching an absorbing state.

b. Verify your value for k by finding T^k and checking that every nonabsorbing state row has a positive probability in at least one absorbing state column.

23. $\begin{pmatrix} 0.60 & 0 & 0.20 & 0.20 \\ 0 & 1 & 0 & 0 \\ 0 & 0.50 & 0.50 & 0 \\ 0.50 & 0 & 0 & 0.50 \end{pmatrix}$

24. $\begin{pmatrix} 0.80 & 0 & 0 & 0.20 & 0 \\ 0 & 0.80 & 0.15 & 0 & 0.05 \\ 0 & 0 & 1 & 0 & 0 \\ 0.90 & 0.10 & 0 & 0 & 0 \\ 0 & 0 & 0 & 0 & 1 \end{pmatrix}$

Rewrite each absorbing Markov transition matrix T in standard form and identify the submatrices R and Q.

25. $\begin{pmatrix} 0.50 & 0.50 & 0 & 0 \\ 0 & 1 & 0 & 0 \\ 0.20 & 0 & 0.80 & 0 \\ 0 & 0 & 1 & 0 \end{pmatrix}$

26. $\begin{pmatrix} 0.60 & 0 & 0.30 & 0.10 \\ 0 & 1 & 0 & 0 \\ 0.10 & 0.10 & 0.80 & 0 \\ 0 & 0 & 0 & 1 \end{pmatrix}$

Find the fundamental matrix for each absorbing transition matrix in standard form. (You may use 🖩 if permitted by your instructor.)

27. $\left(\begin{array}{c|cc} 1 & 0 & 0 \\ \hline 0.30 & 0.30 & 0.40 \\ 0.10 & 0.10 & 0.80 \end{array} \right)$

28. $\left(\begin{array}{cc|cc} 1 & 0 & 0 & 0 \\ 0 & 1 & 0 & 0 \\ \hline 0.10 & 0 & 0.60 & 0.30 \\ 0.05 & 0.05 & 0.10 & 0.80 \end{array} \right)$

29.
$$\left(\begin{array}{c|ccc} 1 & 0 & 0 & 0 \\ \hline 0.10 & 0.70 & 0 & 0.20 \\ 0.10 & 0.10 & 0.80 & 0 \\ 0.20 & 0.40 & 0 & 0.40 \end{array}\right)$$

30.
$$\left(\begin{array}{cc|ccc} 1 & 0 & 0 & 0 & 0 \\ 0 & 1 & 0 & 0 & 0 \\ \hline 0.05 & 0.15 & 0.20 & 0 & 0.60 \\ 0.10 & 0.10 & 0.20 & 0.50 & 0.10 \\ 0.10 & 0 & 0.10 & 0 & 0.80 \end{array}\right)$$

For each given initial state, find the expected number of times in the given state before absorption, and find the probability of absorption in the given final state using the indicated transition matrix.

	Initial State	Expected Number of Times in State	Probability of Absorption into State	Use Transition Matrix from Exercise
31.	N_1	N_1	A_1	27
32.	N_2	N_1	A_1	27
33.	N_1	N_2	A_2	28
34.	N_2	N_2	A_1	28
35.	N_3	N_1	A_1	29
36.	N_2	any N_i	A_1	29
37.	N_1	N_2	A_2	30
38.	N_3	any N_i	A_1	30

Represent each situation as an absorbing Markov chain by constructing a state-transition diagram and a corresponding transition matrix in standard form. Find the fundamental matrix of this transition. Be sure to state your final answers in terms of the original questions.

39. **Computer Obsolescence** A personal computer (PC) market analyst has estimated that each year 60% of the PCs from the previous year are still manufactured, 20% have been upgraded to newer models, and the remainder have been discontinued, while 30% of the new-model PCs from the previous year are still manufactured, 60% undergo further upgrading, and the remainder are discontinued. How many years is a new model expected to be manufactured before it is discontinued?

40. **Car Loans** The Final Federal Credit Corporation classifies its car loans as paid in full (and thus closed out), current, late, overdue, or bad (in which case they are sold to a collection agency). Each month 10% of the current accounts are paid in full, 10% are late, and the rest remain current; 3% of the late accounts are paid in full, 30% improve to be current, 50% become overdue, 7% become bad, and the rest remain late; 2% of the overdue accounts are paid in full, 10% become current, 20% improve to be late, 18% become bad, and the rest remain overdue. What is the probability that an overdue account will (ultimately) be paid in full?

Projects and Essays

The following projects and essays are based on the material in this chapter. There are no right or wrong answers—the results depend only on your imagination and resourcefulness.

1. Many board games using dice can be modeled as Markov chains. Look up the article "Monopoly as a Markov Process," by Robert Asi and Richard Bishop, *Mathematics Magazine* 45 (January 1972), pp. 26–29, and either write a one-page report on what you find or develop a similar model for your favorite game.

2. Markov chains have been used by sociologists to model human mobility among social classes, occupations, and geographical locations. But the assumption that everyone in a particular rank, job, or town has the same probability of moving to another regardless of their past histories is not reasonable, and many models have not been very successful. The "Cornell Mobility Model" attempted to address this shortcoming by adjusting for the amount of time someone had spent in a particular situation. Find out more by reading the article "A Finite Model of Mobility," by Neil Henry, Robert McGinnis, and Heinrich Tegtmeyer, *Journal of Mathematical Sociology* 1 (1971), 107–118, or by finding other papers by Robert McGinnis, and write a one-page report on what you learn.

3. When many divisions of a large company contribute to the development and manufacture of a new product, it may be difficult to make an accurate statement of the costs incurred by each area. Look up the article "The Cost Accounting Problem," by Don O. Koehler, *Mathematics Magazine* 53, 1 (January 1980), pp. 3–12, and write a one-page report describing the Markov chain method of solving these kinds of problems.

4. Imagine you had a magical penny that always knew whether it came up heads (*H*) or tails (*T*) on the last toss and then came up heads or tails on the next with the following transition probabilities:

$$\begin{array}{cc} & \begin{array}{cc} H & T \end{array} \\ \begin{array}{c} H \\ T \end{array} & \begin{pmatrix} \frac{1}{3} & \frac{2}{3} \\ \frac{2}{3} & \frac{1}{3} \end{pmatrix} \end{array}$$

Compare the results of tossing this magical penny to those from an ordinary "fair" coin. Which coin gives results that seem more random?

5. The mixing of dominant and recessive genes in a large plant or animal population can be modeled as a Markov process. Find out more about *Snyder's ratios* by browsing through the genetics literature in your library, searching the Internet, or talking to a friendly biologist, and write a one-page report on what you learn.

6. More complicated Markov chain transition matrices can be created by joining together the types we have studied in this chapter. Use the methods outlined in Exercises 31–40 in Section 2 (pages 534–535) to explore both the short-term and the long-term behavior of the Markov chains with transition matrices

$$\left(\begin{array}{cc|cc} 0 & 1 & 0 & 0 \\ 1 & 0 & 0 & 0 \\ \hline 0 & 0 & 0 & 1 \\ 0 & 0 & 1 & 0 \end{array}\right) \quad \text{and} \quad \left(\begin{array}{cc|cc} 0 & 1 & 0 & 0 \\ 1 & 0 & 0 & 0 \\ \hline 0 & 0 & 0.10 & 0.90 \\ 0 & 0 & 0.30 & 0.70 \end{array}\right)$$

Make up additional examples of your own and write a report on your discoveries.

8

Logic

Logic and valid arguments play fundamental roles in our judicial system.

8.1

Statements and Connectives

Grammar, Logic, and Rhetoric

Grammar, logic, and rhetoric are the first three of the seven liberal arts that dominated education from Roman times to the past century. Rhetoric uses every possible tool for persuasion, and the following are examples of several *invalid* forms of rhetorical reasoning that are continuing traps for the unwary.

Fallacy of Ambiguity. "Tom was mad when he broke the vase. Mad people should be locked away in lunatic asylums. Therefore, Tom should be confined to an asylum." This is not a valid argument because it confuses two different meanings of "mad"—"angry" and "crazy."

Fallacy of Composition. "Every member of the committee is wise. Therefore, the committee will act wisely." This confuses a characteristic of each part with that of the whole: If every part of a television costs less than $25, does the TV itself cost less than $25?

Fallacy of the Complex Question. "Have you stopped partying and gotten down to work yet?" This "loaded" question includes an assumption so that neither a "yes" nor a "no" answer will be accurate for someone who has been working all along.

Fallacy of the non sequitor. "Everyone wants lower taxes. I know about complicated finance because I'm a banker. Therefore, electing me will restore responsible government." The final sentence literally "does not follow" from the others: Each separately seems acceptable, but taken together, the first and second do not lead to the third.

Fallacy of the False Cause. "He said that the stock market would rise yesterday. It did. Therefore, he can predict the market." But did he really know or was he just guessing? Implying a connection is not the same as demonstrating a cause-and-effect relationship.

In this chapter we will explore some of the structures that arguments must have in order to be valid.

Introduction

While language is useful for warning of impending disaster or asking directions, one of its most powerful aspects is its potential for talking someone else into doing what you want done. Persuasion requires conviction, tact, and patience, but its effect may be lost if the justifications contain even the subtlest fallacy. We begin our discussion of symbolic logic as "the mathematics of correct reasoning" by introducing several fundamental ideas and notations.

Statements

A *statement* is a declarative sentence that is either *true* or *false.* Commands ("Don't interrupt"), opinions ("Chocolate tastes good"), questions ("Is today Tuesday?"), and paradoxes ("This sentence is false") are not statements. We do not need to *know* whether a particular statement is true or false—we require only that the statement has the property that it must be one or the other ("There are exactly 1783 pennies in this jar").

We will use lowercase letters (p, q, and so on) for statements and the capital letters T for "true" and F for "false." A *tautology* is a statement that is always true and we reserve the lowercase letter t for such a statement. Similarly, f will represent a *contradiction*, which is a statement that is always false.

A *compound statement* is a combination of statements using one or more of the following logical *connectives:*

Logical Connectives

Word(s)	Type	Symbol	Example
not	negation	$\sim$	$\sim p$
and	conjunction	$\wedge$	$p \wedge q$
or	disjunction	$\vee$	$p \vee q$
if . . . , then . . .	implication	$\rightarrow$	$p \rightarrow q$

We will discuss *implications* in more detail in Section 3 and *universal quantifiers* ("all," "every") and *existential quantifiers* ("there is," "for at least one") in Section 5.

The presence of one of these words in a sentence does not necessarily mean that the sentence expresses a compound statement: "'Mocha madness' is a flavor of Tom and Jerry's ice cream" is a statement, but the "and" is not used as a logical connective.

The *order of operations* for logical connectives is $\sim$, $\wedge$, $\vee$, and then $\rightarrow$, so that negation ("not") is carried out first, then conjunction ("and"), then disjunction ("or"), and finally implication ("if . . . , then . . ."). We shall use parentheses when necessary or just to clarify the meaning. Thus, according to the order of operations,

$$\sim p \vee q \quad \text{means} \quad (\sim p) \vee q \quad \text{rather than} \quad \sim (p \vee q)$$

Since $\sim$ is done before $\vee$

and

$$p \to q \land r \quad \text{means} \quad p \to (q \land r) \quad \text{rather than} \quad (p \to q) \land r$$

Since $\land$ is done before $\to$

EXAMPLE 1   Translating Symbols into English

Let p represent the statement "the home team won" and let q represent the statement "the celebration went on beyond 1 o'clock." Express each symbolic statement as a sentence.

a. $\sim p$ **b.** $p \land q$ **c.** $p \lor \sim q$ **d.** $p \to q$

Solution

a. We could translate $\sim p$ as "it is not true that the home team won," but we prefer the more natural "the home team lost."
b. "The home team won and the celebration went on beyond 1 o'clock."
c. "The home team won or the celebration ended by 1 o'clock."
d. "If the home team won, then the celebration went on beyond 1 o'clock."

■

PRACTICE PROBLEM 1

(*Continuation of Example 1*) Translate $\sim q \to \sim p$.

Solution at the back of the book

EXAMPLE 2   Translating from English Into Symbols

Let p represent the statement "John ordered codfish," and let q represent the statement "John ordered ice cream." Represent each sentence symbolically using p, q, and logical connectives.

a. John did not order ice cream.
b. John ordered codfish or ice cream.
c. John ordered codfish but not ice cream.
d. If John did not order codfish, then he ordered ice cream.

Solution

a. $\sim q$
b. $p \lor q$
c. $p \land \sim q$
d. $\sim p \to q$

■

PRACTICE PROBLEM 2

(*Continuation of Example 2*) Translate into symbols: "John ordered neither codfish nor ice cream."   *Solution at the back of the book*

Truth Values

The *truth value* of a compound statement is either *true* (*T*) or *false* (*F*), depending on the truth or falsity of its components. For instance, the compound statement "the ink is blue and the paper is white" is *false* if the pen has red ink even if the paper actually *is* white (because both parts of a conjunction must be true for the conjunction to be true). Although the calculation of truth values for a compound statement may be complicated and will be explored further in the next section, we can make several general observations about truth values. Negation reverses truth and falsity, so we immediately have

$\sim t$	is the same as	f	t is a statement that is always true
$\sim f$	is the same as	t	f is a statement that is always false
$\sim(\sim p)$	is the same as	p	Double negation is the same as the original

As observed above, for a conjunction to be true, *both* parts of must be true, so we have

$p \wedge t$	is the same as	p	If p is true, both are true, while if p is false, both are false
$p \wedge f$	is the same as	f	If either part of a conjunction is false, then the conjunction is false

For a disjunction, *either* part being true makes it true, so we have:

$p \vee t$	is the same as	t	
$p \vee f$	is the same as	p	If p is true, both are true, while if p is false, both are false

PRACTICE PROBLEM 3

Express "I ain't not going to bed" without the double negative.

Solution at the back of the book

In the next section we will develop and extend the calculation of truth values.

Section Summary

A statement is either true (*T*) or false (*F*). A compound statement is a combination of statements using logical connectives and parentheses (as necessary).

$\sim p$	not *p*
$p \wedge q$	*p* and *q*
$p \vee q$	*p* or *q*
$p \rightarrow q$	if *p*, then *q*

A tautology is a statement that is always true, while a contradiction is always false.

EXERCISES 8.1

Let *p* represent the statement "the yard is large," and let *q* represent "the house is small." Express each symbolic statement as a sentence.

1. $\sim p$ **2.** $\sim q$

3. $p \wedge q$ **4.** $\sim p \vee q$

5. $p \rightarrow q$ **6.** $\sim q \rightarrow \sim p$

Let *p* represent the statement "the cat is purring," and let *q* represent "the dog is barking." Express each symbolic statement as a sentence.

7. $p \vee q$ **8.** $p \wedge \sim q$

9. $\sim p \wedge q$ **10.** $\sim p \vee \sim q$

11. $q \rightarrow p$ **12.** $\sim p \rightarrow \sim q$

Let *p* represent the statement "the pizza is ready," let *q* represent "the pool is cold," and let *r* represent "the music is loud." Express each sentence in symbolic form.

13. The pizza isn't ready.

14. The pool isn't cold or the music is loud.

15. The pizza is ready and the pool is cold.

16. The pizza is not ready and the music is loud.

17. The pool is cold or the music isn't loud.

18. The music is loud and the pool is cold, or the pizza is ready.

19. If the music isn't loud, then the pizza is ready.

20. If it is not the case that the pizza is ready and the pool is cold, then the music is loud.

Explorations and Excursions

The following problems extend and augment the material presented in the text.

Many common sentences do not express what was probably intended. Reformulate each sentence to express the intended statement correctly.

21. This door must remain closed at all times.

22. Let us now review some facts that may be forgotten.

23. All applicants were not selected.

24. An epilepsy support group is forming for persons considering brain surgery on the second Tuesday of each month.

25. I tried to find a book about solving logic problems without success.

8.2 Truth Tables

Circuits

In the two circuits shown below, the light will be "off" when the switches are "open" and will be "on" when the switches are "closed."

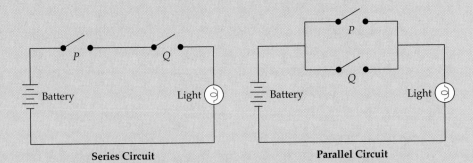

Series Circuit   Parallel Circuit

But the "series" circuit on the left requires both switches *P and Q* to be closed for the light to be on, while the "parallel" circuit on the right requires only switch *P or Q* to be closed. We may show these results in tables for the various possibilities:

Series Circuit

Switch P	Switch Q	Light
Closed	Closed	On
Closed	Open	Off
Open	Closed	Off
Open	Open	Off

← On only if *both* are closed

Parallel Circuit

Switch P	Switch Q	Light
Closed	Closed	On
Closed	Open	On
Open	Closed	On
Open	Open	Off

} On when *either* is closed

The alternatives "closed/open" and "on/off" play the same role for circuits that "true/false" plays in logic, and in this section we will make "truth tables" for logical expressions similar to those above. The observation that logical calculations can be carried out by electrical circuits is the basis for today's electronic computers, whose switches are transistors in integrated circuits etched on silicon wafers. Recent chemical advances indicate that synthetic molecular "switches" may work 100 billion times faster than today's silicon circuits, suggesting the possibility of enormous increases in computing speeds.

Introduction

Symbolic logic has many similarities to the ordinary algebra of real numbers. We have already mentioned that $\sim (\sim p)$ is the same as p, and this bears a striking resemblance to the usual rule for signed numbers that $-(-x) = x$. But before we learned to manipulate algebraic expressions, such as factoring $x^2 - 3x - 4$ into $(x - 4)(x + 1)$, we learned how to evaluate such expressions. In this section, we will "evaluate" compound statements for various truth "values" of their components.

Truth Tables for $\sim$, $\wedge$, and $\vee$

A *truth table* for a compound statement is a list of the truth or falsity of the statement for every possible combination of truth and falsity of its components. We begin with negation:

Negation

p	$\sim p$	
T	F	"Not true" is "false"
F	T	"Not false" is "true"

Because this truth table completely explains the result of negating a statement, it may be taken as the *definition* of negation.

The truth tables for conjunction ("and") and disjunction ("or") require more rows of possibilities because they involve *two* statements, each of which may be true or false:

Conjunction

p	q	$p \wedge q$	
T	T	T	← True only if
T	F	F	*both* are true
F	T	F	
F	F	F	

Disjunction

p	q	$p \vee q$	
T	T	T	True if
T	F	T	*either*
F	T	T	is true
F	F	F	

The disjunction $p \vee q$ is sometimes called the "inclusive disjunction" since it is true when either or *both* of p and q are true. This agrees with the interpretation of "or" in statements such as "Call a doctor if you are in pain or have a temperature," in which the intention is clearly to include the case where *both* are true.*

We will give the truth table for the implication $p \rightarrow q$ in the next section.

---

* The *exclusive disjunction*, as in "Each dinner comes with rice or potatoes," is discussed in the Explorations and Excursions at the end of this section.

Calculating Truth Tables

When making a truth table for a compound statement, we use the "order of operations" to evaluate the various parts of the expression in the correct order. One systematic way of carrying this out is demonstrated in the following example.

EXAMPLE 1 **Constructing a Truth Table**

Construct a truth table for the compound statement $p \lor (\sim p \land q)$.

Solution

Each p and q may be either of two values, so we need $2 \times 2 = 4$ rows in our table. We divide the region under the formula for our compound statement into columns corresponding to the parts we must successively calculate.

p	q	p	$\lor$	$(\sim p$	$\land$	$q)$
T	T					
T	F					
F	T					
F	F					

Under p:
2 T's, then 2 F's

Under q:
alternate T's and F's

The columns under the connectives $\land$ and $\lor$ stand for the conjunction and disjunction, respectively, of the expressions on the two sides of that connective. For example, the column headed $\land$ stands for the expression $(\sim p \land q)$. Working from the inside of the parentheses outward, the order of operations indicates that we should calculate the columns in the order shown below, completing one new column in each numbered step.

1.

p	q	p	$\lor$	$(\sim p$	$\land$	$q)$
T	T			F		
T	F			F		
F	T			T		
F	F			T		

↑
Negating the
p column

2.

p	q	p	$\lor$	$(\sim p$	$\land$	$q)$
T	T			F		T
T	F			F		F
F	T			T		T
F	F			T		F

↑
Copying the
q column

3.

p	q	p	$\vee$	$(\sim p$	$\wedge$	$q)$
T	T			F	F	T
T	F			F	F	F
F	T			T	T	T
F	F			T	F	F

↑
∧ of the $\sim p$ and
the q columns

4.

p	q	p	$\vee$	$(\sim p$	$\wedge$	$q)$
T	T	T		F	F	T
T	F	T		F	F	F
F	T	F		T	T	T
F	F	F		T	F	F

↑
Copying the
p column

5.

p	q	p	$\vee$	$(\sim p$	$\wedge$	$q)$
T	T	T	T	F	F	T
T	F	T	T	F	F	F
F	T	F	T	T	T	T
F	F	F	F	T	F	F

↑
$\vee$ of the p and the
$(\sim p \wedge q)$ columns

The red column is the
$\vee$ of the blue columns

The fifth table is the completed truth table, with the truth values for the entire statement shown in the "final" column under the $\vee$.

Before continuing, be sure that you understand the steps in the construction of this truth table, because from now on we may omit the intermediate steps.

If the truth values in the final column were all T's, the statement would be a *tautology*, while if they were all F's, it would be a *contradiction*. The statement in Example 1 is neither a tautology nor a contradiction.

Keeping only the final column from Example 1, we see that the truth table for the compound statement $p \vee (\sim p \wedge q)$ is the same as that for the disjunction $p \vee q$:

From Example 1

p	q	$p \vee (\sim p \wedge q)$
T	T	T
T	F	T
F	T	T
F	F	F

Disjunction

p	q	$p \vee q$
T	T	T
T	F	T
F	T	T
F	F	F

Different statements
can have the same
truth tables

When we say that two statements have the *same* truth table, we are referring only to their *final* columns.

PRACTICE PROBLEM 1

Construct a truth table for the compound statement $p \vee \sim q$.

Solution at the back of the book

Logical Equivalence

Two statements a and b are *logically equivalent*, written $a \equiv b$, if they have the same truth table. Since we saw from Example 1 that the statement $p \lor (\sim p \land q)$ has the same truth table as the disjunction $p \lor q$, these statements are logically equivalent:

$$p \lor (\sim p \land q) \equiv p \lor q$$

Therefore, whenever we encounter the statement $p \lor (\sim p \land q)$, we may replace it by the simpler statement $p \lor q$. This logical simplification of statements is similar to the algebraic simplification of expressions, such as replacing $2x - (x - 1)$ by the shorter and simpler $x + 1$. We may even replace one statement by a logically equivalent one *within a more complicated statement*, as formalized by the following *substitution principle:*

Substitution Principle

For any statements a and b, let $P(a)$ be a compound statement containing statement a, and let $P(b)$ be the same compound statement but with each instance of statement a replaced by statement b. If a and b are logically equivalent, so are $P(a)$ and $P(b)$; that is,

$$\text{if} \quad a \equiv b, \quad \text{then} \quad P(a) \equiv P(b).$$

We now list some of the basic properties of logical equivalences, each of which can be shown by a truth table.

An *identity* is a quantity that leaves every quantity unchanged when combined with it under a given operation. For real numbers, 0 is the additive identity since $0 + x = x$ for every x, while 1 is the multiplicative identity since $1 \cdot x = x$ for every x. In symbolic logic, the tautology t (which is always true) and the contradiction f (which is always false) act as identities for conjunction and disjunction, respectively, as shown in laws 1 and 5 below.

Identity Laws

1. $p \land t \equiv p$	**4.** $p \lor t \equiv t$
2. $p \land f \equiv f$	**5.** $p \lor f \equiv p$
3. $p \land \sim p \equiv f$	**6.** $p \lor \sim p \equiv t$

Identity laws 1, 2, 4, and 5 were discussed on page 560, and laws 3 and 6 are true since exactly one of p and $\sim p$ can be true, so their conjunction is false and their disjunction is true. We also remarked on page 560 that the negation of a negation is the same as the original statement.

Double Negation Law

$\sim (\sim p) \equiv p$	The negation of a negation is the original

An *idempotent*[*] quantity is one that is unchanged when combined with itself. For real numbers, 0 is the only additive idempotent since $0 + 0 = 0$, and 1 is the only multiplicative idempotent since $1 \cdot 1 = 1$. In symbolic logic, every statement is both a conjunctive and a disjunctive idempotent because repeating the same statement a second time does not alter its meaning.

Idempotent Laws

1. $p \wedge p \equiv p$	A statement "and" itself is just the original statement
2. $p \vee p \equiv p$	A statement "or" itself is just the original statement

The order of the statements make no difference to their combined truth ("the tea is hot and the muffins are buttered" is the same as "the muffins are buttered and the tea is hot"). Such statements are said to "commute."[†]

Commutative Laws

1. $p \wedge q \equiv q \wedge p$	In an "and" statement, the order may be reversed
2. $p \vee q \equiv q \vee p$	In an "or" statement, the order may be reversed

Similarly, a string of conjunctions or disjunctions may be combined (or "associated") in any order.

Associative Laws

1. $p \wedge (q \wedge r) \equiv (p \wedge q) \wedge r$	A series of "and"s may be grouped in any order
2. $p \vee (q \vee r) \equiv (p \vee q) \vee r$	A series of "or"s may be grouped in any order

While ordinary multiplication distributes over addition [for instance, $2 \cdot (3 + 4) = 2 \cdot 3 + 2 \cdot 4$], in symbolic logic both conjunction and

---

[*] From the Latin words *idem* for "same" and *potens* for "strength," suggesting that repetitions do not increase the meaning.

[†] From the Latin word *commutare* meaning "to change with."

disjunction distribute over the other ("dessert" and "coffee or tea" means "dessert and coffee" or "dessert and tea").

Distributive Laws

1. $p \wedge (q \vee r) \equiv (p \wedge q) \vee (p \wedge r)$	An "and" applied to an "or" is two "and"s joined by an "or"
2. $p \vee (q \wedge r) \equiv (p \vee q) \wedge (p \vee r)$	An "or" applied to an "and" is two "or"s joined by an "and"

Unlike ordinary algebra, symbolic logic includes a special fact that essentially says that having redundant information does not provide anything new.

Absorption Laws

1. $p \wedge (p \vee q) \equiv p$	Since (for each law) if p is true, then both sides are true, and if
2. $p \vee (p \wedge q) \equiv p$	p is false, then both sides are false

EXAMPLE 2 **Verifying the First Distributive Law**

Use truth tables to establish the first distributive law:

$$p \wedge (q \vee r) \equiv (p \wedge q) \vee (p \wedge r).$$

Solution

Since each of p, q, and r may be either of two values, we need $2 \times 2 \times 2 = 8$ rows in our table. Rather than building two separate tables (one for each side of the $\equiv$ sign), we shall make one large table that first finds $p \wedge (q \vee r)$ and then $(p \wedge q) \vee (p \wedge r)$. If the final columns for these compound statements are the same, then their logical equivalency will be established. The completed table is shown below.

p	q	r	p	$\wedge$	$(q$	$\vee$	$r)$	$(p$	$\wedge$	$q)$	$\vee$	$(p$	$\wedge$	$r)$	
T	T	T	T	T	T	T	T	T	T	T	T	T	T	T	
T	T	F	T	T	T	T	F	T	T	T	T	T	F	F	
T	F	T	T	T	F	T	T	T	F	F	T	T	T	T	Same final
T	F	F	T	F	F	F	F	T	F	F	F	T	F	F	columns
F	T	T	F	F	T	T	T	F	F	T	F	F	F	T	means
F	T	F	F	F	T	T	F	F	F	T	F	F	F	F	logically
F	F	T	F	F	F	T	T	F	F	F	F	F	F	T	equivalent
F	F	F	F	F	F	F	F	F	F	F	F	F	F	F	
			↑		↑		↑	↑			↑		↑		

Do these columns first.

Since both final columns are the same, we have proved that $\wedge$ distributes over $\vee$: $+ p \wedge (q \vee r) \equiv (p \wedge q) \vee (p \wedge r)$.

■

PRACTICE PROBLEM 2

Verify the second "absorption law" by showing that $p \vee (p \wedge q) \equiv p$.

Solution at the back of the book

Logical Equivalence by Symbolic Manipulation

We now have a second way, besides truth tables, of establishing the logical equivalency of two statements: If one can be transformed into the other using the laws of symbolic logic, then they are equivalent.

EXAMPLE 3   **Using Symbolic Manipulation**

Use symbolic manipulation to show that $p \vee (\sim p \wedge q)$ is logically equivalent to $p \vee q$.

Solution

We begin with the more complicated side, $p \vee (\sim p \wedge q)$, and simplify it by successive logical equivalences until we obtain the other side, $p \vee q$.

$$p \vee (\sim p \wedge q) \equiv (p \vee \sim p) \wedge (p \vee q) \qquad \text{Distributive law 2}$$

$$\equiv (t) \wedge (p \vee q) \qquad \text{Identity law 6}$$

$$\equiv p \vee q \qquad \begin{array}{l}\text{Commutative law 1} \\ \text{and identity law 1}\end{array}$$

Therefore, $p \vee (\sim p \wedge q) \equiv p \vee q$.

■

Notice that symbolic manipulation is very similar to the familiar manipulation of variables that you learned in algebra. By the comment on page 565 (after Example 1), we have already found this same equivalence by truth tables. Both methods will be important in the next sections.

PRACTICE PROBLEM 3

Use the laws of symbolic logic to show that $\sim p \wedge (p \vee q)$ is logically equivalent to $\sim p \wedge q$.

Solution at the back of the book

De Morgan's Laws

Suppose that someone claims he is rich and famous. To prove him wrong, you only need to show *either* that he is not rich *or* that he is not famous (since he is claiming both). In other words, the negation of an "and" is the "or" of the negations, which may be written in symbolic form as:

$$\sim (p \wedge q) \equiv \sim p \vee \sim q$$

Similarly, if a more modest person claims that she is rich *or* famous, to prove her wrong you need to show that she is both not rich *and* not famous. That is, the negation of an "or" is the "and" of the negations:

$$\sim (p \vee q) \equiv \sim p \wedge \sim q$$

These two observations are known as *De Morgan's laws.*[*]

De Morgan's Laws

1. $\sim (p \wedge q) \equiv \sim p \vee \sim q$	To negate an "and," change it to "or" and negate each part
2. $\sim (p \vee q) \equiv \sim p \wedge \sim q$	To negate an "or," change it to "and" and negate each part

De Morgan's laws can be verified by truth tables (see Exercises 39 and 40), and the following exploration shows how a truth table can be done with a graphing calculator.

Graphing Calculator Exploration

Truth tables can be constructed on most graphing calculators by using 1 for *T* and 0 for *F*. We verify the second of De Morgan's laws, $\sim (p \vee q) \equiv \sim p \wedge \sim q$, as follows: Enter the *T T F F* values for *p* in the list L1 as {1,1,0,0} and the *T F T F* values for *q* in the list L2 as {1,0,1,0}. Evaluate $\sim (p \vee q)$ as not(L1 or L2) and $\sim p \wedge \sim q$ as not(L1) and not(L2).

```
L₁
        (1 1 0 0) ←── TTFF
L₂
        (1 0 1 0) ←── TFTF
```

```
~(p∨q) ──→ not(L₁ or L₂)
                      (0 0 0 1) ←──
~p∧~q ──→ not(L₁) and not(          The same
          L₂)                       truth values
                      (0 0 0 1) ←──
```

Since the resulting truth values are the same, the two expressions are logically equivalent and we have shown that $\sim (p \vee q) \equiv \sim p \wedge \sim q$.

[*] After Augustus de Morgan (1806–1871), British mathematician and logician.

EXAMPLE 4   **Negating a Compound Statement**

Negate the compound statement "The cup is blue and the plate is green."

Solution

Although a correct solution could be

It is not true that "the cup is blue and the plate is green."

we can use De Morgan's law that $\sim (p \wedge q) \equiv \sim p \vee \sim q$ to obtain a better sentence for the negation of this conjunction:

The cup is not blue or the plate is not green.

PRACTICE PROBLEM 4

Use De Morgan's laws to negate the compound statement "The spoon is silver or the fork is not gold."   *Solution at the back of the book*

Are Both "And" and "Or" Necessary?

Do we really need both words "and" and "or," or can one be expressed using the other? The answer is that either one can be expressed using the other together with negations, as the following equivalences show.

$$p \wedge q \equiv \sim (\sim p \vee \sim q)$$ From negating both sides of $\sim (p \wedge q) \equiv \sim p \vee \sim q$

$$p \vee q \equiv \sim (\sim p \wedge \sim q)$$ From negating both sides of $\sim (p \vee q) \equiv \sim p \wedge \sim q$

The first equivalence shows that it is logically possible to eliminate the word "and" because it can be expressed in terms of "or" and "not." However, this would mean replacing the statement "he is rich and famous" by the statement "it is not true that he is either not rich or not famous." Both statements are logically equivalent, but it is clearly convenient to retain both "and" and "or" for simplicity and clarity.

Section Summary

A truth table is a complete list of the truth values of a statement for every possible combination of the truth and falsity of its components. Two statements are logically equivalent if they have the same truth table, and logically equivalent statements may be used in place of each other by the substitution principle. Using the laws of symbolic logic, a statement can be transformed into other logically equivalent forms.

Identity laws		
	1. $p \wedge t \equiv p$	4. $p \vee t \equiv t$
	2. $p \wedge f \equiv f$	5. $p \vee f \equiv p$
	3. $p \wedge \sim p \equiv f$	6. $p \vee \sim p \equiv t$

Double negation	$\sim (\sim p) \equiv p$	

Idempotent laws	1. $p \wedge p \equiv p$	2. $p \vee p \equiv p$
Commutative laws	1. $p \wedge q \equiv q \wedge p$	2. $p \vee q \equiv q \vee p$
Associative laws	1. $p \wedge (q \wedge r) \equiv (p \wedge q) \wedge r$	2. $p \vee (q \vee r) \equiv (p \vee q) \vee r$
Distributive laws	1. $p \wedge (q \vee r) \equiv (p \wedge q) \vee (p \wedge r)$	2. $p \vee (q \wedge r) \equiv (p \vee q) \wedge (p \vee r)$
Absorption laws	1. $p \wedge (p \vee q) \equiv p$	2. $p \vee (p \wedge q) \equiv p$
De Morgan's laws	1. $\sim (p \wedge q) \equiv \sim p \vee \sim q$	2. $\sim (p \vee q) \equiv \sim p \wedge \sim q$

EXERCISES 8.2

For each compound statement:

a. Find its truth value given that p is T, q is F, and r is T.
 b. Check your answer using a graphing calculator.

1. $\sim p$ **2.** $\sim q$

3. $p \wedge q$ **4.** $p \vee q$

5. $\sim p \vee r$ **6.** $q \wedge \sim r$

7. $p \wedge (q \vee r)$ **8.** $p \vee (q \wedge r)$

9. $\sim p \vee \sim (q \wedge \sim r)$ **10.** $\sim p \vee (\sim q \vee r)$

For each compound statement:

a. Construct a truth table. Identify any tautologies and contradictions.
 b. Check your answer using a graphing calculator.

11. $p \wedge \sim q$ **12.** $\sim p \vee q$

13. $(\sim p \vee q) \wedge \sim q$ **14.** $\sim p \vee (p \wedge \sim q)$

15. $(p \vee \sim q) \vee \sim (p \wedge q)$ **16.** $\sim (\sim p \vee q) \wedge (p \wedge q)$

17. $\sim (p \vee r) \wedge (\sim q \wedge r)$ **18.** $(p \vee q) \vee \sim (q \wedge \sim r)$

Simplify each statement by symbolic manipulation using the laws of symbolic logic. Identify any tautologies and contradictions.

19. $p \wedge (\sim p \vee q)$ **20.** $p \vee (\sim p \wedge q)$

21. $(p \wedge q) \vee (\sim p \wedge q)$ **22.** $(p \wedge \sim q) \vee (\sim p \wedge q)$

23. $\sim (p \wedge \sim q) \vee p$ **24.** $p \wedge \sim (p \vee \sim q)$

25. Show that the first absorption law and De Morgan's laws lead to the second absorption law.

26. Show that the second absorption law and De Morgan's laws lead to the first absorption law.

Use De Morgan's laws to negate each statement.

27. The coffee is strong and the donuts are hot.

28. The tea is weak or the muffins are ready.

29. The wolf isn't hungry or the sheep are safe.

30. The cat is asleep and the birds aren't in danger.

Verify each of the following properties of logical equivalences by constructing truth tables for both statements and checking that they are the same.

31. $p \wedge t \equiv p$ **32.** $p \vee f \equiv p$

33. $p \vee q \equiv q \vee p$ **34.** $p \wedge q \equiv q \wedge p$

35. $p \wedge (q \wedge r) \equiv (p \wedge q) \wedge r$

36. $p \vee (q \vee r) \equiv (p \vee q) \vee r$

37. $p \vee (q \wedge r) \equiv (p \vee q) \wedge (p \vee r)$

38. $p \wedge (p \vee q) \equiv p$

39. $\sim (p \wedge q) \equiv \sim p \vee \sim q$

40. $\sim (p \vee q) \equiv \sim p \wedge \sim q$

Explorations and Excursions

The following problems extend and augment the material presented in the text.

Exclusive Disjunction $p \veebar q$ interprets "*p* or *q*" to mean "either *p* or *q* but not both" and is thus defined by the truth table

Exclusive Disjunction

p	*q*	$p \veebar q$
T	*T*	*F*
T	*F*	*T*
F	*T*	*T*
F	*F*	*F*

Identify each disjunction as "inclusive" or "exclusive."

41. You may start dinner with soup or salad.

42. The music bonus is a free tape or CD.

43. It's cold! Shut the window or the door.

44. See a doctor when you have a broken arm or leg.

For each compound statement:

a. Construct a truth table. Identify any tautologies and contradictions.

b. Check your answer using a graphing calculator. [*Hint:* The exclusive disjunction is often denoted *xor.*]

45. $p \veebar \sim q$ **46.** $\sim p \veebar q$

47. $\sim p \veebar (p \wedge q)$ **48.** $p \veebar (p \vee \sim q)$

Use truth tables to establish each logical equivalence.

49. $\sim (p \veebar q) \equiv p \veebar \sim q \equiv \sim p \veebar q$

50. $p \veebar q \equiv \sim p \veebar \sim q$

Disjunctive Normal Form How can we find a statement that matches a given truth table? For tables with only one row ending in *T* and all the other rows ending in *F*'s, we can use negation and conjunction to build a statement that is *T* in just that one particular case:

p	*q*	??	
T	*T*	*F*	
T	*F*	*F*	
F	*T*	*T*	← $(\sim p) \wedge q$ is *T* only when *p* is *F* and *q* is *T*
F	*F*	*F*	

Use $\sim$ and $\wedge$ to find a statement for each truth table.

51.

p	*q*	??
T	*T*	*F*
T	*F*	*T*
F	*T*	*F*
F	*F*	*F*

52.

p	*q*	??
T	*T*	*F*
T	*F*	*F*
F	*T*	*F*
F	*F*	*T*

53.

p	*q*	*r*	??
T	*T*	*T*	*F*
T	*T*	*F*	*F*
T	*F*	*T*	*T*
T	*F*	*F*	*F*
F	*T*	*T*	*F*
F	*T*	*F*	*F*
F	*F*	*T*	*F*
F	*F*	*F*	*F*

54.

p	*q*	*r*	??
T	*T*	*T*	*F*
T	*T*	*F*	*F*
T	*F*	*T*	*F*
T	*F*	*F*	*F*
F	*T*	*T*	*F*
F	*T*	*F*	*T*
F	*F*	*T*	*F*
F	*F*	*F*	*F*

Because the disjunction of several statements is true whenever at least one is true, if we have a truth table with several rows ending in *T*'s, we can build statements using $\sim$ and $\wedge$ that are true only for one row and then take their disjunction to obtain a statement for the table. This statement is the "disjunctive normal form" of any statement that has the given truth table.

Find the disjunctive normal form for each truth table.

55.

p	*q*	??
T	*T*	*F*
T	*F*	*F*
F	*T*	*T*
F	*F*	*T*

56.

p	*q*	??
T	*T*	*T*
T	*F*	*T*
F	*T*	*F*
F	*F*	*F*

57.

p	*q*	*r*	??
T	*T*	*T*	*F*
T	*T*	*F*	*F*
T	*F*	*T*	*T*
T	*F*	*F*	*F*
F	*T*	*T*	*T*
F	*T*	*F*	*F*
F	*F*	*T*	*F*
F	*F*	*F*	*F*

58.

p	*q*	*r*	??
T	*T*	*T*	*F*
T	*T*	*F*	*T*
T	*F*	*T*	*F*
T	*F*	*F*	*T*
F	*T*	*T*	*F*
F	*T*	*F*	*T*
F	*F*	*T*	*F*
F	*F*	*F*	*F*

59. Show that the disjunctive normal form of $p \vee q$ is $(p \wedge q) \vee (\sim p \wedge q) \vee (p \wedge \sim q)$.

60. Show that the disjunctive normal form of $p \veebar q$ is $(\sim p \wedge q) \vee (p \wedge \sim q)$.

Implications

"If . . . , then . . ." as a Computer Instruction

Most computer programming languages control the sequence of calculations with special statements that select the next operation on the basis of the values that are currently known. The simplest of these is the "if . . . then . . . else . . ." statement shown in the graphing calculator screen on the left below. This program takes an input value (X) supplied by the user and responds "HELLO" if the value is positive ($X > 0$) and "GOODBYE" otherwise.

```
PROGRAM:IFTHEN
:Input "X? ",X
:If X>0
:Then
:Disp "HELLO"
:Else
:Disp "GOODBYE"
:End
```

Calculator Program

```
prgmIFTHEN
X? 3
HELLO
                Done
X? -2
GOODBYE
                Done
```

Program Input and Output

Do you see how the program on the left gives the output on the right?

Although closely related to this use of "if . . . , then . . ." to control the flow of a program, the logical implication $p \rightarrow q$ (which is read "if p, then q") discussed in this section is a *connective* that forms a new statement from the statements p and q.

Introduction

The application of the scientific method to analyze cause-and-effect mechanisms has been a spectacular success during the last few centuries, both in creating new knowledge and in destroying useless superstition. Although seemingly similar in intent, the logical "if . . . , then . . ." statements discussed in this section do not suppose any underlying cause-and-effect relationship. We view implications simply as logical connectives.

Implications

The implication $p \rightarrow q$ links an *antecedent* to a *consequent:*

$$p \rightarrow q \quad \text{means} \quad \text{if } p, \text{ then } q$$

p is the antecedent
q is the consequent

but makes no claim as to why such a linkage exists. Both of the statements "if you are wicked, then you will be punished" and "if the sky is clear, then the teacup is broken" are equally acceptable as conditional statements, even though the first seems intuitively more reasonable.

How does the truth or falsity of an implication depend on the truth and falsity of its antecedent and consequent? The implication "if you are wicked, then you will be punished" cannot be true if you *are* wicked and yet go unpunished. That is, the implication "if p, then q" must be *false* if p is true and yet q is false, and in all other cases we will say that it is *true*. This leads to the truth values shown below, which serve as a *definition* of the implication $p \rightarrow q$.

Conditional

p	q	$p \rightarrow q$
T	T	T
T	F	F
F	T	T
F	F	T

$\leftarrow$ False only when p is true and q is false

For instance, consider the conditional statement

If I wear my lucky hat, then the Yankees win.   If p, then q

$\underbrace{\hspace{3cm}}_{\text{Antecedent}}$ $\underbrace{\hspace{3cm}}_{\text{Consequent}}$

The only way this could be false is when "I wear my lucky hat" yet "the Yankees lose" (that is, when p is T and q is F). Notice that if "I don't wear my lucky hat," we don't know how the Yankees might do, and so we accept the conditional statement as true by default (even though it is not applicable), as shown in the last two rows of the preceding truth table.*

Since $p \rightarrow q$ is false only when p is true and q is false, and the same is true for $\sim p \vee q$, we have an equivalent way to express implications:

$$p \rightarrow q \equiv \sim p \vee q.$$

---

* This somewhat arbitrary definition of the implication as true in all cases *other* than "true" $\rightarrow$ "false" can also be justified on the basis that experiments in science are designed to *disprove* theories, so an implication should be considered false only under the most restrictive conditions.

Conditional Equivalence

$p \rightarrow q \equiv\; \sim p \vee q$	An implication can be expressed as an "or" with the antecedent negated

Thus "if I wear my lucky hat, then the Yankees will win" can also be expressed "I don't wear my lucky hat or the Yankees win"—both statements have the same logical meaning. When performing symbolic calculations, we will often find it helpful to replace the implication $p \rightarrow q$ by the logically equivalent $\sim p \vee q$.

PRACTICE PROBLEM 1

   a. Use symbolic manipulation to show that $\sim (p \rightarrow q) \equiv p \wedge \sim q$

   b. Use the equivalence in part (a) to negate the conditional statement "if you are at least eighteen, then you can vote."

Solution at the back of the book

From Practice Problem 1a, we have that the negation of an implication can be expressed as follows.

Negating an Implication

$\sim (p \rightarrow q) \equiv p \wedge \sim q$	The negation of an implication can be expressed as an "and" with the consequent negated.

SPREADSHEET EXPLORATION

The spreadsheet below shows the truth table for $\sim p \vee q$ using the spreadsheet functions NOT(...) and OR(... , ...) together with the Boolean values TRUE and FALSE.

C2	▼	=	=OR(NOT(A2),B2)

	A	B	C
1	TRUE	TRUE	TRUE
2	TRUE	FALSE	FALSE
3	FALSE	TRUE	TRUE
4	FALSE	FALSE	TRUE

Because the truth tables for $\sim p \vee q$ and $p \rightarrow q$ are the same, these statements are logically equivalent. Is $p \rightarrow q$ also equivalent to the spreadsheet expression NOT(AND(A1,NOT(B1)))?

There are several other phrases besides "if p, then q" that are commonly used to express the implication $p \rightarrow q$.

Alternative Expressions for "If p, then q"

Form	Example
q if p	The Yankees win if I wear my lucky hat.
q provided p	The Yankees win provided I wear my lucky hat.
q when p	The Yankees win when I wear my lucky hat.
p implies q	Wearing my lucky hat implies that the Yankees win.
p only if q	I am wearing my lucky hat only if the Yankees win.
p is sufficient for q	Wearing my lucky hat is sufficient for the Yankees to win.
q is necessary for p	That the Yankees win is necessary for me to be wearing my hat.

Notice that each of these formulations is wrong only when "I wear my lucky hat" is true and "the Yankees win" is false.

PRACTICE PROBLEM 2   Identify the antecedent and the consequent in each conditional statement.

a. You will pass this course if you study well.

b. I go swimming only if it is hot.   *Solution at the back of the book*

Converse, Inverse, and Contrapositive

For a given conditional statement, switching the antecedent with the consequent or negating them produces three related conditional statements:

Direct statement	$p \rightarrow q$	Original
Converse	$q \rightarrow p$	Switched
Inverse	$\sim p \rightarrow \sim q$	Negated
Contrapositive	$\sim q \rightarrow \sim p$	Switched and negated

Are any of these equivalent to each other? Using the conditional equivalence and the commutative and double negation laws, we have

$$p \rightarrow q \equiv \sim p \vee q \equiv q \vee \sim p \equiv \sim (\sim q) \vee \sim p \equiv \sim q \rightarrow \sim p$$

and

$$q \rightarrow p \equiv \sim q \vee p \equiv p \vee \sim q \equiv \sim (\sim p) \vee \sim q \equiv \sim p \rightarrow \sim q$$

Thus we have established the following logical equivalences:

More Conditional Equivalences

$p \rightarrow q \equiv \sim q \rightarrow \sim p$	The direct is equivalent to the contrapositive
$q \rightarrow p \equiv \sim p \rightarrow \sim q$	The converse is equivalent to the inverse

Note that $p \rightarrow q$ is *not* equivalent to $q \rightarrow p$. (This mistake is often made: "If you work hard, you will get rich. He is rich, so he must have worked hard." This is wrong: Perhaps he received an inheritance or won the lottery.)

EXAMPLE 1   **Finding the Converse, Inverse, and Contrapositive**

State the converse, inverse, and contrapositive of the conditional statement

                If Harold drives a Ford, then Tom has a boat. $p \rightarrow q$

Solution

The antecedent p is "Harold drives a Ford," and the consequent q is "Tom has a boat." Therefore,

Converse:	If Tom has a boat, then Harold drives a Ford.	$q \rightarrow p$
Inverse:	If Harold doesn't drive a Ford, then Tom doesn't have a boat.	$\sim p \rightarrow \sim q$
Contrapositive:	If Tom doesn't have a boat, then Harold doesn't drive a Ford.	$\sim q \rightarrow \sim p$

PRACTICE PROBLEM 3

Which of the answers to Example 1 is logically equivalent to the direct statement?   *Solution at the back of the book*

Biconditional

The biconditional, written $p \leftrightarrow q$ and expressed "p if and only if q," is the double implication that $p \to q$ and $q \to p$:

$$p \leftrightarrow q \equiv (p \to q) \land (q \to p)$$

That is, $p \leftrightarrow q$ means that p and q are either both true or both false:

Biconditional

p	q	$p \leftrightarrow q$
T	T	T
T	F	F
F	T	F
F	F	T

True when p and q are the same.

How does the biconditional $p \leftrightarrow q$ differ from the equivalence $p \equiv q$? The biconditional is a statement about p and q that can be true or false, but "p is equivalent to q" exactly when $p \leftrightarrow q$ is *true*. To put this another way, the biconditional symbol $\leftrightarrow$ may be thought of as $\equiv$ with a truth value T or F saying whether it holds or not. The biconditional is frequently used to define one idea in terms of another: "A positive integer n is even if and only if $n = 2 \cdot k$ for some positive integer k."

Biconditional and the Order of Operations

We evaluate biconditionals *last* of all of the logical connectives, so the order of operations is $\sim$, $\land$, $\lor$, $\to$, and $\leftrightarrow$. For example, $p \leftrightarrow q \land r$ means $p \leftrightarrow (q \land r)$.

Section Summary

The implication $p \to q$ links the antecedent p to the consequent q and is logically equivalent to the disjunction $\sim p \lor q$. The negation $\sim (p \to q)$ is logically equivalent to $p \land \sim q$. The four forms of the conditional implication are

Direct statement	$p \to q$
Converse	$q \to p$
Inverse	$\sim p \to \sim q$
Contrapositive	$\sim q \to \sim p$

The direct and contrapositive are equivalent, as are the converse and inverse.

The biconditional $p \leftrightarrow q$ is the double implication that $p \to q$ and $q \to p$.

EXERCISES 8.3

Write in sentence form the (a) converse, (b) inverse, (c) contrapositive, and (d) negation of the conditional statement. [*Hint:* It may be helpful to rewrite the statement in "if … , then …" form to identify the antecedent and the consequent.]

1. Justice will be done provided the jury is wise.

2. If mice were men, then cowards could be kings.

3. The tooth will be saved when the dentist is skillful.

4. A graceful melody implies a beautiful song.

5. A full bird feeder is sufficient for there to be many birds in the yard.

6. Frequent oil changes are necessary for your motor to last.

Construct a truth table for each statement. Identify any tautologies and contradictions.

7. $p \to \sim q$ 8. $\sim q \to p$

9. $p \to (p \lor q)$ 10. $p \to (p \land q)$

11. $p \land \sim (\sim q \to p)$ 12. $[\sim p \to (p \land q)] \to p$

13. $(p \land q) \to r$ 14. $p \to (q \lor r)$

15. $p \leftrightarrow \sim q$ 16. $p \leftrightarrow (p \lor q)$

Simplify each statement by replacing the conditional $a \to c$ by the equivalent $\sim a \lor c$ and then using the laws of symbolic logic from Section 2 (page 572). Identify any tautologies and contradictions.

17. $(p \to q) \to p$ 18. $(p \to q) \to q$

19. $(p \land q) \to (p \lor q)$ 20. $(p \to q) \land (p \land \sim q)$

Explorations and Excursions

The following problems extend and augment the material presented in the text.

More About the Biconditional   Verify each logical equivalence by showing that the truth tables of both sides are the same.

21. $p \leftrightarrow q \equiv (p \to q) \land (q \to p)$

22. $p \leftrightarrow q \equiv (p \land q) \lor (\sim p \land \sim q)$
    (The disjunctive normal form for $p \leftrightarrow q$)

23. $\sim (p \leftrightarrow q) \equiv \sim p \leftrightarrow q$

24. $p \leftrightarrow q \equiv (p \lor q) \to (p \land q)$

25. $p \leftrightarrow q \equiv (p \to q) \leftrightarrow (q \to p)$

26. $(p \leftrightarrow q) \land (q \leftrightarrow r) \equiv [(p \land r) \leftrightarrow q] \land (p \leftrightarrow r)$

Use symbolic manipulation together with the results of Exercises 21–26 to establish each logical equivalence.

27. $p \leftrightarrow q \equiv q \leftrightarrow p$ 28. $p \leftrightarrow q \equiv \sim p \leftrightarrow \sim q$

29. $\sim (p \leftrightarrow q) \equiv p \leftrightarrow \sim q$ 30. $(p \land q) \leftrightarrow p \equiv p \to q$

Consistent and Contrary Statements   Two statements are *consistent* if they can both be true about the same object: "the car is fast" and "the car is red" are consistent statements. Two statements are *contrary* if they cannot both be true about the same object: "the car is a Jaguar" and "the car is a Porsche" are contrary statements.

Identify each pair of statements as "consistent" or "contrary."

31. She is smart.   She is beautiful.

32. He is smart.   He is stupid.

33. It has hooves.   It has claws.

34. It has wings.   It has feathers.

35. He can sing.   He is a lawyer.

36. She is rich.   She is poor.

37. $x + 2 > 5$ $x + 3 < 4$

38. x is negative. x^2 is positive.

39. The watch is gold.   The watch is fast.

40. The card is blue.   The card is yellow.

Valid Arguments

Fuzzy Logic and Artificial Intelligence

Although people have made many attempts over the centuries to create thinking machines that might be said to possess "artificial intelligence," many now believe that modern computers with their high speeds and vast memories might succeed in the near future. Unfortunately, statements about the real world are often hard to classify as definitely true or false, but many can be rephrased as "probably true" or "possibly false" with a *probability* or *degree of truth* indicated by a number between 0 and 1, where 1 means "absolutely true" and 0 means "absolutely false." Such truth values lead to a "fuzzy" version of logic in which the old *F*'s and *T*'s are replaced by probabilities between 0 and 1. Let us write *P* for the probability that the statement *p* is true and *Q* for the probability that the statement *q* is true. Clearly, $\sim p$ is then true with the "complementary" probability $1 - P$. Because the "fuzzy truth" of a conjunction can be no better than the truths of its parts, we have

$$p \wedge q \text{ is true with probability } \min(P, Q).$$

$\min(P,Q)$ means the *minimum* of the numbers P and Q

Similarly, a disjunction is as true as the better of its parts, so

$$p \vee q \text{ is true with probability } \max(P, Q).$$

$\max(P, Q)$ means the *maximum* of the numbers P and Q

Because $p \to q \equiv \sim p \vee q$, we also have

$$p \to q \text{ is true with probability } \max(1 - P, Q).$$

For example, beginning with the fuzzy truth probabilities $P = 0.8$ and $Q = 0.7$, we obtain the following "fuzzy truth table" using the above formulas:

p	q	$p \wedge q$	$p \vee q$	$p \to q$
0.8	0.7	0.7	0.8	0.7
0.8	0.3	0.3	0.8	0.3
0.2	0.7	0.2	0.7	0.8
0.2	0.3	0.2	0.3	0.8

Because large values approximate T and small values approximate F, these values should remind you of our previous truth tables using T's and F's. Fuzzy logic has had many applications, particularly with "expert systems" that concentrate on a particular area of practical knowledge (such as controlling factory assembly lines or suggesting medical diagnoses). Can you imagine how to use fuzzy logic to develop a "fuzzy set" theory?

Introduction

Although one sense of the word "argument" is "quarrel," in this section we take it to mean simply "discourse meant to persuade." By a "valid argument" we do not demand that the conclusion be a true statement but only that a correct path of reasoning leads from the assumptions to the conclusion. Naturally, a valid argument with true assumptions results in a true conclusion. But a logically valid argument with one or more false assumptions may result in a false conclusion, so the validity of the argument does *not* guarantee the truth of its conclusion.

Valid Argument

An *argument* is a claim that one or more *premises*, when taken together, justify a *conclusion.* We do not ask whether the premises or the conclusion are true, only that *if* the premises are true, *then* the conclusion is also true. That is, a *valid* argument is logically correct independently of the truth or falsity of the premises and the conclusion. A *fallacy* is an argument that is not valid.

Valid Argument

The argument that the premises $P_1, P_2, \ldots, P_n$ imply the conclusion Q is *valid* if and only if $P_1 \wedge P_2 \wedge \ldots \wedge P_n \to Q$ is a tautology.

We will write an argument by listing the premises, drawing a horizontal line, and then stating the conclusion. For example:

If Mary is awake, then the sun is up.	Premise P_1
Jim doesn't have hiccups or Mary is awake.	Premise P_2
If the birds are silent, then the sun isn't up.	Premise P_3
Jim has hiccups.	Premise P_4

Therefore: The birds are singing.   Conclusion Q

One way to determine whether this argument is valid would be to calculate the truth table for $P_1 \wedge P_2 \wedge P_3 \wedge P_4 \rightarrow Q$ and check whether the final column is all T's. However, this method would be tedious, and it has the unpleasant feature that even a minor variation of the argument would require a whole new truth table. Instead, we will develop a basic collection of valid arguments that can be used to check the validity of other arguments. Then we will return to this particular argument.

The Simplest Valid Arguments

As is usual in mathematics, the simplest ideas will be the most fundamental and useful for generalization. Consider first the following argument with just one premise implying the conclusion:

	It is raining.	Premise
Therefore:	It is raining or snowing.	Conclusion
In symbols,	$p \rightarrow (p \vee q)$	p is "it is raining" q is "it is snowing"

We could show that this is a valid argument by constructing its truth table and observing that the final column is all T's (see Exercise 21). Instead, we use symbolic manipulation beginning with the conditional equivalence from page 576 to show that the argument is equivalent to a tautology t:

$$p \rightarrow (p \vee q) \equiv \sim p \vee (p \vee q) \equiv (\sim p \vee p) \vee q \equiv t \vee q \equiv t$$

Since the argument is a tautology, it is valid.

Extension

The argument	$p \rightarrow (p \vee q)$	is valid.	A statement implies itself *or* another statement

Notice that reasoning by extension decreases the quality of the information: It trades the stronger premise "it is raining" for the weaker conclusion "it is raining or snowing."
    Now consider the argument

	It is rainy and dark.	Premise
Therefore:	It is rainy.	Conclusion
In symbols,	$(p \wedge q) \rightarrow p$	p is "it is rainy" q is "it is dark"

Symbolic manipulation (including De Morgan's laws) shows that it is equivalent to a tautology and thus is valid:

$$(p \wedge q) \rightarrow p \equiv \sim (p \wedge q) \vee p \equiv (\sim p \vee \sim q) \vee p \equiv (\sim q \vee \sim p) \vee p \equiv \sim q \vee (\sim p \vee p) \equiv \sim q \vee t \equiv t$$

Simplification

The argument $(p \wedge q) \to p$ is valid.	An "and" implies (either) statement

Reasoning by simplification decreases the amount of information, trading the stronger premise "it is rainy and dark" for the weaker conclusion "it is rainy."

Syllogisms

A *syllogism* (from the Greek word *sullogismo* meaning "a reckoning together") is an argument with two premises, which are sometimes called the "major" and the "minor" premises. Like other forms of arguments, a syllogism is either valid or a fallacy.

EXAMPLE 1   **Determining Whether a Syllogism Is Valid**

Is the following syllogism valid?

If it quacks like a duck, then it is a duck.	Premise P_1
It quacks like a duck.	Premise P_2
Therefore: It is a duck.	Conclusion Q

Solution

Let p be "it quacks like a duck," and let q be "it is a duck." Then this syllogism has the form

$$p \to q$$
$$\underline{ p }$$
$$q$$

Instead of reducing this argument to a tautology, we find an equivalent expression for the premises.

$$(p \to q) \wedge p \equiv (\sim p \vee q) \wedge p \equiv (\sim p \wedge p) \vee (p \wedge q) \equiv f \vee (p \wedge q) \equiv (p \wedge q).$$

By the substitution principle (page 566), this syllogism is equivalent to the "simplification" argument:

$$[(p \to q) \wedge p] \to q \equiv (p \wedge q) \to q$$

Since simplification is valid, so is the argument $(p \to q) \wedge p \to q$. "It is a duck" does follow from the premises.

The form of syllogism shown to be valid in Example 1 is called *modus ponens** and is sometimes called the "law of detachment."

Modus Ponens

The argument	$\begin{array}{c} p \to q \\ \underline{p} \\ q \end{array}$	is valid.	An implication and its antecedent imply its consequent

Replacing p and q by $\sim q$ and $\sim p$, respectively, this valid syllogism becomes

$$\begin{array}{c} \sim q \to \sim p \\ \underline{\sim q} \\ \sim p \end{array} \quad \text{or equivalently} \quad \begin{array}{c} p \to q \\ \underline{\sim q} \\ \sim p \end{array}$$

The contrapositive $\sim q \to \sim p$ is equivalent to $p \to q$

This variation of modus ponens is known as *modus tollens†* and is also called the "law of contraposition" or "indirect reasoning."

Modus Tollens

The argument	$\begin{array}{c} p \to q \\ \underline{\sim q} \\ \sim p \end{array}$	is valid.	An implication and the negation of its consequent imply the negation of its antecedent

Invalid Arguments

Modus ponens and modus tollens combine $p \to q$ with either p or $\sim q$. The other two choices give *invalid* arguments.

Fallacy of the Converse

The argument	$\begin{array}{c} p \to q \\ \underline{q} \\ p \end{array}$	is *not* valid.	An implication and its consequent do *not* imply its antecedent

* From the Latin words *modus* meaning "method" and *ponens* from the verb "to put," suggesting that this method puts in the antecedent so the consequent can be concluded.
† *Tollens* is from the Latin verb meaning "to take away," suggesting that the consequent is taken away (by being negated) so that the opposite of the antecedent must be concluded.

This is a fallacy because the premises $p \to q$ and q are equivalent to just q by absorption. Thus the argument reduces to $q \to p$, which is *not* a tautology. This last implication $q \to p$ is *false* when q is true and p is false, showing that the fallacy of the converse fails to be logically correct when p is false and q is true. Similarly,

Fallacy of the Inverse

The argument	$p \to q$ $\sim p$ ——— $\sim q$	is *not* valid.	An implication and the negation of its antecedent do *not* imply the negation of its consequent

These invalid arguments show that you must be careful about reasoning "backwards" with an implication: The first claims that the *consequent* implies the *antecedent*, and the second claims that the antecedent being false implies that the consequent is false. Both are invalid forms of reasoning.

EXAMPLE 2 **Determining the Validity of a Syllogism**

Is the following syllogism valid?

	If the eggs are fried, then I need toast.	Premise P_1
	The eggs aren't fried.	Premise P_2
Therefore:	I don't need toast.	Conclusion Q

Solution

Let p be "the eggs are fried," and let q be "I need toast." This syllogism has the form of the fallacy of the inverse:

$$p \to q$$
$$\sim p$$
$$\overline{}$$
$$\sim q$$

This syllogism is not valid. ∎

PRACTICE PROBLEM 1

For what truth values of p and q is the "fallacy of the inverse" syllogism false?

Solution at the back of the book

Using double negation, we can recognize $p \vee q$ as $\sim (\sim p) \vee q \equiv \sim p \rightarrow q$, and this observation establishes the following variation on modus ponens as a valid argument.

Disjunctive Syllogism

The argument	$p \vee q$ $\sim p$ $\overline{}$ q	is valid.	An "or" with the negation of one part implies the other part

Our final syllogism links two implications together to form a third. It is called the *hypothetical syllogism* since both the premises and the conclusion are of the hypothetical "if ... , then ... " form. Unlike modus tollens and the disjunctive syllogism, this is not another variant of modus ponens, and it is not reducible to the simplification argument. We defer the proof of its validity until the end of this section.

Hypothetical Syllogism

The argument	$p \rightarrow q$ $q \rightarrow r$ $\overline{}$ $p \rightarrow r$	is valid.	Two implications with the consequent of the first being the antecedent of the second may be combined into one

Analyzing Longer Arguments

We now use syllogisms to *simplify* a longer argument, reducing it in stages until we arrive at a form whose validity we can recognize. We return to the first argument of this section.

EXAMPLE 3   **Analyzing a Longer Argument**

Is the following argument valid?

If Mary is awake, then the sun is up.	Premise P_1
Jim doesn't have hiccups or Mary is awake.	Premise P_2
If the birds are silent, then the sun isn't up.	Premise P_3
Jim has hiccups.	Premise P_4
Therefore: The birds are singing.	Conclusion Q

Solution

We first assign names to each statement:

   m is "Mary is awake"

   s is "the sun is up"   so ~ *s* is "the sun isn't up"

   b is "the birds are singing"   so ~ *b* is "the birds are silent"

   j is "Jim has hiccups"   so ~ *j* is "Jim doesn't have hiccups"

This argument then takes the symbolic form

$$
\begin{array}{ll}
m \rightarrow s & P_1 \\
\sim j \vee m & P_2 \\
\sim b \rightarrow \sim s & P_3 \\
\underline{j} & P_4 \\
b & Q
\end{array}
$$

We simplify the premises in stages, beginning with the simplest premise, *j*, and the other premise involving this letter, ~ *j* $\vee$ *m*, which can be combined using the disjunctive syllogism [since ~ (~ *j*) $\equiv$ *j*]:

$$
\begin{array}{l}
\sim j \vee m \\
\underline{j} \\
m
\end{array}
\qquad
\begin{array}{l}
\text{which reduces} \\
\text{the argument to}
\end{array}
\qquad
\begin{array}{ll}
m \rightarrow s & P_1 \\
m & P_2, P_4 \\
\underline{\sim b \rightarrow \sim s} & P_3 \\
b & Q
\end{array}
$$

In the simplified argument, *m* and *m* $\rightarrow$ *s* can be combined by modus ponens:

$$
\begin{array}{l}
m \rightarrow s \\
\underline{m} \\
s
\end{array}
\qquad
\begin{array}{l}
\text{reducing the} \\
\text{argument further to}
\end{array}
\qquad
\begin{array}{ll}
s & P_1, P_2, P_4 \\
\underline{\sim b \rightarrow \sim s} & P_3 \\
b & Q
\end{array}
$$

In this last argument, replacing ~ *b* $\rightarrow$ ~ *s* by its equivalent *s* $\rightarrow$ *b* (using the substitution principle from page 566) and switching the order of the premises, we see that the argument reduces to modus tollens:

$$
\begin{array}{ll}
s \rightarrow b & P_3 \\
\underline{s} & P_1, P_2, P_4 \\
b & Q
\end{array}
$$

Thus *the original argument is valid* because it can be reduced to a valid argument.

■

PRACTICE PROBLEM 2

Give a second solution of Example 3 by first replacing P_2 with $j \to m$ and P_3 with $s \to b$ (justify these substitutions!), and then showing that the resulting argument is valid.   *Solution at the back of the book*

Deducing Valid Conclusions

Besides verifying that an argument is valid, symbolic logic also enables us to deduce consequences from lists of assumptions, as shown in the following example.

EXAMPLE 4   **Finding a Valid Conclusion**

Supply a valid conclusion for the following premises.

If it is Saturday, then the stock exchange is closed.	Premise P_1
The stock exchange is open or my loan payment is due.	Premise P_2
My loan payment isn't due.	Premise P_3

Solution

We first assign names to each statement used in these premises:

a is "it is Saturday"	so $\sim a$ is "it is not Saturday"
b is "the stock exchange is open"	so $\sim b$ is "the stock exchange is closed"
c is "my loan payment is due"	so $\sim c$ is "my loan payment isn't due"

Then these premises take the symbolic form

$$a \to \sim b \qquad\qquad P_1$$
$$b \lor c \qquad\qquad P_2$$
$$\sim c \qquad\qquad P_3$$

By the disjunctive syllogism, $b \lor c$ and $\sim c$ imply b, and then $a \to \sim b$ and b imply $\sim a$ by modus tollens.

It is not Saturday.

is a valid conclusion from the given premises.   ∎

Proof That the Hypothetical Syllogism Is Valid

The following symbolic calculation establishes the validity of the hypothetical syllogism.

$$[(p \rightarrow q) \wedge (q \rightarrow r)] \rightarrow (p \rightarrow r) \qquad \text{Hypothetical syllogism}$$

$$\equiv\; \sim [(p \rightarrow q) \wedge (q \rightarrow r)] \vee (p \rightarrow r) \qquad \text{Conditional equivalence}$$

$$\equiv\; [\sim (p \rightarrow q) \vee \sim (q \rightarrow r)] \vee (p \rightarrow r) \qquad \text{De Morgan's laws}$$

$$\equiv\; (p \wedge \sim q) \vee)\vee (\sim p \vee r) \qquad \text{Negation of conditionals}$$

$$\equiv\; [(p \wedge \sim q) \vee \sim p] \vee [(q \wedge \sim r) \vee r] \qquad \begin{array}{l}\text{Commutative and}\\\text{associative laws}\end{array}$$

$$\equiv\; [\underbrace{(\sim p \vee p)}_{t} \wedge (\sim p \vee \sim q)] \vee [(q \vee r) \wedge \underbrace{(\sim r \vee r)}_{t}] \qquad \begin{array}{l}\text{Distributive and}\\\text{commutative laws}\end{array}$$

$$\equiv\; \sim p \vee \sim q \vee q \vee r \qquad \begin{array}{l}\text{Identity and}\\\text{associative laws}\end{array}$$

$$\equiv\; (\sim p \vee r) \vee \underbrace{(\sim q \vee q)}_{t} \qquad \begin{array}{l}\text{Commutative and}\\\text{associative laws}\end{array}$$

$$\equiv\; (p \rightarrow r) \vee t \equiv t \qquad \text{Identity law 4}$$

Since the hypothetical syllogism is logically equivalent to a tautology, it is a valid argument. Notice that the calculation removes the tautologies $\sim p \vee p$ and $\sim r \vee r$ to reduce it to the disjunction of the conclusion $p \rightarrow r$ and the tautology $\sim q \vee q$ of the linking statement q.

Section Summary

The argument that the premises $P_1, P_2, \ldots, P_n$ imply the conclusion Q is valid if and only if $P_1 \wedge P_2 \wedge \cdots \wedge P_n \rightarrow Q$ is a tautology. The following arguments are valid:

Valid Arguments

Extension	$p \rightarrow (p \vee q)$
Simplification	$(p \wedge q) \rightarrow p$
Modus ponens	$[(p \rightarrow q) \wedge p] \rightarrow q$
Modus tollens	$[(p \rightarrow q) \wedge \sim q] \rightarrow \sim p$
Disjunctive syllogism	$[(p \vee q) \wedge \sim p] \rightarrow q$
Hypothetical syllogism	$[(p \rightarrow q) \wedge (q \rightarrow r)] \rightarrow (p \rightarrow r)$

The following arguments are not valid:

Fallacies

Fallacy of the converse	$[(p \rightarrow q) \wedge q] \rightarrow p$
Fallacy of the inverse	$[(p \rightarrow q) \wedge \sim p] \rightarrow \sim q$

A longer argument can often be simplified by replacing several premises with their conclusion, and a valid conclusion can similarly be found from a list of premises.

EXERCISES 8.4

Identify the type of each syllogism and whether it is "valid" or a "fallacy."

1. If you like chocolate, you'll love this cake.
You like chocolate.

You'll love this cake.

2. Fred plays third base if Bob pitches.
Fred plays shortstop.

Bob doesn't pitch.

3. He's either studying or watching TV.
He's not watching TV.

He's studying.

4. The dog barks if there is a knock at the door.
The dog barks.

There is a knock at the door.

5. You'll burn your tongue if the soup is hot.
If you burn your tongue, you'll be in a bad mood.

You'll be in a bad mood if the soup is hot.

6. If the book is thick, it must be difficult.
I can't understand it if it is difficult.

If the book is thick, I can't understand it.

7. If it's hot, we'll go to the beach.
We go to the beach.

It's hot.

8. The mystery will be solved if the detective is clever.
The detective is not clever.

The mystery will not be solved.

Analyze each argument and identify it as "valid" or a "fallacy."

9. If it is snowing, the train will be late.
The train is not late or dinner is ruined.
It is snowing.

Dinner is ruined.

10. If it is raining, the road is muddy.
The road is not muddy or the car is stuck.
The car is stuck.

It is raining.

11. If many are sick, the hospitals will be crowded.
If the disease is contagious, many will be sick.
The hospitals are not crowded.

The disease is not contagious.

12. If I can't drive my car, I can't get to work.
I can't drive my car if gas prices go up.
I can get to work.

Gas prices did not go up.

13. I'll get my term paper done if I stay up all night and the computer doesn't break.
My term paper isn't done.

I didn't stay up all night or the computer broke.

14. If the batter concentrates, it will be a triple play or a home run.
It wasn't a triple play and it wasn't a home run.

The batter didn't concentrate.

15. I can graduate and get a job if I work hard.
I can't graduate or I am a success or I can't get a job.
I am not a success.

I don't work hard.

16. If the stock market is up, then the investors are confident.
The investors are not confident or the investors buy new stock offerings.
The investors are taking risks if they buy new stock offerings.
The stock market is up.

The investors are taking risks.

Supply a valid conclusion for each list of premises.

17. If I have chicken for dinner, then I have ice cream for dessert.
I don't have ice cream for dessert or I go to bed early.

18. The flower is yellow or the leaf is green.
The spider is poisonous if the leaf is green.
The flower is red.

19. I go swimming and I have fun if I go to the beach.
I do not go swimming or I do not have fun or I am sunburned.
I go to the beach.

20. If I go to the mountains, I hike and canoe.
I do not hike or I do not canoe or I camp.
I do not camp.

Explorations and Excursions

The following problems extend and augment the material presented in the text.

Although symbolic manipulations give simple and direct proofs of the validity of the arguments presented in this section, it is also a worthwhile undertaking to establish the same results by truth tables.

Construct a complete truth table for each argument and verify that it is a tautology and thus is valid.

21. $p \to (p \lor q)$ (extension)

22. $(p \land q) \to p$ (simplification)

23. $[(p \to q) \land p] \to q$ (modus ponens)

24. $[(p \to q) \land \sim q] \to \sim p$ (modus tollens)

25. $[(p \lor q) \land \sim p] \to q$ (disjunctive syllogism)

26. $[(p \to q) \land (q \to r)] \to (p \to r)$ (hypothetical syllogism)

27. $[(p \to \sim q) \land (q \lor r) \land (\sim r)] \to \sim p$ (See Example 4 on page 589.)

28. $[(p \to q) \land (\sim r \lor p) \land (\sim s \to \sim q) \land r] \to s$ (See Example 3 on pages 587–588.)

Construct a complete truth table for each argument and verify that it is *not* a tautology and thus is a fallacy.

29. $[(p \to q) \land q] \to p$ (fallacy of the converse)

30. $[(p \to q) \land \sim p] \to \sim q$ (fallacy of the inverse)

8.5   Quantifiers and Euler Diagrams

APPLICATION PREVIEW

The Dog Walking Ordinance*

The following transcript of a Borough Council meeting in England illustrates the difficulties of expressing a simple idea in precise and unambiguous language.

---

* From *The Reader Over Your Shoulder* by Robert Graves and Alan Hodge (New York, 1943), as reprinted on pages 1890–1891 of Ernest Nagel, "Symbolic Notation, Haddock's Eyes and the Dog-Walking Ordinance," on pages 1878–1900 of James R. Newman, *The World of Mathematics*, vol. III (New York: Simon and Schuster, 1956).

From the Minutes of a Borough Council Meeting:

Councillor Trafford took exception to the proposed notice at the entrance of South Park: "No dogs must be brought to this Park except on a lead." He pointed out that this order would not prevent an owner from releasing his pets, or pet, from a lead when once safely inside the Park.

The Chairman (Colonel Vine): What alternative wording would you propose, Councillor?

Councillor Trafford: "Dogs are not allowed in this Park without leads."

Councillor Hogg: Mr. Chairman, I object. The order should be addressed to the owners, not to the dogs.

Councillor Trafford: That is a nice point. Very well then: "Owners of dogs are not allowed in this Park unless they keep them on leads."

Councillor Hogg: Mr. Chairman, I object. Strictly speaking, this would prevent me as a dog-owner from leaving my dog in the back-garden at home and walking with Mrs. Hogg across the Park.

Councillor Trafford: Mr. Chairman, I suggest that our legalistic friend be asked to redraft the notice himself.

Councillor Hogg: Mr. Chairman, since Councillor Trafford finds it so difficult to improve on my original wording, I accept. "Nobody without his dog on a lead is allowed in this Park."

Councillor Trafford: Mr. Chairman, I object. Strictly speaking, this notice would prevent me, as a citizen, who owns no dog, from walking in the Park without first acquiring one.

Councillor Hogg (with some warmth): Very simply, then: "Dogs must be led in this Park."

Councillor Trafford: Mr. Chairman, I object: this reads as if it were a general injunction to the Borough to lead their dogs into the Park.

Councillor Hogg interposed a remark for which he was called to order; upon his withdrawing it, it was directed to be expunged from the Minutes.

The Chairman: Councillor Trafford, Councillor Hogg has had three tries; you have had only two . . .

Councillor Trafford: "All dogs must be kept on leads in this Park."

The Chairman: I see Councillor Hogg rising quite rightly to raise another objection. May I anticipate him with another amendment: "All dogs in this Park must be kept on the lead."

This draft was put to the vote and carried unanimously, with two abstentions.

Introduction

If you get into a conversation with someone, you might hear statements such as "Rich people are snobs," "Midwesterners are friendly," and "There's someone out there who's right for you." In this section we will consider statements that say that *everyone* in a certain class has a certain property (*universal* statements) or that *there is at least one* person in a class with a certain property (*existential* statements). We begin by considering statements whose truth or falsity depends on the individual to whom they are applied.

Open Statements

Consider the sentence "x is a registered Democrat," where x may be any citizen of the United States. We denote the sentence as $p(x)$, which is read "p of x," since its truth or falsity depends on the particular citizen x to whom it is applied. When x is replaced by a particular citizen, then $p(x)$ will be a statement, since it will then be true or false (depending on the individual chosen).

In general, an *open statement* $p(x)$ is a statement about x, where x may be any member of a specified *universal set* U of allowable objects. An open statement $p(x)$ is not a statement, since its truth value cannot be determined until x is specified.

EXAMPLE 1 **Finding Truth and Falsity Using an Open Statement**

For the universal set

$$U = \{\text{table, lamp, taxi, cat, horse, crow}\}$$

and the open statement

$$p(x) = \text{"the } x \text{ has four legs"}$$

find the truth or falsity of

$$p(\text{table}) \quad \text{and} \quad p(\text{crow}).$$

Solution

Although $p(x)$ is neither true nor false (since x is not specified), we have that

$$p(\text{table}) \text{ is true} \qquad \text{"The table has four legs" is true}$$

and

$$p(\text{crow}) \text{ is false} \qquad \text{"The crow has four legs" is false}$$

Truth Sets

The *truth set* of an open statement $p(x)$ is the set of all x in U such that $p(x)$ is true. We denote the truth set of $p(x)$ by P. For the U and $p(x)$ in Example 1, the truth set is

$$P = \{\text{table, cat, horse}\} \qquad \text{\small Set of } x \text{ in } U \text{ with four legs}$$

In general, if $p(x)$ is a tautology, then P is the universal set, $P = U$, while if $p(x)$ is a contradiction, then P is the null set, $P = \emptyset$. The truth set of the negation $\sim p(x)$ is the complement P^c of the truth set of $p(x)$. For the preceding example, the truth set of $\sim p(x)$ is

$$P^c = \{\text{lamp, taxi, crow}\} \qquad \text{\small Set of } x \text{ in } U \text{ not with four legs}$$

PRACTICE PROBLEM 1

Find (a) the truth set Q of the open statement $q(x) = $ "the x can give you a ride" where $U = \{\text{table, lamp, taxi, cat, horse, crow}\}$, and (b) the truth set of $\sim q(x)$. *Solution at the back of the book*

Quantifiers

We will use the symbol $\forall$ to mean "for all" and the usual set notation symbol $\in$ to mean "is an element of." Again letting $p(x) = $ "the x has four legs," the statement that all elements of the set $S = \{\text{table, horse}\}$ have four legs may be written as

$$\forall x \in S, p(x) \qquad \text{\small For all } x \text{ in \{table, horse\},} \atop \text{\small the } x \text{ has four legs}$$

Universal Quantifier

Let S be a subset of U, and let $p(x)$ be an open statement for x in U. Then the statement "for all $x \in S, p(x)$" is written

$$\forall x \in S, p(x) \qquad \text{\small For all } x \text{ in } S, p \text{ of } x$$

This statement is true if and only if $p(x)$ is true for *all* x in S.

Other phrases for the universal quantifier $\forall$ are "for each" and "for every."

Clearly, the truth set P of $p(x)$ has the property that

$$\forall x \in P, p(x) \qquad \text{\small } p(x) \text{ holds for all } x \atop \text{\small in its truth set}$$

Since the truth set P is the *largest* set of x's in U for which $p(x)$ is true, the statement "$\forall x \in S, p(x)$" is true if and only if S is a subset of P.

Be careful! "$\forall x \in S, p(x)$" is a statement about the truth of $p(x)$ for each x in S, and is *not* a claim that S contains every possible x for which $p(x)$ is true.

We use the symbol $\exists$ to mean "there exists."

Existential Quantifier

Let S be a subset of U, and let $p(x)$ be an open statement for x in U. Then the statement "there exists an $x \in S$ such that $p(x)$" is written

$$\exists x \in S, p(x)$$

There exists an x in S such that p of x

This statement is true if and only if $p(x)$ is true for *at least one x in S.*

Other phrases for the existential quantifier $\exists$ are "there is an," "for at least one," and "for some." Since the truth set P of $p(x)$ contains every x in U for which $p(x)$ is true, the statement "$\exists x \in S, p(x)$" is true if and only if P and S have at least one element in common (thus the intersection of P and S is nonempty: $P \cap S \neq \varnothing$).

EXAMPLE 2   Using Quantifiers

Let U be the set of colors

$$U = \{\text{violet, blue, green, yellow, orange, pink, red}\}$$

with subsets

$$A = \{\text{violet, yellow}\} \quad \text{and} \quad B = \{\text{blue, orange, red}\}.$$

Let $p(x)$ and $q(x)$ be the open statements

$$p(x) = \text{"the word } x \text{ has more than five letters"}$$

$$q(x) = \text{"the word } x \text{ contains the letter } n.\text{"}$$

Determine the truth values of

a. $\forall x \in A, p(x)$ **b.** $\forall x \in B, p(x)$ **c.** $\exists x \in A, q(x)$ **d.** $\exists x \in B, q(x)$

Solution

a. "$\forall x \in A, p(x)$" means that all words in $A = \{\text{violet, yellow}\}$ have more than five letters, which is clearly *true.*

b. "$\forall x \in B, p(x)$" means that all words in $B = \{\text{blue, orange, red}\}$ have more than five letters, which is clearly *false.*

c. "$\exists x \in A, q(x)$" means that there exists a word in $A = \{\text{violet, yellow}\}$ that contains the letter n, which is clearly *false.*

d. "$\exists x \in B, q(x)$" means that there exists a word in $B = \{\text{blue, orange, red}\}$ that contains the letter n, which is clearly *true.*

We could also have found these answers using truth sets. For instance, the truth set for $p(x)$ is P = {violet, yellow, orange}, and since A = {violet, yellow} is a subset of P, the statement "$\forall x \in A$, $p(x)$" in part (a) is *true*, just as we found above.

PRACTICE PROBLEM 2

Let $r(x)$ be the open statement "the word x contains the letter o" with the same sets U, A, and B used in Example 2. Determine the truth values of: **a.** $\forall x \in A$, $r(x)$ **b.** $\exists x \in B$, $r(x)$.   *Solution at the back of the book*

De Morgan's Laws

The statement "$\forall x \in S$, $p(x)$" is not true if we can identify even one *single* x in S for which $p(x)$ is false. Similarly, "$\exists x \in S$, $p(x)$" is not true if it happens that $p(x)$ is false for *every* x in S. These observations, known as De Morgan's laws, are generalizations of the identically named laws on page 570.

De Morgan's Laws

1. $\sim [\forall x \in S, p(x)] \equiv \exists x \in S, \sim p(x)$	To negate an $\forall$, change it to an $\exists$ and negate the inside statement
2. $\sim [\exists x \in S, p(x)] \equiv \forall x \in S, \sim p(x)$	To negate an $\exists$, change it to an $\forall$ and negate the inside statement

That is, the negation of a universally quantified statement is the existence of its negation, and the negation of an existentially quantified statement is the universality of its negation.

EXAMPLE 3   **Negating Quantified Statements**

Let U = {penny, nickel, dime, quarter} with subsets A = {penny, nickel, dime} and B = {dime, quarter}. Let $p(x)$ be the open statement "x is worth less than fifteen cents." Express each statement in sentence form, use De Morgan's laws to state the negation in symbolic and sentence form, and determine the truth value of each.

a. $\forall x \in A$, $p(x)$ **b.** $\exists x \in B$, $\sim p(x)$

Solution

a. "$\forall x \in A$, $p(x)$" means "every coin in the set {penny, nickel, dime} is worth less than fifteen cents," which is clearly *true*.

By De Morgan's law 1, the negation of "$\forall x \in A, p(x)$" is "$\exists x \in A,$ $\sim p(x)$," which means "there is a coin in {penny, nickel, dime} that is worth fifteen or more cents." This statement is clearly *false*, as we would expect for the negation of a true statement.

b. "$\exists x \in B, \sim p(x)$" means "there is a coin in the set {dime, quarter} that is worth fifteen or more cents," which is clearly *true*.

By De Morgan's law 2, the negation of "$\exists x \in B, \sim p(x)$" is "$\forall x \in B,$ $p(x)$" [using the double negation $\sim (\sim p(x)) \equiv p(x)$]. This means "every coin in the set {dime, quarter} is worth less than fifteen cents," which is clearly *false*, as we would expect for the negation of a true statement.

∎

PRACTICE PROBLEM 3

For the same sets U and A used in Example 3, express "$\exists x \in A,$ $\sim p(x)$" in sentence form, use De Morgan's laws to state its negation in symbolic and sentence form, and determine the truth value of each.

Solution at the back of the book

Quantified Syllogisms and Euler Diagrams

A syllogism involving "all," "some," or "none" is called a *quantified syllogism*, and the validity of such an argument can often be established visually by using *Euler diagrams,** as shown in the following examples. We begin with a syllogism due to Aristotle (384–322 B.C.), who was the first to systematically investigate logic and its relationship to mathematics.

EXAMPLE 4 **Using an Euler Diagram**

Is the following syllogism valid?

All men are mortal.

Socrates is a man.
_____

Therefore: Socrates is mortal.

Solution

Since the first premise states that all men are mortal, we represent "all men" as a region within a larger region representing "all mortals."

An Euler diagram

---

* Leonhard Euler (1707–1783), a prolific Swiss mathematician, first used his diagrams more than a century before the formal beginnings of modern set theory and Venn diagrams (see Section 1 of Chapter 5).

If we represent Socrates by a dot •, where should this dot be placed? By the second premise, this dot must be inside the region representing "all men."

But then the dot representing Socrates is clearly inside the region representing "all mortals," showing that "Socrates is mortal," and thus the syllogism is valid.

When drawing Euler diagrams, we must draw each region to reflect *all* possible positions, rather than just those favorable to the conclusion.

EXAMPLE 5   **More about Euler Diagrams**

Is the following syllogism valid?

> Some home owners are in debt.
>
> Charles is in debt.
> _____
Therefore: Charles is a home owner.

Solution

The regions for "home owners" and those who are "in debt" overlap by the first premise, but we are *not* told that one lies completely within the other. So we must begin by drawing

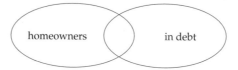

If we represent Charles by a dot •, we must place this dot within the "in debt" region. But where?

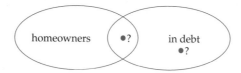

Since it is possible to place the dot within the "in debt" region but not within the "home owners" region, this syllogism is not valid.

Section Summary

An open statement $p(x)$ is a statement about x, where x may be any member of a specified universal set U of allowable objects. The truth set P of $p(x)$ is the set of all x in U such that $p(x)$ is true.

Universal quantifier $\forall$	$\forall x \in S, p(x)$	$p(x)$ is true for *all* x in S
Existential quantifier $\exists$	$\exists x \in S, p(x)$	$p(x)$ is true for at *least one* x in S

$$\sim [\forall x \in S, p(x)] \equiv \exists x \in S, \sim p(x)$$
$$\sim [\exists x \in S, p(x)] \equiv \forall x \in S, \sim p(x)$$

De Morgan's laws

The validity of a quantified syllogism can often be demonstrated by an Euler diagram showing the relationship between the truth sets of quantified statements.

EXERCISES 8.5

Let U = {apple, broccoli, grape, pear, potato} with subsets A = {apple, pear} and B = {broccoli, grape, potato}. Let $p(x)$ be the open statement "x is a fruit." Express each symbolic statement as a sentence and identify it as "true" or "false."

1. $p(\text{apple})$

2. $\forall x \in A, p(x)$

3. $\forall x \in B, p(x)$

4. $\exists x \in A, \sim p(x)$

5. $\exists x \in B, \sim p(x)$

Let U = {coffee, tea, milk, cookies, cake} with subsets A = {milk, cookies} and B = {coffee, tea, milk}. Let $p(x)$ be the open statement "x is a beverage." Express each symbolic statement as a sentence and identify it as "true" or "false."

6. $\sim p(\text{tea})$

7. $\exists x \in A, p(x)$

8. $\forall x \in B, p(x)$

9. $\forall x \in A, p(x)$

10. $\exists x \in B, \sim p(x)$

Let U = {1, 2, 3, 4, 5, 10, 16, 24, 49, 56} with subsets A = {2, 10, 16, 24} and B = {16, 49}. Let $p(x)$ be the open statement "x is an even integer," and let $q(x)$ be the open statement "x is a perfect square."

11. Find the truth set P of $p(x)$.

12. Find the truth set Q of $q(x)$.

Determine the truth value of each statement [with U, A, B, $p(x)$, and $q(x)$ as defined for Exercises 11 and 12].

13. $\exists x \in A, p(x)$

14. $\exists x \in A, q(x)$

15. $\forall x \in B, p(x)$

16. $\forall x \in B, q(x)$

17. $\exists x \in A, p(x) \wedge q(x)$

18. $\exists x \in B, p(x) \wedge q(x)$

19. $\forall x \in B, p(x) \vee q(x)$

20. $\forall x \in A, \sim q(x)$

Let U = {Colorado, Missouri, Nevada, New York, Sacramento, Tennessee} with subsets A = {Colorado, Nevada, Sacramento} and B = {Missouri, Tennessee}. Let $p(x)$ be the open statement "x is a state," and let $q(x)$ be the open statement "x is a major river."

21. Find the truth set P of $p(x)$.

22. Find the truth set Q of $q(x)$.

Determine the truth value of each statement [with U, A, B, $p(x)$, and $q(x)$ as defined for Exercises 21 and 22].

23. $\exists x \in A, p(x)$

24. $\exists x \in A, q(x)$

25. $\forall x \in B, p(x) \wedge q(x)$

26. $\forall x \in A, p(x) \vee \sim q(x)$

27. $\exists x \in A, \sim p(x) \wedge q(x)$

28. $\exists x \in B, \sim p(x) \wedge q(x)$

29. $\forall x \in A, \sim p(x) \vee q(x)$

30. $\forall x \in B, p(x) \to q(x)$

Negate each statement using De Morgan's laws.

31. $\exists x \in A, \sim p(x)$

32. $\forall x \in B, \sim q(x)$

33. $\forall x \in A, p(x) \wedge q(x)$

34. $\exists x \in B, p(x) \vee q(x)$

35. $\exists x \in A, p(x) \to q(x)$

36. $\forall x \in B, \sim p(x) \wedge \sim q(x)$

37. Every rich person is a snob.

38. All midwesterners are friendly.

39. There is a quadrilateral that is not a square.

40. There is a positive integer that is neither odd nor even.

Use Euler diagrams to identify each argument as "valid" or a "fallacy."

41. All pears are ripe.
Try this pear.
_____
It is ripe.

42. All pears are ripe.
This is ripe.
_____
This is a pear.

43. Some apples are rotten.
Try this apple.
_____
It is rotten.

44. Some apples are rotten.
All the apples are green.
_____
This apple is rotten and green.

45. All hurricanes are dangerous.
Floyd is a hurricane.
_____
Floyd is dangerous.

46. Some storms are hurricanes.
All hurricanes are dangerous.
_____
This storm is dangerous.

47. All business majors take mathematics.
Jim takes mathematics.
_____
Jim is a business major.

48. All business majors take mathematics.
Jim is a business major.
_____
Jim takes mathematics.

49. All sports cars are fast.
All fast cars are expensive.
_____
All sports cars are expensive.

50. Some cold days are snowy.
Today is not snowy.
_____
Today is not cold.

Explorations and Excursions

The following problems extend and augment the material presented in the text.

More About Sets and Logic   Let U be a universal set for the open statements $p(x)$ and $q(x)$, let P be the truth set for $p(x)$, and let Q be the truth set for $q(x)$. Justify each of the following facts relating set theory to logic.

51. The truth set of $p(x) \wedge q(x)$ is $P \cap Q$.

52. The truth set of $p(x) \vee q(x)$ is $P \cup Q$.

53. The truth set of a tautology is U.

54. The truth set of a contradiction is $\varnothing$ (the null set).

55. The truth set of $\sim p(x)$ is P^c.

56. $P \subseteq Q \leftrightarrow \forall x, p(x) \to q(x)$ is true.

57. $P = Q \leftrightarrow \forall x, p(x) \leftrightarrow q(x)$ is true.

58. The truth set of $\sim p(x) \wedge q(x)$ is $P^c \cap Q$.

59. The truth set of $p(x) \vee \sim q(x)$ is $P \cup Q^c$.

60. Show that De Morgan's laws for symbolic logic (see page 570) imply De Morgan's laws for sets (see Exercises 57 and 58 on page 394).

Chapter Summary with Hints and Suggestions

Reading the text and doing the exercises in this chapter have helped you to master the following skills, which are listed by section (in case you need to review them) and are keyed to particular Review Exercises. Answers for all Review Exercises are given at the back of the book, and full solutions can be found in the Student Solutions Manual.

8.1 STATEMENTS AND CONNECTIVES

- Express a symbolic statement as a sentence. *(Review Exercises 1–4.)*

$$\sim p \qquad \text{not } p$$

$$p \wedge q \qquad p \text{ and } q$$

$$p \vee q \qquad p \text{ or } q$$

$$p \rightarrow q \qquad \text{if } p, \text{ then } q$$

- Express a sentence in symbolic form. *(Review Exercises 5–10.)*

8.2 TRUTH TABLES

- Find the truth value of a compound statement using the correct order of operations. *(Review Exercises 11–12.)*

$$(\ ), \sim, \wedge, \vee, \text{ and then } \rightarrow$$

- Construct a truth table for a compound statement. *(Review Exercises 13–14.)*
- Simplify a compound statement by symbolic manipulation using the laws of symbolic logic. *(Review Exercises 15–18.)*

$$\sim(\sim p) \equiv p$$

$$p \vee q \equiv q \vee p \qquad\qquad p \wedge q \equiv q \wedge p$$

$$p \vee (q \vee r) \equiv (p \vee q) \vee r \qquad p \wedge (q \wedge r) \equiv (p \wedge q) \wedge r$$

$$p \vee (q \wedge r) \equiv (p \vee q) \wedge (p \vee r)$$

$$p \wedge (q \vee r) \equiv (p \wedge q) \vee (p \wedge r)$$

$$p \vee (p \wedge q) \equiv p \qquad\qquad p \wedge (p \vee q) \equiv p$$

- Use De Morgan's laws to express the negation of a conjunction as a disjunction or the negation of a conjunction as a disjunction. *(Review Exercises 19–20.)*

$$\sim(p \wedge q) \equiv \sim p \vee \sim q \qquad \sim(p \vee q) \equiv \sim p \wedge \sim q$$

8.3 IMPLICATIONS

- State the converse, inverse, contrapositive, and negation of a conditional statement. *(Review Exercises 21–22.)*

$$p \rightarrow q \qquad \text{Direct statement}$$

$$p \wedge \sim q \qquad \text{Negation}$$

$$q \rightarrow p \qquad \text{Converse}$$

$$\sim p \rightarrow \sim q \qquad \text{Inverse}$$

$$\sim q \rightarrow \sim p \qquad \text{Contrapositive}$$

- Construct a truth table for a compound statement involving the conditional or biconditional connective. *(Review Exercises 23–26.)*

$$p \rightarrow q \qquad p \leftrightarrow q$$

- Simplify a compound statement involving the conditional connective by using the laws of symbolic logic. *(Review Exercises 27–30.)*

$$p \rightarrow q \equiv \sim p \vee q$$

8.4 VALID ARGUMENTS

- Identify the type of a syllogism and whether it is "valid" or a "fallacy." *(Review Exercises 31–34.)*

$[(p \to q) \wedge p] \to q$	Modus ponens
$[(p \to q) \wedge \sim q] \to \sim p$	Modus tollens
$[(p \vee q) \wedge \sim p] \to q$	Disjunctive syllogism
$[(p \to q) \wedge (q \to r)] \to (p \to r)$	Hypothetical syllogism
$[(p \to q) \wedge q] \to p$	Fallacy of the converse
$[(p \to q) \wedge \sim p] \to \sim q$	Fallacy of the inverse

- Analyze an argument and identify it as "valid" or a "fallacy." *(Review Exercises 35–38.)*

- Supply a valid conclusion for a list of premises. *(Review Exercises 39–40.)*

8.5 QUANTIFIERS AND EULER DIAGRAMS

- Find the truth value of an open statement $p(x)$ for a particular choice of $x \in U$. *(Review Exercises 41–42.)*

- Find the truth set of an open statement. *(Review Exercises 43–44.)*

$$P = \{\text{all } x \in U \text{ such that } p(x) \text{ is true}\}$$

- Find the truth value of a quantified statement. *(Review Exercises 45–46.)*

$$\forall x \in S, p(x) \qquad \exists x \in S, p(x)$$

- Use De Morgan's laws to negate a quantified statement. *(Review Exercises 47–48.)*

$$\sim [\forall x \in S, p(x)] \equiv \exists x \in S, \sim p(x)$$

$$\sim [\exists x \in S, p(x)] \equiv \forall x \in S, \sim p(x)$$

- Determine the validity of a quantified syllogism using an Euler diagram. *(Review Exercises 49–50.)*

HINTS AND SUGGESTIONS

- **Overview:** Although useful conclusions require correct premises, the validity of logical arguments does not depend on the truth of the premises. Truth values lead to truth table definitions of the logical connectives, and logical equivalences and tautologies lead to valid arguments.

- Truth tables are useful for small calculations, but involved compound statements can be analyzed more easily using the laws of symbolic logic.

- The four variations of the implication $p \to q$ are not all equivalent: The converse and the inverse are equivalent, as are the direct statement and the contrapositive.

- Logically equivalent statements are interchangeable, but it is usually best to replace a complicated statement by a simpler version when analyzing a statement or argument.

- The implication $p \to q$ is equivalent to $\sim p \vee q$, so the negation $\sim (p \to q)$ is equivalent to $p \wedge \sim q$ by De Morgan's laws.

- Even if $P_1 \wedge P_2 \wedge \ldots \wedge P_n \to Q$ is true for a particular choice of truth values for $P_1, P_2, \ldots, P_n$ and Q, it is *not* a valid argument unless it is true for *every* possible choice of truth values.

- Notice the technical differences in the usage and meaning of certain terms: A *statement* can be a *tautology* (always true) or a *contradiction* (always false), or something between (sometimes true and sometimes false depending on the truth or falsity of its components). An *argument* (a statement in the form of an implication) will be either *valid* (always true) or a *fallacy* (all other possibilities, so at least sometimes false). That is, tautology is analogous to a valid argument but contradiction is more extreme than a fallacy.

- When drawing an Euler diagram, make every effort to discredit the argument before declaring it valid.

- **Practice for Test:** Review Exercises 2, 6, 11, 14, 17, 20, 21, 24, 28, 32, 37, 39, 43, 46, 48, 49.

Review Exercises for Chapter 8   *Practice test exercise numbers are in blue.*

8.1   Statements and Connectives

Let p represent the statement "roses are red," and let q represent "violets are blue." Express each symbolic statement as a sentence.

1. $p \wedge q$ 2. $p \vee q$

3. $\sim q$ 4. $\sim p \rightarrow q$

Let p represent the statement "it is summer," let q represent "the leaves are green," and let r represent "it is snowing." Express each sentence in symbolic form.

5. It is summer and the leaves are green.

6. It is summer and it is not snowing.

7. The leaves are green or it is not summer.

8. It is not snowing or the leaves are green and it is summer.

9. If it is snowing, then it is not summer.

10. If it is summer and the leaves are green, then it is not snowing.

8.2   Truth Tables

For each compound statement:

a. Find its truth value given that p is T, q is T, and r is F.

b. Check your answer using a graphing calculator.

11. $p \wedge (\sim q \vee r)$ 12. $(p \vee \sim q) \wedge r$

For each compound statement:

a. Construct a truth table. Identify any tautologies and contradictions.

b. Check your answer using a graphing calculator.

13. $(p \vee \sim q) \wedge \sim p$ 14. $\sim (p \wedge r) \vee (\sim q \vee r)$

Simplify each statement using the laws of symbolic logic. Identify any tautologies and contradictions.

15. $(p \vee \sim q) \wedge (p \vee q)$ 16. $(\sim p \vee q) \wedge (p \wedge \sim q)$

17. $p \wedge \sim (p \vee q)$ 18. $\sim (p \wedge q) \vee p$

Use De Morgan's laws to negate each statement.

19. The glasses are broken and the watch won't run.

20. His ankle is fine or his knee is bruised.

8.3   Implications

Write in sentence form the (a) converse, (b) inverse, (c) contrapositive, and (d) negation of the conditional statement. [*Hint:* It may be helpful to rewrite the statement in "if ... , then ..." form to identify the antecedent and the consequent.]

21. The frame is ornate if the painting is expensive.

22. The freezer works properly only if the ice cream is hard.

Construct a truth table for each statement. Identify any tautologies and contradictions.

23. $\sim p \rightarrow \sim q$ 24. $[(p \rightarrow q) \wedge p] \rightarrow q$

25. $(p \vee q) \rightarrow (q \wedge r)$ 26. $\sim p \leftrightarrow q$

Simplify each statement using the laws of symbolic logic. Identify any tautologies and contradictions.

27. $(p \vee q) \rightarrow p$ 28. $(p \wedge q) \rightarrow p$

29. $(p \rightarrow q) \rightarrow \sim p$ 30. $p \rightarrow \sim p$

8.4   Valid Arguments

Identify the type of each syllogism and whether it is "valid" or a "fallacy."

31. Freshmen must take English composition.
    Tim is a freshman.
    _____
    Tim takes English composition.

32. The band sounds great when Justin plays bass.
    The band was awful.
    _____
    Justin didn't play bass.

33. She's talking on the phone or visiting her sister.
    She's not visiting her sister.
    _____
    She's talking on the phone.

34. If it's cloudy, we'll go fishing.
    It's clear.
    _____
    We're not going fishing.

Analyze each argument and identify it as "valid" or a "fallacy."

35. If the virus is virulent, then many will die.
Not many will die or it will be an economic disaster.
The virus is virulent.

It will be an economic disaster.

36. Many old people will freeze this winter if home heating oil is too expensive.
If oil prices go up, home heating oil will be too expensive.
Oil prices have not gone up.

Not many old people will freeze this winter.

37. If I win the lottery or get a job, then I can afford a car.
I can't afford a car or I won't take the bus.
I ride the bus.

I didn't win the lottery and I didn't get a job.

38. If interest rates are down, industry can invest to increase productivity.
Industry cannot invest to increase productivity or unemployment will decrease.
If unemployment decreases, then housing starts will increase.
Housing starts have decreased.

Interest rates are up.

Supply a valid conclusion for each list of premises.

39. If I have soup for lunch, then I have cookies for tea.
I don't have cookies for tea or I stay late at work.

40. The book is heavy if it is thick.
The book is light or I won't buy it.
I bought the book.

8.5   Quantifiers and Euler Diagrams

Let $U = \{$Brahms, Cézanne, Debussy, Dürer, Pachelbel, Vermeer$\}$ with subsets $A = \{$Debussy, Dürer$\}$ and $B = \{$Brahms, Cézanne$\}$. Let $p(x)$ be the open statement "x is a famous composer," and let $q(x)$ be the open statement "x is a famous Frenchman."

Find the truth value of each statement.

41. p(Brahms)   **42.** q(Vermeer)

43. Find the truth set P of $p(x)$.

44. Find the truth set Q of $q(x)$.

Determine the truth value of each statement.

45. $\forall x \in A, p(x)$ 46. $\exists x \in B, q(x)$

Negate each statement using De Morgan's laws.

47. All great composers have names beginning with the letter B.

48. There is a logic problem that I cannot solve.

Use Euler diagrams to identify each quantified argument as "valid" or a "fallacy."

49. All snowy days are cold.
Today is not cold.

Today is not snowy.

50. Some sports cars are expensive.
All sports cars are fast.

All fast cars are expensive.

Projects and Essays

The following projects and essays are based on Chapter 8. There are no right or wrong answers—the results depend only on your imagination and resourcefulness.

1. Locate and read the article on "Symbolic Logic" by John E. Pfeiffer in the December 1960 issue of *Scientific American,* and write a one-page report on either (a) one of his examples or (b) how computer advances since 1960 have (or haven't) fulfilled some of his predictions.

2. Besides *Alice's Adventures in Wonderland* and *Through the Looking Glass,* written under his pen name "Lewis Carroll," Charles Dodgson (1832–1898) also wrote *The Game of Logic.* Locate a reprint of this classic or some excerpts included in a logic book, and write a one-page analysis of one of his examples.

3. Although our discussion of the conditional explicitly rejected any cause-and-effect linkage between the antecedent and the consequent, such an approach can be challenged on the basis that there is a legitimate need for "meaning" in logic statements and analysis. Investigate the differences between "implication in formal meaning" and "implication in material meaning," perhaps by starting with the article "Symbolic Logic" by Alfred Tarski as reprinted in *The World of Mathematics,* vol. 3, by James R. Newman (New York: Simon and Schuster, 1956), pp. 1901–1931, and write a one-page report on your findings.

4. The problem of separating logically valid facts from conclusions that we might subconsciously draw from the content of the statements has been addressed by the philosopher H. P. Grice in his theory of "conversational implicature." Find out more about Grice and his work, and write a one-page report on some of his examples.

5. Closely read a credit card or car loan agreement and find examples of the legal phrase "and/or." Write a report on how this language clarifies possible confusions between the inclusive and exclusive meanings of the disjunction.

Appendix

Normal Probabilities Using Tables

This appendix, showing how to find normal probabilities from tables, replaces pages 494–499 (through Practice Problem 3) of Chapter 6 (Statistics) for readers who do not have graphing calculators that calculate normal probabilities.

z-Scores

The *standard normal distribution* is the normal distribution with mean $\mu = 0$ and standard deviation $\sigma = 1$ (the word "standard" indicates mean zero and standard deviation one). The letter z is traditionally used to denote the variable for this special distribution. Since $\mu = 0$, the highest point on the curve is at $z = 0$, and the curve is symmetric about this value.

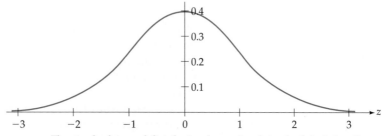

The standard normal distribution (mean 0 and standard deviation 1)

To change any x-value for a normal distribution with mean μ and standard deviation σ into the corresponding z-score for the *standard normal distribution*, subtract the mean and then divide by the standard deviation.

z-Score

$z =$	Standardizing an x-value

The z-score shows how many standard deviations the value is away from its mean.

EXAMPLE 1 Finding a z-Score

Convert each x-value into a z-score using the given values for μ and σ.

a. $x = 8$ with $\mu = 4$ and $\sigma = 2$ b. $x = 8$ with $\mu = 12$ and $\sigma = 4$

Solution

a. $z = \dfrac{8 - 4}{2} = 2$ b. $z = \dfrac{8 - 12}{4} = -1$

In part (a) the z-score $z = 2$ means that the original 8 is *two* standard deviations to the right of its mean. In part (b) the z-score $z = -1$ means that the 8 in that distribution is one standard deviation to the *left* of *its* mean.

■

PRACTICE PROBLEM 1

Convert $x = 20$ with $\mu = 15$ and $\sigma = 1$ into a z-score. Is this value far from the mean?   *Solution at the back of the book*

The *central limit theorems* developed by Carl F. Gauss (1777–1855) and other mathematicians during the nineteenth century proved that the errors in observed values, the means of random samples, and many other statistical quantities were normally distributed. Extensive tables of the probabilities of collections of z-scores were laboriously calculated, and the results were worth the effort because they could be used for *any* normally distributed x-values by first converting them to z-scores.

On page 614 of this appendix is a brief table for the normal distribution, giving the probability that a standard normal random variable takes a value between 0 and any given positive number. For example, to find $P(0 \leq Z \leq 1.24)$ (in words, the probability that a standard normal random variable has a value between 0 and 1.24), we locate in the table the row headed **1.2** and the column headed **0.04** (the second decimal place), and $P(0 \leq Z \leq 1.24)$ is the number where this row and column intersect, **0.3925**.

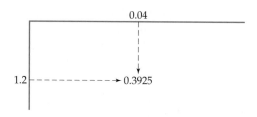

In the table on
page 614, the
row headed **1.2**
and the column
headed **0.04** inter-
sect at the table
value **0.3925**

Therefore,

$$P(0 \leq Z \leq 1.24) \approx 0.3925$$

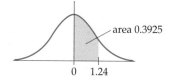

area 0.3925

0   1.24

Probabilities for other intervals can be found by adding and subtracting such areas and using the symmetry of the normal curve, as
the following examples illustrate.

EXAMPLE 2   **Finding a Probability for a Normal
Random Variable**

The heights of American men are approximately normally distributed
with mean $\mu = 68.1$ inches and standard deviation $\sigma = 2.7$ inches.*
Find the proportion of men who are between 5 feet 9 inches and 6 feet
tall.

Solution

Converting to inches, we want the probability of a man's height being
between 69 inches and 72 inches. We must convert these numbers to
the corresponding z-scores for a *standard* normal random variable by
subtracting the mean and dividing by the standard deviation:

$$x = 69 \text{ corresponds to }\quad z = \frac{69 - 68.1}{2.7} \approx 0.33$$

Using $z = \frac{x - \mu}{\sigma}$ with
$\mu = 68.1$ and $\sigma = 2.7$

$$x = 72 \text{ corresponds to }\quad z = \frac{72 - 68.1}{2.7} \approx 1.44$$

Using these values, we then want $P(0.33 \leq Z \leq 1.44)$, which is equivalent to the shaded area in the first of the following graphs, which is

* These and other data in this appendix are from *Handbook of Human Factors* by Gavriel
Salvendy (New York: John Wiley-Interscience, 1987).

equal to the difference between the next two areas on the right. These two areas (or probabilities) are found from the table on page 614, with the calculations shown below.

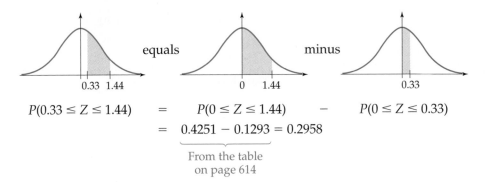

$$P(0.33 \leq Z \leq 1.44) \quad = \quad P(0 \leq Z \leq 1.44) \quad - \quad P(0 \leq Z \leq 0.33)$$

$$= \quad 0.4251 - 0.1293 = 0.2958$$

From the table
on page 614

Therefore, about 30% of American men are between 5 feet 9 inches and 6 feet tall.

PRACTICE PROBLEM 2

The weights of American women are approximately normally distributed with mean 134.7 pounds and standard deviation 30.4 pounds. Find the proportion of women who weigh between 130 and 150 pounds. *Solution at the back of the book*

The Normal and Binominal Distributions

As we saw on page 450, the binomial distribution too has a kind of "bell" shape, with a peak at the expected value $\mu = np$ and falling on both sides to very small probabilities several standard deviations $\sigma = \sqrt{np(1 - p)}$ away.

$n = 20, p = 0.4$ $n = 15, p = 0.5$ $n = 25, p = 0.7$

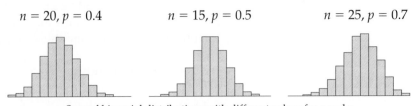

Several binomial distributions with different values for n and p

In the eighteenth century, Abraham de Moivre (1667–1754) and Pierre-Simon Laplace (1749–1827) discovered and proved that for any choice of p between 0 and 1, the binomial distribution *approaches* the normal distribution as n becomes large. This fundamental fact is known as the *de Moivre–Laplace theorem*.

We may use the de Moivre–Laplace theorem to approximate binomial distributions by the normal distribution.

Normal Approximation to the Binomial

Let X be a binomial random variable with parameters n and p. If $np > 5$ and $n(1 - p) > 5$, then the distribution of X is approximately normal with mean $\mu = np$ and standard deviation $\sigma =$

To have the same width for a "slice" of the normal distribution as for the binomial, we adopt the convention that the binomial probability $P(X = x)$ corresponds to the area under the normal distribution curve from $x - \frac{1}{2}$ to $x + \frac{1}{2}$. This is called the *continuous correction.*

$P(X = x)$ is approximately   Area from $x - \frac{1}{2}$ to $x + \frac{1}{2}$

EXAMPLE 3   **Normal Approximation of a Binomial Probability**

Estimate $P(X = 12)$ for a binomial random variable X with $n = 25$ and $p = 0.6$ using the corresponding normal distribution.

Solution

Since $np = (25)(0.6) = 15$ and $n(1 - p) = (25)(0.4) = 10$ are both greater than 5, we may use the normal distribution with $\mu = np = (25)(0.6) = 15$ and $\sigma = \sqrt{np(1 - p)} = \sqrt{(25)(0.6)(0.4)} = \sqrt{6} \approx 2.45$. By the continuous correction we interpret the event $X = 12$ as $11.5 \leq X \leq 12.5$ to include numbers that would round to 12. Converting these x-values into z-scores, we find that

$$x = 11.5 \text{ corresponds to } \quad z = \frac{11.5 - 15}{2.45} \approx -1.43$$

$$x = 12.5 \text{ corresponds to } \quad z = \frac{12.5 - 15}{2.45} \approx -1.02$$

Using $z = \dfrac{x - \mu}{\sigma}$ with $\mu = 15$ and $\sigma = 2.45$

The probability $P(-1.43 \le Z \le -1.02)$ is represented below by the shaded area in the first graph, which, by symmetry, is equivalent to the second graph, which in turn is equivalent to the difference between the third and fourth graphs. The calculation with the probabilities found from the normal table is shown below.

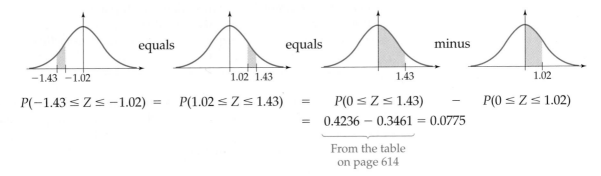

$$P(-1.43 \le Z \le -1.02) \; = \; P(1.02 \le Z \le 1.43) \; = \; P(0 \le Z \le 1.43) \; - \; P(0 \le Z \le 1.02)$$
$$= \; \underbrace{0.4236 - 0.3461}_{\substack{\text{From the table} \\ \text{on page 614}}} = 0.0775$$

The required probability is (about) 0.078. This is a good approximation to the actual value of $_{25}C_{12}(0.6)^{12}(1 - 0.6)^{13} \approx 0.0759667$. ∎

EXAMPLE 4   **Management MBAs**

At a major Los Angeles accounting firm, 73% of the managers have MBA degrees. In a random sample of 40 of these managers, what is the probability that between 27 and 32 will have MBAs?

Solution

Because each manager either has or does not have an MBA, presumably independently of each other, the question asks for the probability that a binomial random variable X with $n = 40$ and $p = 0.73$ satisfies $27 \le X \le 32$. Since $np = (40)(0.73) = 29.2$ and $n(1 - p) = (40)(0.27) = 10.8$ are both greater than 5, this probability can be approximated as the area under the normal distribution curve with mean $\mu = np = (40)(0.73) = 29.2$ and standard deviation $\sigma = \sqrt{np(1 - p)} = \sqrt{(40)(0.73)(0.27)} \approx 2.81$ from $26\frac{1}{2}$ to $32\frac{1}{2}$ (again using the continuous correction to include values that would round to between 27 and 32). We convert the x-values to z-scores:

$x = 26.5$ corresponds to $\quad z = \dfrac{26.5 - 29.2}{2.81} \approx -0.96 \qquad$ Using $z = \dfrac{x - \mu}{\sigma}$

$x = 32.5$ corresponds to $\quad z = \dfrac{32.5 - 29.2}{2.81} \approx 1.17 \qquad$ with $\mu = 29.2$ and $\sigma = 2.81$

The probability $P(-0.96 \leq Z \leq 1.17)$ is represented by the shaded area in the first graph, which is equivalent to the sum of the next two areas, with the calculation shown below.

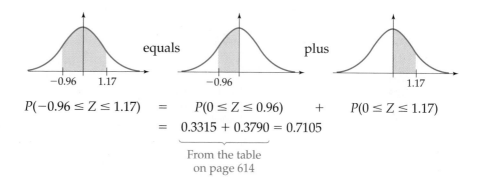

$$P(-0.96 \leq Z \leq 1.17) \quad = \quad P(0 \leq Z \leq 0.96) \quad + \quad P(0 \leq Z \leq 1.17)$$
$$= \quad 0.3315 + 0.3790 = 0.7105$$

From the table
on page 614

The probability that the number of MBAs will be between 27 and 32 is 0.711, or about 71%. (The exact answer to this problem, from summing the binomial distribution from 27 to 32, is 71.5%, so the normal approximation is very accurate.)

PRACTICE PROBLEM 3

A brand of imported VCR is known to have defective tape rewind mechanisms in 8% of the units imported last April. If Jerry's Discount Electronics received a shipment of 80 of these VCRs, what is the probability that 10 or more are defective?   *Solution at the back of the book*

After completing Practice Problem 3, you should return to page 499 to read the Section Summary and do the Exercises in Section 6.4.

**Area under the standard
normal distribution from 0 to z**

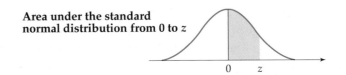

x	0.00	0.01	0.02	0.03	0.04	0.05	0.06	0.07	0.08	0.09
0.0	0.0000	0.0040	0.0080	0.0120	0.0160	0.0199	0.0239	0.0279	0.0319	0.0359
0.1	0.0398	0.0438	0.0478	0.0517	0.0557	0.0596	0.0636	0.0675	0.0714	0.0754
0.2	0.0793	0.0832	0.0871	0.0910	0.0948	0.0987	0.1026	0.1064	0.1103	0.1141
0.3	0.1179	0.1217	0.1255	0.1293	0.1331	0.1368	0.1406	0.1443	0.1480	0.1517
0.4	0.1554	0.1591	0.1628	0.1664	0.1700	0.1736	0.1772	0.1808	0.1844	0.1879
0.5	0.1915	0.1950	0.1985	0.2019	0.2054	0.2088	0.2123	0.2157	0.2190	0.2224
0.6	0.2258	0.2291	0.2324	0.2357	0.2389	0.2422	0.2454	0.2486	0.2518	0.2549
0.7	0.2580	0.2612	0.2642	0.2673	0.2704	0.2734	0.2764	0.2794	0.2823	0.2852
0.8	0.2881	0.2910	0.2939	0.2967	0.2996	0.3023	0.3051	0.3078	0.3106	0.3133
0.9	0.3159	0.3186	0.3212	0.3238	0.3264	0.3289	0.3315	0.3340	0.3365	0.3389
1.0	0.3413	0.3438	0.3461	0.3485	0.3508	0.3531	0.3554	0.3577	0.3599	0.3621
1.1	0.3643	0.3665	0.3686	0.3708	0.3729	0.3749	0.3770	0.3790	0.3810	0.3820
1.2	0.3849	0.3869	0.3888	0.3907	0.3925	0.3944	0.3962	0.3980	0.3997	0.4015
1.3	0.4032	0.4049	0.4066	0.4082	0.4099	0.4115	0.4131	0.4147	0.4162	0.4177
1.4	0.4192	0.4207	0.4222	0.4236	0.4251	0.4265	0.4279	0.4292	0.4306	0.4319
1.5	0.4332	0.4345	0.4357	0.4370	0.4382	0.4394	0.4406	0.4418	0.4429	0.4441
1.6	0.4452	0.4463	0.4474	0.4484	0.4495	0.4505	0.4515	0.4525	0.4535	0.4545
1.7	0.4554	0.4564	0.4573	0.4582	0.4591	0.4599	0.4608	0.4616	0.4625	0.4633
1.8	0.4641	0.4649	0.4656	0.4664	0.4671	0.4678	0.4686	0.4693	0.4699	0.4706
1.9	0.4713	0.4719	0.4726	0.4732	0.4738	0.4744	0.4750	0.4756	0.4761	0.4767
2.0	0.4772	0.4778	0.4783	0.4788	0.4793	0.4798	0.4803	0.4808	0.4812	0.4817
2.1	0.4821	0.4826	0.4830	0.4834	0.4838	0.4842	0.4846	0.4850	0.4854	0.4857
2.2	0.4861	0.4864	0.4868	0.4871	0.4875	0.4878	0.4881	0.4884	0.4887	0.4890
2.3	0.4893	0.4896	0.4898	0.4901	0.4904	0.4906	0.4909	0.4911	0.4913	0.4916
2.4	0.4918	0.4920	0.4922	0.4925	0.4927	0.4929	0.4931	0.4932	0.4934	0.4936
2.5	0.4938	0.4940	0.4941	0.4943	0.4945	0.4946	0.4948	0.4949	0.4951	0.4952
2.6	0.4953	0.4955	0.4956	0.4957	0.4959	0.4960	0.4961	0.4962	0.4963	0.4964
2.7	0.4965	0.4966	0.4967	0.4968	0.4969	0.4970	0.4971	0.4972	0.4973	0.4974
2.8	0.4974	0.4975	0.4976	0.4977	0.4977	0.4978	0.4979	0.4979	0.4980	0.4981
2.9	0.4981	0.4982	0.4982	0.4983	0.4984	0.4984	0.4985	0.4985	0.4986	0.4986
3.0	0.4987	0.4987	0.4987	0.4988	0.4988	0.4989	0.4989	0.4989	0.4990	0.4990
3.1	0.4990	0.4991	0.4991	0.4991	0.4992	0.4992	0.4992	0.4992	0.4993	0.4993
3.2	0.4993	0.4993	0.4994	0.4994	0.4994	0.4994	0.4994	0.4995	0.4995	0.4995
3.3	0.4995	0.4995	0.4995	0.4996	0.4996	0.4996	0.4996	0.4996	0.4996	0.4997
3.4	0.4997	0.4997	0.4997	0.4997	0.4997	0.4997	0.4997	0.4997	0.4997	0.4998
3.5	0.4998	0.4998	0.4998	0.4998	0.4998	0.4998	0.4998	0.4998	0.4998	0.4998

Solutions to Practice Problems

SECTION 1.1

1. $-1,000,000$ [the negative sign makes it less than (to the left of) the positive number $\frac{1}{100}$]

2. **a.** $\{x \mid x \geq -7\}$
b. the set of all x such that x is less than -1

3. **a.** S_1
b. S_4

4. $m = \dfrac{7-1}{4-2} = \dfrac{6}{2} = 3$ From points $(2, 1)$ and $(4, 7)$
$y - 1 = 3(x - 2)$ Using the point–slope form with
$y - 1 = 3x - 6$ $(x_1, y_1) = (2, 1)$
$y = 3x - 5$

5. $x = -2$

6. $x - \dfrac{y}{3} = 2$

     $-\dfrac{y}{3} = -x + 2$ Subtracting x from each side
     $y = 3x - 6$ Multiplying each side by -3

Slope is $m = 3$ and y-intercept is $(0, -6)$.

SECTION 1.2

1. **a.** $\dfrac{x^5 \cdot x}{x^2} = \dfrac{x^6}{x^2} = x^4$
b. $\left[(x^3)^2\right]^2 = x^{3 \cdot 2 \cdot 2} = x^{12}$

2. **a.** $2^0 = 1$
b. $2^{-4} = \dfrac{1}{2^4} = \dfrac{1}{16}$

3. $\left(\dfrac{2}{3}\right)^{-2} = \left(\dfrac{3}{2}\right)^2 = \dfrac{9}{4}$

4. **a.** $(-27)^{1/3} = \sqrt[3]{-27} = -3$
b. $\left(\dfrac{16}{81}\right)^{1/4} = \sqrt[4]{\dfrac{16}{81}} = \dfrac{2}{3}$

5. **a.** $16^{3/2} = \left(\sqrt{16}\right)^3 = 4^3 = 64$
b. $(-8)^{2/3} = \left(\sqrt[3]{-8}\right)^2 = (-2)^2 = 4$

6. **a.** $25^{-3/2} = \dfrac{1}{25^{3/2}} = \dfrac{1}{\left(\sqrt{25}\right)^3} = \dfrac{1}{5^3} = \dfrac{1}{125}$
b. $\left(\dfrac{1}{4}\right)^{-1/2} = \left(\dfrac{4}{1}\right)^{1/2} = \sqrt{4} = 2$
c. $5^{1.3} \approx 8.103$

SECTION 1.3

1. Domain: $\{x \mid x \leq 0 \text{ or } x \geq 3\}$, Range: $\{y \mid y \geq 0\}$

2. **a.** $g(27) = \sqrt{27 - 2} = \sqrt{25} = 5$
b. Domain: $\{z \mid z \geq 2\}$
c. Range: $\{y \mid y \geq 0\}$

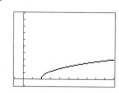

$y_1 = \sqrt{x - 2}$ on $[-1, 10]$ by $[-1, 10]$

3. $D(x) = 25 + 0.05x$

4. $9x - 3x^2 = -30$
     $-3x^2 + 9x + 30 = 0$
     $-3(x^2 - 3x - 10) = 0$
     $-3(x - 5)(x + 2) = 0$
     $x = 5, x = -2$ or from

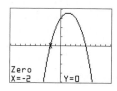

SECTION 1.4

1. $2x^3 - 4x^2 - 48x = 0$
     $2x(x^2 - 2x - 24) = 0$
     $2x(x + 4)(x - 6) = 0$
     $x = 0, x = -4, x = 6$

2. Domain: $\{x \mid x \neq 0,\ x \neq -10\}$
　　Range: $\{y \mid y > 0 \text{ or } y \leq -7200\}$

3. a. $f(g(x)) = [g(x)]^2 + 1 = \left(\sqrt[3]{x}\right)^2 + 1$ or $x^{2/3} + 1$

　　b. $g(f(x)) = \sqrt[3]{f(x)} = \sqrt[3]{x^2 + 1}$ or $(x^2 + 1)^{1/3}$

SECTION 1.5

1. $10{,}000(1 + 0.07)^{30} = 10{,}000 \cdot 1.07^{30} \approx 76{,}122.55$
　　The value will be \$76,122.55.

2. $50{,}000(1 - 0.20)^4 = 50{,}000(0.8)^4 = \$20{,}480$

SECTION 1.6

1. $\log 10{,}000 = 4$　　　　　　　(Since $10^4 = 10{,}000$)

2. $\ln 8.34 \approx 2.121$　　　　　　(Using a calculator)

SECTION 2.1

1. $I = (50{,}000)(0.198)(92/360) = 2530$, so the Banker's rule interest is \$2530. A 3-month loan would have interest $I = (50{,}000)(0.198)(3/12) = 2475$, which means the Banker's rule gives the lender a $2530 - 2475 = \$55$ advantage.

2. $PV = \dfrac{5000}{1 + (4)(0.12)} = \dfrac{5000}{1.48} \approx 3378.38$

3. $r_s = \dfrac{0.06}{1 - (0.06)(3)} \approx 0.0732.$ The effective simple

　　interest rate is 7.32%. $r_s = \dfrac{0.06}{1 - (0.06)(5)} \approx 0.0857.$
　　The effective simple interest rate is 8.57%. Can you think of an intuitive reason for the effective rate to be higher if the term is longer? [*Hint:* Think of how much earlier the lender gets the money or of how much less the borrower really gets.]

SECTION 2.2

1. $A = 3500\left(1 + \dfrac{0.051}{6}\right)^{(6)(8)} \approx 5254.26.$
　　The amount due is \$5254.26.

2. $2P = P\left(1 + \dfrac{0.06}{4}\right)^{4t}$, which simplifies to
　　$2 = (1 + 0.06/4)^{4t}$. Taking logarithms:
　　$\log 2 = \log(1 + 0.06/4)^{4t} = 4t \log(1 + 0.06/4)$ so

$4t = \dfrac{\log 2}{\log(1 + 0.06/4)} \approx 46.6$ (quarters!).
Rounding up gives 47 quarters, or $11\frac{3}{4}$ years.

3. $r_e = \left(1 + \dfrac{0.062}{12}\right)^{12} - 1 \approx 0.0638.$
The effective rate is 6.38%.

SECTION 2.3

1. $A = 40\,\dfrac{(1 + 0.05/52)^{(52)(35)} - 1}{0.05/52} \approx 197{,}590.27.$ The final balance is \$197,590.27. Because the deposits total $\$40 \times 52 \times 35 = \$72{,}800$, the final balance contains \$124,790.27 interest.

2. $P = 18{,}000\,\dfrac{0.045/26}{(1 + 0.045/26)^{(26)(3)} - 1} \approx 215.742.$
\$215.75 should be deposited every other week. (Compare double this amount, \$431.50, to the \$467.95 monthly payment found in Example 2 on page 131.)

3. $12t = \dfrac{\log\left(\dfrac{18{,}000}{250}\dfrac{0.045}{12} + 1\right)}{\log\left(1 + \dfrac{0.045}{12}\right)} \approx 63.9$ months,

which rounds up to 64 months. It will take 5 years 4 months.

SECTION 2.4

1. $PV = 850\,\dfrac{1 - (1 + 0.0753/12)^{-(12)(20)}}{0.0753/12} \approx 105{,}272.468.$
The present value is \$105,272.47.

2. $P = 150{,}000\,\dfrac{0.086/12}{1 - (1 + 0.086/12)^{-(12)(25)}} \approx 1217.97.$ The required payment is \$1217.97 each month. The borrower will pay a total of $\$1217.97 \times 12 \times 25 = \$365{,}391.$

3. The monthly payment to amortize this loan is
$P = 12{,}000\,\dfrac{0.047/12}{1 - (1 + 0.047/12)^{-(12)(4)}} \approx 274.724,$
which rounds up to \$274.73. The remaining payments form a 3-year annuity at 4.7%, and the present value of this annuity is
$PV = 274.73\,\dfrac{1 - (1 + 0.047/12)^{-(12)(3)}}{0.047/12} \approx 9207.884.$
The amount still owed is \$9207.88.

SECTION 3.1

1. Let x be the number of pennies in the jar and y be the number of nickels. Then the first statement may be expressed as "$x + y = 80$" and the second as "$0.01x + 0.05y = 1.60$" since each penny is worth $0.01 and each nickel $0.05 (if you want to write the second statement in cents rather than dollars, you would have "$x + 5y = 160$"). The situation may be represented as

$$\begin{cases} x + y = 80 \\ 0.01x + 0.05y = 1.60 \end{cases}$$

2. Since the first equation can be solved for y as $y = 10 - 2x$, we can substitute $10 - 2x$ for y in the second equation: $x + 2(10 - 2x) = 8$. Multiplying out and collecting like terms, $x + 20 - 4x = 8$ so $-3x = -12$ so that $x = 4$. Substituting $x = 4$ into $y = 10 - 2x$ gives $y = 10 - 2(4) = 10 - 8 = 2$. The solution is $x = 4$, $y = 2$. There are several other ways of solving this problem by the substitution method, and all reach the same conclusion.

3. $\begin{cases} x + y = 100 \\ x - y = -20 \end{cases}$ $\xrightarrow[\text{to first}]{\text{Add second}}$ $\begin{cases} 2x + 0y = 80 \\ x - y = -20 \end{cases}$ $\xrightarrow[\text{by 2}]{\text{Divide first}}$

$\begin{cases} 1x + 0y = 40 \\ x - y = -20 \end{cases}$ $\xrightarrow[\text{from second}]{\text{Subtract first}}$ $\begin{cases} 1x + 0y = 40 \\ 0x - y = -60 \end{cases}$ $\xrightarrow[\text{by} -1]{\text{Multiply second}}$

$\begin{cases} 1x + 0y = 40 \\ 0x + 1y = 60 \end{cases}$

The solution is $x = 40$, $y = 60$. There are many other possible sequences of equivalent systems that solve this problem, and all reach the same conclusion.

SECTION 3.2

1. **a.** $\begin{pmatrix} 2 & -1 & 14 \\ 1 & 3 & 21 \end{pmatrix}$

   b. $\begin{cases} 3x + 2y = 35 \\ x + 3y = 21 \end{cases}$

2. **a.** $\begin{pmatrix} 6 & 3 & 42 \\ 1 & -3 & 21 \end{pmatrix} R_1' = 3R_1$

   b. $\begin{pmatrix} 7 & 0 & 63 \\ 1 & -3 & 21 \end{pmatrix} R_1' = R_1 + R_2$

   c. $\begin{pmatrix} 1 & 0 & 9 \\ 1 & -3 & 21 \end{pmatrix} R_1' = \frac{1}{7}R_1$

   d. No: you can multiply only by a *nonzero* number.
   e. No: Multiplying row 1 by 5 must still give row 1 (not row 2).

3. Continuing from the solutions to Practice Problem 2, $\begin{pmatrix} 1 & 0 & 9 \\ 0 & -3 & 12 \end{pmatrix} R_2' = R_2 - R_1$ and then $\begin{pmatrix} 1 & 0 & 9 \\ 0 & 1 & -4 \end{pmatrix} R_2' = -\frac{1}{3}R_2$. The solution is $x = 9$, $y = -4$ and the system of equations is independent and consistent. There are many other sequences of row operations to reduce this augmented matrix and all reach the same conclusion.

4. For $x = -5$, $y = -20$, the first equation becomes $6(-5) - 3(-20) = -30 + 60 = 30$ and the second becomes $-8(-5) + 4(-20) = 40 - 80 = -40$ as needed. For $x = 15$, $y = 20$, the first equation becomes $6(15) - 3(20) = 90 - 60 = 30$ and the second becomes $-8(15) + 4(20) = -120 + 80 = -40$ as needed.

SECTION 3.3

1. $x_1 = -2t$
   $x_2 = t$
   $x_3 = 3$
   $x_4 = 4$

2. We found in Example 3 that the corresponding augmented matrix $\begin{pmatrix} 5 & 5 & 0 & 5 & 50 \\ 2 & 3 & 1 & 0 & 17 \\ 2 & 2 & 1 & -1 & 9 \\ 2 & 3 & 1 & 1 & 22 \end{pmatrix}$ row-reduces

   to $\begin{pmatrix} 1 & 0 & 0 & 0 & 2 \\ 0 & 1 & 0 & 0 & 3 \\ 0 & 0 & 1 & 0 & 4 \\ 0 & 0 & 0 & 1 & 5 \end{pmatrix}$. The system is independent

   (there are no zero rows) and consistent (there is no row of zeros ending in a 1). The solution, read from the last column, is $x_1 = 2$, $x_2 = 3$, $x_3 = 4$, $x_4 = 5$. We check this solution by substitution into the original equations:

   $$5(2) + 5(3) + 0(4) + 5(5) = 50$$
   $$2(2) + 3(3) + 1(4) + 0(5) = 17$$
   $$2(2) + 2(3) + 1(4) - 1(5) = 9$$
   $$2(2) + 3(3) + 1(4) + 1(5) = 22$$

   It checks!

SECTION 3.4

1. $(1 \quad 2) \cdot \begin{pmatrix} 3 \\ 4 \end{pmatrix} = (1 \cdot 3 + 2 \cdot 4) = (11)$

2. $\begin{pmatrix} 1.00 & 1.50 & 0.75 \\ 1.25 & 1.75 & 0.50 \end{pmatrix} \begin{pmatrix} 4 \\ 2 \\ 5 \end{pmatrix}$

$$= \begin{pmatrix} 1.00 \cdot 4 + 1.50 \cdot 2 + 0.75 \cdot 5 \\ 1.25 \cdot 4 + 1.75 \cdot 2 + 0.50 \cdot 5 \end{pmatrix} = \begin{pmatrix} 10.75 \\ 11.00 \end{pmatrix}$$

The prices are $10.75 at McBurger and $11 at BurgerQueen.

SECTION 3.5

1. $A^{-1}A = \begin{pmatrix} 1 & 2 & -2 \\ -1 & 0 & 1 \\ 0 & -1 & 1 \end{pmatrix} \begin{pmatrix} 1 & 0 & 2 \\ 1 & 1 & 1 \\ 1 & 1 & 2 \end{pmatrix}$

$$= \begin{pmatrix} 1+2-2 & 0+2-2 & 2+2-4 \\ -1+0+1 & 0+0+1 & -2+0+2 \\ 0-1+1 & 0-1+1 & 0-1+2 \end{pmatrix}$$

$$= \begin{pmatrix} 1 & 0 & 0 \\ 0 & 1 & 0 \\ 0 & 0 & 1 \end{pmatrix} = I$$

2. Using $X = A^{-1}B$:

$$\begin{pmatrix} x_1 \\ x_2 \\ x_3 \end{pmatrix} = \begin{pmatrix} 1 & 2 & -2 \\ -1 & 0 & 1 \\ 0 & -1 & 1 \end{pmatrix} \cdot \begin{pmatrix} -5 \\ 10 \\ 0 \end{pmatrix} = \begin{pmatrix} 15 \\ 5 \\ -10 \end{pmatrix}$$

so $\begin{cases} x_1 = 15 \\ x_2 = 5 \\ x_3 = -10 \end{cases}$

SECTION 4.1

1. The boundary is the line $3x - 5y = 60$. From $x = 60/3 = 20$, the x-intercept is $(20, 0)$. From $y = 60/(-5) = -12$, the y-intercept is $(0, -12)$. Because $3 \cdot 0 - 5 \cdot 0$ is ≤ 60, the origin $(0, 0)$ is on the correct side of the boundary line.

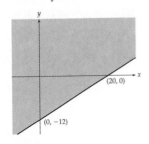

2. The boundaries are the lines $x + 2y = 20$ [with intercepts $(20, 0)$ and $(0, 10)$], $x + y = 10$ [with intercepts $(10, 0)$ and $(0, 10)$], and $x = 10$ [a vertical line with x-intercept $(10, 0)$ and no y-intercept]. The origin $(0, 0)$ is on the correct side of $x + 2y \leq 20$ and $x \leq 10$ but not on the correct side of $x + y \geq 10$. This *is* a feasible system of linear inequalities.

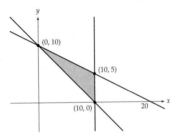

3. Two of the three vertices of the region are known from the x- and y-intercepts of the boundary lines in the previous sketch for Practice Problem 2. The third is the intersection of the lines $x + 2y = 20$ and $x = 10$. Substituting $x = 10$ into the first equation, $(10) + 2y = 20$ means $2y = 20 - 10 = 10$ so $y = 10/2 = 5$ and this vertex is $(10, 5)$. The vertices of the region are $(10, 0)$, $(10, 5)$, and $(0, 10)$. The region *is* bounded.

SECTION 4.2

1.

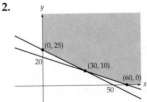

Vertex	$P = 5x + 2y$
$(0, 0)$	0
$(20, 0)$	100
$(8, 36)$	112
$(0, 36)$	72

The maximum value is 112 when $x = 8$ and $y = 36$.

2.

Vertex	$C = 5x + 11y$
$(60, 0)$	300
$(30, 10)$	260
$(0, 25)$	275

The minimum value is 260 when $x = 30$ and $y = 10$.

SECTION 4.3

1. The matrix form of the problem is

$$\text{Maximize } P = (9 \quad -1 \quad 10 \quad 12)\begin{pmatrix} x_1 \\ x_2 \\ x_3 \\ x_4 \end{pmatrix}$$

Subject to
$$\begin{cases} \begin{pmatrix} 1 & 1 & 2 & 2 \\ 2 & 2 & 1 & 1 \\ 1 & 2 & 2 & 1 \end{pmatrix}\begin{pmatrix} x_1 \\ x_2 \\ x_3 \\ x_4 \end{pmatrix} \leq \begin{pmatrix} 16 \\ 20 \\ 18 \end{pmatrix} \\[2em] \text{and } \begin{pmatrix} x_1 \\ x_2 \\ x_3 \\ x_4 \end{pmatrix} \geq 0 \end{cases}$$

↑
Standard since
$$b = \begin{pmatrix} 16 \\ 20 \\ 18 \end{pmatrix} \geq 0$$

The simplex tableau is

	x_1	x_2	x_3	x_4	s_1	s_2	s_3	
s_1	1	1	2	2	1	0	0	16
s_2	2	2	1	1	0	1	0	20
s_3	1	2	2	1	0	0	1	18
P	−9	1	−10	−12	0	0	0	0

2. a. The smallest negative entry in the bottom row is
−6, so the pivot column is column 3. The ratios
are $\frac{4}{1} = 4$ and $\frac{9}{3} = 3$, so the smallest nonnegative
ratio is 3 and the pivot row is row 2. The pivot
element is the 3 in column 3 and row 2 of the
simplex tableau.

	x_1	x_2	x_3	x_4	s_1	s_2			
s_1	−2	0	0	$\frac{1}{3}$	1	$-\frac{1}{3}$	1	$R_1^{new} = R_1 - (1)R_{pivot}^{new}$
x_3	1	2	1	$\frac{2}{3}$	0	$\frac{1}{3}$	3	$R_{pivot}^{new} = R_{pivot}/3$
P	1	9	0	2	0	2	18	$R_3^{new} = R_3 - (-6)R_{pivot}^{new}$

b. The smallest negative entry in the bottom row is
−4, so the pivot column is column 3. The first
row may not be considered for the pivot row because the pivot column entry is negative. The
second row may not be considered for the pivot
row because the pivot column entry is zero.
There is no pivot row for this simplex tableau,
and therefore *there is no pivot element.* This means

that the original linear programming problem
has *no* solution (see page 301).

3. The basic variables take the values at the right ends
of their rows, so $x_2 = 6$, $s_2 = 6$, and $x_1 = 4$. The nonbasic variables are zero, so $s_1 = 0$ and $s_3 = 0$. The
value of the objective function appears in the bottom right corner, so $P = 62$. *Solution:* The maximum
is 62 when $x_1 = 4$ and $x_2 = 6$.

SECTION 4.4

1. In matrix form, this problem is

$$\text{Minimize } C = (11 \quad 9 \quad 7)\begin{pmatrix} y_1 \\ y_2 \\ y_3 \end{pmatrix}$$

Subject to
$$\begin{cases} \begin{pmatrix} 2 & 1 & 1 \\ 1 & 1 & 0 \\ 0 & 1 & 1 \\ 1 & 1 & 3 \end{pmatrix}\begin{pmatrix} y_1 \\ y_2 \\ y_3 \end{pmatrix} \geq \begin{pmatrix} 4 \\ 7 \\ 5 \\ 6 \end{pmatrix} \\[2em] \text{and } \begin{pmatrix} y_1 \\ y_2 \\ y_3 \end{pmatrix} \geq 0 \end{cases}$$

and the dual maximum problem is

$$\text{Maximize } P = (4 \quad 7 \quad 5 \quad 6)\begin{pmatrix} x_1 \\ x_2 \\ x_3 \\ x_4 \end{pmatrix}$$

Subject to
$$\begin{cases} \begin{pmatrix} 2 & 1 & 0 & 1 \\ 1 & 1 & 1 & 1 \\ 1 & 0 & 1 & 3 \end{pmatrix}\begin{pmatrix} x_1 \\ x_2 \\ x_3 \\ x_4 \end{pmatrix} \leq \begin{pmatrix} 11 \\ 9 \\ 7 \end{pmatrix} \\[2em] \text{and } \begin{pmatrix} x_1 \\ x_2 \\ x_3 \\ x_4 \end{pmatrix} \geq 0 \end{cases}$$

The initial simplex tableau is

	x_1	x_2	x_3	x_4	s_1	s_2	s_3	
s_1	2	1	0	1	1	0	0	11
s_2	1	1	1	1	0	1	0	9
s_3	1	0	1	3	0	0	1	7
P	−4	−7	−5	−6	0	0	0	0

2. The bottom row of the final tableau for the dual maximum problem displays the values for the slack variables ($t_1 = 0$, $t_2 = 0$, $t_3 = 0$), the variables ($y_1 = 7$, $y_2 = 3$, $y_3 = 1$, $y_4 = 0$), and the objective function ($C = 3390$). The minimum value is 3390 when $y_1 = 7$, $y_2 = 3$, $y_3 = 1$, and $y_4 = 0$.

SECTION 4.5

1. a. The dual pivot row is row 1 because -10 is the smallest negative entry in the rightmost column (omitting the bottom row). The dual pivot column is column 2 because the ratio $\frac{-6}{-1} = 6$ is greater than the ratio $\frac{-5}{-1} = 5$ for the first column and the ratio $\frac{8}{-2} = -4$ for the fourth column; the other columns may not be considered since their dual pivot row entries are zero or positive, and the rightmost column is never considered. The dual pivot element is the -1 in row 1 and column 2.

b. The tableau does not have a dual pivot element. The dual pivot row is row 4 because the smallest negative entry in the rightmost column (omitting the bottom row) is -20. The other entries in row 4 are either zero or positive, so there is no dual pivot column. (This means that the constraints are infeasible. The fourth row represents the inequality $2x_1 + x_2 + x_3 + 4x_4 \leq -20$ and this is impossible because the variables are nonnegative. The problem has no solution.)

2. Rewriting the second inequality as $-2x_1 - x_2 - 2x_3 \leq -10$, this problem may be written in matrix form as

Maximize $P = (4 \quad 1 \quad 3)\begin{pmatrix} x_1 \\ x_2 \\ x_3 \end{pmatrix}$

Subject to $\begin{cases} \begin{pmatrix} 1 & 2 & 1 \\ -2 & -1 & -2 \end{pmatrix}\begin{pmatrix} x_1 \\ x_2 \\ x_3 \end{pmatrix} \leq \begin{pmatrix} 50 \\ -10 \end{pmatrix} \\ \\ \text{and } \begin{pmatrix} x_1 \\ x_2 \\ x_3 \end{pmatrix} \geq 0 \end{cases}$

The initial simplex tableau is

	x_1	x_2	x_3	s_1	s_2	
s_1	1	2	1	1	0	50
s_2	-2	-1	-2	0	1	-10
P	-4	-1	-3	0	0	0

This is not feasible because $s_2 = -10$. Pivoting on the dual pivot element in row 2 and column 1, the tableau becomes feasible:

	x_1	x_2	x_3	s_1	s_2	
s_1	0	3/2	0	1	1/2	45
x_1	1	1/2	1	0	-1/2	5
P	0	1	1	0	-2	20

Pivoting on the (regular) pivot element in column 5 and row 1, the tableau becomes optimal:

	x_1	x_2	x_3	s_1	s_2	
s_2	0	3	0	2	1	90
x_1	1	2	1	1	0	50
P	0	7	1	4	0	200

This is the final tableau because it is both feasible and optimal. The maximum is $P = 200$ when $x_1 = 50$ and $x_2 = 0$.

SECTION 4.6

1. Since the 10, 20, and 5 are all nonnegative, this maximum problem is in normal form except for the order of the constraints. In normal form, the problem is

Maximize $P = 5x_1 + 6x_2 - 14x_3 - 8x_4$

Subject to $\begin{cases} x_1 + x_2 - x_3 - x_4 \leq 20 \\ x_1 + x_2 - 2x_3 + 2x_4 \geq 10 \\ x_2 - 3x_3 - 2x_4 \geq 5 \\ x_1 \geq 0, x_2 \geq 0, x_3 \geq 0, x_4 \geq 0 \end{cases}$

The slack variable s_1 from the first constraint can be used in the initial basis, but the second and third require the introduction of artificial variables a_2 and a_3:

	x_1	x_2	x_3	x_4	s_1	s_2	s_3	a_2	a_3	
s_1	1	1	-1	-1	1	0	0	0	0	20
a_2	1	1	-2	2	0	-1	0	1	0	10
a_3	0	1	-3	-2	0	0	-1	0	1	5
P	$-M-5$	$-2M-6$	$5M+14$	8	0	M	M	0	0	$-15M$

2. We need one artificial variable a_2 for the second constraint.

	x_1	x_2	x_3	s_1	s_2	a_2	
s_1	1	2	1	1	0	0	50
a_2	2	1	2	0	-1	1	10
P	$-2M-4$	$-M-1$	$-2M-3$	0	M	0	$-10M$

Pivoting on the 2 in column 1 and row 2, we obtain

	x_1	x_2	x_3	s_1	s_2	a_2	
s_1	0	3/2	0	1	1/2	$-1/2$	45
x_1	1	1/2	1	0	$-1/2$	1/2	5
P	0	1	1	0	-2	$M+2$	20

Pivoting on the 1/2 in column 5 and row 1, we reach the final tableau:

	x_1	x_2	x_3	s_1	s_2	a_2	
s_2	0	3	0	2	1	-1	90
x_1	1	2	1	1	0	0	50
P	0	7	1	4	0	M	200

Because the extended problem has a solution and the artificial variable is zero, the solution of the original problem is: The maximum is $P = 200$ when $x_1 = 50$, $x_2 = 0$, and $x_3 = 0$.

SECTION 5.1

1. Let D be the set of businesses offering dental insurance, and let V be those offering vision insurance. Then

$$n(D \cup V) = n(D) + n(V) - n(D \cap V)$$
$$= 150 + 150 - 100 = 200$$

Two hundred of the three hundred businesses surveyed offer dental or vision insurance.

2. Since $26 \cdot 26 \cdot 26 \cdot 26 \cdot 10 \cdot 10 \cdot 10 \cdot 10 = 4{,}569{,}760{,}000$,

$$\underbrace{\qquad}_{\text{Four letters}} \underbrace{\qquad}_{\text{Four digits}}$$

there are 4,569,760,000 different passwords.

SECTION 5.2

1. There are $6! = 6 \cdot 5 \cdot 4 \cdot 3 \cdot 2 \cdot 1 = 720$ different orders.

2. Since the winners must be selected in order (first, second, and third place), there are ${}_{35}P_3 = 35 \cdot 34 \cdot 33 = 39{,}270$ different ways of choosing the winners.

3. Since the courses can be taken in any order, there are ${}_{10}C_6 = \frac{10 \cdot 9 \cdot 8 \cdot 7 \cdot 6 \cdot 5}{6 \cdot 5 \cdot 4 \cdot 3 \cdot 2 \cdot 1} = 210$ different ways to fulfill a minor in computer science.

SECTION 5.3

1. "No heads" becomes $0 = \{(T,T)\}$, "one head" becomes $1 = \{(H,T), (T,H)\}$, and "two heads" becomes $2 = \{(H,H)\}$.

2. $P(A) = \frac{1}{4}$, $P(B) = \frac{1}{2}$, and $P(C) = \frac{1}{4}$.

3. In three rolls there are $6 \cdot 6 \cdot 6 = 6^3 = 216$ possible outcomes, each with probability $\frac{1}{216}$. If the event D represents at least one six (in three rolls), then D^c (*no* sixes in three rolls) contains $5 \cdot 5 \cdot 5 = 5^3 = 125$ possible outcomes, each having probability $\frac{1}{216}$. Using the summation formula, $P(D^c) = \frac{125}{216} \approx 0.579$. The probability of at least one six in three rolls of a die is then $P(D) = 1 - P(D^c) \approx 1 - 0.579 = 0.421$. Therefore, betting on at least one six in *three* rolls would win only about 42% of the time.

SECTION 5.4

1. If H^c is the event that your parents don't arrive home soon (that is, that you go on foot), then $P(H^c) = 0.50$ and $P(A \text{ given } H^c) = 0.60$. Then

$$P(A \text{ and } H^c) = P(A \text{ given } H^c) \cdot P(H^c)$$
$$= (0.60) \cdot (0.50) = 0.30$$

The probability of arriving on time and on foot is 30%.

2. Since the probability that any one salesperson does not make the sale is 0.20, the probability that all five do not is

$$\underbrace{0.20 \cdot 0.20 \cdot 0.20 \cdot 0.20 \cdot 0.20}_{\text{5 salespeople}} = 0.20^5 = 0.00032.$$

SECTION 5.5

Using the notation from the solution to Example 1 (page 436), we have

$P(I$ given $V)$

$= \dfrac{P(V \text{ given } I) \cdot P(I)}{P(V \text{ given } D) \cdot P(D) + P(V \text{ given } R) \cdot P(R) + P(V \text{ given } I) \cdot P(I)}$

$= \dfrac{(0.90) \cdot (0.25)}{(0.70) \cdot (0.45) + (0.80) \cdot (0.30) + (0.90) \cdot (0.25)} \approx 0.288$

The probability that a voter in the last election was an Independent is about 29%.

SECTION 5.6

1. $E(X) = 3000 \cdot \frac{1}{400} + 1000 \cdot \frac{2}{400} + 100 \cdot \frac{10}{400} + 0 \cdot \frac{387}{400} = 15$
   The expected value is $15.00.

2. ${}_6C_3\left(\frac{1}{2}\right)^3\left(\frac{1}{2}\right)^3 = \frac{6 \cdot 5 \cdot 4}{3 \cdot 2 \cdot 1}\left(\frac{1}{2}\right)^6 = 20 \cdot \frac{1}{64} = \frac{5}{16}$

   $\mu = n \cdot p = 6 \cdot \frac{1}{2} = 3$

   $\sigma = \sqrt{np(1-p)} = \sqrt{6 \cdot \frac{1}{2} \cdot \frac{1}{2}} \approx 1.22$

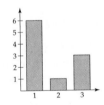

SECTION 6.1

1. **a.** Nominal **b.** Ordinal **c.** Ratio

2.
Vehicle	Tally	Frequency
1	ⴵⴵⴵ I	6
2	I	1
3	III	3

3. Choosing the stem to be the units digit and the leaf to be the first decimal place, we obtain

Stem	Leaf
0	
1	
2	5
3	7
4	6, 4
5	4, 6, 7
6	2, 0, 4, 2, 7
7	8, 6
8	1
9	

SECTION 6.2

1. Since 20 occurs three times and no other value occurs this often, the mode is 20.

2. The median of nominal data is not defined (if the integers represent different colors, then 5.5 would be half way between two of them).

3. The mode is 1, the median is 2, and the mean is

   $$\bar{x} = \tfrac{1}{5}(1 + 1 + 2 + 3 + 8) = \tfrac{1}{5}(15) = 3$$

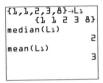

SECTION 6.3

1. The largest value is 30 and the smallest is 10, so the range is $30 - 10 = 20$.

2. The minimum is 8 and the maximum is 29. Since there are eleven values, the median is 16, the sixth value. From the first five values, {8, 10, 13, 14, 15}, the first quartile is 13, and from the last five values, {17, 19, 23, 24, 29}, the third quartile is 23. From this five-point summary, the box-and-whisker plot is

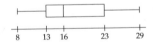

The tightest clustering is the quarter of the values between 13 and 16.

3. First, find the mean: $\bar{x} = \tfrac{1}{8}(2 + 6 + 7 + 8 + 9 + 15 + 16 + 17) = \tfrac{1}{8}(80) = 10$. Then

$$s = \sqrt{\dfrac{(2-10)^2 + \cdots + (17-10)^2}{8-1}}$$

$$= \sqrt{\dfrac{64 + 16 + 9 + 4 + 1 + 25 + 36 + 49}{7}}$$

$$= \sqrt{\dfrac{204}{7}} \approx 5.4$$

SECTION 6.4

1.
```
normalcdf(130,15
0,134.7,30.4)
          .2540533807
```

About 25% of American women weigh between 130 and 150 pounds. (For a solution using tables, see page A11.)

2. $z = \dfrac{20 - 15}{1} = 5$. Since more than 99% of the z-scores are within 3 of the mean, this value is *very* far to the right of the mean!

3. Since $np = (80)(0.08) = 6.4 > 5$ and $n(1 - p) = (80)(0.92) = 73.6 > 5$, we may use the normal distribution with $\mu = np = (80)(0.08)$ and $\sigma = \sqrt{np(1 - p)} = \sqrt{(80)(0.08)(0.92)}$ to approximate $P(X \geq 10)$ as the area under the normal curve from 9.5 to 80.5 (corresponding to all 80 being defective).

The probability is (about) 0.10. (For a solution using tables, see page A11.)

```
normalcdf(9.5,80
.5,80*.08,√(80*.
08*.92))
          .1007041942
```

SECTION 7.1

1. The probability of the transition $G \rightarrow G$ is 0.20.

2. Only matrix (a) is stochastic. Matrix (b) is not stochastic because the first row does not sum to 1, and matrix (c) is not stochastic because one of the entries is negative.

3. The identity matrix sends each state to itself, so every state is absorbing.

SECTION 7.2

1. $\left(\frac{1}{2} \quad \frac{1}{2}\right) \cdot \begin{pmatrix} \frac{1}{4} & \frac{3}{4} \\ \frac{3}{4} & \frac{1}{4} \end{pmatrix} = \left(\frac{1}{2} \cdot \frac{1}{4} + \frac{1}{2} \cdot \frac{3}{4} \quad \frac{1}{2} \cdot \frac{3}{4} + \frac{1}{2} \cdot \frac{1}{4}\right)$

$= \left(\frac{4}{8} \quad \frac{4}{8}\right) = \left(\frac{1}{2} \quad \frac{1}{2}\right)$.

Therefore, $\left(\frac{1}{2} \quad \frac{1}{2}\right)$ *is* a steady-state distribution for the ergodic transition matrix $T = \begin{pmatrix} \frac{1}{4} & \frac{3}{4} \\ \frac{3}{4} & \frac{1}{4} \end{pmatrix}$.

2. $\left(\frac{3}{5} \quad \frac{2}{5}\right) \cdot \begin{pmatrix} \frac{1}{3} & \frac{2}{3} \\ 1 & 0 \end{pmatrix} = \left(\frac{3}{5} \cdot \frac{1}{3} + \frac{2}{5} \cdot 1 \quad \frac{3}{5} \cdot \frac{2}{3} + \frac{2}{5} \cdot 0\right) = \left(\frac{3}{5} \quad \frac{2}{5}\right)$.

Therefore, $\left(\frac{3}{5} \quad \frac{2}{5}\right)$ *is* the steady-state distribution for the regular transition matrix $T = \begin{pmatrix} \frac{1}{3} & \frac{2}{3} \\ 1 & 0 \end{pmatrix}$.

SECTION 7.3

1. Both $\quad \begin{array}{c} \\ W \\ F \\ G \end{array} \begin{matrix} W & F & G \\ \begin{pmatrix} 1 & 0 & 0 \\ 0 & 0.99 & 0.01 \\ 0.02 & 0 & 0.98 \end{pmatrix} \end{matrix}$

and $\quad \begin{array}{c} \\ W \\ G \\ F \end{array} \begin{matrix} W & G & F \\ \begin{pmatrix} 1 & 0 & 0 \\ 0.02 & 0.98 & 0 \\ 0 & 0.01 & 0.99 \end{pmatrix} \end{matrix}$

are standard forms of the transition matrix.

2. The expected times are 100 years in the factory storage pool and then 50 years in the ground water (from the fundamental matrix, interpreting the columns by their headings).

3. 0.54 (from the entry in row 1 and column 2 of $(I - Q)^{-1} \cdot R$ in the Graphing Calculator Exploration).

4. The expected lifetime is $4 + 1 + 1 = 6$ years, and the probability that it will become part of an agribusiness conglomerate is 0.11, or 11%.

SECTION 8.1

1. "If the celebration ended by 1 o'clock, then the home team lost."

2. This could be translated as either $\sim (p \vee q)$ or $\sim p \wedge \sim q$.

3. "I ain't not going to bed" is really $\sim(\sim$ "I am going to bed"), which is just "I am going to bed."

SECTION 8.2

1. (1)

p	q	$p \vee \sim q$
T	T	
T	F	
F	T	
F	F	

(2)

p	q	$p \vee \sim q$
T	T	T
T	F	T
F	T	F
F	F	F

(3)

p	q	p	$\vee$	$\sim q$
T	T	T		F
T	F	T		T
F	T	F		F
F	F	F		T

(4)

p	q	p	$\vee$	$\sim q$
T	T	T	T	F
T	F	T	T	T
F	T	F	F	F
F	F	F	T	T

2. (1)

p	q	p	$\vee$	$(p$	$\wedge$	$q)$
T	T					
T	F					
F	T					
F	F					

(2)

p	q	p	$\vee$	$(p$	$\wedge$	$q)$
T	T	T	T			T
T	F	T	T			F
F	T	F	F			T
F	F	F	F			F

(3)

p	q	p	$\vee$	$(p$	$\wedge$	$q)$
T	T	T	T	T	T	T
T	F	T	T	T	F	F
F	T	F	F	F	F	T
F	F	F	F	F	F	F

(4)

p	q	p	$\vee$	$(p$	$\wedge$	$q)$
T	T	T	T	T	T	T
T	F	T	T	T	F	F
F	T	F	F	F	F	T
F	F	F	F	F	F	F

The column under $\vee$ is the same as the column for p, so $p \vee (p \wedge q) \equiv p$.

3. $\sim p \wedge (p \vee q) \equiv (\sim p \wedge p) \vee (\sim p \wedge q)$ Distributive law 1

$\equiv (f) \vee (\sim p \wedge q)$ Commutative law 1 and Identity Law 3

$\equiv \sim p \wedge q$ Commutative law 2 and Identity Law 5

4. "The spoon is not silver and the fork is gold" by using De Morgan's law that $\sim (p \vee q) \equiv \sim p \wedge \sim q$ and the double negation $\sim (\sim q) \equiv q$.

SECTION 8.3

1. a. $\sim (p \to q) \equiv \sim (\sim p \vee q)$ Replacing $p \to q$ by the equivalent $\sim p \vee q$

$\equiv p \wedge \sim q$ Using De Morgan's law 2

b. You are at least eighteen and you cannot vote.

$$p \qquad \wedge \qquad \sim q$$

2. a. If you study well, then you will pass this course.

     Antecedent   Consequent

b. If I go swimming, then it is hot.

     Antecedent   Consequent

3. The contrapositive, "If Tom doesn't have a boat, then Harold doesn't drive a Ford," is logically equivalent to the direct statement.

SECTION 8.4

1. Since $(p \to q) \wedge \sim p \equiv (\sim p \vee q) \wedge \sim p \equiv \sim p$ by the absorption laws, this syllogism is equivalent to $\sim p \to \sim q$, which is false when p is false (so that $\sim p$ is true) and q is true (so that $\sim q$ is false).

2. Since $\sim j \vee m \equiv j \to m$ and $\sim b \to \sim s \equiv s \to b$, the argument becomes

$m \to s$	P_1
$j \to m$	P_2
$s \to b$	P_3
j	P_4
b	Q

Changing the order, this is the same as

$$(j \to m) \wedge (m \to s) \wedge (s \to b) \wedge j \to b$$

Implies $(j \to b)$ by the hypothetical syllogism

Since $[(j \to b) \wedge j] \to b$ is valid by modus ponens, the argument of Example 3 is valid.

SECTION 8.5

1. a. $Q = \{\text{taxi, horse}\}$, since $q(x)$ is true only when x is replaced by "taxi" or "horse."

b. $Q^c = \{\text{table, lamp, cat, crow}\}$, since the truth set of $\sim q(x)$ is the complement of the truth set of $q(x)$.

2. a. True, since the letter o appears in each element of $A = \{\text{violet, yellow}\}$.

b. True, since the letter o appears in at least one element, "orange," of $B = \{\text{blue, orange, red}\}$.

3. $\exists x \in A, \sim p(x)$ means "there is a coin in the set {penny, nickel, dime} that is worth fifteen or more cents," which is clearly false.

By De Morgan's law 2, the negation is "$\forall x \in A$, $p(x)$" [using the double negation $\sim (\sim p(x)) \equiv p(x)$], which means "every coin in the set {penny, nickel, dime} is worth less than fifteen cents." This statement is true, as we would expect for the negation of a false statement.

APPENDIX

1. $z = \dfrac{20 - 15}{1} = 5$. Since more than 99% of the z-scores are within 3 of the mean, this value is *very* far to the right of the mean!

2. We first change the weights into z-scores using $z = \dfrac{x - \mu}{\sigma}$ with $\mu = 134.7$ and $\sigma = 30.4$:

$$x = 130 \text{ corresponds to } z = \frac{130 - 134.7}{30.4} \approx -0.15$$

$$x = 150 \text{ corresponds to } z = \frac{150 - 134.7}{30.4} \approx 0.50$$

Using these values, we then want $P(-0.15 \leq Z \leq 0.50)$, which is equivalent to the shaded area shown in the graph on the left below, which is equal to the *sum* of the following two shaded areas. The two areas are found from the table on page 614, with the calculation shown below.

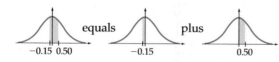

$$P(-0.15 \leq Z \leq 0.50) = P(0 \leq Z \leq 0.15) + P(0 \leq Z \leq 0.50)$$
$$= 0.0596 + 0.1915 = 0.2511$$

From the table
on page 614

Therefore, about 25% of American women weigh between 130 and 150 pounds.

3. Since $np = (80)(0.08) = 6.4$ and $n(1 - p) = (80)(0.92) = 73.6$ are each greater than 5, we may use the normal distribution with $\mu = np = (80)(0.08) \approx 6.4$ and $\sigma = \sqrt{np(1 - p)} = \sqrt{(80)(0.08)(0.92)} \approx 2.43$ to approximate $P(X \geq 10)$ as the area under the normal curve from 9.5 to 80.5 (corresponding to all 80 being defective, and again including rounding). Converting the x-values into z-scores using $z = \dfrac{x - \mu}{\sigma}$ with $\mu = 6.4$ and $\sigma = 2.43$:

$$x = 9.5 \text{ corresponds to } z = \frac{9.5 - 6.4}{2.43} \approx 1.28$$

$$x = 80.5 \text{ corresponds to } z = \frac{80.5 - 6.4}{2.43} \approx 30.49$$

The probability $P(1.28 \leq Z \leq 30.49)$ is represented by the shaded area on the left, which is equivalent to the difference between the two areas on the right with the calculation shown below.

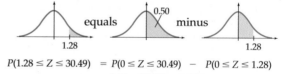

$$P(1.28 \leq Z \leq 30.49) = P(0 \leq Z \leq 30.49) - P(0 \leq Z \leq 1.28)$$
$$= 0.50 - 0.3997 = 0.1003$$

From the table
on page 614

The probability of at least ten defective VCRs is (about) 0.10, or 10%.

Answers to Selected Exercises

1. $\{x \mid 0 \le x < 6\}$ **3.** $\{x \mid x \le 2\}$ **5. a.** Increase by 15 units

b. Decrease by 10 units   **7.** $m = -2$ **9.** $m = \dfrac{1}{3}$ **11.** $m = 0$ **13.** Slope is undefined.

15. $m = 3, (0, -4)$ **17.** $m = -\dfrac{1}{2}, (0, 0)$ **19.** $m = 0, (0, 4)$

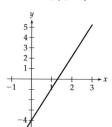 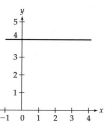 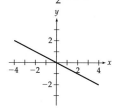

21. Slope and y-intercept do not exist.   **23.** $m = \dfrac{2}{3}, (0, -4)$ **25.** $m = -1, (0, 0)$

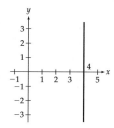 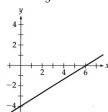 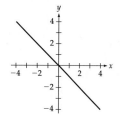

27. $m = 1, (0, 0)$ **29.** $m = \dfrac{1}{3}, \left(0, \dfrac{2}{3}\right)$ **31.** $m = \dfrac{2}{3}, (0, -1)$

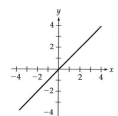 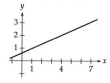 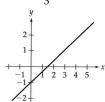

33.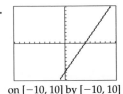

on $[-10, 10]$ by $[-10, 10]$

35.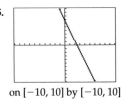

on $[-10, 10]$ by $[-10, 10]$

37.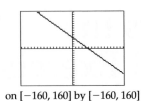

on $[-160, 160]$ by $[-160, 160]$

39. $y = -2.25x + 3$ **41.** $y = 5x + 3$ **43.** $y = -4$ **45.** $x = 1.5$ **47.** $y = -2x + 13$ **49.** $y = -1$

51. $y = -2x + 1$ **53.** $y = \dfrac{3}{2}x - 2$ **55.** $y = -x + 5, y = -x - 5, y = x + 5, y = x - 5$

57. Substituting $(0, b)$ into $y - y_1 = m(x - x_1)$ gives $y - b = m(x - 0)$, or $y = mx + b$.   **59.** $(-b/m, 0), m \neq 0$

61. a.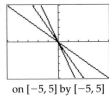

on $[-5, 5]$ by $[-5, 5]$

b.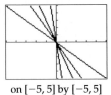

on $[-5, 5]$ by $[-5, 5]$

63. Low: $[0, 8)$; average: $[8, 20)$; high: $[20, 40)$; critical $[40, \infty)$ **65. a.** 3 minutes 38.28 seconds   **b.** the year 2033

67. a. $y = 4x + 2$ **b.** \$10 million   **c.** \$22 million   **69. a.** $y = \dfrac{9}{5}x + 32$ **b.** $68°$ **71. a.** $V = 50,000 - 2200t$

b. \$39,000   **c.**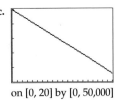

on $[0, 20]$ by $[0, 50,000]$

73. a.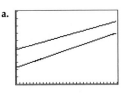

b. Men: 28.1 years;
women: 26.4 years

c. Men: 28.7 years; women: 27.2 years   **75. b.**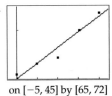

on $[-5, 45]$ by $[65, 72]$

$y_1 = 0.158x + 65.06$ **c.** 76.9 years

EXERCISES 1.2 page 28

1. 64   **3.** $\dfrac{1}{16}$ **5.** 8   **7.** $\dfrac{8}{5}$ **9.** $\dfrac{1}{32}$ **11.** $\dfrac{8}{27}$ **13.** 1   **15.** $\dfrac{4}{9}$ **17.** 5   **19.** 125   **21.** 8   **23.** 4   **25.** -32

27. $\dfrac{125}{216}$ **29.** $\dfrac{9}{25}$ **31.** $\dfrac{1}{4}$ **33.** $\dfrac{1}{2}$ **35.** $\dfrac{1}{8}$ **37.** $\dfrac{1}{4}$ **39.** $-\dfrac{1}{2}$ **41.** $\dfrac{1}{4}$ **43.** $\dfrac{4}{5}$ **45.** $\dfrac{64}{125}$ **47.** -243 **49.** 2.14

51. 274.37   **53.** -128 **55.** 6.25   **57.** 0.5   **59.** 0.4   **61.** 0.977 (rounded)   **63.** 2.720 (rounded)   **65.** x^{10}

67. z^{27} **69.** x^8 **71.** w^5 **73.** y^5/x **75.** $27y^4$ **77.** $u^2v^2w^2$ **79.** 25.6 ft   **81.** Costs will be multiplied by 2.3.

83.

on [0, 5] by [0, 3]

Capacity can be multiplied by about 3.2.   **85.** 125 beats per minute

87.

on [0, 200] by [0, 150]

Heart rate decreases more slowly as body weight increases.

89. About 42.6 thousand work-hours, or 42,600 work-hours, rounded to the nearest hundred hours

91. a. About 32 times more ground motion   **b.** About 8 times more ground motion   **93.** About 312 mph

95.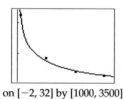

on [0, 100] by [0, 4]

$x \approx 18.2$. Therefore, the land area must be increased by a factor of more than 18 to double the number of species.

97. b.

on [−2, 32] by [1000, 3500]

$y = 3261x^{-0.267}$ **c.** 1147 work-hours

EXERCISES 1.3 page 44

1. Yes   **3.** No   **5.** No   **7.** No   **9.** Domain: $\{x \mid x \le 0 \text{ or } x \ge 1\}$; Range: $\{y \mid y \ge -1\}$

11. a. $f(10) = 3$ **b.** $\{x \mid x \ge 1\}$ **c.** $\{y \mid y \ge 0\}$ **13. a.** $h(-5) = -1$ **b.** $\{z \mid z \ne -4\}$ **c.** $\{y \mid y \ne 0\}$

15. a. $h(81) = 3$ **b.** $\{x \mid x \ge 0\}$ **c.** $\{y \mid y \ge 0\}$ **17. a.** $f(-8) = 4$ **b.** $\mathbb{R}$ **c.** $\{y \mid y \ge 0\}$

19. a. $f(0) = 2$ **b.** $\{x \mid -2 \le x \le 2\}$ **c.** $\{y \mid 0 \le y \le 2\}$ **21. a.** $f(-25) = 5$ **b.** $\{x \mid x \le 0\}$ **c.** $\{y \mid y \ge 0\}$

23. 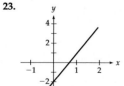   **25.** 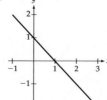   **27.** 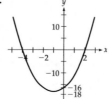   **29.**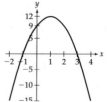

31. a. (20, 100)   **b.**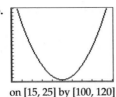

33. a. $(-40, -200)$ **b.**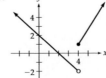

on [15, 25] by [100, 120]

on $[-45, -35]$ by
$[-220, -200]$

35. $x = 7, x = -1$ **37.** $x = 3, x = -5$ **39.** $x = 4, x = 5$ **41.** $x = 0, x = 10$ **43.** $x = 5, x = -5$ **45.** $x = -3$

47. $x = 1, x = 2$ **49.** No solutions   **51.** No solutions   **53.** $x = -4, x = 5$ **55.** $x = 4, x = 5$ **57.** $x = -3$

59. No (real) solutions   **61.** $x = 1.14, x = -2.64$ **63. a.** The slopes are all 2, but the y-intercepts differ.

b. $y = 2x - 8$ **65.** $C(x) = 4x + 20$ **67.** $P(x) = 15x + 500$ **69. a.** 17.7 lb/in^2 **b.** 15,765 lb/in^2

71. 132 ft   **73. a.** 400   **b.** 5200   **75.** About 208 mph   **77.** 2.92 seconds

79. a. Break even at 40 and 200 units.   **b.** Profit maximized at 120 units. Maximum profit is $12,800.

81. a. Break even at 20 and 80 units.   **b.** Profit maximized at 50 units. Maximum profit is $1800.

83. a. 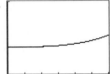   **b.** About 40% (from 39.8)   **c.** About 60% (from 60.4)

85. $v = \dfrac{c}{w + a} - b$ **87. a.** Shifted 4 units to the right; vertex: (4, 0)   **b.** Shifted 3 units to the left; vertex: $(-3, 0)$

c. Shifted a units to the right, with vertex $(a, 0)$. A plus sign means a shift to the left.

89. a. Shifted 3 units to the right and 2 units up; vertex: (3, 2)   **b.** Shifted 2 units to the left and 5 units down; vertex: $(-2, -5)$

c. Shifted a units to the right and b units up, with vertex (a, b). A negative a or b reverses the direction.

EXERCISES 1.4 page 61

1. Domain: $\{x \mid x > 0 \text{ or } x < -4\}$; Range: $\{y \mid y > 0 \text{ or } y < -2\}$ **3. a.** $f(-3) = 1$ **b.** $\{x \mid x \neq -4\}$ **c.** $\{y \mid y \neq 0\}$

5. a. $f(-1) = -\dfrac{1}{2}$ **b.** $\{x \mid x \neq 1\}$ **c.** $\{y \mid y \leq 0 \text{ or } y \geq 4\}$

7. a. $f(2) = 1$ **b.** $\{x \mid x \neq 0, x \neq -4\}$ **c.** $\{y \mid y > 0 \text{ or } y \leq -3\}$ **9. a.** $g(-5) = 3$ **b.** $\mathbb{R}$ **c.** $\{y \mid y \geq 0\}$

11. $x = 0, x = -3, x = 1$ **13.** $x = 0, x = 2, x = -2$ **15.** $x = 0, x = 3$ **17.** $x = 0, x = 5$ **19.** $x = 0, x = 3$

21. $x = -2, x = 0, x = 4$ **23.** $x = -1, x = 0, x = 3$ **25.** $x = 0, x = 3$ **27.** $x = -5, x = 0$ **29.** $x = 0, x = 1$

31. $x \approx -1.79, x = 0, x \approx 2.79$ **33.** 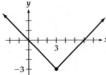   **35.**

37.

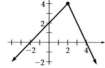

39. Polynomial   **41.** Piecewise linear   **43.** Polynomial   **45.** Rational

47. Piecewise linear   **49.** Polynomial   **51.** None (not a polynomial because of the fractional exponent)

53. a. y_3 **b.** y_1 **c.**

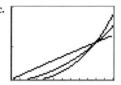

d. $(10, 1000)$.

55. a.

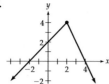

$y = \text{INT}(x)$ on $[-5, 5]$ by $[-5, 5]$ Note that each line segment in this graph includes its left endpoint but excludes its right endpoint, so it should be drawn like ●——○.

b. Domain: $\mathbb{R}$; Range: $\{\ldots, -3, -2, -1, 0, 1, 2, 3, \ldots\}$, that is, the set of all integers.

57. a. $(7x - 1)^5$ **b.** $7x^5 - 1$ **59. a.** $\dfrac{1}{x^2 + 1}$ **b.** $\left(\dfrac{1}{x}\right)^2 + 1$ **61. a.** $(\sqrt{x} - 1)^3 - (\sqrt{x} - 1)^2$ **b.** $\sqrt{x^3 - x^2} - 1$

63. a. $\dfrac{(x^2 - x)^3 - 1}{(x^2 - x)^3 + 1}$ **b.** $\left(\dfrac{x^3 - 1}{x^3 + 1}\right)^2 - \dfrac{x^3 - 1}{x^3 + 1}$ **65. a.** $f(g(x)) = acx + ad + b$ **b.** Yes

67. a. 2.70481   **b.** 2.71815   **c.** 2.71828   **d.** Yes, 2.71828

69. Shifted left 3 units and up 6 units

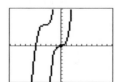

71. a. \$300   **b.** \$500   **c.** \$2000   **d.**

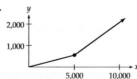

73. $R(v(t)) = 2(60 + 3t)^{0.3}$, $R(v(10)) \approx 7.714$ million dollars   **75. a.**

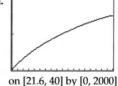

b. About $x = 27.9$ mpg

on $[21.6, 40]$ by $[0, 2000]$

EXERCISES 1.5 page 72

1. 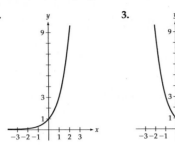   **3.**   **5.** 5.697 (rounded to three decimal places)

7. a. e^x **b.** e^x **c.** e^x **d.** e^x **e.** e^x will exceed any power of x for large enough x.

9. $1,096,000 (approx.)   **11.** $1,989,300 (approx.)   **13.** $101,257   **15. a.** $4463   **b.** $20,156

17. 7.1 billion   **19. a.** 0.53 (the chances are better than 50–50)   **b.** 0.70 (quite likely)

21. a. 0.267 or 26.7%   **b.** 0.012 or 1.2%   **23. a.** 1.3 mg   **b.** 0.84 mg   **25.** 208

27. a. About 153°   **b.** About 123°   **29.** 38   **31.** 6.5%

33. b. Texas, Florida, New York   **c.** Florida, Texas, New York   **d.** about 2016 (from $x \approx 26$ years after 1990)

35. By about 25%

EXERCISES 1.6 page 87

1. a. 5   **b.** -2 **c.** $\frac{1}{2}$ **3. a.** 5   **b.** -1 **c.** $\frac{1}{3}$ **5. a.** 0   **b.** 1   **c.** $\frac{2}{3}$ **7. a.** 2   **b.** -1 **c.** $\frac{1}{2}$

9. a. 1.348   **b.** 3.105   **11.** $\ln x$ **13.** $2 \ln x$ or $\ln x^2$ **15.** $\ln x$ **17.** $3x$ **19.** $7x$

21.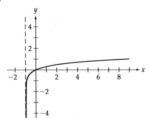

23.   Domain: $\{x \mid x > 1 \text{ or } x < -1\}$
       Range: $\mathbb{R}$

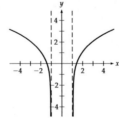

25. 7.27 years. It will have doubled in 8 years.   **27.** 33.35 years. It will have doubled in 34 years.

29. 29.23 years. It will have doubled in 30 years.   **31.** About 31,400 years   **33.** About 1.7 million years

35. 5 years (from 4.67)   **37.** 13 years   **39.** 228 million years   **41. a.** $\log_b b^x = x$, property 8 of logarithms

b. Follows directly from (a)   **c.** Using the change-of-base formula, cancellation

CHAPTER 1 REVIEW EXERCISES page 91

1. $\{x \mid 2 < x \le 5\}$ **2.** $\{x \mid -2 \le x < 0\}$ **3.** $\{x \mid x \ge 100\}$

4. $\{x \mid x \le 6\}$ **5.** Hurricane: $[74, \infty)$; storm: $[55, 74)$; gale: $[38, 55)$; small craft warning: $[21, 38)$

6. a. $(0, \infty)$ **b.** $(-\infty, 0)$ **c.** $[0, \infty)$ **d.** $(-\infty, 0]$ **7.** $y = 2x - 5$ **8.** $y = -3x + 3$ **9.** $x = 2$ **10.** $y = 3$

11. $y = -2x + 1$ **12.** $y = 3x - 5$ **13.** $y = 2x - 1$ **14.** $y = -\dfrac{1}{2}x + 1$ **15. a.** $V = 25{,}000 - 3000t$ **b.** \$13,000

16. a. $V = 78{,}000 - 5000t$ **b.** \$38,000

17. b.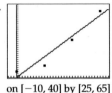
on $[-2, 22]$ by $[17, 27]$

The regression line $y_1 = -0.396x + 25.68$ fits the data well.

c. 13.8 million tons in the year 2005 [from $y_1(30)$]; 11.8 million tons in the year 2010 [from $y_1(35)$]

18. b.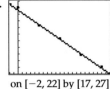
on $[-10, 40]$ by $[25, 65]$

The regression line $y_1 = 1.03x + 26.3$ fits the data reasonably well.

c. 73 to 1 in the year 2005 [from $y_1(45) = 72.65$]; 78 to 1 in the year 2010 [from $y_1(50) = 77.8$]

19. 36 **20.** $\dfrac{3}{4}$ **21.** 8 **22.** 10 **23.** $\dfrac{1}{27}$ **24.** $\dfrac{1}{1000}$ **25.** $\dfrac{9}{4}$ **26.** $\dfrac{64}{27}$ **27.** 13.97 **28.** 112.32

29. a. $f(11) = 2$ **b.** $\{x \mid x \geq 7\}$ **c.** $\{y \mid y \geq 0\}$ **30. a.** $g(-1) = \dfrac{1}{2}$ **b.** $\{t \mid t \neq -3\}$ **c.** $\{y \mid y \neq 0\}$

31. a. $h(16) = \dfrac{1}{8}$ **b.** $\{w \mid w > 0\}$ **c.** $\{y \mid y > 0\}$ **32. a.** $w(8) = \dfrac{1}{16}$ **b.** $\{z \mid z \neq 0\}$ **c.** $\{y \mid y > 0\}$ **33.** Yes

34. No **35.** 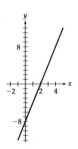 **36.** 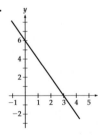 **37.** 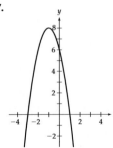 **38.**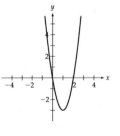

39. $x = 0, x = -3$ **40.** $x = 5, x = -1$ **41.** $x = -2, x = 1$ **42.** $x = 1, x = -1$

43. a. Vertex: $(5, -50)$ **b.**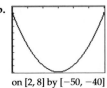
on $[2, 8]$ by $[-50, -40]$

44. a. Vertex: $(-7, -64)$ **b.**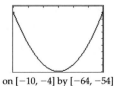
on $[-10, -4]$ by $[-64, -54]$

45. $C(x) = 45 + 0.12x$ **46.** $I(t) = 800t$ **47.** $T(x) = 70 - \dfrac{x}{300}$

48. $C(t) = 25 + 0.58t$; in about 9 years from 1997, so in about 2006.

49. a. Break even at 15 and 65 units. **b.** Profit maximized at 40 units. Maximum profit: $1250.

50. a. Break even at 150 and 450 units. **b.** Profit maximized at 300 units. Maximum profit: $67,500.

51. a. $f(-1) = 1$ **b.** $\{x \mid x \neq 0 \text{ and } x \neq 2\}$ **c.** $\{y \mid y > 0 \text{ or } y \leq -3\}$

52. a. $f(-8) = \dfrac{1}{2}$ **b.** $\{x \mid x \neq 0 \text{ and } x \neq -4\}$ **c.** $\{y \mid y > 0 \text{ or } y \leq -4\}$ **53. a.** $g(-4) = 0$ **b.** $\mathbb{R}$ **c.** $\{y \mid y \geq -2\}$

54. a. $g(-5) = -10$ **b.** $\mathbb{R}$ **c.** $\{y \mid y \leq 0\}$ **55.** $x = 0, x = 1, x = -3$ **56.** $x = 0, x = 2, x = -4$ **57.** $x = 0, x = 5$

58. $x = 0, x = 2$ **59.** **60.** **61.**

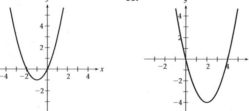
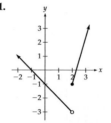

62.

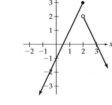

63. a. $f(g(x)) = \left(\dfrac{1}{x}\right)^2 + 1 = \dfrac{1}{x^2} + 1$ **b.** $g(f(x)) = \dfrac{1}{x^2 + 1}$

64. a. $f(g(x)) = \sqrt{5x - 4}$ **b.** $g(f(x)) = 5\sqrt{x} - 4$ **65. a.** $f(g(x)) = \dfrac{x^3 + 1}{x^3 - 1}$ **b.** $g(f(x)) = \left(\dfrac{x + 1}{x - 1}\right)^3$

66. a. $f(g(x)) = |x + 2|$ **b.** $g(f(x)) = |x| + 2$ **67. a.** $f(g(x)) = x$ **b.** $g(f(x)) = x$ **68. a.** $f(g(x)) = 5$

b. $g(f(x)) = 12$ **69.** $A(p(t)) = 2(18 + 2t)^{0.15}$, $A(p(4)) \approx \$3.26$ million **70. a.** $x = -1, x = 0, x = 3$

b.

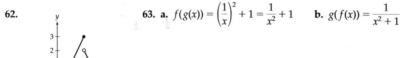

71. a. $x = -3, x = 0, x = 1$ **b.**

on $[-5, 5]$ by $[-5, 5]$ on $[-5, 5]$ by $[-5, 5]$

72. a. The points suggest a parabolic (quadratic) curve. **b.**

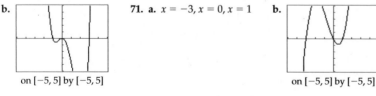

c. $3.6 million, $4.8 million

on $[0.5, 5.5]$ by $[1.5, 3]$

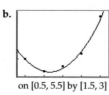

73. 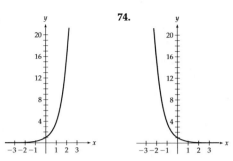   **74.**   **75.** $15.3 million

76. $3.49 million   **77. a.** 800,000 $(0.8)^t$ **b.** $327,680   **78. a.** 5.4 $(0.88)^t$ **b.** $1.5 million

79. a. In about 2004 (from $x \approx 8.8$)   **b.** In about 2050 (from $x \approx 55$)

80. 4096 megabits (which is enough to hold the text of sixteen 16-volume encyclopedias on one chip)

81. a. 3   **b.** -3 **c.** 3   **d.** $\frac{1}{4}$ **82. a.** $\frac{1}{2}$ **b.** 8   **c.** -1 **d.** $\frac{3}{2}$ **83.** $f(x) = \ln x$ **84.** $f(x) = 2x - 1$

85. 4.2 years. It will have doubled in 5 years.   **86.** In about 13 years

87. 22.7 years. It will have doubled in 23 years.   **88.** 29.2 years. It will have doubled in 30 years.

89. About 50.7 million years   **90.** About 1.85 million years

EXERCISES 2.1 page 107

1. $1050   **3.** $3120   **5.** $298.57   **7.** $15.36   **9.** $15.70   **11.** $2550   **13.** $7751.58   **15.** $3250.26   **17.** 6.4%

19. $2500   **21.** 4 years   **23.** $7500   **25.** $9450.90   **27.** $4669.41   **29.** 6 years   **31.** 8 years 9 months   **33.** 20 years

35. 5.48%   **37.** No; the $1000 is the same as the interest, so you would be agreeing to pay $1000 of interest on a loan of $0.

39. $950.40   **41.** 50%   **43.** 58.25%   **45.** 17.89%

EXERCISES 2.2 page 123

1. a. $32,383.87   **b.** $33,120.59   **c.** $33,362.60   **3. a.** $30,352.10   **b.** $30,874.78   **c.** $31,046.39

5. a. $16,408.93   **b.** $16,488.53   **c.** $16,504.89   **7. a.** $11,308.73   **b.** $10,926.92   **c.** $10,803.21

9. a. $12,119.55   **b.** $11,966.45   **c.** $11,917.53   **11. a.** $8106.29   **b.** $8057.40   **c.** $8047.38

13. 17 years and 9 months   **15.** 8 years and 9 months   **17.** 5 years and 11 months   **19.** About 8 years; 8.04 rounds up to 9 years

21. About 9.11 years; 8 years and 41 weeks   **23.** About 11.80 years; 11.36 years   **25.** 4.39%   **27.** 8.75%   **29.** 9.90%

31. $6827.69   **33.** $14,467.34   **35.** 30 years and 6 months   **37.** 17.45%

39. The bond gives the greater return (6.80%; the CD returns only 6.63%).   **41.** 32.89 years

43. People's State Bank offers the higher effective yield (4.27%; Statewide Federal's effective rate is just 4.19%).

EXERCISES 2.3 page 136

1. $20,724.67   **3.** $68,852.94   **5.** $514,647.37   **7.** $32.20   **9.** $172.81   **11.** $345.85   **13.** 24 years

15. 22 years and 6 months   **17.** 7 years and 2 months   **19.** 8.40%   **21.** 12.20%   **23.** 6.70%

25. Joe will have $316,781.40, and Jill will have $364,548.28.   **27.** $4001.27   **29.** $6.39

31. 1 year and 9 months   **33.** 5.65%

43. a. 242   **b.** 242 [part (b) is just part (a) "backward"]   **c.** 333,333. The sum on page 129 has the correct value of $10,819.57

EXERCISES 2.4 page 147

1. $105,353.72   **3.** $95,896.47   **5.** $212,572.07   **7.** $584.59   **9.** $116   **11.** $872.41   **13.** $85,934.02   **15.** $2719.70

17. $10,284.16   **19.** $4,306,638   **21.** $14,204,375   **23.** $70,777.95   **25.** $468,407.35; John should buy $500,000 of life insurance.

27. a. $3334.58   **b.** It saves $334.25 each month.   **c.** The longer mortgage costs an extra $2,640,702.24.   **29.** $31.53

31. $115.66   **33.** $41,640.59   **35.** 11.9%   **37.** $2531.32

47. The amortization table uses an annual payment of $7518.83 and the unpaid balance after 3 years is $38,879.49, a difference of 4¢ compared to Exercise 16. The final payment is $7518.87, a correction of 4¢ from the others in the table.

55. This unpaid balance formula gives slightly different answers because the payments are not rounded to the upper penny before continuing the calculation.

CHAPTER 2 REVIEW EXERCISES page 152

1. $217.50   **2.** $52.90   **3.** $2250   **4.** $336.88   **5.** $33.86   **6.** $56.63   **7.** $16.53   **8.** $188.08   **9.** $10,212.75

10. $1491.53   **11.** $2138.56   **12.** $4288.61   **13.** 6.8%   **14.** $1800   **15.** 1 year and 6 months   **16.** 4 years

17. 5 years   **18.** $4500   **19.** $9394.08   **20.** $47,267.91   **21.** 42.86%   **22.** 21.43%   **23.** 6.6%

24. $500,000   **25.** 11.1%; $1110   **26.** $31,535.24   **27.** $264,247.79   **28.** $370,893.09   **29.** $10,749.24

30. No; it would only be worth $112 billion.   **31.** $10,198.43   **32.** $18,563.50   **33.** $71,056.01   **34.** $12,702.50

35. $8466.50   **36.** 18 years and 6 months   **37.** 4 years and 2 months   **38.** 7.35 years   **39.** 3 years   **40.** 18 years

41. 9 years (9.01 years)   **42.** 12 years; 11 years and 7 months   **43.** 6 years; 5 years and 44 weeks   **44.** 6 years (6.12 years)

45. 7 years (7.27 years)   **46.** 13.92% if quarterly and 14.17% if continuously   **47.** 28.19%   **48.** 106.64%   **49.** 13.51%

50. 2.94%   **51.** $96,304.25   **52.** $19,541.30   **53.** $33,322.04   **54.** $935,211.12   **55.** $715,609.58   **56.** $1559.49

57. $205.98   **58.** $106.28   **59.** $168.76   **60.** $19.51   **61.** 26 years   **62.** 24 years and 6 months

63. 5 years and 1 month   **64.** 15 years and 7 months   **65.** 13 years and 6 months   **66.** 9.80%   **67.** 2.97%

68. 9.42%   **69.** 6.58%   **70.** 3.00% with 365 days per year; 3.12% with 360 days per year   **71.** $319,583.90

72. $20,923,408   **73.** $30,147,921   **74.** $6,773,662.45   **75.** $158,891   **76.** $106.12   **77.** $46.14   **78.** $423.17

79. $1180.05   **80.** $548.25; do not confuse paying off a current debt with accumulating money in the future.

81. $75,148.14   **82.** $142,742.00   **83.** $5449.20   **84.** $523,692.95   **85.** $143,927.90

EXERCISES 3.1 page 170

1. $\begin{cases} x + y = 18 \\ x - y = 2 \end{cases}$ **3.** $\begin{cases} x - y = 6 \\ x + y = 40 \end{cases}$ **5.** $\begin{cases} x + y = 30 \\ x + 5y = 70 \end{cases}$ **7.** $\begin{cases} x + y = 100 \\ 10x + 5y = 650 \end{cases}$ **9.** $\begin{cases} x + y = 225 \\ x - 2y = 0 \end{cases}$

11. The solution is $x = 4, y = 2$. The equations are independent and consistent.

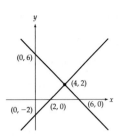

13. The solution is $x = 3, y = 2$. The equations are independent and consistent.

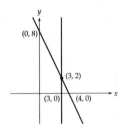

15. The solution is $x = 2, y = -2$. The equations are independent and consistent.

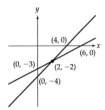

17. There is no solution. The equations are independent and inconsistent.

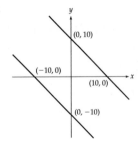

19. There are infinitely many solutions that may be parameterized as $x = 10 - t, y = t$. The equations are dependent.

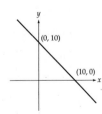

21. $x = 4, y = 3$. The equations are independent and consistent.

23. $x = 5, y = 10$. The equations are independent and consistent.

25. $x = 10, y = -10$. The equations are independent and consistent.

27. $x = 4, y = 9$. The equations are independent and consistent.

29. There are infinitely many solutions that may be parameterized as $x = 10 + t, y = t$. The equations are dependent.

31. $x = 3, y = 8$. The equations are independent and consistent.

33. $x = 4, y = 3$. The equations are independent and consistent.

35. $x = 8, y = 3$. The equations are independent and consistent.

37. $x = 15, y = 6$. The equations are independent and consistent.

39. There is no solution. The equations are independent and inconsistent.

41. There are 34 nickels and 26 dimes in the jar.

43. The retired couple should invest $2000 in the money market account and $8000 in the stock mutual fund.

45. The concession stand sold 1800 sodas and 1200 hot dogs.

47. The federal tax is $4900 and the state tax is $900.

49. The required calcium and phosphorus can be provided by just 5 tablets of supplement B (with none of supplement A) each day.

51. The equations are inconsistent and dependent.

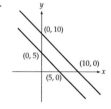

53. The equations are consistent and independent. The solution is $x = 15, y = 4$.

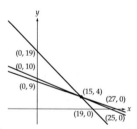

55. The equations are inconsistent and independent.

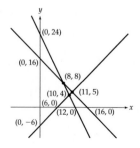

EXERCISES 3.2 page 185

1. $3 \times 2; 1; 6$ **3.** $3 \times 4; 3; -6$ **5.** $4 \times 4; 1; 1; 1; 1; 0$ **7.** $1 \times 4; 6$ **9.** $5 \times 1; 8; 6$ **11.** $\begin{pmatrix} 1 & 2 & 2 \\ 3 & 4 & 12 \end{pmatrix}$

13. $\begin{pmatrix} -4 & 3 & 84 \\ 5 & -2 & 70 \end{pmatrix}$ **15.** $\begin{pmatrix} 3 & -2 & 24 \\ 1 & 0 & 6 \end{pmatrix}$ **17.** $\begin{pmatrix} 5 & -15 & 30 \\ -4 & 12 & 24 \end{pmatrix}$ **19.** $\begin{pmatrix} 1 & 0 & 20 \\ 0 & 1 & 30 \end{pmatrix}$ **21.** $\begin{cases} x + y = 9 \\ y = 4 \end{cases}$

23. $\begin{cases} -4x + 3y = -60 \\ x - 2y = 20 \end{cases}$ **25.** $\begin{cases} x - 3y = -70 \\ x + y = 10 \end{cases}$ **27.** $\begin{cases} 2x + y = 6 \\ x + 2y = -6 \end{cases}$ **29.** $\begin{cases} 20x - 15y = 60 \\ -16x + 12y = -48 \end{cases}$

31. $\begin{pmatrix} 5 & 6 & 30 \\ 3 & 4 & 24 \end{pmatrix} \begin{matrix} R_1' = R_2 \\ R_2' = R_1 \end{matrix}$ **33.** $\begin{pmatrix} 2 & 2 & -4 \\ 6 & 5 & 60 \end{pmatrix} R_1' = R_1 - R_2$ **35.** $\begin{pmatrix} 5 & 6 & 30 \\ 5 & 10 & 90 \end{pmatrix} R_2' = 5R_2$

37. $\begin{pmatrix} 6 & 6 & -12 \\ 0 & 1 & -72 \end{pmatrix} R_2' = R_1 - R_2$ **39.** $\begin{pmatrix} 1 & -2 & -42 \\ 0 & 1 & 15 \end{pmatrix} R_2' = \frac{1}{8}R_2$

41. $x = 7, y = -3$. The equations are independent and consistent.

43. No solution. The equations are independent and inconsistent.

45. No solution. The equations are independent and inconsistent.

47. $x = 3 - 2t, y = t$ (infinitely many solutions). The equations are dependent.

49. $x = t, y = -3$ (infinitely many solutions). The equations are dependent.

51. $x = 3, y = 2$. The equations are independent and consistent.

53. $x = 3, y = 1$. The equations are independent and consistent.

55. $x = 1, y = 2$. The equations are independent and consistent.

57. $x = 3, y = 2$. The equations are independent and consistent.

59. $x = 9, y = 2$. The equations are independent and consistent.

61. No solution. The equations are independent and inconsistent.

63. $x = 8, y = -15$. The equations are independent and consistent.

65. $x = 4, y = 15$. The equations are independent and consistent.

67. $x = 9 + 3t, y = t$ (infinitely many solutions). The equations are dependent.

69. $x = 21, y = -4$. The equations are independent and consistent.

71. The commodities speculator invested $10,000 in soybean futures and $5000 in corn futures.

73. The older brother receives $4.8 million and the younger brother receives $2.4 million, leaving $4.8 million for their sister.

75. There are 175 nickels and 112 quarters in the jar.

77. The dietician probably wants positive whole numbers for the solution, so the possibilities are

Cans of NutraDrink:	12	8	4	0
Tablets of VitaPills:	0	5	10	15

79. The campaign manager should use 7 TV ads and 30 radio ads.

81. $x = 18, y = 16$. The equations are independent and consistent.

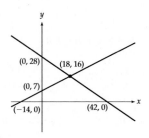

83. No solution. The equations are independent and inconsistent.

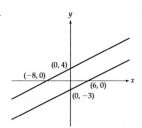

85. No solution. The equations are independent and inconsistent.

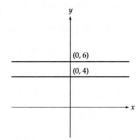

87. $x = 14 - \frac{7}{2}t, y = t$ (infinitely many solutions). The equations are dependent.

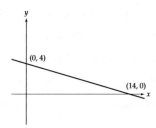

89. $x = t, y = 6$ (infinitely many solutions). The equations are dependent.

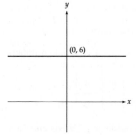

91. $\begin{pmatrix} 3 & -10 & -65 \\ -4 & 13 & 84 \end{pmatrix} \to \begin{pmatrix} -4 & 13 & 84 \\ 3 & -10 & -65 \end{pmatrix} \begin{matrix} R_1' = R_2 \\ R_2' = R_1 \end{matrix} \to \begin{pmatrix} 3 & -10 & -65 \\ -4 & 13 & 84 \end{pmatrix} \begin{matrix} R_1' = R_2 \\ R_2' = R_1 \end{matrix}$

93. $\begin{pmatrix} 3 & -10 & -65 \\ -4 & 13 & 84 \end{pmatrix} \to \begin{pmatrix} 15 & -50 & -325 \\ -4 & 13 & 84 \end{pmatrix} R_1' = 5R_1 \to \begin{pmatrix} 3 & -10 & -65 \\ -4 & 13 & 84 \end{pmatrix} R_1' = \tfrac{1}{5}R_1$

95. $\begin{pmatrix} 3 & -10 & -65 \\ -4 & 13 & 84 \end{pmatrix} \to \begin{pmatrix} 3 & -10 & -65 \\ -1 & 13/4 & 21 \end{pmatrix} R_2' = \tfrac{1}{4}R_2 \to \begin{pmatrix} 3 & -10 & -65 \\ -4 & 13 & 84 \end{pmatrix} R_2' = 4R_2$

97. $\begin{pmatrix} 3 & -10 & -65 \\ -4 & 13 & 84 \end{pmatrix} \to \begin{pmatrix} -1 & 3 & 19 \\ -4 & 13 & 84 \end{pmatrix} R_1' = R_1 + R_2 \to \begin{pmatrix} 3 & -10 & -65 \\ -4 & 13 & 84 \end{pmatrix} R_1' = R_1 - R_2$

99. $\begin{pmatrix} 3 & -10 & -65 \\ -4 & 13 & 84 \end{pmatrix} \to \begin{pmatrix} 7 & -23 & -149 \\ -4 & 13 & 84 \end{pmatrix} R_1' = R_1 - R_2 \to \begin{pmatrix} 3 & -10 & -65 \\ -4 & 13 & 84 \end{pmatrix} R_1' = R_1 + R_2$

EXERCISES 3.3 page 201

1. $\begin{pmatrix} 1 & 1 & 1 & 4 \\ 1 & 2 & 1 & 3 \\ 1 & 2 & 2 & 5 \end{pmatrix}$ **3.** $\begin{pmatrix} 2 & -1 & 2 & 11 \\ -1 & 1 & -3 & -12 \\ 2 & -2 & 7 & 27 \end{pmatrix}$ **5.** $\begin{pmatrix} 2 & 1 & 5 & 4 & 5 & 2 \\ 1 & 1 & 3 & 3 & 3 & -1 \end{pmatrix}$ **7.** $\begin{pmatrix} 6 & 3 & 5 & 8 \\ 1 & 2 & 2 & 1 \\ 4 & 3 & 4 & 5 \\ 5 & 1 & 3 & 7 \end{pmatrix}$

9. $\begin{pmatrix} 3 & 4 & 2 & 4 & 12 \\ 1 & 2 & 1 & 1 & 4 \\ 4 & 5 & 2 & 5 & 14 \\ 6 & 6 & 1 & 6 & 15 \end{pmatrix}$ **11.** $\begin{cases} 4x_1 + 3x_2 + 2x_3 = 11 \\ 3x_1 + 3x_2 + x_3 = 6 \\ x_1 - 2x_2 + 3x_3 = 13 \end{cases}$ **13.** $\begin{cases} 2x_1 + x_2 + x_3 = 7 \\ 2x_1 + 2x_2 + x_3 = 6 \\ 3x_1 + 3x_2 + 2x_3 = 10 \end{cases}$

15. $\begin{cases} 8x_1 + 3x_2 - 2x_3 + 19x_4 = 15 \\ 3x_1 + x_2 - x_3 + 7x_4 = 6 \end{cases}$ **17.** $\begin{cases} 2x_1 + 3x_2 + 2x_3 = 5 \\ 3x_1 + 5x_2 + 3x_3 = 8 \\ x_1 + 2x_2 + 2x_3 = 2 \\ 4x_1 + 7x_2 + 5x_3 = 9 \end{cases}$ **19.** $\begin{cases} 3x_1 + 3x_2 + 5x_3 + 4x_4 = 11 \\ 2x_1 + 2x_2 + 3x_3 + 3x_4 = 9 \\ 2x_1 + x_2 + 2x_3 + 2x_4 = 7 \\ 3x_1 + 2x_2 + 3x_3 + 3x_4 = 11 \end{cases}$

21. The system of equations is independent and consistent, with solution $x_1 = 4$, $x_2 = 5$, $x_3 = -4$.

23. The system of equations is independent and consistent, with solution $x_1 = 2$, $x_2 = -1$, $x_3 = 3$, $x_4 = 1$.

25. The system of equations is independent and inconsistent, with no solution.

27. The system of equations is dependent and consistent, with solution $x_1 = -5 + t$, $x_2 = 5 - t$, $x_3 = t$.

29. The system of equations is dependent and consistent, with solution $x_1 = 8 + t_1 - t_2$, $x_2 = t_1$, $x_3 = 4 + t_2$, $x_4 = t_2$.

31. $\begin{pmatrix} 1 & 0 & 0 & 1 \\ 0 & 1 & 0 & 2 \\ 0 & 0 & 1 & 3 \end{pmatrix}$ **33.** $\begin{pmatrix} 1 & 0 & 0 & 1 \\ 0 & 1 & 0 & 2 \\ 0 & 0 & 1 & 3 \end{pmatrix}$ **35.** $\begin{pmatrix} 1 & 2 & 0 & 3 \\ 0 & 0 & 1 & -3 \\ 0 & 0 & 0 & 0 \end{pmatrix}$ **37.** $\begin{pmatrix} 1 & 0 & 0 & -2 & 0 \\ 0 & 1 & 0 & -1 & 0 \\ 0 & 0 & 1 & 1 & 0 \\ 0 & 0 & 0 & 0 & 1 \end{pmatrix}$ **39.** $\begin{pmatrix} 1 & 0 & 0 & 0 & 2 \\ 0 & 1 & 0 & 0 & 2 \\ 0 & 0 & 1 & 0 & 3 \\ 0 & 0 & 0 & 1 & 1 \end{pmatrix}$

41. The system of equations is independent and consistent, with solution $x_1 = 1$, $x_2 = -2$, $x_3 = 3$.

43. The system of equations is independent and consistent, with solution $x_1 = 1$, $x_2 = -2$, $x_3 = 3$.

45. The system of equations is independent and consistent, with solution $x_1 = 1$, $x_2 = 1$, $x_3 = 1$, $x_4 = 1$.

47. The system of equations is independent and inconsistent, with no solution.

49. The system of equations is dependent and consistent, with solution $x_1 = 3 + 7t$, $x_2 = 4 - 10t$, $x_3 = t$.

51. The system of equations is independent and consistent, with solution $x_1 = 2$, $x_2 = 1$, $x_3 = -2$, $x_4 = 3$.

53. The system of equations is independent and consistent, with solution $x_1 = -1, x_2 = 1, x_3 = 2, x_4 = -2$.

55. The system of equations is independent and consistent, with solution $x_1 = 1, x_2 = 2, x_3 = -2, x_4 = -1, x_5 = 1$.

57. The system of equations is dependent and inconsistent, with no solution.

59. The system of equations is dependent and consistent, with solution $x_1 = 2 + t_1 - t_2, x_2 = 1 - 2t_1 - t_2, x_3 = t_1, x_4 = t_2$.

61. The gardener should use 2 bags of GrowRite, 1 bag of MiracleMix, and 3 bags of GreatGreen.

63. Letting Q be the number of quarters, there are $200 + 3Q$ nickels and $500 - 4Q$ dimes. Because the number of each can never be negative, there can be no more than 125 quarters in the jar.

65. a. The federal tax is \$80,510, the state tax is \$16,490, and the city tax is \$2490.
  b. The effective combined tax rate is (about) 55.3%.

67. a. Letting B be the number of blouses and S be the number of skirts, the shop can make $60 + \frac{6}{5}B - \frac{4}{5}S$ scarves and $60 - \frac{9}{10}B - \frac{2}{5}S$ dresses (where none of these quantities are negative). **b.** 76 scarves and 38 dresses

69. Letting R be the number of radio ads, the number of TV ads is $20 - \frac{1}{5}R$, and the number of newspaper ads is 60. The promotional director may choose R to be 0, 5, 10, 15, 20, . . . , 85, 90, 95, or 100 because these are the only values for R that will result in a whole number of TV ads, while ensuring that neither value is negative.

71. $\begin{cases} x_1 + 2x_2 = 3 \\ x_2 = 1 \end{cases}$ **73.** $\begin{cases} x_1 + x_2 + x_3 = 4 \\ x_2 + x_3 = 3 \\ x_3 = 2 \end{cases}$ **75.** All are equivalent to $\begin{pmatrix} 1 & 0 & 1 \\ 0 & 1 & 1 \end{pmatrix}$ and so to each other.

77. All are equivalent to $\begin{pmatrix} 1 & 0 & 0 & 1 \\ 0 & 1 & 0 & 1 \\ 0 & 0 & 1 & 2 \end{pmatrix}$ and so to each other. **79.** $\begin{pmatrix} 1 & 0 & 1 \\ 0 & 1 & 1 \end{pmatrix}$ **81.** $x_1 = 1, x_2 = 1$

83. $x_1 = 1, x_2 = 1$ **85.** $x_1 = 1, x_2 = 1, x_3 = 2$ **87.** $x_1 = 3, x_2 = 2, x_3 = 4, x_4 = 1$ **89.** $x_1 = 3, x_2 = -1, x_3 = 2$

EXERCISES 3.4 page 218

1. $\begin{pmatrix} 1 & 4 & 7 \\ 2 & 5 & 8 \\ 3 & 6 & 9 \end{pmatrix}$ **3.** $\begin{pmatrix} 3 & 18 & 24 \\ 12 & 6 & 21 \\ 27 & 15 & 9 \end{pmatrix}$ **5.** $\begin{pmatrix} -9 & -8 & -7 \\ -6 & -5 & -4 \\ -3 & -2 & -1 \end{pmatrix}$ **7.** $\begin{pmatrix} 2 & 8 & 11 \\ 8 & 7 & 13 \\ 16 & 13 & 12 \end{pmatrix}$ **9.** $\begin{pmatrix} -1 & 4 & 5 \\ 0 & -4 & 1 \\ 2 & -3 & -7 \end{pmatrix}$

11. (6) **13.** $\begin{pmatrix} 3 & 4 \\ 6 & 8 \end{pmatrix}$ **15.** $\begin{pmatrix} -2 \\ 5 \end{pmatrix}$ **17.** $\begin{pmatrix} 6 & 6 \\ 7 & 7 \end{pmatrix}$ **19.** $\begin{pmatrix} 11 & 7 & 11 \\ 6 & 7 & 6 \\ -1 & 3 & -1 \end{pmatrix}$ **21.** $\begin{pmatrix} 3 & -1 & 4 \\ 2 & 1 & 3 \end{pmatrix}$ **23.** $\begin{pmatrix} 2 & 8 \\ 4 & 1 \\ 5 & -4 \end{pmatrix}$

25. $\begin{pmatrix} -1 & -4 & -1 \\ -2 & 6 & -1 \end{pmatrix}$ **27.** $\begin{pmatrix} 7 & 1 & 4 \\ 0 & -1 & 0 \\ 4 & 2 & 2 \end{pmatrix}$ **29.** $\begin{pmatrix} 5 & 5 & 6 \\ 7 & -7 & 4 \end{pmatrix}$ **31.** $\begin{pmatrix} 1 & 5 & 4 \\ 1 & 1 & 1 \\ 2 & 3 & 3 \end{pmatrix}\begin{pmatrix} x_1 \\ x_2 \\ x_3 \end{pmatrix} = \begin{pmatrix} 6 \\ 4 \\ 9 \end{pmatrix}$

33. $\begin{pmatrix} 4 & 3 & -1 \\ 3 & 3 & 2 \\ 2 & 1 & -3 \end{pmatrix}\begin{pmatrix} x_1 \\ x_2 \\ x_3 \end{pmatrix} = \begin{pmatrix} 2 \\ 9 \\ -6 \end{pmatrix}$ **35.** $\begin{pmatrix} 5 & 2 & -4 & 1 & 5 \\ 3 & 1 & -3 & 1 & 3 \end{pmatrix}\begin{pmatrix} x_1 \\ x_2 \\ x_3 \\ x_4 \\ x_5 \end{pmatrix} = \begin{pmatrix} 7 \\ 5 \end{pmatrix}$ **37.** $\begin{cases} 5x_1 + 9x_2 + 9x_3 = 11 \\ 4x_1 + 7x_2 + 6x_3 = 9 \\ 3x_1 + 5x_2 + 3x_3 = 8 \\ 4x_1 + 7x_2 + 5x_3 = 10 \end{cases}$

39. $\begin{cases} 5x_1 + 4x_2 + 7x_3 + 6x_4 = 18 \\ 2x_1 + 2x_2 + 3x_3 + 3x_4 = 9 \\ 4x_1 + 3x_2 + 5x_3 + 5x_4 = 16 \\ 3x_1 + 2x_2 + 3x_3 + 3x_4 = 11 \end{cases}$ **41.** $\begin{pmatrix} 1 & 0 & 0 & 0 \\ 0 & 0 & 0 & 1 \\ 0 & 0 & 1 & 0 \\ 0 & 1 & 0 & 0 \end{pmatrix}$ **43.** $\begin{pmatrix} 1 & 0 & 0 & 0 \\ 0 & 1 & 0 & 0 \\ 0 & 0 & 1 & 0 \\ 0 & 0 & 0 & 3 \end{pmatrix}$ **45.** $\begin{pmatrix} 1 & 0 & -1 & 0 \\ 0 & 1 & 0 & 0 \\ 0 & 0 & 1 & 0 \\ 0 & 0 & 0 & 1 \end{pmatrix}$

47. $\begin{pmatrix} 1 & 0 & 0 & 0 \\ 0 & 1 & 0 & 0 \\ 0 & 0 & 1 & -2 \\ 0 & 0 & 0 & 1 \end{pmatrix}$ **49.** $\begin{pmatrix} 1 & 0 & -3 & 0 \\ 0 & 1 & -2 & 0 \\ 0 & 0 & 1 & 0 \\ 0 & 0 & -1 & 1 \end{pmatrix}$

51. Let P be the "price" matrix of selling prices and C be the "commission" matrix so that $C = 0.15P$. Choosing P to be a 3×3 matrix with the rows representing the manufacturers SlumberKing, DreamOn, and RestEasy in that order and the columns representing the models "economy," "best," and "deluxe" in that order, $P = \begin{pmatrix} 300 & 350 & 500 \\ 350 & 400 & 550 \\ 400 & 500 & 700 \end{pmatrix}$ and $C = \begin{pmatrix} 45.00 & 52.50 & 75.00 \\ 52.50 & 60.00 & 82.50 \\ 60.00 & 75.00 & 105.00 \end{pmatrix}$.

53. Let D be the "dealer invoice" matrix and S be the "sticker price" matrix so that the "markup" matrix M is $M = S - D$. Choosing D to be a 2×4 matrix with the rows representing the sales lots in Oakdale and Roanoke in that order and the columns representing the vehicle models "sedan," "station wagon," "van," and "pickup truck" in that order, $D = \begin{pmatrix} 15,000 & 19,000 & 23,000 & 25,000 \\ 15,000 & 19,000 & 23,000 & 25,000 \end{pmatrix}$ and $S = \begin{pmatrix} 18,900 & 22,900 & 26,900 & 29,900 \\ 19,900 & 21,900 & 27,900 & 28,900 \end{pmatrix}$ so then $M = \begin{pmatrix} 3900 & 3900 & 3900 & 4900 \\ 4900 & 2900 & 4900 & 3900 \end{pmatrix}$.

55. Let L be the "labor costs" matrix and M be the "materials cost" matrix so that the "total cost" matrix T is $T = L + M$. Choosing L to be a 2×3 matrix of values in pennies with the rows representing the countries Costa Rica and Honduras in that order and the columns representing the apparel items "shorts," "tee-shirts," and "caps" in that order, $L = \begin{pmatrix} 75 & 25 & 45 \\ 80 & 20 & 55 \end{pmatrix}$ and $M = \begin{pmatrix} 160 & 95 & 115 \\ 150 & 80 & 110 \end{pmatrix}$ so then $T = \begin{pmatrix} 235 & 120 & 160 \\ 230 & 100 & 165 \end{pmatrix}$.

57. Let P be the "sale price" row matrix with values in dollars and N be the "number of items" column matrix so that the "total cost" matrix C is $C = P \cdot N$. Choosing the columns of P and the rows of N to represent the items "bottles of soda," "bottles of pickles," "packages of hot dogs," and "bags of chips" in that order, $P = (0.89 \quad 1.29 \quad 2.39 \quad 1.69)$ and $N = \begin{pmatrix} 12 \\ 2 \\ 3 \\ 4 \end{pmatrix}$ so then $C = (27.19)$. The total cost of these items at these prices is $27.19.

59. Let T be the "time" matrix and L be the "labor hourly cost" matrix so that the "production cost" matrix C is $C = T \cdot L$. Choosing T to be a 3×3 matrix with the rows representing the furniture items "table," "chair," and "desk" in that order and the columns representing the manufacturing steps "cutting and milling," "assembly," and "finishing" in that order, $T = \begin{pmatrix} 2 & 1 & 2 \\ 1.5 & 1 & 0.5 \\ 3 & 2 & 3 \end{pmatrix}$, and choosing L to be a 3×2 matrix with the rows representing the manufacturing steps "cutting and milling," "assembly," and "finishing" in that order and the columns representing the factory locations "Wytheville" and "Andersen" in that order, $L = \begin{pmatrix} 9 & 10 \\ 14 & 13 \\ 13 & 12 \end{pmatrix}$, so then $C = \begin{pmatrix} 58 & 57 \\ 34 & 34 \\ 94 & 92 \end{pmatrix}$. For these choices of T and L, the rows of C represent the furniture items "table," "chair," and "desk" in that order, and the columns represent the factory locations "Wytheville" and "Andersen" in that order.

61. $A^t \cdot A = \begin{pmatrix} 17 & 22 & 27 \\ 22 & 29 & 36 \\ 27 & 36 & 45 \end{pmatrix}$ and $A \cdot A^t = \begin{pmatrix} 14 & 32 \\ 32 & 77 \end{pmatrix}$; both are symmetric.

63. $A^t \cdot A = \begin{pmatrix} 1 & -1 & 2 \\ -1 & 5 & -4 \\ 2 & -4 & 6 \end{pmatrix}$ and $A \cdot A^t = \begin{pmatrix} 6 & -4 & 2 \\ -4 & 5 & -1 \\ 2 & -1 & 1 \end{pmatrix}$; both are symmetric.

67. $f(g(x)) = \dfrac{8x-1}{5x}$ and $F \cdot G = \begin{pmatrix} 8 & -1 \\ 5 & 0 \end{pmatrix}$ **69.** $f(g(x)) = \dfrac{4x-1}{2x+1}$ and $F \cdot G = \begin{pmatrix} 4 & -1 \\ 2 & 1 \end{pmatrix}$ **81.** $\begin{pmatrix} 1 & \frac{1}{2} & \frac{1}{2} \\ 0 & \frac{1}{2} & \frac{3}{2} \\ 0 & \frac{1}{2} & \frac{1}{2} \end{pmatrix}$

83. $\begin{pmatrix} 1 & 0 & -2 \\ 0 & 1 & 4 \\ 0 & 0 & -1 \end{pmatrix}$ **85.** $(2)(\frac{1}{2})(-1) = -1$ **87.** $(3)(\frac{-19}{3})(\frac{-5}{19}) = 5$ **89.** 1 (switched rows 1 and 2)

91. -5 (multiplied second row by 5)   **93.** -1 (added row 1 to row 2 five times)   **95.** -1 (transposed the matrix)

EXERCISES 3.5 page 233

1. $\begin{pmatrix} 1 & 2 \\ -1 & -1 \end{pmatrix}\begin{pmatrix} -1 & -2 \\ 1 & 1 \end{pmatrix} = \begin{pmatrix} 1 & 0 \\ 0 & 1 \end{pmatrix}$ so this pair of matrices is a matrix and its inverse.

3. $\begin{pmatrix} 1 & 1 & 0 \\ 2 & 1 & 1 \\ 1 & 0 & 0 \end{pmatrix}\begin{pmatrix} 0 & 0 & 1 \\ 1 & 0 & -1 \\ -1 & 1 & -1 \end{pmatrix} = \begin{pmatrix} 1 & 0 & 0 \\ 0 & 1 & 0 \\ 0 & 0 & 1 \end{pmatrix}$ so this pair of matrices is a matrix and its inverse.

5. $\begin{pmatrix} 4 & 6 & 3 \\ 3 & 4 & 1 \\ 5 & 7 & 3 \end{pmatrix}\begin{pmatrix} -5 & -3 & 6 \\ 4 & 3 & -5 \\ -1 & -2 & 2 \end{pmatrix} = \begin{pmatrix} 1 & 0 & 0 \\ 0 & 1 & 0 \\ 0 & 0 & 1 \end{pmatrix}$ so this pair of matrices is a matrix and its inverse.

7. $\begin{pmatrix} 10 & -4 & -7 \\ -7 & 3 & 5 \\ 4 & -1 & -3 \end{pmatrix}\begin{pmatrix} 4 & 5 & -1 \\ 1 & 2 & 1 \\ 5 & 6 & 2 \end{pmatrix} = \begin{pmatrix} 1 & 0 & -28 \\ 0 & 1 & 20 \\ 0 & 0 & -11 \end{pmatrix}$ so this pair of matrices is not a matrix and its inverse.

9. $\begin{pmatrix} 2 & 0 & 1 & 0 \\ 1 & 1 & 1 & 0 \\ -2 & 0 & -1 & 1 \\ 1 & 0 & 0 & 1 \end{pmatrix}\begin{pmatrix} -1 & 0 & -1 & 1 \\ -2 & 1 & -1 & 1 \\ 3 & 0 & 2 & -2 \\ 1 & 0 & 1 & 0 \end{pmatrix} = \begin{pmatrix} 1 & 0 & 0 & 0 \\ 0 & 1 & 0 & 0 \\ 0 & 0 & 1 & 0 \\ 0 & 0 & 0 & 1 \end{pmatrix}$ so this pair of matrices is a matrix and its inverse.

11. $\begin{pmatrix} 1 & -3 \\ 0 & 1 \end{pmatrix}$ **13.** $\begin{pmatrix} -1 & 2 \\ 6 & -11 \end{pmatrix}$ **15.** $\begin{pmatrix} 0 & 1 & -2 \\ 1 & -1 & 2 \\ 0 & -1 & 3 \end{pmatrix}$ **17.** Singular matrix   **19.** $\begin{pmatrix} 1 & 0 & 0 & -1 \\ 0 & 1 & 0 & 0 \\ -1 & 0 & 1 & 1 \\ 0 & -1 & 0 & 1 \end{pmatrix}$

21. $\begin{pmatrix} 1 & -3 & 1 \\ 0 & 2 & -1 \\ -1 & 0 & 1 \end{pmatrix}$ **23.** $\begin{pmatrix} 1 & 2 & 0 \\ -2 & -2 & 1 \\ -5 & -7 & 2 \end{pmatrix}$ **25.** $\begin{pmatrix} 3 & -5 & 1 & 1 \\ -1 & 1 & 0 & 0 \\ -4 & 5 & -1 & 1 \\ 0 & 1 & 0 & -1 \end{pmatrix}$ **27.** Singular matrix

29. $\begin{pmatrix} 0 & -2 & 0 & 1 & 0 \\ 3 & 3 & -1 & -1 & 1 \\ 1 & 6 & -1 & -3 & 2 \\ -5 & -5 & 2 & 2 & -2 \\ 0 & -4 & 1 & 2 & -2 \end{pmatrix}$ **31.** $\begin{pmatrix} x_1 \\ x_2 \end{pmatrix} = \begin{pmatrix} -1 & 2 \\ 6 & -11 \end{pmatrix}\begin{pmatrix} 9 \\ 5 \end{pmatrix} = \begin{pmatrix} 1 \\ -1 \end{pmatrix}$ so $\begin{cases} x_1 = 1 \\ x_2 = -1 \end{cases}$

33. $\begin{pmatrix} x_1 \\ x_2 \\ x_3 \end{pmatrix} = \begin{pmatrix} 0 & 1 & -2 \\ 1 & -1 & 2 \\ 0 & -1 & 3 \end{pmatrix}\begin{pmatrix} 2 \\ 5 \\ 2 \end{pmatrix} = \begin{pmatrix} 1 \\ 1 \\ 1 \end{pmatrix}$ so $\begin{cases} x_1 = 1 \\ x_2 = 1 \\ x_3 = 1 \end{cases}$ **35.** $x_1 = 1, x_2 = 1, x_3 = 1$ **37.** $x_1 = 4, x_2 = -1, x_3 = -2$

39. $x_1 = 2, x_2 = -1, x_3 = 2, x_4 = 1$ **41.** $x_1 = 1, x_2 = -2, x_3 = 1, x_4 = 2$ **43.** $x_1 = 1, x_2 = -2, x_3 = 2, x_4 = -1, x_5 = 1$

45. $x_1 = 1, x_2 = -2, x_3 = 1, x_4 = -2, x_5 = 1$ **47.** $x_1 = 2, x_2 = 4, x_3 = 1$ **49.** $x_1 = -5, x_2 = 10, x_3 = -20, x_4 = 15$

51. The multiplex sold 150 adult tickets and 350 child tickets for Film No. 1; 200 adult tickets and 200 child tickets for Film No. 2; 250 adult tickets and 200 child tickets for Film No. 3; 400 adult tickets and 100 child tickets for Film No. 4; and 600 adult tickets and no child tickets for Film No. 5.

53. The red jar contains 250 pennies, 150 nickels, and 100 dimes; the green jar contains 150 pennies, 350 nickels, and 200 dimes; and the blue jar contains 325 pennies, 115 nickels, and 160 dimes.

55. Billy needs 4 drops of Supplement No. 1, no drops of Supplement No. 2, 5 drops of Supplement No. 3, and 2 drops of Supplement No. 4; Susie needs 2 drops of Supplement No. 1, 2 drops of Supplement No. 2, 3 drops of Supplement No. 3, and 6 drops of Supplement No. 4; and Jimmy needs 3 drops of Supplement No. 1, 1 drop of Supplement No. 2, no drops of Supplement No. 3, and 8 drops of Supplement No. 4.

57. Mr. and Mrs. Jordan should invest $100,000 in the stock fund, $150,000 in the money market fund, and $50,000 in the bond fund; Mr. and Mrs. French should invest $78,300 in the stock fund, $51,100 in the money market fund, and $105,500 in the bond fund; and Mrs. Daimen should invest $90,000 in the stock fund, $105,000 in the money market fund, and $75,000 in the bond fund.

59. The mass transit manager should assign 120 subway cars, 50 buses, and 10 jitneys to Brighton; 100 subway cars, 30 buses, and 20 jitneys to Conway; 100 subway cars, 40 buses, and 10 jitneys to Longwood; and 110 subway cars, 40 buses, and 15 jitneys to Oakley.

61. $X = (A + I)^{-1} \cdot C; \ X = \begin{pmatrix} 60 \\ 64 \\ -12 \end{pmatrix}$ **63.** $X = (A + B)^{-1} \cdot (C + D); \ X = \begin{pmatrix} -60 \\ 100 \\ 180 \end{pmatrix}$

EXERCISES 3.6 page 248

1. $\begin{pmatrix} 0.30 & 0.50 \\ 0.40 & 0.10 \end{pmatrix}$ for sectors A and L **3.** $\begin{pmatrix} 0.20 & 0.10 & 0.15 \\ 0.10 & 0.30 & 0.20 \\ 0.40 & 0.15 & 0.10 \end{pmatrix}$ for sectors C, E, and L

5.

7.

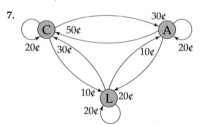

9. $Y = \begin{pmatrix} 71 \\ 37 \end{pmatrix}$ **11.** $Y = \begin{pmatrix} 89 \\ 118 \\ 101 \end{pmatrix}$ **13.** $X = \begin{pmatrix} 150 \\ 120 \end{pmatrix}$ **15.** $X = \begin{pmatrix} 80 \\ 60 \\ 100 \end{pmatrix}$

17. Heavy industry production must be $300 million and light industry production must be $280 million.

19. The current excess production from these sectors is $115 million (made up of $20 million from heavy industry, $50 million from light industry, and $45 million from the railroads). Each $10 million increase in heavy industry production raises the excess production by $1 million (but because the heavy industry production consumes both light industry and railroad production, this increase is composed of an $8 million increase in heavy industry production together with decreases of $5 million from light industry and $2 million from the railroads). When the heavy industry production level reaches $200 million, all light industry excess production will be consumed by the heavy industry sector, and no further expansion will be possible without expanding the light industry production.

21. $x = 5, y = -4$ **23.** $x_1 = 2, x_2 = 3, x_3 = 4$ **25.** Inconsistent, yet "almost" satisfied by $x = 5, y = 5$

27. Inconsistent, yet "almost" satisfied by $x_1 = 1, x_2 = 3, x_3 = 3$

29. Inconsistent, yet "almost" satisfied by $x_1 = 5, x_2 = 3, x_3 = 2, x_4 = 1$

31. $y = 9x + 34$ **33.** $y = -29x + 92$ **35.** $y = 9x + 165$ **37.** His sales in the fifth month may be expected to be $910,000.

39. At 79¢ per eight-ounce bag, the manufacturer can expect 1711 sales per 20,000 customers.

CHAPTER 3 REVIEW EXERCISES page 254

1. Let x be the number of 30-day advance sale tickets and let y be the number of full-fare tickets: $\begin{cases} x + y = 30 \\ 79x + 159y = 3970 \end{cases}$

2. Let x be the number of cows and let y be the number of horses: $\begin{cases} x + y = 420 \\ x - 2y = 0 \end{cases}$

3. The solution is $x = 13, y = 5$. The equations are independent and consistent.

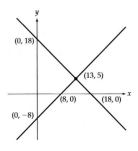

4. There are infinitely many solutions that may be parameterized as $x = 18 + 2t, y = t$. The equations are dependent.

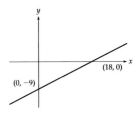

5. $x = 14, y = 30$. The equations are independent and consistent.

6. No solution. The equations are independent and inconsistent.

7. $x = 9, y = 3$. The equations are independent and consistent.

8. $x = -15, y = 24$. The equations are independent and consistent.

9. He has 10 rosebushes and 15 tomato plants in his garden.

10. The fraternity should use 2 cars and 2 vans for the trip to Orlando.   **11.** $3 \times 3; 5; 6; 4$ **12.** $4 \times 4; 7; 10; 13$

13. $\begin{pmatrix} 3 & 4 & 12 \\ 3 & 6 & 6 \end{pmatrix} R_2' = 3R_2$ **14.** $\begin{pmatrix} 1 & -1 & 8 \\ 2 & -3 & 6 \end{pmatrix} R_1' = R_1 + R_2$

15. $x = 15, y = 16$. The equations are independent and consistent.

16. $x = 20, y = 30$. The equations are independent and consistent.

17. No solution. The equations are independent and inconsistent.

18. $x = 3 + \frac{3}{4}t, y = t$ (infinitely many solutions). The equations are dependent.

19. The pharmacist filled 63 prescriptions for antibiotics and 29 prescriptions for cough suppressants.

20. The price of each old *Life* magazine is $6, and each old copy of *The New Yorker* costs $5.

21. The system of equations is independent and consistent, with solution $x_1 = 3, x_2 = -3, x_3 = 6$.

22. The system of equations is dependent and consistent, with solution $x_1 = 4 - t, x_2 = t, x_3 = 2$.

23. $\begin{pmatrix} 1 & 0 & 0 & 4 \\ 0 & 1 & 0 & -3 \\ 0 & 0 & 1 & 2 \end{pmatrix}$ **24.** $\begin{pmatrix} 1 & 0 & 0 & 1 & 0 \\ 0 & 1 & 0 & -1 & 0 \\ 0 & 0 & 1 & 0 & 0 \\ 0 & 0 & 0 & 0 & 1 \end{pmatrix}$

25. The system of equations is independent and consistent, with solution $x_1 = 4$, $x_2 = 3$, $x_3 = 1$, $x_4 = 2$.

26. The system of equations is dependent and consistent, with solution $x_1 = 2 - t$, $x_2 = 3 - t$, $x_3 = t$, $x_4 = 4$.

27. The system of equations is independent and inconsistent, and there is no solution.

28. The system of equations is dependent and inconsistent, and there is no solution.

29. The Nyack Nursery should raise 48 dahlias, 328 chrysanthemums, and 48 daisies.

30. Each hot dog costs $1. If the mirror house is $1.50 and a soda is $1, then a go-kart ride costs $4.50.   **31.** $\begin{pmatrix} 4 & -6 \\ 6 & -4 \end{pmatrix}$

32. $\begin{pmatrix} 1 & 2 & 3 & 4 \\ -4 & -3 & -2 & -1 \\ 1 & 2 & 3 & 4 \\ -4 & -3 & -2 & -1 \end{pmatrix}$ **33.** $\begin{pmatrix} 15 & 25 \\ 5 & 2 \end{pmatrix}$ **34.** $\begin{pmatrix} 12 & 3 & -2 \\ -2 & 12 & 11 \\ 9 & 3 & 1 \end{pmatrix}$ **35.** $\begin{pmatrix} 1 & 4 & 1 \\ 2 & 8 & 3 \\ 1 & 5 & 2 \end{pmatrix}\begin{pmatrix} x_1 \\ x_2 \\ x_3 \end{pmatrix} = \begin{pmatrix} 15 \\ 26 \\ 17 \end{pmatrix}$

36. $\begin{pmatrix} 2 & 3 & -1 & 1 \\ 5 & 4 & 1 & 2 \\ 2 & 1 & 1 & 1 \end{pmatrix}\begin{pmatrix} x_1 \\ x_2 \\ x_3 \\ x_4 \end{pmatrix} = \begin{pmatrix} 20 \\ 35 \\ 12 \end{pmatrix}$ **37.** $\begin{pmatrix} 1 & -1 & 0 \\ 0 & 1 & 0 \\ 0 & 0 & 1 \end{pmatrix}$ **38.** $\begin{pmatrix} 0 & 1 & 0 \\ 1 & 0 & -1 \\ 0 & -2 & 1 \end{pmatrix}$

39. Let T be the "this year" matrix and L be the "last year" matrix so that the "growth" matrix G is $G = T - L$. Choosing T to be a 3×2 matrix with the rows representing the grandchildren Thomas, Richard, and Harriet in that order and the columns representing their heights and weights in that order,

$$T = \begin{pmatrix} 61 & 90 \\ 54 & 75 \\ 47 & 60 \end{pmatrix} \text{ and } L = \begin{pmatrix} 58 & 80 \\ 52 & 70 \\ 46 & 55 \end{pmatrix} \text{ so then } G = \begin{pmatrix} 3 & 10 \\ 2 & 5 \\ 1 & 5 \end{pmatrix}.$$

40. Let N be the "number of items needed" column matrix and P be the "price" matrix so that the "cost of her order" matrix C is $C = P \cdot N$. Choosing the rows of N to represent the items "jacket," "blouse," "skirt," and "slacks" in that order, N is the

4×1 matrix $N = \begin{pmatrix} 200 \\ 300 \\ 250 \\ 175 \end{pmatrix}$. Then the columns of P must also represent the items "jacket," "blouse," "skirt," and "slacks" in that

order, so P is the 2×4 matrix $P = \begin{pmatrix} 195 & 85 & 145 & 130 \\ 190 & 90 & 150 & 125 \end{pmatrix}$ and the rows represent the prices of the East Coast designer and the

Italian team in that order. Then $C = \begin{pmatrix} 123{,}500 \\ 124{,}375 \end{pmatrix}$. The cost of her order from the East Coast designer is $123,500, and from the

Italian team it is $124,375.

41. $\begin{pmatrix} 1 & 2 & 3 \\ 1 & 1 & 1 \\ 0 & 1 & 3 \end{pmatrix}\begin{pmatrix} -2 & 3 & 1 \\ 3 & -3 & -2 \\ -1 & 1 & 1 \end{pmatrix} = \begin{pmatrix} 1 & 0 & 0 \\ 0 & 1 & 0 \\ 0 & 0 & 1 \end{pmatrix}$ so this pair of matrices is a matrix and its inverse.

42. $\begin{pmatrix} -3 & 0 & 1 \\ 1 & 3 & 1 \\ -3 & 2 & 2 \end{pmatrix}\begin{pmatrix} -4 & -2 & 3 \\ 5 & 3 & -4 \\ -11 & -6 & 8 \end{pmatrix} = \begin{pmatrix} 1 & 0 & -1 \\ 0 & 1 & -1 \\ 0 & 0 & -1 \end{pmatrix}$ so this pair of matrices is not a matrix and its inverse.

43. $\begin{pmatrix} 3 & 0 & -1 \\ -2 & 0 & 1 \\ -6 & 1 & 3 \end{pmatrix}$ **44.** Singular matrix   **45.** $\begin{pmatrix} x_1 \\ x_2 \end{pmatrix} = \begin{pmatrix} 1 & -4 \\ -3 & 13 \end{pmatrix}\begin{pmatrix} 33 \\ 8 \end{pmatrix} = \begin{pmatrix} 1 \\ 5 \end{pmatrix}$ so $\begin{cases} x_1 = 1 \\ x_2 = 5 \end{cases}$

46. $\begin{pmatrix} x_1 \\ x_2 \end{pmatrix} = \begin{pmatrix} -5 & 8 \\ 2 & -3 \end{pmatrix}\begin{pmatrix} 25 \\ 16 \end{pmatrix} = \begin{pmatrix} 3 \\ 2 \end{pmatrix}$ so $\begin{cases} x_1 = 3 \\ x_2 = 2 \end{cases}$ **47.** $\begin{pmatrix} x_1 \\ x_2 \\ x_3 \end{pmatrix} = \begin{pmatrix} 1 & 1 & -3 \\ 0 & 1 & -1 \\ -2 & -3 & 8 \end{pmatrix}\begin{pmatrix} 11 \\ 7 \\ 5 \end{pmatrix} = \begin{pmatrix} 3 \\ 2 \\ -3 \end{pmatrix}$ so $\begin{cases} x_1 = 3 \\ x_2 = 2 \\ x_3 = -3 \end{cases}$

48. $\begin{pmatrix} x_1 \\ x_2 \\ x_3 \\ x_4 \end{pmatrix} = \begin{pmatrix} 1 & 0 & 0 & -1 \\ 0 & 1 & -2 & 0 \\ 0 & -2 & 6 & -1 \\ -1 & 0 & -1 & 2 \end{pmatrix}\begin{pmatrix} 7 \\ 10 \\ 4 \\ 6 \end{pmatrix} = \begin{pmatrix} 1 \\ 2 \\ -2 \\ 1 \end{pmatrix}$ so $\begin{cases} x_1 = 1 \\ x_2 = 2 \\ x_3 = -2 \\ x_4 = 1 \end{cases}$

49. The Kingman store can display 7 living room suites (2 in the window and 5 more only on the showroom floor) and 8 bedroom suites (3 in the window and 5 more only on the showroom floor); the Prescott store can display 11 living room suites (3 in the window and 8 more only on the showroom floor) and 10 bedroom suites (3 in the window and 7 more only on the showroom floor); and the Holbrook store can display 9 living room suites (3 in the window and 6 more only on the showroom floor) and 12 bedroom suites (2 in the window and 10 more only on the showroom floor).

50. Mr. Dahlman's taxes are $17,100 (federal), $7600 (state), and $3600 (city); Mrs. Farrell's taxes are $8550 (federal), $3800 (state), and $1800 (city); Ms. Mazlin's taxes are $13,680 (federal), $6080 (state), and $2880 (city); and Mr. Seidner's taxes are $25,650 (federal), $11,400 (state), and $5400 (city).

51. $\begin{pmatrix} 0.15 & 0.25 \\ 0.30 & 0.20 \end{pmatrix}$ for sectors A and B   **52.** $\begin{pmatrix} 0.10 & 0.20 & 0.30 & 0 \\ 0.20 & 0.10 & 0 & 0.20 \\ 0.30 & 0 & 0.10 & 0.30 \\ 0 & 0.20 & 0.30 & 0.10 \end{pmatrix}$ for sectors A, B, C, and D

53.

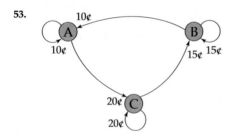

54.

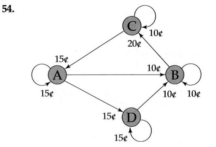

55. $Y = \begin{pmatrix} 371 \\ 449 \\ 419 \end{pmatrix}$ **56.** $Y = \begin{pmatrix} 170 \\ 115 \\ 160 \\ 143 \end{pmatrix}$ **57.** $X = \begin{pmatrix} 364 \\ 436 \end{pmatrix}$ **58.** $X = \begin{pmatrix} 457 \\ 605 \\ 529 \end{pmatrix}$

59. Each division must produce $4 million.

60. The other sectors of the economy can use $841 million of domestic oil and $1948 million of foreign oil, but all the military protection budget is consumed in the production of this output.

61. $x = 4, y = 3$ **62.** $x = -3, y = 8$ **63.** Inconsistent, yet "almost" satisfied by $x = 5, y = 3$

64. Inconsistent, yet "almost" satisfied by $x = 2, y = -1$ **65.** $y = 7x + 9$ **66.** $y = 73x + 14$ **67.** $y = 13x + 13$

68. $y = 19x - 52$ **69.** They can expect sales of $4405.   **70.** She can expect 12 minor accidents.

EXERCISES 4.1 page 275

1. b   **3.** c   **5.** a   **7.** d   **9.** b

11. The vertices are (0, 0), (40, 0), and (0, 20).
The region is bounded.

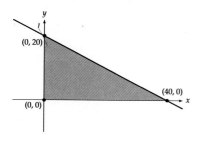

13. The vertices are (0, 0), (10, 0), (10, 30), and (0, 10).
The region is bounded.

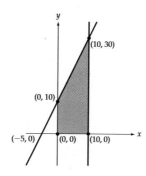

15. The vertices are (0, 0), (6, 0), (4, 2), and (0, 4).
The region is bounded.

17. The vertices are (4, 0) and (0, 10).
The region is unbounded.

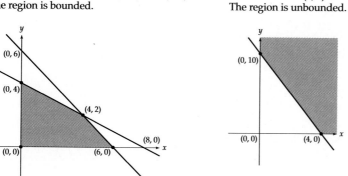

19. The vertices are (3, 4) and (0, 8).
The region is unbounded.

21. The vertices are (8, 0), (2, 6), and (0, 12).
The region is unbounded.

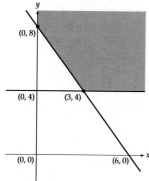

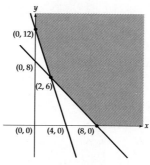

23. The vertices are (0, 0), (15, 0), (15, 5), (10, 10), and (0, 10). The region is bounded.

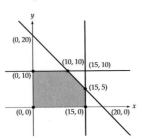

25. The vertices are (30, 0), (40, 0), (0, 80), and (0, 10). The region is bounded.

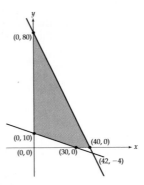

27. The vertices are (15, 0) and (0, 30). The region is unbounded.

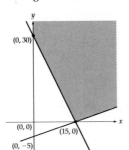

29. The vertices are (0, 0), (9, 0), (8, 2), (3, 7), and (0, 8). The region is bounded.

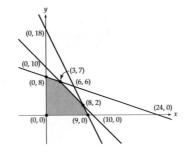

31. Let x be the number of goats and let y be the number of llamas.

$$\begin{cases} 2x + 5y \leq 400 & \text{(Land)} \\ 100x + 80y \leq 13{,}200 & \text{(Money)} \\ x \geq 0, y \geq 0 & \text{(Nonnegativity)} \end{cases}$$

The vertices are (0, 0), (132, 0), (100, 40), and (0, 80).

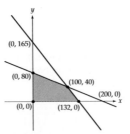

33. Let x be the number of dinghies and let y be the number of rowboats.

$$\begin{cases} 2x + 3y \leq 120 & \text{(Metal work)} \\ 2x + 2y \leq 100 & \text{(Painting)} \\ x \geq 0, y \geq 0 & \text{(Nonnegativity)} \end{cases}$$

The vertices are (0, 0), (50, 0), (30, 20), and (0, 40).

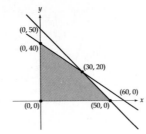

35. Let x be the number of servings of SugarSnaks and let y be the number of bags of Gobbl'Ems.

$$\begin{cases} 5x + 8y \le 80 & \text{(Fat)} \\ 125x + 250y \le 2250 & \text{(Calories)} \\ x \ge 0, y \ge 0 & \text{(Nonnegativity)} \end{cases}$$

The vertices are $(0, 0)$, $(16, 0)$, $(8, 5)$, and $(0, 9)$.

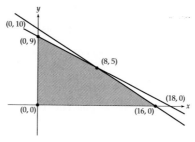

37. Let x be the number of hours the Ohio factory operates and let y be the number of hours the Pennsylvania factory operates.

$$\begin{cases} 4x + 4y \le 64 & \text{(Sulfur dioxide)} \\ 5x + 3y \le 60 & \text{(Particulates)} \\ x \ge 0, y \ge 0 & \text{(Nonnegativity)} \end{cases}$$

The vertices are $(0, 0)$, $(12, 0)$, $(6, 10)$, and $(0, 16)$.

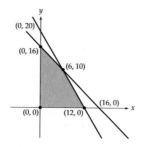

39. Let x be the amount of money (in millions of dollars) invested in stock funds and let y be the amount of money (in millions of dollars) invested in bond funds.

$$\begin{cases} x + y \le 8 & \text{(Money)} \\ x \le y & \text{(Limit risk)} \\ x \ge 0, y \ge 0 & \text{(Nonnegativity)} \end{cases}$$

The vertices are $(0, 0)$, $(4, 4)$, and $(0, 8)$.

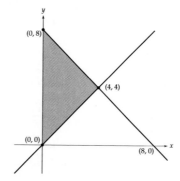

EXERCISES 4.2 page 290

1. The maximum is 45 (when $x = 15$, $y = 15$).   **3.** The minimum is 30 (when $x = 10$, $y = 0$).

5. The maximum is 15 (when $x = 15$, $y = 0$).   **7.** The minimum is 60 (when $x = 0$, $y = 60$).

9. There is no maximum because the region is unbounded in the positive y direction and P increases for increasing y values.

11. The maximum is 400 (when $x = 0$, $y = 10$).   **13.** The minimum is 150 (when $x = 10$, $y = 0$).

15. The maximum is 160 (when $x = 15$, $y = 20$).   **17.** The minimum is 232 (when $x = 6$, $y = 16$).

19. The maximum is 140 (when $x = 0$, $y = 20$).   **21.** The minimum is 180 (when $x = 30$, $y = 0$).

23. The maximum is 230 (when $x = 40$, $y = 10$).   **25.** The maximum is 1080 (when $x = 0$, $y = 90$).

27. The minimum is 885 (when $x = 33, y = 9$).   **29.** The minimum is 1545 (when $x = 45, y = 220$).

31. The rancher should raise 100 goats and 40 llamas to obtain the greatest possible profit of $9600.

33. The company should manufacture 32 prams and 9 yawls to obtain the greatest possible profit of $6420.

35. The county should operate the Norton incinerator 4 hours each day and the Wiseburg incinerator 6 hours each day to obtain the least cost of $620.

37. Each bunny should receive 30 handfuls of greens and 16 drops of supplement each week to minimize the owner's costs at 76¢.

39. No grassland and 2400 acres of forest should be reclaimed this year to raise the greatest possible amount of $360,000 in long-term leases for use in next year's reclamation efforts.

EXERCISES 4.3 page 314

1.

	x_1	x_2	x_3	s_1	s_2	
s_1	3	2	4	1	0	12
s_2	6	1	5	0	1	15
P	-8	-9	-7	0	0	0

3.

	x_1	x_2	s_1	s_2	s_3	
s_1	4	3	1	0	0	12
s_2	5	2	0	1	0	20
s_3	1	6	0	0	1	12
P	-13	-7	0	0	0	0

5.

	x_1	x_2	x_3	x_4	s_1	s_2	
s_1	2	1	1	3	1	0	6
s_2	1	4	-2	1	0	1	8
P	-5	2	-10	5	0	0	0

7.

	x_1	x_2	x_3	s_1	s_2	s_3	
s_1	8	1	4	1	0	0	32
s_2	3	5	7	0	1	0	30
s_3	6	2	9	0	0	1	28
P	-10	-20	-15	0	0	0	0

9.

	x_1	x_2	x_3	s_1	s_2	s_3	s_4	
s_1	1	2	3	1	0	0	0	45
s_2	6	5	4	0	1	0	0	40
s_3	7	8	9	0	0	1	0	63
s_4	12	11	10	0	0	0	1	60
P	-90	-80	-100	0	0	0	0	0

11. The smallest negative entry in the bottom row is -9, so the pivot column is column 3. The ratios are $\frac{4}{1} = 4$, $\frac{5}{1} = 5$, and $\frac{6}{3} = 2$, so the pivot row is row 3. The pivot element is the 3 in column 3 and row 3 of the simplex tableau.

	x_1	x_2	x_3	s_1	s_2	s_3		
s_1	5/3	1/3	0	1	0	$-1/3$	2	$R_1^{new} = R_1 - R_{pivot}^{new}$
s_2	2/3	4/3	0	0	1	$-1/3$	3	$R_2^{new} = R_2 - R_{pivot}^{new}$
x_3	1/3	$-1/3$	1	0	0	1/3	2	$R_{pivot}^{new} = R_{pivot}/3$
P	-4	-11	0	0	0	3	18	$R_4^{new} = R_4 + 9R_{pivot}^{new}$

13. The smallest negative entry in the bottom row is -8, so the pivot column is column 3. The ratios are $\frac{3}{1} = 3$, $\frac{4}{1} = 4$, and (omit), so the pivot row is row 1. The pivot element is the 1 in column 3 and row 1 of the simplex tableau.

	x_1	x_2	x_3	s_1	s_2	s_3		
x_3	1	0	1	1	0	0	3	$R_{pivot}^{new} = R_{pivot}$
s_2	-1	1	0	-1	1	0	1	$R_2^{new} = R_2 - R_{pivot}^{new}$
s_3	1	1	0	0	0	1	5	$R_3^{new} = R_3$
P	2	-7	0	8	0	0	24	$R_4^{new} = R_4 + 8R_{pivot}^{new}$

15. The smallest negative entry in the bottom row is -5, so the pivot column is column 2. The ratios are $\frac{12}{2} = 6$ and $\frac{8}{1} = 8$, so the pivot row is row 1. The pivot element is the 2 in column 2 and row 1 of the simplex tableau.

	x_1	x_2	x_3	x_4	s_1	s_2	
x_2	2	1	3	1	1/2	0	6
s_2	1	0	-1	0	$-1/2$	1	2
P	6	0	21	2	5/2	0	30

$R_{pivot}^{new} = R_{pivot}/2$
$R_2^{new} = R_2 - R_{pivot}^{new}$
$R_3^{new} = R_3 + 5R_{pivot}^{new}$

17. No pivot column because there are no negative entries in the bottom row. The solution has been found: The maximum is 90 when $x_1 = 10$, $x_2 = 15$, $x_3 = 0$, and $x_4 = 0$ (and $s_1 = 10$, $s_2 = 0$, and $s_3 = 0$).

19. The smallest negative entry in the bottom row is -10 so the pivot column is column 5. Since all of the entries in the pivot column are zero or negative, there is no pivot row. There is no maximum value.

21. The maximum is 28 when $x_1 = 0$, $x_2 = 14$.   **23.** The maximum is 30 when $x_1 = 0$, $x_2 = 0$, $x_3 = 6$, $x_4 = 0$.

25. The maximum is 2000 when $x_1 = 0$, $x_2 = 50$, $x_3 = 0$.

27. There is no maximum (the second pivot column is column 2 but there is no pivot row).

29. The maximum is 300 when $x_1 = 0$, $x_2 = 20$, $x_3 = 0$, $x_4 = 40$.   **31.** The maximum is 16 when $x_1 = 3$, $x_2 = 2$.

33. The maximum is 12,600 when $x_1 = 0$, $x_2 = 160$, $x_3 = 60$.   **35.** The maximum is 30 when $x_1 = 5$, $x_2 = 5$, $x_3 = 5$.

37. The maximum is 2000 when $x_1 = 0$, $x_2 = 24$, $x_3 = 32$.   **39.** The maximum is 460 when $x_1 = 15$, $x_2 = 10$, $x_3 = 20$.

41. The shop should rebuild no carburetors, 35 fuel pumps, and 20 alternators for a greatest possible profit of $690.

43. The recycling center should accept 300 crates of paper products and 500 crates of glass bottles to raise the greatest possible amount of $59 each week.

45. The company should process no agates, 49 trays of onyxes, and 7 trays of garnets each day to obtain the greatest possible profit of $581.

47. The politician should run 3 daytime ads, 7 prime time ads, and no late night ads to reach 47,000 voters, the most possible.

49. The farmer should plant 30 acres of corn, 90 acres of peanuts, and 120 acres of soybeans for the greatest possible profit of $43,500.

51. a. $5x + 2y = 70$ means $x = 14 - \frac{2}{5}y$. Substitution in the second equation gives $4\left(14 - \frac{2}{5}y\right) + 3y = 84$ so $\frac{7}{5}y = 28$. Then $y = 20$ and $x = 14 - \frac{2}{5}(20) = 6$.

b. Pivoting gives

	x	y	
	1	2/5	14
	0	7/5	28

The first row represents the equation $x + \frac{2}{5}y = 14$, which is equivalent to $x = 14 - \frac{2}{5}y$.

The second row represents the equation $\frac{7}{5}y = 28$, which we found before on the way to obtaining $y = 20$.

53. Yes.

55. For Exercise 19: Pivoting at [column 3, row 1], [column 4, row 2], [column 5, row 3], and [column 6, row 4], the simplex tableau becomes

	x_1	x_2	s_1	s_2	s_3	s_4	
s_1	0	1	1	0	0	0	30
s_2	0	1	0	1	0	0	20
s_3	−1	1	0	0	1	0	10
s_4	0	1	0	0	0	1	25
P	−10	−15	0	0	0	0	0

That is,

Maximize $P = 10x_1 + 15x_2$

Subject to $\begin{cases} x_2 \le 30 \\ x_2 \le 20 \\ -x_1 + x_2 \le 10 \\ x_2 \le 25 \\ x_1 \ge 0, x_2 \ge 0 \end{cases}$

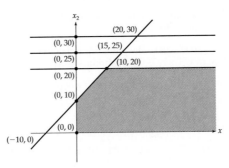

The region is unbounded to the right and the objective function increases in this direction. There is no maximum value.

For Exercise 20: Pivoting at [column 3, row 1], [column 4, row 2], [column 5, row 3], and [column 6, row 4], the simplex tableau becomes

	x_1	x_2	s_1	s_2	s_3	s_4	
s_1	1	−1	1	0	0	0	15
s_2	1	0	0	1	0	0	20
s_3	1	−1	0	0	1	0	10
s_4	1	0	0	0	0	1	25
P	−20	−15	0	0	0	0	0

That is,

Maximize $P = 20x_1 + 15x_2$

Subject to $\begin{cases} x_1 - x_2 \le 15 \\ x_1 \le 20 \\ x_1 - x_2 \le 10 \\ x_1 \le 25 \\ x_1 \ge 0, x_2 \ge 0 \end{cases}$

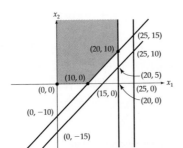

The region is unbounded moving upward and the objective function increases in this direction. There is no maximum value.

57. a. The final tableau is

	x_1	x_2	x_3	x_4	s_1	s_2	s_3	
x_3	0	0	1	0	0	0	1	1
s_1	0	−8	0	30	1	−2	3	3
x_1	1	−24	0	6	0	2	1	1
P	0	2	0	21/2	0	3/2	5/4	5/4

The maximum is $\frac{5}{4}$ when $x_1 = 1$, $x_2 = 0$, $x_3 = 1$, $x_4 = 0$.

b. The final tableau is

	x_1	x_2	x_3	x_4	s_1	s_2	s_3	
x_1	1	−6	0	6	0	2	1	1
x_3	0	0	1	0	0	0	1	1
s_1	0	−1/2	0	15/2	1	−1/2	3/4	3/4
P	0	1/2	0	21/2	0	3/2	5/4	5/4

The maximum is $\frac{5}{4}$ when $x_1 = 1$, $x_2 = 0$, $x_3 = 1$, $x_4 = 0$.

59. No.   **61.** The maximum is 100 when $x_1 = 100$, $x_2 = 0$.

63. Maximize $P = (1 \quad 10 \quad 100 \quad 1000) \begin{pmatrix} x_1 \\ x_2 \\ x_3 \\ x_4 \end{pmatrix}$

Subject to $\begin{pmatrix} 0 & 0 & 0 & 1 \\ 0 & 0 & 1 & 20 \\ 0 & 1 & 20 & 200 \\ 1 & 20 & 200 & 2000 \end{pmatrix} \begin{pmatrix} x_1 \\ x_2 \\ x_3 \\ x_4 \end{pmatrix} \le \begin{pmatrix} 1 \\ 100 \\ 10{,}000 \\ 1{,}000{,}000 \end{pmatrix}$ and $\begin{pmatrix} x_1 \\ x_2 \\ x_3 \\ x_4 \end{pmatrix} \ge 0$.

The maximum is 1,000,000 when $x_1 = 1{,}000{,}000$, $x_2 = 0$, $x_3 = 0$, $x_4 = 0$.

EXERCISES 4.4 page 335

1. Minimize $C = (60 \quad 100 \quad 300) \begin{pmatrix} y_1 \\ y_2 \\ y_3 \end{pmatrix}$ subject to $\begin{pmatrix} 1 & 2 & 3 \\ 4 & 5 & 6 \end{pmatrix} \begin{pmatrix} y_1 \\ y_2 \\ y_3 \end{pmatrix} \ge \begin{pmatrix} 180 \\ 120 \end{pmatrix}$ and $\begin{pmatrix} y_1 \\ y_2 \\ y_3 \end{pmatrix} \ge 0$. The dual problem is maximize

$P = (180 \quad 120) \begin{pmatrix} x_1 \\ x_2 \end{pmatrix}$ subject to $\begin{pmatrix} 1 & 4 \\ 2 & 5 \\ 3 & 6 \end{pmatrix} \begin{pmatrix} x_1 \\ x_2 \end{pmatrix} \le \begin{pmatrix} 60 \\ 100 \\ 300 \end{pmatrix}$ and $\begin{pmatrix} x_1 \\ x_2 \end{pmatrix} \ge 0$ with initial simplex tableau

	x_1	x_2	s_1	s_2	s_3	
s_1	1	4	1	0	0	60
s_2	2	5	0	1	0	100
s_3	3	6	0	0	1	300
P	−180	−120	0	0	0	0

3. Minimize $C = (3 \quad 20)\begin{pmatrix} y_1 \\ y_2 \end{pmatrix}$ subject to $\begin{pmatrix} 3 & 2 \\ 1 & 4 \\ 3 & 4 \end{pmatrix}\begin{pmatrix} y_1 \\ y_2 \end{pmatrix} \geq \begin{pmatrix} 150 \\ 100 \\ 228 \end{pmatrix}$ and $\begin{pmatrix} y_1 \\ y_2 \end{pmatrix} \geq 0$. The dual problem is maximize

$P = (150 \quad 100 \quad 228)\begin{pmatrix} x_1 \\ x_2 \\ x_3 \end{pmatrix}$ subject to $\begin{pmatrix} 3 & 1 & 3 \\ 2 & 4 & 4 \end{pmatrix}\begin{pmatrix} x_1 \\ x_2 \\ x_3 \end{pmatrix} \leq \begin{pmatrix} 3 \\ 20 \end{pmatrix}$ and $\begin{pmatrix} x_1 \\ x_2 \\ x_3 \end{pmatrix} \geq 0$ with initial simplex tableau

	x_1	x_2	x_3	s_1	s_2	
s_1	3	1	3	1	0	3
s_2	2	4	4	0	1	20
P	-150	-100	-228	0	0	0

5. Minimize $C = (84 \quad 21)\begin{pmatrix} y_1 \\ y_2 \end{pmatrix}$ subject to $\begin{pmatrix} 3 & 1 \\ 4 & -1 \end{pmatrix}\begin{pmatrix} y_1 \\ y_2 \end{pmatrix} \geq \begin{pmatrix} 21 \\ 0 \end{pmatrix}$ and $\begin{pmatrix} y_1 \\ y_2 \end{pmatrix} \geq 0$. The dual problem is maximize

$P = (21 \quad 0)\begin{pmatrix} x_1 \\ x_2 \end{pmatrix}$ subject to $\begin{pmatrix} 3 & 4 \\ 1 & -1 \end{pmatrix}\begin{pmatrix} x_1 \\ x_2 \end{pmatrix} \leq \begin{pmatrix} 84 \\ 21 \end{pmatrix}$ and $\begin{pmatrix} x_1 \\ x_2 \end{pmatrix} \geq 0$ with initial simplex tableau

	x_1	x_2	s_1	s_2	
s_1	3	4	1	0	84
s_2	1	-1	0	1	21
P	-21	0	0	0	0

7. Minimize $C = (15 \quad 20 \quad 5)\begin{pmatrix} y_1 \\ y_2 \\ y_3 \end{pmatrix}$ subject to $\begin{pmatrix} -1 & 1 & 2 \\ 1 & 2 & 1 \end{pmatrix}\begin{pmatrix} y_1 \\ y_2 \\ y_3 \end{pmatrix} \geq \begin{pmatrix} -30 \\ 30 \end{pmatrix}$ and $\begin{pmatrix} y_1 \\ y_2 \\ y_3 \end{pmatrix} \geq 0$. The dual problem is maximize

$P = (-30 \quad 30)\begin{pmatrix} x_1 \\ x_2 \end{pmatrix}$ subject to $\begin{pmatrix} -1 & 1 \\ 1 & 2 \\ 2 & 1 \end{pmatrix}\begin{pmatrix} x_1 \\ x_2 \end{pmatrix} \leq \begin{pmatrix} 15 \\ 20 \\ 5 \end{pmatrix}$ and $\begin{pmatrix} x_1 \\ x_2 \end{pmatrix} \geq 0$ with initial simplex tableau

	x_1	x_2	s_1	s_2	s_3	
s_1	-1	1	1	0	0	15
s_2	1	2	0	1	0	20
s_3	2	1	0	0	1	5
P	30	-30	0	0	0	0

9. Minimize $C = (105 \quad 40)\begin{pmatrix} y_1 \\ y_2 \end{pmatrix}$ subject to $\begin{pmatrix} 7 & 5 \\ 3 & 1 \\ 5 & 2 \end{pmatrix}\begin{pmatrix} y_1 \\ y_2 \end{pmatrix} \geq \begin{pmatrix} 70 \\ 45 \\ 80 \end{pmatrix}$ and $\begin{pmatrix} y_1 \\ y_2 \end{pmatrix} \geq 0$. The dual problem is maximize

$P = (70 \quad 45 \quad 80)\begin{pmatrix} x_1 \\ x_2 \\ x_3 \end{pmatrix}$ subject to $\begin{pmatrix} 7 & 3 & 5 \\ 5 & 1 & 2 \end{pmatrix}\begin{pmatrix} x_1 \\ x_2 \\ x_3 \end{pmatrix} \leq \begin{pmatrix} 105 \\ 40 \end{pmatrix}$ and $\begin{pmatrix} x_1 \\ x_2 \\ x_3 \end{pmatrix} \geq 0$ with initial simplex tableau

	x_1	x_2	x_3	s_1	s_2	
s_1	7	3	5	1	0	105
s_2	5	1	2	0	1	40
P	-70	-45	-80	0	0	0

11. The minimum is 900 when $y_1 = 0$, $y_2 = 40$, $y_3 = 10$.   **13.** The minimum is 43 when $y_1 = 7$, $y_2 = 3$.

15. The minimum is 300 when $y_1 = 20$, $y_2 = 0$.   **17.** The minimum is 45 when $y_1 = 0$, $y_2 = 15$.

19. The minimum is 10 when $y_1 = 0$, $y_2 = 10$.   **21.** The minimum is 4950 when $y_1 = 30$, $y_2 = 45$.

23. The minimum is 98 when $y_1 = 2$, $y_2 = 2$, $y_3 = 0$.   **25.** There is no minimum.

27. The minimum is 90 when $y_1 = 0$, $y_2 = 10$, $y_3 = 30$, $y_4 = 0$.   **29.** The minimum is 2376 when $y_1 = 6$, $y_2 = 12$, $y_3 = 6$.

31. The athlete should use 8 Bulk-Up Bars and 4 cans of Power Drink to receive the needed fat and protein at the least possible cost of $5.64.

33. The office manager should buy 45 packages from Jack's Office Supplies and 30 packages from John's Discount to restock the store room at the least possible cost of $1230.

35. The project engineer should use 350 heavy-duty dump truck loads of dirt, 300 heavy-duty dump truck loads of crushed rock, no regular dump truck loads of dirt, and 150 regular dump truck loads of crushed rock to get the project finished on time at the least possible cost of $66,000.

37. The farmer should buy 8000 pounds of Miracle Mix and 18,000 pounds of the store brand to meet the fertilizer needs of his field at the least possible cost of $2640.

39. The Kentucky warehouse should ship 200 cartons to Kansas, 200 cartons to Texas, and none to Oregon, and the Utah warehouse should ship none to Kansas, 100 cartons to Texas, and 100 cartons to Oregon to incur the smallest possible shipping cost of $1600.

45. The final tableau shows that the solutions of the minimum and maximum problems are both 3390 with the variables taking the following values.

Minimum problem:	$y_1 = 7$	$y_2 = 3$	$y_3 = 1$	$y_4 = 0$	$t_1 = 0$	$t_2 = 0$	$t_3 = 0$
Maximum problem:	$s_1 = 0$	$s_2 = 0$	$s_3 = 0$	$s_4 = 75$	$x_1 = 60$	$x_2 = 45$	$x_3 = 30$

Thus one of each pair x_1 and t_1, x_2 and t_2, x_3 and t_3, s_1 and y_1, s_2 and y_2, s_3 and y_3, and s_4 and y_4 is zero.

EXERCISES 4.5 page 350

1. Maximize $P = (15 \quad 20 \quad 18)\begin{pmatrix} x_1 \\ x_2 \\ x_3 \end{pmatrix}$ subject to $\begin{pmatrix} 3 & 2 & 8 \\ -5 & -1 & -6 \end{pmatrix}\begin{pmatrix} x_1 \\ x_2 \\ x_3 \end{pmatrix} \leq \begin{pmatrix} 96 \\ -30 \end{pmatrix}$ and $\begin{pmatrix} x_1 \\ x_2 \\ x_3 \end{pmatrix} \geq 0$ with initial simplex tableau

	x_1	x_2	x_3	s_1	s_2	
s_1	3	2	8	1	0	96
s_2	-5	-1	-6	0	1	-30
P	-15	-20	-18	0	0	0

The dual pivot element is the -1 in row 2 and column 2 of the simplex tableau.

3. Maximize $P = (6 \quad 4 \quad 6)\begin{pmatrix} x_1 \\ x_2 \\ x_3 \end{pmatrix}$ subject to $\begin{pmatrix} -2 & -1 & -3 \\ -1 & -1 & -2 \end{pmatrix}\begin{pmatrix} x_1 \\ x_2 \\ x_3 \end{pmatrix} \leq \begin{pmatrix} -30 \\ -20 \end{pmatrix}$ and $\begin{pmatrix} x_1 \\ x_2 \\ x_3 \end{pmatrix} \geq 0$ with initial simplex tableau

	x_1	x_2	x_3	s_1	s_2	
s_1	-2	-1	-3	1	0	-30
s_2	-1	-1	-2	0	1	-20
P	-6	-4	-6	0	0	0

The dual pivot element is the -1 in row 1 and column 2 of the simplex tableau.

5. Maximize $P = \begin{pmatrix} 20 & 30 & 10 \end{pmatrix} \begin{pmatrix} x_1 \\ x_2 \\ x_3 \end{pmatrix}$ subject to $\begin{pmatrix} -1 & -1 & -1 \\ 1 & 2 & 3 \\ 1 & 2 & 1 \end{pmatrix} \begin{pmatrix} x_1 \\ x_2 \\ x_3 \end{pmatrix} \le \begin{pmatrix} -8 \\ 30 \\ 18 \end{pmatrix}$ and $\begin{pmatrix} x_1 \\ x_2 \\ x_3 \end{pmatrix} \ge 0$ with initial simplex tableau

	x_1	x_2	x_3	s_1	s_2	s_3	
s_1	-1	-1	-1	1	0	0	-8
s_2	1	2	3	0	1	0	30
s_3	1	2	1	0	0	1	18
P	-20	-30	-10	0	0	0	0

The dual pivot element is the -1 in row 1 and column 2 of the simplex tableau.

7. Maximize $P = \begin{pmatrix} 3 & 2 & 5 & 4 \end{pmatrix} \begin{pmatrix} x_1 \\ x_2 \\ x_3 \\ x_4 \end{pmatrix}$ subject to $\begin{pmatrix} -1 & -1 & -1 & -1 \\ -2 & -3 & -2 & -1 \\ 4 & 2 & 1 & 2 \end{pmatrix} \begin{pmatrix} x_1 \\ x_2 \\ x_3 \\ x_4 \end{pmatrix} \le \begin{pmatrix} -30 \\ -20 \\ 80 \end{pmatrix}$ and $\begin{pmatrix} x_1 \\ x_2 \\ x_3 \\ x_4 \end{pmatrix} \ge 0$ with initial simplex tableau

	x_1	x_2	x_3	x_4	s_1	s_2	s_3	
s_1	-1	-1	-1	-1	1	0	0	-30
s_2	-2	-3	-2	-1	0	1	0	-20
s_3	4	2	1	2	0	0	1	80
P	-3	-2	-5	-4	0	0	0	0

The dual pivot element is the -1 in row 1 and column 3 of the simplex tableau.

9. Maximize $P = \begin{pmatrix} 2 & 2 & 1 \end{pmatrix} \begin{pmatrix} x_1 \\ x_2 \\ x_3 \end{pmatrix}$ subject to $\begin{pmatrix} -1 & -2 & -1 \\ -2 & -1 & -1 \\ 5 & 3 & 2 \\ 3 & 1 & 2 \end{pmatrix} \begin{pmatrix} x_1 \\ x_2 \\ x_3 \end{pmatrix} \le \begin{pmatrix} -40 \\ -50 \\ 120 \\ 150 \end{pmatrix}$ and $\begin{pmatrix} x_1 \\ x_2 \\ x_3 \end{pmatrix} \ge 0$ with initial simplex tableau

	x_1	x_2	x_3	s_1	s_2	s_3	s_4	
s_1	-1	-2	-1	1	0	0	0	-40
s_2	-2	-1	-1	0	1	0	0	-50
s_3	5	3	2	0	0	1	0	120
s_4	3	1	2	0	0	0	1	150
P	-2	-2	-1	0	0	0	0	0

The dual pivot element is the -1 in row 2 and column 2 of the simplex tableau.

11. The maximum is 80 when $x_1 = 10, x_2 = 40$.   **13.** The maximum is 1200 when $x_1 = 0, x_2 = 60, x_3 = 0$.

15. There is no maximum.   **17.** The maximum is 270 when $x_1 = 2, x_2 = 20, x_3 = 0$.

19. The maximum is 13 when $x_1 = 8, x_2 = 5$.   **21.** There is no maximum.

23. The maximum is 720 when $x_1 = 0, x_2 = 0, x_3 = 60, x_4 = 0$.

25. The maximum is 808 when $x_1 = 0, x_2 = 14, x_3 = 0, x_4 = 24$.

27. The maximum is 30 when $x_1 = 10, x_2 = 40, x_3 = 0$.   **29.** The maximum is 240 when $x_1 = 0, x_2 = 0, x_3 = 0, x_4 = 30$.

31. The retired couple should invest \$15,000 in certificates of deposit and \$5000 in treasury bonds.

33. The manager should place 5 newspaper ads, 15 radio commercials, and 5 TV spots to reach 295,000 potential customers, the greatest possible number.

35. The furniture shop should manufacture 20 desks, 10 tables, and 80 chairs to obtain the greatest possible profit of $7620.

37. The farmer should grow 100 acres of wheat, 300 acres of barley, and 100 acres of oats to obtain the greatest possible profit of $21,500.

39. The electric power plant should purchase 40,000 tons of low-sulfur coal and 50,000 tons of high-sulfur coal to obtain the most energy of 2,300,000 BTUs.

41. a. The minimum of C is 12 when $x = 4$, $y = 0$, while the maximum of P is -12, also when $x = 4$, $y = 0$. Yes.

b. The simplex tableaux for the dual of this problem are

	x_1	x_2	s_1	s_2	
s_1	-1	2	1	0	3
s_2	-1	1	0	1	4
P	10	-8	0	0	0

and then

	x_1	x_2	s_1	s_2	
x_2	$-1/2$	1	$1/2$	0	3/2
s_2	$-1/2$	0	$-1/2$	1	5/2
P	6	0	4	0	12

c. The simplex tableaux for this nonstandard problem are

	x_1	x_2	s_1	s_2	
s_1	1	1	1	0	10
s_2	-2	-1	0	1	-8
P	3	4	0	0	0

and then

	x_1	x_2	s_1	s_2	
s_1	0	1/2	1	1/2	6
x_1	1	1/2	0	$-1/2$	4
P	0	5/2	0	3/2	-12

d. Yes; yes and yes

47. The maximum is 230 when $x_1 = 0$, $x_2 = 20$, $x_3 = 20$, $x_4 = 10$.   **49.** The maximum is 12 when $x_1 = 0$, $x_2 = 4$.

EXERCISES 4.6 page 370

1. Already in normal form; the initial simplex tableau is

	x_1	x_2	x_3	s_1	s_2	a_2	
s_1	3	2	8	1	0	0	96
a_2	5	1	6	0	-1	1	30
P	$-5M - 15$	$-M - 20$	$-6M - 18$	0	M	0	$-30M$

3. Already in normal form; the initial simplex tableau is

	x_1	x_2	x_3	s_1	s_2	a_1	a_2	
a_1	2	1	3	-1	0	1	0	30
a_2	1	1	2	0	-1	0	1	20
P	$-3M - 6$	$-2M - 4$	$-5M - 6$	M	M	0	0	$-50M$

5. For the normal form

Maximize $P = 20x_1 + 30x_2 + 10x_3$

Subject to $\begin{cases} x_1 + 2x_2 + 3x_3 \le 30 \\ x_1 + 2x_2 + x_3 \le 18 \\ x_1 + x_2 + x_3 \ge 8 \\ x_1 \ge 0, x_2 \ge 0, x_3 \ge 0 \end{cases}$

the initial simplex tableau is

	x_1	x_2	x_3	s_1	s_2	s_3	a_3	
s_1	1	2	3	1	0	0	0	30
s_2	1	2	1	0	1	0	0	18
a_3	1	1	1	0	0	-1	1	8
P	$-M - 20$	$-M - 30$	$-M - 10$	0	0	M	0	$-8M$

7. Already in normal form; the initial simplex tableau is

	x_1	x_2	x_3	x_4	s_1	s_2	a_1	a_2	a_3	
a_1	1	1	1	1	-1	0	1	0	0	30
a_2	2	3	2	1	0	-1	0	1	0	20
a_3	4	2	1	2	0	0	0	0	1	80
P	$-7M - 3$	$-6M - 2$	$-4M - 5$	$-4M - 4$	M	M	0	0	0	$-130M$

9. For the normal form

Maximize $P = 2x_1 + 2x_2 + x_3$

Subject to $\begin{cases} 5x_1 + 3x_2 + 2x_3 \le 120 \\ 3x_1 + x_2 + 2x_3 \le 150 \\ x_1 + 2x_2 + x_3 \ge 40 \\ 2x_1 + x_2 + x_3 \ge 50 \\ x_1 \ge 0, x_2 \ge 0, x_3 \ge 0 \end{cases}$

the initial simplex tableau is

	x_1	x_2	x_3	s_1	s_2	s_3	s_4	a_3	a_4	
s_1	5	3	2	1	0	0	0	0	0	120
s_2	3	1	2	0	1	0	0	0	0	150
a_3	1	2	1	0	0	-1	0	1	0	40
a_4	2	1	1	0	0	0	-1	0	1	50
P	$-3M - 2$	$-3M - 2$	$-2M - 1$	0	0	M	M	0	0	$-90M$

11. The maximum is 80 when $x_1 = 10, x_2 = 40$.

13. The maximum is 1200 when $x_1 = 0, x_2 = 60, x_3 = 0$.

15. There is no maximum.

17. The maximum is 270 when $x_1 = 2, x_2 = 20, x_3 = 0$.

19. The maximum is 13 when $x_1 = 8, x_2 = 5$.

21. There is no maximum.

23. The maximum is 480 when $x_1 = 0, x_2 = 40, x_3 = 20, x_4 = 0$.

25. The maximum is 808 when $x_1 = 0, x_2 = 14, x_3 = 0, x_4 = 24$.

27. The maximum is 30 when $x_1 = 10, x_2 = 40, x_3 = 0$.

29. The maximum is 240 when $x_1 = 0, x_2 = 0, x_3 = 0, x_4 = 30$.

31. The retired couple should invest $15,000 in certificates of deposit and $5000 in Treasury bonds.

33. The manager should place 5 newspaper ads, 15 radio commercials, and 5 TV spots to reach 295,000 potential customers, the greatest possible number.

35. The furniture shop should manufacture 20 desks, 10 tables, and 80 chairs to obtain the greatest possible profit of $7620.

37. The farmer should grow 100 acres of wheat, 300 acres of barley, and 100 acres of oats to obtain the greatest possible profit of $21,500.

39. The electric power plant should purchase 40,000 tons of low-sulfur coal and 50,000 tons of high-sulfur coal to obtain the most energy of 2,300,000 million BTUs.

41. (a) The minimum of C is 12 when $x = 4, y = 0$, while the maximum of P is -12, also when $x = 4, y = 0$. Yes.

47. The maximum is 230 when $x_1 = 0, x_2 = 20, x_3 = 20, x_4 = 10$.

49. The maximum is 12 when $x_1 = 0, x_2 = 4$.

CHAPTER 4 REVIEW EXERCISES page 379

1.

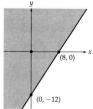

2.

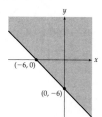

3. The vertices are $(0, 0)$, $(20, 0)$, and $(0, 20)$. The region is bounded.

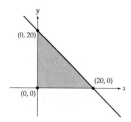

4. The vertices are $(10, 0)$ and $(0, 5)$. The region is unbounded.

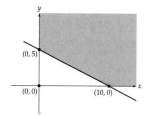

5. The vertices are $(0, 0)$, $(10, 0)$, $(5, 10)$, and $(0, 15)$. The region is bounded.

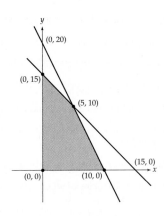

6. The vertices are $(8, 0)$, $(4, 4)$, and $(0, 12)$. The region is unbounded.

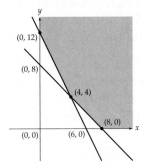

7. The vertices are $(2, 0)$, $(8, 0)$, $(4, 6)$, and $(2, 8)$. The region is bounded.

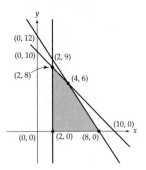

8. The vertices are $(0, 0)$, $(9, 0)$, $(9, 1)$, $(3, 7)$, and $(0, 4)$. The region is bounded.

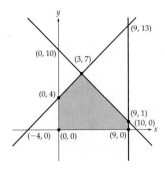

9. Let x be the number of Irish setters and let y be the number of Labrador retrievers.

$$\begin{cases} \frac{1}{2}x + \frac{3}{4}y \le 6 & \text{(Time in hours)} \\ 8x + 8y \le 80 & \text{(Dog treats)} \\ x \ge 0, y \ge 0 & \text{(Nonnegativity)} \end{cases}$$

The vertices are $(0, 0)$, $(10, 0)$, $(6, 4)$, and $(0, 8)$.

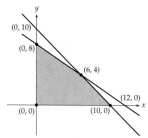

10. Let x be the number of pounds of SongBird brand bird seed and let y be the number of pounds of MeadowMix brand bird seed.

$$\begin{cases} 2x + 4y \ge 104 & \text{(Sunflower hearts)} \\ 3x + 2y \ge 84 & \text{(Crushed peanuts)} \\ x \ge 0, y \ge 0 & \text{(Nonnegativity)} \end{cases}$$

The vertices are $(52, 0)$, $(16, 18)$, and $(0, 42)$.

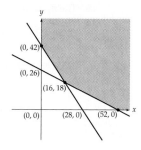

11. The maximum is 40 when $x = 0$, $y = 10$.   **12.** The minimum is 10 when $x = 0$, $y = 5$.

13. There is no maximum because the region is unbounded to the right and upward, and P increases in these directions.

14. The minimum is 60 when $x = 30$, $y = 0$.   **15.** The maximum is 240 when $x = 12$, $y = 0$.

16. The minimum is 2400 when $x = 48$, $y = 24$.   **17.** The maximum is 350 when $x = 4$, $y = 9$.

18. The minimum is 140 when $x = 20$, $y = 0$.

19. The shop should make 24 wall clock cases and 15 mantle clock cases to obtain the greatest possible profit of $9000.

20. The cat's nutritional needs can be met by 30 ounces of canned food and 4 ounces of dry food at a least cost of $1.82.

21.

	x_1	x_2	x_3	s_1	s_2	
s_1	5	2	3	1	0	30
s_2	2	3	4	0	1	24
P	-4	-10	-9	0	0	0

22.

	x_1	x_2	s_1	s_2	s_3	
s_1	1	1	1	0	0	10
s_2	2	1	0	1	0	14
s_3	1	-1	0	0	1	4
P	-5	-7	0	0	0	0

23. The smallest negative entry in the bottom row is -8, so the pivot column is column 3. The ratios are $\frac{6}{1} = 6$, $\frac{8}{2} = 4$, and (omit), so the pivot row is row 2. The pivot element is the 2 in column 3 and row 2 of the simplex tableau.

	x_1	x_2	x_3	s_1	s_2	s_3		
s_1	$-1/2$	$1/2$	0	1	$-1/2$	0	2	$R_1^{new} = R_1 - R_{pivot}^{new}$
x_3	$3/2$	$1/2$	1	0	$1/2$	0	4	$R_{pivot}^{new} = R_{pivot}/2$
s_3	$11/2$	$5/2$	0	0	$3/2$	1	18	$R_3^{new} = R_3 + 3R_{pivot}^{new}$
P	6	8	0	0	4	0	32	$R_4^{new} = R_4 + 8R_{pivot}^{new}$

24. No pivot column because there are no negative entries in the bottom row. The solution has been found: The maximum is 150 when $x_1 = 0$ and $x_2 = 10$ (and $s_1 = 8$, $s_2 = 42$, and $s_3 = 0$).

25. The maximum is 96 when $x_1 = 0$, $x_2 = 12$, $x_3 = 0$.   **26.** The maximum is 90 when $x_1 = 0$, $x_2 = 9$, $x_3 = 0$.

27. The maximum is 37 when $x_1 = 2$, $x_2 = 3$, $x_3 = 0$, $x_4 = 0$.   **28.** The maximum is 61 when $x_1 = 8$, $x_2 = 0$, $x_3 = 7$.

29. The publisher should print no trade edition copies, no book club edition copies, and 250,000 paperback edition copies for the greatest possible profit of $1,000,000.

30. The plastics factory should produce 20 batches of toy racing cars, no toy jet airplanes, and 10 batches of toy speed boats for the greatest possible profit of $3500.

31. Minimize $C = (10 \quad 15 \quad 20) \begin{pmatrix} y_1 \\ y_2 \\ y_3 \end{pmatrix}$ subject to $\begin{pmatrix} 2 & -1 & 1 \\ 1 & 1 & 3 \end{pmatrix} \begin{pmatrix} y_1 \\ y_2 \\ y_3 \end{pmatrix} \geq \begin{pmatrix} 40 \\ 30 \end{pmatrix}$ and $\begin{pmatrix} y_1 \\ y_2 \\ y_3 \end{pmatrix} \geq 0$ is a standard problem because $\begin{pmatrix} 10 \\ 15 \\ 20 \end{pmatrix} \geq 0$.

The dual problem is maximize $P = (40 \quad 30) \begin{pmatrix} x_1 \\ x_2 \end{pmatrix}$ subject to $\begin{pmatrix} 2 & 1 \\ -1 & 1 \\ 1 & 3 \end{pmatrix} \begin{pmatrix} x_1 \\ x_2 \end{pmatrix} \leq \begin{pmatrix} 10 \\ 15 \\ 20 \end{pmatrix}$ and $\begin{pmatrix} x_1 \\ x_2 \end{pmatrix} \geq 0$ with initial simplex tableau

	x_1	x_2	s_1	s_2	s_3	
s_1	2	1	1	0	0	10
s_2	-1	1	0	1	0	15
s_3	1	3	0	0	1	20
P	-40	-30	0	0	0	0

32. Minimize $C = (30 \quad 20) \begin{pmatrix} y_1 \\ y_2 \end{pmatrix}$ subject to $\begin{pmatrix} 5 & 2 \\ 3 & 4 \\ -5 & -4 \end{pmatrix} \begin{pmatrix} y_1 \\ y_2 \end{pmatrix} \geq \begin{pmatrix} 210 \\ 252 \\ -380 \end{pmatrix}$ and $\begin{pmatrix} y_1 \\ y_2 \end{pmatrix} \geq 0$ is a standard problem because $\begin{pmatrix} 30 \\ 20 \end{pmatrix} \geq 0$. The

dual problem is maximize $P = (210 \quad 252 \quad -380) \begin{pmatrix} x_1 \\ x_2 \\ x_3 \end{pmatrix}$ subject to $\begin{pmatrix} 5 & 3 & -5 \\ 2 & 4 & -4 \end{pmatrix} \begin{pmatrix} x_1 \\ x_2 \\ x_3 \end{pmatrix} \leq \begin{pmatrix} 30 \\ 20 \end{pmatrix}$ and $\begin{pmatrix} x_1 \\ x_2 \\ x_3 \end{pmatrix} \geq 0$ with initial simplex

tableau
	x_1	x_2	x_3	s_1	s_2	
s_1	5	3	-5	1	0	30
s_2	2	4	-4	0	1	20
P	-210	-252	380	0	0	0

33. Minimize $C = (42 \quad 36)\begin{pmatrix} y_1 \\ y_2 \end{pmatrix}$ subject to $\begin{pmatrix} 7 & 4 \\ 1 & 1 \end{pmatrix}\begin{pmatrix} y_1 \\ y_2 \end{pmatrix} \geq \begin{pmatrix} 84 \\ 18 \end{pmatrix}$ and $\begin{pmatrix} y_1 \\ y_2 \end{pmatrix} \geq 0$ is a standard problem because $\begin{pmatrix} 42 \\ 36 \end{pmatrix} \geq 0$. The dual

problem is maximize $P = (84 \quad 18)\begin{pmatrix} x_1 \\ x_2 \end{pmatrix}$ subject to $\begin{pmatrix} 7 & 1 \\ 4 & 1 \end{pmatrix}\begin{pmatrix} x_1 \\ x_2 \end{pmatrix} \leq \begin{pmatrix} 42 \\ 36 \end{pmatrix}$ and $\begin{pmatrix} x_1 \\ x_2 \end{pmatrix} \geq 0$ with initial simplex tableau

	x_1	x_2	s_1	s_2	
s_1	7	1	1	0	42
s_2	4	1	0	1	36
P	-84	-18	0	0	0

34. Minimize $C = (130 \quad 40 \quad 98)\begin{pmatrix} y_1 \\ y_2 \\ y_3 \end{pmatrix}$ subject to $\begin{pmatrix} 3 & 1 & 2 \\ 4 & 1 & 2 \\ 3 & 1 & 3 \end{pmatrix}\begin{pmatrix} y_1 \\ y_2 \\ y_3 \end{pmatrix} \geq \begin{pmatrix} 51 \\ 60 \\ 57 \end{pmatrix}$ and $\begin{pmatrix} y_1 \\ y_2 \\ y_3 \end{pmatrix} \geq 0$ is a standard problem because $\begin{pmatrix} 130 \\ 40 \\ 98 \end{pmatrix} \geq 0$.

The dual problem is maximize $P = (51 \quad 60 \quad 57)\begin{pmatrix} x_1 \\ x_2 \\ x_3 \end{pmatrix}$ subject to $\begin{pmatrix} 3 & 4 & 3 \\ 1 & 1 & 1 \\ 2 & 2 & 3 \end{pmatrix}\begin{pmatrix} x_1 \\ x_2 \\ x_3 \end{pmatrix} \leq \begin{pmatrix} 130 \\ 40 \\ 98 \end{pmatrix}$ and $\begin{pmatrix} x_1 \\ x_2 \\ x_3 \end{pmatrix} \geq 0$ with initial simplex

		x_1	x_2	x_3	s_1	s_2	s_3	
	s_1	3	4	3	1	0	0	130
tableau	s_2	1	1	1	0	1	0	40
	s_3	2	2	3	0	0	1	98
	P	-51	-60	-57	0	0	0	0

35. The minimum is 192 when $y_1 = 8$, $y_2 = 5$, $y_3 = 0$.   **36.** The minimum is 510 when $y_1 = 11$, $y_2 = 5$.

37. There is no solution.   **38.** The minimum is 60 when $y_1 = 0$, $y_2 = 12$, $y_3 = 0$.

39. The student should buy no single pens, 3 packages of ink cartridges, and 3 "writer's combo" packages to spend the least possible amount of $6.24.

40. The pasta company should purchase 18 small-capacity machines and 8 large-capacity machines to expand its linguini production at the least possible cost of $138,000.

41. Maximize $P = (80 \quad 30)\begin{pmatrix} x_1 \\ x_2 \end{pmatrix}$ subject to $\begin{pmatrix} 4 & 1 \\ 2 & 3 \\ -1 & -1 \end{pmatrix}\begin{pmatrix} x_1 \\ x_2 \end{pmatrix} \leq \begin{pmatrix} 40 \\ 60 \\ -10 \end{pmatrix}$ and $\begin{pmatrix} x_1 \\ x_2 \end{pmatrix} \geq 0$ with initial simplex tableau

	x_1	x_2	s_1	s_2	s_3	
s_1	4	1	1	0	0	40
s_2	2	3	0	1	0	60
s_3	-1	-1	0	0	1	-10
P	-80	-30	0	0	0	0

The dual pivot element is the -1 in row 3 and column 1 of the simplex tableau.

42. Maximize $P = (4 \quad 8 \quad 6 \quad 10) \begin{pmatrix} x_1 \\ x_2 \\ x_3 \\ x_4 \end{pmatrix}$ subject to $\begin{pmatrix} 2 & 3 & -3 & 1 \\ -7 & 5 & -1 & -5 \end{pmatrix} \begin{pmatrix} x_1 \\ x_2 \\ x_3 \\ x_4 \end{pmatrix} \leq \begin{pmatrix} 30 \\ -35 \end{pmatrix}$ and $\begin{pmatrix} x_1 \\ x_2 \\ x_3 \\ x_4 \end{pmatrix} \geq 0$ with initial simplex tableau

	x_1	x_2	x_3	x_4	s_1	s_2	
s_1	2	3	-3	1	1	0	30
s_2	-7	5	-1	-5	0	1	-35
P	-4	-8	-6	-10	0	0	0

The dual pivot element is the -1 in row 2 and column 3 of the simplex tableau.

43. The maximum is 40 when $x_1 = 0$, $x_2 = 10$.   **44.** The maximum is 280 when $x_1 = 35$, $x_2 = 0$, $x_3 = 0$.

45. The maximum is 48 when $x_1 = 8$, $x_2 = 0$.   **46.** The maximum is 300 when $x_1 = 25$, $x_2 = 0$, $x_3 = 0$.

47. The maximum is 59 when $x_1 = 4$, $x_2 = 5$.   **48.** The maximum is 300 when $x_1 = 0$, $x_2 = 0$, $x_3 = 30$.

49. The sawmill should produce 6 thousand board-feet of rough-cut lumber and 4 thousand board-feet of finished-grade boards each day to obtain the greatest possible profit of $1120.

50. The money manager should invest $100,000 in U.S. bonds and $50,000 in Canadian stocks (and nothing in U.S. stocks or in Canadian bonds) to obtain the greatest possible return of $10,500.

51. Already in normal form; the initial simplex tableau is

	x_1	x_2	s_1	s_2	s_3	a_3	
s_1	4	1	1	0	0	0	40
s_2	2	3	0	1	0	0	60
a_3	1	1	0	0	-1	1	10
P	$-M - 80$	$-M - 30$	0	0	M	0	$-10M$

52. For the normal form

Maximize $P = 4x_1 + 8x_2 + 6x_3 + 10x_4$

Subject to $\begin{cases} 2x_1 + 3x_2 - 3x_3 + x_4 \leq 30 \\ 7x_1 - 5x_2 + x_3 + 5x_4 \geq 35 \\ x_1 \geq 0, x_2 \geq 0, x_3 \geq 0, x_4 \geq 0 \end{cases}$

the initial simplex tableau is

	x_1	x_2	x_3	x_4	s_1	s_2	a_2	
s_1	2	3	-3	1	1	0	0	30
a_2	7	-5	1	5	0	-1	1	35
P	$-7M - 4$	$5M - 8$	$-M - 6$	$-5M - 10$	0	M	0	$-35M$

53. The maximum is 40 when $x_1 = 0$, $x_2 = 10$.

54. The maximum is 280 when $x_1 = 35$, $x_2 = 0$, $x_3 = 0$.

55. The maximum is 48 when $x_1 = 8$, $x_2 = 0$.

56. The maximum is 300 when $x_1 = 25$, $x_2 = 0$, $x_3 = 0$.

57. The maximum is 59 when $x_1 = 4$, $x_2 = 5$.

58. The maximum is 300 when $x_1 = 0$, $x_2 = 0$, $x_3 = 30$.

59. The sawmill should produce 6 thousand board-feet of rough-cut lumber and 4 thousand board-feet of finished-grade boards each day to obtain the greatest possible profit of $1120.

60. The money manager should invest $100,000 in U.S. bonds and $50,000 in Canadian stocks (and nothing in U.S. stocks and in Canadian bonds) to obtain the greatest possible return of $10,500.

EXERCISES 5.1 page 393

1. 27 **3.** 60 **5.** 44 **7.** 31 **9.** 17 **11.** $A \cap B^c \cap C^c$ **13.** $A^c \cap B^c \cap C$ **15.** $A \cap B \cap C^c$ **17.** $A \cap B \cap C$

19.

A   B   C

1 2 3 4   1 2 3 4   1 2 3 4

Therefore, 12 permits.

21.

Windbreakers   Ski Jackets   Overcoats

Red   Blue   Red   Blue   Red   Blue

Therefore, 6 kinds.

23. No: 26 lines each leading to 10 more is impractical for a tree. **25.** 17 **27.** 200

29. $25 \cdot 9 \cdot 9 = 2025$ **31.** $15^8 = 2,562,890,625$ **33.** $10^4 = 10,000$ **35.** $3 \cdot 4 \cdot 4 \cdot 5 = 240$

37.

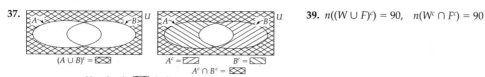

$(A \cup B)^c = \boxtimes$

$A^c = \boxbackslash$ $B^c = \boxbackslash$

$A^c \cap B^c = \boxtimes$

Note that the $\boxtimes$ shadings agree.

39. $n((W \cup F)^c) = 90$, $n(W^c \cap F^c) = 90$

EXERCISES 5.2 page 406

1. a. 2 **b.** 720 **3.** 720 **5.** 39,800 **7.** 120 **9.** 8 **11. a.** 17,160 **b.** 154,440 **c.** 1,235,520

13. a. 10 **b.** 30,240 **c.** 3,628,800 **15.** 15 **17.** 35 **19. a.** 330 **b.** 462 **c.** 462 **d.** 330

21. a. 12 **b.** 66 **c.** 924 **d.** 12 **e.** 1 **23. a.** $\dfrac{n!}{n} = \dfrac{n(n-1) \cdot \ldots \cdot 1}{n} = (n-1) \cdot \ldots \cdot 1 = (n-1)!$

b. $\dfrac{n!}{n} = (n-1)!$ with $n = 1$ gives $\dfrac{1!}{1} = (1-1)!$, which simplifies to $1 = 0!$

25. $x^3 + 3x^2y + 3xy^2 + y^3$ **27.** $a^5 + 10a^4 + 40a^3 + 80a^2 + 80a + 32$ **29.** $w^4 - 12w^3 + 54w^2 - 108w + 81$

31. $x^5 + 10x^4y + 40x^3y^2 + 80x^2y^3 + 80xy^4 + 32y^5$ **33.** $16x^4 + 96x^3y + 216x^2y^2 + 216xy^3 + 81y^4$

35. Pascal's triangle with the next two lines is

$$
\begin{array}{ccccccccccccc}
 & & & & & & 1 & & & & & & \\
 & & & & & 1 & & 1 & & & & & \\
 & & & & 1 & & 2 & & 1 & & & & \\
 & & & 1 & & 3 & & 3 & & 1 & & & \\
 & & 1 & & 4 & & 6 & & 4 & & 1 & & \\
 & 1 & & 5 & & 10 & & 10 & & 5 & & 1 & \\
1 & & 6 & & 15 & & 20 & & 15 & & 6 & & 1
\end{array}
$$

37. $x^{20} + 20x^{19}y + 190x^{18}y^2$ **39.** $2^8 = 256$ subsets

41. Taking $x = 1$ and $y = -1$, the binomial theorem becomes

$$(1 - 1)^n = {_nC_0}\, 1^n + {_nC_1} 1^{n-1}(-1) + {_nC_2} 1^{n-2}(-1)^2 + {_nC_3}\, 1^{n-3}(-1)^3 + \cdots + {_nC_n}\,(-1)^n \text{ so that } 0 = {_nC_0} - {_nC_1} + {_nC_2} - {_nC_3} + \cdots \pm {_nC_n}.$$

43. $16 = 1 + 4 + 6 + 4 + 1$. For 4 people there are 16 possible committees: 1 of size 0, 4 of size 1, 6 of size 2, 4 of size 3, and 1 of size 4.

45. $8! = 40{,}320$ **47.** ${_{12}P_4} = 11{,}880$ **49.** $8 \cdot {_9P_6} = 483{,}840$

51. $36 \cdot 35 \cdot 34 \cdot 33 = 1{,}413{,}720$; $34 \cdot 33 \cdot 32 \cdot 31 = 1{,}113{,}024$ **53.** ${_6P_2} = 30$ **55.** ${_{12}C_5} = 792$

57. ${_{13}C_5} = 1287$; $4 \cdot 1287 = 5148$ **59.** $({_{10}C_4})({_{12}C_4}) = 103{,}950$ **61.** ${_{12}C_{10}} = {_{12}C_2} = 66$

63. ${_{100}C_5} = 75{,}287{,}520$; $\frac{100 \cdot 98 \cdot 96 \cdot 94 \cdot 92}{5!} = 67{,}800{,}320$ **65.** $2^6 = 64$; and ${_6C_2} = 15$ **67.** $2^{10} - 1 = 103$

69.

n	$n!$	$\sqrt{2\pi n}\, n^n\, e^{-n}$	Ratio	% Error
1	1	0.922	1.084	8.44
2	2	1.919	1.042	4.22
5	120	118.02	1.017	1.68
10	3,628,800	3,598,695.6	1.008	0.84
50	$3.041 \cdot 10^{64}$	$3.036 \cdot 10^{64}$	1.002	0.17

71. a. $\dfrac{(2n)!}{n!\,n!} \approx \dfrac{\sqrt{2\pi 2n}\,(2n)^{2n} e^{-2n}}{\left(\sqrt{2\pi n}\, n^n e^{-n}\right)^2} = \dfrac{2\sqrt{\pi n}\,2^{2n} n^{2n} e^{-2n}}{2\pi n \cdot n^{2n} e^{-2n}} = \dfrac{\sqrt{\pi n}\,2^{2n}}{\pi n} = \dfrac{2^{2n}}{\sqrt{n\pi}}$

b. $\left(\begin{matrix}\text{Number of committees} \\ \text{of size } n\end{matrix}\right) \approx \left(\begin{matrix}\text{Total number} \\ \text{of committees}\end{matrix}\right)\Big/ \sqrt{n\pi}$

EXERCISES 5.3 page 420

1. $\{(A, B), (A, C), (A, D), (B, C), (B, D), (C, D)\}$; $\{(A, B), (A, D), (B, C), (C, D)\}$

3. $\{(S_1, J_1), (S_1, J_2), (S_2, J_1), (S_2, J_2), (S_3, J_1), (S_3, J_2), (S_4, J_1), (S_4, J_2)\}$

5. $\{(R, R), (R, G), (R, B), (G, R), (G, G), (G, B), (B, R), (B, G), (B, B)\}$; $\{(R, G), (R, B), (G, R), (G, B), (B, R), (B, G)\}$

7. a. $\{3\}$ **b.** $\{2, 3, 4, 6\}$ **9. a.** $\{(6, 2), (6, 4), (6, 6), (2, 6), (4, 6)\}$ **b.** $\emptyset$ **11. a.** 0.80 **b.** 1

13. a. 0.20 **b.** 0.15 **15.** $1/({_{12}C_3}) = \frac{1}{220}$, $1/({_{12}P_3}) = \frac{1}{1320}$ **17.** $P(R) = \frac{6}{10} = \frac{3}{5}$, $P(B) = \frac{4}{10} = \frac{2}{5}$

19. $P(6) = \frac{1}{2}$, $P(8) = \frac{1}{4}$, $P(12) = \frac{1}{4}$ **21. a.** $\frac{1}{20}$ **b.** $\frac{19}{20}$ **23.** 32%

25. $63\% + 48\% - 15\% + 10\% = 106\%$ when it should add to 100%.   **27. a.** $\dfrac{_4C_3}{_{12}C_3} = \dfrac{1}{55}$ **b.** $\dfrac{_4C_3}{_{12}C_3} + \dfrac{_8C_3}{_{12}C_3} = \dfrac{3}{11}$

29. $1 - \dfrac{_9C_3}{_{10}C_3} = \dfrac{3}{10}$ **31.** $\dfrac{_{48}C_1}{_{52}C_5} \approx 0.000018$ **33.** $\dfrac{7 \cdot 6 \cdot 5 \cdot 4 \cdot 3}{7^5} \approx 0.15$ **35.** $\dfrac{_{90}C_{10}}{_{100}C_{10}} \approx 0.33$ **37.** $1 - \dfrac{_{98}C_2}{_{100}C_2} \approx 0.04$

39. $\dfrac{_4C_1}{_6C_1} = \dfrac{2}{3}$ **41.** $P(H) = \frac{1}{2}$ and $P(T) = \frac{1}{2}$ give $\frac{1}{2}:\frac{1}{2}$ or $1:1$.   **43.** $P(E^c) = \frac{m-n}{m}$ so the odds are $\frac{n}{m} : \frac{m-n}{m}$, or $n:(m - n)$.

45. Since the probabilities must add to 1, we divide the odds by $n + m$, so $P(E) = \frac{n}{n+m}$.   **47.** $\frac{7}{11}$ **49.** $\frac{7}{9}$

EXERCISES 5.4 page 432

1. a. $\frac{0.2}{0.4} = \frac{1}{2}$ **b.** $\frac{0.2}{0.6} = \frac{1}{3}$ **3.** Both parts require first finding $P(A \cap B) = 0.3$.   **a.** $\frac{3}{5}$ **b.** $\frac{3}{4}$

5. $\frac{2}{3}$ **7.** $\frac{1}{3}$ **9.** $\frac{1}{3}$ **11.** $\frac{0.38}{0.68} \approx 0.56$ **13.** $0.95 \cdot 0.20 + 0.70 \cdot 0.80 = 0.75$ **15.** 0.16

17. Yes $[P(A) \cdot P(B) = \frac{1}{2} \cdot \frac{1}{2}$ and $P(A$ and $B) = \frac{1}{4}]$ **19.** No $[P(A) \cdot P(B) = \frac{1}{2} \cdot \frac{3}{36} = \frac{1}{24}$ and $P(A$ and $B) = \frac{2}{36} = \frac{1}{18}]$

21. $\left(\frac{1}{6}\right)^3 = \frac{1}{216}$ **23.** $\frac{0.9}{1 - 0.1 \cdot 0.3} \approx 0.928$, so about 93%   **25. a.** $\left(\frac{1}{6}\right)^3 = \frac{1}{216}$ **b.** $\frac{6}{216} = \frac{1}{36}$ **c.** $\frac{35}{36}$

27. $P(A$ and $B^c) = P(A \cap B^c) = P(A) - P(A \cap B) = P(A) - P(A) \cdot P(B) = P(A)[1 - P(B)] = P(A) \cdot P(B^c)$

29. $P(A$ and $B) = \frac{1}{4} = \frac{1}{2} \cdot \frac{1}{2} = P(A) \cdot P(B)$

$P(A$ and $C) = \frac{1}{4} = \frac{1}{2} \cdot \frac{1}{2} = P(A) \cdot P(C)$

$P(B$ and $C) = \frac{1}{4} = \frac{1}{2} \cdot \frac{1}{2} = P(B) \cdot P(C)$

$P(A$ and B and $C) = \frac{1}{4}$ but $P(A) \cdot P(B) \cdot P(C) = \frac{1}{2} \cdot \frac{1}{2} \cdot \frac{1}{2} = \frac{1}{8}$

EXERCISES 5.5 page 438

1. $\frac{0.55 \cdot 0.60}{0.55 \cdot 0.60 + 0.65 \cdot 0.40} \approx 0.559$, so about 56%   **3.** $\frac{0.95 \cdot 0.001}{0.95 \cdot 0.001 + 0.05 \cdot 0.999} \approx 0.019$, so about 2%

5. $\frac{0.02 \cdot 0.3}{0.03 \cdot 0.5 + 0.02 \cdot 0.3 + 0.01 \cdot 0.2} \approx 0.261$, so about 26%   **7.** $\frac{0.90 \cdot \frac{30}{300,000,000}}{0.90 \cdot \frac{30}{300,000,000} + 0.0005 \cdot \frac{299,999,970}{300,000,000}} \approx 0.00018$, so about 0.02%

9. $\frac{0.05 \cdot 0.99}{0.05 \cdot 0.99 + 0.95 \cdot 0.01} \approx 0.839$, so about 84%   **11.** $\frac{0.70 \cdot 0.2}{0.90 \cdot 0.5 + 0.85 \cdot 0.3 + 0.70 \cdot 0.2} \approx 0.166$, so about 17%

13. $\frac{0.75 \cdot 0.02}{0.75 \cdot 0.02 + 0.3 \cdot 0.98} \approx 0.049$, so about 5%   **15.** $\frac{0.006}{0.009 + 0.01 + 0.008 + 0.0075 + 0.006} \approx 0.148$, so about 15%

EXERCISES 5.6 page 452

1.

x	0	1	2	3
$P(X = x)$	$\frac{1}{8}$	$\frac{3}{8}$	$\frac{3}{8}$	$\frac{1}{8}$

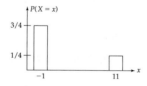

3.

x	-1	11
$P(X = x)$	$\frac{3}{4}$	$\frac{1}{4}$

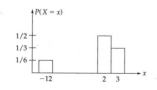

5.

x	-12	2	3
$P(X = x)$	$\frac{1}{6}$	$\frac{1}{2}$	$\frac{1}{3}$

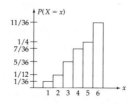

7.

x	1	2	3	4	5	6
$P(X = x)$	$\frac{1}{36}$	$\frac{1}{12}$	$\frac{5}{36}$	$\frac{7}{36}$	$\frac{1}{4}$	$\frac{11}{36}$

9. $\mu = 0 \cdot \frac{1}{8} + 1 \cdot \frac{3}{8} + 2 \cdot \frac{3}{8} + 3 \cdot \frac{1}{8} = \frac{3}{2}$ **11.** $\mu = 2, \sigma = \sqrt{(11 - 2)^2 \cdot \frac{1}{4} + (-1 - 2)^2 \cdot \frac{3}{4}} = \sqrt{27} \approx 5.20$

13. $\mu = 0, \sigma = \sqrt{29} \approx 5.39$ **15.** $\mu = \frac{161}{36} \approx 4.5$ **17.** $\mu = 3\frac{1}{4}$ **19.** $\sigma = 1, \sigma = 10$

21. $\mu = 20 \cdot \frac{1}{2} = 10$

$\sigma = \sqrt{10 \cdot \frac{1}{2} \cdot \frac{1}{2}} = \sqrt{5} \approx 2.24$

23. $\mu = 16$

$\sigma = \sqrt{3.2} \approx 1.79$

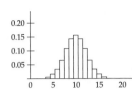

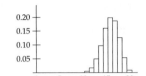

25. 0.78125, or about 78%   **27.** $_8C_4\left(\frac{1}{2}\right)^8 = \frac{70}{256} \approx 0.273$, or about 27%

29. Most likely: 2 heads, with probability 0.329 (approx.)
Least likely: 6 heads, with probability 0.0014 (approx.)

31. Most likely: 1 six, with probability 0.372 (approx.), $\mu = \frac{4}{3}$ **33.** \$14,000   **35.** $\mu = 0.2, \sigma = \sqrt{10 \cdot 0.02 \cdot 0.98} \approx 0.443$

37. $P(7 \le X \le 10) \approx 0.172$ **39.** 0.972 (compared to 0.90 for transmitting a single digit)   **41.** $\left(_{48}C_{13}/_{52}C_{13}\right)^3 \approx 0.028$, or about 3%

43. a. $_3C_2\left(\frac{3}{5}\right)^2\left(\frac{2}{5}\right)^1 + _3C_3\left(\frac{3}{5}\right)^3\left(\frac{2}{5}\right)^0 \approx 0.648$, or about 65%   **b.** 0.682, or about 68%   **c.** 0.710, or about 71%

45. By the definition of expected value.

47. The expected value of a sum is the sum of the expected values (from Exercise 46) and the formula for the expected value of a binomial random variable with parameters $n = 1$ and p (from Exercise 45).

49. (1) Definition of variance   **(2)** $E(X + Y) = E(X) + E(Y)$, algebra, and independence   **(3–5)** Algebra
(6) Definition of variance, probabilities sum to 1, definition of expected value
(7) $E(X - E(X)) = E(X) - E(X) = 0$, and similarly for Y

51. (1) Definition of variance   **(2)** Separating the sum into two parts   **(3)** Dropping a nonnegative quantity
(4) Since $(x - \mu) \ge (k\sigma)^2$

53. Reversing sides and dividing by σ^2 and k^2 **55.** Algebra   **57.** $P(\mu - 5\sigma < X < \mu + 5\sigma) > 1 - \frac{1}{5^2} = 0.96$

CHAPTER 5 REVIEW EXERCISES page 458

1. a. 17   **b.** 32   **c.** 21   **d.** 15   **2.** 110   **3. a.** $26^3 = 17{,}576$ **b.** $26 \cdot 25 \cdot 24 = 15{,}600$

4. a. 120   **b.** 20   **5.** $_{20}C_4 = 4845, _{20}P_4 = 116{,}280$ **6.** $_{13}C_5 = 1287$

7. a. $x^{10} + _{10}C_1x^9y^1 + _{10}C_2x^8y^2 = x^{10} + 10x^9y^1 + 45x^8y^2$ **b.** $8a^4 - 32a^3 + 24a^2 - 8a + 1$

8. $\{(C_1, S_1, H_1), (C_1, S_1, H_2), (C_1, S_2, H_1), (C_1, S_2, H_2), (C_2, S_1, H_1), (C_2, S_1, H_2), (C_2, S_2, H_1), (C_2, S_2, H_2)\}$

9. a. $\{(B, B), (B, Y), (B, R), (Y, B), (Y, Y), (Y, R), (R, B), (R, Y), (R, R)\}$

   b. $\{(B, Y), (B, R), (Y, B), (Y, R), (R, B), (R, Y)\}$

10. a. $\{(H, H), (H, T), (T, H)\}$ **b.** $\{(H, T), (T, H), (T, T)\}$ **c.** $\{(H, T), (T, H)\}$

11. a. $\frac{1}{4}$ **b.** $\frac{3}{4}$ **c.** $\frac{1}{2}$ **12.** $1/_{15}C_2 = \frac{1}{105}, 1/_{15}P_2 = \frac{1}{210}$ **13.** $P(1) = P(3) = P(5) = P(7) = \frac{1}{4}$

14. $P(1) = \frac{1}{2}, P(2) = P(4) = P(6) = \frac{1}{6}$ **15.** $P(R) = \frac{1}{6}, P(G) = \frac{1}{3}, P(B) = \frac{1}{2}$ **16.** 0.45

17. a. $\frac{1}{3}$ **b.** $\frac{1}{2}$ **18.** $_5C_4/_{40}C_5 \approx 0.00000760$

19. a. $_{12}C_5/_{52}C_5 \approx 0.000305$, or about 0.03%   **b.** $_{40}C_5/_{52}C_5 \approx 0.253$, or about 25%

20. $_{39}C_{13}/_{52}C_{13} \approx 0.0128$, or about 1%   **b.** $_{40}C_{13}/_{52}C_{13} \approx 0.0189$, or about 2%

21. $_{28}C_5/_{30}C_5 \approx 0.690$, or about 69%   **22.** $_{48}C_2/_{50}C_4 \approx 0.0049$, or about 0.5%; $_{48}C_4/_{50}C_4 \approx 0.8449$, or about 84.5%

23. $_{28}C_1/_{30}C_3 = \frac{1}{145} \approx 0.00690$, or about 0.7%   **24.** $_6C_2/_{10}C_2 = \frac{1}{3}$

25. a. $P(A \text{ given } B) = \frac{3}{4}$ **b.** $P(B \text{ given } A) = \frac{3}{5}$ **26.** $\frac{1}{3}$ **27.** $\frac{1}{7}$

28. 0.00235, or about 0.2%   **29.** $\frac{3}{16}$ **30.** 0.32   **31.** $0.8 \cdot 0.6 + 0.9 \cdot 0.4 = 0.84$ **32.** 0.983 (approx.)

33. a. No $\left(\frac{3}{4} \cdot \frac{3}{4} \neq \frac{1}{2}\right)$ **b.** Yes $\left(\frac{1}{2} \cdot \frac{1}{2} = \frac{1}{4}\right)$ **34.** 0.9999   **35. a.** $\frac{1}{32}$ **b.** $\frac{1}{16}$ **c.** $\frac{1}{16}$

36. $\frac{2}{3}$ **37.** $\frac{0.04 \cdot 0.25}{0.04 \cdot 0.25 + 0.03 \cdot 0.35 + 0.02 \cdot 0.40} \approx 0.351$, or about 35%   **38.** 0.247, or about 25%

39.

Value of X	Outcomes
$X = 5$	(H, H, H, H, H)
$X = 1$	$(H, T, T, T, T), (T, H, T, T, T), (T, T, H, T, T), (T, T, T, H, T), (T, T, T, T, H)$
$X = 0$	(T, T, T, T, T)

40.

x	34	-2
$P(X = x)$	$\frac{1}{4}$	$\frac{3}{4}$

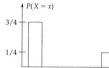

41. $\mu = 7$, $\sigma = \sqrt{243} \approx 15.6$

42.

x	0	1	2
$P(X = x)$	$\frac{5}{18}$	$\frac{5}{9}$	$\frac{1}{6}$

43. $\mu = \frac{8}{9}$ **44.** $\mu = \$3.10$

45.

x	0	1	2	3	4
$P(X = x)$	$\frac{1}{625}$	$\frac{16}{625}$	$\frac{96}{625}$	$\frac{256}{625}$	$\frac{256}{625}$

$\mu = \frac{16}{5} = 3\frac{1}{5}$ $\sigma = \frac{4}{5}$

46. 0.711 (approx.), or about 71%   **47.** \$360   **48.** 2   **49.** $\mu = 12 \cdot 0.01 = 0.12$, $\sigma = \sqrt{12 \cdot 0.01 \cdot 0.99} \approx 0.345$

50. 0.859, or about 86%   **51.** 0.633, or about 63%

EXERCISES 6.1 page 472

1. Nominal   **3.** Ordinal   **5.** Ordinal   **7.** Ordinal   **9.** Ratio

11.

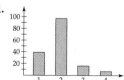

1 = Never married
2 = Married
3 = Widowed
4 = Divorced

13.

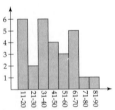

1 = Business executives
2 = Real estate developers
3 = Lawyers
4 = Doctors
5 = Certified public accountants
6 = Retirees

15.

Stem	Leaf
0	0, 9, 0, 0
1	8, 9, 2
2	2, 9, 3, 7, 1, 9, 9, 5, 7, 5, 1, 1, 5, 4
3	4, 8, 5, 2, 2
4	8, 2, 3
5	0

17.

Stem	Leaf
2	2, 8, 6, 6, 4
3	1, 4, 3, 7, 8, 4, 3, 6, 4, 5, 2, 7, 5, 7
4	8, 1, 2, 6
5	4

19.

Stem	Leaf
4	6, 0, 9, 2
5	2, 9, 2, 3, 5, 6, 6, 9, 4, 2, 8, 9
6	8, 6, 5, 0, 6, 1, 3
7	6, 4, 0, 6, 4, 8, 7, 1, 5

21. Using 6 classes of width 8 beginning at 12.5:

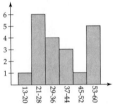

23. Using 8 classes of width 10 beginning at 10.5:

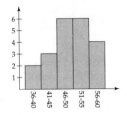

25. Using 11 classes of width 32 beginning at 218.5:

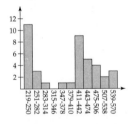

27. Using 5 classes of width 5 beginning at 35.5:

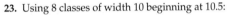

29. Using 7 classes of width 3 beginning at 8.5:

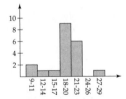

EXERCISES 6.2 page 481

1. Mode = 9, median = 10, $\bar{x} = 11$ **3.** Mode = 7 and 14 (bimodal), median = 9, $\bar{x} = 10$

5. Mode = 19, median = 15, $\bar{x} = 15.2$ **7.** Mode = 14, median = 13, $\bar{x} = 12$ **9.** Mode = 19, median = 18, $\bar{x} = 16$

11. Mode = 12 and 20 (bimodal), median = 15, $\bar{x}$ = 15.75, each in minutes

13. Mode = 10, median = 10, $\bar{x}$ = 11.1, each in hours   **15.** Mode = 9, median = 9, $\bar{x}$ = 10, each in weeks

19. $\bar{x} \approx$ 44.95 dollars

EXERCISES 6.3 page 489

1. 62   **3.** 20   **5.** 56   **7.** Minimum = 3
First quartile = 10
Median = 14
Third quartile = 21
Maximum = 23

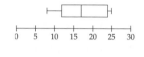

9. Minimum = 8
First quartile = 10
Median = 13
Third quartile = 17
Maximum = 26

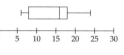

11. Minimum = 8
First quartile = 12
Median = 17
Third quartile = 24
Maximum = 25

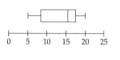

13. Minimum = 6
First quartile = 8
Median = 16
Third quartile = 18
Maximum = 24

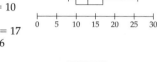

15. Minimum = 5
First quartile = 8.5
Median = 15.5
Third quartile = 17.5
Maximum = 20

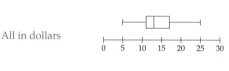

17. $s \approx$ 5.79 (from $\bar{x}$ = 11)   **19.** $s \approx$ 8.63 (from $\bar{x}$ = 13)

21. Range = 20
Minimum = 5
First quartile = 11
Median = 13   } All in dollars
Third quartile = 17
Maximum = 25
$s \approx$ 4.64

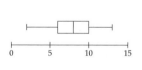

23. Range = 11
Minimum = 2
First quartile = 6   } All in
Median = 8   numbers
Third quartile = 10   of people
Maximum = 13
$s \approx$ 2.55

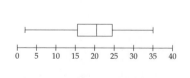

25. Range = 33
Minimum = 2
First quartile = 15.5
Median = 20.5   } All in weeks
Third quartile = 24.5
Maximum = 35
$s \approx$ 8.19

29. $s = \sqrt{\dfrac{739 - \frac{1}{5}(55)^2}{5 - 1}} = \sqrt{33.5} \approx 5.79$ (same answer as Exercise 17)

EXERCISES 6.4 page 500

(*Note:* Answers may differ depending on rounding and on whether graphing calculators or tables were used.)

1. 0.6827   **3.** 0.9973   **5.** 0.3413   **7.** 0.0674   **9.** 0.3361   **11.** $z = 1$ **13.** $z = 0$ **15.** $z = -1$ **17.** $z = -5$

19. $z = 2$ **21.** 0.0298 [from $\mu = 14$, $\sigma \approx 2.05$, after checking that $np > 5$ and $n(1 - p) > 5$]

23. 0.1883 [from $\mu = 14$, $\sigma \approx 2.05$, after checking that $np > 5$ and $n(1 - p) > 5$]

25. 0.0036 [from $\mu = 14$, $\sigma \approx 2.05$, after checking that $np > 5$ and $n(1 - p) > 5$]

27. 0.4029 [from $\mu = 14$, $\sigma \approx 2.05$, after checking that $np > 5$ and $n(1 - p) > 5$]

29. 0.9992 [from $\mu = 14$, $\sigma \approx 2.05$, after checking that $np > 5$ and $n(1 - p) > 5$]   **31.** About 0.89, or 89%

33. About 0.25, or 25% using $P(x \geq 101)$. [Alt. answers: 28% for $P(x > 100.5)$ or 31% for $P(x > 100)$]

35. About 0.31, or 31%   **37.** About 0.24, or 24% [from $\mu = 564$, $\sigma \approx 15$, after checking that $np > 5$ and $n(1 - p) > 5$]

39. About 0.26, or 26% [from $\mu = 80$, $\sigma \approx 6.93$, after checking that $np > 5$ and $n(1 - p) > 5$]

CHAPTER 6 REVIEW EXERCISES page 503

1. Nominal   **2.** Ordinal   **3.** Interval   **4.** Ratio

5.

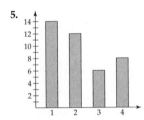

1 = Army
2 = Navy
3 = Marines
4 = Air Force

6.

1 = Handguns
2 = Other firearms
3 = Knives
4 = Blunt instruments
5 = Hands

7.

Stem	Leaf
2	9
3	6, 8, 3
4	7, 3, 5, 0, 6
5	9, 9, 5, 6, 7, 0, 8, 4, 5
6	0
7	
8	7

8.

Stem	Leaf
0	4, 9, 3
1	4, 8, 3, 9, 7
2	5, 1, 7, 6
3	6, 9, 1, 6, 0, 8, 2, 7
4	8, 5, 9, 1
5	7, 2, 8, 1, 3
6	0

9. Using 8 classes of width 4 beginning at 33.5:

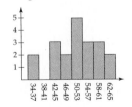

10. Using 10 classes of width 79 beginning at 697.5:

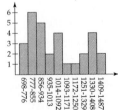

11. Mode = 11, median = 10, $\bar{x} = 9$ **12.** Mode = 6 and 13 (bimodal), median = 11, $\bar{x} = 10$

13. Mode = 13, median = 12.5, $\bar{x} = 11$ **14.** Mode = 8, median = 8, $\bar{x} = 8.50$ (all in dollars)

15. Mode = 4, median = 20, $\bar{x} = 17.4$ (all in numbers of calls)   **16.** 18

17. Minimum = 4
First quartile = 6
Median = 10
Third quartile = 14.5
Maximum = 20

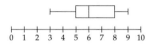

18. $s = 4$

19. Range = 6
Minimum = 3
First quartile = 5
Median = 6
Third quartile = 8
Maximum = 9
$s \approx 1.96$
} All in minutes

20. Range = 40
Minimum = 114
First quartile = 124
Median = 131
Third quartile = 134
Maximum = 154
$s \approx 9.74$
} All in thousands of dollars

21. 0.8186
22. 0.3023
} Answers may differ slightly

23. $z = 2$ **24.** $z = -2$ **25.** 0.0151 [from $\mu = 8.75$, $\sigma \approx 2.38$, after checking that $np > 5$ and $n(1 - p) > 5$]

26. 0.683 [from $\mu = 8.75$, $\sigma \approx 2.38$, after checking that $np > 5$ and $n(1 - p) > 5$]   **27.** About 0.66, or 66%

28. About 0.157, or 16%   **29.** About 0.634, or 63% [from $\mu = 6.9$, $\sigma \approx 2.61$, after checking that $np > 5$ and $n(1 - p) > 5$]

30. About 0.382, or 38% [from $\mu = 48$, $\sigma \approx 5.71$, after checking that $np > 5$ and $n(1 - p) > 5$]

EXERCISES 7.1 page 519

1. $\begin{array}{c} \\ A \\ B \\ C \end{array} \begin{array}{ccc} A & B & C \\ \begin{pmatrix} 0 & 1 & 0 \\ 0 & 0 & 1 \\ 1 & 0 & 0 \end{pmatrix} \end{array}$, oscillating

3. $\begin{array}{c} \\ D \\ E \end{array} \begin{array}{cc} D & E \\ \begin{pmatrix} \frac{2}{5} & \frac{3}{5} \\ 1 & 0 \end{pmatrix} \end{array}$, mixing

5. $\begin{array}{c} \\ F \\ G \end{array} \begin{array}{cc} F & G \\ \begin{pmatrix} \frac{2}{5} & \frac{3}{5} \\ 0 & 1 \end{pmatrix} \end{array}$, absorbing

7. Oscillating

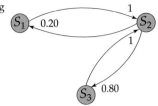

9. Mixing

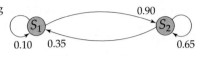

11. Absorbing

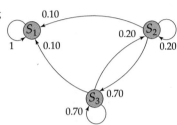

13. $D_1 = \left(\frac{7}{20} \quad \frac{13}{20}\right), D_2 = \left(\frac{33}{80} \quad \frac{47}{80}\right)$ **15.** $D_1 = (0.35 \quad 0.25 \quad 0.40), D_2 = (0.31 \quad 0.17 \quad 0.52)$ **17.** 5   **19.** $2\frac{1}{2}$

21. The expected number of minutes that a green light for traffic on Main Street will be green before changing to red is 4 minutes.

23. Next week, 71% will be eating cereal; the week after, 71.8% will be eating cereal.

25. Tomorrow, 25% will be fishing; and the day after, 52.5% will be golfing.

27. In two years, approximately 75.4% of the portfolio will be "secure." Because "bankrupt" is an absorbing state for all the other states, ultimately every company in the portfolio will become bankrupt.

29. The expected number of weekends that the laundry room will be manageable, given that it is now manageable, is 2 weekends.

EXERCISES 7.2 page 532

1. Ergodic   **3.** Regular   **5.** Neither   **7.** Regular   **9.** (0.50 0.50)

11. (0.40 0.60)   **13.** (0.20 0.70 0.10)   **15.** (0.10 0.85 0.05)   **17.** 2   **19.** 3

21. The voter records lead to a steady-state distribution with 60% of the population voting, which is consistent with the national average (the surveys lead to a steady-state distribution with 80% of the population voting, which is not consistent with the national average).

23. One million of the three million commuters will be driving alone.

25. The long-term market share for DentiMint toothpaste will be 24%.

27. 45% + 30% = 75% of the broker's auto policies are rated satisfactory or better.

29. Yes; a long-term moving probability of $\frac{1}{3}$ is consistent with the national average of about 34.3%.

EXERCISES 7.3 page 545

1. 2 transitions   **3.** 3 transitions   **5.** $k = 2$ **7.** $k = 3$

9.
$$
\begin{array}{c}
 \\
S_2 \\
S_1 \\
S_3
\end{array}
\begin{array}{ccc}
S_2 & S_1 & S_3 \\
\left(\begin{array}{c|cc}
1 & 0 & 0 \\
\hline
0.02 & 0.88 & 0.10 \\
0.16 & 0.04 & 0.80
\end{array}\right)
\end{array}
\text{ and other answers are possible.}
$$

11.
$$\begin{array}{c c} & \begin{array}{c c c c c} S_2 & S_4 & S_3 & S_1 & S_5 \end{array} \\ \begin{array}{c} S_2 \\ S_4 \\ S_3 \\ S_1 \\ S_5 \end{array} & \left(\begin{array}{c c | c c c} 1 & 0 & 0 & 0 & 0 \\ 0 & 1 & 0 & 0 & 0 \\ \hline 0.20 & 0.30 & 0.50 & 0 & 0 \\ 0 & 0 & 0.20 & 0.80 & 0 \\ 0 & 0 & 0.05 & 0.20 & 0.75 \end{array}\right) \end{array}$$ Other answers are possible.

13. $\begin{pmatrix} 2 & 1 \\ 1 & 3 \end{pmatrix}$ **15.** $\begin{pmatrix} 2 & 3 \\ 1 & 4 \end{pmatrix}$ **17.** $\begin{pmatrix} 2 & 2 & 3 \\ 0 & 10 & 0 \\ 1 & 1 & 4 \end{pmatrix}$ **19.** $\begin{pmatrix} 2 & 2 & 2 \\ 1 & 6 & 1 \\ 1 & 2 & 3 \end{pmatrix}$

21. 2; 1 **23.** 1; 0.70 **25.** 10; 1 **27.** 2; 0.70 **29.** $1 + 6 + 1 = 8$; 0.40

31. a. 30% of the town's original population survived.

b. The expected time for the disease to run its course was 2 months.

33. With 0.60 probability, a resident requiring assistance will ultimately be transferred to a nursing home.

35. a. A policy that has just been renewed can be expected to continue for $4 + 2 = 6$ periods, so $6 \times 5 = 30$ years.

b. A policy that is renewed with increased coverage will close with a death benefit payment with probability 0.53.

37. A tooth needing a filling can be expected to need 4 fillings before being extracted.

39. Since 80% of the companies receiving advice become successful and 76% of those receiving loans become successful, the trust's claim is justified.

CHAPTER 7 REVIEW EXERCISES page 551

1. $\begin{array}{c c} & \begin{array}{c c} A & B \end{array} \\ \begin{array}{c} A \\ B \end{array} & \begin{pmatrix} \frac{2}{3} & \frac{1}{3} \\ 1 & 0 \end{pmatrix} \end{array}$, mixing **2.** $\begin{array}{c c} & \begin{array}{c c c} X & Y & Z \end{array} \\ \begin{array}{c} X \\ Y \\ Z \end{array} & \begin{pmatrix} 0 & 0 & 1 \\ 1 & 0 & 0 \\ 0 & 1 & 0 \end{pmatrix} \end{array}$, oscillating

3. Absorbing **4.** Mixing

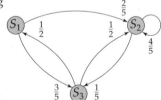

5. $D_1 = (0.35 \quad 0.65)$, $D_2 = (0.325 \quad 0.675)$ **6.** $D_1 = (0.60 \quad 0.30 \quad 0.10)$, $D_2 = (0.19 \quad 0.42 \quad 0.39)$ **7.** 20 **8.** 5

9. A working cab can be expected to remain in service for 100 days before breaking down.

10. Of these 875 students, 130 will be at the second-year level after 2 years.

11. Ergodic **12.** Neither **13.** Regular **14.** Regular

15. $(0.25 \quad 0.75)$ **16.** $(0.15 \quad 0.50 \quad 0.35)$ **17.** 2 transitions **18.** 2 transitions

19. Of the manufacturer's 2660 dealers, $0.40 \cdot 2660 = 1064$ have service departments rated "excellent" by their customers.

20. Out of the next 25 years, 10 will have terrific weather for growing corn, if the old timers are right.

21. 2 transitions **22.** 3 transitions **23.** 3 transitions **24.** 3 transitions

25.
$$
\begin{array}{c c}
 & \begin{array}{cccc} S_2 & S_1 & S_3 & S_4 \end{array} \\
\begin{array}{c} S_2 \\ S_1 \\ S_3 \\ S_4 \end{array} &
\left(\begin{array}{cc|cc}
1 & 0 & 0 & 0 \\
0.50 & 0.50 & 0 & 0 \\
0 & 0.20 & 0.80 & 0 \\
0 & 0 & 1 & 0
\end{array}\right)
\end{array}
$$
Other answers are possible.

26.
$$
\begin{array}{c c}
 & \begin{array}{cccc} S_2 & S_4 & S_3 & S_1 \end{array} \\
\begin{array}{c} S_2 \\ S_4 \\ S_3 \\ S_1 \end{array} &
\left(\begin{array}{cc|cc}
1 & 0 & 0 & 0 \\
0 & 1 & 0 & 0 \\
0.10 & 0 & 0.80 & 0.10 \\
0 & 0.10 & 0.30 & 0.60
\end{array}\right)
\end{array}
$$
Other answers are possible.

27. $\begin{pmatrix} 2 & 4 \\ 1 & 7 \end{pmatrix}$ **28.** $\begin{pmatrix} 4 & 6 \\ 2 & 8 \end{pmatrix}$ **29.** $\begin{pmatrix} 6 & 0 & 2 \\ 3 & 5 & 1 \\ 4 & 0 & 3 \end{pmatrix}$ **30.** $\begin{pmatrix} 2 & 0 & 6 \\ 1 & 2 & 4 \\ 1 & 0 & 8 \end{pmatrix}$ **31.** 2; 1   **32.** 1; 1

33. 6; 0.30   **34.** 8; 0.60   **35.** 4; 1   **36.** $3 + 5 + 1 = 9$; 1   **37.** 0; 0.30   **38.** $1 + 0 + 8 = 9$; 0.85

39. A new model PC can be expected to be manufactured for $3 + 4 = 7$ years before it is discontinued.

40. With probability 0.39 an overdue account will (ultimately) be paid in full.

EXERCISES 8.1 page 561

1. The yard is small.   **3.** The yard is large and the house is small.

5. If the yard is large, then the house is small.   **7.** The cat is purring or the dog is barking.

9. The cat is not purring and the dog is barking.   **11.** If the dog is barking, then the cat is purring.

13. $\sim p$ **15.** $p \wedge q$ **17.** $q \vee \sim r$ **19.** $\sim r \to p$

EXERCISES 8.2 page 572

1. F **3.** F **5.** T **7.** T **9.** T

11.

p	q	p	$\wedge$	$\sim q$
T	T	T	F	F
T	F	T	T	T
F	T	F	F	F
F	F	F	F	T

13.

p	q	$(\sim p$	$\vee$	$q)$	$\wedge$	$\sim q$
T	T	F	T	T	F	F
T	F	F	F	F	F	T
F	T	T	T	T	F	F
F	F	T	T	F	T	T

15.

p	q	$(p$	$\vee$	$\sim q)$	$\vee$	$\sim$	$(p \wedge q)$
T	T	T	T	F	T	F	T
T	F	T	T	T	T	T	F
F	T	F	F	F	T	T	F
F	F	F	T	T	T	T	F

so $(p \vee \sim q) \vee \sim (p \wedge q)$ is a tautology.

17.

p	q	r	~	(p ∨ r)	∧	(~q	∧	r)
T	T	T	F	T	F	F	F	T
T	T	F	F	T	F	F	F	F
T	F	T	F	T	F	T	T	T
T	F	F	F	T	F	T	F	F
F	T	T	F	T	F	F	F	T
F	T	F	T	F	F	F	F	F
F	F	T	F	T	F	T	T	T
F	F	F	T	F	F	T	F	F

so ~ (p ∨ r) ∧ (~ q ∧ r) is a contradiction.

19. p ∧ q   **21.** q   **23.** t, so ~ (p ∧ ~ q) ∨ p is a tautology.   **27.** The coffee is weak or the donuts are cold.

29. The wolf is hungry and the sheep are in danger.   **41.** Exclusive   **43.** Inclusive

45.

p	q	p	⊻	~q
T	T	T	T	F
T	F	T	F	T
F	T	F	F	F
F	F	F	T	T

47.

p	q	~p	⊻	(p ∧ q)
T	T	F	T	T
T	F	F	F	F
F	T	T	T	F
F	F	T	T	F

51. p ∧ ~ q   **53.** p ∧ ~ q ∧ r   **55.** (~ p ∧ q) ∨ (~ p ∧ ~ q)   **57.** (p ∧ ~ q ∧ r) ∨ (~ p ∧ q ∧ r)

EXERCISES 8.3 page 580

1. (*direct*)   If the jury is wise, then justice is done.
(*converse*)   If justice is done, then the jury is wise.
(*inverse*)   If the jury is not wise, then justice is not done.
(*contrapositive*)   If justice is not done, then the jury is not wise.
(*negation*)   The jury is wise and justice is not done.

3. (*direct*)   If the dentist is skillful, then the tooth is saved.
(*converse*)   If the tooth is saved, then the dentist is skillful.
(*inverse*)   If the dentist is not skillful, then the tooth is not saved.
(*contrapositive*)   If the tooth is not saved, then the dentist is not skillful.
(*negation*)   The dentist is skillful and the tooth is not saved.

5. (*direct*)   If the bird feeder is full, then there are many birds in the yard.
(*converse*)   If there are many birds in the yard, then the bird feeder is full.
(*inverse*)   If the bird feeder is not full, then there aren't many birds in the yard.
(*contrapositive*)   If there aren't many birds in the yard, then the bird feeder is not full.
(*negation*)   The bird feeder is full and there aren't many birds in the yard.

7.

p	q	p	→	~q
T	T	T	F	F
T	F	T	T	T
F	T	F	T	F
F	F	F	T	T

9.

p	q	p	→	(p ∨ q)
T	T	T	T	T
T	F	T	T	T
F	T	F	T	T
F	F	F	T	F

so p → (p ∨ q) is a tautology.

11.

p	q	p	∧	~	(~ q	→	p)
T	T	T	F	F	F	T	T
T	F	T	F	F	T	T	T
F	T	F	F	F	F	T	F
F	F	F	F	T	T	F	F

so p ∧ ~ (~ q → p) is a contradiction.

13.

p	q	r	(p ∧ q)	→	r
T	T	T	T	T	T
T	T	F	T	F	F
T	F	T	F	T	T
T	F	F	F	T	F
F	T	T	F	T	T
F	T	F	F	T	F
F	F	T	F	T	T
F	F	F	F	T	F

15.

p	q	p	↔	~ q
T	T	T	F	F
T	F	T	T	T
F	T	F	T	F
F	F	F	F	T

17. p   **19.** t, so (p ∧ q) → (p ∨ q) is a tautology.   **31.** Consistent

33. Contrary   **35.** Consistent   **37.** Contrary   **39.** Consistent

EXERCISES 8.4 page 591

1. Modus ponens; valid   **3.** Disjunctive syllogism; valid   **5.** Hypothetical syllogism; valid

7. Fallacy of the converse; fallacy   **9.** Valid   **11.** Valid   **13.** Valid   **15.** Valid

17. If I have chicken for dinner, then I go to bed early.   **19.** I am sunburned.

EXERCISES 8.5 page 600

1. An apple is a fruit. True.   **3.** Everything in {broccoli, grape, potato} is a fruit. False.

5. There is at least one thing in {broccoli, grape, potato} that is not a fruit. True.

7. There is at least one thing in {milk, cookies} that is a beverage. True.

9. Everything in {milk, cookies} is a beverage. False.

11. $P = \{2, 4, 10, 16, 24, 56\}$ **13.** T   **15.** T   **17.** T   **19.** T

21. $P = \{$Colorado, Missouri, Nevada, New York, Tennessee$\}$ **23.** T   **25.** T   **27.** T   **29.** F

31. $\forall x \in A, p(x)$ **33.** $\exists x \in A, \sim (p(x) \wedge q(x))$ or $\exists x \in A, \sim p(x) \vee \sim q(x)$

35. $\forall x \in A, p(x) \wedge \sim q(x)$ **37.** There is at least one rich person who is not a snob.

39. All quadrilaterals are squares.   **41.** Valid   **43.** Fallacy   **45.** Valid   **47.** Fallacy   **49.** Valid

CHAPTER 8 REVIEW EXERCISES page 604

1. Roses are red and violets are blue.   **2.** Roses are red or violets are blue.

3. Violets are not blue.   **4.** If roses are not red, then violets are blue.

5. $p \wedge q$ **6.** $p \wedge {\sim} r$ **7.** $q \vee {\sim} p$ **8.** ${\sim} r \vee (q \wedge p)$ **9.** $r \rightarrow {\sim} p$

10. $(p \wedge q) \rightarrow {\sim} r$ **11.** F **12.** F

13.

p	q	$(p$	$\vee$	${\sim} q)$	$\wedge$	${\sim} p$
T	T	T	T	F	F	F
T	F	T	T	T	F	F
F	T	F	F	F	F	T
F	F	F	T	T	T	T

14.

p	q	r	${\sim}$	$(p$	$\wedge$	$r)$	$\vee$	$({\sim} q$	$\vee$	$r)$
T	T	T	F	T	T	T	T	F	T	T
T	T	F	T	T	F	F	T	F	F	F
T	F	T	F	T	T	T	T	T	T	T
T	F	F	T	T	F	F	T	T	T	F
F	T	T	T	F	F	T	T	F	T	T
F	T	F	T	F	F	F	T	F	F	F
F	F	T	T	F	F	T	T	T	T	T
F	F	F	T	F	F	F	T	T	T	F

so ${\sim} (p \wedge r) \vee ({\sim} q \vee r)$ is a tautology.

15. p **16.** f, so $({\sim} p \vee q) \wedge (p \wedge {\sim} q)$ is a contradiction.

17. f, so $p \wedge {\sim} (p \vee q)$ is a contradiction.   **18.** t, so ${\sim} (p \wedge q) \vee p$ is a tautology.

19. The glasses aren't broken or the watch works.   **20.** His ankle is twisted and his knee is fine.

21. (*direct*)   If the painting is expensive, then the frame is ornate.
(*converse*)   If the frame is ornate, then the painting is expensive.
(*inverse*)   If the painting is not expensive, then the frame is not ornate.
(*contrapositive*)   If the frame is not ornate, then the painting is not expensive.
(*negation*)   The painting is expensive and the frame is not ornate.

22. (*direct*)   If the freezer works properly, then the ice cream is hard.
(*converse*)   If the ice cream is hard, then the freezer works properly.
(*inverse*)   If the freezer does not work properly, then the ice cream is not hard.
(*contrapositive*)   If the ice cream is not hard, then the freezer does not work properly.
(*negation*)   The freezer works properly and the ice cream is not hard.

23.

p	q	${\sim} p$	$\rightarrow$	${\sim} q$
T	T	F	T	F
T	F	F	T	T
F	T	T	F	F
F	F	T	T	T

24.

p	q	$[(p \rightarrow q)$	$\wedge$	$p]$	$\rightarrow$	q
T	T	T	T	T	T	T
T	F	F	F	T	T	F
F	T	T	F	F	T	T
F	F	T	F	F	T	F

so $[(p \rightarrow q) \wedge p] \rightarrow q$ is a tautology.

25.

p	q	r	$(p \vee q)$	$\rightarrow$	$(q \wedge r)$
T	T	T	T	T	T
T	T	F	T	F	F
T	F	T	T	F	F
T	F	F	T	F	F
F	T	T	T	T	T
F	T	F	T	F	F
F	F	T	F	T	F
F	F	F	F	T	F

26.

p	q	$\sim p$	$\leftrightarrow$	q
T	T	F	F	T
T	F	F	T	F
F	T	T	T	T
F	F	T	F	F

27. $q \rightarrow p$ **28.** t, so $(p \wedge q) \rightarrow p$ is a tautology.   **29.** $p \rightarrow \sim q$ and also $q \rightarrow \sim p$

30. $\sim p$ **31.** Modus ponens; valid   **32.** Modus tollens; valid

33. Disjunctive syllogism; valid   **34.** Fallacy of the inverse; fallacy   **35.** Valid

36. Fallacy   **37.** Valid   **38.** Valid   **39.** If I have soup for lunch, then I stay late at work.

40. The book is thin.   **41.** T **42.** F **43.** $P = \{$Brahms, Debussy, Pachelbel$\}$

44. $Q = \{$Cézanne, Debussy$\}$ **45.** F **46.** T

47. There is a great composer whose name does not begin with the letter B.

48. I can solve every logic problem.   **49.** Valid   **50.** Fallacy

INDEX

FUNCTIONS

$$m = \frac{y_2 - y_1}{x_2 - x_1} \qquad\qquad y = mx + b \qquad\qquad ax + by = c$$

$$x^0 = 1 \qquad x^{-n} = \frac{1}{x^n} \qquad x^{\frac{1}{n}} = \sqrt[n]{x} \qquad x^m \cdot x^n = x^{m+n} \qquad (x^m)^n = x^{m \cdot n} \qquad \frac{x^m}{x^n} = x^{m-n}$$

$$y = ax^2 + bx + c \quad \text{has vertex at} \quad x = \frac{-b}{2a} \quad \text{and } x\text{-intercepts at} \quad x = \frac{-b \pm \sqrt{b^2 - 4ac}}{2a}$$

$$|x| = \begin{cases} x & \text{if } x \geq 0 \\ -x & \text{if } x < 0 \end{cases} \qquad\qquad\qquad \begin{aligned} y = \log x & \quad \text{means} \quad x = 10^y \\ y = \ln x & \quad \text{means} \quad x = e^y \end{aligned}$$

$$\log 1 = 0 \qquad \log(M \cdot N) = \log M + \log N \qquad \log\left(\frac{M}{N}\right) = \log M - \log N \qquad \log(M^N) = N \cdot \log M$$

FINANCE

Simple Interest:
$$I = Prt \qquad\qquad A = P(1 + rt) \qquad\qquad r_s = \frac{r}{1 - rt}$$

Compound Interest:
$$\begin{aligned} A &= P(1 + r)^t \\ A &= P(1 + r/m)^{mt} \\ A &= Pe^{rt} \end{aligned} \qquad \begin{aligned} r_e &= (1 + r/m)^m - 1 \\ r_e &= e^r - 1 \end{aligned}$$

Rule of 72:
$$\left(\begin{array}{c} \text{Doubling} \\ \text{Time} \end{array}\right) \approx \frac{72}{r \times 100}$$

Annuities and Amortization:
$$A = P\frac{(1 + r/m)^{mt} - 1}{r/m} \qquad mt = \frac{\log\left(\frac{A}{P}\frac{r}{m} + 1\right)}{\log(1 + r/m)} \qquad PV = P\frac{1 - (1 + r/m)^{-mt}}{r/m} \qquad P = D\frac{r/m}{1 - (1 + r/m)^{-mt}}$$

MATRICES

$$\begin{cases} ax + by = h \\ cx + dy = k \end{cases} \xrightarrow{\text{augmented matrix}} \begin{pmatrix} a & b & h \\ c & d & k \end{pmatrix} \qquad\qquad A \cdot A^{-1} = I = A^{-1} \cdot A \text{ and } (A|I) \xrightarrow{\text{row reduction}} (I|A^{-1})$$

$$\xrightarrow{\text{row reduction may give}} \begin{pmatrix} 1 & 0 & p \\ 0 & 1 & q \end{pmatrix} \qquad\qquad A^{-1} \text{ solves } A \cdot X = B \text{ as } X = A^{-1} \cdot B$$

LINEAR PROGRAMMING

Maximize $P = c^t X$

Subject to $\begin{cases} AX \leq b \\ X \geq 0 \end{cases}$ $\xrightarrow{\text{simplex tableau}}$

	X	S	
S	A	I	b
P	$-c^t$	0	0

$dial \updownarrow problems$

Minimize $C = b^t Y$

Subject to $\begin{cases} A^t Y \geq c \\ Y \geq 0 \end{cases}$

Pivot Operation:
$$\begin{cases} R_{pivot}^{new} = R_{pivot}/PivotElement \\ R_{other}^{new} = R_{other} - \left(\begin{array}{c} Pivot \\ Column \\ Entry \end{array}\right) R_{pivot}^{new} \end{cases}$$